Organometallics in Synthesis

Fourth Manual

Organometallics in Synthesis

Fourth Manual

Edited by

Bruce H. Lipshutz
University of California, Santa Barbara, California, USA

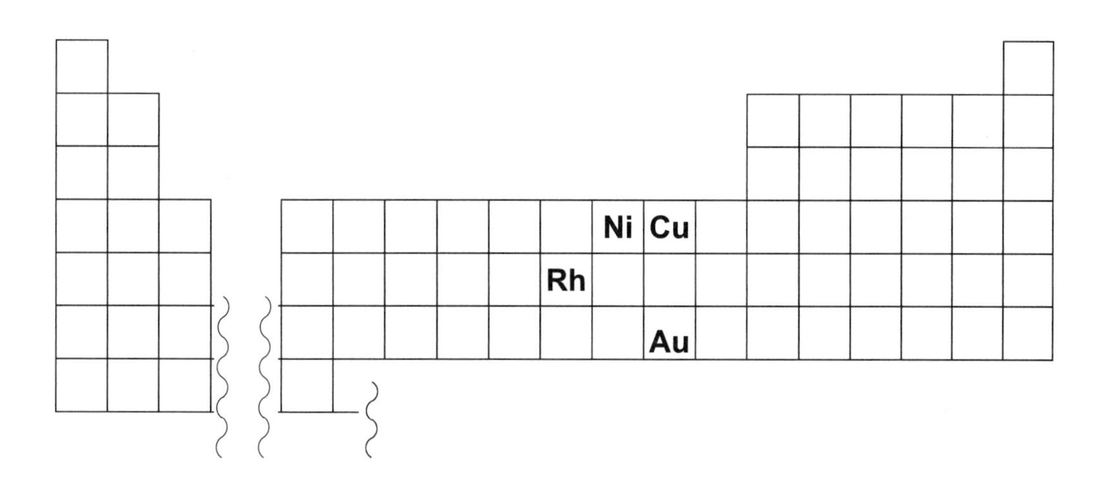

Published by John Wiley & Sons, Inc., Hoboken, New Jersey

Published simultaneously in Canada

Library of Congress Cataloging-in-Publication Data

Organometallics in synthesis : fourth manual / edited by Bruce H. Lipshutz, University of California, Santa Barbara, California, USA. — Fourth edition.
pages cm
Includes bibliographical references and index.
ISBN 978-1-118-48882-9 (pbk.)
1. Organic compounds—Synthesis. 2. Organometallic compounds. I. Lipshutz, Bruce H.
QD262.O745 2013b
547'.05—dc23 2013009447

Printed in the United States of America.

10 9 8 7 6 5 4 3 2 1

Contents

Contributors .... **vii**

Preface .... **ix**

Foreword .... **xii**
Prof. Dr. Reinhard W. Hoffmann

I Organocopper Chemistry **1**
Bruce H. Lipshutz

II Organorhodium Chemistry **135**
Iwao Ojima. Alexandra A. Athan, Stephen J. Chaterpaul, Joseph J. Kaloko, and Yu-Han Gary Teng

III Organonickel Chemistry **317**
John Montgomery

IV Organogold Chemistry **426**
Norbert Krause

Index **547**

Contributors

Chapter I

BRUCE H. LIPSHUTZ
Department of Chemistry & Biochemistry, University of California, Santa Barbara, CA, USA

Chapter II

IWAO OJIMA
ALEXANDRA A. ATHAN
STEPHEN J. CHATERPAUL
JOSEPH J. KALOKO
YU-HAN GARY TENG
Department of Chemistry, State University of New York, Stony Brook, NY, USA

Chapter III

JOHN MONTGOMERY
Department of Chemistry, University of Michigan, Ann Arbor, MI, USA

Chapter IV

NORBERT KRAUSE
Organic Chemistry, Dortmund University of Technology, Dortmund, Germany

Preface

When the first edition of *Organometallics in Synthesis, A Manual* appeared back in 1994, it offered the community a rare, atypical source of experimental "inside information." It was a stand-alone resource of how to's and why's, written by card-carrying organic chemists with a heavy accent on organic synthesis where the C-C, C-H, and C-heteroatom bonds being made are mediated by metals that serve as either reagents or catalysts. The second edition, published in 2002, was expanded by four metals (from 7 to 11) and surpassed 1200 pages in length. And yet, it was essentially a new book. Now, more than a decade later, the third edition has just appeared. Increasing the length of this monograph, however, was no longer an option from a publishing perspective. So, to do justice to the evolving technologies associated with metals discussed in prior editions, as well as to add metals that had not been covered previously, the Manual was divided, as was the work associated with bringing each to fruition. Thus, in the third edition, Manfred Schlosser oversaw inclusion of chapters focused on organoalkali reagents, as well as those highlighting organo-silicon, -magnesium, -zinc, -iron, and -palladium chemistry, each written, again, by a "who's who" in these prominent areas.

With this fourth edition, the torch has been passed. Chapters can be found on organo-rhodium, -copper, -nickel, and -gold chemistry, where the theme is unquestionably on catalysis. And although this opus is equally split between two precious (Rh, Au) and two base (Ni, Cu) metals, the discussion is heavily, and at times exclusively, devoted to the use of each metal in catalytic quantities. Price is no longer the sole driving force; indeed, environmental considerations have become a major topic in the planning and execution of organic synthesis, regardless of the metal involved.

Three of the four contributing authors in this 4th edition are new to this series; hence, this is the first coverage in the *Manual* on rhodium, nickel, and gold chemistry. As for copper, this chapter has been completely redone; there is no redundancy whatsoever insofar as prior discussions in earlier editions are concerned. Thus, this monograph compliments the third edition beautifully. And although it may show a different name below as editor, it is likely to continue to be viewed as another worthy addition to the *Schlosser Manuals.*

Bruce H. Lipshutz
May 2013

Foreword

OMCOS Chemistry: What's the Value?

Analysis and synthesis define the main activities of chemistry that shape the material world around us. Synthesis is the only means to secure new substances and materials when non-natural products are to be studied and to be used. Present-day synthesis of complex organic chemicals is subject to challenging demands: It should be efficient and short, it should be safe and environmentally acceptable, and it should be resource efficient and proceed in high yield, being economically feasible. These concepts imply that the methods by which we do synthesis today need to be constantly challenged as to how well they meet these criteria.

Of the operations to be carried out in synthesis, those that construct the molecular skeleton are the most important and demanding. That is precisely where the development of new methods has brought substantial progress during the last 150 years, and it is in particular the application of organometallic reagents to synthesis that is indicative of the progress made. Imagine the state of synthetic methodology in which there were no organometallic reagents: You are left with certain variants of the Aldol- and Michael-additions, the alkylation of active methylene compounds, the Friedel-Crafts acylation, and the Diels-Alder addition to build the molecular skeleton of your target structures. This is hardly anything more than what nature uses in biosynthesis. Such a restricted arsenal of methods might fascinate some advocates of green chemistry, but it would not be sufficient to address the current practices and needs of medicinal chemistry. It is thus evident that adapting organometallic reagents to the needs of organic synthesis marked the progress in synthetic methodology over the last century; it enabled organic synthesis to fulfill most of the tasks in an acceptable manner, and it constitutes one of the major cultural achievements of chemistry.

Given that situation, one may hold that a *Manual* of organometallics in synthesis might be more of a historic exercise than an opus that meets the needs of today's chemists. The value of such a *Manual*, however, becomes evident when one takes the viewpoint of those that apply these methods in synthesis. Those doing actual synthesis of complex target molecules are focused on the synthesis and do not want to lose time or materials by applying methods with which they are not familiar; that is, in general, chemists doing synthesis are highly conservative regarding the methods they apply.

For instance, a survey of (arbitrarily chosen) 41 syntheses of complex natural products carried out in 1981 revealed that more than 70% of the skeleton bond-forming steps fell to enolate and Grignard reactions, the Wittig reaction and cuprate reactions, or what we would call classic organometallic reactions. In the late 1970s, a golden era of organometallics in organic synthesis started, providing the synthetic chemists with olefin metathesis; the Heck, Negishi, Suzuki, and Sonogashira reactions; and many related cross-couplings. Yet, when surveying 58 syntheses of complex natural products published in 2006, these new methods amounted to (already or merely!) 20% of the skeleton bond-forming reactions used, whereas the classic organometallic reactions mentioned lost only 15% of their usage. This underscores the statement that synthetic chemists are slow in taking up new (and, hopefully, more advantageous) methods as long as the older, established ones do not do too badly . The motivation to apply any method (whether old or new) depends on the practitioner's level of confidence in the scope and reliability of the method. Although a time-consuming, thorough literature search could provide this information, chemists tend to get around this and to stick with the familiar methods as long as justifiable. It is clear that it is the work of a small number of daring chemists applying newly described methods in synthesis that provided the basis for a *Manual* in which scope, usefulness, and reliability of a method can be documented. The merit of such a *Manual* is invaluable as it gives the majority of synthetic chemists the information to make a reasonable judgment as to which method might best serve their immediate synthetic needs. As the present *Manual* covers the long established organometallic methods side by side with the methods of younger vintage, it allows the synthetic chemists to make objective choices regarding the methods to be applied. The availability of such a *Manual* should reduce the reluctance of chemists to apply methods with which they are not (yet) familiar. Therefore, the fourth edition of *Organometallics in Synthesis, A Manual* is timely and a well-taken endeavor.

PROF. DR. REINHARD W. HOFFMANN

Chapter One

Organocopper Chemistry

Bruce H. Lipshutz

Department of Chemistry & Biochemistry
University of California, Santa Barbara, USA

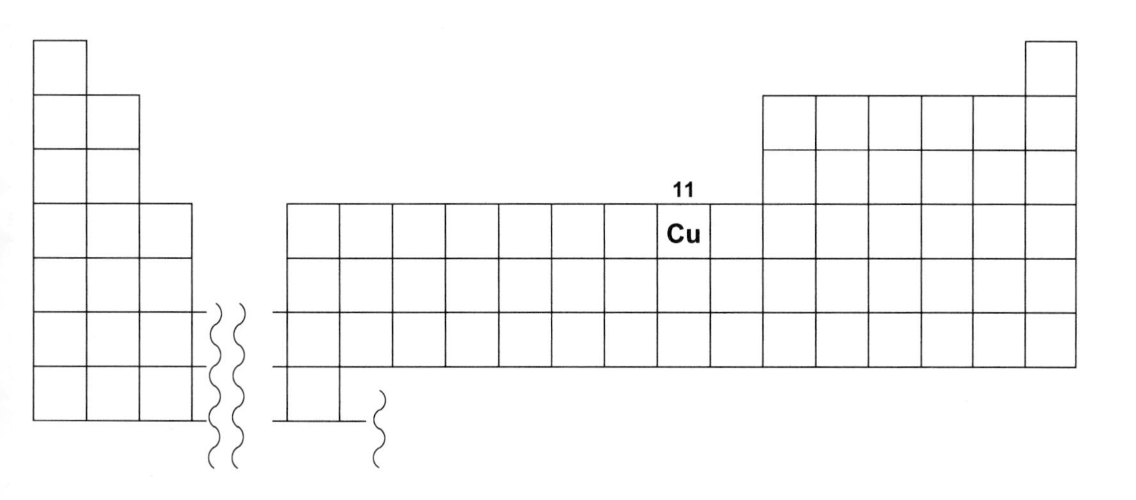

Contents

1.	**Introduction**	3
2.	**Click Chemistry: Just Add Copper**	4
3.	**Cu-Catalyzed Aminations and Amidations**	16
	3.1 Aminations	16
	3.2 Amidations	34
4.	**Cu-Catalyzed *O*- and *S*-Arylations**	38
	4.1 Diaryl Ether Formation	39
	4.2 Diaryl Thioether Formation	46
5.	**Asymmetric Cu-Catalyzed Borylations**	49
6.	**Oxidations of Organocopper Complexes**	58
	6.1 Biaryl Syntheses	58
	6.2 Aminations	63
7.	**1,2-Additions to Imines**	64
8.	**Cu-Catalyzed Asymmetric C-C Bond Formation**	70
	8.1 Copper Enolates	70
	8.2 Conjugate Additions	71
	8.3 Asymmetric Allylic Alkylations (AAA)	85
9.	**Copper Hydride Chemistry**	94
10.	**Miscellaneous Processes**	113
	10.1 C-H Activation	113
	10.2 Protodecarboxylation	115
	10.3 Cu-Catalyzed Carbometallation	116
11.	**Conclusions and Outlook**	118
12.	**Acknowledgments**	118
13.	**References**	119

1. Introduction

Synthetic Chemistry of Cu(I): Still in Focus and "Hot"

Catalysis. It was the buzzword in the 1990s, and certainly insofar as organocopper chemistry is concerned, there has been no letup on this front in the new millennium. Very good reasons exist for this emphasis, both in terms of development of new methodologies as well as in applications. The key driver is usually the cost of waste disposal; whether copper is contained in an aqueous workup mixture, or in amounts greater than the ppm level allowed in pharmaceuticals, it requires attention. Starting with as little copper as possible in a reaction just makes sense, notwithstanding its "base" metal status. So, whereas the accent in previous versions of the *Manual* was on reagents, this chapter is organized by reaction type and focuses heavily on processes that are catalytic in copper(I). Nonetheless, attention is also directed to traditional albeit stoichiometric copper chemistry, acknowledging the fundamentals that began with Henry Gilman more than 60 years ago[1] and still are very much valued by the synthetic community today.

Within the catalysis manifold, remarkable advances have been brought to light in both asymmetric as well as achiral synthesis. As already witnessed with precious metal-based technologies, such as asymmetric hydrogenations, the metal is important, but the ligands "rule." Research over the past decade has led to several mono- and (mostly) bidentate nonracemic ligands that possess, and translate, their extraordinary innate facial biases as their derived copper complexes to a host of substrate types. Several newly discovered species now present themselves to the practitioner, where enantioselectivities are routinely in excess of 90%. On the other hand, chiral upgrades often allow industrial chemists to realize the desired high levels of enantiometric excesses (ee's) needed in pharma, thereby dettaching any real significance to the "magic" barrier of 90% ee. Nonetheless, the challenges extended by nature to chemists to achieve as close to 100% stereocontrol have not in the past, nor are they likely in the future, to be ignored. Of course, stereocontrol is not the only element that counts in such catalysis; indeed, with up-front costs for copper essentially nonexistent from the perspective of economics, it is turnover number (TON) that may dictate usage. The gap between what academicians may see as "catalytic amounts" (usually 1–5 mol%) and the required low loadings for an industrial process can be hard to fill. Hopefully, some of the progress made as highlighted in this contribution will entice our industrial colleagues.

A wealth of achiral copper chemistry is also covered; in fact, it is still the lion's share of reports in this field. Whether copper(I) in the form of its salts (CuX), perhaps (*in situ*) derived organocopper species (RCu), or even a cuprate ($R_2Cu^-M^+$), there is no denying the textbook status of this metal in organic synthesis. Nonetheless, both silver and especially gold chemistry are making astounding advances. But of the group 11 metals, copper still offers the widest variety of synthetically valued chemoselectivities. And while questions regarding mechanisms, aggregation state(s), and structure of organocopper complexes remain, tremendous progress has been made on these fronts as well. The spirited debate throughout the 1990s on the existence, or not, of "higher order" cuprates provided new incentives for computational, structural, spectroscopic, and physical organic chemistry that has since appeared and has helped to advance the field.

As noted in earlier editions of this *Manual*, and as is true herein, this chapter is written for the practitioner who faces an ever expanding literature on organometallics. Notwithstanding the emphasis now squarely placed on catalysis, many of the same practical questions arise as highlighted previously, choices of copper salt, solvent, precursor reagent, stoichiometry, and additives, and today, there have been new variables that can play major roles, such as the choice of (*e.g.*, nonracemic) ligand, potential for heterogeneous catalysis, and the option to employ microwave assistance. Thus, with additional insights from several colleagues who have shared their experiences in organocopper chemistry, many of the "secrets to success" are again revealed in this single source. Unfortunately, however, the field is too broad for this opus to be comprehensive; indeed, tough decisions had to be made as to coverage. Thus, there are entire areas even within Cu(I)-catalyzed chemistry that are not included (*e.g.*, cyclopropanations, Diels-Alder constructions, etc.), and surely others that may cause the reader to wonder "What about …?". The explanation is simple: Each author had a page limitation to his chapter, and it was recommended (mainly as a result of technical matters associated with binding) that the editor not violate his own rules!

2. Click Chemistry: Just Add Copper

Reviews. Diez-Gonzalez, S. *Curr. Org. Chem.* **2011**, *15*, 2830; Diez-Gonzalez, S. *Catal. Sci Technol.* **2011**, *1*, 166; Cantillo, D.; Avalos, M.; Babiano, R.; Cintas, P.; Jimenez, J. L.; Palacios, J. C. *Org. Biomol. Chem.* **2011**, *9*, 2952; Elchinger, P-H.; Faugeras, P-A.; Boens, B.; Brouillette, F.; Montplaisir, D.; Zerrouki, R.; Lucas, R. *Polymers*, **2011**, *3*, 1607; Becer, C. R.; Hoogenboom, R.; Schubert, U. S. *Angew. Chem., Int. Ed.* **2009**, *48*, 4900 (*metal-free*); Meldal, M.; Tornoe, C. W. *Chem. Rev.* **2008**, *108*, 2952; Lutz, J.-F. *Angew. Chem., Int. Ed.* **2007**, *46*, 1018; Fokin, V. V.; Wu, P. *Aldrichimica Acta* **2007**, *40*, 7; Bock, B. D.; Hiemstra, H.; van Maarseveen, J. H. *Eur. J. Org. Chem.* **2006**, 51.

The 2002 papers by the groups of Sharpless[2] at Scripps (La Jolla, CA) and Meldal[3] at the Carlsberg Laboratory (Denmark) highlighting the remarkable acceleration of Huisgen cycloadditions between organic azides[4] and terminal alkynes by Cu(I) have led to an avalanche of renewed interest in, and usage of, copper(I) in synthesis. The facility with which these two relatively high-energy educts "click" to form heteroaromatic 1,2,3-triazoles is truly impressive, and the community at large is

R—≡—H + R'N_3 → (Huisgen cycloaddition, Δ) 1,4-triazole + 1,5-triazole; Cu(I)-catalysis → 1,4-triazole

using this chemistry in both routine and highly innovative ways. Importantly, these otherwise thermally driven cycloadditions, which often lead to mixtures of 1,4- and 1,5-disubstituted triazoles,[5] are fully controlled in the presence of Cu(I) to afford only the 1,4-regioisomer[6] (while Ru leads to the corresponding 1,5-isomer).[7] Mechanistic details have not been fully elucidated, but significant progress has been made, most notably by Finn[8a] and Fokin,[8b] and more recently using density functional theory (DFT) calculations.[9] The data suggest involvement of copper(I) acetylides, shown to be associated within a dimeric array. Electron-withdrawing groups on the alkyne increase reactivity. A key point for the synthetic practitioner here is the acid/base chemistry that must ensue *en route* to the acetylide, implying strong potential influence of base in the medium. Indeed, it is now well accepted that selected bases can dramatically influence rates of click reactions.[6, 10] Curiously, however, the nature of the copper species selected can make a difference in the outcome, with the field narrowing in on two approaches: 1) the original Sharpless protocol using $CuSO_4$, a copper(II) salt that is readily reduced in aqueous *t*-butanol by excess sodium ascorbate (*i.e.*, the inexpensive Na salt of vitamin C).[2] Usually, *ca.* 1% $CuSO_4$ is employed in the presence of excess ascorbate, and with unhindered partners, cycloadditions occur at ambient temperatures in high yields; 2) CuI in organic solvent. Each of these is represented by the two procedures below. A third alternative involving *in situ* oxidation of Cu(0) is also available[8b] and, although considered of equal efficiency, is less frequently used.

17-ethynyl estradiol + N_3CH$_2$CH(OH)CH$_2$OH → (10% $CuSO_4 \cdot 5H_2O$, sodium ascorbate, H_2O/*t*-BuOH, rt, 12 h) triazole product (94%)

Copper(I)-catalyzed synthesis of 1,4-disubstituted 1,2,3-triazoles; general procedure. *2'S-17-[1-(2',3'-Dihydroxypropyl)-1H-[1,2,3]-triazol-4-yl]-estradiol*[2]

17-Ethynyl estradiol (888 mg, 3 mmol) and (*S*)-3-azidopropane-1,2-diol (352 mg, 3 mmol) were suspended in 12 mL of a 1:1 water/*t*-butanol mixture. Sodium ascorbate (0.3 mmol, 300 μL of freshly prepared 1-M solution in water) was added, followed by copper(II) sulfate pentahydrate (7.5 mg, 0.03 mmol, in 100 μL of water). The heterogeneous mixture was stirred vigorously overnight, at which point it cleared and thin layer chromatography (TLC) analysis indicated complete consumption of the reactants. The reaction mixture was diluted with 50 mL of water and cooled in ice, and the white precipitate was collected by filtration. After being washed with cold water (2 × 25 mL), the precipitate was dried under vacuum to afford 1.17 g (94%) of pure product as an off-white powder; mp 228–230 °C.

CuI-DIPEA-catalyzed click chemistry.[11]

AcO OAc AcO O AcO O N3 OAc — $NHC_{14}H_{29}$ alkyne; CuI, DIPEA, PhMe, rt → triazole product $NHC_{14}H_{29}$ (85%)

The azide (20 mg, 0.047 mmol), alkyne (12.7 mg, 0.047 mmol, 1 equiv), and CuI (0.9 mg, 0.004 mmol, 0.1 equiv) were dissolved in toluene (500 μL) in a glass vial (15.5 × 50 mm). To this mixture was added diisopropylethylamine (8.3 μL, 0.047 mmol, 1 equiv) and the vial was capped. After stirring for 18 h at room temperature (RT), the crude product was filtered over Celite (Sigma-Aldrich, St. Louis, MO) and purified by flash chromatography on silica gel using as eluent *n*-hexane/EtOAc (from 1:1 to 1:2). The product (28 mg) was isolated in 85% yield as a single 1,4-regioisomer as an amorphous solid; HR-MALDI-FTMS calcd. for $C_{33}H_{54}N_4O_{11}Na$ $[M + Na]^+$, 705.3681; found 705.3681.

Click chemistry with copper metal. 2,2-Bis((4-phenyl-1H-1,2,3-triazol-1-yl)-methyl)-propane-1,3-diol[8b]

HO, N_3, N_3, HO + H—≡—Ph; copper metal, H_2O/*t*-BuOH, rt, 24 h → bis-triazole product (98%)

Phenylacetylene (2.04 g, 20 mmol) and 2,2-*bis*(azidomethyl)-propane-1,3-diol (1.86 g, 10 mmol) were dissolved in a 1:2 *t*-butyl alcohol/water mixture (50 mL). About 1 g of copper metal turnings was added, and the reaction mixture was stirred for 24 h; after which time, TLC analysis indicated complete consumption of starting materials. Copper was removed, and the white product was filtered off, washed with water, and dried to yield 3.85 g (98%) of pure *bis*-triazole product; mp 211–212 °C; ESIMS *m/z*: 391.2 $(M + H^+)$ 413.2 $(M + Na^+)$.

Stabilized Cu(I) in the form of its *N*-heterocyclic carbene (NHC) complex, *e.g.*, (SIMes)CuBr (SIMes = *N,N'*-*bis*(2,4,6-trimethylphenyl)-4,5-dihydroimidazol-2-ylidene), and the cyclohexyl analog $[(ICy)_2Cu]PF_6$, catalyzes click reactions very well in aqueous *t*-butanol, and even better in water alone.[12] Low conversions were noted in nonaqueous solvents such as tetrahydrofuran (THF), *t*-BuOH, and dichloromethane (DCM). Starting from an alkyl bromide, triazoles could be smoothly generated by *in situ* conversion to the corresponding azide (aqueous NaN_3) followed by copper-catalyzed cycloaddition. This is but one example of the potential for combining several steps in a single flask that culminates with a click reaction (*vide infra*). The alternative use of $CuBr(Ph_3P)_3$ (0.5 mol%) in these 3-component couplings with NaN_3 (1.3 equiv) at room temperature is also best carried out in water as solvent.[13]

NaN_3 (1.05 equiv), 5% (SIMes)CuBr, water, rt, 30 min; (92%); SIMes

In general, as suggested by the above examples, outstanding compatibility exists among azides, 1-alkynes, and copper(I) along with the product triazoles with a vast array of other functional groups that may be present in either educt. Steric factors do exert an effect on rates. In most situations, the beneficial impact of a trialkylamine base (*e.g.*, *i*-Pr_2NEt and 1,4-diazabicyclo[2.2.2]octane [DABCO]) or others (*e.g.*, 1,8-diazabicyclo[5.4.0]undec-7-ene [DBU] and 2,6-lutidine) is used to great advantage, in particular when less soluble CuI in toluene (or even CH_3CN) is the catalyst. A recent study suggests that in the presence of catalytic amounts of HOAc, together with *i*-Pr_2NEt (1:1), rapid quenching of the intermediate copper species occurs leading to enhanced reaction rates.[14] The counterion in the Cu(I) salt (*e.g.*, Br, I, and OAc), or the Cu(II) precursor to catalytically active Cu(I) (*e.g.*, -OAc and -NO_3) can exert influence. Microwave assistance can reduce reaction times from hours to minutes, although yields are mostly unaffected by this mainly thermal phenomenon. Tandem events in one-pot include initial azide formation by halide substitution resulting in a net three-component coupling all under microwave irradiation (procedure below).

SO_2N_3; **1**; 10% CuI, 2,6-lutidine, $CHCl_3$, 0 °C, 12 h; **2** (90%)

Remarkable is the *in situ* conversion of amines into azides by Cu(II)-catalyzed diazo transfer in a mixed aqueous environment (using trifluoromethanesulfonylazide, TfN_3), followed by click cyclization.[15] Ligands such as *tris*(benzyltriazolylmethyl)amine (TBTA) and clickphine (below) have been used to accelerate these copper-catalyzed cycloadditions.[16] Recently, conditions have been found that lead to *N*-sulfonyl-4-substituted-1,2,3-triazoles using sulfonylazides (*e.g.*, **1** to **2**), avoiding products from competing ring-opening α-diazoimino tautomers.[17a] Reactions are best run in chloroform at 0 °C using catalytic CuI, giving yields in the moderate-to-high range. In a mixed aqueous solvent environment, or water alone, the corresponding *N*-tosylated amides are formed in good yields, rather than the corresponding triazoles.[17b] Sequential displacement of α-tosyloxy ketones by azide ion and subsequent one-pot cyclization in a polyethylene glycol (PEG)/water mixture at room temperature leads to carbonyl-containing triazole derivatives.[18]

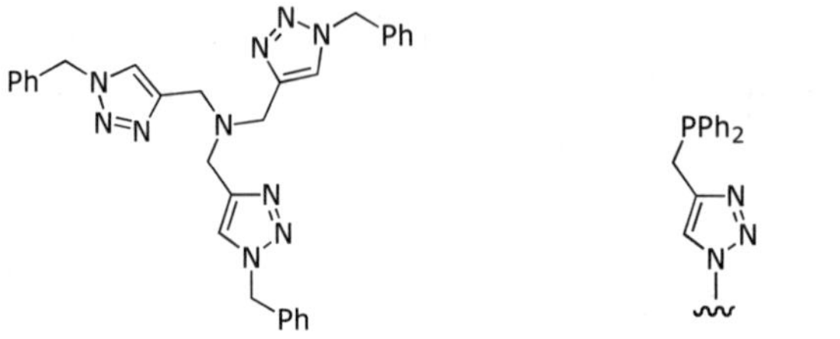

ligand TBTA clickphine ligand

Sequential one-pot process for diazo transfer and azide-alkyne cycloaddition using $CuSO_4$ and sodium ascorbate[15]

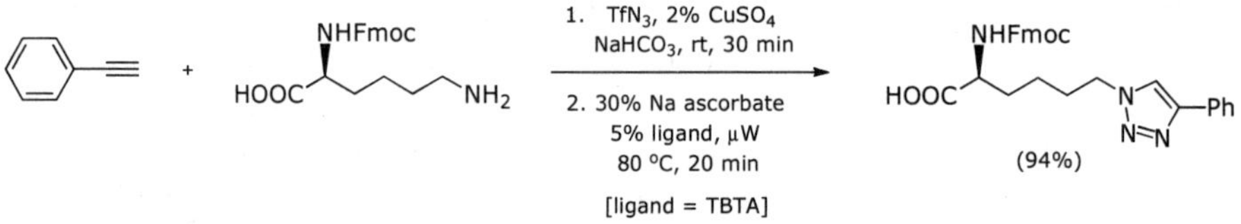

1. TfN_3, 2% $CuSO_4$, $NaHCO_3$, rt, 30 min
2. 30% Na ascorbate, 5% ligand, μW, 80 °C, 20 min

[ligand = TBTA]

(94%)

Triflyl azide (TfN_3) was freshly prepared prior to each reaction. NaN_3 (6 equiv per substrate amine) was dissolved in a minimum volume of water (solubility of NaN_3 in water is approximately 0.4 g/mL). At 0 °C, an equal volume of DCM was added and triflic anhydride (Tf_2O; 3 equiv) was added dropwise to the vigorously stirred solution. After stirring for 2 h at 0 °C, the aqueous phase was once extracted with DCM. The combined organic phases were washed with sat. $NaHCO_3$ solution and used without further purification. The amine, Fmoc-Lys-OH (81 mg, 0.22 mmol), $CuSO_4$ (2 mol%), and $NaHCO_3$ (1 equiv) were dissolved/suspended in water (equal volume relative to DCM used for TfN_3). The TfN_3 solution was added, followed by addition of methanol until the mixture became homogeneous. The reaction was stirred at RT (*ca.* 30 min) until TLC showed complete consumption of the amine. Phenylacetylene (24 μL, 0.22 mmol) was added, then ligand TBTA (5 mol%), and then sodium ascorbate (30 mol%), and the reaction was heated to 80 °C in the microwave until complete loss of azide (≤30 min). The reaction mixture was then diluted with water and the organics extracted. Solvents were removed under reduced pressure. After flash chromatography ($CHCl_3$/MeOH/AcOH 96:3:1), the product (103 mg, 94%) was isolated as a white powder. ESIMS (MeOH): *m/z* calcd. for 495.2 $[M-H]^-$; found 495.3.

General procedure for microwave-assisted, three-component coupling reactions. 4-Phenyl-1-(3,4,5-trimethoxybenzyl)-1H-1,2,3-triazole[19]

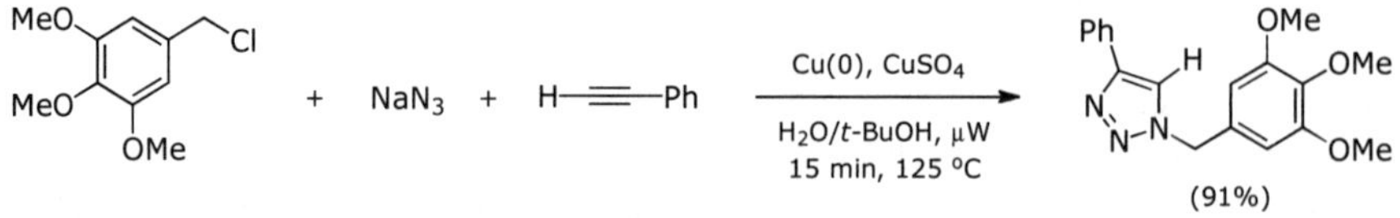

+ NaN_3 + H—≡—Ph; Cu(0), $CuSO_4$; H_2O/*t*-BuOH, μW, 15 min, 125 °C

(91%)

The benzylic halide (1.0 mmol), phenylacetylene (1.1 mmol), and sodium azide (1.1 mmol) were suspended in a 1:1 mixture of water and *t*-BuOH (1.5 mL each) in a 10-mL glass vial equipped with a small magnetic stirring bar. To this was added copper wire (50 mg) and copper sulfate solution (200 μL, 1 N), and the vial was tightly sealed with an aluminum/Teflon crimp top. The mixture was then irradiated using an irradiation power of 100 W (CEM Discover instrument; Matthews, NC). After completion of the reaction, the vial was cooled to 50 °C and then diluted with water (20 mL) and filtered.

The residue was washed with cold water (20 mL), 0.25-N HCl (10 mL), and petroleum ether (50 mL) to furnish the product triazole in 91% yield; EIMS [M^+]: 325 (100%).

Although aqueous *t*-butanol is commonly employed according to the original recipe, click cyclizations between 1-alkynes and aryl or aliphatic azides can be realized "on water" at room temperature.[20] Copper(I) bromide (5 mol%) in the presence of thioanisole (50 mol%) leads for most substrates within minutes (or hours with selected cases, such as with azidothymidine [AZT], below) to good isolated yields of triazoles. Good functional group tolerance is evident, especially given the "green" conditions used (*i.e.*, water as the medium, and no additional energy input).

General procedure for copper-catalyzed cycloadditions of aliphatic and aryl azides with alkynes to triazoles "on water"[20]

Boc-NH-CH(CH2Ph)-C(O)-NH-CH2-C≡CH + AZT → (cat. CuBr, PhSMe, H_2O, rt, 6 h) → triazole (88%)

Water (1-2 mL), alkyne (0.6 mmol), azide (0.5 mmol), CuBr (0.05 mmol, 7.5 mg), and thioanisole (0.25 mmol, 31 mg) were added to a flask with a stir bar, and the mixture was stirred at RT without exclusion of air. After total consumption of the starting azide (by TLC), the resulting solution was poured into a water/EtOAc mixture. After extraction of the aqueous phase with EtOAc, the combined organic layers were dried over anhydrous magnesium sulfate and then filtered. The solvent was removed by rotary evaporation, and the crude product was purified on a short silica gel column using EtOAc/petroleum ether (v/v, 10:1 to 1:1) as eluent to give the product triazole.

The potential for sequential copper-catalyzed processes can also be illustrated in the case of formation of fully substituted 1,2,3-triazoles.[21] In this sequence, the same copper catalyst is promoting two distinct types of catalysis: [3+2]-cycloaddition and arylation *via* "C-H activation." Each reaction type tolerates both electron-rich and -poor substrates, as well as steric hindrance, adding noteworthy breadth to this scheme. A 4-component sequence using NaN_3, rather than an alkyl azide, is shown below. The diamine DMEDA (*N,N'*-dimethylethylenediamine) is used to stabilize the copper catalyst.

Ph-C≡CH + $PhCH_2N_3$ → (1. cat. CuI, DMF, 60 °C; 2. I-C6H4-F, LiO-*t*-Bu, DMF, 140 °C) → 1-Bn-4-Ph-5-(4-fluorophenyl)-1,2,3-triazole (79%)

Representative procedure; synthesis of a 1,4,5-trisubstituted-triazole[21]

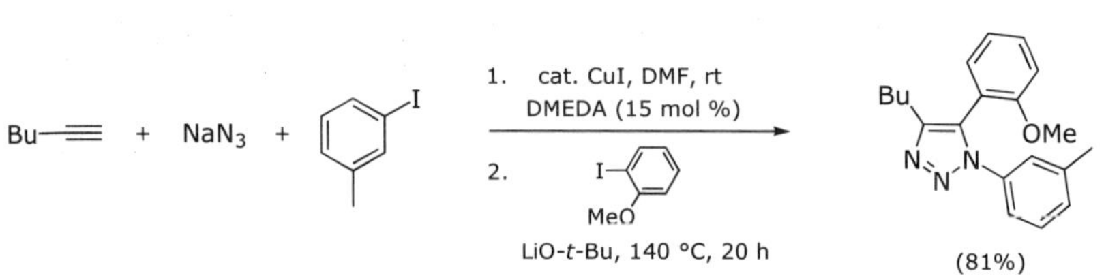

To a suspension of CuI (19 mg, 0.10 mmol, 10 mol%) and NaN_3 (69 mg, 1.05 mmol) in *N,N*-dimethylformamide (DMF) (3 mL) was added 1-hexyne (82 mg, 1.00 mmol), 3-iodotoluene (234 mg, 1.00 mmol), and *N,N'*-dimethylethylenediamine (13 mg, 0.15 mmol, 15 mol%), and the mixture was stirred under N_2 at RT for 2 h. Then, LiO-*t*-Bu (160 mg, 2.00 mmol), 2-iodoanisole (702 mg, 3.00 mmol), and DMF (2.0 mL) were added and the resulting suspension was stirred under N_2 at 140 °C for 20 h. Upon cooling to RT, Et_2O (50 mL) and H_2O (50 mL) were added, and the separated aqueous layer was extracted with Et_2O (2 × 75 mL). The combined organic layers were washed with sat. aq. NH_4Cl (50 mL), H_2O (50 mL), and brine (50 mL), and then they were dried over anhydrous Na_2SO_4. Concentration *in vacuo* gave a residue that was purified by column chromatography on silica gel (*n*-hexane/EtOAc 12/1 to 10/1) to yield the product as an off-white solid (260 mg, 81%).

Several heterogeneous catalysts have been shown to effect related multicomponent couplings. These include cross-linked polymeric ionic liquid material-supported copper (Cu-CPSIL), silica-dispersed CuO (CuO/SiO_2), and imidazolium-loaded Merrifield resin-supported copper (Cu-PSIL),[22] all of which can be used in water at room temperature to arrive at 1,4-disubstituted-1,2,3-triazoles from alkyl halides, NaN_3, and terminal alkynes. Each can be filtered and reused several times with minimal loss of efficacy. Multistep flow synthesis,[23] specifically including generation of underused vinyl azides and their subsequent click conversions to vinyl triazoles, has also been reported.[24]

Access to regio-defined trisubstituted triazoles (*i.e.*, with substituents at the 5-position), using an acetylenic halide (in particular, bromide), has been reported.[25] The resulting derivative allows for subsequent manipulation, including metal-halogen exchange, assuming compatibility within the remaining portion(s) of the educt. Mechanistically, it remains unclear how the sequence proceeds, but in light of the exclusive isomer formed, a copper acetylide intermediate is likely.

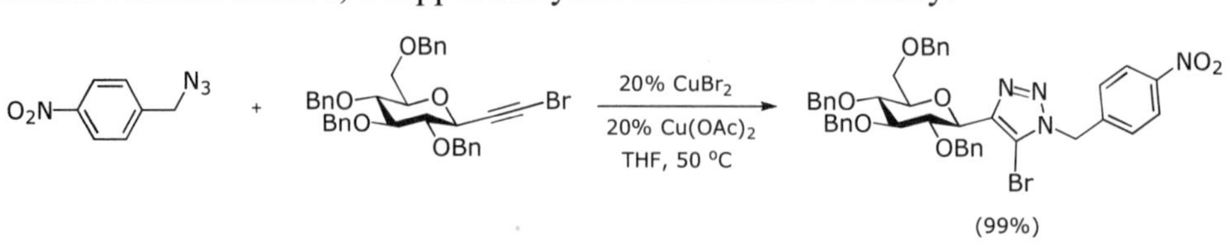

N-Unsubstituted 1,2,3-triazoles are also readily prepared using trimethylsilylazide (TMS-N_3) as a precursor to *in situ* generated HN_3 in the presence of methanol (or water).[26] Using CuI and nonactivated terminal alkynes, high yields of 4-substituted products result, as shown below. Cu powder is also useful for these cycloadditions, while other salts with group 11 metals (*e.g.*, $AuCl_3$ and AgCl) were totally ineffective, as was $ZnCl_2$. Noteworthy is that bulky acetylenes (*e.g.*, TIPS-acetylene) and conjugated networks (*e.g.*, an enyne) clicked under standard conditions (DMF/MeOH, 9:1, 0.5 M, 100 °C, 10-24 h).

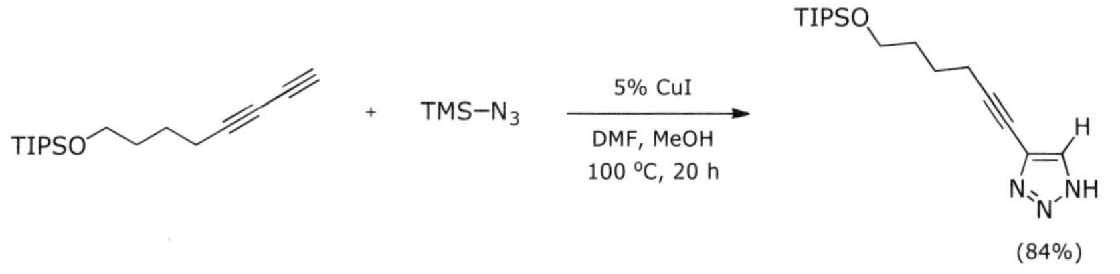

TIPSO–(CH₂)₄–C≡C–C≡CH + TMS–N_3 → (5% CuI; DMF, MeOH; 100 °C, 20 h) → TIPSO–(CH₂)₄–C≡C–triazole (N-H) (84%)

Representative procedure for the synthesis of N-unsubstituted 1,2,3-triazoles[26]

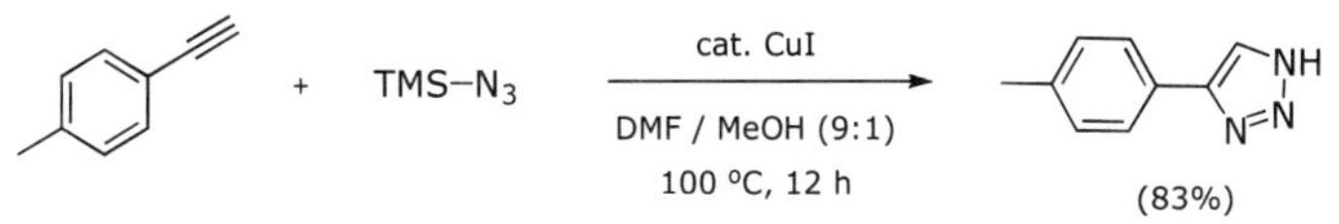

p-tolylacetylene + TMS–N_3 → (cat. CuI; DMF / MeOH (9:1); 100 °C, 12 h) → 4-(p-tolyl)-1H-1,2,3-triazole (83%)

Trimethylsilylazide (0.1 mL, 0.75 mmol) was added to a DMF and MeOH solution (1 mL, 9:1) of CuI (4.8 mg, 0.025 mmol) and *p*-tolylacetylene (58 mg, 0.5 mmol) under Ar in a pressure vial. The reaction mixture was stirred at 100 °C for 12 h in a tightly capped 5-mL microvial. After completion, the mixture was cooled to RT and filtered through a short pad of Florisil and concentrated *in vacuo*. The residue was purified by silica gel column chromatography (*n*-hexane/EtOAc 10:1 to 2:1) to afford 66 mg of 4-(*p*-tolyl)-1H-1,2,3-triazole (83%).

N-Allylated 1,2,3-triazoles, where the allyl moiety is positioned at either *N*-1 or *N*-2, can be generated regiospecifically by virtue of a Pd-Cu bimetallic-catalyzed, three-component coupling.[27] Terminal alkynes and *in situ* formed HN_3 (see procedure above) undergo Huisgen cycloaddition, as expected when Cu(I) is present. However, with both catalytic Pd(0) and an allylic carbamate in the pot, *N*-deallylation/re-allylation can afford either regioisomer (Scheme 1-1). The combination of $Pd_2(dba)_3$•$CHCl_3$/$CuCl(PPh_3)$/$P(OPh)_3$ leads exclusively to 2-allylated products, while switching the copper/phosphine source to $CuBr_2$/PPh_3 inverts the product ratio entirely favoring *N*-1 allylated, 1,4-disubstituted triazoles. The presence of $P(OPh)_3$ in the former cyclization is crucial for regiocontrol. Nonpolar solvents decreased yields, while more polar media gave mixtures. For the latter, the absence of phosphine led to no reaction (recovery of alkyne); other sources of copper (*e.g.*, CuBr and $CuCl_2$) produced a mixture of triazoles.

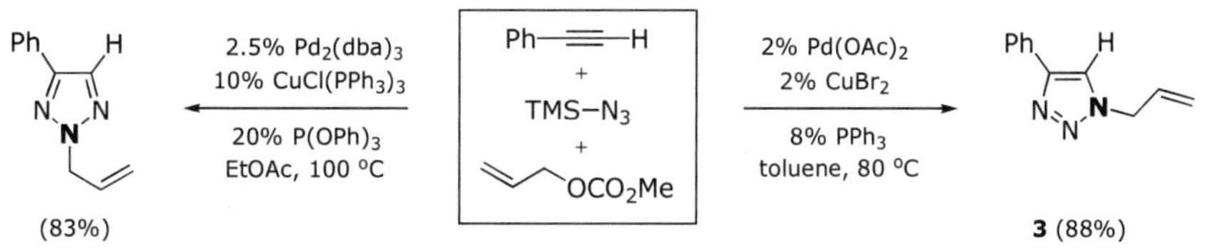

Ph–C≡C–H + TMS–N_3 + allyl–OCO_2Me

← (2.5% $Pd_2(dba)_3$; 10% $CuCl(PPh_3)_3$; 20% $P(OPh)_3$; EtOAc, 100 °C): 2-allyl-4-phenyl-2H-1,2,3-triazole (83%)

→ (2% $Pd(OAc)_2$; 2% $CuBr_2$; 8% PPh_3; toluene, 80 °C): 1-allyl-4-phenyl-1H-1,2,3-triazole **3** (88%)

Scheme 1-1

Representative procedure for the synthesis of 1-allyltriazoles (**3**)[27]

To a toluene solution (1 mL) of $Pd(OAc)_2$ (2.3 mg. 0.01 mmol), PPh_3 (10.5 mg, 0.04 mmol), and $CuBr_2$ (2.2 mg, 0.01 mmol) were added phenylacetylene (55 μL, 0.5 mmol), allylmethylcarbonate (68 μL, 0.6 mmol), and $TMSN_3$ (80 μL, 0.6 mmol) under an Ar atmosphere. The reaction mixture was stirred at 80 °C for 3 h in a tightly capped 5 mL microvial. After completion, the mixture was cooled to RT and filtered through a short pad of Florisil (U.S. Silica Co., Frederick, MD) with Et_2O (~100 mL) and concentrated *in vacuo*. The residue was purified by silica gel column chromatography (*n*-hexane/EtOAc 20:1 to 2:1) to afford 81.3 mg of 1-allyl-4-phenyl-1H-[1,2,3]-triazole (88%).

Applications of click chemistry are already far too numerous to acknowledge by even offering an example from each area herein. Some of these areas include dendrimers, polypeptides, polymeric materials, conformationally restricted macrocycles, new ligand designs, and carbohydrates. One representative procedure is provided below. A related study describes sugar-based silica gels for use as hydrophilic interaction chromatography (HILIC) for separation of monosaccharides.[28]

Heterogeneous polysaccharide click chemistry[29]

Reactions were performed in dried glass tubes sealed with plugs that contain an activated drying agent. 5-Hexynoic acid (3.4 mmol) and tartaric acid (0.17 mmol) were mixed in glass vials, and known amounts of paper samples (17 mg) were introduced into the mixture. The vials were sealed with screw caps, and the reactions were run for 6 h at 110 °C. After cooling, the filter papers were Soxhlet extracted using DCM and water, respectively. The samples were dried prior to further analysis. The cellulose paper was added to Wang's probe **4** (25 mg, 0.1 mmol) in water/methanol/EtOAc (1:1:1, v/v, 5 mL). Next, a freshly prepared solution of sodium ascorbate (20 μL, 0.02 mmol, 1 M) in water and a 7.5% solution of copper(II) sulfate pentahydrate in water (17 μL, 0.005 mmol) were added. The heterogeneous mixture was stirred vigorously overnight in the dark at RT. Finally, the cellulose paper was extensively washed with cold water and Soxhlet extracted using DCM and water. The synthesized probes were analyzed using ^1H and ^{13}C nuclear magnetic resonance (NMR) spectroscopy. The cellulose samples were analyzed directly (Perkin-Elmer Spectrum One FT-IR; Waltham, MA). The fluorescence of the derivatized cellulose samples was analyzed using a Leica (Allendale, NJ) fluorescence microscope with an excitation/emission filter cub (filter cub A): excitation (λ_{ex}) =340–380 nm and emission (λ_{em}) = >425 nm.

Solid-phase click chemistry in organic solvents (*e.g.*, DMF and THF) works well, seemingly independent of the support. Resins such as polystyrene and polyethylene glycol (PEGA) attest to tolerance to both nonpolar and polar types. As outlined in the original Meldal *et al.* papers,[3] novel triazole-containing polypeptides are realizable based on heterogeneous 4-hydroxymethylbenzoic acid (HMBA)-$PEGA_{800}$–bound acetylene substrates that "click" in THF with excess CuI/Hünig's base at room temperature.[3] Conversions were high (>95%), and all amino acids examined participated offering considerable prospects for synthesis of large libraries of novel products.

Solid phase click chemistry. General procedure to peptidotriazoles[3]

CuI, DIPEA
THF, rt, 16 h

NaOH
H_2O

(>95%)

1-(2-Deoxy-1-thiophenyl-α-D-galactopyranos-2-yl)-1*H*-[1,2,3]-triazol-4-ylcarbonyl-Phe-Gly-Phe-Gly-OH. DIPEA (50 equiv), CuI (2 equiv), and phenyl-3,4,6-tri-*O*-acetyl-2-azide-2-deoxy-1-thio-α-D-galacto-pyranoside (2 equiv) were added to resin-bound alkyne (5 mg of resin, *ca.* 2 µmol swollen in 200 µL of THF) and reacted for 16 h at 25 °C. The resin was then washed with THF, water, and THF again. A sample was cleaved using aqueous NaOH, and the product was analyzed by high-profile liquid chromatography (HPLC) and mass spectrometry (MS). Conversion was >95% with 75–99% purity. ESIMS: *m/z* calcd. for (MH^+) $C_{37}H_{42}N_7O_{10}S^+$: 776.3; found: 776.8.

Insofar as heterogeneous Cu(I)-catalyzed click chemistry is concerned, new catalysts continue to appear. Nanoparticles of copper can be generated and used in aqueous solution, from which crystalline products often precipitate. Inclusion of copper (as CuCl) into Lewis acidic zeolites, using in particular "Cu(I)-USY" (pore size 6–8 Å), is one such catalyst employed in toluene.[30] Other solvents such as DCM, CH_3OH, CH_3CN, and benzene were not recommended, and yields of 1,4-disubstituted triazoles can be highly variable.

Nanoparticles of Cu(0) powder that undergo oxidation in the presence of an amine hydrochloride lead to active Cu(I).[31] The amine is presumed to play several roles: reduction of Cu(II) to Cu(I); stabilization of Cu(I) as ligand; and enhancement of solubility of copper in organic media. Nanosize clusters also catalyze click reactions, in this case in H_2O/*t*-BuOH at 25 °C without salts.[32] A simplified route to copper nanoparticles (CuNPs) relies on $CuCl_2$, lithium metal, and catalytic amounts of 4,4'-di-*t*-butylbiphenyl (DTBB) in THF at ambient temperatures.[33] They exist as a range of nanospheres mainly between 1 and 6 nm, as analyzed by transmission electron microscopy (TEM). Terminal alkynes and a variety of azides can be cyclized in THF (88–98% isolated yields) between room and refluxing temperatures. Cycloadditions take place within 10–30 minutes, and simple filtration suffices to remove the catalyst. Recycling of CuNPs, however, is not an option.

$$EtO_2CCH_2N_3 + Ph\text{—}\equiv \xrightarrow[\text{THF, Et}_3\text{N, }\Delta\text{, 30 min}]{\text{10\% CuNPs}} \text{1-(CH}_2\text{CO}_2\text{Et)-4-Ph-1,2,3-triazole (88\%)}$$

An extensive study has been made on the catalyst "copper-in-charcoal" (Cu/C; Sigma-Aldrich #70910-7), where $Cu(NO_3)_2$ has been impregnated into the pores of commercially available activated charcoal.[34] The solid support relies on readily available wood (and not coconut) charcoal (Aldrich, catalog #242276) of 100 mesh size. Larger size particles with less surface area do not lead to active catalyst, while finer grades (*i.e.*, higher mesh sizes) are too difficult to handle to be practical. Only $Cu(NO_3)_2$ can be used as the source of copper, where mounting, based on literature precedent (albeit at much higher temperatures), reportedly relies on loss of oxides of nitrogen $[(NO)_x]$ resulting in both Cu_2O and CuO within the charcoal matrix. Other copper salts possessing alternative counterions (halides, acetate, etc.) are apparently readily washed out during processing to Cu/C. Nanoparticle distribution of copper aggregates are formed by simply mixing $Cu(NO_3)_2$ and charcoal in water, followed by ultrasonication, distillation, and finally drying. The free-flowing charcoal, in general, is best stored away from light, air, and moisture, although its use as a catalyst specifically for click reactions requires no such precautions; *i.e.*, it can be placed in a container (vial, bottle, etc.) on the shelf. More recently, a streamlined procedure has been developed that eliminates the distillation step. Thus, by mixing $Cu(NO_3)_2$ + charcoal in water, ultrasonication of the mixture in a standard bath overnight, and filtration, the resulting (wet, or dried) Cu/C is active. The modified procedure for the preparation of Cu/C is as follows.

Simplified preparation of Cu/C[34]

$$Cu(NO_3)_2{\bullet}3H_2O + \text{charcoal} \xrightarrow[\text{2. drying}]{\text{1. H}_2\text{O, ultrasound}} \text{"Cu/C"}$$

Darco KB-activated carbon (50.0 g, 100 mesh, 25% H_2O content) was added to a 500-mL round-bottom flask containing a stir bar. A solution of $Cu(NO_3)_2{\bullet}3H_2O$ (Acros Organics [Geel, Belgium], 11.114 g, 46.0 mmol) in deionized H_2O (100 mL) was added to the flask and deionized H_2O (100 mL) was further added to wash down the sides of the flask. The flask was loosely capped and stirred in air for 30 min, and then submerged in an ultrasonic bath for 7 h. Subsequent filtration and washing (H_2O, then toluene) followed by air drying (3 h) by vacuum suction yielded *ca.* 85 g of "wet" Cu/C. The catalyst can be used at this stage, or further dried *in vacuo* at 120 °C overnight, to yield 44 g of "dry" Cu/C.

Click reactions mediated by Cu/C require no special precautions with respect to handling; reaction partners can be weighed out in air, as can the catalyst. Without additives of any kind (*e.g.*, reducing agents, base, ligands, etc.), mild heating to 60 °C is enough to complete the conversion into triazoles. Significant rate enhancements are to be expected when Et_3N is present (only 1 equiv), used "out of the bottle," and most reactions with unhindered educts take place at ambient temperatures. Solvents ranging from nonpolar (*e.g.*, toluene) to 100% water give similar results. Most examples to date, however, have been run in undistilled dioxane (although toluene is, in fact, the preferred solvent), open to air. Microwave heating to 150 °C reduces the time requirements for typical cycloadditions in the absence of base from a few hours (at 60 °C) to <5 minutes with no loss in regioselectivity (*i.e.*, only 1,4-triazoles are formed). Higher-than-average molecular weight products can be made using this catalyst (*e.g.*, **7**). Yields in almost all cases reported exceed 90%. Cu/C can be recovered by filtration and recycled several times without erosion of efficacy. Control experiments and ICP-AES (inductively coupled plasma-atomic emission spectroscopy) suggest that little bleeding of copper into solution is occurring.

Procedure for Cu/C-catalyzed "click" reaction[34]

6

Cu/C, TEA, dioxane
60 °C, 30 min

5

7; MW = 1029 (87%)

Cu/C (50 mg, 1.01 mmol/g, *ca.* 0.05 mmol) is added to a clean 10-mL flask fitted with a stir bar and septum. Dioxane (2 mL) is added slowly to the sidewalls of the flask, rinsing the catalyst down. While the heterogeneous mixture is stirred, triethylamine (0.153 mL, 1.1 mmol), azide **5** (0.377 g, 1.0 mmol), and alkyne **6** (0.783 g, 1.2 mmol) are added. The flask is stirred at 60 °C, and the reaction progress is monitored by TLC until complete consumption of azide has occurred. The mixture is filtered through a pad of Celite to remove the catalyst, and the filter cake is further washed with EtOAc to ensure complete transfer. The volatiles are removed *in vacuo* to give the crude triazole, which was further purified *via* flash chromatography with 5:1 hexanes/EtOAc as eluent yielding 0.891 g of colorless oil (87%).

Yet here is another approach that, in this case, avoids solvent altogether, and leads to rapid and ligand-free cycloaddition reactions run in a planetary ball mill (*e.g.*, see http://www.fritsch-milling.com/products/milling/planetary-mills/pulverisette-7-classic-line/). High conversions and selectivities are to be expected using $Cu(OAc)_2$ in catalytic amounts (5 mol%), with reaction times on the order of only 10 minutes.[35]

Bottom-line comments. Copper-catalyzed click chemistry is experimentally too easy and works too well in most cases not to be strongly considered as a means of "stitching" together just about any terminal alkyne and azide. The regiospecificity for 1,4-disubstituted heteroaromatic triazoles is complemented by the corresponding ruthenium-catalyzed cycloadditions to afford 1,5-adducts. And with options for heterogeneous processes, microwave-assisted thermal rate enhancements, and solvent-free conditions, click chemistry has become, and rightfully so, a tremendously powerful tool in synthesis.

3. Cu-Catalyzed Aminations and Amidations

Reviews. Zhang, M. *Synthesis* **2011**, 3408; Sadig, J. E. R.; Willis, M. C. *Synthesis* **2011**, 1; Rao, H; Fu, H. *Synlett* **2011**, 745; Das, P.; Sharma, D.; Kumar, M.; Singh, B. *Curr. Org. Chem.* **2010**, *14*, 754; Monnier, F.; Taillefer, M. *Angew. Chem., Int. Ed.* **2009**, *48*, 6954; Evano, G.; Blanchard, N.; Toumi, M. *Chem. Rev.* **2008**, *108*, 3054; Ma, D.; Cai, Q. *Acc. Chem. Res.* **2008**, *41*, 1450; Beletskaya, I. P.; Cheprakov, A. V. *Coord. Chem. Rev.* **2004**, *248*, 2337.

3.1 Aminations

Modern amination reactions of both aromatic and heteroaromatic rings have become some of the most valuable tools in synthesis, especially to the medicinal chemist. Problems associated with traditional Ullmann-like couplings to form di- and tri-arylamines, such as stoichiometric levels of copper in solution, have been largely solved using more recent transition-metal–based catalysis. While palladium-catalyzed processes offer considerable promise and have been enthusiastically embraced by industry,[36] extensive efforts to update copper as an inexpensive, viable alternative have, indeed, led to significant advances. And as with palladium catalysis, much of the story is focused on associated ligands,[37] although the nature of the base, and/or the solvent, surely plays an important role. Many combinations of substrate/amine are possible; that is, educts can be aryl or heteroaryl halides or pseudohalides, while amines can be 1° or 2°, aryl, diaryl, alkyl, dialkyl, heteroaromatic, etc. In addition, derivatives of amines (*e.g.*, amides) can also serve as nucleophilic partners. Akin to developments with palladium catalysis, studies on aryl halides as substrates began with iodides and have more recently concentrated on bromides. Some preliminary work on aryl chlorides using catalytic Cu_2O/NaO-*t*-Bu in hot *N*-methyl-2-pyrrolidone (NMP) have led to reasonable yields of aniline products.[38] However, no general methodology yet exists for such aminations of aryl chlorides based on Cu(I), although this is likely to appear in the not-too-distant future. Preliminary work employing Cu nanoparticles that catalyze cross-couplings of activated chlorides with imidazoles looks especially promising.[39] Studies on the origin of copper(I) catalysis initiating from copper(II) precursors, using ultraviolet (UV)-vis spectroscopy and ^1H NMR analyses on phenanthroline complexes, have shown that reduction occurs *in situ*, with both the amine and the base being required for the proposed β-hydride elimination sequence.[40]

Table 1-1 lists many of the more successful combinations involving aryl iodides or bromides, highlighting the variety of both (substituted) amines and recommended ligands. Some trends are clear: Usually, highly polar solvents are involved together with carbonate or phosphate bases. Ligands are both noticeably variable (*e.g.*, among amino acids) and often structurally unrelated, and they can be even quite large and/or complex (*e.g.*, dendrimers). Aryl iodides are highly prone toward substitution by nitrogen, whether primary, secondary, or aryl. Mild heating, from 40 °C to 80 °C, is often used depending on substituents in the halide, while anilines take somewhat more vigorous conditions. CuI is commonly used as catalyst, as in the examples below. The best ratios of ligand to Cu tend to be 2:1 or greater.

Table 1-1. Aminations/Amidations of aryl halides.

***N*-Nucleophile**	**Base/Solvent**	**Ligands**	**References**
Anilines	K_2CO_3 / DMSO	L-proline	[41]
	K_3PO_4 / DMF	DPP	[42]
	Cs_2CO_3 / toluene	PPh_3	[43]
	KO-*t*-Bu / toluene	2,6-diphenylpyridine	[44, 45]
	Cs_2CO_3/DMF	2,2'-bijmidazole	[46]
	Cs_2CO_3/DMSO	8-hydroxy quino, lin-N-oxide	[47]
	Cs_2CO_3/DMF	3-acetylcoumarin	[48]
	K_2CO_3 / NMP	2-[dimethylamino]methyl]benzene thiol	[49]
1° amines	K_2CO_3 / DMSO	L-proline	[41]
	K_3PO_4 / DMF	DPP	[42]
	Cs_2CO_3 / DMF	2-isobutyroylcyclohexanone	[50]
	K_3PO_4 / H_2O	deanol	[51]
	Cs_2CO_3 / DMSO	*rac*-BINOL	[52]
	K_2CO_3 / DMSO	proline	[53]
	KO-*t*-Bu / -	phosphinidenecyclobutene	[54]
	K_3PO_4 / *i* - PrOH	ethylene glycol	[55]
	H_2O, or solvent free	-	[56]
2° amines	K_2CO_3 / DMSO	L-proline	[41]
	K_3PO_4 / DMF	DPP	[42]
	Cs_2CO_3 / DMSO	*rac*-BINOL	[52]
	K_2CO_3 / DMSO	proline	[53]
	KO-*t*-Bu / -	phosphinidenecyclobutene	[54]
	K_3PO_4 / *i*-PrOH	ethylene glycol	[55]

amides	K_3PO_4 / DMF, dioxane	*trans*-1,2-diaminocyclohexanes	[57–59]
	KF, Al_2O_3 / toluene	1,10-phenanthroline	[60]
	Cs_2CO_3 / CH_3CN	Chxn-Py-Al	[61]
	K_3PO_4 / toluene	1,10-phenanthroline	[62]
indoles	K_2CO_3 / DMSO	L-proline, *N,N*-dimethylglycine	[63]
amino acids	K_3PO_4 / H_2O	dcanol	[64]
amino alcohols	NaOH / DMSO-H_2O	-	[65]
hydrazides	Cs_2CO_3 / DMF	1,10-phenanthroline	[66]
imidazole	K_2CO_3 / DMSO	L-proline	[63]
	Cs_2CO_3 / CH_3CN	Chxn-Py-Al	[61]
pyrroles	K_2CO_3, Cs_2CO_3 / DMSO	N,N-dimethylglycine	[63]
pyrazoles	K_2CO_3 / DMSO	L-proline, *N,N*-dimethylglycine	[63]
	Cs_2CO_3 / DMSO	*rac*-BINOL	[52]
	K_2CO_3 / DMSO	proline	[53]
carbazoles	Cs_2CO_3 / DMSO	L-proline	[63]

General procedure for the coupling reaction of aryl iodides with amines catalyzed by CuI and L-proline[41]

(R)-aryl–I + HNR_1R_2 → (R)-aryl–NR_1R_2 (20% CuI, K_2CO_3; 40% L-proline; DMSO, 40-90 °C)

A mixture of aryl iodide (2 mmol), amine (3 mmol), K_2CO_3 (4 mmol), CuI (0.2 mmol), and L-proline (0.4 mmol) in 4 mL of dimethylsulfoxide (DMSO) was heated (40–90 °C, depending on partners). The cooled mixture was partitioned between water and EtOAc. The organic layer was separated, and the aqueous layer was extracted with EtOAc. The combined organic layers were washed with brine, dried over anhydrous Na_2SO_4, and concentrated *in vacuo*. The residual oil was loaded on a silica gel column and eluted with 1:10 to 1:8 EtOAc/petroleum ether to afford the desired product.

While the example above relies on L-proline as ligand, switching to 1,3-diketone **8** has been found to allow for CuI-catalyzed aminations of iodides at ambient temperatures.[50, 67] The mild conditions are unusual and noteworthy, and although 20 mol% **8** is involved, several functional groups are tolerated (*e.g.*, COOH, cyclopropyl, Br, and OH). Intramolecular aminations are also possible at room temperature in minutes under these conditions even with a bromide (Eqn. 1-1). In general, however, aryl bromides require heating to 90 °C to effect aminations with 1° or 2° amines.

NH_2 Br
5% CuI, 20% **8**
2 Cs_2CO_3, DMF
rt, 30 min
(no ligand: 3 h)
N H
(90%)
(Eqn. 1-1)

Room temperature coupling of aryl iodides with amines. (3,4-Difluorophenyl)-[2-(1-methyl-pyrrolidin-2-yl)-ethyl]amine[50]

F F I + H_2N N
5% CuI, 20% **8**
Cs_2CO_3, DMF, rt
[**8** = O O]
F F H N N
(98%)

An oven-dried Schlenk tube equipped with a Teflon valve was charged with a magnetic stir bar, CuI (10 mg, 0.05 mmol, 5 mol%), Cs_2CO_3 (650 mg, 2 mmol), and any remaining solids (amine and/or aryl halide). The tube was evacuated and backfilled with argon (repeated 3 times). Under argon, the amine (2-(1-methyl-pyrrolidin-2-yl)-ethylamine; 192 mg, 1.5 mmol), aryl halide (1,2-difluoro-4-iodobenzene; 240 mg, 1.0 mmol), and DMF (0.5 mL) were added by syringe. Finally, ligand **8** (34 mg, 0.2 mmol, 20 mol%) was added *via* syringe, the tube was sealed, and the mixture was allowed to stir under argon at RT for 1 h. Upon completion of the reaction, the mixture was diluted with EtOAc and passed through a fritted glass filter to remove the inorganic salts, and the solvent was removed with the aid of a rotary evaporator. The residue was purified by column chromatography on silica gel using 50:1 DCM (saturated with NH_3)/MeOH to afford a yellow oil; yield: 235 mg (98%).

Several excellent procedures now exist for animations of aryl bromides. CuI, again, is the favored copper salt, as CuBr, $CuCl_2$, and $CuSO_4$ all gave inferior results. Using racemic diphenylpyrrolidine-2-phosphonate (DPP) (**9**) as ligand in DMF (90–110 °C), yields are somewhat higher than those obtained in dioxane or toluene. A small percentage (*ca.* 2%) of water relative to DMF (v/v) also enhances product yields. Unfortunately, DPP is not commercially available; it is made from l-pyrroline trimer and diphenylphosphite.[68]

O $P(OPh)_2$ NH
9
⟹ N N N + O $HP(OPh)_2$

Representative procedure for aminations using DPP. 1-(4-(4-Bromophenylamino)-phenyl)ethanone[42]

10% CuI, 20% DPP
K_3PO_4, DMF, 100 °C
(73%)
DPP = $P(OPh)_2$•HCl

A flask was charged with CuI (40 mg, 0.2 mmol), diphenylpyrrolidine-2-phosphonate hydrochloride (136 mg, 0.4 mmol), and potassium phosphate (552 mg, 4 mmol), evacuated, and then backfilled with nitrogen at low temperature. Aryl halide (3 mmol), amine (3 mmol), and DMF (3 mL, containing 2% H_2O [v/v]) were added to the flask under nitrogen. The flask was immersed in an oil bath, and the reaction mixture was stirred at 100 °C for 24 h. The reaction mixture was cooled to RT, EtOAc (10 mL) was added, the resulting suspension was filtered, the filtrate was concentrated, and the residue was purified by column chromatography on silica gel (hexanes/EtOAc 20:1-8:1) to provide the desired product as a white solid (73%); mp 114–116 °C, HREIMS *m/z* calcd. for $C_{14}H_{12}BrNO$: 289.0102, found: 289.0109.

Primary amines couple well with aryl bromides in the presence of 5–20 mol% of phenolic ligand **10** (next page), with CuI (5 mol%) in the pot.[69] The *ortho*-substitution pattern in *N,N*-diethylsalicylamide (**10**), a commercially available material (Aldrich #644234), is especially important: 2-picolinamide afforded greatly reduced yields. Bases such as K_2CO_3 and K_3PO_4 are equally effective, but DABCO and DBU are not recommended. DMF is the solvent of choice, as other media (including dioxane, toluene, Et_3N, and 1,2-dimethoxyethane [DME]) gave inferior yields in the model study conducted (Eqn. 1-2). A free NH_2 group in the aryl bromide is tolerated, and *N*-arylation is chemoselectively carried out in the presence of an unprotected alcohol (*e.g.*, 4-aminobutanol). Other compatible functionality includes nitro, nitrile, ketone, and thioether. These aminations can also be effected under solvent-free conditions at 100 °C in good yields, where the reactants are absorbed onto the solids present in the mixture. Secondary amines, however, are not amenable to this methodology.

5% CuI, 20% **10**
2 K_3PO_4, DMF
90 °C, 20 h
(91%)
(Eqn. 1-2)

Cu-Catalyzed amination of functionalized aryl bromides using ligand **10**. *3,4(Methylenedioxy)-N-furfurylaniline*[69]

5% CuI, 20% **10**

K_3PO_4, DMF, 90 °C

10 =

(87%)

CuI (10 mg, 0.05 mmol), *N,N*-diethylsalicylamide (39 mg, 0.20 mmol), and K_3PO_4 (425 mg, 2.0 mmol) were added to a screw-capped test tube with a Teflon-lined septum. The tube was then evacuated and backfilled with argon (three cycles). Aryl bromide (4-bromo-1,2-(methylenedioxy)benzene; 120 μL, 1.0 mmol), furfurylamine (132 μL, 1.5 mmol), and DMF (0.5 mL) were added by syringe at RT. The reaction mixture was stirred at 90 °C for 22 h and then allowed to cool to RT. EtOAc (~2 mL), water (~10 mL), ammonium hydroxide (~0.5 mL), and dodecane (227 μL, GC standard) were added. The organic phase was analyzed by gas chromatography (GC) or GC-MS. The reaction mixture was further extracted with EtOAc (4 × 10 mL). The combined organic phases were washed with brine and dried over anhydrous Na_2SO_4. Solvent was removed *in vacuo*, and the residue was purified by flash column chromatography on silica gel (hexane/EtOAc 8:1) to afford the desired product as a colorless oil (187 mg, 87%); R_f= 0.5 (hexane/EtOAc 5:1); EIMS *m/z* (relative intensity) 217 (M^+, 55), 136 (50), 81 (100).

A recent finding indicates that aminations can be run at room temperature in the presence of organic ionic bases, in particular, tetraalkylphosphonium salts (**11**).[70] The recipe calls for 10 mol% of CuI and 20 mol% of *N,N*-dimethylglycine as ligand, together in DMSO as solvent. The favored base in couplings with aryl bromides is tetrabutylphosphonium malonate (TBPM; shown below). Using NH_3 in dioxane, the same approach leads directly to primary amines, yet another welcomed addition to the many alternatives available (*vide infra*). The difference between bases within the phosphonium, as well as the corresponding tetraalkylammonium series, is attributed to their ionization abilities, supported by electrical conductivity measurements (in DMF).

cat. CuI, TBPM

cat. *N,N*-dimethylglycine

DMSO, rt

(70%) (96%) (82%) (95%)

Organic Bases = $[R_4P]^{\oplus} A^{\ominus}$

TBPM = $Bu_4P^{\oplus}$malonate$^{\ominus}$

11; R = Me, *n*-Bu

$A^{\ominus}$ = OAc, adipate, malonate

Representative procedure for amination at RT[70]

A mixture of CuI (9.5 mg, 0.05 mmol, 10 mol%), *N,N*-dimethylglycine (10.3 mg, 0.1 mmol, 20 mol%), TBPM (465 mg, 0.75 mmol), and the (solid) aryl bromide were added to a vacuum tube filled with Ar. The tube was evacuated and backfilled with argon (repeated 3 times). Under argon, the amine, aryl bromide (if liquid), and DMSO (0.5 mL) were added by syringe. The tube was sealed, and the mixture was allowed to stir under argon at ambient temperature (25 ± 1 °C) for 24 h. Upon completion of the reaction, the mixture was diluted with EtOAc. The solvent was removed *in vacuo*. The residue was purified by column chromatography on silica gel, and the product was dried under high vacuum for at least 0.5 h before it was weighed and characterized by NMR spectroscopy.

Effective use of ligands can impart chemoselective amination over *O*-arylation in aminoalcohol couplings with aryl iodides. Thus, diketone ligand **8** is particularly useful in the presence of CuI (5%) for directing *N*-arylation.[71] While a 1,2-relationship in the nucleophile can be problematic as a result of likely 5-membered ring chelation, more distal positioning affords highly favored *N*-substituted products (usually 18 to >50:1 over *O*-substitution). Yields tend be very good, and early indications suggest good functional group tolerance (*e.g.*, ketals, 2° amides, and 3° basic nitrogen). Anilines, however, are problematic and require palladium catalysis to form diarylamines. Switching to a phenanthroline ligand inverts the coupling, giving aryl ethers with equally impressive selectivities and efficiencies.

5% CuI, 20% **8**
2 Cs_2CO_3, DMF, rt
83% (18:1 *N*- *vs.* *O*-)

Aminations of pyridyl systems, as well as related heteroaromatics (*e.g.*, bromopyrimidines), have been achieved in moderate-to-good yields under ligandless conditions using Cu powder in the presence of CsOAc in DMSO.[72] Both iodides and bromides serve equally well; both primary and secondary amines can be used; and the procedure is insensitive to both air and moisture.

Cu (10 mol %), CsOAc (2 equiv)
DMSO, 90 °C, 24 h
(82%)

General procedure for C-N bond formation. Synthesis of N-benzylpyridin-3-amine[72]

After cooling of an oven-dried tube to room temperature under Ar, it was charged with copper powder (3.3 mg, 0.05 mmol) and CsOAc (196 mg, 1.0 mmol). 3-Bromopyridine (50 mL, 0.5 mmol) and benzylamine (84 mL, 0.75 mmol) were added followed by dry DMSO (0.5 mL). The tube was sealed, and the mixture was heated to 90 °C. After stirring at this temperature for 24 h, the heterogeneous mixture was cooled to RT and diluted with EtOAc (10 mL). The resulting solution was filtered through a pad of silica gel and concentrated to give the crude product. Purification by silica gel chromatography (1:1 pentane/ethyl acetate) gave *N*-benzylpyridin-3-amine as a white solid: 91 mg (98%). The identity and purity of the product was confirmed by ^{1}H and ^{13}C NMR spectroscopic analyses.

Other procedures also of recent vintage provide entries to primary aryl amines.[70, 73] One that introduces the NH_2 residue directly *via* aqueous ammonia, rather than an ammonia surrogate (*e.g.*, amides, carbamates, imines, amidines, etc.), applies to both aryl iodides and bromides. Copper iodide (10 mol%) in warm DMF makes up the conditions, absent any special equipment for controlling pressure.[73] Simple dicarbonyl compounds such as **12** (0.4 equiv) serve as ligands, although there is an as yet undetermined relationship between those that afford good results and those that lead to low levels of conversion. While the source (and oxidation state) of copper is not crucial, the base is important, with Cs_2CO_3 giving best results. A curious biphasic mixture is formed upon heating in this medium; other solvents (DMSO, CH_3CN, water) were not nearly as effective.

NH_3•H_2O, cat. $Cu(acac)_2$; Ligand **12**, Cs_2CO_3; DMF, 90 °C, 24 h

(92%) (84%) (88%)

12 =

General procedure for copper-catalyzed amination reactions[73]

After standard cycles of evacuation and back-filling with dry and pure nitrogen, a Schlenk tube equipped with a magnetic stirring bar was charged with $Cu(acac)_2$ (0.1 equiv), Cs_2CO_3 (2 equiv), and the aryl halide (2 mmol, 1 equiv), if a solid. The tube was evacuated, and then backfilled with nitrogen. If a liquid, the aryl halide was added under a stream of nitrogen by syringe at RT, followed by 2,4-pentadione (0.4 equiv), anhydrous and degassed DMF (4.0 mL), and 600 μL of ammonia solution (28%). The tube was sealed under a positive pressure of nitrogen, stirred, and heated to 70 °C or 90 °C for 24 h. After cooling to RT, the mixture was diluted with DCM and filtered through a plug of Celite, the filter cake being further washed with DCM. The filtrate was washed twice with water. The combined aqueous phases were extracted with DCM (5 times). The organic layers were combined, dried over anhydrous Na_2SO_4, filtered, and concentrated *in vacuo* to yield the crude product, which was purified by silica gel chromatography eluting with mixtures of cyclohexanes and EtOAc. The products were characterized by NMR and HRMS, and the data were compared with those from authentic commercial products.

Additional methods of late for the direct introduction of the NH_2 moiety using aqueous ammonia are also available, with each made possible by the choice of ligand.[74] Aryl bromides and iodides (but not chlorides) react in hot (120 °C) ammonium hydroxide in the presence of CuI (10 mol%) to afford anilines in good yields, so long as a piperazine ligand was present (shown below); in its absence, no amination took place. Copper was also shown to be essential, and other sources (*e.g.*, CuCl and $CuSO_4$) were less effective. Both electron-rich and -poor aryl halides readily react, although *ortho*-alkyl substitution is not well tolerated in the educt. Functional groups such as a conjugated ketone, nitro residue, and chloride were unaffected by these conditions.

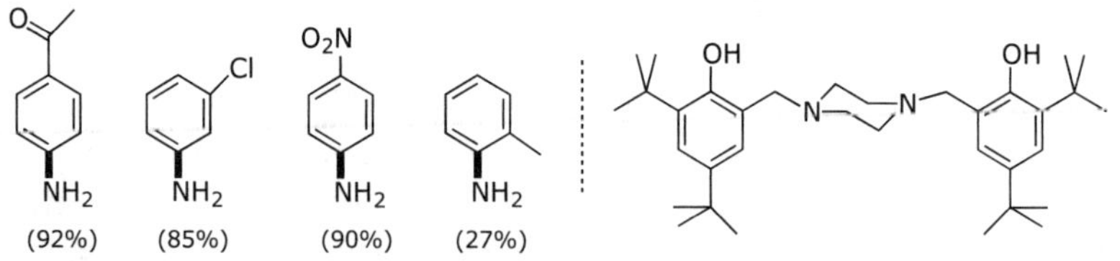

(92%) (85%) (90%) (27%)

D-Glucosamine (10 mol%), as the hydrochloride salt, has also been found to function as a ligand for CuI-catalyzed aminations of aryl iodides and bromides, which were studied in an effort to use greener, more eco-friendly conditions.[75] Couplings can best be conducted in a mixed aqueous acetone solvent system at 90 °C (over 12–30 hours), with K_2CO_3 (2 equiv) as base (as opposed to the weaker base NaOAc or stronger base NaOMe). Other sources of copper led to couplings but in inferior yields (*e.g.*, CuBr, Cu_2O, $CuCl_2$, $Cu(OAc)_2$, etc.). As in the previous method, *ortho*-substitution may be problematic.

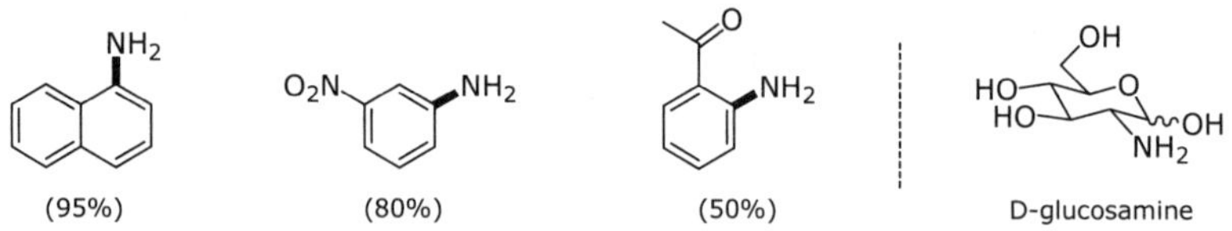

(95%) (80%) (50%) D-glucosamine

By using a 2-carboxylic acid-quinoline-N-oxide ligand (shown below) complexed with CuI, aminations of aryl iodides and bromides can be carried out in DMSO at 50 °C with the former, and at 80 °C with the latter, substrates.[76] Yields are typically >90%, regardless of the electronic nature of the aryl halide. Neither DMF nor toluene is an effective alternative solvent for this coupling. Use of less expensive K_2CO_3, as opposed to Cs_2CO_3, is another virtue. Some representative examples, as well as a typical procedure, follow.

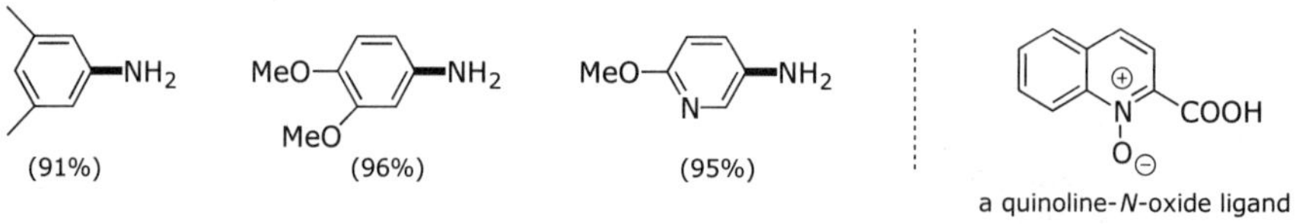

(91%) (96%) (95%)

a quinoline-*N*-oxide ligand

General procedure for the coupling of aryl and heteroaryl bromides with aqueous ammonia[76]

A mixture of aryl iodide (1 mmol), aqueous ammonia (28%, 0.3 mL, 5.0 mmol), CuI (38 mg, 0.2 mmol), 2-carboxylic acid-quinoline-*N*-oxide (75.7 mg, 0.4 mmol), and K_2CO_3 (346 mg, 2.5 mmol) in 2 mL of DMSO was heated at 80 °C for 23 h. The cooled mixture was then partitioned between water and EtOAc. The organic layer was separated, and the aqueous layer was extracted with EtOAc (3 × 10 mL). The combined organic layers were washed with brine, dried over anhydrous Na_2SO_4, and concentrated under vacuum. The residue was purified by chromatography on silicon gel with petroleum ether and EtOAc as eluent to provide the primary aryl amine.

Another method for direct amination of aryl rings relies on arylboronic acids rather than on aryl halides, using ammonia (25% NH_3 in water) and Cu_2O-catalysis. These are run in MeOH at ambient temperatures in the absence of a ligand, a base, or other additives.[77] Air is crucial for catalysis; increasing the temperature to even 40 °C lowered the yield as a result of a decrease in solubility of NH_3. Yields of derived anilines are uniformly high, most between 80% and 93%.

B(OH)2 → cat. Cu_2O, $NH_3 \bullet H_2O$ (5 equiv), MeOH, air, rt → NH2 (92%) | NH2 (92%) | NH2 (84%)

General procedure: copper-catalyzed primary aromatic amines[77]

A round-bottom flask containing a magnetic stir bar was charged with an aromatic boronic acid (1 mmol), methanol (2 mL), 25% aqueous ammonia (5 mmol), and Cu_2O (0.1 mmol, 15 mg). The flask was not sealed, and the mixture was allowed to stir under an atmosphere of air at RT until complete (as monitored by TLC). The mixture was then filtered, and the solvent of the filtrate was removed *via* rotary evaporation. The residue was purified by column chromatography on silica gel to provide the desired product.

A facile one-pot, CuI-catalyzed amidation/hydrolysis leads to net Ar-NH_2 bond formation using the ammonia equivalent in crystalline 2,2,2-trifluoroacetamide.[78] Aryl iodides react in DMF at 45 °C, while bromides require dioxane at 75 °C. DMEDA (*N,N'*-dimethylethylenediamine) is the ligand of choice. Yields are consistently good, although *ortho*-substitution is not tolerated. Both electron-rich and -poor substrates can be used with iodides under the milder conditions. Simple addition of methanol and water to the reaction mixture with continued stirring leads to the desired corresponding anilines.

Br, CN → 1. cat. CuI, H_2NCOCF_3, DMEDA, dioxane, 75 °C; 2. MeOH, water → NH_2, CN (95%) | NH_2, NO_2 (93%) [from bromide] | NH_2, CO_2Et (99%) [from iodide]

General procedure for the one-pot synthesis of primary aryl amines[78]

An oven-dried Schlenk tube was charged with CuI (10 mg), base (2.0 mmol; K_3PO_4 for Ar-I or K_2CO_3 for Ar-Br), and 4-Å molecular sieves (500 mg). The tube was evacuated and backfilled with Ar. The aryl halide (1.0 mmol), trifluoroacetamide (170 mg), DMEDA (12 μL), and solvent (1.0 mL, DMF for Ar-I, or dioxane for Ar-Br) were added under Ar. The reaction mixture was stirred for 24–48 h at 45 °C (for Ar-I) or 75 °C (for Ar-Br). Subsequently, to the reaction mixture was added CH_3OH/H_2O (3.0 mL/3.0 mL) under Ar. The reaction was further stirred for 5–12 h. The resulting suspension was cooled to RT, concentrated *in vacuo*, and extracted with EtOAc (20 mL × 3). The combined organic layers were concentrated *in vacuo*, and the residue was purified by column chromatography to afford the product.

A similar amidation/hydrolysis approach using amidine hydrochlorides, likewise, produces anilines in one pot.[79] Aryl bromides tend to take slightly higher temperatures (120 °C) than iodides (110 °C) to form the C-N bond, which is best achieved in DMF using 10% CuI, 20% L-proline, and Cs_2CO_3 as base. DMSO in place of DMF was not effective; L-proline as ligand was the most active; and other bases (K_2CO_3, K_3PO_4) gave inferior yields, although K_2CO_3 was synthetically useful. The amidines studied include acetamidine, benzamidine, and butyramidine; all gave comparable yields, although the two bulkier cases showed slightly lower reactivities. Water produced during the reaction (between the amidine salt and base) presumably is responsible for secondary conversion of the initial amidine adduct into the isolated primary amine.

General procedure for the copper-catalyzed synthesis of primary arylamines via amidation/hydrolysis[79]

R–C6H4–X [X = Br, I] + H2N–C(R')=NH•HCl —(cat. CuI, cat. L-Proline; Cs2CO3, DMF, 110-120 °C)→ R–C6H4–NH2 (64-92%)

A flask was charged with CuI (19 mg, 0.1 mmol), L-proline (23 mg, 0.2 mmol), and Cs_2CO_3 (2 or 3 mmol) in DMF (2 mL), an aryl halide (1 mmol), and amidine hydrochloride (1.2–2 mmol) was added to the flask at RT under nitrogen. The mixture was stirred at 110–120 °C. After the coupling reaction (10 h), the resulting solution was cooled to RT and diluted with DCM (5 mL). The solution was filtered, and the inorganic salts were removed. The filtrate was concentrated *via* rotary evaporation, and the residue was purified by column chromatography on silica gel using petroleum ether/EtOAc (10:1 to 1:1) as eluent to provide the desired product.

An extensive study on CuI-catalyzed aminations using several heteroaromatic amines (pyrazole, indole, carbazole, pyrrole, and imidazole) has led to efficient procedures using a variety of substituted aryl bromides (and iodides),[63] as well as arylboronic acids.[80] Selected amino acids as ligands, such as glycines (*e.g.*, **13**, below) and L-proline in DMSO appear to offer considerable generality as to substrates, some functional group compatibility, opportunities for intramolecularity, and good yields of isolated products. Compounds **14**–**17** are representative of these aminations. Reactions tend to take place at temperatures in the 60–90 °C range, although couplings of electron-rich aryl bromides with pyrazole or imidazole require 110 °C.

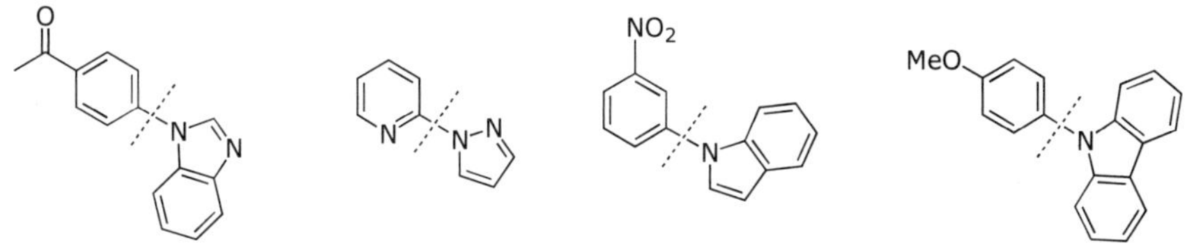

14 (75%) **15** (94%) **16** (93%) **17** (93%)

Mechanistic possibilities that acknowledge the bidentate nature of the ligands include a four-coordinate Cu(III) species **18** that is subject to halide displacement by an amine (followed by reductive elimination, "R. E."), and a Cu(I) π-complex **19**, activated toward nucleophilic halide substitution. A modified phenanthroline ligand (*i.e.*, the 4,7-dimethoxy derivative; **20**, R' = OMe) has been found to be especially useful for *N*-arylation of hindered imidazoles and/or aryl halides, using a combination of catalytic Cu_2O/**20**/PEG in *n*-butyrylnitrile at 110 °C.[81] Likewise, pyrimidinediol ligand **21** effects coupling with imidazoles and benzimidazoles in the presence of catalytic CuBr/TBAF at 145–150 °C in the *absence* of solvent.[82]

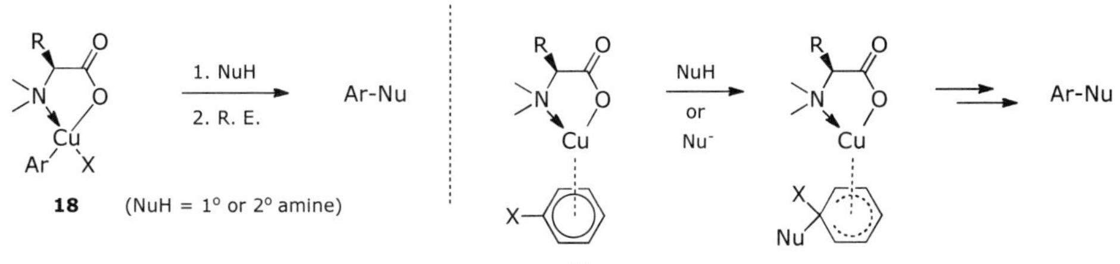

1. NuH; 2. R. E. → Ar-Nu

18 (NuH = 1° or 2° amine)

NuH or Nu^- → Ar-Nu

19

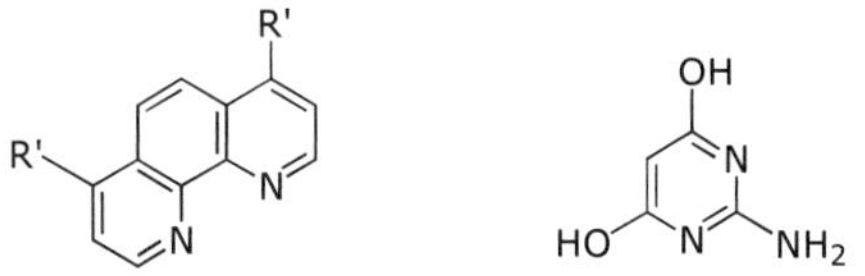

20 **21**

General procedure for coupling aryl halides with pyrazole and imidazole catalyzed by CuI and an amino acid[63]

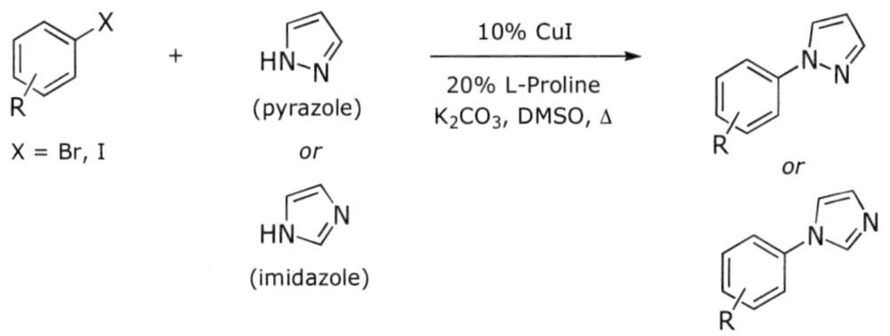

X = Br, I; (pyrazole) *or* (imidazole); 10% CuI, 20% L-Proline, K_2CO_3, DMSO, Δ

A resealabe tube was charged with CuI (0.2 mmol), L-proline or *N,N*-dimethylglycine (0.4 mmol), K_2CO_3 (4 mmol), aryl halide (2.2 mmol), and azole (2 mmol); evacuated; and backfilled with nitrogen. To this mixture was added DMSO (3 mL) by syringe at RT under nitrogen. The mixture was heated (pyrazoles: 75–110 °C;

imidazoles: 60–110 °C) before cooling and then partitioning between water and EtOAc. The organic layer was separated, and the aqueous layer was extracted with EtOAc. The combined organic layers were washed with brine, dried over anhydrous Na_2SO_4, and concentrated *in vacuo*. The residual oil was loaded on a silica gel column and eluted with 1:10 to 1:8 EtOAc/petroleum ether to afford the coupling product.

General procedure for coupling aryl halides with indole, pyrrole, or carbazole catalyzed by CuI and an amino acid[63]

X = Br, I; (indole) or (pyrrole); 10% CuI, 20% **13**, DMSO, Δ; K_2CO_3 (X = I), Cs_2CO_3 (X = Br); [**13** = Me_2N–CH_2–COOH]

A resealable tube was charged with CuI (0.2 mmol), L-proline or *N,N*-dimethylglycine (0.4 mmol), and K_2CO_3 (4 mmol) for aryl iodides or Cs_2CO_3 (4 mmol) for aryl bromides, aryl halide (2.2 mmol), and *N*-containing heterocycle (2 mmol), evacuated; and backfilled with nitrogen. To this mixture was added DMSO (3 mL) by syringe at RT under nitrogen. The mixture was heated (indoles: 75–90 °C; pyrroles: 90–110 °C; carbazoles: 90 °C) until the reaction was complete; after which, it was worked up and the product was isolated as in the procedure above.

Pyrazoles react with aryl bromides in refluxing acetonitrile over three hours, in the presence of Cs_2CO_3, when exposed to CuI complexed to two phosphines associated with the new butadienyl-phosphine ligand **22**, for which an X-ray structure determination has been made.[83] Preparation of **22** using a Wittig olefination favors the *Z*-isomer (yields of isomers: 75% *Z*, 25% *E*). Extensive mechanistic studies indicate that the initial CuI is coordinated to two ligands **22**, one of which is then replaced by the pyrrazole ring. Subsequent oxidative addition to the aryl halide leads to the cross-coupled product *via* a likely Cu(III) intermediate. The prospects for use as well in diaryl ether formation are also encouraging.

General procedure for copper-catalyzed coupling reactions using ligand **22** *(1.5 mmol scale)*[83]

cat. L_nCu, Cs_2CO_3; CH_3CN, Δ; (97%); L = **22**

After standard cycles of evacuation and backfilling with dry and pure nitrogen, an oven-dried Schlenk tube equipped with a magnetic stirring bar was charged with the Cu- ligand complex as catalyst (122.8 mg, 0.15 mmol), Cs_2CO_3 (978 mg, 3 mmol), the nucleophile (2.25 mmol; 1.5 equiv), and the aryl halide (1.5 mmol), if a solid. If a liquid, the aryl halide was added under a stream of nitrogen by syringe at RT, followed by anhydrous CH_3CN (1.5 mL). The tube was sealed under a positive pressure of nitrogen, stirred, and heated to 82 °C for the required time period. After cooling to RT, the mixture was diluted with DCM (~20 mL) and filtered through a plug of Celite, the filter cake being further washed with DCM (~5 mL). The filtrate was washed twice with water (~10 mL × 2). The aqueous phases were twice extracted with DCM (~10 mL). The organic layers were combined, dried over anhydrous Na_2SO_4, filtered, and concentrated *in vacuo* to yield the crude product, which was purified by silica gel chromatography with hexanes/DCM as eluent.

Procedures that rely on co-catalytic systems involving copper have appeared for effecting aminations with heteroaromatic amines, most of which involve iron/copper combinations (*e.g.*, $Fe(acac)_3/CuO$[84] and $Fe_2O_3/Cu(acac)_2$[85]). An alternative bimetallic catalyst that includes manganese in the form of MnF_2/CuI leads to C-N couplings at a moderate 60 °C, and it takes place in water only, although only aryl iodides are the coupling partners.[86] The ligand recommended is *trans*-1,2-diaminocyclohexane (20 mol%). Included in this study are pyrazole, benzpyrazole, and indole, among others.

MnF_2 (30 mol %)
CuI (10 mol %)
KOH (2 equiv)
H_2O, 60 °C, 24 h
ligand
(91%)
H_2N NH_2
(ligand)

General procedure for N-arylation of nitrogen nucleophiles[86]

The *N*-nucleophile (1.47 mmol), CuI (Sigma-Aldrich, 99.999% purity, 0.147 mmol), MnF_2 (Sigma-Aldrich, 98% purity, 0.441 mmol), KOH (2.94 mmol), the aryl halide (2.21 mmol), *trans*-1,2-diaminocyclohexane (0.294 mmol), and water (0.75 mL) were added to a reaction vial and a screw cap was fitted to it. The reaction mixture was stirred under air in a closed system at 60 °C for 24 h. After cooling to RT, the mixture was diluted with DCM and filtered through a pad of Celite. The combined organic extracts were dried with anhydrous Na_2SO_4, and the solvent was removed under reduced pressure. The crude product was purified by silica-gel column chromatography to afford the *N*-arylated product. The identity and purity of known products was confirmed by ^1H and ^{13}C NMR spectroscopic analyses.

A new procedure for *N*-arylation of imidazole that obviates added ligands entirely has been described, where Ullmann-type coupling is achieved under very mild conditions (<40 °C in DMF).[87] Although aryl iodides are needed, and the procedure calls for 20 mol% CuI, considerable variety within the aryl substrate can be tolerated: a nitro group, bromide, ester, amide, and aldehyde all smoothly couple. Noteworthy is the observation that neither a free phenolic OH nor an anilino NH_2 group within the iodide competes with imidazole.

20 mol % CuI, K_3PO_4
DMF, 35-40 °C, 40 h
(73%)

Representative procedure. 4-(4-(1H-Imidazol-1-yl)phenyl)-2-methylbut-3-yn-2-ol[87]

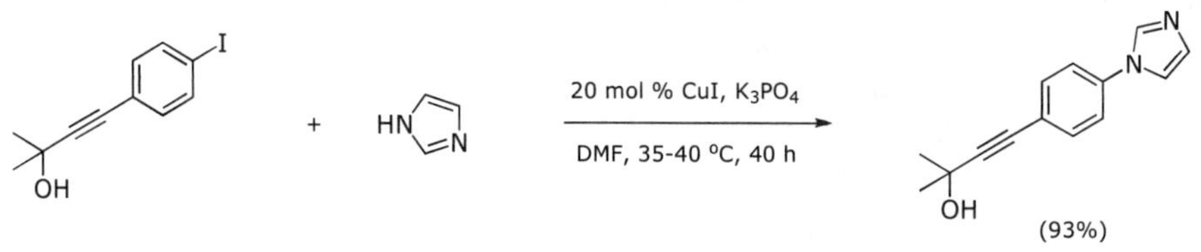

20 mol % CuI, K_3PO_4
DMF, 35-40 °C, 40 h
(93%)

A flame-dried Schlenk tube with a magnetic stirring bar was charged with CuI (38.4 mg, 0.2 mmol), K_3PO_4 (0.422 g, 2.0 mmol), a nitrogen-containing heterocycle (1.4 mmol), aryl halide (1.0 mmol), and DMF (1 mL) under a N_2 atmosphere. A rubber septum was replaced with a glass stopper, and the system was then evacuated twice and backfilled with N_2. After being stirred at 35–40 °C for 40 h, the reaction mixture was diluted with EtOAc (2–3 mL), filtered through a plug of silica gel, and washed with 10–20 mL of EtOAc. The filtrate was concentrated *in vacuo*, and the resulting residue was purified by column chromatography on silica gel with EtOAc/petroleum ether (1/1) to give a white solid (93% yield). HRESIMS calcd. for $[C_{11}H_8N_2+K]^+$: 207.0325, found: 207.0325.

An updated protocol now accomplishes the same goal but reduces the amount of copper salt needed (CuBr; 5 mol%) in DMSO at 60 °C and applies to aryl bromides. In this case, however, a ligand is needed: an acylpyridinyl species, identified as the ligand of choice.[88]

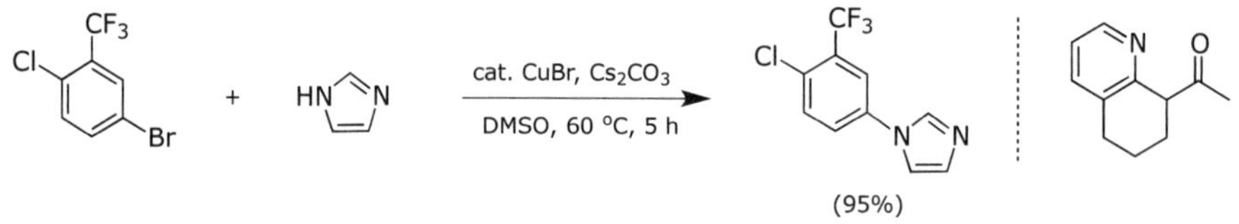

cat. CuBr, Cs_2CO_3
DMSO, 60 °C, 5 h
(95%)

General procedure for aminations of aryl bromides with imidazole (as above).[88]

A sealed tube equipped with a Teflon valve was charged with a magnetic stir bar, CuBr (7 mg, 0.05 mmol, 5 mol%), Cs_2CO_3 (650 mg, 2 mmol) and any remaining solids (azole or aryl halide). The tube was evacuated and backfilled with N_2. Under a counter flow of N_2, imidazole (1.2 mmol), DMSO (1.0 mL), and ligand (18 mg, 0.10 mmol, 10 mol%) were added by syringe. The tube was evacuated and again backfilled with N_2 and sealed. The mixture was heated to 60 °C for the required time period. After cooling to rom temperature the mixture was diluted with EtOAc (25 mL), passed through a fritted glass filter to remove the inorganic salts and the filtrate was washed by H_2O, brine, then removed under vacuum. The residue was purified by column chromatography (silica gel) and the product was dried under high vacuum (at least 1 h).

Applications of Cu-mediated aminations are plentiful and suggest that the virtues of this technology are being recognized. As examples, *N*-substituted nicotinamides such as **23** were prepared using proline (20%) as ligand (10% CuI, K_2CO_3, DMSO, 120 °C, 15 h).[53] A highly active GPIIb/IIIa receptor antagonist, SB-214857 (**24**), was made (without racemization) by way of intramolecular amination using catalytic CuI/K_2CO_3 in DMF at 90 °C[89]; likewise, intramolecular animations were key 5- and 6-membered ring-forming steps in syntheses of antitumor agents (+)-yatakemycin and the duocarmycins, which proceeded *via* intermediates **25** and **26**, respectively.[90, 91]

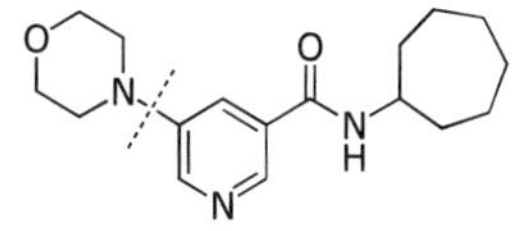

23 (80%) 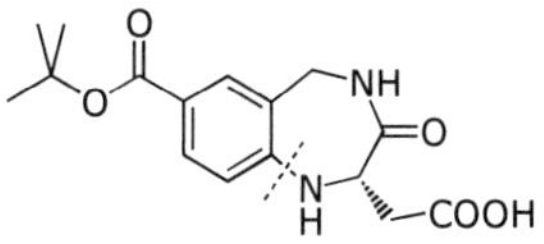 **24** (67%) 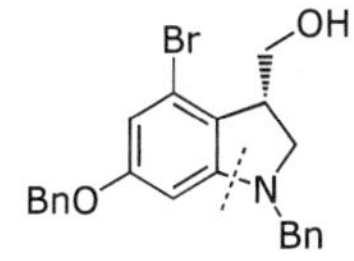 **25** (>67%) 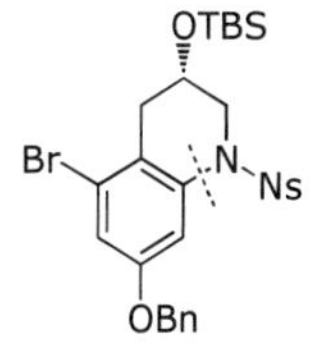 **26** (83%)

As important and popular as direct nucleophilic displacements by nitrogen are for aminations of aryl halides, the umpolung approach to amines using electrophilic sources of nitrogen broadens considerably the available options. Early work showed that both Grignards[92, 93] (Eqn. 1-3) and organozinc reagents (zincates R_3ZnM, M = Li, MgBr) could be used with appropriately matched sources of "R_2N^+"; the former reacting with sulfonyloxime **27** by way of catalysis with CuCN, followed by imine hydrolysis (and derivatization if desired) to afford an ammonium salt **28**. More recently, as alternatives to oxime derivative **27** (reactions with Grignards), *N*-chloroamines (reactions with arylboronic acids),[94] and either *O*-methylhydroxylamine **29** or oxime derivative **30** (with zinc reagents),[95] as well as *O*-benzoyl hydroxylamines **31** have been found to be excellent coupling partners with various diorganozinc species (R_2Zn) when catalyzed by copper (Scheme 1-2).[96, 97] The precursors are easily prepared [$(PhCO_2)_2$, K_2HPO_4, DMF, RT] on a gram scale.[98, 99] Only 0.6 equivalents of R_2Zn are needed. $(CuOTf)_2{\bullet}C_6H_6$ is the commercial source of catalytically active Cu(I), although $CuCl_2$ works as well.

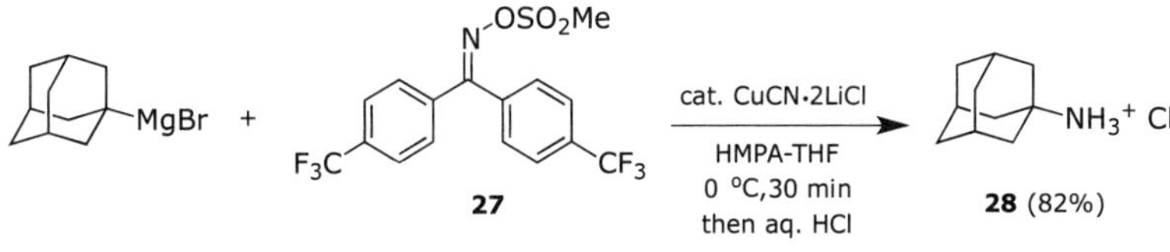

cat. CuCN•2LiCl; HMPA-THF, 0 °C, 30 min, then aq. HCl (Eqn. 1-3)

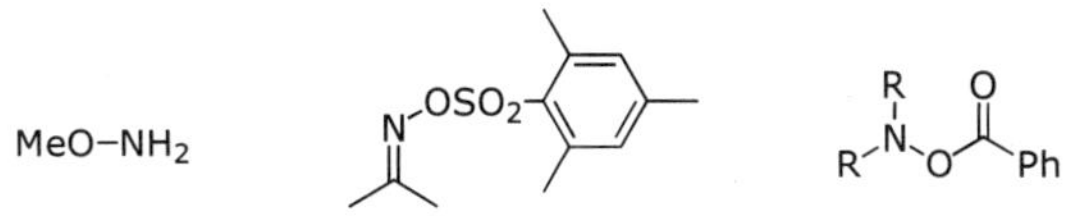

27 **28** (82%)

$MeO{-}NH_2$

29 **30** **31**

$(CuOTf)_2{\bullet}C_6H_6$ (1.25 mol %); THF, rt, ≤ 1 h

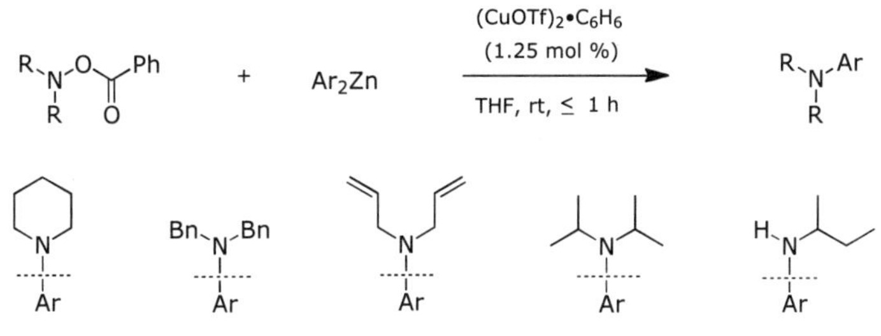

Scheme 1-2

Reactions proceed at ambient temperatures within an hour in THF. Functionalized aryl zinc reagents containing an ester, CN, Cl, and NO_2 groups, prepared by aryl halide metalation with *i*-PrMgBr followed by metathesis with $ZnCl_2$ (0.5 equiv), react smoothly to give the corresponding anilines. The scope of amines that can be fashioned is broad; two are shown below.[100] Steric factors appear not to be of major consequence. No additives are needed (*e.g.*, ligands for Cu). Although only briefly studied, prospects for using Grignards + catalytic $ZnCl_2$ (10 mol%), along with catalytic copper (2.5 mol%), look promising (Eqn. 1-4).

+ PhMgBr —(cat. $ZnCl_2$, cat. CuOTf; THF, rt)→ (69%) (Eqn. 1-4)

General procedure for the copper-catalyzed amination of functionalized diarylzinc reagents. N-(2-Nitrophenyl)-dibenzylamine[100]

+ —(1.25% $[CuOTf]_2$•C_6H_6; THF, rt)→ (96%)

An oven-dried, round-bottom flask equipped with magnetic stir bar was charged with 1-iodo-2-nitrobenzene (0.137 g, 0.55 mmol) and anhydrous THF (3.0 mL), and the solution was cooled to –34 °C. A THF solution of phenylmagnesium bromide (0.60 mL, 1.0 M) was slowly added along the edges of the flask, and the reaction mixture was stirred at –35 °C for 10 min. A THF solution of zinc(II) chloride (1.0 mL, 0.28 M) was slowly added along the edges of the flask, and the reaction mixture was stirred at –35 °C for 10 min. A second oven-dried, round-bottom flask equipped with a magnetic stir bar was charged with an *N,N*-dialkyl-*O*-benzoylhydroxylamine derivative (0.79 g, 0.25 mmol), copper(I) trifluoromethane-sulfonate benzene complex ($[CuOTf]_2$•C_6H_6, 2 mg, 0.0031 mmol,), and anhydrous THF (2.0 mL). The contents of the first flask were added to the second flask, and the reaction mixture was stirred at RT for 1 h. The reaction mixture was diluted with Et_2O (10 mL), and the organic layer was washed with a saturated sodium bicarbonate solution (2 × 10 mL), dried over anhydrous $MgSO_4$, and the solvent removed under reduced pressure. The crude product was purified by flash chromatography to afford *N*-(2-nitrophenyl)-dibenzylamine (0.076 g, 0.24 mmol, 96%) as a yellow viscous oil after flash chromatography with 15% EtOAc/hexanes. The desired product was determined to be of ≥95% purity by ^1H NMR spectroscopy and GLC or supercritical fluid chromatographic (SFC) analysis.

Several reports of late have described aromatic C-N bond formations using aryl boronic acids and copper catalysis. Initial reports from DuPont on this topic in the late 1990s relied on stoichiometric $Cu(OAc)_2$, but more recently, it has been shown that reduced levels of copper (5–10%) can be used, together with myristic acid (10–20 mol%), forming diarylamines (Eqn. 1-5).[101] Representative products, and yields, are shown below (**32**–**35**).

Ar-NH$_2$ (1 equiv) + Ar'-B(OH)$_2$ (1.5 equiv) → [cat. Cu(OAc)$_2$, cat. myristic acid; 2,6-lutidine (1 equiv), air, toluene] Ar-NH-Ar' (Eqn. 1-5)

Limitations were noted, such as reduced yields with *ortho*-substitution (*e.g.*, **34**) and with *para*-halo substitution (*e.g.*, *p*-$ClC_6H_4B(OH)_2$ + $PhNH_2$ to <10% product). Allylamines can also be used, albeit moderate efficiencies are to be expected (50–64%).

32 (77%) **33** (62%) **34** (50%) **35** (50%)

N-Arylation of imidazoles using aryl boronic acids has been of particular interest. While a Cu(II) catalyst was found (*i.e.*, $[Cu(OH)\cdot TMEDA]_2Cl_2$) such that catalysis can be carried out in water, simple salts (*e.g.*, CuCl) can effect the coupling in methanol at reflux.[102] Both CuBr and $CuClO_4$ gave equally impressive results (*e.g.*, **36** and **37**).

36 (98%) **37** (98%)

A heterogeneous source of copper, the well-characterized, strongly basic, copper-exchanged fluorapatite (CuFAP), is formed as the (mainly) Cu(II) species from $Ca_{10}(PO_4)_6(F)_2$ upon treatment with $Cu(OAc)_2$.[103–105] CuFAP mediates displacement of all aryl halides (F, Cl, Br, I) in DMF (X = F, Cl) or DMSO (X = Br, I), as shown below (Eqn. 1-6). Less basic CHAP [from $Ca_{10}(PO_4)_6(OH)_2$ + $Cu(OAc)_2$], $Cu(OAc)_2$, CuI alone, or Cu powder are inactive. Interestingly, in 3-chloro-4-fluoronitrobenzene, the C-F bond reacts selectively with imidazole in high yield (85%, isolated). Polymer-supported copper (I)[106] and copper (II)[107] on polystyrene have both been shown useful for arylations of anilines and heteroaromatics with boronic acids.

Ar−X + imidazole → [CuFAP, K_2CO_3; DMF, 120 °C or DMSO, 110 °C] Ar-imidazole (Eqn. 1-6)

X= Br, I 85-98% (isolated)
X= F, Cl 65-100% (GC)

3-chloro-4-fluoronitrobenzene + imidazole → [CuFAP; DMF, K_2CO_3, 120 °C] 1-(2-chloro-4-nitrophenyl)imidazole (85%)

Typical procedure for N-arylation of imidazole with chloroarenes. 1-(2-Chloro-4-nitrophenyl)-1H-imidazole[103]

Chlorobenzene (0.1 mL, 1 mmol), imidazole (81 mg, 1.2 mmol), potassium carbonate (276 mg, 2 mmol), potassium *t*-butoxide (11 mg, 10 mol% for electron-rich chloroarenes only), and the catalyst (CuFAP, 100 mg) were stirred in DMF (4 mL) at 120 °C for 36 h. The catalyst was filtered and the filtrate was quenched with aqueous sodium hydrogen carbonate, and the product was extracted with EtOAc. The combined organic extracts were dried over anhydrous sodium sulfate and filtered. Solvent was evaporated under reduced pressure and concentrated *in vacuo* to give the crude product. The crude product was purified by column chromatography on silica gel (hexane/EtOAc, 70/30) to afford *N*-phenylimidazole (115 mg, 80%); mp 181 °C.

Nanoparticles of CuO can be fashioned as catalysts for C-N bond formations, where both the size and the morphologies can be controlled and play pivotal roles in catalyst activity.[108] Fabrication of CuO nanocrystals can be accomplished by dehydration of $Cu(OH)_2$ nanowires at 180 °C over 12 hours, followed by additional heating at 250 °C to promote crystallization. Related techniques can arrive at the corresponding nanospheres, nanocubes, nanorods, etc. Alternatively, a cost-effective method to arrive at CuO nanoparticles relies on the precipitation method starting with $Cu(OAc)_2$.[109] This catalyst can be used under ligandless conditions in dimethylacetamide (DMAc) at 120 °C to form *N*-arylated benzimidazoles (see examples below). Another route to spherical granules of CuO begins with the thermal decomposition of $Cu(NO_3)_2$, followed by immobilization into a polymeric matrix of sodium alginate.[110] Final calcination forms pebble-type materials of high porosity that mediate aminations of iodobenzene with both aryl and alkyl amines at 110 °C either in DMSO or under solvent-free conditions.

+ Ar-X — CuO nanoparticles, K_2CO_3 / DMAc, 120 °C, *ca.* 24 h →

CHO (X = F: 67%; Cl: 77%; Br: 80%)

COOH (X = F: 75%)

(X = Br; 88%)

3.2 Amidations

Amidations of aromatics (the Goldberg reaction)[111] have also generated considerable interest, including detailed mechanistic studies that address longstanding questions about ligands, the copper oxidation state, and the nature of intermediates involved. As with traditional aminations (Ullmann couplings), these were previously characterized by high temperature and polar solvent requirements, often along with stoichiometric amounts of copper and with activated substrates. Advances in ligand design have led to both inter- and intramolecular procedures that result in *N*-arylations of amides (1° and 2°), lactams, and related derivatives (*e.g.*, oxazolidinones and a 2H-pyridazin-3-one as in **38**). Leading cases call for catalytic Cu_2O/chelating Schiff base ligand **39** or **40** in DMF or CH_3CN,[61] or catalytic CuI/diamine ligand **41** (or, in some cases, simpler *N,N'*-dimethylethylenediamine) in toluene or dioxane at reflux.[57, 58] An alternative based on CuI/1,10-phenanthroline, assisted by KF/Al_2O_3, also works well on aryl iodides in refluxing toluene.[60] The former method has been used to make the sedative amphenidone (**43**).[61] Use of 3 Å molecular sieves in the mixture minimized amide hydrolysis (<1%) and competitive *O*-arylation of the iodide by water. *N*-Arylation (*vs.* *O*-arylation) of 2- and 4-pyridones is strongly favored (>99:1), done in hot DMSO, using 4,7-dimethoxyphenanthroline (**20**, R' = OMe) and 2,2,6,6-tetramethylheptane-3,5-dione (**42**) as ligands, respectively.

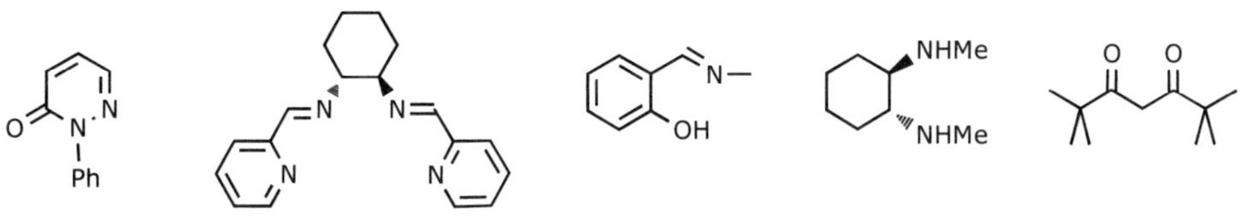

38 **39** (Chxn-Py-Al) **40** (salox) **41** **42**

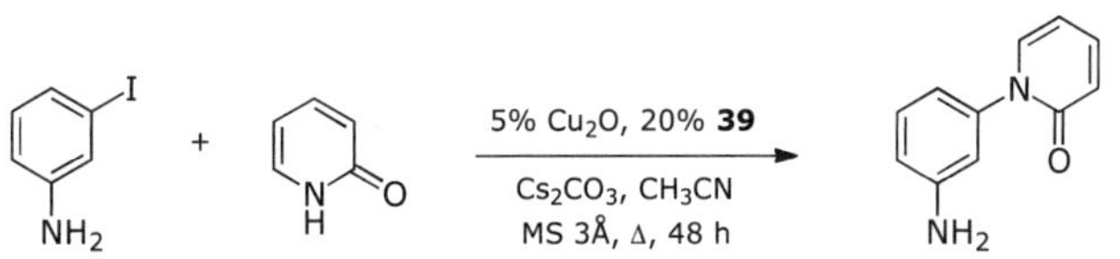

5% Cu_2O, 20% **39**; Cs_2CO_3, CH_3CN; MS 3Å, Δ, 48 h

43 (82%)

The ligand *trans-N,N'*-dimethyl-1,2-cyclohexanediamine (**41**) is broadly applicable to amidations of aryl iodides by 1° and 2° amides. In addition to toluene and dioxane, THF and DMF are also acceptable media. No special handling of reagents or reaction mixtures (other than an inert atmosphere) is required. There is a lack of competing side reactions, such as homocoupling and reduction, which is not characteristic of related Pd-catalyzed couplings. Also, these diamine ligands (*N,N'*-dimethylethylenediamine and **41**) are of low molecular weight, implying economic advantages on a cost/mole basis. Other copper salts afford comparable yields (CuCl, Cu_2O), as does copper powder, and a catalyst loading as low as 0.2% can be used. Both K_3PO_4 and Cs_2CO_3 can be employed with equal success, although the former is less costly. Functional groups not amenable to the corresponding Pd-catalyzed process, *e.g.*, a free NH_2 (**44**, **45**) and an allyl ester (**46**), are tolerated. Particularly useful may be the observation that an aliphatic amino residue does not compete with amidation in the presence of ligand **41**, while switching to the CuI/ethylene glycol combination reverses chemoselectivity favoring the more basic nitrogen. A kinetic study implicates a monomeric copper(I) amidate as the reactive species, the dimerization of which is prevented by higher concentrations of the diamine ligand (relative to copper).

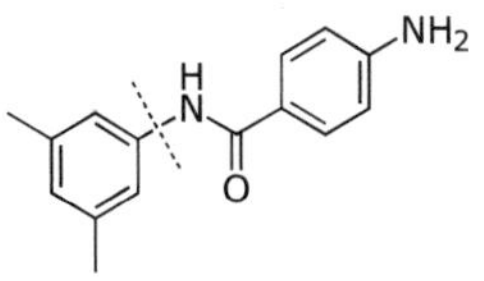

44 (98%) 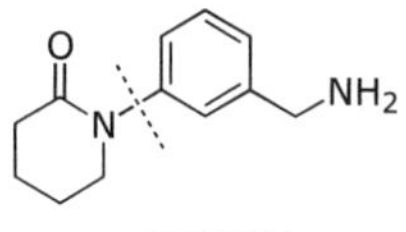 **45** (95%) 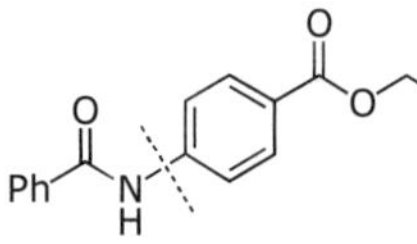 **46** (86%)

Aryl bromides, not surprisingly, are less reactive partners, but they can still be coupled in high isolated yields.[112] For these cases, K_2CO_3 is the best base, used in refluxing toluene or dioxane. Some representative examples can be seen in products **47–49**.

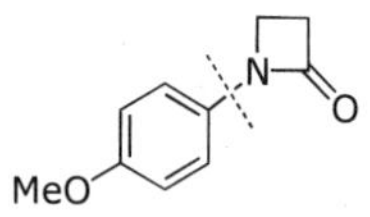

47 (90%) 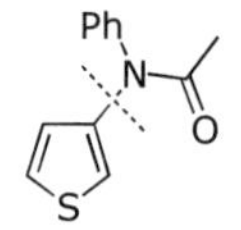 **48** (81%) 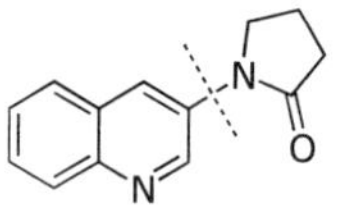 **49** (98%)

Cu-Catalyzed amidation. trans-N-(4-Dimethylaminophenyl)-3-phenylpropenamide[112]

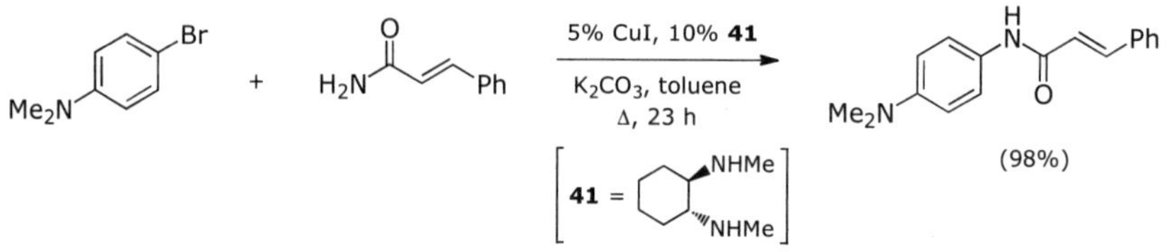

5% CuI, 10% **41**; K_2CO_3, toluene, Δ, 23 h

(98%)

A Schlenk tube was charged with CuI (9.6 mg, 0.050 mmol, 5.0 mol%), 4-dimethylamino)-1-bromobenzene (210 mg, 1.00 mmol), *trans*-cinnamamide (178 mg, 1.21 mmol), and K_2CO_3 (280 mg, 2.03 mmo,); briefly evacuated; and then backfilled with Ar. Racemic *trans-N,N'*-dimethyl-1,2-cyclohexanediamine (16 μL, 0.10 mmol, 10 mol%) and toluene (1.0 mL) were added under Ar. The Schlenk tube was sealed with a Teflon valve, and the reaction mixture was stirred at 110 °C for 23 h. The resulting bright yellow suspension was allowed to reach RT and then filtered through a 0.5 × 1 cm pad of silica gel eluting with 1:1 EtOAc/DCM (10 mL). The filtrate was concentrated *in vacuo*, and the residue was dissolved in DCM (10 mL) and purified by flash chromatography on silica gel (2 × 20 cm, EtOAc/DCM 1:4, 15 mL fractions). Fractions 10–20 provided 261 mg (98%) of the desired product as a bright yellow solid; mp 171–173 °C.

Amidations of vinyl iodides have also been used to advantage in synthesis. Enamides such as **50–52** are representative of the procedures available (Scheme 1-3).[113–115] By proper choice of nucleophile, annulations to either pyrroles **54** (15 examples) or pyrazoles **55** (10 examples) can be induced with CuI-catalysis after an initial amidation of enyne **53**.[116] Subsequent *5-endo-dig* cyclization produces the heteroaromatic products (Scheme 1-4). Several total syntheses have successfully applied this approach, as examples, to construction of key enamide antibiotic CJ-15,801 (**56**), and the naturally occurring cyclopeptide alkaloid paliurine F (**57**), where this methodology served as a tool for macrocyclization.[117, 118]

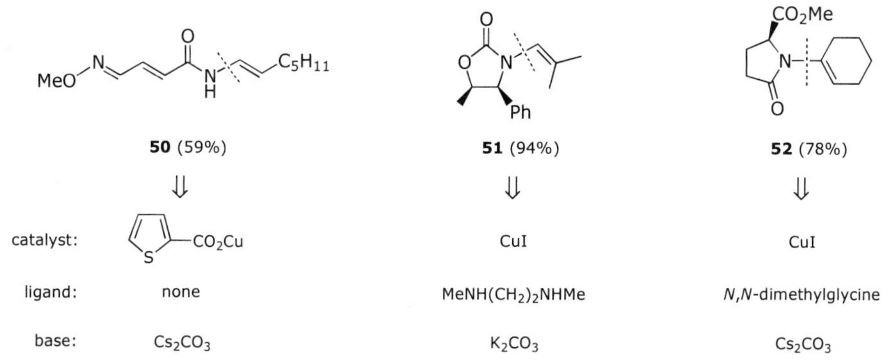

50 (59%) ⇓ catalyst: thiophene-2-CO_2Cu; ligand: none; base: Cs_2CO_3

51 (94%) ⇓ catalyst: CuI; ligand: $MeNH(CH_2)_2NHMe$; base: K_2CO_3

52 (78%) ⇓ catalyst: CuI; ligand: *N*,*N*-dimethylglycine; base: Cs_2CO_3

Scheme 1-3

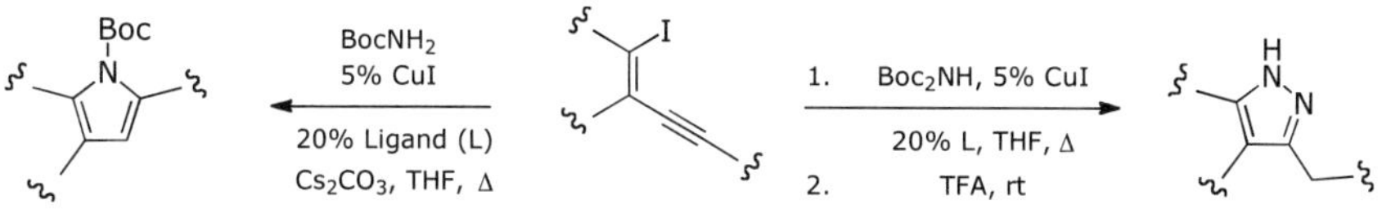

54 (68-95%) ← $BocNH_2$, 5% CuI, 20% Ligand (L), Cs_2CO_3, THF, Δ — **53** — 1. Boc_2NH, 5% CuI, 20% L, THF, Δ; 2. TFA, rt → **55** (66-93%)

Ligand = $MeNH(CH_2)_2NHMe$

Scheme 1-4

cat. Cu(I), ligand, Rb_2CO_3, DMA (90%)

[Cu(I) = $Cu(CH_3CN)_4PF_6$; ligand = 3,4,7,8-tetramethyl-1,10-phenanthroline]

56 (CJ-15,801)

cat. CuI, Ligand, Cs_2CO_3, THF (70%)

3 steps: R = Boc → R = *iso*-leu-*N*,*N*-dimethylleucine

(**57**) paliurine F

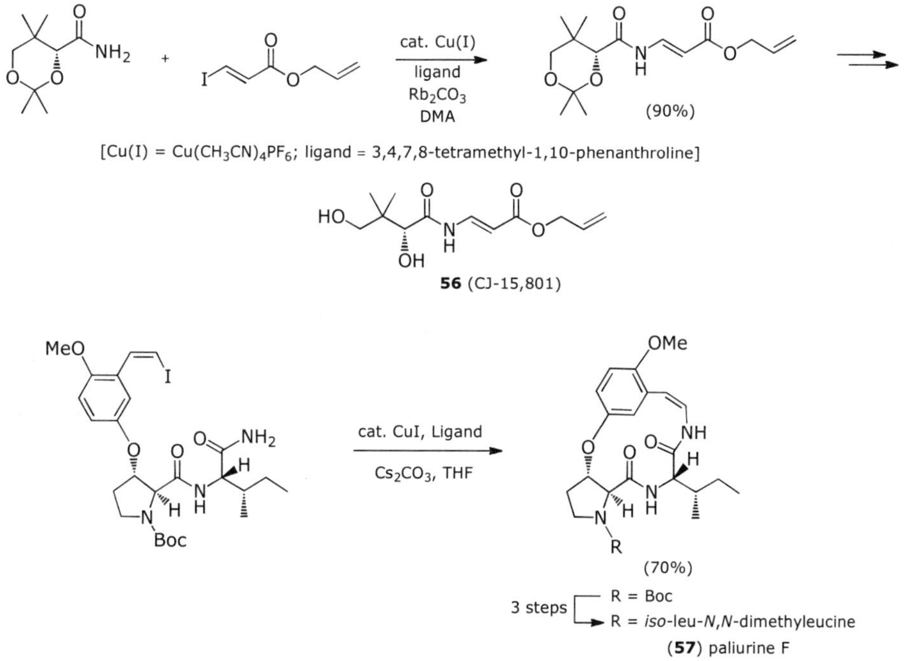

Enamides are also readily available through a copper-catalyzed amidation of vinyltrifluoroborate salts.[119] A mixture of solvents including DMSO together with DCM (1:1) is recommended, given the limited solubility of amides in this chlorinated medium alone. Interestingly, the coupling occurs in a "ligandless" environment, in an oxygen atmosphere at 40 °C: *E*-enamides are favored, and *Z*-configured products appear to be far less stable. Two equivalents of the borate were used in most examples, although it was shown that with increased levels, higher yields can be realized.

General procedure for enamide formation using a trifluoroborate salt[119]

cat. $Cu(OAc)_2$
$DMSO/CH_2Cl_2$ (1:1)
4Å MS, O_2, 40 °C
(80%)

DCM (1.0 mL) and DMSO (1.0 mL) were added to $Cu(OAc)_2$ (0.05 mmol), amide (0.5 mmol), potassium alkenyltrifluoroborate salt (1.0 mmol), and 4-Å molecular sieves (0.38 g). The suspension was stirred for 20 h at 40 °C under an atmosphere of oxygen. The reaction mixture was filtered through a plug of Celite, which was then washed with EtOAc. The crude products were purified by silica gel chromatography using EtOAc/*n*-hexanes and 1% v/v Et_3N.

Bottom line comments. There is now a sufficient number of procedures available such that it is likely that a direct C-N bond between most primary or secondary amines, whether alkyl or aryl, can be made to an aryl or heteroaryl bromide or iodide. Fewer approaches to amidation exist, but significant progress is being made. Several methods for this type of amination give good yields of products; nonetheless, most rely on 10% CuI under homogeneous conditions, which may give rise to separation and/or waste disposal issues for larger scale applications. Heterogeneous catalysis may provide an alternative. The umpolung approach using electrophilic nitrogen provides another avenue well worth consideration in the course of retrosynthetic analysis. Interestingly, just how much copper is truly needed to effect aminations has recently come into question. The observation that ppm levels (or "homeopathic amounts") are sufficient to catalyze certain C-N bond formations that are often reported, wherein 5 mol% CuX or more is required, suggests that there exists a yet-to-be-understood interplay between levels of ligand and temperature and that with new insights, remarkable discoveries in such catalysis may be on the horizon.[120]

4. Cu-Catalyzed *O*- and *S*-Arylation Reactions

Reviews. Monnier, F.; Taillefer, M. *Angew. Chem., Int. Ed.* **2009**, *48*, 6954; Evano, G.; Blanchard, N.; Toumi, M. *Chem. Rev.* **2008**, *108*, 3054; Frlan, R.; Kikelj, D. *Synthesis* **2006**, 2271; Beletskaya, I. P.; Cheprakov, A. V. *Coord. Chem. Rev.* **2004**, *248*, 2337; Ley, S. V.; Thomas, A. W. *Angew. Chem., Int. Ed.* **2003**, *42*, 5400; Sawyer, J. S. *Tetrahedron*, **2000**, *56*, 5045.

4.1 Diaryl Ether Formation

As the reviews rightly remind us, over 100 years have passed since Fritz Ullmann published on reactions between aryl halides and phenols, mediated by copper, to arrive at diaryl ethers.[121] Beginning in the 1980s, increasing attention began to be paid to this functional group, as several important biologically active natural products possessing a diaryl ether were identified, such as verbenachalcone (**58**),[122] bouvardin (**59**),[123] and the especially well-known antibiotic vancomycin.[124]

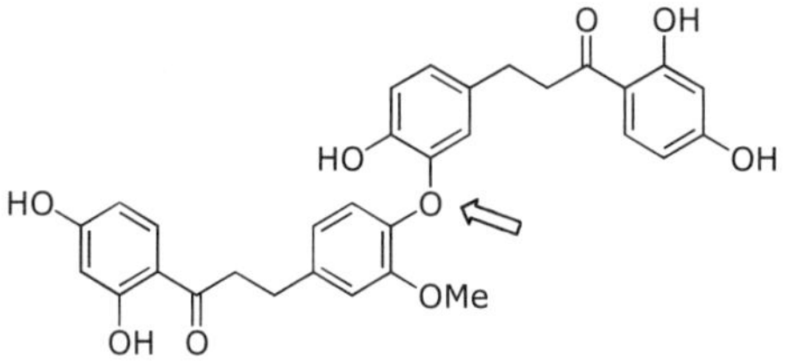

58 (verbenachalcone)

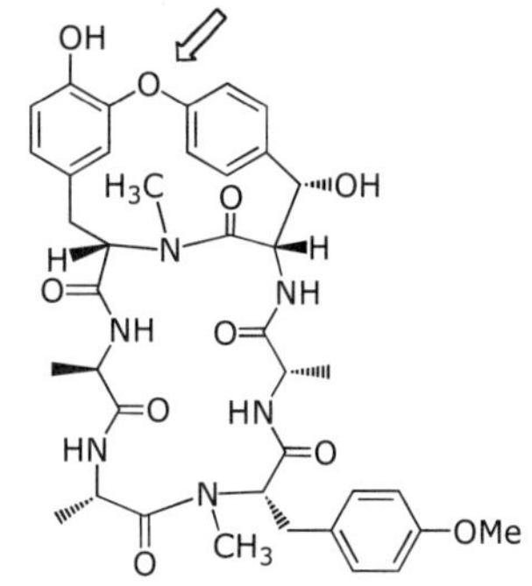

59 (bouvardin)

Intermolecular processes, many of which involve copper catalysis under homogeneous conditions, usually vary widely as to ligand used, although in some cases, no additional chelating species need be present. As with aminations, there are several ligands that have been identified, some being common to both types of reaction, and even beyond (*i.e.*, can be applied to amination, etherification, and *S*-arylation). Representative ligands include 1-naphthoic acid (**60**),[125] tripod species **61**,[126] dione **43**,[127] 2-aminopyridine (**62**),[128] phosphazene P_4-*t*-Bu (**63**),[129] and neocuproine (**64**).[130] Other ligands that also serve well for both *O*- and *N*-arylations include DPP (**9**),[42] *N,N*-dimethylglycine (**13**),[131] Chxn-Py-Al (**39**),[132] salox (**40**),[132] and 8-hydroxyquinoline (**65**).[128] Aryl iodides and bromides are the substrates most often studied. No general procedure for *O*-arylation of aryl chlorides has been reported as yet. The source of copper, as with aminations, appears not to have a major influence. While use of CuI is commonplace for *N*-arylations, methods for diaryl etherifications also recommend (in addition to CuI) CuCl, Cu_2O, and $(CuOTf)_2$•C_6H_6. Cs_2CO_3 is the base most frequently employed (in excess), presumably due to the greater solubility of the corresponding phenoxide salts. Most procedures use 5–10% of a copper salt in DMF, toluene, or dioxane and occasionally mixed media (*e.g.*, DMF/dioxane) arrived at empirically. Ligand **65** is best matched to DME or diglyme.[128] Catalyst loadings well below 1 mol% have been quoted for $(CuOTf)_2$•C_6H_6 in toluene at reflux, although yields are higher at the 5 mol% Cu level. While aryl iodides can be coupled with phenols used in excess without added ligand, bromides may require 1-naphthoic acid (2 equiv). Good functional group compatibility in the halide is to be expected. Similar results have been obtained on iodides using catalyst **61**/Cs_2CO_3 in hot dioxane,[126] where 1,1,1-*tris*(hydroxymethyl)-ethane is the inexpensive and commercially available tridentate *O*-donor ligand.

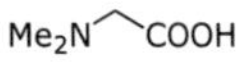

13 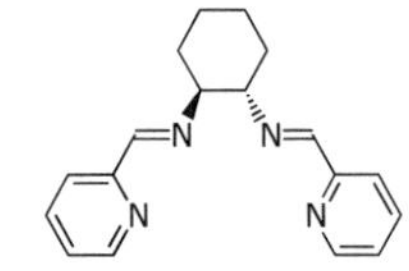 39 (Chxn-Py-Al) 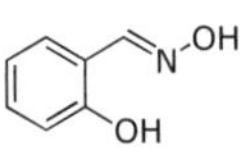 40 (salox) 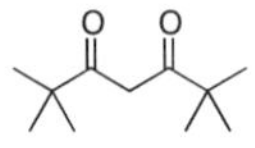 42 (TMDH) 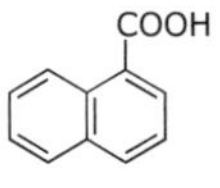 60 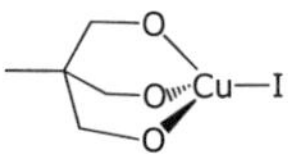 61

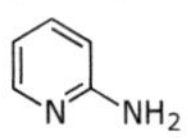

62 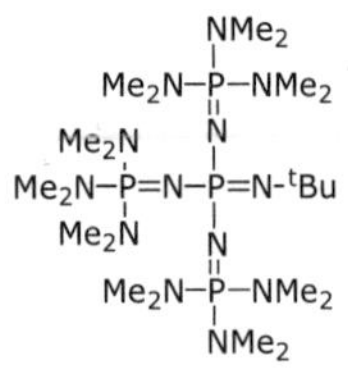 63 (phosphazene P_4-*t*-Bu) 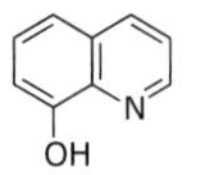 65 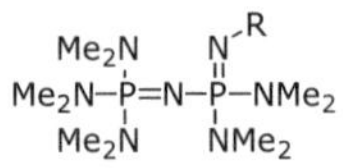 66 (phosphazene) 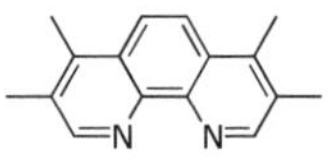 67

Copper-catalyzed couplings of phenols with aryl halides. 2-(3',4'-Dimethylphenoxy)benzoic acid[125]

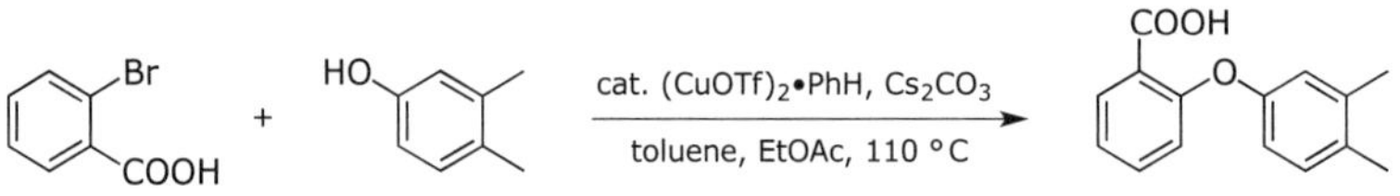

cat. $(CuOTf)_2$•PhH, Cs_2CO_3; toluene, EtOAc, 110 °C (77%)

The aryl halide (2-bromobenzoic acid; 2.5 mmol), phenol (3,4-dimethylphenol; 5.0 mmol), Cs_2CO_3 (5.0 mmol), $(CuOTf)_2$•PhH (0.063 mmol, 5.0 mol% Cu), EtOAc (0.125 mmol, 5.0 mol%), and toluene (2.0 mL) were added to an oven-dried test tube that was then sealed with a septum, purged with Ar, and heated to 110 °C under Ar until the aryl halide was consumed as determined by GC analysis. The reaction mixture was then allowed to cool to RT, diluted with Et_2O, and washed sequentially with 5% aqueous NaOH, water, and brine. The organic layer was dried over anhydrous $MgSO_4$ and concentrated under vacuum to give the crude product. Purification by flash column chromatography on silica gel (20% Et_2O/pentane) afforded an analytically pure product as a white solid (466 mg, 77% yield), mp 98 °C.

Couplings of aryl bromides to phenols have benefited from use of ligands **13** and **42**.[133] With 10% CuI/Cs_2CO_3 and glycine ligand **13** (30%), several diaryl ethers such as **68–70** can be fashioned at 90 °C in dioxane. With *ortho*-substitution in the halides, a higher temperature gives a better yield (see **68**). Both activated and deactivated bromides can be used. Not unexpectedly, a bromide reacts exclusively over an aryl chloride.

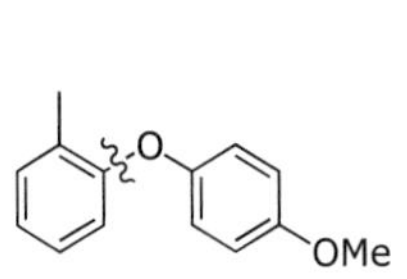

68 59% at 90 °C 75% at 105 °C

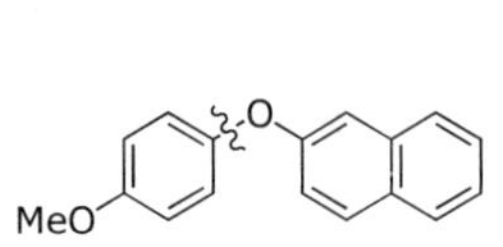

69 (83%)

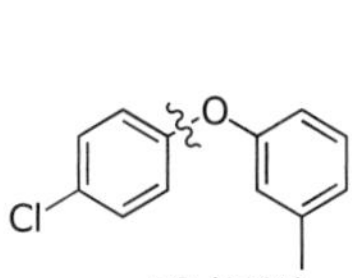

70 (90%)

Using ligand **42** (TMHD = 2,2,6,6-tetramethylheptane-3,5-dione), electron-rich bromides (*e.g.*, *p*-bromoanisole) react with an electron-poor phenol (*e.g.*, *p*-fluorophenol) under catalysis by CuCl in NMP at 130 °C with Cs_2CO_3 as base.[134] Other ligands also examined gave inferior levels of conversion (2-aminopyridine: 43%; 1,10-phenanthroline: 68%; 8-hydroxyquinoline: 82%). The corresponding pentanedione was less efficient than TMHD. CuCl is soluble under these conditions, suggesting that the ligands' role is not simply to provide homogeneous Cu(I). The analogous Cu(II) complex, Cu(II) *bis*-(2,2,6,6-TMHD), an item of commerce,[135] was found to be a less effective and more costly catalyst.

cat. CuCl, TMHD
Cs_2CO_3, NMP
130 °C, 10 h
(82%)

General procedure for diaryl ether formation[134]

cat. CuCl, Cs_2CO_3
Ligand **42** (TMHD)
NMP, 120-130 °C
(77-85%)

Cesium carbonate (7.36 g, 22.6 mmol) was added to the phenol (22.6 mmol) in NMP (19 mL). The slurry was degassed and filled with nitrogen three times. The aryl halide (ArX; X = Br, I; 11.3 mmol) and 2,2,6,6-tetramethylheptane-3,5-dione (1.1 mmol) were added followed by copper(I) chloride (5.6 mmol). The reaction mixture was degassed and filled with nitrogen three times and then heated to 120 °C under nitrogen. Reactions were monitored by HPLC and stopped when >98% conversion (by peak area) was obtained. The reaction mixture was cooled to RT and diluted with MTBE (25 mL). The slurry was filtered and the filter cake washed with methyl *tert*-butyl ether (MTBE) (25 mL). A 1-mL sample of the combined filtrates was diluted 100-fold for assay yield. Combined filtrates were washed subsequently with 2 N HCl (35 mL), 0.6 N HCl (35 mL), 2 M NaOH (30 mL), and 10% NaCl (30 mL). The resulting organic layer was dried over anhydrous $MgSO_4$ and concentrated *in vacuo*. The crude product mixture was purified by column chromatography or recrystalization.

In just the past two to three years alone, aryl ether formation has become even more popular, and many new procedures beyond those already discussed above have appeared. Table 6-2 summarizes these. It is interesting to point out how each highlights a different structural motif with respect to the ligand involved, and how it presumably coordinates to copper, thereby catalyzing C-O bond formation. Most are unlike those already discussed above. Nonetheless, these couplings remain challenging from the standpoint that each requires refluxing conditions of at least 82 °C, and more often >100 °C. In the absence of a ligand, diaryl ethers can be formed from aryliodides under microwave conditions in NMP at 170 °C in *ca.* 30 minutes.[136]

Table 6-2. Recent methods in Cu-catalyzed aryl ether formation.

Alcohol	Aryl Halide	Cu Source	Ligand	Conditions	Yields	References
aliphatic phenolic	bromides iodides	$Cu(CH_3CN)_4^-$ BF_4^-	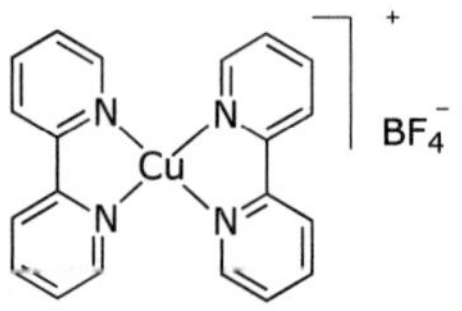 $[Cu(bipy)_2]^+$ BF_4^-	K_2CO_3 DMF 90-110 °C	~80–90%	[137]
aliphatic phenolic	bromides iodides	CuI	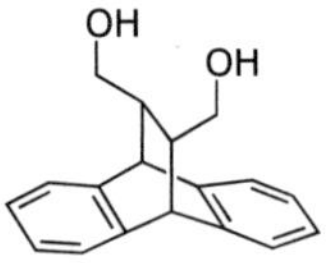	Cs_2CO_3 CH_3CN Δ	50–90%	[138]
phenolic	bromides iodides	$Cu(OTf)_2$	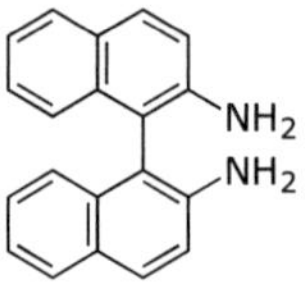 (BINAM)	Cs_2CO_3 dioxane Δ	60–84%	[139]
aliphatic	bromides iodides	CuI	BINAM	Cs_2CO_3 dioxane Δ	60–96%	[140]
aliphatic	bromides	CuI	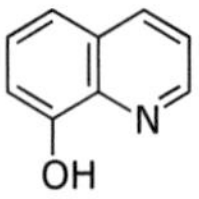	K_3PO_4 ROH 110 °C	>80%	[141]
phenolic	bromides iodides	CuI	 CONHMe·HCl	K_3PO_4 DMF 110 °C	79–96%	[142]
phenolic	chlorides iodides	CuBr	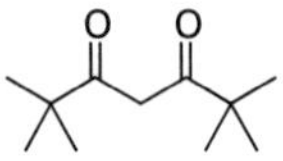 (TMHD)	Cs_2CO_3 DMF 140 °C	53–99%	[143]
phenolic	iodides	CuI	Ligand- and additive-free	K_3PO_4 Bu_4NBr DMF Δ	40–90%	[144]
phenolic	bromides	CuCl	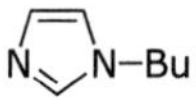 N-Bu imidazole	K_2CO_3 toluene 120-140 °C	54–99%	[145]

Diaryl ethers can also be fashioned using heterogeneous sources of copper. One recent contribution starts with silica gel (100–200 mesh), which after activation with acid, washing, and drying is silylated in dry toluene to give the functionalized support.[146]

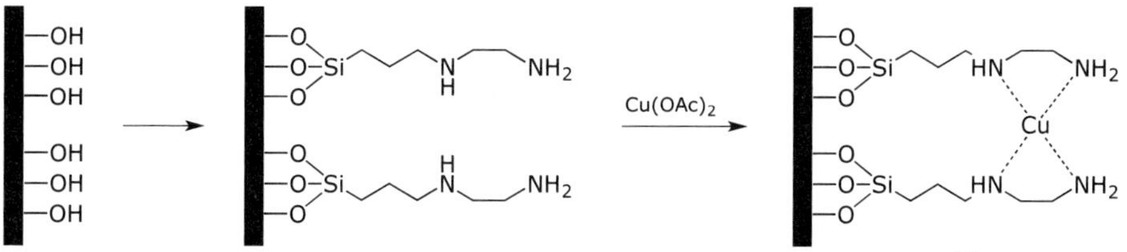

Complexation with $Cu(OAc)_2$ leads to a heterogeneous catalyst **71** that mediates coupling in DMSO at 130 °C. KF (2 equiv) was determined to be the most effective base, with comparisons being made with K_2CO_3, K_3PO_4, and even Cs_2CO_3, among others. No reaction was noted in several solvents (*e.g.*, dioxane, toluene, CH_3CN, and ethylene glycol) at reflux temperatures. The catalyst could be recycled 10 times without loss in activity, after a series of washings and oven drying. Noteworthy are the cases of activated aryl chlorides that led to diaryl ethers in good yields; *e.g.*, as illustrated below.

Typical procedure for the Ullmann diaryl etherification using heterogeneous catalyst **71**[146]

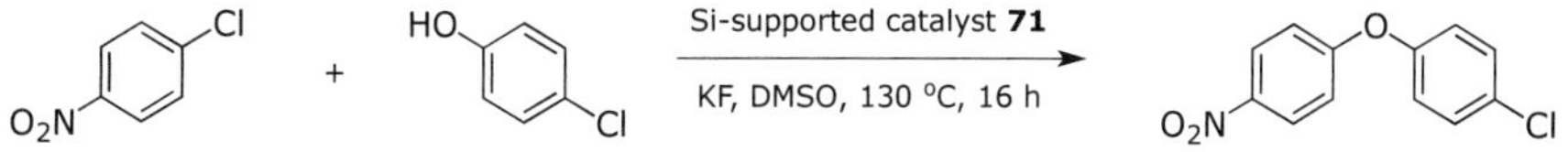

Under a nitrogen atmosphere, an oven-dried, round-bottomed flask was charged with silica-supported copper(II) (**71**, 125 mg, containing 0.05 mmol of Cu), KF (116 mg, 2.0 mmol) aryl halide (1.0 mmol), phenol (1.0 mmol), and DMSO (4 mL). The reaction mixture was refluxed at 130 °C for 16 h. After cooling to RT, the reaction mixture was vacuum filtered using a sintered glass funnel and then washed with DCM (2 × 5 mL). The combined organic extracts were dried over anhydrous Na_2SO_4, filtered, and concentrated *in vacuo*, and the residue was purified by flash chromatography on silica gel to give the desired cross-coupling product.

The bipy unit in designed sol gel ligand **74** chelates copper upon introduction of CuI.[147] This form of immobilized Cu(I) efficiently catalyzes diaryl ether formation in DMF at 110–120 °C, using aryl iodides, and *in situ* prepared aryl iodides from aryl bromides in the presence of KI (2 equiv; *e.g.*, **72** + **73**, below). Bridged silsesquioxane catalyst **74** was prepared from 4,4'-diamino-2,2'-bipyridine in two steps, and it was found to be recyclable without loss in reactivity.

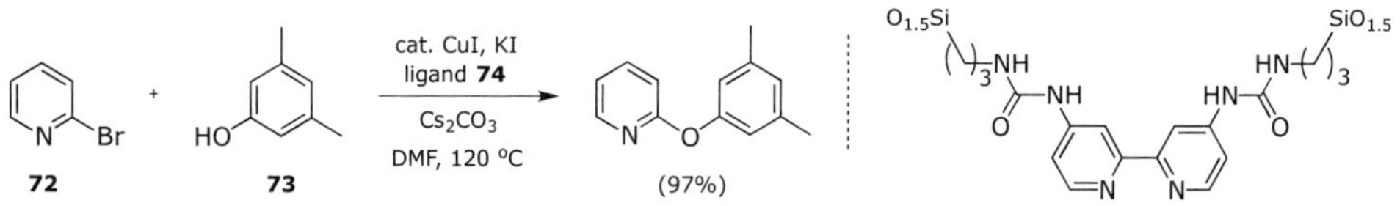

Another approach to heterogeneous Cu-catalyzed diaryl ethers calls for use of copper-in-charcoal (Cu/C), preferably with a 10–15 wt% loading of the metal into the support.[148] All diaryl ether formations using Cu/C (5 mol% *vs.* substrate) have been run at high temperatures (180–200 °C) using either conventional heating in a Teflon resealable round-bottom flask or pressure vial (Fig. 1-1A and B) or in a standard microwave irradiation tube (Fig. 1-1C).[148]

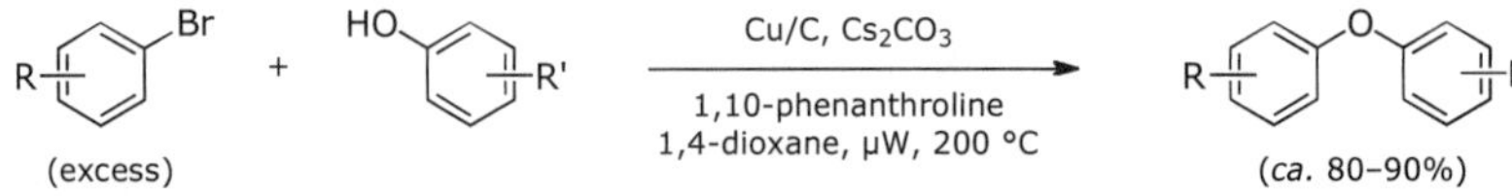

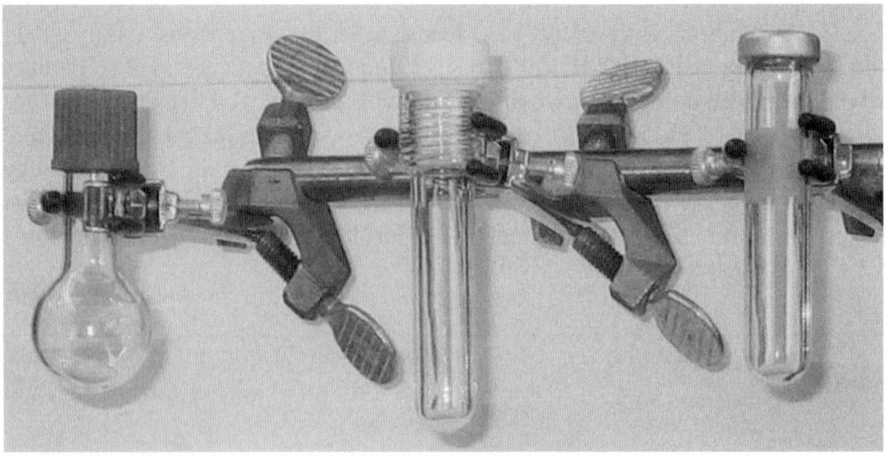

R–C6H4–Br (excess) + HO–C6H4–R' → (Cu/C, Cs_2CO_3; 1,10-phenanthroline; 1,4-dioxane, μW, 200 °C) → R–C6H4–O–C6H4–R' (*ca.* 80–90%)

Figure 1-1A **Figure 1-1B** **Figure 1-1C**

The higher loading in Cu/C allows for less mass of charcoal being used, a factor that aids with stirring especially within a microwave reactor. Both the base (Cs_2CO_3) and the ligand (1,10-phenanthroline) are crucial for successful couplings on aryl bromides, all run in dioxane over *ca.* an hour. Control experiments unequivocally established that Cu is leached into solution, and that each of the three components (phenol, base, and ligand) is needed for leaching to occur. Nonetheless, by adjusting stoichiometry such that the phenol is the limiting species, the presumed ligand-solubilized cesium phenoxide is lost upon etherification. Hence, copper no longer remains in solution prior to workup. Aromatic and heteroaromatic bromides can be used, *ortho*-substitution in either partner (or both) is tolerated, and several functional groups can be present in the halide (*e.g.*, CHO, NO_2, and ketone). Recent work indicates that there is no special "microwave effect" due to any selective catalyst heating; oil bath heating gives the same results.[149]

Procedure for microwave-assisted, Cu/C-catalyzed Ullmann diaryl ether coupling[148]

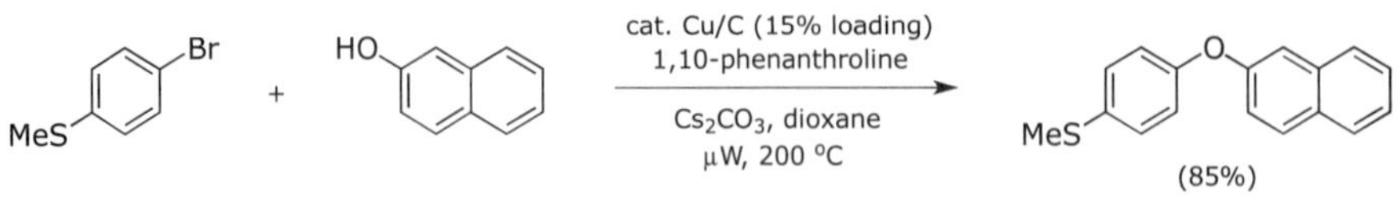

4-MeS–C6H4–Br + 2-naphthol → (cat. Cu/C (15% loading), 1,10-phenanthroline; Cs_2CO_3, dioxane, μW, 200 °C) → 4-MeS–C6H4–O–(2-naphthyl) (85%)

Cu/C (60 mg, 0.10 mmol, 1.8 mmol/g), 1,10-phenanthroline (180.2 mg, 1.00 mmol), cesium carbonate (781.9 mg, 2.4 mmol), 2-naphthol (288.3 mg, 2.0 mmol), and 4-bromothioanisole (609.3 mg, 3.0 mmol) were added to an oven-dried Biotage (Uppsala, Sweden) 2–5-mL process vial under argon at RT. Dry dioxane (3.0 mL) was added by syringe and the slurry stirred at RT for 30 min. The resulting heterogeneous mixture was heated in the microwave with the following settings: *Temperature:* 200 °C, *Time:* 4800 s, *Fixed hold time:* On, *Sample Absorption:* Normal, *Prestirring:* 10 s. After cooling to RT, the crude reaction mixture was filtered through Celite and the filter cake was further rinsed with EtOAc. Volatiles were removed on a rotary evaporator, and the crude product was purified on silica gel by flash chromatography (4/1 hexanes/toluene), $R_f = 0.35$, to yield 452 mg (85%) as a colorless solid; mp 53–54 °C.

Nanoparticles of CuO, prepared from pH-adjusted aqueous solutions of $Cu(NO_3)_2{\bullet}H_2O$, catalyze cross-couplings of both aryl bromides and iodides under basic conditions (KOH or Cs_2CO_3, respectively) in DMSO at 110 °C.[150] Steric effects due to substituents in the *ortho*-position reduce yields. Transmission electron microscopy (TEM) revealed particles on the order of 28–29 nm. Re-isolation of the catalyst by centrifugation allows for reuse; after five additional recycles, similar yields were observed as expected, as TEM images of the reclaimed catalyst revealed no change in particle shape or size.

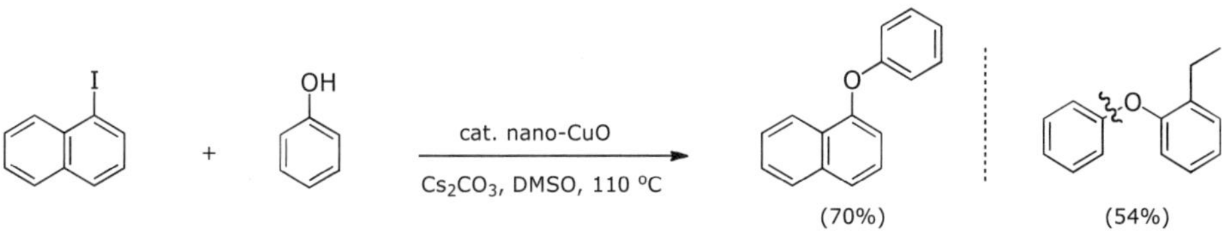

cat. nano-CuO

Cs_2CO_3, DMSO, 110 °C

(70%) (54%)

In addition to silica-based catalyst **71**, sol gel ligand **74**-complexed copper, and Cu/C, a fourth option for heterogeneous, copper-catalyzed ether formations is use of the copper-containing perovskite: $La_{0.9}Ce_{0.1}CO_{0.6}Cu_{0.4}O_3$ (or "LaCu").[151] Two other perovskites, $YBa_2Cu_3O_7$ and $LaFe_{0.57}Cu_{0.38}Pd_{0.05}O_3$, were found to be less active. Reactions using 1% Cu are run in refluxing toluene containing Cs_2CO_3 (1.4 equiv) over 48 hours. EtOAc (5%) is added to increase the solubility of copper and to significantly improve yields in what appears to be a "release-and-capture" mechanism for the active metal involved. Iodides, bromides, and some electron-poor chlorides led to the corresponding ethers in good isolated yields. Some representative examples are shown below (**75**–**77**; Fig. 1-2). Control reactions demonstrated that copper in LaCu is essential for activity.

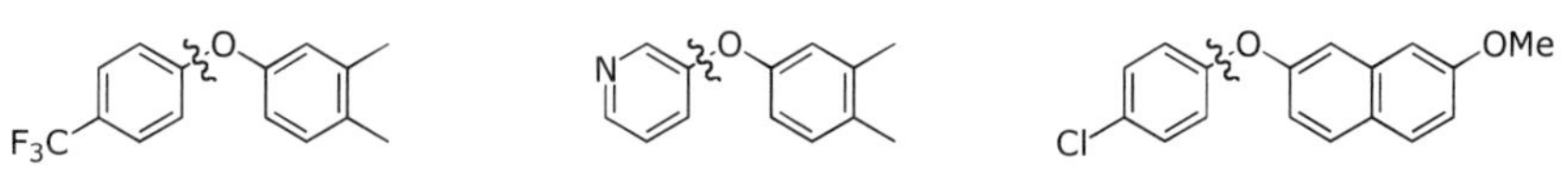

75 (90%) [from the chloride]

76 (74%) [from the bromide]

77 (82%) [from the iodide]

Figure 1-2. Diaryl ethers made using "LaCu" as catalyst

General procedure for $La_{0.9}Ce_{0.1}Co_{0.6}Cu_{0.4}O_3$-catalyzed Ullmann couplings of phenols[151]

$La_{0.9}Ce_{0.1}Co_{0.6}Cu_{0.4}O_3$ (4.4 mg, 5 mol%) was added to a mixture of aryl halide (0.71 mmol), phenol (1.00 mmol), Cs_2CO_3 (325 mg, 1.00 mmol), and 1 drop of EtOAc in toluene (1 mL) in a microwave vial. The tube was flushed with Ar, sealed, and heated in an oil bath at 110 °C for 48 h. The reaction mixture was filtered through a Chem Elut (Agilent Technologies, Santa Clara, CA) CE1005 drying column, eluting with Et_2O

(3 × 10 mL). The combined filtrates were concentrated *in vacuo* and purified by column chromatography to afford the desired product.

Attachment of 1,5-pentanediol to a Merrifield resin (benzylic chloride) under basic conditions (NaH, DMF, 80 °C) followed by conversion to the bromide and then esterification with *p*-iodobenzoic acid leads to coupling precursor **78**.[152] Many examples of diaryl ether formation have been done with phenolic derivatives using catalytic CuCl, TMHD (2,2,6,6-tetramethylheptane-3,5-dione) in a mixture of DMF and CH_3CN at 120 °C. Tolerance to *ortho*-substitution is seen, except in selected cases (*e.g.*, 2,6-di-*t*-Bu, or dichlorophenols). Cleavage from the resin is achieved by transesterification with methanolic KCN, or *via* saponification (KOH, IPA/H_2O, 2:1, 90 °C). Several other esters, besides benzoates, could also be attached to the solid support (*e.g.*, hydroxycinnamates) with which unsymmetrical diaryl ethers could also be generated using similar conditions (albeit with mainly aryl iodides as coupling partners).

pentanediol

resin

78

1. cat. CuCl, Cs_2CO_3 TMHD DMF, CH_3CN, 120 °C
2. KCN, MeOH, 90 °C, 12 h

(75%)

ligand = TMHD = **42** =

4.2 Diaryl Thioether Formation

Related thioethers can also be prepared using the same perovskite catalyst, LaCu, although conditions for its use needed further optimization. While the one study reported was limited,[151] prospects for "tailoring" metals and level of oxygenation in these industrial materials as catalysts look encouraging. Typical thioethers generated by the methodology are illustrated below.

Ar–X + HS-Ar-R → Ar-S-Ar-R

2.5% LaCu
NaO-*t*-Bu (2 equiv)
10% neocuproine, toluene
110 °C, 48 h

(64%) [from the iodide]

(78%) [from the bromide]

(99%) [from the bromide]

L =

64 (neocuproine)

Catalytic amounts of both CuBr and CuI are effective for cross-couplings of aryl iodides with aryl thiols in refluxing toluene.[153] The former is done in the presence of Schweisinger's phosphazene base P_2-Et (2 equiv; Aldrich catalog #79417), and while P_1-*t*-Bu is also acceptable, other bases such as Cs_2CO_3, Et_3N, Hünig's base, and *N,N*-dimethylaminopyridine afforded no diaryl thioether. DBU could be used in place of P_2-Et, although reaction times increased. Under these conditions, chemoselectivity is achieved for thioether over diaryl ether formation (Eqn. 1-6).

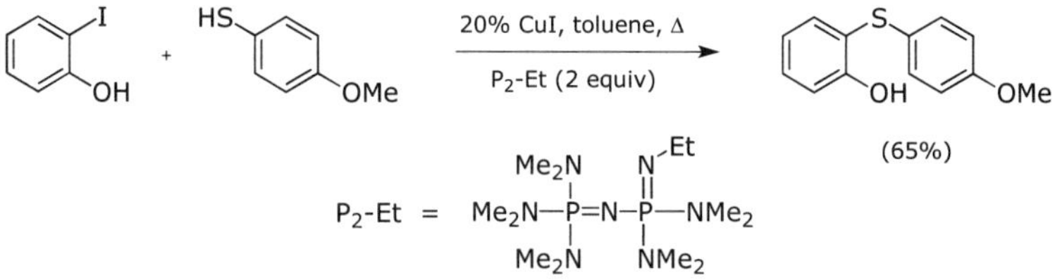

20% CuI, toluene, Δ; P_2-Et (2 equiv) (65%) (Eqn. 1-6)

P_2-Et = Me_2N–P(NMe_2)$_2$=N–P(=NEt)(NMe_2)$_2$

Higher yields were noted in this case with stoichiometric CuBr (95% *vs.* 65%). Representative products are shown in **79–81**, highlighting functional group compatibility and steric issues. Activated bromides also smoothly couple (*e.g.*, **81**).

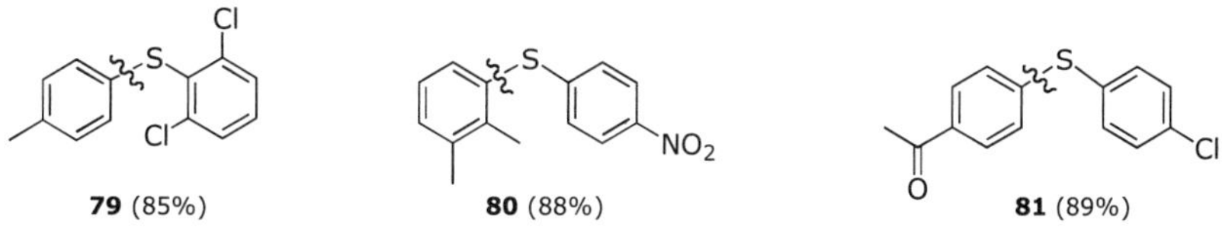

79 (85%) **80** (88%) **81** (89%)

Similar results are realized using a catalytic CuI/NaO-*t*-Bu/neocuproine (**64**) mixture with aryl iodides and, thus, a far less expensive base.[154] Cs_2CO_3 here, too, led to poor results (<50% conversion), but KO-*t*-Bu is an acceptable replacement for its sodium counterpart, as is K_3PO_4 (Scheme 1-5). Bromides are not good coupling partners. Noteworthy is the observation that both aryl and alkyl mercaptans can be used as nucleophiles (*e.g.*, **82–84**).

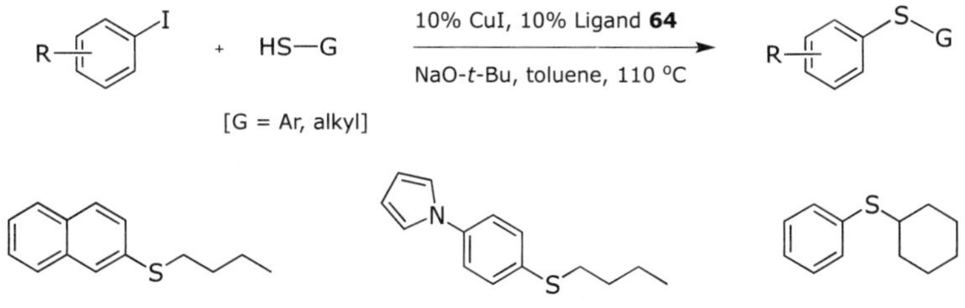

10% CuI, 10% Ligand **64**; NaO-*t*-Bu, toluene, 110 °C

[G = Ar, alkyl]

82 (95%) **83** (95%) **84** (77%)

Scheme 1-5

Copper-catalyzed coupling of thiophenols with aryl iodides; 2,6-dimethylphenyl sulfide[154]

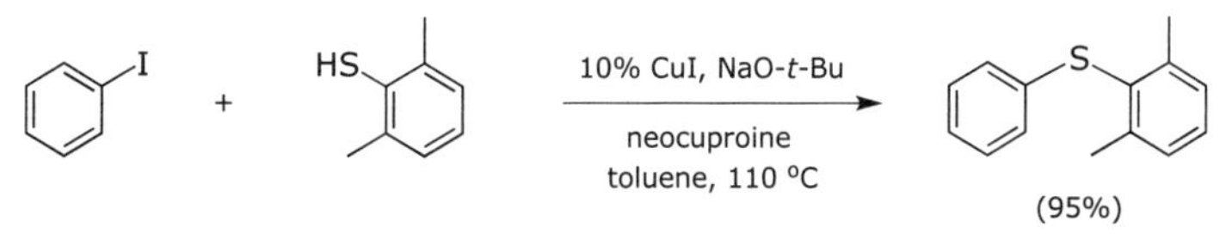

10% CuI, NaO-*t*-Bu; neocuproine; toluene, 110 °C (95%)

In an argon-filled glove box, a Pyrex glass tube (2.5 cm in diameter) equipped with a Teflon stir bar was charged with sodium *t*-butoxide (Acros Organics catalog #25255; 3.0 mmol), CuI (10 mol% with respect to the aryl iodide), and neocuproine (10 mol% with respect to aryl iodide). The tube was then sealed with a rubber septum, taken out of the glove box and thiophenol (2.2 mmol), and the aryl iodide (2.00 mmol) and toluene (6.0 mL) were injected into the tube through the septum. The contents were then stirred at 110 °C for 24 h. The reaction mixture was then cooled to RT and filtered to remove any insoluble residues. The filtrate was concentrated *in vacuo*, and the residue was purified by flash column chromatography (hexane) on silica gel to obtain the analytically pure product as a clear oil (409 mg, 95%). Anal. calcd. for $C_{14}H_{14}S$: C, 78.45, H, 6.58; S, 14.96; found: C, 78.58; H, 6.71; S, 14.98. Due to the stench of thiols, all glassware and syringes used were washed with bleach to reduce the odor.

Alternatively, boronic acids react under the influence of Cu-catalysis with electrophilic *N*-thioimides under mild conditions and in the absence of base (Eqn. 1-7).[155] Various copper carboxylates can be used (*e.g.*, CuMeSal, CuTC, and CuOAc) but not salts such as CuCl, CuCN, or Cu_2O. A presumed Cu(I)/Cu(III) cycle is put forth, where the thioimide is oxidatively added to the Cu(I) catalyst, and the $RB(OH)_2$ supplies the "R" group on the metal (with loss of imide-$B(OH)_2$) prior to reductive elimination. Several biaryl thioethers were made in this way (e.g., **85**, **86**) in THF (or dioxane), along with aryl alkenyl derivative **87**. All bases examined (K_2CO_3, NaOH, TBAF, pyridine, Et_3N) inhibited the reaction.

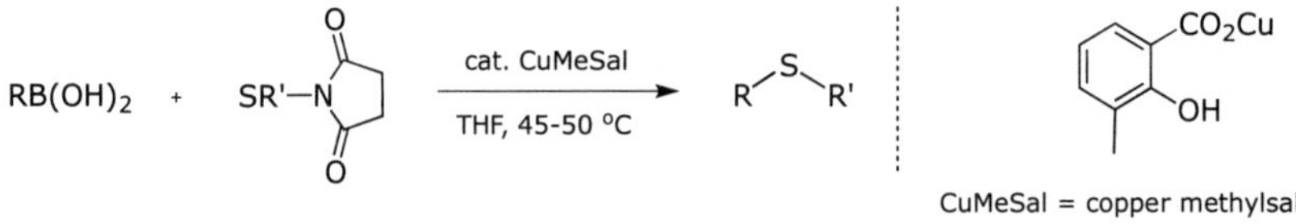

$RB(OH)_2$ + SR'–N(succinimide) —(cat. CuMeSal; THF, 45-50 °C)→ R–S–R' (Eqn. 1-7)

CuMeSal = copper methylsalicylate

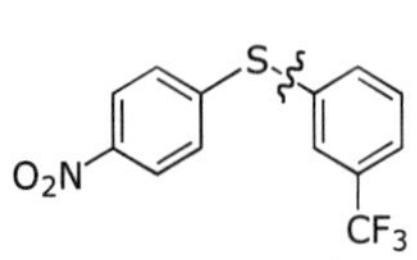

85 (69%) 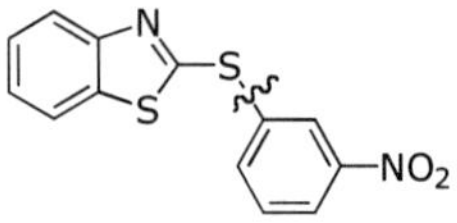 **86** (56%) 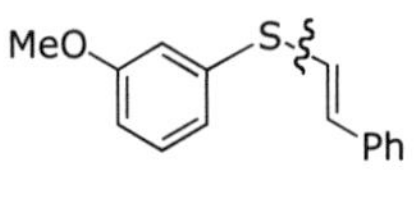 **87** (72%)

Bottom line comments. An impressive array of excellent modern procedures now exists, developed over the past decade, that offer considerable choices as to coupling partners, ligands, solvents, homo- *vs.* heterogeneous processes, etc. for making diaryl ethers and related bonds. And the good news is that all rely on catalysis with a base metal: copper. They are also convincing; that is, there is reason to seriously consider a copper-promoted, diaryl ether-forming step somewhat later in a synthesis, as examples illustrating functional group tolerance continue to appear. More advances are likely, as greater insight regarding ligand effects are likely to broaden the scope of reactive substrates (*e.g.*, chlorides and tosylates) and, in principle, allow for couplings to even take place in water at room temperature.

5. Asymmetric Cu-Catalyzed Borylations

In just the past few years alone, a wealth of new procedures has been generated using copper catalysis for introducing carbon-boron bonds, notably with absolute stereocontrol at C_{sp3} centers. Asymmetric 1,4-additions to acyclic enones of a presumed active copper boryl complex, derived from *bis*(pinacolato)diboron (B_2pin_2) and containing the preferred ligand (diphenylphosphino)ferrocenyl]ethyldicyclohexylphosphine (JOSIPHOS; 3 mol%), proceed in typically >90% ee.[156] Crucial for conversion is the presence of MeOH (1 equiv) in a THF medium at room temperature, which quenches the intermediate copper enolate to form L_nCuOMe for reentry into the catalytic cycle, upon subsequent reaction with pinacolborane (pinBH).

cat. CuCl, cat. Na-O-*t*-Bu, B_2pin_2
cat. JOSIPHOS, MeOH (1 equiv), THF, rt

(95%) 90% ee
(72%) 92% ee
(94%) 97% ee

Analogous conjugate additions to β-substituted cyclic enones can be achieved with B_2pin_2 but under different conditions of copper catalysis, including solvent: DMSO; copper salt: $CuPF_6(CH_3CN)_4$ (5–10 mol%); and ligand: (*R,R*)-QuinoxP*.[157] These asymmetric reactions are particularly impressive in that the substrates are β,β-disubstituted systems, and yet they react selectively even at room temperature in the absence of MeOH. In fact, in these cases, use of MeOH was detrimental to the yield. On the other hand, the additive LiO-*t*-Bu was found to be advantageous. Alternatively, copper (from *in situ*–formed CuO-*t*-Bu) complexed by (*R,S*)-Taniaphos in THF (+1 equiv MeOH) also leads to high levels of induction using similar educts.[158]

(n = 0-2) + B_2pin_2
cat. $CuPF_6(CH_3CN)_4$
Li-O-*t*-Bu (1.5 mol %)
Ligand, DMSO, rt
(>80%) 70-98% ee

Ligand = (*R,R*)-QuinoxP*
(91%) 94% ee
(85%) 95% ee
(95%) 90% ee [using Taniaphos]

General procedure for the asymmetric copper-catalyzed conjugate boration of β-substituted cyclic enones[157]

DMSO (500 μL) was added to a mixture of *bis*(pinacolato)-diboron (155 mg, 0.6 mmol) cyclic enone (0.4 mmole), $CuPF_6(CH_3CN)_4$ (7.5 mg, 0.02 mmol), and QuinoxP* (8 mg, 0.024 mmol), and the mixture was stirred for 10 min at RT. A THF solution of 1 M lithium *t*-butoxide (30 μL, 0.03 mmol) was then added, and the solution was stirred for 12 h. The reaction mixture was diluted with EtOAc (5 mL), water (1 mL) was added, and the mixture was stirred for 5 min. After phase separation, the organic layer was separated and dried over anhydrous Na_2SO_4. The inorganic salts were removed by filtration through cotton, and the filtrate was evaporated to dryness *in vacuo*. The residue was purified by neutral silica gel column chromatography (hexane/EtOAc 4:1), giving the product.

General procedure for the asymmetric β-boration of β-unsubstituted cyclic enones with (R,S)-Taniaphos[158]

To a oven-dried Schlenk tube equipped with a stir bar were added CuCl (0.010 mmol, 1.5 mg), NaO-*t*-Bu (0.015 mmol, 1.4 mg), (*R*,*S*)-Taniaphos (0.020 mmol, 13.8 mg), and THF (0.4 mL) under nitrogen. After the mixture had been stirred at RT for 30 min, *bis*(pinacolato)diboron (0.11 mmol, 140 mg) dissolved in THF (0.30 mL) was added. The reaction mixture was stirred for 10 min. The cyclic enone (0.5 mmol) was then added followed by MeOH (1 mmol, 0.04 mL). The reaction tube was washed with THF (0.3 mL) and sealed. The reaction mixture was stirred for 24 h; after which it was filtered through a pad of Celite and concentrated *in vacuo*. The product was purified by silica gel chromatography.

Related 1,4-additions on mono-β-substituted enoates lead to products **88** under the influence of a different ligand, nonracemic QUINAP (ee's up to 79%).[159] The alternative series of ligands, Quinazolinap, has been prepared and screened, but it was found not to be as effective.

Ligand	product ee(%)
QUINAP	79
Quinazolinap (R = *i*-Pr, R' = H)	42

An alternative method for copper-catalyzed asymmetric conjugate addition of pinacolborane to enoates relies on a newly engineered, atypical, 6-membered NHC-Cu(I) complex containing a rigid imidazoquinazoline core (**89a**; R = Me). This salt, prepared in two steps (>90% overall), is shelf-stable and is smoothly converted into its corresponding copper carbene (**89b**) upon treatment with CuCl and KO-*t*-Bu in DCM.[160] The combination of this catalyst (1 mol%) together with B_2Pin_2 (1.1 equiv) and NaO-*t*-Bu (0.3 equiv) in ether at –55 °C afforded the β-borylated esters in ee's between 82% and 96%, with yields in all cases >90%. Key to these reactions is the inclusion of MeOH in the pot (2 equiv), presumably as a low-temperature quenching agent of excess B_2Pin_2 that prevents unwanted reactions upon warming to room temperature. When run at 0 °C, a catalyst loading dropped from 1 to 0.01 mol% was still effective, thus, increasing the TON from 100 to 10,000.

1. NH⁻Li⁺ (aryl amide), THF, Δ; 2. NH_4BF_4, $(EtO)_3CH$, Δ → **89a** (R= Me, *t*-Bu); CuCl, KO-*t*-Bu, THF, DCM, 0° to rt, 18 h → **89b** (R= Me, *t*-Bu)

Representative procedure for β-borylation of enoates[160]

(93%) 90% ee (91%) 96% ee (95%) 96% ee

The α,β-unsaturated ester (0.20 mmol) and *bis*(pinacolato)diborane (56 mg, 0.22 mmol) were dissolved in Et_2O or toluene (1 mL). NaO-*t*-Bu (6 mg, 0.060 mmol) was added to the reaction mixture. The reaction mixture was cooled to –55 °C, and MeOH (18 µL, 0.4 mmol) was added. After 5 min, the 6-NHC copper catalyst was added (1.1 mg, 0.0020 mmol or 3.3 mg, 0.0060 mmol). After complete consumption of the α,β-unsaturated ester, the reaction mixture was filtered through silica gel and washed with Et_2O. The filtrate was concentrated on a rotary evaporator. The resulting residue was purified by column chromatography (hexane : EtOAc = 20 : 1) to afford the desired products. Note: The order of addition was critical to reproducibly achieve high yields and ee's.

While the procedure above works well with β-mono-substituted enoates, generation of a quaternary carbon center at the same β-site requires a different catalyst, in this case, a chiral, C_1-symmetric, 5-membered NHC-ligated source of Cu(I), **89c**.[161] Using 2.6 equivalents of NaO-*t*-Bu relative to CuCl (5 mol%) in THF at –78 °C, asymmetric addition takes place with both β-aryl/alkyl and β-alkyl/alkyl enoates in typically ≥90% ee's. The limitations noted include the case with an *o*-methyl group on the β-aryl ring, and a β-isopropyl residue that prevents addition of the Bpin moiety. The example below is illustrative of this protocol.

Representative procedure for NHC–Cu-catalyzed enantioselective boron conjugate additions. Preparation of the desired NHC–Cu-O-t-Bu[161]

conditions → (80%) 93% ee

89c precursor imidazolium salt

conditions: NHC presursor salt (5 mol %), CuCl (5 mol %), NaO-*t*-Bu (13 mol %), B_2Pin_2 (1.1 equiv), MeOH (1.2 equiv), THF, -78 °C, 24 h; quenched by 30% HCl in MeOH

In an oven-dried vial (6 × 1 cm) equipped with a stir bar, imidazolinium tetrafluoroborate salt (22 mg, 0.033 mmol), NaO-*t*-Bu (8.3 mg, 0.086 mmol), and CuCl (3.3 mg, 0.033 mmol) were placed and THF (1.0 mL) was added in the glove box. After the solution was allowed to stir for 2 h at 22 °C under a dry N_2 atmosphere, it was filtered through a short plug of flame-dried Celite and rinsed with 1.0 mL of THF.

An appropriate portion of the solution of NHC–Cu-O-*t*-Bu (0.017 mmol in 1.0 mL of THF) was placed in a separate oven-dried vial (6 × 1 cm), and the resulting solution was charged with $B_2(Pin)_2$ (93 mg, 0.37 mmol). The vessel was removed from the glove box, placed in a fume hood, and allowed to cool to –78 °C. (*E*)-Ethyl 3-phenylbut-2-enoate (63 mg, 0.33 mmol) and MeOH (16 uL, 0.40 mmol) were added, and the mixture was allowed to stir for 24 h at –78 °C; after which, the reaction was quenched by the addition of 30% HCl in MeOH (1 mL). The resulting mixture was subsequently allowed to warm to RT, H_2O (3 mL) was added, and the solution was neutralized through addition of a saturated solution of aq. $NaHCO_3$. The layers were separated, and the aqueous portion was washed with Et_2O (5 mL × 3). The combined organic layers were dried over anhydrous $MgSO_4$ and filtered. The volatiles were removed *in vacuo*, and the resulting light yellow oil was purified by silica gel chromatography (hexanes/Et_2O : 5/1) to afford 86 mg (0.27 mmol, 80% yield) of (*S*)-ethyl 3-phenyl-3-(4,4,5,5-tetramethyl-1,3,2-dioxaborolan-2-yl)butanoate as a colorless oil. Specific rotation: $[\alpha]_D^{20}$ –13.6 (c 1.70, $CHCl_3$) for a sample with 96.4:3.6 er. Enantiomeric purity was determined by HPLC analysis in comparison with authentic racemic material; 96.4:3.6 er; chiralpak AD–H column (25 cm × 0.46 cm), 99.8/0.2 hexanes/*i*-PrOH, 0.5 mL/min, 220 nm).

β,β-Disubstituted unsaturated thioesters react under similar conditions; yet they require somewhat higher temperatures (–50 °C instead of –78 °C); nonetheless, resultant ee's are actually higher than those found with enoates.[161]

EtS, Ph — *conditions* (as above) → EtS, BPin, Ph

(87%) 98% ee

Nonracemic pinacolborane derivatives are also readily available from styrene precursors *via* copper-catalyzed hydroboration.[162] The regiochemical outcome is virtually 100% controlled, with boron resulting at the benzylic site. Using *P*-chiral tangphos (see below; 3.3 mol%) as ligand, ee's are typically >80%. Other bidentate phosphines, such as *p*-Tol-(1,1′-binaphthalene-2,2′-diyl)bis(diphenylphosphine) (*p*-Tol-BINAP), MeO-BIPHEP, JOSIPHOS, and Me- 1,2-*bis*-phospholanobenzene (Me-DUPHOS), led to ee's of 64%, 70%, 70%, and 83%, respectively, with styrene as the model educt. Reaction temperatures between RT and 40 °C were used. Note that the combination of reagents leads to *in situ*–generated CuH, which is postulated to effect initial hydridocupration followed by transmetalation with pinacolborane (pinBH, 1.2 equiv). The pinBH is also the source of hydride that initially forms the catalyst L_nCuH once CuCl and NaO-*t*-Bu have metathesized to CuO-*t*-Bu. *E*-Disubstituted examples were successful as well, although *Z*-alkenes (*e.g.*, indene) are not reactive.

CuCl + NaO-*t*-Bu —toluene→ [CuO-*t*-Bu] —pinBH, L_n*→ L_n*CuH —Ar-CH=CH$_2$→ Ar-CH(CuL$_n$*)-CH$_3$ —pinBH→ Ar-CH(Bpin)-CH$_3$

(*S,S,R,R*)-tangphos
[Ligand L_n]

(95%) 82% ee

(90%) 91% ee

(61%) 88% ee

General procedure for copper-catalyzed asymmetric hydroboration.[162]

A mixture of CuCl (0.015 mmol, 1.5 mg), NaO-*t*-Bu (0.03 mmol, 3.0 mg), and (*S,S,R,R*)-tangphos (0.0165 mmol, 4.7 mg) in anhydrous toluene (0.2 mL) was stirred for 10 min in a Schlenk tube under an atmosphere of nitrogen. Pinacolborane (0.6 mmol, 89.8 μL) was added to the reaction mixture, and stirring was continued for 10 min at RT. Styrene or a styrene derivative (0.5 mmol) and tetradecane (0.25 mmol) as an internal standard were added in toluene (0.1 mL), and the reaction tube was washed with toluene (0.2 mL), sealed, and monitored by TLC and GC. Upon completion, the mixture was filtered through a pad of Celite and concentrated *in vacuo*. The product was purified by chromatography on silica gel.

Remarkably, the B(pin) moiety can be introduced at the β- rather than at the α-site (*vide supra*) of a styrene or styrene-like substrate using a nonracemic copper-NHC complex, derived from **89d**, in variable overall yields (51–98%).[163] The reactions are run in THF at –50 °C over 48 hours in the presence of MeOH (2 equiv). The same B_2Pin_2 + CuCl + KO-*t*-Bu combination is used to generate the active Cu-B species.

89d

(75%) 89% ee
[after oxidative work-up]

(98%) 89% ee

An updated procedure has been developed that applies to α-substituted styrenes; hence, while the B(Pin) moiety is directed to the β-site, a new stereogenic center is created at the α-carbon.[164] Evaluation of several C_1-symmetric NHC ligand precursors led to the chiral bidentate, naphthyl-substituted array **89e**. Reactions performed in THF at –50 °C in the presence of MeOH (2 equiv) tended to give the best results in terms of ee's, although these can be highly variable (*ca.* 60–92%), as can the yields (37% to >98%), although the latter reflect the extent of conversion rather than the reaction eficiency. Flexibility as to substitution on the aromatic ring is extensive, while the hydrocarbon residue at the α-site can be varied as well (R = Me, Et, *i*-Pr, *i*-Bu) with larger groups leading to better ee's. Note that the new stereocenter derives from secondary protonation of an intermediate quaternary C-Cu bond formed *via* initial stereocontrolled addition of Cu-B to the alkene. Some representative examples follow.

Ar(R)C=CH₂ → CuCl (7.5 mol %), salt **89e** (7.5 mol %); NaO-*t*-Bu (0.8 equiv), B_2Pin_2 (1.1 equiv); MeOH (2.0 equiv), THF, -50 °C, 2 d → Ar(R)CH–CH₂BPin

89e

(60%) 75% ee (90%) 90% ee (57%) 93% ee

Regiocontrolled net hydroboration of terminal alkynes can also be realized due to the nature of NHC-Cu complex employed. Thus, with an *N*-arylated NHC (*e.g.*, IMes or IPr), α-vinylboronates are formed, while with a bulky *N*-alkyl NHC-Cu, the β-isomer prevails. Substrates such as propargyl ethers, amines, and a variety of aryl-substituted alkynes have all been studied.[165]

R'—≡ → *conditions* → R'C(BPin)=CH₂ or R'CH=CHBPin

conditions: NHC-Cu complex (5 mol %), NaO-*t*-Bu (5 mol %), B_2Pin_2 (1.0 equiv), MeOH (1.1 equiv), THF, -50 °C, 24 h

R = Me; IMesCuCl
R = *i*-Pr; IPrCuCl
R = *Ad*; IAdCuCl

BocNH (93%), BnO (76%) — (using IMesCuCl) —; CF₃ (70%) (using IPrCuCl); CF₃ (80%) (using IAdCuCl)

An asymmetric allylic alkylation (AAA) route to substituted allylic boronates relies on the influence of phosphoramidite ligand **90**, together with 2 mol% copper thiophenecarboxylate and a Grignard reagent in DCM at –78 °C.[166] These S_N2' displacements occur with *ca.* 7:1 regioselectivity using vinylboronate **91** derived from 2,2-dimethylpropanediol. Other borates, on occasion, give better regiocontrol, but the ee's were highest when derived from this allylic chloride precursor, **91**. A one-pot process through to homoallylic alcohol **92** of essentially identical ee is suggestive of virtually complete chirality transfer in the ternary transition state (*i.e.*, aldehyde, boronate, and BF_3).

91 → cat. CuTC, EtMgBr; Ligand **90**, CH_2Cl_2, -78 °C → 95.5% ee

RCHO, $BF_3·OEt_2$, -78 °C

90 **92**; >90% ee

Several alternative routes to nonracemic allylic boronates now exist, including use of *Z*-allylic carbonates that react with the combination catalytic (*R,R*)-QuinoxP* (**90**)-complexed Cu(I)/B_2Pin_2 (in THF at 0 °C) to afford (*S*)-products in usually >90% ee's. The corresponding *E*-educts give ee's that are not synthetically useful.[167]

R, OCO$_2$Me → cat. Cu-O-*t*-Bu, cat. QuinoxP*; B$_2$pin$_2$, THF, 0 °C, 48 h → BPin, R; ee's ≥90%

93 (*R,R*)-QuinoxP*

A stereo*divergent* approach that addresses both *Z*- and *E*-allylic carbonate conversions to *S*- and *R*-allylic boronates, respectively, has appeared.[168] It relies on the analog of **89e**, the bidentate NHC precursor salt **89f**, which varies by replacement of the naphthyl residue a 2,6-diisopropylphenyl ring. Unexpectedly, the Cu(II) salt $Cu(OTf)_2$ was identified as optimal, and DME (as opposed to THF) led to the best results. Products from potentially competitive S_N2 delivery of the BPin moiety were not observed. Importantly, opportunities to apply this technology for generation of quaternary centers also exist (albeit by switching from **89f** to the corresponding IMes analog that replaces the diisopropylphenyl ring), although an alkyl residue larger than methyl affords reduced levels of enantioselectivity. In most cases, ee's are ≥90%.

R, OCO$_2$Me → BPin, R

conditions

R, OCO$_2$Me → BPin, R

89f

conditions: NHC presursor salt **89f** (6 mol %), Cu(OTf)$_2$ (5 mol %), NaOMe (0.8 equiv), B$_2$Pin$_2$ (2.0 equiv), DME, -30 °C, 24 h

Representative experimental procedure for enantioselective Cu-catalyzed substitution reaction of (E)-3-cyclohexylbut-2-en-1-yl methyl carbonate with B_2Pin_2. (R)-2-(2-Cyclohexylbut-3-en-2-yl)-4,4,5,5-tetramethyl-1,3,2-dioxaborolane[168a]

OCO$_2$Me → *conditions* → BPin

(>98%) 96% ee

In an N_2-filled glove box, an oven-dried vial (4 mL, 17 × 38 mm) with magnetic stir bar was charged with the appropriate imidazolinium salt (6.0 mg, 0.012 mmol, 6.0 mol%), $Cu(OTf)_2$ (3.7 mg, 0.010 mmol, 5.0 mol%), NaOMe (8.8 mg, 0.16 mmol, 80 mol%), and DME (1.0 mL) under an N_2 atmosphere. The mixture was sealed with a cap (phenolic open top cap with red PTFE/white silicone) and allowed to stir for 30 min. The color of the solution was clear blue. *Bis*-(pinacolato)diboron (102 mg, 0.400 mmol, 2.0 equiv) was added to the solution. The color of the solution immediately turned dark brown. The vial was resealed with the same type of cap. After 30 min, the vial was removed from the glove box and was cooled to –30 °C, stirred for 10 min, and then (*E*)-3-cyclohexylbut-2-en-1-yl methyl carbonate (42.5 mg, 0.200 mmol, 1.0 equiv) was added neat by syringe. The reaction was allowed to stir 24 h at –30 °C. The solution was quenched by the addition of THF (reagent grade), allowed to warm to RT, and passed then through a short plug of Celite and silica gel, eluting with Et_2O (3 × 2 mL). The filtrate was concentrated *in vacuo* to provide a dark brown oil, which was purified by silica gel chromatography (0% →1% Et_2O in hexanes) to afford the allylic boronate, (*R*)-2-(2-cyclohexylbut-3-en-2-yl)-4,4,5,5-tetramethyl-1,3,2-dioxaborolane, as a colorless oil (30.8 mg, 0.200 mmol, >98% yield); $[\alpha]_D^{20}$ –12.6 (c = 1.00, $CHCl_3$). The er 98:2 was confirmed by oxidation to the alcohol (*S*)-2-cyclohexylbut-3-en-2-ol.

Yet another advance in the preparation of nonracemic allylboronates, in this case a stereo*convergent* method, has been reported that uses the same novel 6-membered NHC-copper complex **89b**, albeit with R = *t*-Bu, applied to asymmetric 1,4-additions of B(pin) to enoates (*vide supra*).[169] Several distinguishing features set this study apart from the other procedures described herein, beyond the differences in the NHC ligand associated with the copper catalyst, and the need for only 1 mol% of **89b**. First, the substrates do not contain the typical carbonate leaving groups, or even related leaving groups such as acetate, halide, or pseudo-halide. Rather, a *m*-nitrophenyl or 3,5-dimethylphenyl *ether* serves admirably in this context, and in fact, it reacts much faster (minutes) relative to the corresponding methyl carbonates (hours). Second, the B(pin) residue is exclusively delivered at a 3-monosubstituted, over 3,3-disubstituted, allylic ether. Last, but likely to be viewed as most significant, is that both *E*- and *Z*-isomers of the starting allylic ethers afford the same stereo-outcome associated with the allylic boronate. Hence, mixtures of *E/Z* isomeric educts obtained, *e.g.*, by olefin cross-metathesis, are of no consequence in this chemistry.

TBSO (12:1 *E:Z*) (<1:30 *E:Z*) NO_2 — conditions → TBSO BPin 83% (*S*) 91% (*S*)

Br (3.3:1 *E:Z*) NO_2 — conditions → Br BPin 92% (*S*)

conditions: **89b** (R = *t*-Bu; 1 mol %), B_2Pin_2 (1.1 equiv), NaO-*t*-Bu (0.3 equiv), MeOH (2 equiv), Et_2O, -55 °C, 14 h

Typical reaction conditions for allylic substitution reactions[169]

Allylic aryl ethers (0.20 mmol) and *bis*-(pinacolato)diboron (56 mg, 0.22 mmol) were dissolved in Et_2O (1 mL). Na-O-*t*-Bu (6 mg, 0.060 mmol) was added to the reaction mixture. The reaction mixture was then cooled to –55 °C and MeOH (18 μL, 0.40 mmol) was added. After 5 min, the 6-NHC copper catalyst **89b**, R = *t*-Bu (1.2 mg, 0.0020 mmol) was added. After complete consumption of the allylic aryl ether, the reaction mixture was filtered through silica gel and washed with Et_2O. The filtrate was concentrated under rotary evaporator. The resulting residue was purified by column chromatography (hexane:Et_2O) to afford the desired product. Absolute configurations were determined by comparison with known data.

Readily available *N-t*-butanesulfinylimine derivatives of aldehydes react with B_2pin_2 in the presence of the NHC-copper catalyst (ICy)Cu-O-*t*-Bu (5 mol%) to give high ratios of diastereomeric α-amino boronate esters.[170] Benzene is the solvent of choice, as lower yields are observed for the reaction below, run in toluene (69%), THF (50%), or dioxane (62%). Both alkyl and aryl aldimines participate to give dr's >95:5. The utility of this highly diastereoselective process was demonstrated by its application to a synthesis of bortezomib (Velcade), a U.S. Food & Drug Administration (FDA)-approved and clinically useful protease inhibitor.[171]

General procedure for the addition of bis(pinacolato)diboron to aryl N-tert-butanesulfinyl imines[170]

+ B_2pin_2 — cat. (ICy)Cu-O-*t*-Bu, toluene, 0 °C → (52-61%) dr > 95:5

Reactions were set up in a glove box. To a Schlenk tube containing a stir bar and *bis*(pinacolato)diboron (0.508 g, 2.00 mmol, 2.0 equiv) in toluene (2.0 mL) was added the sulfinylimine (1.00 mmol, 1.0 equiv) in 2.0 mL of toluene. (ICy)CuO-*t*-Bu (36.6 mg, 0.10 mmol, 0.10 equiv) was dissolved in toluene (2.0 mL) and added to the reaction vessel. The Schlenk tube was sealed and the mixture placed in a precooled bath at 0 °C with stirring at that temperature for 28–48 h. The reaction mixture was then removed from the Schlenk flask and diluted and extracted with EtOAc (10 mL). The mixture was then concentrated under reduced pressure. The products were isolated by *rapid* silica gel chromatography on deactivated silica gel (35% w/w) using EtOAc/DCM mixtures and were visualized with CMA stain.

bortezomib ⇒ [*via* hydrolysis with HCl/MeOH of the sulfinylimine]

Bottom line comments. Access to valued nonracemic, boron-containing intermediates for use in synthesis, or even as a functional group within final targets, is rapidly growing by virtue of new methodologies based on copper catalysis. The levels of stereocontrol already realized are in the 90% ee category, and considering that these approaches are of recent vintage, further refinements within this hot topic are likely to be forthcoming.

6. Oxidations of Organocopper Complexes

Reviews. Kozlowski, M. C.; Morgan, B. J.; Linton, E. C. *Chem. Soc. Rev.* **2009**, *38*, 3193; Surry, D. S.; Spring, D. R. *Chem. Soc. Rev.* **2006**, *35*, 218.

6.1 Biaryl Syntheses

Exposure of organocuprates to an excess of oxidating agent usually leads to C-C bond formation between the two carbon-based ligands on copper, although carbon-heteroatom bonds can also be made. Several oxidizing agents are available that readily induce this reductive elimination following conversion to a Cu(II) and, possibly, maybe even likely, a Cu(III) species. Historically, ground state oxygen has been used extensively as the oxidizing agent,[172, 173] as have transition metal salts[174] as well as mono- or polynitro-substituted benzenes.[175] More recent alternatives include benzamide derivative **94**,[176] which offers the advantage of ease of product separation due to the basic tertiary amine, and variously substituted *p*-benzoquinones (**95**), with the tetra-methyl derivative (R = CH_3) of particular note.[177] Amide **94** is readily prepared in large quantities and has excellent shelf life at ambient temperatures. These are reactions stoichiometric in copper; catalytic process, however, are starting to accrue (vide infra).

Common oxidizing agents:

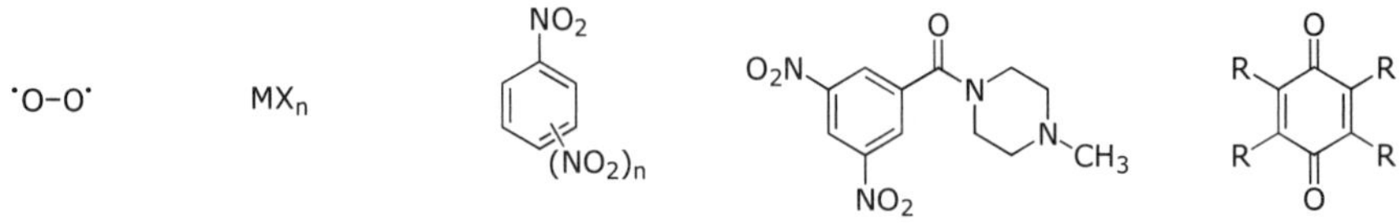

$^{\bullet}O\text{-}O^{\bullet}$ — oxygen

MX_n — transition metal salts

nitrobenzenes (n = 0,1)

94

95

Most of the recent cuprate oxidations have used cyanocuprates as precursors to targeted biaryls. The prescription follows from earlier work involving induced oxidative decomposition of a preformed "higher order" (HO) cyanocuprate (*i.e.*, a Cu(I) dianionic species, $R_2Cu(CN)Li_2$).[178] Based on a considerable body of spectroscopic and theoretical data, the initially formed lower order cuprate RCu(CN)Li undergoes cleavage of the Cu-CN bond upon reaction with the second equivalent of RLi to generate a lower order Gilman cuprate, $[R\text{-}Cu\text{-}R']^{-}Li^{+}$, with cyanide as part of the cluster bound to lithium.[179] Upon treatment with an oxidizing agent, three possible products can result: R-R, R-R', and R'-R' (Scheme 1-6). However, when two aryl halides within a single molecule are both lithiated and treated with 0.5 equivalents of CuCN, selectivity for intramolecular cross-coupling (*i.e.*, ring formation) is observed.[180] The linear, 180° arrangement of the groups attached to copper in the intermediate diarylcuprate, along with the relatively long (*ca.* 1.9 Å) carbon-copper bonds, are favorable considerations that have been applied to otherwise challenging medium-sized ring formations.[176] Aryl halide precursors can lead to diastereomeric products reflecting newly created axial chirality, ratios being a function of ring size and reaction parameters. Under milder, kinetic conditions, products such as **96** and **97** are formed in good yields. Subsequent heating to 120–150 °C equilibrates the initially formed mix of isomers (*e.g.*, **96**).[180a] This chemistry is highly amenable to solid-phase synthesis on polystyrene beads.

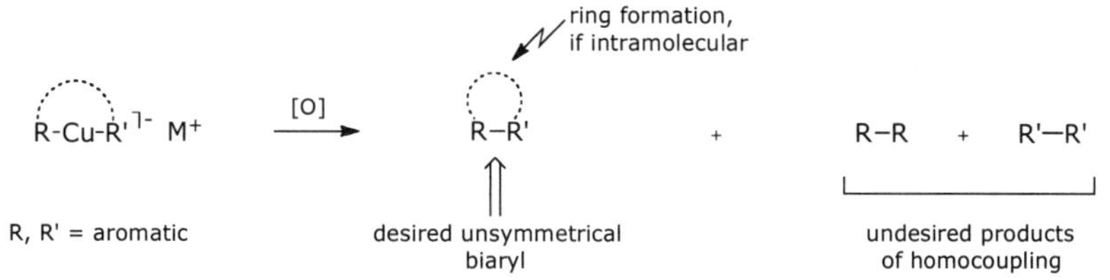

R-Cu-R']⁻ M+ —[O]→ R–R' + R–R + R'–R'

ring formation, if intramolecular

R, R' = aromatic

desired unsymmetrical biaryl

undesired products of homocoupling

Scheme 1-6

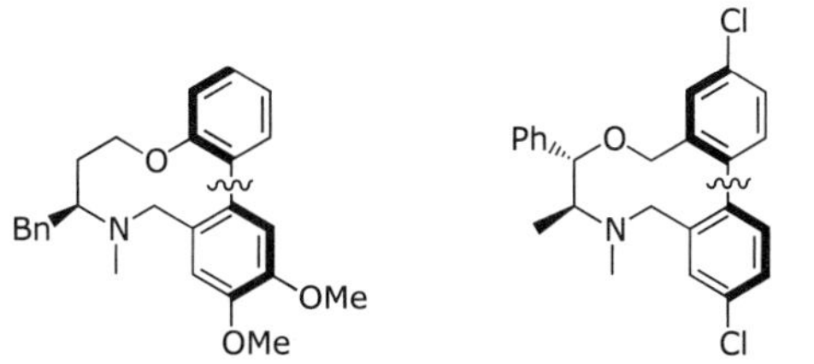

96; 16:1 dr (92%)
(1:11 dr thermodynamic)

97; 6:1 dr (94%)
(1:2 dr thermodynamic)

Preparation of (M,3S)-17-bromo-5-methyl-3-phenyl-2-oxa-5-azatricyclo-[11.4.0.^{7,12}]-heptadeca-7,9,11,13,15,17-hexane[180a]

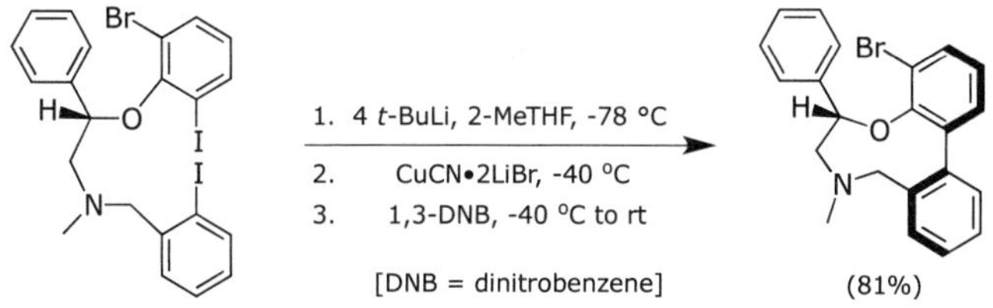

1. 4 *t*-BuLi, 2-MeTHF, -78 °C
2. CuCN•2LiBr, -40 °C
3. 1,3-DNB, -40 °C to rt

[DNB = dinitrobenzene]

(81%)

An oven-dried, three-necked, round-bottomed flask containing the cyclization precursor and equipped with a magnetic stirring bar, a septum, a low-temperature thermometer, and a vacuum/gas inlet (greased joints) was evacuated and twice heated with a heat gun over 30 min. The flask was filled with Ar and charged with the diiodide and then anhydrous 2-methyltetrahydrofuran (2-MeTHF, 0.05–0.15 M). The solution was cooled to –78 °C, and *t*-BuLi (4.0 equiv) was added dropwise. After addition the reaction was stirred for 30 min to give a clear yellow solution of the dilithium intermediate. A freshly prepared solution of CuCN•2LiBr (1 M) in 2-MeTHF (1.0 equiv) was added dropwise, and then the mixture was stirred for 2 h allowing the reaction temperature to slowly rise to –40 °C. At this temperature, a freshly prepared solution of 1,3-dinitrobenzene (1 M) in 2-MeTHF (4.0 equiv) was added to give a dark brown-black solution. The solution was allowed to warm to RT over 1–12 h. The black solution was quenched with a 1-M solution of HCl in MeOH to give a clear yellow-brown solution that was stirred for 30 min. A solution of 10% NH_4OH in saturated aqueous NH_4Cl was added, and the resulting two-phase mixture was stirred for 40 min, then separated, and the aqueous phase was extracted with Et_2O (3 times). The combined organic layers were washed (saturated aqueous NaCl), dried (anhydrous K_2CO_3), filtered, concentrated *in vacuo*, and chromatographed to give a colorless oil (81%); kinetic dr: 1:>50 (*P*:*M*); R_f 0.48 (SiO_2; hexane/EtOAc/Et_3N 90:9:1); *m/z* (ESI) 394 (MH^+, 100%); HRMS calcd. for $C_{22}H_{21}BrNO$ (MH^+) 394.0806; found 394.0815.

This approach to biaryls using diarylcuprate oxidations, proceeding by way of *t*-BuLi-induced metal-halogen exchange to lithiated aromatics, has been successfully applied to natural products total synthesis. Syntheses of lignans such as interiotherin A[181] and angeloylgomisin R,[181] gomisins O and E (of general structure **98**),[182] and perylenequinones calphostin A[183] and phleichrome (**99**),[183b] all rely on such methodology. Unnatural products, including 10-membered ring cyclophanes composed of thiophenes as in **100**, and both nonaphenylenes and dodecaphenylenes, have also been realized by electron-transfer oxidations of preformed cyanocuprates.[184]

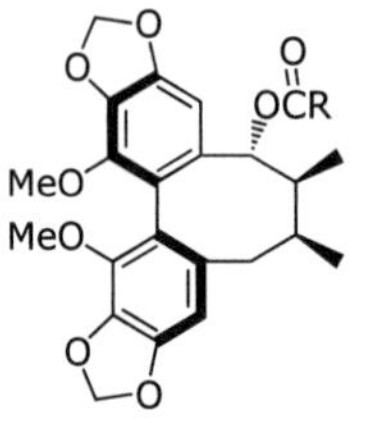

98
dibenzocyclooctadiene lignan natural products

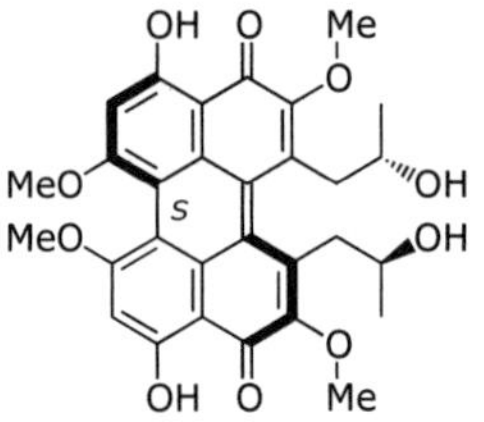

99
phleichrome

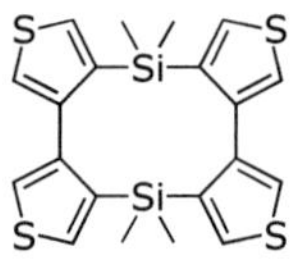

100
silicon-bridged cyclophane

p-Benzoquinones are effective oxidizing agents, and the tetramethyl-derivative (duroquinone, **95**, R = Me; 1.5 equiv) in particular, appears to offer remarkable generality for intermolecular homocouplings of simpler aryl bromides.[185] No by-product resulting from aromatic-to-CN coupling was seen. Yields based on either CuI or CuSCN were substantially inferior. For especially hindered cases (*e.g.*, mesitylene bromide), excess *t*-BuLi (2.5 equiv relative to Ar-Br) was needed to ultimately afford the tetra-*ortho*-substituted biaryl. Nitrogen-containing substrates (*e.g.*, 2-pyridyl, 4-cyanophenyl, and 4-Me_2N-C_6H_4) gave no biaryl product.

General procedure for the synthesis of biaryls using electron-transfer oxidation. Preparation of 4,4'-dichlorobiphenyl[185]

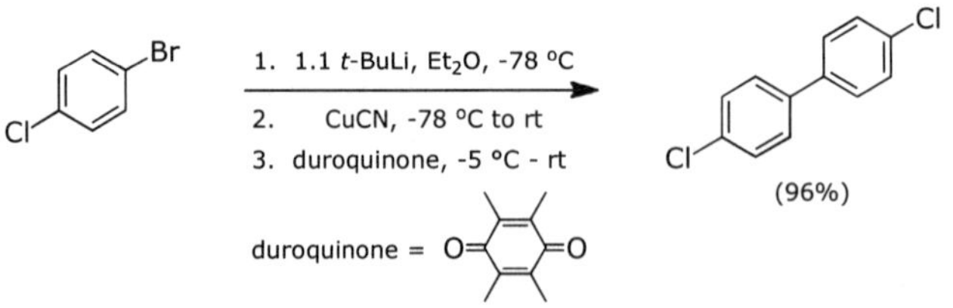

1. 1.1 *t*-BuLi, Et_2O, -78 °C
2. CuCN, -78 °C to rt
3. duroquinone, -5 °C - rt

(96%)

duroquinone =

To a solution of 4-chlorobromobenzene (382 mg, 2.0 mmol) in ether (120 mL) was added 1.5 mL (2.2 mmol, 1.1 equiv) of *t*-BuLi (1.49 M in pentane) at –78 °C. The mixture was stirred at this temperature for 1.5 h, and then CuCN (90 mg, 1.0 mmol) was added. After addition, the cooling bath was removed and the mixture was vigorously stirred until all the CuCN dissolved. To the clear pale yellow solution of cuprate was added solid tetramethyl-*p*-quinone (duroquinone; 492 mg, 3.0 mmol, 1.5 equiv) at –5 °C giving rise to a deep bluish-green solution of durosemiquinone anion radical. The mixture was stirred at RT for 3 h to complete the reaction. Water was then added; the deep bluish-green color disappeared to give a yellow solution. The organic layer was separated, and the aqueous layer was extracted with ether. The combined organics were dried over anhydrous $MgSO_4$; after which, the solvent was evaporated *in vacuo*. The crude product was purified by column chromatography on silica gel with hexane/benzene (3:1) as eluent to afford 214 mg (96%) of 4,4′-dichlorobiphenyl.

Binaphthyl couplings can also be accomplished using copper complexes of nonracemic diamine ligands, such as the 1,5-diaza-*cis*-decalin Cu(II) derivative **101**, and the CuCl/(*S*)-Phbox ligand **102**.[186, 187] Substituted 2-naphthols undergo oxidative dimerization to give variable yields and ee's of 1,1'-binaphthyl-2,2'-diol (BINOL)-like products. Such couplings have been used successfully *en route* to perylenequinones (*i.e.*, natural products with extended oxidized aromatics, such as pleichrome, **99**) and nigerone (Scheme 1-7).[188] Copper(I) chloride or iodide in acetonitrile, a solvent known to stabilize Cu(I), leads to the highest ee's and yields; other sources of copper(I), *e.g.*, -OTf, -CN, or Cu(II) salts with BF_4, OAc, NO_3, Br, Cl, or SO_4 were less effective on one or both outcomes.

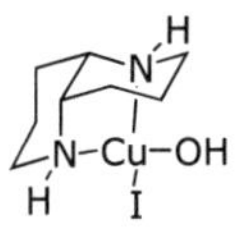

101 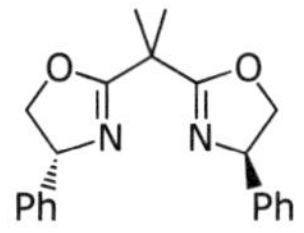 **102**

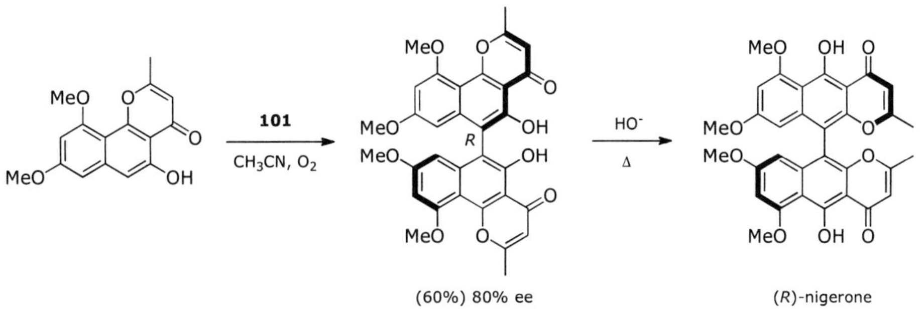

Scheme 1-7

Nonracemic biaryl ligands have been constructed using an intramolecular Ullmann coupling,[189] where axial chirality is induced with high fidelity from central chirality in the tether to the biaryl bond (*ca.* 70% yield; Eqn. 1-8).[190a] As with tunaphos ligands,[190] *bis*-phosphines **103** have varying dihedral angles as a function of methylene units (n in **103**) in the tether, a feature that has far-reaching implications for stereoinduction, especially in asymmetric hydrogenations.[191] Thus, with n = 0, the angle is computed (Chem 3D MM2) to be only 64.8°, while with n = 1 and n = 2, it goes up to 80.0° and 88.8°, respectively. By way of comparison,[190a] MeO-BIPHEP and BINAP are 83.2° and 86.6°, respectively.

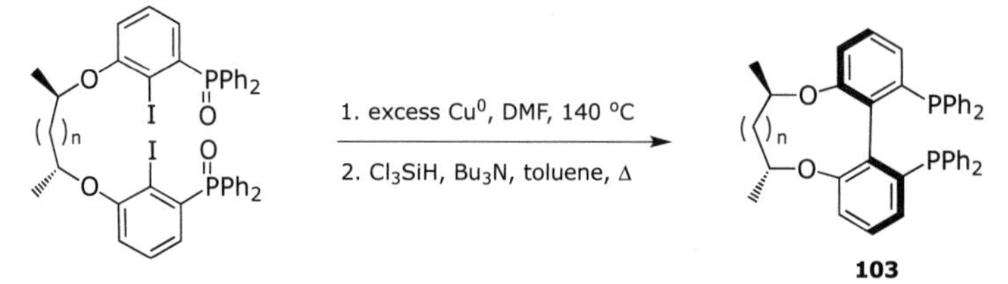

1. excess Cu^0, DMF, 140 °C
2. Cl_3SiH, Bu_3N, toluene, Δ

(Eqn. 1-8)

103

A newly introduced method for a one-pot unsymmetrical biaryl construction takes advantage of an initial selective iodination of one of the coupling partners, followed by copper-catalyzed cross-coupling.[192] Since the mechanism by which an arene undergoes iodination varies by the nature of the arene, the iodination step needed to be worked out for each type of arene (*i.e.*, electron-rich, electron-poor, both 5- and 6-membered aromatics and heteroaromatics, etc.). Common to each set of conditions is the use of molecular iodine as the oxidant. Moreover, CuI complexed by phenanthroline could be used in all cases. The base and solvent, therefore, were two key parameters that need to be varied. Reaction temperatures in all cases are between 100 °C and 135 °C, and reaction times can vary from hours to, often, days. Yields in most cases tend to be moderate. The method does not apply to two electron-rich arenes. Also, arenes with protons that are of pK_a greater than that of DMSO (*i.e.*, **37**) are not amenable as they fail to form the intermediate copper species required for the

[*arylation of an electron-rich arene*]

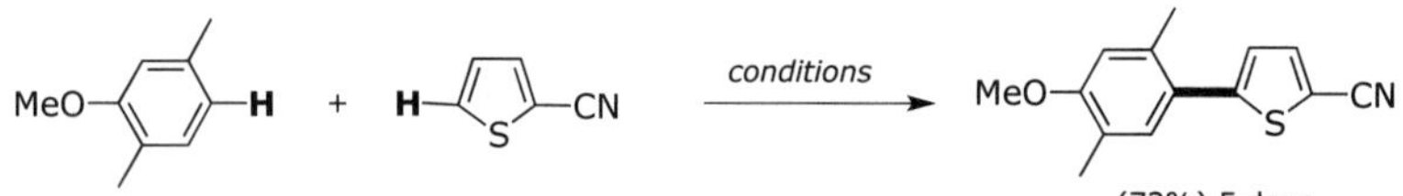

(72%) 5 days

conditions: 10% CuI/phenanthroline, I_2 (1.3 equiv), pyr. (0.5 equiv), K_3PO_4 (3.5 equiv), *o*-dichlorobenzne, 130 °C

[*arylation of an electron-poor arene*]

(50%) 4 days

conditions: 10% CuI/phenanthroline, I_2 (1.6 equiv), LiO-*t*-Bu (3.0 equiv), K_3PO_4 (2.0 equiv), dioxane, 120 °C

[*arylation of 5-membered ring heterocycles*]

(60%) 2 days

conditions: 10% CuI/phenanthroline, I_2 (1.4 equiv), pyr. (0.5 equiv), K_3PO_4 (3.5 equiv), dioxane, 120 °C

[*arylation of pyridines*]

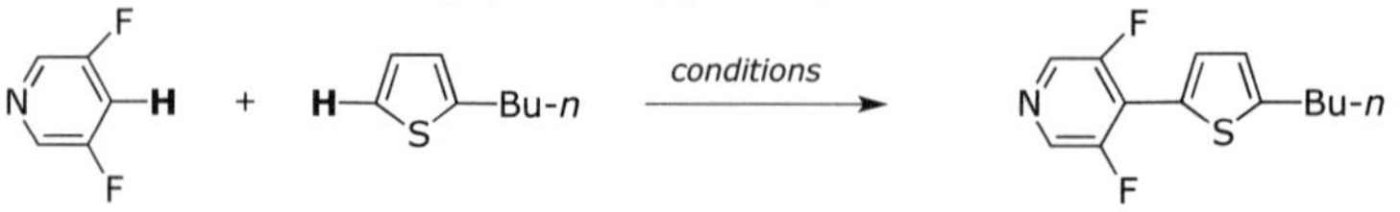

(79%) 7 days

conditions: 10% CuI/phenanthroline, I_2 (1.3 equiv), K_3PO_4 (4.0 equiv), *o*-dichlorobenzene, 120 °C

coupling. Among the dozens of examples cited, one representative biaryl associated with each of the four types of substitution patterns is highlighted above.

6.2 Aminations

Oxidations of Gilman-type cuprates that contain both an amido and an aryl ligand on copper ultimately lead (upon reductive elimination) to a net aryl amination (see Section 3.1 for direct aryl aminations).[193] Aromatics are metalated using Grignard reagents, either by direct deprotonation or metal-halogen exchange. Conversion to a solubilized aryl copper species (using 2 equiv LiCl) is then followed by the addition of an *N*-lithiated amide to form cuprate **104**. Treatment of **104** with chloranil (**105**) gives modest-to-good yields of aryl amines, improved with *bis*[2-(*N,N*-dimethylamino)ethyl]ether in the pot (Scheme 1-8). Noteworthy is that $LiN(TMS)_2$ (LiHMDS) can be used in this cuprate formation/oxidation sequence, and subsequent treatment of the initial adduct with tetra-*n*-butylammonium fluoride (TBAF) affords primary anilines. *tert*-Butyldimethylsilyl (TBS)-protected lithiated anilides as cuprate precursors, by the same rule, give *N*-TBS-protected products available for desilylation to secondary diarylamines. Heterocyclic Grignards are also participants. Highly hindered amides are not problematic (see **106**). Representative amines prepared by this protocol are shown below.

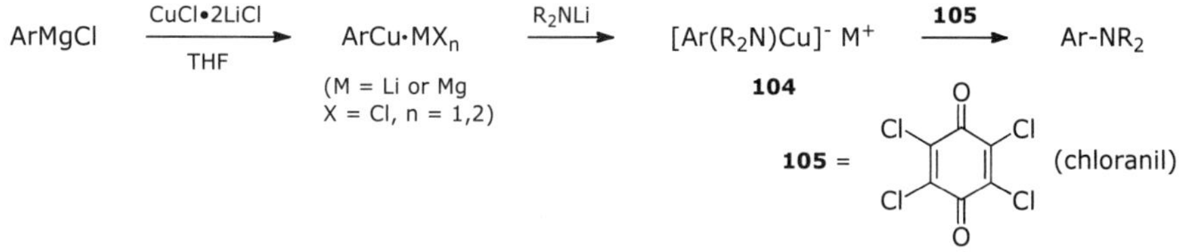

ArMgCl —(CuCl•2LiCl, THF)→ ArCu•MX_n (M = Li or Mg, X = Cl, n = 1,2) —(R_2NLi)→ $[Ar(R_2N)Cu]^- M^+$ **104** —(**105**)→ Ar-NR_2

105 = (chloranil)

Scheme 1-8

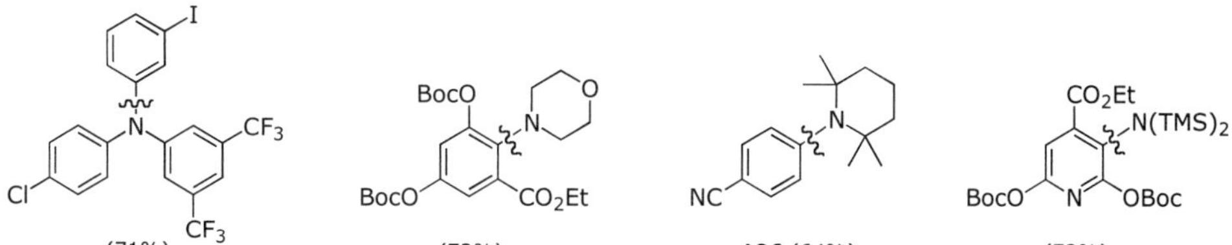

(71%) (72%) **106** (64%) (72%)

Typical procedure. Metalation, cuprate formation, oxidation; net amination[193a]

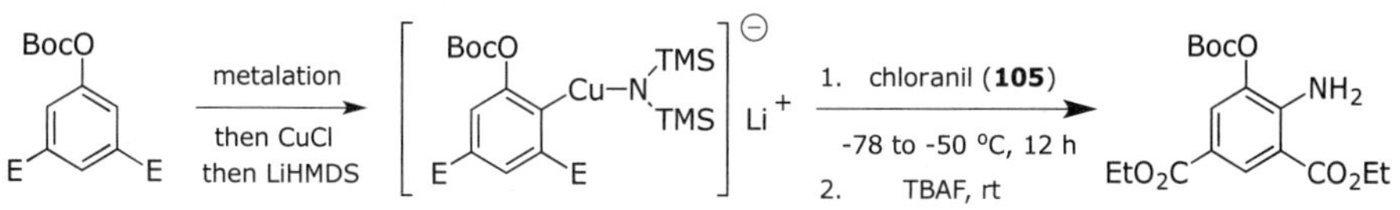

metalation, then CuCl, then LiHMDS → 1. chloranil (**105**), -78 to -50 °C, 12 h; 2. TBAF, rt (72%)

A dry, argon-flushed Schlenk flask, equipped with a magnetic stirring bar and a septum, was charged with a solution of the diester (338 mg, 1.0 mmol) in dry THF (3 mL). After cooling the mixture to 0 °C, TMPMgCl•LiCl (0.92 mL, 1.2 M in THF,

1.1 mmol; TMP = tetramethylpiperidide) was added dropwise and the mixture was stirred for 1 h to afford the corresponding Grignard reagent. This reagent was then added dropwise to a solution of CuCl•2LiCl (1.2 mL, 1.0 M in THF, 1.2 mmol) and *bis*-[2-*N*,*N*-dimethylamino)ethyl]ether (192 mg, 1.2 mmol) at –50 °C under argon, and the mixture was stirred for 45 min. LiHMDS (2.0 mL, 1.0 M in THF, 2.0 mmol) was added dropwise to the resulting aryl cuprate, and the mixture was further stirred for 90 min at –50 °C. The reaction mixture was cooled to –78 °C, and then a solution of chloranil (295 mg, 1.2 mmol) in dry THF (7 mL) was added slowly over a period of 45 min. The reaction was then allowed to reach –50 °C and stirred for 12 h. Et_2O (10 mL) was poured into the crude reaction mixture, which was then filtered through Celite, and the residue was washed with Et_2O (*ca.* 100 mL). The organic phase was washed with 2 × 10-mL portions of aqueous NH_4OH (2.0 M) and extracted with Et_2O. The combined organic layers were dried over anhydrous $MgSO_4$, filtered, and concentrated under reduced pressure. This crude material was dissolved in Et_2O (3 mL); after which TBAF (2 mL, 1.0 M in THF, 2 mmol) was added in one portion. The mixture was stirred at RT for 10 min, poured into EtOAc (10 mL), and washed with water (3 × 10 mL). The combined organic extracts were dried (anhydrous $MgSO_4$), filtered, and concentrated *in vacuo*. Purification by flash chromatography (pentane/Et_2O; 4:1) yielded the aromatic amine (254 mg, 72%) as a white crystalline solid (mp 86.7–88.1 °C).

Bottom line comments: Cuprate oxidation as an entry to biaryls offers a unique opportunity to combine two metalated (*i.e.*, anionic) species to form a single C-C bond, as opposed to a more traditional "nucleophile + electrophile" approach. This methodology becomes all the more attractive with the recent advent of new conditions for aromatic metalations in the presence of a wide range of functionality, including electrophilic centers (*e.g.*, esters, nitriles, etc.). Together with the latest advances in oxidizing agents, a better appreciation for cuprate structure, and the documented opportunities for controlling atropisomerism, this often overlooked technology may find increased usage going forward. The downside, of course, is that copper is still needed in stoichiometric amounts.

7. 1,2-Additions to Imines

Reviews. Friestad, G. K.; Mathies, A. K. *Tetrahedron* **2007**, *63*, 2541; Riant, O.; Hannedouche, J. *Org. Biomol. Chem.* **2007**, *5*, 873.

Nonracemic nitrogen-based scaffolds are commonly accessed *via* enantioselective additions to imine derivatives. As the number of copper-complexing ligands continues to grow, so does the number of methods that provides entry to amine derivatives from such 1,2-additions.[194] Copper can play various roles in assisting this mode of reaction, including service as the reactive organometallic formed by way of transmetalation from another metal species (*e.g.*, allylic borates),[195] as the Lewis acidic site of *N*-complexation, or both. At issue in all situations is the residue attached to nitrogen of the imine; most of the literature where copper is concerned suggests that both aryl sulfonyl (-SO_2Ar) or phosphinoyl (-$P(O)Ar_2$) moieties are especially useful, although on occasion other groups can function in a similar capacity when matched with appropriate nonracemic copper chelating ligands.

Propargylic amines have been consistent targets for copper-based methodologies. Preformed imines derived from aryl aldehydyes and anilines or substituted anilines react with terminal acetylenes using $\leq$10 mol% CuOTf or $Cu(OTf)_2$ in the presence of biaryl ligands such as **107** and **108**, typically best run in toluene at room temperature.[196] The extent of induction tends to be in the range ee ~80–85%.

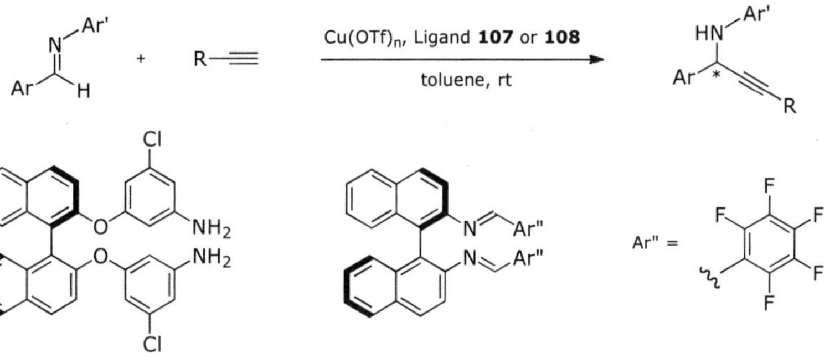

107 (with $Cu(OTf)_2$) **108** (with CuOTf)

Related routes to propargylic amines include 3-component systems, with *in situ* imine formation using dibenzyl- or diallyl-amine (Scheme 1-9).[197] Copper(I) bromide seems to be a good source of the eventually formed catalyst. With ligands such as QUINAP and PINAP (**109**), reactions can be performed in toluene at ambient temperatures. Acetylene itself does not lead to the expected high enantioselectivities usually seen with monosubstituted alkynes. Examples have been studied where silicon serves as a bulky proton (*i.e.*, protected equivalents; *e.g.*, TMS, TBS, or triisopropylsilyl [TIPS]); only the TMS cases give good ee's, depending on the ligand.

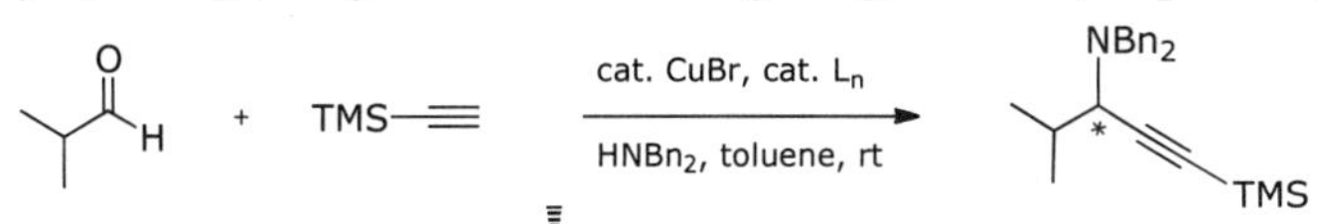

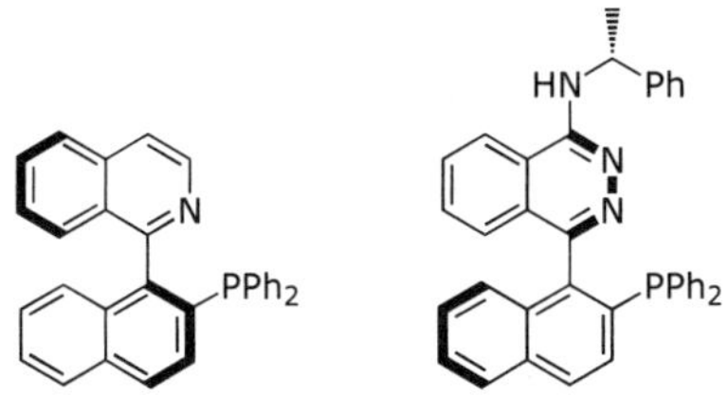

L_n	yield(%)	ee(%)
(*R*)-QUINAP	87	96 (*S*)
(*R*,*P*)-PINAP	84	98 (*R*)

(*S*)-QUINAP **109**; (*R*,*P*)-PINAP

Scheme 1-9

Procedure for formation of a nonracemic propargylamine[197a]

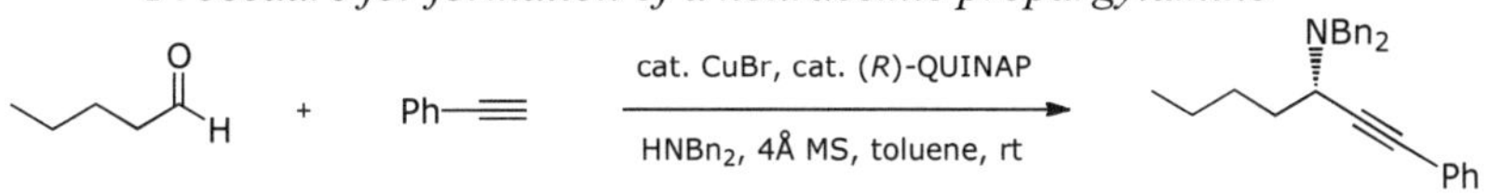

(85%) 83% ee

A dry, Ar-flushed, 10-mL flask equipped with a magnetic stirrer and a septum was charged with copper(I) bromide (3.6 mg, 0.025 mmol) and (*R*)-QUINAP (12.1 mg, 0.028 mmol). Dry toluene (2 mL) was added and the mixture was stirred at RT for 30 min. Molecular sieves (4 Å, 0.3 g) were added followed by phenylacetylene (51 mg, 0.5 mmol), valeraldehyde (43 mg, 0.5 mmol), and dibenzylamine (99 mg, 0.5 mmol). The reaction mixture was stirred for 48 h at RT. The molecular sieves were removed by filtration and washed with Et_2O. The crude product was concentrated *in vacuo* and purified by chromatography on silica gel (pentane/Et_2O 99:1) yielding the desired propargylamine (156 mg, 85%, 83% ee) as a colorless oil. $[\alpha]^{20}_D = -239$ (c = 1.00, $CHCl_3$); HREIMS: *m/z* calcd. for $C_{27}H_{33}N$: 366.2222; found: 366.2215 [M^+ - H]; Anal. calcd. for $C_{27}H_{33}N$: C 88.24, H 7.95, N 3.81; found: C 87.85, H 7.84, N 3.73; HPLC (OD-H, 99% *n*-heptane/1% IPA, 0.2 mL/min): t_R = 45.4 (+), 53.8 min (–).

Several aniline-derived imines also make for excellent substrates toward various acetylenes to produce propargylic amines.[198] Here, $CuPF_6$ is ligated by a C-2 symmetric Pybox ligand **110**, and it mediates delivery in excellent ee's. Reactions are performed in $CHCl_3$, where coordination of the imine nitrogen to copper leads to several transition state-stabilizing interactions (*e.g.*, C-H•••π and π•••π) that account for the enantioselectivity. Noteworthy is the case of *p*-methoxyphenyl (PMP), which is subject to potential facile oxidative removal from the initial amine adduct.

PhCHO + PMP-NH_2 + [aryl acetylene] —10% (L_n)$CuPF_6$, $CHCl_3$→ [propargylamine, (*S*)] (97%) 99% ee

Ligand = L_n = **110**; *i*-Pr-Pybox-diPh

Aldehydes converted to their corresponding sulfonylimines make for excellent substrates subject to Cu-complexation/nucleophilic attack.[199] For example, tosyl derivatives RCH=NTs react with glycine ester derivative **111**; either *syn*- or *anti*-diamine products can be fashioned in high yields and selectivities. Ligand **112** leads to *anti*-selectivity, while ligand **113** bearing electron-withdrawing aryl residues reverses the diastereoselectivity. The copper salt $CuClO_4$ serves as a catalyst precursor (11 mol%).

111 + RCH=NTs —cat. $CuClO_4$, cat. L_n, Et_3N, 4 Å MS, THF, -78 °C→ *syn* + *anti*

L_n = [ferrocenyl oxazoline, PAr_2]

112; Ar = 4-MeO-C_6H_4 ⟹ *anti*-diamine

113; Ar = 3,5-$F_2C_6H_3$ ⟹ *syn*-diamine

Another example of a 1,2-addition to a tosylimine derivative mediated by a CuOTf-derived complex with *bis-N*-oxide ligand **114** is the (aza)Henry reaction.[200] An enantioselective version can be done with ketoimines of both aromatic and aliphatic ketones. The solvent is unusual in PhOEt, and both excess CH_3NO_2 and molecular sieves (4 Å) are needed. The reactions take (up to 10) days to reach completion, but the yields can be good and ee's tend to be over 85%.

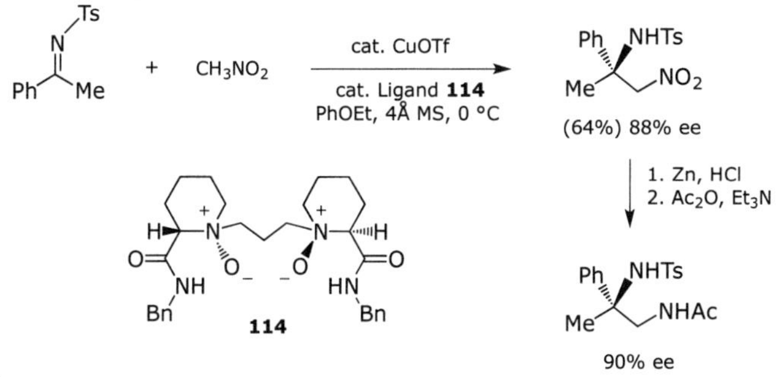

Other *N*-sulfonyl-containing imines, such as the 2-thienylaldimine derivative **115**, and the 8-quinolyl analog **116** (see procedure below), are valuable for copper-catalyzed vinylogous Mannich reactions,[201] and direct Mannich reactions (below), respectively. The former is effected in DCM at room temperature with the Fesulphos ligand 1-dinaphthyl analog **117**, while the latter employs the parent (diphenyl) system (**118**). A catalytic silver salt ($AgClO_4$) is part of the recipe, which at –40 °C in the presence of "diene" silyloxyfuran **119**, gives butenolides in good dr's and ee's >89% (*e.g.*, **120**). Upon treatment with Mg/MeOH, the 2-thienyl (2-Th) moiety is cleaved, a nice feature that after hydrogenation, leads to lactam **121**.

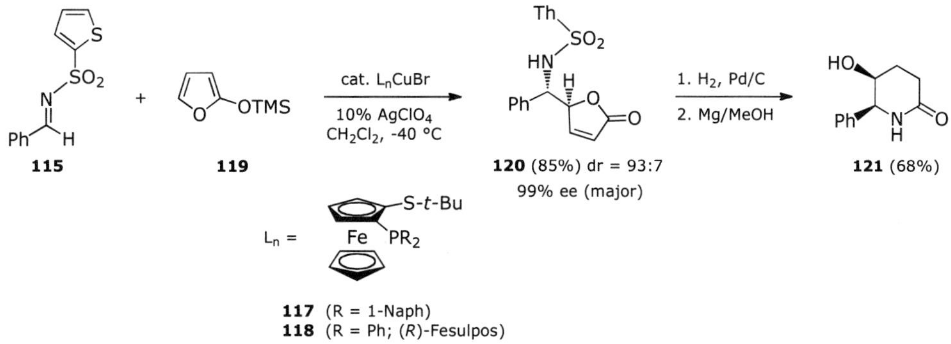

The addition of a glycine Schiff's base (**122**) to the corresponding aldimine **116** using ligand **118** (complexed to copper, using $Cu(CH_3CN)_4PF_6$) gives vicinal diamines in differentiated/protected form.[202] The highly favored *anti*-isomer is controlled by nitrogen in the quinoline, as the sterically equivalent naphthylsulfonyl residue gave a complex reaction mixture. Other copper salts were slightly less effective (*e.g.*, CuOTf: 77%, 91% ee; *anti:syn* = 94:6).

General procedure for the asymmetric Mannich reaction[202]

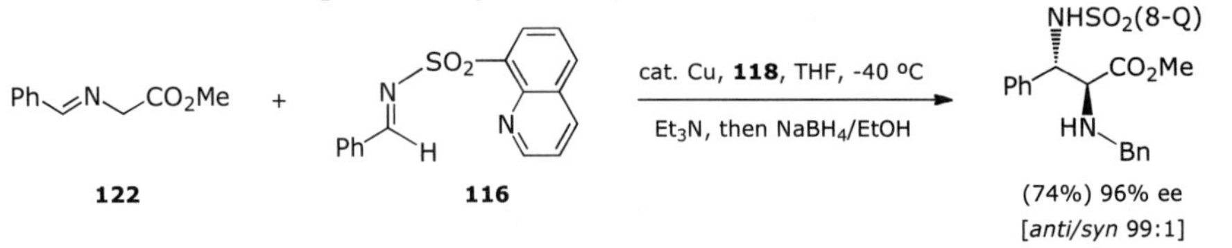

122 + 116 → cat. Cu, **118**, THF, -40 °C; Et_3N, then $NaBH_4$/EtOH → (74%) 96% ee [*anti/syn* 99:1]

To a solution of ligand (*R*)-Fesulphos (**118**; 2.8 mg, 0.082 mmol, 5 mol%) and $Cu(CH_3CN)_4PF_6$ (3.4 mg, 0.075 mmol, 5 mol%) in THF (0.5 mL), at the optimal temperature (−40 °C or −78 °C), were successively added a solution of the glycinate imine (0.15 mmol) in THF (1.0 mL), Et_3N (3 μL, 0.16 mmol, 10 mol%), and a solution of the corresponding *N*-sulfonylimine (0.165 mmol, 1.1 equiv) in THF (1.0 mL). Upon consumption of the starting material (TLC monitoring), the reaction mixture was cooled to 0 °C before being treated with EtOH (2 mL) and $NaBH_4$ (5.7 mg, 0.15 mmol). The mixture was allowed to reach RT and stirred at this temperature for 30 min before it was quenched with saturated aq. NH_4Cl (2 mL) and extracted with EtOAc (3 × 30 mL). The combined organic phase was washed with brine, dried (anhydrous $MgSO_4$), and concentrated *in vacuo*. The residue was analyzed by ^1H NMR to determine the diastereomeric ratio, and then purified by flash chromatography.

Considerable effort has been made in developing *N*-diphenylphosphinoyl (Dpp) ald- and keto-imines as electrophilic partners in copper-catalyzed 1,2-additions (Scheme 1-10).[203] Thioamides and aldimines (*e.g.*, **123**) in the presence of a soft Lewis acidic Cu(I) salt complexed by *bis*-phosphine **124**, together with a hard base in the form of a lithium phenoxide, react to afford products in good-to-excellent ee's. Activation resulting from a likely Cu-S interaction increases the acidity of an α-proton and, hence, more facile deprotonation. With alkylnitriles as nucleophiles generated *via* an initial decarboxylative process (*e.g.*, from **125**), a DTBM-SEGPHOS-coordinated copper complex arrives at C-C bond formation in usually >80% ee.

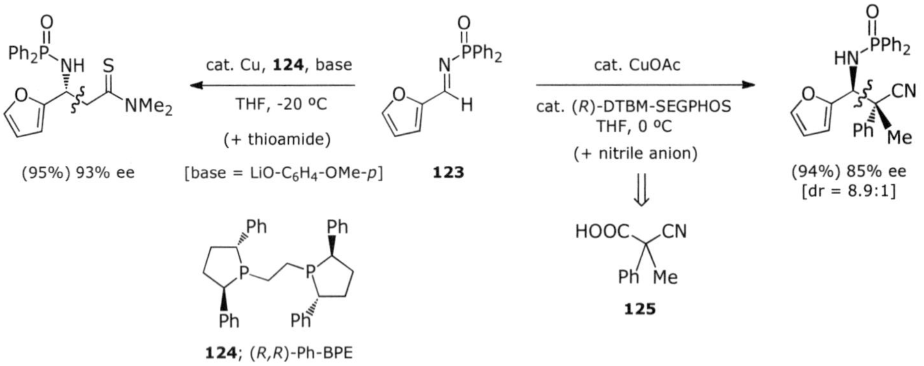

(95%) 93% ee ← cat. Cu, **124**, base; THF, -20 °C; (+ thioamide); [base = LiO-C_6H_4-OMe-*p*] — **123** — cat. CuOAc; cat. (*R*)-DTBM-SEGPHOS; THF, 0 °C; (+ nitrile anion) → (94%) 85% ee [dr = 8.9:1]

124; (*R,R*)-Ph-BPE

125

Scheme 1-10

The 1,2-addition of dialkylzinc reagents, in particular Et_2Zn, to aryl imines is stereocontrolled by a ligated form of copper bearing a hemi-labile, monophosphine oxide (Me-DUPHOS(O), **126**).[204] Nonracemic secondary carbon centers bearing a protected amine are created in >90% ee's and in good yields. With alkyl-substituted imines, however, moisture is a problem; hence, sulfinic acid derivatives **127** serve as imine precursors *via* β-elimination induced by excess Et_2Zn. Extensive work has been done on the synthesis of such ligands that contain a nonracemic phospholane, a labile heteroatom-coordinating group, and a linker between these sites.

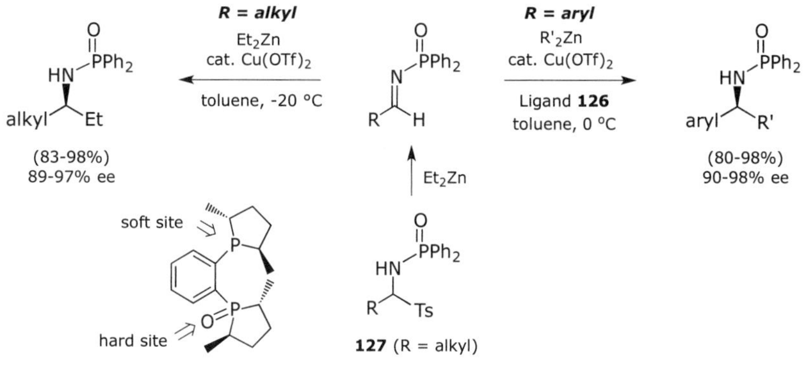

126; Me-DUPHOS(O)

A related *in situ* imine formation/1,2-addition by a ligated copper-containing nucleophile has been found in the form of *N*-formyl imines.[205] Precursor dialkylzinc reagents (*e.g.*, Et_2Zn, *i*-Pr_2Zn, and *n*-Bu_2Zn) add in the presence of 2 mol% $Cu(OTf)_2$ and phosphoramidite ligand **128** (4 mol%) to afford adducts **129** in good yields and ee's. Introduction of a methyl group using far less reactive Me_2Zn required higher temperatures (–30 °C up to –10 °C) at which the ee suffered (27% to 10%). Switching to Me_3Al at –30 °C in *i*-Pr_2O as solvent raised the ee of the (enantiomeric) product to 86% using the same (*S,R,R*) ligand system.

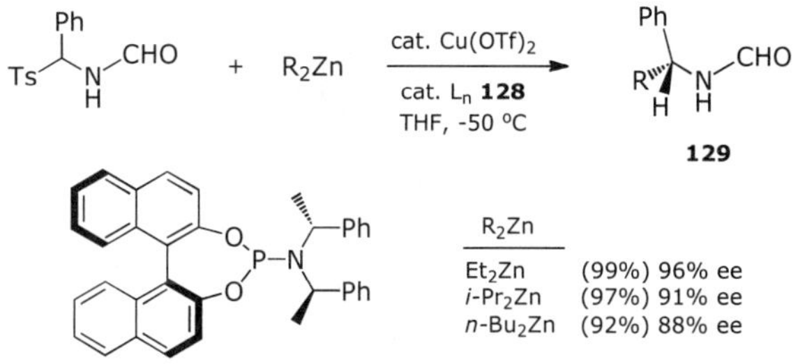

(*S,R,R*)-**128**

R_2Zn	
Et_2Zn	(99%) 96% ee
i-Pr_2Zn	(97%) 91% ee
n-Bu_2Zn	(92%) 88% ee

Bottom line comments. Traditional thinking that carbon residues covalently attached to copper rarely add in a 1,2-sense to carbonyl groups, let alone to imine derivatives, needs to be adjusted. While the fundamental ingredients (*i.e.*, a C=N moiety and copper) have not changed, other reaction variables have as follows: the substituent on nitrogen, the ligands on the metal, and the catalytic nature of the (asymmetric) processes that thereby introduce additional, stoichiometric metals into the equation. The representative contributions and associated procedures in this section point to yet another avenue by which copper can assist in synthesis and, in particular, of amines and derivatives.

8. Cu-Catalyzed Asymmetric C-C Bond Formation

Overview. Das, J. P.; Marek, I. *Chem Commun.* **2011**, *47*, 4593; Thaler, T.; Knochel, P. *Angew. Chem., Int. Ed.* **2009**, *48*, 645.

Asymmetric catalysis by copper(I) is blossoming; indeed, its coverage is worthy of an entire monograph. In the recent (2008) thematic issue of *Chemical Reviews* dedicated to "Coinage Metals in Organic Synthesis," 7 out of 12 articles on copper chemistry featured the word "enantioselective" or "asymmetric" in the title and/or introduction. Among the topics covered in these latest reviews, several are traditional (*e.g.*, conjugate addition and allylic alkylation), while others are less commonly associated with organocopper chemistry, let alone copper catalysis (*e.g.*, 1,2-additions to imines). Nonetheless, as "matches" between copper and both new and existing nonracemic ligands have been identified, remarkable discoveries in synthetic methodology have been forthcoming, and yes, as with so many advances in catalysis, in general, it really is all about the ligand.

8.1 Copper(I) enolates

Although Cu(I) is infrequently regarded as a potent Lewis acid relative to its Cu(II) oxidation state, it certainly can function in this regard, perhaps most notably when introduced as its triflate (*i.e.*, CuOTf). This precursor is highly prone, therefore, to complexation by (nonracemic) ligands, the results from which can be useful upon introduction of selected substrates. For example, racemic β-keto esters **130** are converted *via* intramolecular *syn*-carbocupration to chiral products **131** in typically high ee's (Scheme 1-11).[206] Pre-complexation of copper in a "cooperative" sense with ligand **132** bearing an internal Lewis basic (nitrogen) site, and a urea that supplies a H-bonding donor, is suggestive of an eventual ordered transition state (**133**) involving a copper enolate. The intermediary of vinylcopper species **134** may offer interesting opportunities for subsequent C-C bond formations *via in situ* trapping by electrophiles other than H^+. A β-keto amide was also shown to participate.

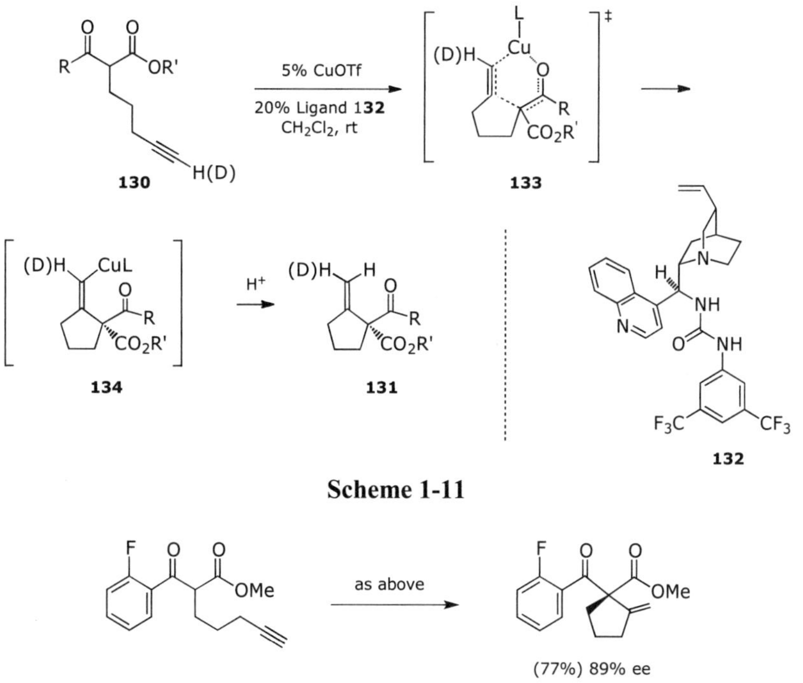

Scheme 1-11

A copper enolate is also regarded as an intermediate in a hetero-Diels-Alder–like sequence between a ketone and an analog of Danishefsky's diene, modified at silicon (*i.e.*, $(OEt)_3$ in place of Me_3; Scheme 1-12).[207] A fluoride-initiated (tetrabutylammonium difluorotriphenylsilicate [TBAT]) cleavage of the starting silyl enol ether leads to a nonracemically ligated (with Walphos) copper enolate, which condenses and then closes to the observed dihydropyranones in modest ee's. Noteworthy in this scheme is generation of the quaternary center.

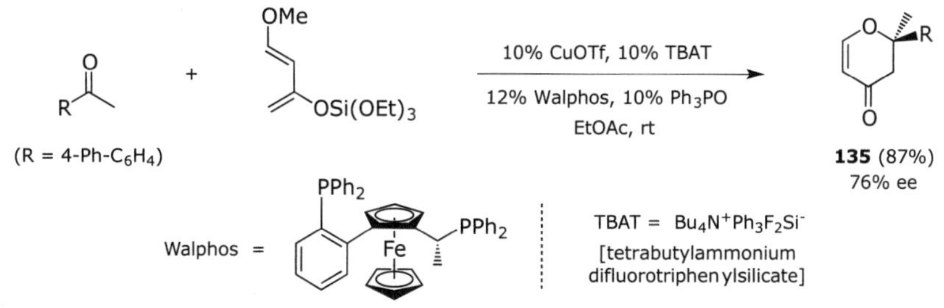

Scheme 1-12

8.2 Conjugate Additions

Reviews. Jerphagnon, T.; Pizzuti, M. G.; Minnaard, A. J.; Feringa, B. L. *Chem. Soc. Rev.* **2009**, *38*, 1039; Lopez, F.; Minnaard, A. J.; Feringa, B. L. *Acct. Chem. Res.* **2007**, *40*, 179.

There are many noteworthy, even spectacular, achievements on this front. Not surprisingly, these are intimately tied to either existing ligand matches to an organometallic or design of a new ligand scaffold. While each focuses on Cu(I) catalysis, often the source of this metal is in the form of Cu(II): usually, $Cu(OTf)_2$. Sources of Cu(I) tend to be the halides ($CuBr{\bullet}Me_2S$, CuCl, CuI), but also the Liebeskind 2-thiophenecarboxylate (*i.e.*, CuTC)[208] is a viable option that has grown in usage. Nucleophiles used in (super)stoichiometric amounts include both highly functional group-tolerant organometallics (*e.g.*, organozinc reagents), as well as Grignards, aluminum reagents,[209] and allylic silanes.[210] Many of these combinations are listed in Table 1-3.

Several recent reviews, in addition to those listed above, cover asymmetric 1,4-additons and related reactions. These can be found in the thematic issue of *Chemical Reviews* (August, 2008) dedicated to "Coinage Metals in Organic Synthesis."

Table 1-3. Representative recent copper-catalyzed asymmetric 1,4-additions.

Organometallic	Ligand	Product	ee (%)	References
Grignard reagent-based				
RMgX	*(R)*-(+)-Tol-BINAP	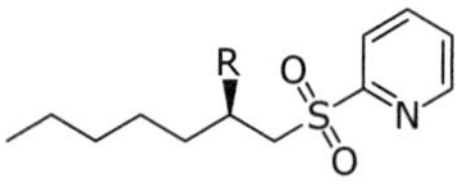	>87	[211]
RMgX	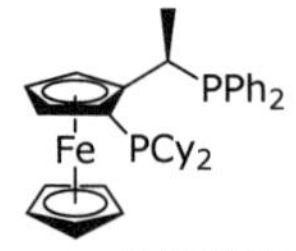 "reversed" JOSIPHOS **143**	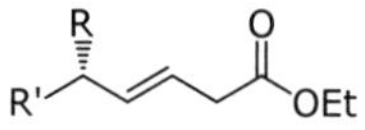	73–96	[212]
RMgX	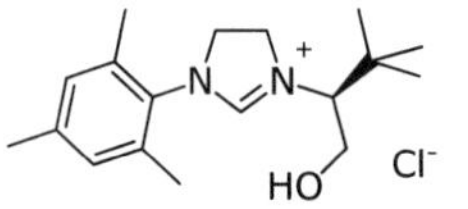 **142**	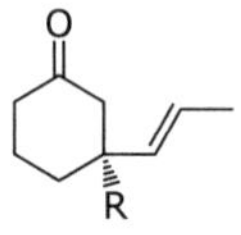	66–99	[213] [218]
RMgX	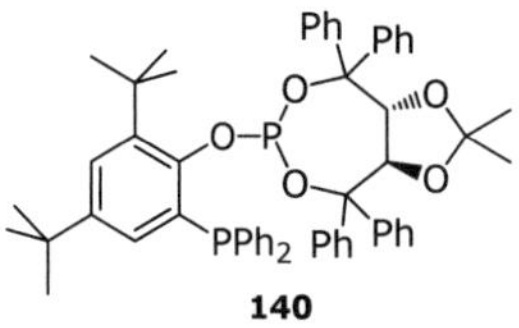 **140**	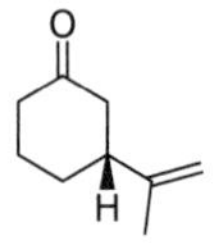	92	[214]
RMgX	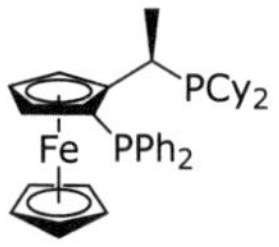 JOSIPHOS (**136**)	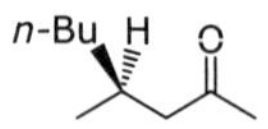	93	[215]

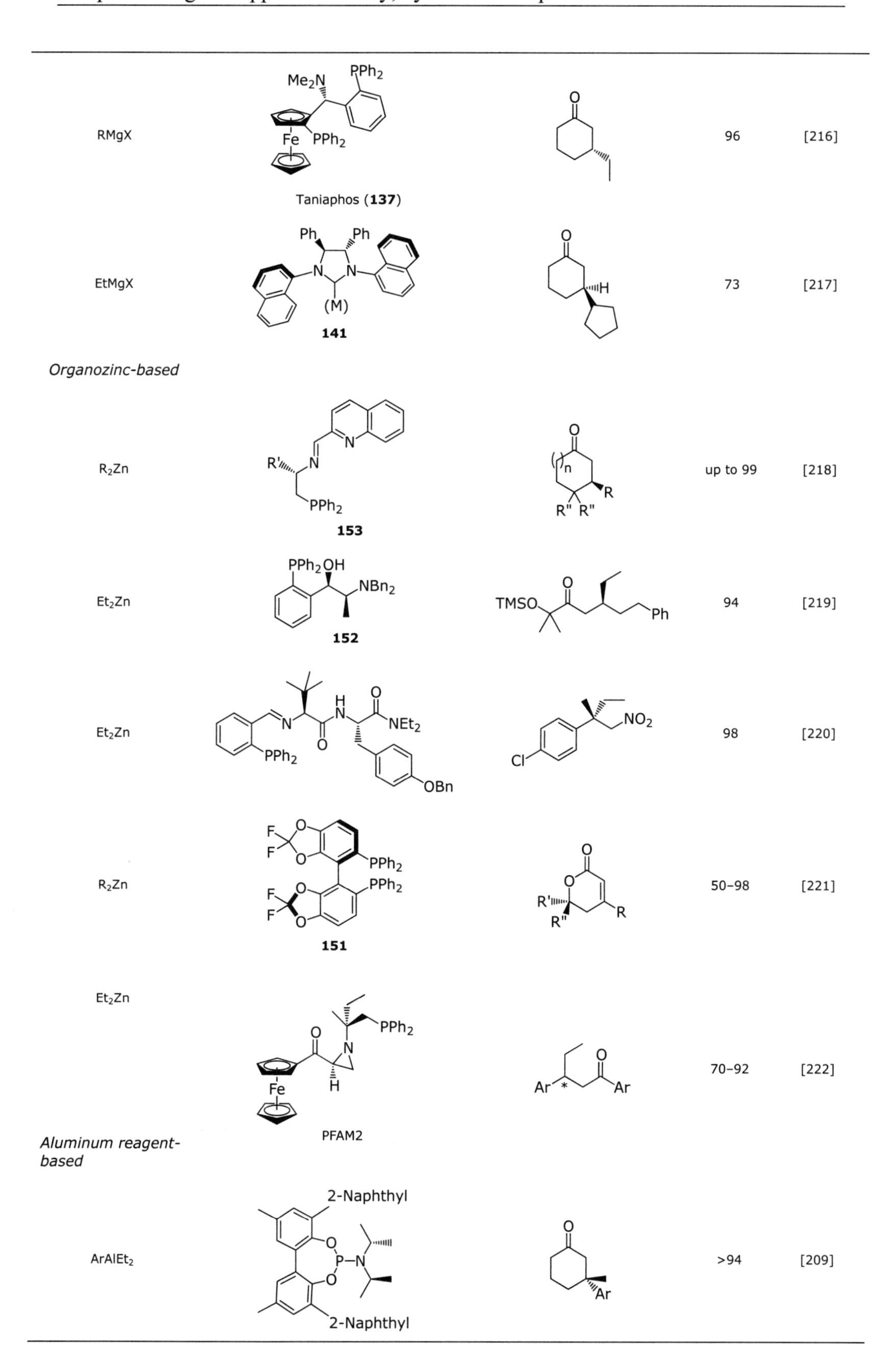
RMgX
Me2N
PPh2
Fe
PPh2
Taniaphos (137)
O
96
[216]
EtMgX
Ph
Ph
N
N
(M)
141
O
H
73
[217]
Organozinc-based
R2Zn
N
R'
N
PPh2
153
O
n
R
R" R"
up to 99
[218]
Et2Zn
PPh2 OH
NBn2
152
TMSO
O
Ph
94
[219]
Et2Zn
H
N
O
N
NEt2
O
PPh2
OBn
NO2
Cl
98
[220]
R2Zn
F
F
O
O
PPh2
PPh2
F
F
O
O
151
O
O
R'
R"
R
50–98
[221]
Et2Zn
O
PPh2
N
Fe
H
PFAM2
O
Ar
*
Ar
70–92
[222]
Aluminum reagent-based
ArAlEt2
2-Naphthyl
O
P–N
O
2-Naphthyl
O
Ar
>94
[209]

Silicon-based

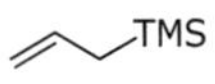 TMS	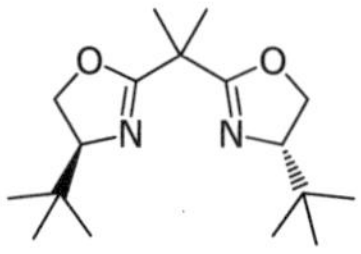	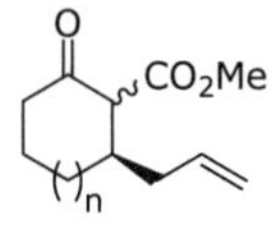 CO_2Me, n	50->98	[210]

Reactions of Grignard Reagents

Most of the procedures of late rely on RMgX as a source of the carbon nucleophile. Several classes of substrates are amenable to copper-catalyzed conjugate additions (CAa), including acyclic and cyclic enones, enoates, thioesters, sulfones, and even enals. While individual processes have been fine-tuned to the point where impressive (*i.e.*, usually >90%) ee's can now be realized with considerable generality, variations in particular with respect to the ligand clearly suggest an element of unpredictability. A good case in point involves conjugated enones, where *acyclic* systems respond best to (6 mol%) JOSIPHOS-complexed, presumably *in situ*–formed, magnesiocuprates, R_2CuMgX (Scheme 1-13).[215] Other substitution patterns on this ferrocenyl-based scaffold gave significantly lower ee's. Taniaphos (**137**)[223] was surprisingly ineffective (1% ee), considering that it works well in several cases of *cyclic* enones (*vide infra*). Solvent and temperature differences were also noted: with acyclic cases, methyl *t*-butyl ether (MTBE) at –75 °C; and with cyclic enones, Et_2O at 0 °C. Regiochemistry can vary depending on the educt: acyclic examples are sensitive to the nature of substituents in the educt, while cycloalkenones show variations as a result of R in RMgX. Interestingly, alteration of the amino residue on Taniaphos (*e.g.*, **138**) led to inverted chirality in the product ketone, indicating that central chirality in the ligand side-chain (rather than planar chirality, which is the same in all cases) controls the stereochemical outcome.

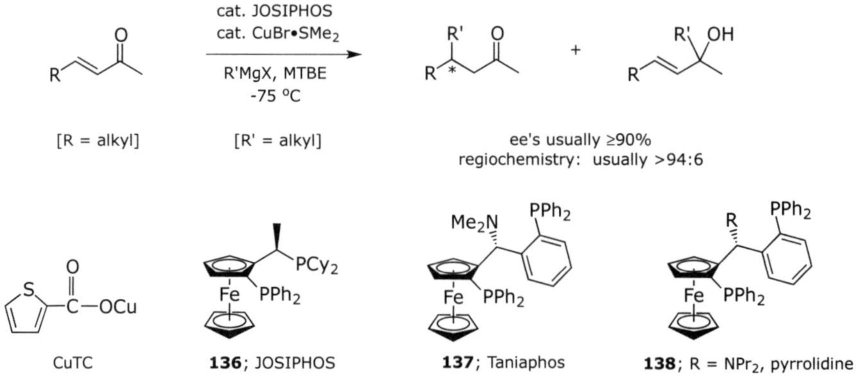

cat. JOSIPHOS
cat. $CuBr{\bullet}SMe_2$
R'MgX, MTBE
-75 °C

[R = alkyl] [R' = alkyl]

ee's usually ≥90%
regiochemistry: usually >94:6

CuTC **136**; JOSIPHOS **137**; Taniaphos **138**; R = NPr_2, pyrrolidine

Scheme 1-13

General procedure for asymmetric conjugate addition of RMgX to cyclic enones[216]

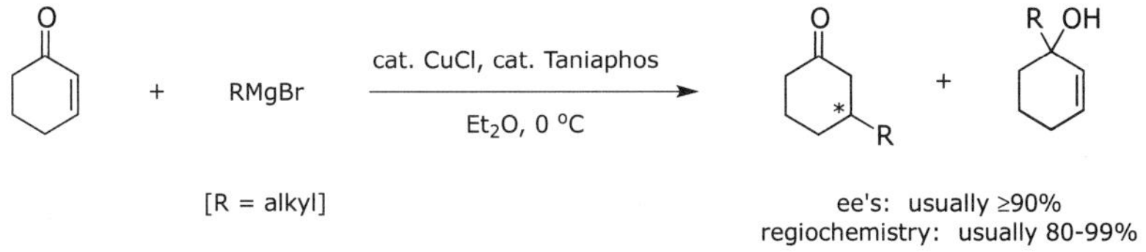

cat. CuCl, cat. Taniaphos

Et_2O, 0 °C

[R = alkyl]

ee's: usually ≥90%
regiochemistry: usually 80-99%

In a Schlenk tube equipped with septum and stirring bar, a mixture of CuCl (12.5 μmol) and the ligand (15 μmol) was dissolved in Et_2O (2.5 mL). After stirring under argon at RT for 30 min, the enone (0.25 mmol) was added. After additional stirring for 10 min, the corresponding Grignard reagent (as a solution in Et_2O, 0.29 mmol) was added dropwise to the resulting mixture over 5 min at 0 °C. After stirring under argon for 15 min, aqueous NH_4Cl solution (1 M, 1 mL) was added to the reaction mixture. The organic phase was separated, filtered over a short silica column, and subjected to ee determination (capillary GC). CuBr•SMe_2 was used instead of CuCl in reactions performed at low temperature. To avoid undesirable tandem aldol reactions, MeOH was used instead of aqueous NH_4Cl solution in the work-up involving cyclopentenone as substrate.

The recent introduction of 4,5-bis(diphenylhydroxymethyl)-2,2-dimethyldioxolane (TADDOL)-derived phosphine-phosphite ligands **139** and **140**, albeit bearing a completely different non-ferrocenyl framework, can also be used for Grignard additions to cyclohexenone.[214] The complexed salt CuBr•SMe_2, as opposed to CuCl, gave better ee's given its greater solubility in 2-methyltetrahydrofuron (2-MeTHF) at –78 °C. Regioselectivities tend to be *ca.* 9:1, 1,4- over 1,2-addition. THF itself gave far inferior ee's. Some results are shown below.

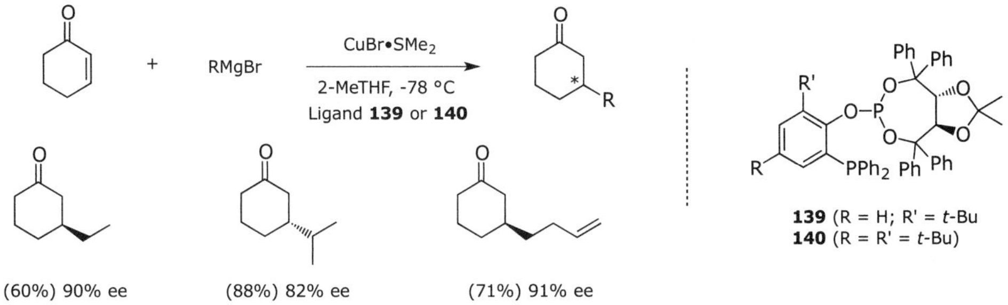

CuBr•SMe_2

2-MeTHF, -78 °C
Ligand **139** or **140**

(60%) 90% ee (88%) 82% ee (71%) 91% ee

139 (R = H; R' = *t*-Bu)
140 (R = R' = *t*-Bu)

All-carbon quaternary centers at the β-position within a cycloalkanone result from asymmetric copper-catalyzed Grignard additions employing yet another ligand system: a nonracemic NHC.[217] While the arrangement of the phenyl substituents in the selected carbene complex **141** is fixed, the naphthyl residues on nitrogen are bulky and may orient (for steric reasons) in such a way as to transmit chiral information to the reaction site(s). Et_2O was the preferred solvent (no ee in THF!) at 0 °C; $Cu(OTf)_2$ (3 mol%) provides the metal. Carbene generation was conveniently carried out using a portion of the Grignard present. Curiously, under controlled amounts of RMgBr, which favors species RCu•NHC, a very low ee (2%) was observed. Thus, with excess RMgBr in the pot, the major species may be derived from a cuprate $R_2CuMgBr$, giving rise to species containing "(NHC)CuR_2," a higher order–like reagent. Other organometallics, *e.g.*, R_2Zn and R_3Al, did not give acceptable ee's.

cat. $Cu(OTf)_2$
4% Ligand from **142**
1.2 RMgBr, Et_2O

R = Et (80% ee; *R*)
R = *n*-Bu (77% ee; *R*)
R = butenyl (90% ee; *S*)
R = *i*-Bu (96% ee; *S*)

NHC Ligand **141**

NHC Ligand precursor **142**

The influence of NHC ligand precursor **142** (with Cl^- as the counterion, as shown below), together with RMgBr and catalytic $Cu(OTf)_2$, is vividly seen in reactions with doubly unsaturated cyclic ketones.[213] Thus, while both zinc- and aluminum reagents direct addition of copper in a 1,6-manner under the influence of a phosphoramidite ligand, **142** directs the *n*-alkyl Grignard-derived copper reagent in a 1,4-mode. Other NHC ligands were found to be 1,6-directing, highlighting the importance of the OH group in **142**.

cat. $Cu(OTf)_2$
MeMgBr, cat. **142**
CH_2Cl_2, -10 °C

(60%) 92% ee

precursor **142**

(60%) 93% ee

(44%) 92% ee

General procedure for the enantioselective 1,4-conjugate addition of a Grignard reagent to a dienone[213]

In a dried Schlenk tube equipped with septum and stirring bar under nitrogen, $Cu(OTf)_2$ (10.8 mg, 6 mol%) and ligand **142** (14.6 mg, 9 mol%) were dissolved in dry DCM (1.5 mL). The mixture was cooled to –10 °C, and the Grignard reagent (2 equiv) was added. After stirring for an additional 5 min, a solution of dienone (0.5 mmol) in dry DCM (5 mL) was added *via* syringe pump over 15 min. The reaction mixture was stirred 1 h at –10 °C, and aqueous NH_4Cl solution (1 M, 0.5 mL) was added. The mixture was warmed to RT, and 5 mL of the NH_4Cl solution and 5 mL of DCM were added and the layers separated. After extraction with DCM (2 × 5 mL), the combined organic extracts were dried and evaporated under vacuum. The residual material was dissolved in DCM (5 mL) under argon and DBU (1 equiv) was added. The mixture was stirred at RT for 1 h and then hydrolyzed with a solution of HCl (1 M). The layers were separated, and after extraction with DCM (2 × 5 mL), the combined organic extracts were dried and evaporated *in vacuo*. Flash chromatography (pentane/diethyl ether, 90:10) yielded the desired compound.

To drive a copper-catalyzed Grignard addition in the 1,6-direction, in the case of dienoates (e.g., **144**), a JOSIPHOS-ligated species can be used.[212] The parent, JOSIPHOS, as well as Taniaphos, led to very low levels of conversion. However, the "reversed" JOSIPHOS; *i.e.*, a switch between substituents on each phosphorus (**143**, *cf.* **136** in Scheme 1-13) gave excellent results: regioselectivity >95:5; yields 73–88%; ee's 73–96%; all β,γ-unsaturated (kinetic) products.

CO_2Et
144
cat. $CuBr{\bullet}SMe_2$, R'MgBr
cat. Ligand **143**
CH_2Cl_2, -70 °C
R', H, R, CO_2Et
PGO, CO_2Et
PG = TBDPS (82%) 73% ee
PG = Bn (69%) 90% ee
CO_2Et
(77%) 93% ee
PPh_2
Fe PCy_2
143 ("reversed" JOSIPHOS)

Unsaturated esters (*i.e.,* enoates) are amenable to copper-catalyzed addition of RMgBr, the induction of chirality also being controlled by the same ligands (JOSIPHOS **136** and "reversed" JOSIPHOS, **143**) used for dienoates, **144** (*vide supra*).[224] Hindered esters (*e.g.*, *i*-Pr) are not required, as competing 1,2-addition was not observed with either methyl or ethyl esters. High yields and ee's are to be expected. Catalyst loadings down to 0.2 mol% gave results comparable to those obtained using 1 mol%. Purification of the reaction solvent (MTBE) is not required. Catalysts **145a** and **145b**, both *bis*-phosphine-complexes with CuBr, could be reisolated from a reaction mixture upon dilution with pentane and reused with equal success.

F_3C, CO_2Me
cat. **145a**, EtMgBr
MTBE, -75 °C
F_3C, CO_2Me
(99%) 96% ee
R', R', P, Br, Fe, P, Cu, R, R, 2
145
a: R = Cy, R' = Ph
b: R = Ph, R' = Cy
CO_2Et
(86%) 98% ee [using **145a**]
CO_2Me
(86%) 95% ee [using **145b**]
Ph, CO_2Me
(84%) 87% ee

A similar study using a CuI•*p*-Tol-BINAP complex (1:1.5 ratio of CuI:ligand) as catalyst for 1,4-additions of Grignards to enoates has also appeared.[225] While there are several similarities (*e.g.*, MTBE as solvent, and roughly comparable yields and ee's), there are differences: here, reaction temperatures are at –40 °C (*vs.* –78 °C); the enantiomer of the ligand was shown to give the enantiomeric product; and *Z*-enoates gave higher ee's than their corresponding *E*-isomers. Also of note is the use of a secondary alkyl Grignard reagent, *e.g.*, *i*-PrMgBr, which works well. On the other hand, MeMgBr leads to a poor-yielding reaction, albeit in high ee.

Asymmetric 1,4-addition of a Grignard reagent to an enoate[225]

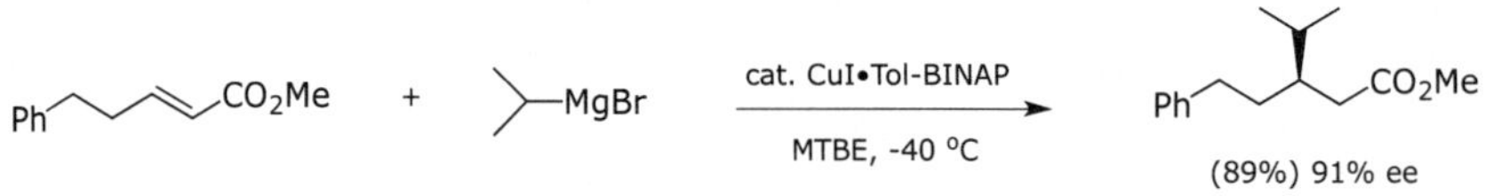

In a Schlenk tube equipped with septum and stirring bar, (*R*)-*p*-Tol-BINAP (5.1 mg, 0.0075 mmol) and CuI (0.95 mg, 0.005 mmol) were added to *t*-BuOMe (1.0 mL) and stirred under nitrogen at RT until a yellow suspension was observed. The mixture was then cooled to –40 °C and EtMgBr (Aldrich, 3.0 M in Et_2O, 2.5 mmol) was added. After stirring for 5 min, a solution of unsaturated ester (0.5 mmol) in *t*-BuOMe (0.25 mL) was added dropwise over 1 h by syringe pump. After stirring at –40 °C for 2–3 h, MeOH (0.5 mL) and aqueous NH_4Cl (1 M, 2 mL) were sequentially added and the mixture was warmed to RT. After extraction with Et_2O (3 × 5 mL), the combined organic phases were dried and concentrated *in vacuo* to a yellow oil that was purified by flash chromatograph (1:99 Et_2O/pentane) to yield the desired product. Racemic 1,4-addition products were obtained by the reaction of the enoates with the corresponding Grignard reagents (–20 °C, Et_2O) and CuI (100 mol%).

To improve, yield-wise, the important delivery of a methyl group to α,β-unsaturated esters, conditions have been found that vary slightly from those above: doubling the catalyst loading to 2% CuI, 3% *p*-Tol-BINAP, and increasing the reaction temperature from –40 °C to –20 °C.[226] These subtle changes avoid competing 1,2-addition and lead to very successful couplings; examples are below. All cases involve methyl esters; no variations were studied.

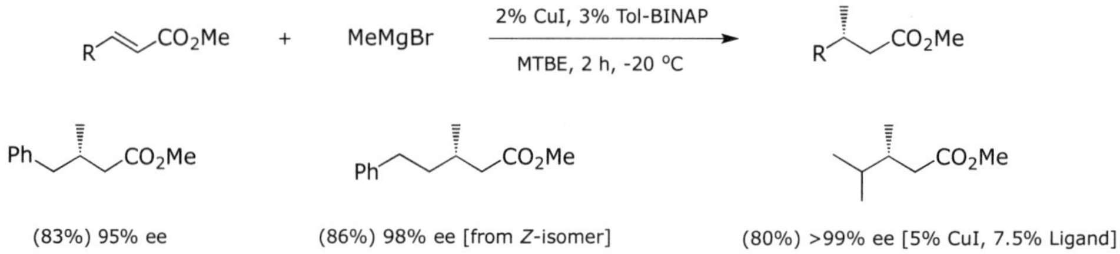

Stereocontrol in copper-catalyzed additions of methyl Grignard can be realized using unsaturated thioesters as Michael acceptors.[227] Initially formed magnesio enolates can be quenched directly to give the derived conjugate adducts. Asymmetric conjugate addition of a MeMgBr/JOSIPHOS complex takes place in MTBE at –75 °C overnight. Alternatively, as highly reactive intermediates, they are subject to second-stage aldol reactions. Hence, subsequent addition of an aryl or branched alkyl aldehyde generates three contiguous asymmetric centers within minutes at this low temperature, with dr's usually >20:1. Stereocontrol derives from both the initial facial selectivity of the nonracemic copper complex as well as the preferred enolate geometry (**146**) that also takes advantage of the larger phenyl *vs.* methyl residue at the γ-site. Final treatment of the mixture with carbonate ultimately affords the corresponding methyl esters.

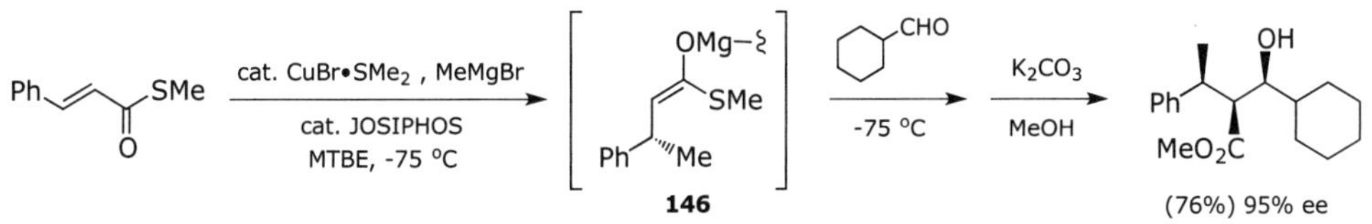

Asymmetric copper-catalyzed conjugate addition of RMgX to α,β-unsaturated thioesters[227]

$CuBr•SMe_2$ (5 mol%, 1.04 mg) and ligand **136** or **147** (6 mol%, 3.57 mg) were dissolved in *t*-BuOMe (0.9 mL) and stirred at RT for 30 min under nitrogen. The mixture was cooled to –75 °C, and the Grignard reagent (0.12 mmol, in Et_2O) was added dropwise. After stirring for 10 min, a solution of thioester (0.1 mmol) in *t*-BuOMe (0.1 mL) was added *via* a syringe pump over 1–2 h. The reaction mixture was stirred at –75 °C for 2–5 h, then quenched by the addition of MeOH, and allowed to warm to RT. Saturated aqueous NH_4Cl solution was then added, the phases were separated, and the aqueous layer was extracted with Et_2O. The combined organic phases were dried over anhydrous $MgSO_4$, concentrated under reduced pressure, and purified by flash chromatography to afford the desired 1,4-addition adduct.

Related work along these lines includes conversion of the conjugate methyl adduct (a thioester) into the reduced aldehyde, followed by Wittig olefination to the next thioenoate **148** (Scheme 1-14).[228] Either *syn*- or *anti*-1,3-dimethylated products can be formed by the choice of chirality within the JOSIPHOS ligand. Noteworthy is the implied conclusion that the existing stereocenter has little-to-no influence on the second asymmetric 1,4-addition. Both methyl and ethyl thioesters can be used interchangeably. Other Grignards also work well, although hindered cases (*i*-Pr, *i*-Bu) give low ee's.

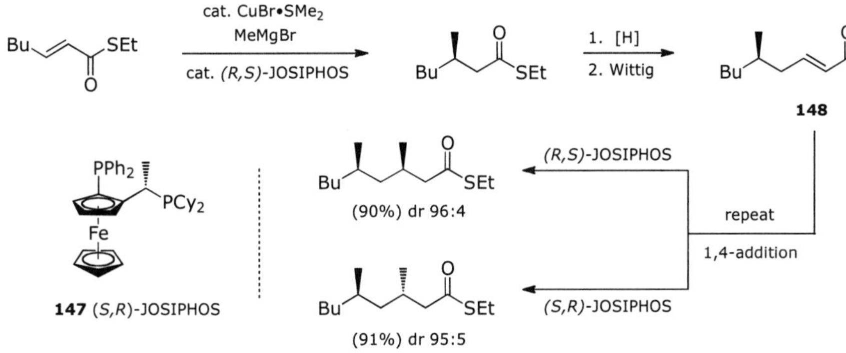

Scheme 1-14

α,β-Unsaturated-β-substituted-2-pyridylsulfones (**149**), by contrast to enones and enoates, are also susceptible to 1,4-addition of Grignards, in this case using a CuI•*p*-Tol-BINAP catalyst.[211] Other copper halide salts work equally well, although CuCN gave lower ee's. Several examples of alkylmagnesium bromides led to ee's between 87% and 94%, while PhMgBr gave an 80% yield of racemic material. The 2-pyridyl (2-py) group is required, as reported previously with related asymmetric reductions of CuH on β,β-disubstituted pyridylsulfones; the *p*-tolylsulfone case affords the product **150** in only 31% ee.

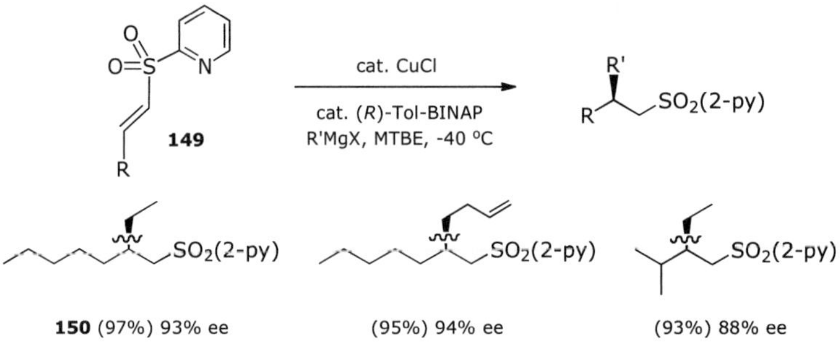

General procedure for the conjugate addition of Grignard reagents to α,β-unsaturated sulfones **149**[211]

CuCl (12.5 μmol, 1.24 mg, 5 mol%) and (*R*)-(+)-*p*-Tol-BINAP (15.0 μmol, 10.18 mg, 6 mol%) were dissolved in *t*-BuOMe (4 mL) under a nitrogen atmosphere. The mixture was stirred for 15 min and cooled to −40 °C and the Grignard reagent (1.2 equiv) then added. After stirring for 15 min, the substrate (0.25 mmol, dissolved in 0.5 mL *t*-BuOMe) was added over 5 h and the mixture stirred overnight. Aqueous saturated NH_4Cl (2 mL) was added, and the mixture was warmed to RT, diluted with Et_2O, and the layers separated. The aqueous layer was extracted with DCM (3 × 10 mL) and the combined organic layers were dried with anhydrous Na_2SO_4, filtered, and the solvent evaporated *in vacuo*. The crude product was purified by flash chromatography on silica (pentane/Et_2O 4:1 to 1:1) to afford the product as colorless-to-pale-yellow oils in good-to-excellent yields (88–97%).

A preliminary study on the asymmetric 1,4-addition of a variety of RMgBr to enals suggests that the potential exists for this type of reaction to be fine-tuned by further advances in ligand design.[229] Using (*R*)-(Tol-BINAP) together with copper thiophenecarboxylate (CuTC) in Et_2O at −78 °C, several enals and Grignards were screened for both regio- and stereo-selectivity. The presence of TMS-Cl (1.3 equiv) was found to be important to achieve the best results, although ratios of 1,4 to 1,2-adducts were still modest, as were the ee's.

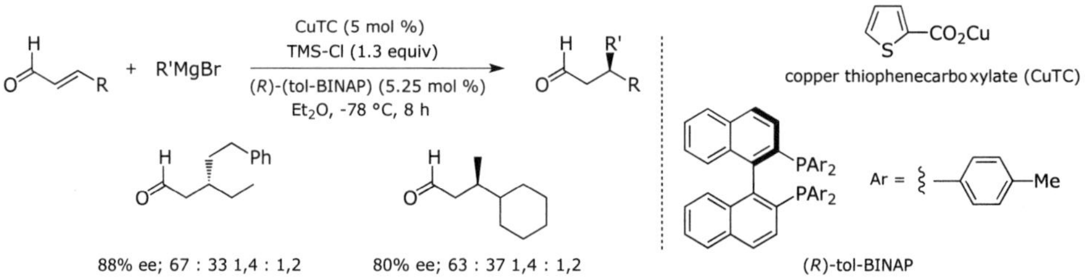

Reactions of Organozinc Reagents

Asymmetric Michael additions of zinc reagents, as with Grignards, are effected with a variety of ligands, some of which are also useful in other types of copper-catalyzed reactions (*e.g.* phosphoramidite **128** and DIFLUORPHOS **151**), while others are focused mainly on C-C bond constructions involving zinc (*e.g.*, tridentate ligand **152**, *N,N,P*-tridentate ligand **153**, and amino acid-based phosphines **154a** and **154b**).

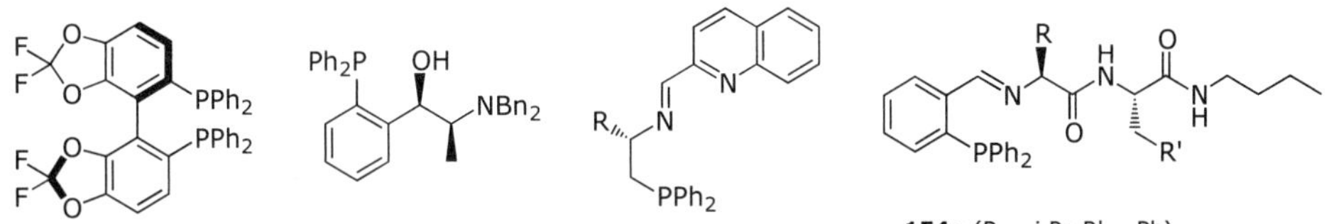

151 [(*R*)-DIFLUORPHOS] **152** **153** (R = *i*-Pr, *t*-Bu) **154a** (R = *i*-Pr, R' = Ph) **154b** (R = *t*-Bu, R' = C_6H_4-*p*-O-*t*-Bu)

Ligand **152** was designed to accommodate a diorganozinc reagent as part of the calculated Cu(III) transition state involved in 1,4-additions to acyclic enones.[230] Using catalytic $Cu(OTf)_2$ (3 mol%), a slight excess of ligand **152** (3.6 mol%), and excess R_2Zn (1.5 equiv), 1,4-additon in DCM occurs smoothly at 0 °C to give the desired adduct with very high levels of enantio-induction (usually $\geq$98%). Other solvents such as THF, Et_2O, and especially DME gave lower ee's. Use of copper salts with oxygen-based counterions is also important; reactions with copper halides led to reduced levels of enantiocontrol in a model case [CuI (93% ee), CuBr (71% ee), and CuBr•SMe_2 (56% ee)]. Enoates were inert to these conditions, and cyclic enones as substrates are less responsive ee-wise, suggesting an important role for an *s-cis* conformation in the transition state.

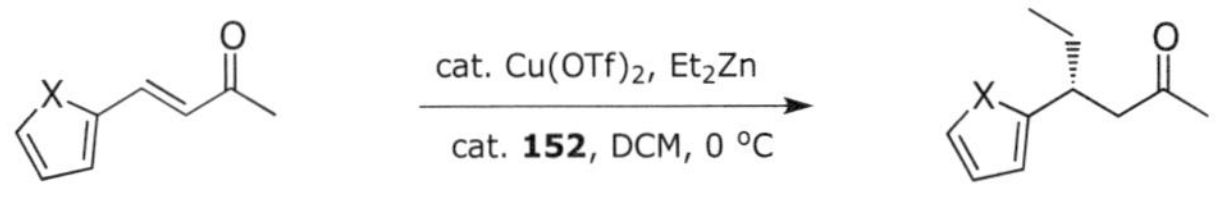

cat. $Cu(OTf)_2$, Et_2Zn / cat. **152**, DCM, 0 °C

X = O (89%) 98.9% ee
X = S (90%) 98.1% ee

Typical procedure for copper-catalyzed enantioselective conjugate additions[230]

In a flame-dried Schlenk tube were placed a copper salt (0.03 mmol), ligand **152** (18.6 mg, 0.036 mmol), and DCM (5 mL). The resulting mixture was stirred at RT for 1 h, and then cooled in an ice bath followed by the addition of a substrate (1 mmol). After stirring for 15 min, Et_2Zn (1.0 M in hexanes, 1.5 mL, 1.5 mmol) was added dropwise over 3 min. After stirring at this temperature for a period of time (see Tables in publication), the reaction was quenched by the addition of saturated aqueous NH_4Cl (2 mL). The resulting mixture was extracted with Et_2O (3 × 5 mL), and the combined organic layer was dried over anhydrous Na_2SO_4 and concentrated under reduced pressure to afford crude product. The material was purified by silica gel chromatography (hexane/Et_2O 100:0 to 80:20) to afford the 1,4-addition product.

The same conditions used above (*i.e.*, with ligand **152**) apply to acyclic enones bearing a silyloxy group α' to the enone carbonyl, as in **155** (Scheme 1-15).[220] Asymmetric addition of R_2Zn, where R = Me, Et, *i*-Pr, and *n*-Bu, affords an initial product that can be oxidatively converted into the derived acids, or reduced with hydride followed by oxidative cleavage to an aldehyde.

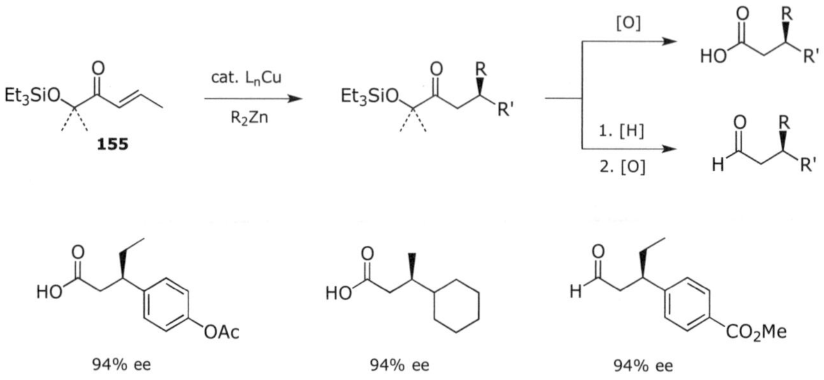

Scheme 1-15

Allenic esters readily react with dialkylzinc reagents in the presence of DIFLUORPHOS-complexed copper (from $Cu(OAc)_2$) in THF at –20 °C.[223] While this initial adduct bears no new central chirality, the intermediate nonracemic copper/zinc enolate can then add in a stereocontrolled 1,2-fashion to unsymmetrical ketones. Ring closure to the resulting δ-lactone completes the sequence. Both 4 Å molecular sieves and 20 mol% Lewis basic $Ph_2S{=}O$ (or DMSO, hexamethylphosphoramide [HMPA]) are required to direct attack at the γ-position rather than the otherwise reversible aldol event at the α-site, thereby facilitating conversion to cyclic products.

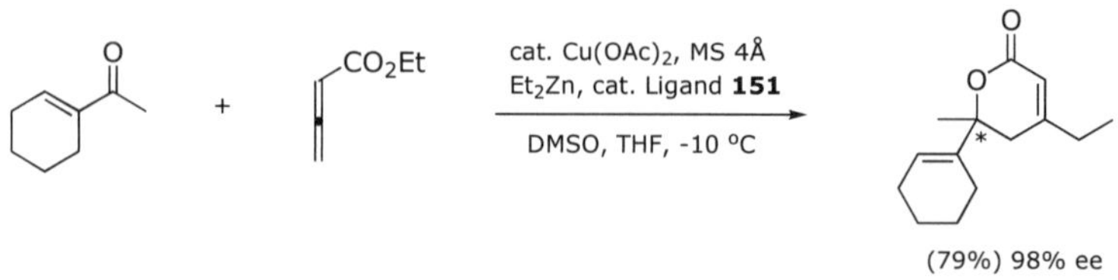

Cyclic enones (mainly 5- and 6-membered rings) react with dialkylzinc reagents Me_2Zn or Et_2Zn in the presence of 0.1 mol% $Cu(OTf)_2$ modified by the tri-coordinate species **153**.[218] Both rates and enantioselectivity require the 2-quinolyl moiety. The corresponding pyridyl ligand was not as effective (55% ee *vs.* >95% ee). Acyclic enones (*e.g.*, chalcone) under identical conditions gave low ee's (with either ligand).

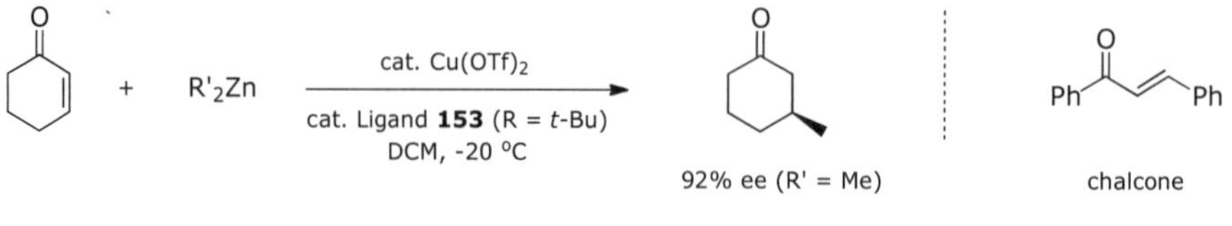

Furanones and pyranones are Michael acceptors, reacting with copper-chelated catalysts derived from ligands **154a** and **154b**.[223] More specifically, substrates such as **156** react with R_2Zn (R = alkyl) in toluene containing a 10 mol% catalyst, so long as benzaldehyde is present to trap the enolate *via* aldol condensation to give species **157**. A reverse aldol is required to give the targeted products in high ee's.

cat. $Cu(OTf)_2$
cat. Ligand **154a**
Et_2Zn, PhCHO
toluene, -30 °C
K_2CO_3
toluene, Δ
156 **157** (74%) >98% ee

Pyranone **158**, likewise, reacts with high facial selectivity to afford adduct **159**.[223] In these cases, THF is essential for minimizing oxygen-copper chelation; in toluene, the ee dropped to ≤64%. With these substrates, PhCHO is not needed.

cat. $Cu(OTf)_2$
cat. Ligand **154a**
Me_2Zn, THF, - 30 °C
158 **159** (58%) >98% ee

Asymmetric 1,4-addition to alkylidene Meldrum's acids creates benzylic quaternary centers, using the catalytic combination of $Cu(OTf)_2/R_2Zn$ in DME.[231] Phosphoramidite ligand **128** (10 mol%) controlled the sense of chiral induction, and in most cases examined, with ee's in the 75–95% range. Alkylzinc reagents (Me, Et, *n*-Bu) gave the best results both yield and ee-wise.

General procedure for the conjugate addition of R_2Zn to alkylidene Meldrum's acids[231]

cat. $Cu(OTf)_2$
cat. Ligand **128**
Bu_2Zn, DME
-40 °C to rt, 48 h
(97%) 97% ee

Reactions were typically carried out using 0.24 mmol of substrate. In a glove box, $Cu(OTf)_2$ (5 mol%) and the nonracemic phosphoramidite ligand (10 mol%) were charged in a flame-dried resealable Schlenk tube. DME (1 mL) was then added to the Schlenk tube to wash down any residual solids. The reaction mixture was allowed to stir at RT for 15 min outside the glove box, and then cooled to –40 °C. In the dry box, the R_2Zn solution (2.0 equiv) was transferred to a round-bottom flask equipped with a septum. This solution was added to the Schlenk tube dropwise and the resulting solution stirred for 5 min. A solution of freshly crystallized alkylidene Meldrum's acid (1.0 equiv) in DME (1 mL) was then added dropwise. Finally, DME (0.4 mL) was added to wash down the remaining solid on the sides of the Schlenk tube. The reaction mixture was allowed to warm slowly to RT. After 48 h of stirring, 5% HCl and ether were added to the reaction mixture. The layers were partitioned and the aqueous layer extracted three times with Et_2O. The combined organic layers were washed twice with water, dried over anhydrous $MgSO_4$, filtered, and concentrated *in vacuo*. The crude residue was purified by flash chromatography on silica gel using hexanes or petroleum ether (35–60 °C) in EtOAc to yield the desired product. HPLC using a chiral column OD, OD-H, or AD-H was used to measure the ee's of the products. The racemates were prepared using EtMgBr (2.5 equiv) in THF at RT.

Reactions of Organoaluminum Reagents

In addition to copper-catalyzed reactions of Grignard and organozinc reagents (*vide supra*), mixed organoaluminum reagents offer yet another approach to substrate elaboration, in particular with respect to enantioselective bond formation arriving at quaternary centers. Focus of late has been on delivery of alkenyl residues, especially their introduction into challenging β-substituted cyclic enones. An initial study using a dimethyl(alkenyl)alane has made good progress on 3-methylcyclohexenone, employing CuTC and SimplePhos in mainly MTBE and/or Et_2O at –30 °C.[232] Some competitive methyl transfer from the alane typically occurs, and yields can be highly variable; ee's can be >90%. Unfortunately, other groups at the β-site (*e.g.*, Ph) and other cyclic enones (*e.g.*, 3-methylcyclopentenone) did not afford consistent levels of stereoinduction.

Me_2Al (2 equiv); CuTC (10 mol %), SimplePhos (11 mol %), MTBE, Et_2O, pentane, hexane, -30 °C, 18 h

SimplePhos

An alternative solution to the same problem uses mixed vinylalanes that contain silicon in the α-position.[233] Since these derive from a diisobutylaluminum (DIBAL)-based hydroalumination of silylalkynes, the resulting species are diisobutyl(alkenyl)-alanes. These undergo transmetalation in the presence of an NHC-Cu complex, derived from $CuCl_2$ and, importantly, a bidentate, dimeric silver NHC precursor (illustrated below). Silyl moieties can be TMS or TBS, although the latter, bulkier group in the starting alane requires room temperature for 1,4-addition to ensue. Both 5- and 6-membered enones can be used; representative examples are shown below. The resulting vinylsilane in each product can be easily converted to alternative functionality, such as a ketone (*via* oxidation with MCPBA), a halide (*e.g.*, an iodide, with NIS), and an alkene (*via* protodesilylation with TFA).[234]

(n = 0, 1) (R = Me, Ph) + i-Bu_2Al (2 equiv)

$CuCl_2 \bullet H_2O$ (5 mol %)
Ag-NHC precursor (2.5 mol %)
THF, 0 *C or rt,12 h

(48%) 78% ee

(70%) >96% ee

(71%) 90% ee

dimeric Ag-NHC precursor

Representative procedure for NHC–Cu-catalyzed conjugate addition of silyl substituted vinylaluminum reagents to unsaturated enones[233]

An oven-dried, 4-dram vial was charged with the NHC–Ag complex (2.70 mg, 5.00 μmol) and $CuCl_2 \bullet 2H_2O$ (0.850 mg, 5.00 μmol) weighed under an N_2 atmosphere in a glove box. The vial was sealed with a cap with septum. Tetrahydrofuran (1.0 mL) was added through a syringe to the vial, and the resulting blue solution was allowed to stir for 5 min. The aluminum reagent ((*E*-di-*i*-butyl(1-trimethylsilyl)pent-1-en-1-yl)aluminum) (160 μL, 2.00 mmol, 1.25 M) (CAUTION! Flammable!) and 3-methylcyclopentenone (9.90 μL, 0.100 mmol) were added through a syringe, sequentially, resulting in a brown solution. The solution was allowed to stir at RT for 15 min; after which, the reaction was quenched upon addition of a saturated aqueous solution of sodium potassium tartrate (2 mL). The solution was washed with Et_2O (2 × 1 mL) and passed through a short plug of silica gel (2 × 1 cm) eluted with Et_2O. The organic layer was dried over anhydrous $MgSO_4$, filtered, and concentrated *in vacuo* to provide a yellow oil that was purified by silica gel chromatography (5% Et_2O/hexanes to 10% Et_2O/hexanes) to afford 18.1 mg of (*R,Z*)-3-methyl-3-(1-(trimethylsilyl)pent-1-en-1-yl)cyclopentanone as a clear oil (0.076 mmol, 76% yield).

8.3 Asymmetric Allylic Alkylation (AAA)

Review. Alexakis, A.; Backvall, J. E.; Krause, N.; Pamis, O.; Dieguez, M. *Chem. Rev.* **2008**, *108*, 2796, and references therein.

In just the past few years alone, tremendous progress has been made in applying copper chemistry to the area of AAA. Although organopalladium catalysis is well established in this regard,[235] along with several other transition metals serving in related capacities (*e.g.*, Mo,[236] W,[237] Rh,[238] Ru,[239] and Ir[240]), most are responsive to lower pK_a, softer nucleophiles (*e.g.*, malonates). By contrast, copper catalysis offers opportunities to use nonstabilized and far more basic organometallics (*e.g.*, RMgX, RZnX, $RAlR'_2$, etc.), while producing products reflecting high levels of stereo-, regio-, and chemo-selectivity. Success, at least defined by ee's, in these recent ventures, not surprisingly, is largely commensurate with matching educt with ligand, and while the initial source of copper can be influential, it is usually quickly determined: halides, OTf, BF_4, PF_6, CN, and thiophenecarboxylate (TC) are the main choices. Solvent rarely varies from DCM or THF, and subambient temperatures often at –78 °C are common. Selection of the ligand tends to fall within two main structural classes: (1) ferrocenyl-based networks; JOSIPHOS, Taniaphos, etc., and (2) phosphoramidites, which are often thought of as BINOL-derived, but this is certainly not a requirement (*e.g.*, TADDOL-based cases), and more recent alternatives such as NHCs are becoming popular. The leaving group is likely to be either chloride or bromide when RMgX is involved, while a phosphate can be an excellent oxygen-based alternative.[241]

Reactions of Grignard Reagents

Grignard reagents have played a leading role in copper-catalyzed AAA chemistry, although both alkylzinc[242] and vinylic alanes[241] are also of current interest. And while there now exists a large body of information supporting the use of "RMgX" in these processes, most procedures are actually based on "RMgBr." In other words, in addition to potential complications associated with Schlenk equilibria, the halide ion is rarely studied in these methodological developments. Not all commercial sources of RMgX are bromide-derived (*e.g.*, cyclohexylmagnesium chloride and *n*-BuMgCl), and Grignards not of commercial origin may be influenced (*e.g.*, of mixed halide content) by the presence of newly formed salts (*e.g.*, R-I → RLi → "RMgX").

While allylic chlorides are generally useful starting materials (*e.g.*, **160**), the simple S_N2' introduction of a methyl group can be challenging relative to other alkyl residues, presumably due to its lower reactivity and reduced steric demands.[243] Nonetheless, for mainly cinnamyl derivatives, catalytic CuBr along with MeMgBr affords high ee's using a (*S,S,S*)-phosphoramidite ligand **161** (R = OMe). The key experimental observation was control of the rate of Grignard addition: slower addition gave a higher γ:α regioselectivity and best ee's. Gradual introduction of MeMgBr enhances the amount of "MeCu•MgBr," a more selective reagent relative to cuprate $Me_2CuMgBr$ that would be formed with catalytic CuBr + stoichiometric MeMgBr in the pot at the same time.

3% CuBr
3% **161** (R = OMe)
MeMgBr, DCM, -78 °C

160

9:1 γ:α ; 91% ee

161 (R = H, OMe)

Reactions of analogous β-disubstituted arrays were better used in the presence of ligand **161**, R = H; the potential coordination of the methoxy residues in **161**, R = OMe, lowers both the regioselectivity and ee.[244] Copper thiophenecarboxylate (CuTC; 3 mol%) as precursor along with a variety of slowly introduced alkyl Grignards (*e.g.*, Et, *n*-Pr, *n*-pentyl, etc.) give good S_N2' : S_N2 ratios (>79:21), yields (usually >80%), and ee's (≥93%).

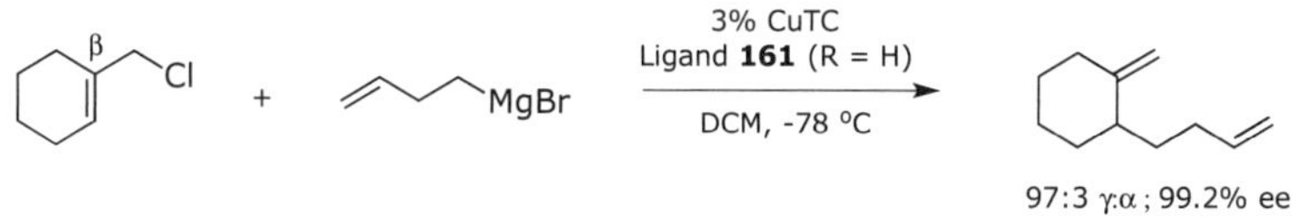

Both *E*- and *Z*-dichlorobutenes as individual substrates represent yet another application of AAA (Aldrich #215082), where the enantio-enriched product of an intial S_N2' addition by RMgX can be further used.[245] Although complete S_N2' addition takes place, ee's vary with ligand. The corresponding *trans*-dibromide **162**, also a commercially available material (Aldrich #D39207), used as received, gave results comparable to or better than those using the corresponding (and, likewise, commercially available) dichloride. Attempts to replace RMgX by R_2Zn were unsuccessful, both with regard to lower ee's (<52%) and greatly reduced reactivity.

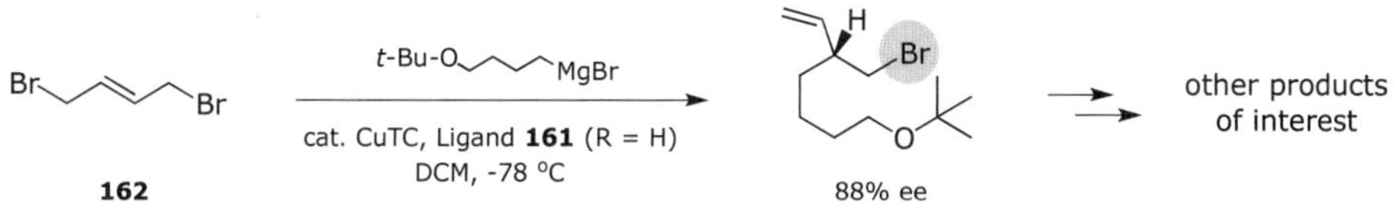

The ferrocene-based ligand Taniaphos (**137**) has been found particularly effective in copper-catalyzed AAA using Grignards and allylic bromides.[246] Here, $CuBr•SMe_2$ is the precursor of choice (1–5 mol%), used in DCM at –78 °C. Cinnamyl systems typically react in high yields and excellent ee's. Catalyst loading of only 1% is sufficient. Reactions in MTBE (*t*-BuOMe) as solvent were not as selective.

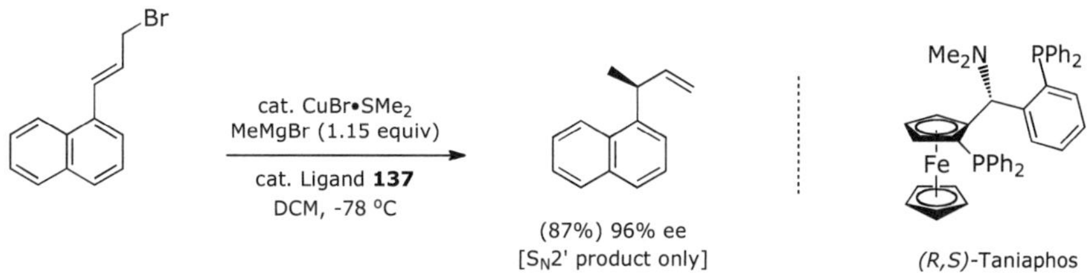

The same catalyst system works well in hetero-allylic asymmetric alkylations (h-AAA; Scheme 1-16).[247] Substrates such as enol esters **163** provide entry to nonracemic esters of allylic alcohols. Remarkably, competing 1,2-addition and/or acyl transfer were not issues; yields are good (80–99%). In these cases, catalyst loading can go as low as 0.8%, and ee's are mostly >95%. Additional chemoselectivity has been noted in the case of cinnamyl ester **163**, where the desired S_N2' AAA takes place without competing Cu-catalyzed 1,4-addition to the enoate. This sets the stage for a subsequent metathesis (GH-2 = Grubbs-Hoveyda second-generation catalyst) *en route* to butenolide **164**.

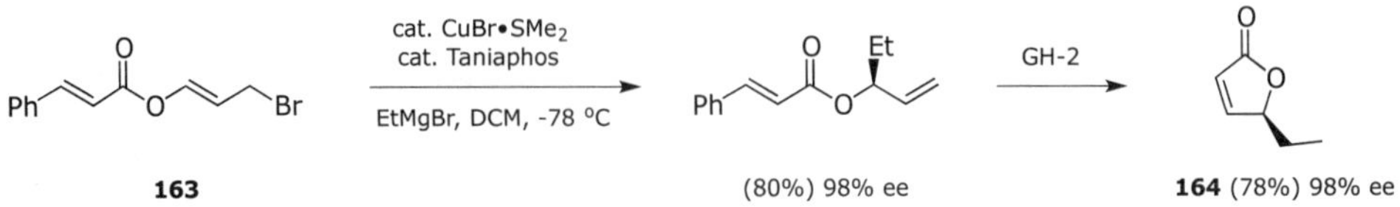

Ph–CH=CH–C(=O)–O–CH₂–CH=CH–CH₂Br **163** → cat. $CuBr{\bullet}SMe_2$, cat. Taniaphos, EtMgBr, DCM, -78 °C → (80%) 98% ee → GH-2 → **164** (78%) 98% ee

Scheme 1-16

Copper-catalyzed asymmetric synthesis of nonracemic allylic esters. Preparation of diene **165**[247]

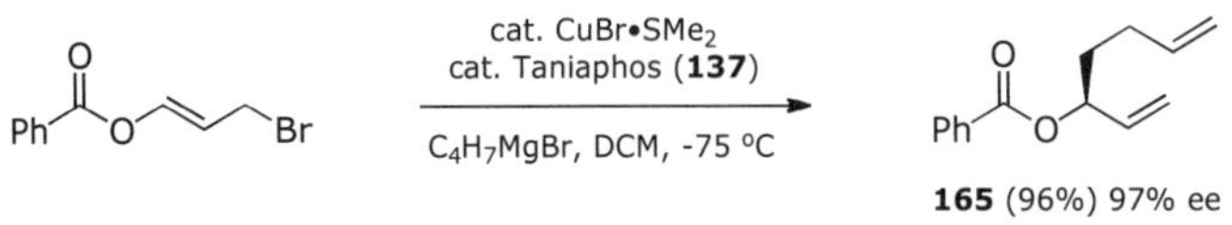

cat. $CuBr{\bullet}SMe_2$, cat. Taniaphos (**137**), C_4H_7MgBr, DCM, -75 °C

165 (96%) 97% ee

Butenylmagnesium bromide (1.1 M, 1.37 mL, 1.51 mmol) was added dropwise over 2 min to a stirred and cooled (–73 °C) solution of the ester (182.0 mg, 0.755 mmol), $CuBr{\bullet}SMe_2$ (8.0 mg, 50 μmol) and *(R,S)*-Taniaphos (30.0 mg, 43.6 μmol) in DCM (3.5 mL) under a nitrogen atmosphere. Stirring was continued for 20 h at –73 °C before the reaction was quenched with MeOH (5 mL). The reaction mixture was removed from the cooling bath and sat. aqueous NH_4Cl (*ca.* 5 mL) was added. The mixture was partitioned between DCM (5 mL) and water. The organic layer was dried over anhydrous $MgSO_4$, filtered, and the solvent evaporated *in vacuo*. Flash chromatography of the dark orange residue over silica gel using 2% Et_2O/pentane gave (+)-*(S)*-benzoic acid-1-vinylpent-4-enyl ester **165** (156.3 mg, 96%) as a clear colorless oil; HRMS *m/z* calcd. for $C_{14}H_{16}O_2$ 216.1150, found 216.1160; $[\alpha]_D^{20}$ = +27.2 (c 0.29, $CHCl_3$). HPLC analysis indicated an enantiomeric excess of 97% [Chiralcel OB-H column; flow: 0.5 mL/min; heptane/IPA: 99.75:0.25; λ = 220 nm; major enantiomer t_R = 15.82 min; minor enantiomer t_R = 18.76 min]. The absolute stereochemistry is based on the conversion to known material.

Regiochemical, as well as stereochemical, issues surround the use of halocrotonates, where no less than four modes of potential addition by an organometallic reagent may take place (see below). To direct the addition of a methyl copper reagent to the S_N2' site, the combination of $CuBr_2{\bullet}SMe_2$, MeMgBr, and the nonracemic ligand (*R,R*)-Taniaphos in DCM at –78 °C leads not only to a good isolated yield of the desired α-stereocenter but also to the product in 99% ee.[248] While the benzyl crotonate worked well, the corresponding thioenoate led to significant by-product formation resulting from competitive 1,4-addition followed by intramolecular displacement to form a disubstituted cyclopropane. Other nonracemic ligands, including JOSIPHOS and "reversed" JOSIPHOS, afforded unacceptable levels of S_N2-derived side products (*ca.* 27%).

1,4-
MeMgBr
S_N2'
1,2-
S_N2' S_N2
(R,R)-Taniaphos

$CuBr•SMe_2$ (1 mol %)
MeMgBr (1.2 equiv)
DCM, -78 °C, 2 h
(81-90%) ≥98% ee

Prospects for dynamic kinetic asymmetric transformation (DYKAT) in allylic substitution reactions using copper-catalyzed Grignard additions to racemic cyclohexenyl allylic bromides look promising.[249] Optimization identified not only bromide over other leaving groups (OAc, Cl, OTf, etc.) but also DCM as solvent (rather than THF or Et_2O) and phosphoramidite ligand **161** (R = H). Copper thiophenecarboxylate (CuTC, 7.5 mol%) was again preferred, and reactions at low temperature gave ee's ≥90% for primary RMgBr (1.2 equiv). A σ-to-π-allylcopper interconversion is likely to account for the opportunity to obtain good yields (79–95%), where equililbration is slow relative to reductive elimination of a presumed favored, nonracemic η^1-Cu(III) intermediate.

t-Bu-O~~~MgBr
cat. CuTC, Ligand **161** (R = H)
DCM, -78 °C, 1 h
(91%) 90% ee

Reactions of Boranes and Boronates

Allylic phosphates have become especially attractive and popular electrophilic coupling partners toward nonracemically ligated copper species capable of delivering their covalently bound residues in an enantiocontrolled S_N2' sense. Remarkably, the technologies have advanced to the state where carbon sp^3 (alkyl), sp^2 (vinyl, aryl), and even sp (alkynyl) residues on boron are all transferrable *via* copper in the desired fashion, and where both yields and ee's (or dr's), for the most part, are impressive.

Alkyl groups associated with products of alkene hydroboratrion with 9-BBN in THF, followed by conversion to the corresponding borate with K-O-*t*-Bu, can be efficiently transmetalated to copper in the presence of CuOAc (10 mol%). Introduction of a *Z*-allylic phosphate leads, upon heating to 60 °C for eight hours, the racemic product of alkyl-allyl coupling in good isolated yields and of *E*-olefin geometry.[250] *E*-Allylic phosphates, on the other hand, led to unacceptable *E/Z* mixtures of olefinic products. Other copper salts, including CuCl and, surprisingly, $Cu(OAc)_2$, were less effective in terms of yields. No ligand on copper is required, and in the absence of the *t*-butoxide salt forming the intermediate borate complex, no coupling took place. Other leaving groups, such as acetate or carbonate, were found to be inhibitory.

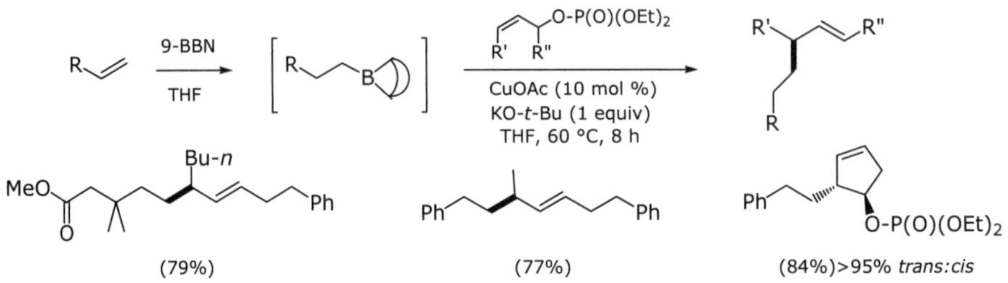

> *Typical procedure for hydroboration/allyl–alkyl coupling*[250]
>
> In a glove box, $(9\text{-BBN-H})_2$ (91.5 mg, 0.375 mmol) was placed in a vial containing a magnetic stirring bar. The vial was sealed with a cap equipped with a Teflon-coated silicon rubber septum, and the vial was removed from the glove box. THF (0.3 mL) and the alkene (0.9 mmol) were sequentially added, and the mixture was stirred at 60 °C for 1 h to prepare the alkylborane.
>
> In a glove box, CuOAc (6.1 mg, 0.05 mmol) was placed in another vial. K-O-*t*-Bu (1 M in THF, 0.5 mL, 0.5 mmol) was added to the alkylborane at 25 °C, prepared in advance, and the mixture was stirred at 25 °C for 5 min to produce the corresponding alkylborate. Next, the alkylborate was then transferred to the vial containing the Cu salt. Finally, the allylic phosphate (0.5 mmol) was added. After stirring at 60 °C for 8 h, DCM was added to the mixture, which was then filtered through a short plug of silica gel, which was washed with diethyl ether. After the solvent was removed under reduced pressure, flash chromatography on silica gel (hexane) provided the desired product.

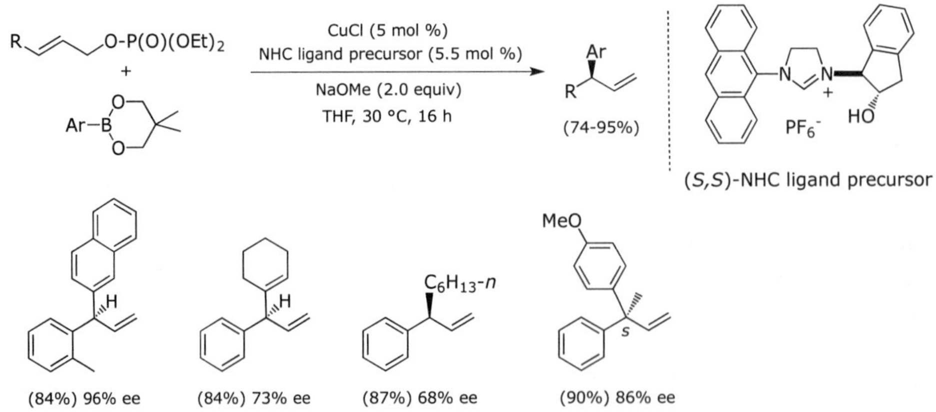

Both tertiary and quaternary carbon centers can be prepared with high regio- and enantioselectivity by reacting allylic phosphates with aryl/alkenyl-boronates (of the neopentylglycol ester type) in the presence of a nonracemic copper-complexed NHC.[251] The choice of NHC ligand precursor (shown below) was not unexpectedly crucial, as was the alkoxide base (NaOMe over either KO-*t*-Bu or NaO-*t*-Bu). These S_N2' reactions, run in THF at 30 °C, work best with cinnamyl-like substrates together with arylboronates, giving ee's in the 89–94% range. β-Alkyl-substituted allylic phosphates lead to reduced levels of stereo-induction (<70%). Extension of this procedure to β,β-disubstituted allylic phosphates generates the corresponding quaternary centers with good stereocontrol (86–90% ee's). Use of an alkenylboronate (*e.g.*, cyclohexenyl) led to a lower level of enantiocontrol. The regiochemistry in almost all cases is usually >98% favoring S_N2' over S_N2.

General procedure for AAA reactions using aryl/alkenylboronates[251]

A solution of CuCl (1.5 mg, 15 μmol), the NHC ligand precursor (8.7 mg, 17 μmol), and NaOMe (32.4 mg, 0.600 mmol) in THF (0.40 mL) was stirred for 15 min at RT. The organoboronic acid neopentylglycol ester (0.600 mmol) was added and the mixture stirred for 5 min at RT. The allylic phosphate (0.300 mmol) was then added with additional THF (0.20 mL), and the resulting mixture was stirred for 16 h at 30 °C. After dilution with EtOAc, the mixture was passed through a pad of silica gel with EtOAc, and the solvent was then removed under vacuum. The residue was purified by silica gel preparative TLC using EtOAc/hexane to afford the desired product. For ee analysis, the product was converted into the corresponding primary alcohol *via* a hydroboration–oxidation sequence.

An allenyl moiety, originating as its pinacolborane derivative, can also be delivered in an S_N2' sense to allylic phosphates.[252] Here again, a sulfonate NHC salt (**i**, R = *i*-Pr; below) must be used to control both the regiochemistry of addition (*i.e.*, S_N2') as well as the level of chiral induction (typically >90% ee's). In related NHC precursor **ii** lacking the sulfonate group, the *opposite* mode of attack is strongly favored (*i.e.*, >98% S_N2). NaOMe is needed to assist with activation of the borate toward transmetallation, and catalytic CuCl (10 mol%) is the source of copper. Both aryl- and non–aryl-substituted allylic phosphates are amenable. Noteworthy here is the option to generate quaternary carbon centers using β,β-disubstituted allylic arrays, although a change to the less sterically demanding NHC ligand **i** (R = Me) must be made. Representative examples follow below.

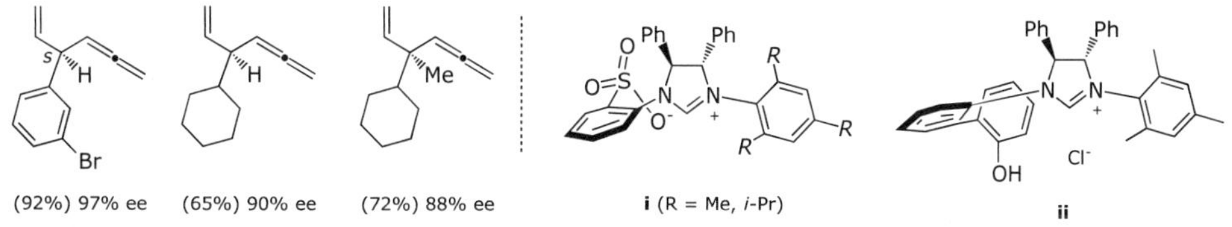

(92%) 97% ee (65%) 90% ee (72%) 88% ee

i (R = Me, *i*-Pr)

ii

conditions: (allene)BPin (1.5 equiv), CuCl (10 mol %), NHC precursor (11 mol %), THF, rt

Reactions of Mixed Alanes

Related reactions that generate quaternary carbon centers from β,β-disubstituted allylic phosphates can be carried out with mixed aluminum reagents $ArAlEt_2$, derived *in situ* from ArLi and Et_2AlCl. Exposure of a β,β-disubstituted allylic phosphate to $ArAlEt_2$/catalytic $CuCl_2{\bullet}2H_2O$ chelated by one of three dimeric Ag-NHC complexes (shown below) leads to S_N2'' substitution with excellent regiocontrol.[253] Aryl as well as heteroaromatics (*e.g.*, 2- and 3-furanyl and thienyl systems), can be selectively transferred from aluminum to carbon. Starting with α-substituted cinnamyl phosphates, unsymmetrical diarylated derivatives can be formed; see the examples below.

R(R')C=CH–CH$_2$–O-P(O)(OEt)$_2$ + Ar–AlEt$_2$ → [$CuCl_2$•$2H_2O$ (1 mol %), NHC ligand precursor (0.5-1.0 mol %), THF, -30 °C, 1-3 h] → R(Ar)(G)C–CH=CH$_2$ (G = aryl, heteroaryl)

(81%) 92% ee (NHC, R = Ph; R' = Me) | (90%) >96% ee (NHC, R = Ph; R' = *i*-Pr) | (92%) 85% ee (NHC, R = Ph; R' = Me) | (71%) 89% ee (NHC, R = H; R' = Me)

dimeric Ag-NHC precursor (R = H, Ph) (R' = Me, *i*-Pr)

Representative procedure for Cu-catalyzed enantioselective allylic substitution reactions of allylic phosphates by diethylaryl/heteroarylaluminum reagents[253]

In an N_2-filled glove box, an oven-dried vial (8 mL, 17 × 60 mm) with a magnetic stir bar is charged with an NHC-Ag(I) complex (7.5 × 10^{-4} mmol) and sealed with a septum before removal from the glove box. To the vial under a N_2 atmosphere are added THF (0.5 mL) and a solution of $CuCl_2$•$2H_2O$ (0.02 M in THF, 75 μL, 1.50 × 10^{-3} mmol). The light blue solution is allowed to stir at RT for 30 min, and a solution of the allylic phosphate (0.150 mmol) in THF (0.5 mL) is added through a syringe. After stirring for 10 min, the reaction mixture is allowed to cool to –78 °C. A solution of the mixed arylalane reagent, $ArAlEt_2$ (0.622 M in pentane, 723 μL, 0.450 mmol), is added slowly through a syringe. The vial is transferred to a –30 °C Cryocool. After 1 h, the reaction is cooled to –78 °C and quenched by the addition of a saturated aqueous solution of Rochelle's salt (potassium sodium tartrate, 2 mL). The aqueous layer is washed with Et_2O (3 × 1 mL), and the combined organic layers are passed through a short plug of anhydrous $MgSO_4$ and silica gel. The filtrate is concentrated under reduced pressure to provide a colorless oily residue, which is purified by silica gel column chromatography (30:1 pentane:Et_2O) to afford the desired product.

Mixed vinylalanes, derived from hydroalumination reactions of alkyl-substituted terminal alkynes, can be used to arrive at 1,4-dienic products bearing an enantioenriched stereocenter at their C-3 position.[254] Nonracemically NHC-complexed copper(I) is again used catalytically to control regioselectivity and induce asymmetry in couplings with β,β-disubstituted allylic phosphates; these species are formed from the same type of NHC precursors illustrated in the examples above (in these cases studied, R = H, Ph; R' = Me; and 2,6-diisopropyl). Substituents on the educt can be aryl/substituted aryl, alkyl, carboalkoxy, and silyl; in all cases, the regiochemistry is >98% S_N2', and the yields are high (82–97%). The ee's of the resulting 1,4-dienes are typically >90%. Arylalkynes require a Ni-catalyzed hydroalumination that minimizes alkyne deprotonation/competitive S_N2' addition leading to products of acetylene transfer that are tough to separate from the desired dienic adducts. The procedure is essentially that used for the reactions with mixed alkynylalanes below.

CF_3 O-P(O)(OEt)$_2$ + n-C_6H_{13} Al(i-Bu)$_2$ (1.5 equiv)

$CuCl_2$•H_2O (1 mol %)
NHC ligand precursor (2.5 mol %)
THF, -15 °C, 3 h

CF_3 C_6H_{13}-n
(96%) 96% ee
(NHC, R = Ph; R' = Me)

CF_3 C_6H_{13}-n (96%) 96% ee
$PhMe_2Si$ Bu-t (>98%) 82% ee
Br (89%) 98% ee
NO_2 (92%) 96% ee

Although an acetylide group is routinely viewed (and used) as a "dummy" ligand in organocuprate chemistry, it can be smoothly transferred when in the form of its NHC complex in reactions with allylic phosphates.[255] The precursor mixed alkynylalanes bearing two isobutyl residues; *i.e.*, (alkynyl-Al(i-Bu)$_2$, selectively releases the alkynyl residue to the NHC-Cu formed *in situ*, which subsequently undergoes S_N2' addition to an allylic educt. Various combinations of β,β-disubstituted allylic phosphates and substituted alkynes can be coupled with outstanding regioselectivity (>98% S_N2'). The precursor NHC ligands of choice in this process are, again, the Ag-based dimeric species seen above, with $CuCl_2$•$2H_2O$ as the source of copper, and THF as solvent.

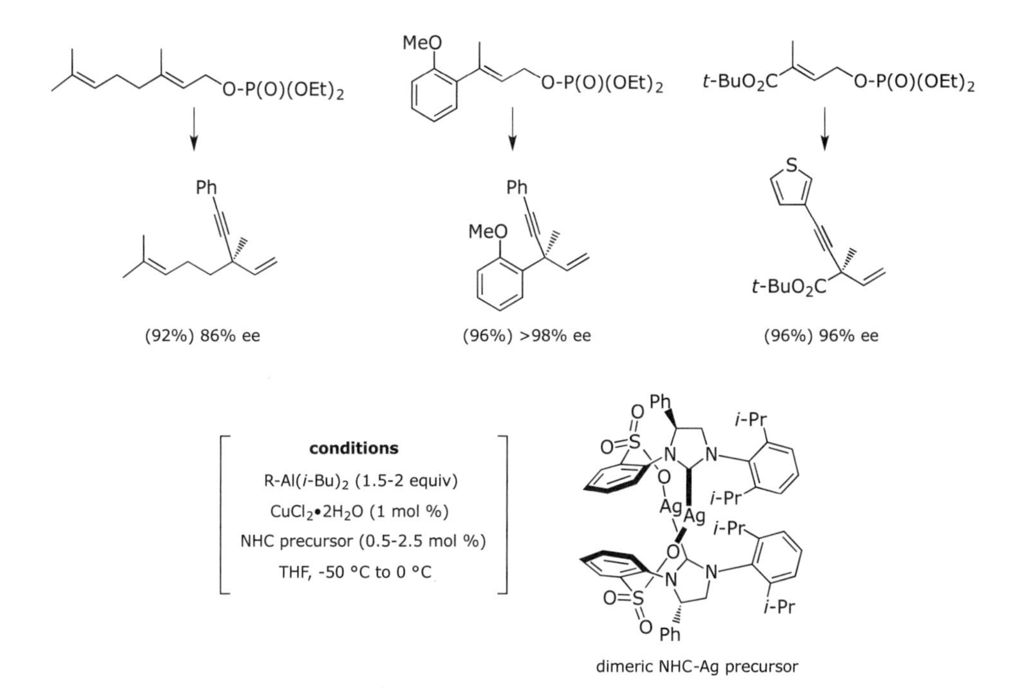

Representative procedure for catalytic enantioselective addition of alkynylaluminum reagents to allylic phosphates[255]

An oven-dried, 4 mL vial equipped with a stir bar was charged with the chiral NHC-Ag complex (2.70 mg, 2.50 μmol) and $CuCl_2$•$2H_2O$ (0.85 mg, 5.00 μmol) in a glove box under an N_2 atmosphere. The vial was sealed with a septum and cap and wrapped with electrical tape before removal from the glove box. THF (0.50 mL) was added at RT through a syringe, and the resulting blue solution was allowed to stir for 5 min prior to the addition of diisobutyl(phenylethynyl)aluminum (160 μL, 0.200 mmol) (FLAMMABLE! USE CAUTION!) at −78 °C resulting in a brown solution. (*E*)-Diethyl 3-phenylbut-2-enyl phosphate (28.4 mg, 0.100 mmol) was added through a syringe as a solution in THF (0.5 mL). The mixture was sealed with a cap and transferred to a Cryocool (−30 °C) and was allowed to stir at this temperature for 6 h. The reaction was quenched upon the addition of a saturated solution of sodium potassium tartrate (1.0 mL) stirring under N_2. After allowing the mixture to warm to RT, the aqueous solution was washed with Et_2O (3 × 1 mL), and the organic layers were passed through a short plug of silica gel (3 × 1 cm) eluting with Et_2O (8.0 mL). The organic layer was concentrated *in vacuo* to a yellow oil, and then purified by silica gel chromatography (100% hexanes) to afford the desired product in the indicated yield.

9. Copper Hydride Chemistry

Reviews. Lipshutz, B. H. *Synlett* **2009**, 509; Deutsch, C.; Lipshutz, B. H.; Krause, N. *Chem. Rev.* **2008**, *108*, 2916; Diez-Gonzalez, S.; Nolan, S. P. *Acct. Chem. Res.* **2008**, *41*, 349; Rendler, S.; Oestreich, M. *Angew. Chem., Int. Ed.* **2007**, *46*, 498; Riant, O.; Mostefaï, N.; Courmarcel, J. *Synthesis* **2004**, 2943.

Although copper hydride enjoys a rich history dating back to the mid-1800s,[256] modern usage of CuH in synthesis is well recognized to have begun with "Stryker's reagent" (SR) in 1988.[257] That the phosphine-stabilized hexamer, $[(Ph_3P)CuH]_6$,[258] can be used catalytically, in particular with inexpensive and environmentally unoffensive silanes, has shifted the spotlight insofar as development of new methodologies is concerned toward asymmetric uses of these net hydrosilylation reactions. The early recipe of Stryker, calling for catalytic amounts of CuCl together with NaO-*t*-Bu to generate CuO-*t*-Bu,[257] is still used frequently, especially since (extremely air-sensitive) CuO-*t*-Bu is no longer commercially available. Key C-H bond constructions that rely on $(Ph_3P)CuH$ can be found in many syntheses; as examples: (1) (±)-*trans*-kumausyne,[259] proceeding through enone **166**; (2) intermediate **167** *en route* to (+)-pinnatoxin A,[260] and **168**, of brevetoxin A,[261] among many others. Nonetheless, advances have been made of late such that the precursor Cu-O bond no longer need be formed *in situ* from these salts; that is,

166

cat. SR | $PhSiH_3$

(63%)[212]

167

cat. SR

1,4-reduced product

(64%)[213]

168

cat. SR | Me_2PhSiH

1,4-reduced product

(>82%)[214]

$Cu(OAc)_2{\bullet}H_2O$ has been found to be an attractive replacement that streamlines the procedure to CuH (a reagent that is tolerant of water or alcohols).[262] Usually, the phosphine is present prior to the addition of the silane to ensure stability of the CuH through complexation (Eqn. 1-9). Considerable effort is also being made to evaluate *N*- heterocyclic carbene (NHC) ligands as a means of avoiding phosphines altogether

$$\text{Cu–X} + \text{M–H} \xrightarrow{\text{ligand (L)}} \text{(L)CuH} + \text{MX} \qquad \text{(Eqn. 1-9)}$$

X = Cl, OAc; M = H, Si, Sn; L = phosphine, NHC

(*vide infra*).[263] A brief summary of several combinations that generate CuH is shown in Table 1-4. Polymethylhydrosiloxane (PMHS)[264] is often listed as a 29mer, and as a stoichiometric source of hydride can vary widely in content between vendors.

Table 1-4. Methods for generation of CuH

Source of Cu(I)	Additive	Hydride	References
CuCl	NaO-*t*-Bu	H_2	[265]
		Bu_3SnH	[266]
		$PhMe_2SiH$	[267, 268]
		$PhSiH_3$	[266]
		PMHS	[269]
		TMDS	[270]
$CuX(PPh_3)$	–	$PhMe_2SiH$	[268]
$Cu(OAc)_2{\bullet}H_2O$	–	PMHS	[271, 272]
$CuCl_2{\bullet}2H_2O$	–	PMHS	[273]
$CuNO_3(P(3,5\text{-xylyl})_3)_2$	–	H_2	[191, 274]
CuO-*t*-Bu	–	PMHS	[275]
CuF_2	–	$PhSiH_3$/PMHS/DEMS	[276, 277]

PMHS = –O–Si(Me)(H)–O–   TMDS = H–Si(Me)₂–O–Si(Me)₂–H   DEMS = H–Si(Me)(OEt)₂

Material from Lancaster (Lancaster, CA; catalog #L14561) appears to give reproducible CuH chemistry, while PMHS from Acros Organics can lead to different (vastly inferior) results. PMHS is used as received from the vendor, although it should be handled and stored under argon in a multiply septumed bottle to maximize lifetime. Tetramethyldisiloxane (TMDS; Alfa catalog #12934) is another inexpensive silane that, on occasion, is a superior source of hydride (*vide infra*). Also available from Alfa Aesar (Ward Hill, MA) are Fleming's silane ($PhMe_2SiH$; catalog #A17901), phenylsilane ($PhSiH_3$; catalog #L04558), and diethoxymethylsilane (DEMS; catalog #A10153) that are occasionally used.

With several sources of mild hydride available, the focus of current CuH chemistry continues to be on catalytic asymmetric processes.[278] Thus, both 1,2- and 1,4-additions of hydride to a variety of electrophilic centers have been developed, where the level of chiral induction derives from the innate bias of the ligand-metal complex. Such interactions, specifically regarding the electronic effect of the P-Cu-P bite angle, have been the subject of a recent theoretical DFT study.[279] Much of the success realized to date with asymmetrically ligated CuH has come from a relatively small subset of either biaryl or ferrocenyl *bis*-phosphines. Most notably, selected biaryls of the BIPHEP (**169**)[280] and SEGPHOS (**170**)[281] series deserve heightened attention not only for their remarkable discriminatory structural features but also because of the high TONs now possible as their CuH complexes. The same is true for certain chelators in the JOSIPHOS series.[282] The biaryl ligands illustrated in Figure 1-3, and the ferrocenyl derivatives shown in Figure 1-5, are also commercially available both in quantity and in enantiomerically pure form. They are stable solids that need be stored to the exclusion of air (*e.g.*, in a glove box) in order to avoid phosphine oxidation. Dipyridyl analogs (*e.g.*, **171**), used in conjunction with CuF_2 and $PhSiH_3$ in toluene, also display good stability.[283] Indeed, their influence on asymmetric hydrosilylations of simple aryl ketones is such that these reactions can be carried out even under an oxidating environment (*i.e.*, in air), as had been observed previously in 1,2-additions.[284] TONs with **171**, likewise, are impressive and can be as high as 100,000:1 in substrate-to-ligand (S/L), with ee's usually >87% (Eqn. 1-10). Two ligands in this bipyridyl series, the parent Ar = phenyl and the xylyl analog shown (**171**), are commercially available in both antipodal forms.

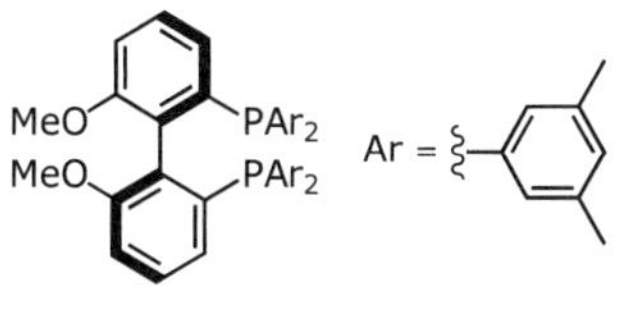

169 [(*R*)-Xyl-MeO-BIPHEP][282]
Roche

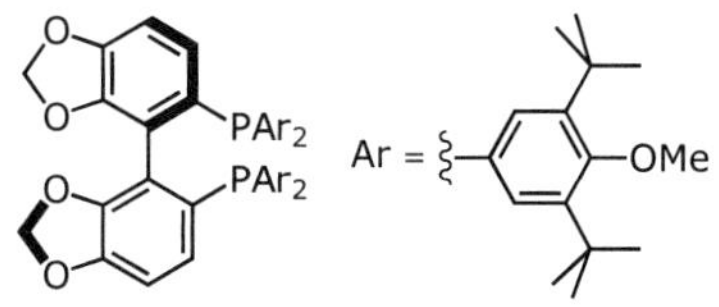

170 [(*R*)-(-)-DTBM-SEGPHOS][283]
Takasago
(Strem Cat.# 15-0066)

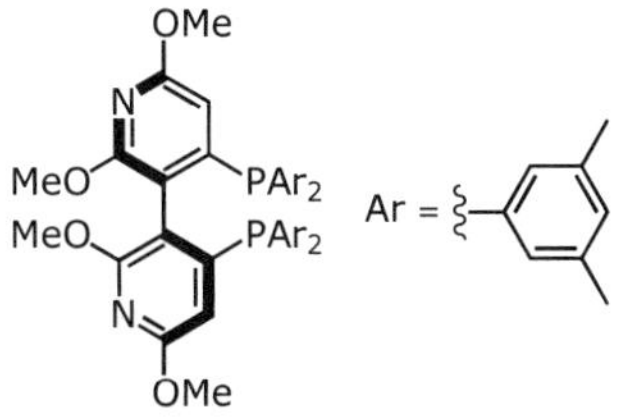

171 [(*S*)-Xyl-P-Phos]
Chan[284]
(Strem Cat.#15-5201)

Figure 1-3. Representative biarylphosphine ligands used in CuH chemistry.

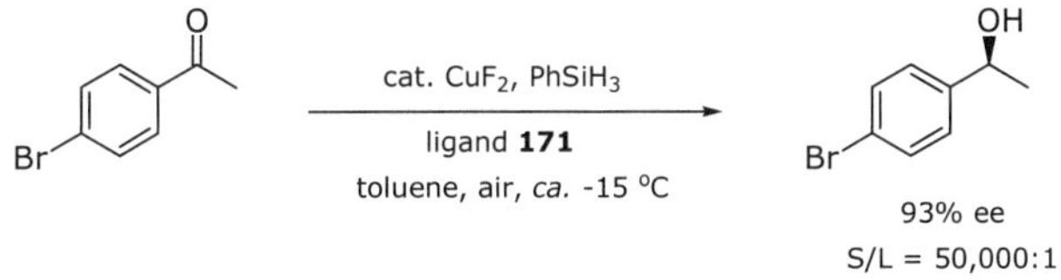

cat. CuF_2, $PhSiH_3$; ligand **171**; toluene, air, *ca.* -15 °C — 93% ee, S/L = 50,000:1 (Eqn. 1-10)

Both ligand **171** and its parent, (*S*)-P-Phos (Ar = Ph), as their complexes with *in situ*–generated CuH, have been studied in asymmetric hydrosilylations of halogen-containing aryl ketones.[285] That is, a variety of ketones bearing either chloride or bromide in the α-, β-, or γ-positions undergo 1,2-carbonyl reduction in toluene at –20 °C, using phenylsilane as the source of hydride. Enantiomeric excesses typically exceed 90%, as illustrated by the examples below.

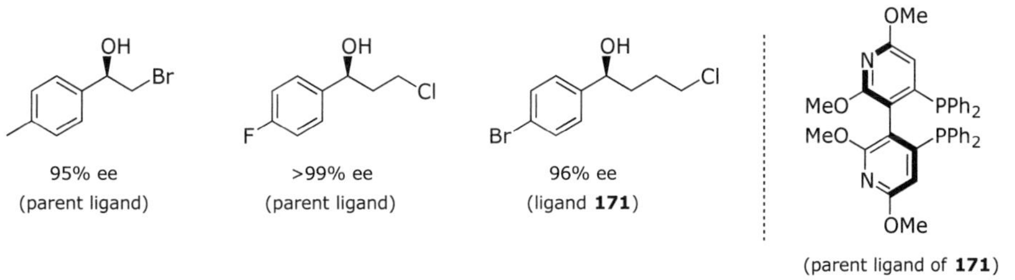

95% ee (parent ligand); >99% ee (parent ligand); 96% ee (ligand **171**)

(parent ligand of **171**)

Both BIPHEP **169** and SEGPHOS **170** form highly kinetically reactive CuH complexes. [*(R)*-(-)-DTBM-SEGPHOS]CuH (**172**) has excellent shelf life at room temperature when protected from air during handling and storage.[286] It is made from $Cu(OAc)_2{\cdot}H_2O$ and excess PMHS in toluene, where each milliliter of a preformed 0.001 M solution used per millimole of substrate translates into S/L = 1000:1.

Preparation of [(R)-DTBM-SEGPHOS]CuH in a Bottle (0.001 M)[286]

$Cu(OAc)_2{\bullet}H_2O$ + PMHS —(toluene, rt; (*R*)-DTBM-SEGPHOS)→ "CuH in a Bottle" [Ar = 3,5-di-*t*-Bu-4-OMe-phenyl]

An oven-dried, poly-coated amber glass bottle equipped with a stir bar was purged under argon and brought into the glove box. $Cu(OAc)_2{\bullet}H_2O$ (10 mg, 0.05 mmol) and (*R*)-DTBM-SEGPHOS (59 mg, 0.05 mmol) were added followed by dry toluene (44 mL), and the reaction mixture was allowed to stir for 2 h at RT. PMHS (6 mL) was added dropwise, and the mixture was allowed to stir for 30 min. The amber bottle was then sealed using a standard (Aldrich) Sure/Seal, or Oxford Sure/Seal, storage valve-cap, and preferably stored at 4 °C.

Hydrosilylations by complexed CuH have been applied to several substrate types (Scheme 1-17). As illustrated by the following examples, the stereochemical outcomes from both 1,2-additions (to aryl ketones[287] and aryl imines[288]) and 1,4-conjugate additions (cyclic ketones,[269] β-aryl[289] and/or β-silyl enoates,[290] and unsaturated lactones)[273] can be controlled by these ligand-accelerated reactions. One of the key "tricks" to this chemistry is to take advantage of the tolerance of CuH complexes to alcohols and water.[273] In fact, several methods rely on the presence of a bulky alcohol (*e.g.*, *t*-BuOH) to significantly enhance reaction rates. It takes relatively little added alcohol (volume-wise) to accelerate the hydrosilylation, usually on the order of 1–3 equivalents. The role of this additive is usually ascribed to the more rapid quenching of an intermediate copper alkoxide or enolate, which necessarily generates a copper alkoxide, an ideal precursor to rapid reformation of CuH in the presence of excess silane.[273] Thus, the rate increase is presumably due to

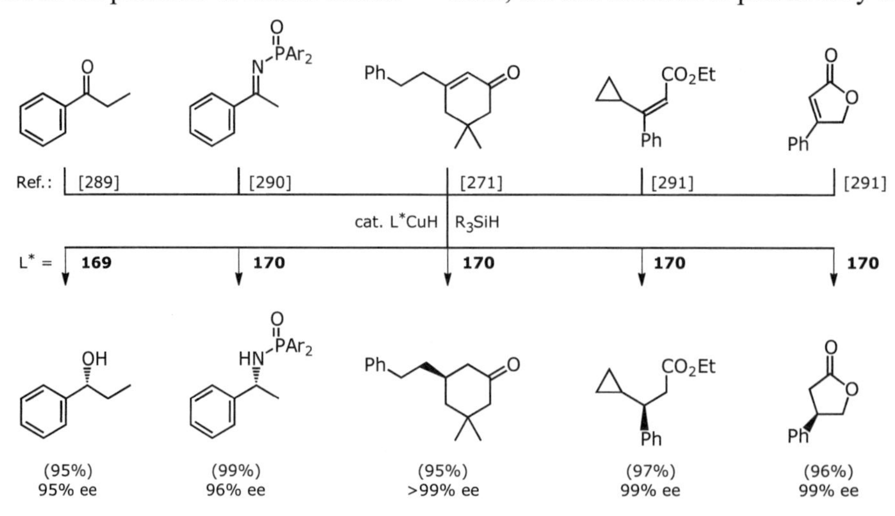

Scheme 1-17. Representative substrate types amenable to CuH-catalyzed hydrosilylations.

bypassing a slower metathesis step between Cu-O and Si-H bonds that is otherwise essential in the catalytic cycle for regenerating CuH.

Early in these studies with DTBM-SEGPHOS-ligated CuH, **172**, the source of catalytic copper hydride (out of convenience) was Stryker's reagent itself [*i.e.*, the hexamer of $(Ph_3P)CuH$].[257] In cases run at ambient temperatures, the "background" (and hence, competing) achiral addition of $(Ph_3P)CuH$ led to lower product ee's. Thus, catalytic asymmetric hydrosilylations by CuH at room temperature should best be run using either [CuCl + NaO-*t*-Bu]- or $[Cu(OAc)_2 \cdot H_2O]$-based recipes. At temperatures of 0 °C or lower, ee's are not likely to suffer from Ph_3P present in solution. Nevertheless, Stryker's reagent has a limited shelf life. Even a fresh bottle of reagent should be checked by ^{1}H NMR for hydride (δ 3.50, in C_6D_6).[270, 291]

All of the examples shown (Scheme 6-17), except for aryl imines, have S/L ratios of >1000:1 (imines: ≤100:1). For selected educts such as acetophenone, the TON is >100,000 using either CuH complexed by ligand **169** or **170**,[286] while for isophorone, it is ≥275,000:1 with (DTBM-SEGPHOS)CuH (procedure below).[269] 1,2-Additions are temperature sensitive in that ee's improve as reactions are cooled toward –78 °C.[287] The variation in ee's between temperatures can easily be >10%, and this is a limitation for larger scale applications. Moreover, at lower temperatures, there may be substrate solubility issues. Solvents other than toluene can be used (*e.g.*, THF and DCM), although their effect on ee's has not been established for many types of substrates. 1,4-Additions appear to be far less temperature-dependent. Dialkyl ketones are also reactive, but ee's are usually far too low to be useful[292]; these have been improved greatly (*vide infra*). Several heteroaromatic ketones have been studied using (DTBM-SEGPHOS)CuH; at low temperatures (–35 °C to –78 °C), good ee's and high yields have been obtained (Fig. 1-4).[274, 293, 294]

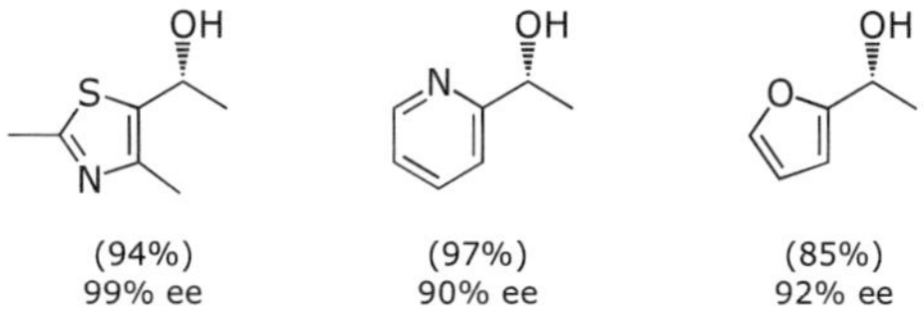

(94%) 99% ee | (97%) 90% ee | (85%) 92% ee

Figure 1-4. Products of asymmetric hydrosilylation of heteroaromatic ketones.

Representative procedure for the asymmetric hydrosilylation of cyclic enones; isophorone (S/L = 1,000:1)[269]

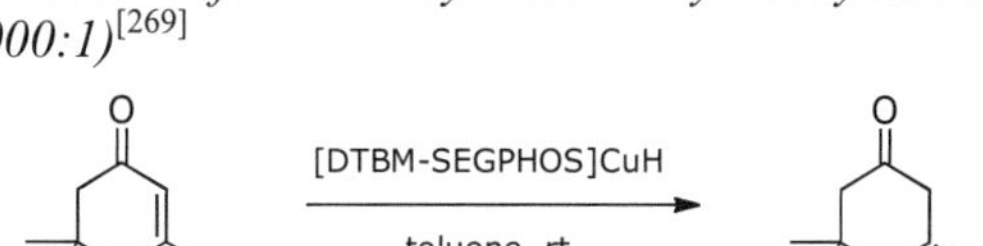

[DTBM-SEGPHOS]CuH; toluene, rt; (S/L = 1000:1) — (90%) >98%ee

To a 5-mL, round-bottom flask, flame dried and purged with Ar, was added $(Ph_3P)CuH$ (4 mg, 0.012 mmol) and *(R)*-DTBM-SEGPHOS (2.8 mg, 0.0024 mmol). Et_2O (2.5 mL) was added, and the solution was cooled to –35 °C. PMHS (312 μL, 4.8 mmol) was introduced *via* syringe and then isophorone (neat, 325 μL, 2.4 mmol) was added. The mixture was stirred at –35 °C for 12 h until the reaction was complete by TLC (20% Et_2O/ligroin). The mixture was quenched by pouring into 3 M NaOH and then diluted with Et_2O/H_2O and stirred for 2 h at RT. The aqueous layer was extracted twice with Et_2O, and the combined organic layers were washed with brine, dried over anhydrous $MgSO_4$, filtered, and concentrated by rotary evaporation with caution due to slight volatility. The residue was purified by flash chromatography (10% Et_2O/ligroin) to afford the product ketone (302 mg, 90%) as a clear oil. TLC: R_f = 0.5 (20% Et_2O/ligroin). Analysis by GC (Chiraldex B-DM 75) indicated an *ee* of 98%; $[\alpha]_D = -24.5°$ (c = 1.1, $CHCl_3$).

The sense of chirality in these hydrosilylations is such that *(R)*-(-)-DTBM-SEGPHOS predictably delivers hydride to aryl alkyl ketones from the *si* face to give alcohols of *R* absolute stereochemistry, which is also true for imines. Stereocontrol at a β-site due to asymmetric 1,4-additions is controlled by either the nature of the geometrical isomer (*i.e.*, *E* or *Z*) or the axial chirality of the ligand. Thus, by switching from the (*R*)- to the (*S*)-enantiomer of the ligand, or from the *E*- to the *Z*-activated olefin isomer, the observed central chirality can be inverted.

Acyclic enones are surprisingly poor substrates for DTBM-SEGPHOS, as are β,β-dialkyl-substituted enoates. Fortunately, the di-*t*-butyl analog of JOSIPHOS, (*R,S*)-PPF-P(*t*-Bu)$_2$ (**173**; Figure 1-5) as its CuH complex leads to outstanding ee's for these two cases, as represented by the conversions of **174** and **175** into their 1,4-adducts.[295] A new nonracemic ferrocene-based phosphine-phosphoramidite ligand (PPFAPhos; see below), which contains (S_c)-central, (R_p)-planar, and (R_a)-axial chirality, used to complex CuH formed from $CuF(Ph_3P)_3$•2MeOH (5 mol%), has been found to reduce enoates in THF at room temperature.[296] PMHS is the hydride source, and *t*-butanol (4 equiv) is present. Bulky esters (*e.g.*, *i*-Pr or *t*-Bu) tended to limit the extent of conversion, while ee's, in general, were in the 80% to >99% range.

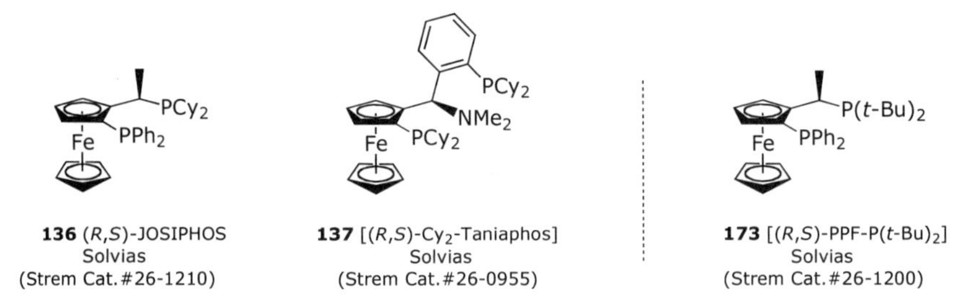

Figure 1-5. JOSIPHOS and derived ligands.

O N P O Fe PPh2

(S_c,R_p,R_a)-PPFAPhos

EtO O Ph **174** → cat. CuH, 1.1 *t*-BuOH / cat. (*R,S*)-PPF-P(*t*-Bu)$_2$ / toluene, 0 °C, PMHS → EtO O Ph (98%); 99% ee

Represenative procedure for the asymmetric hydrosilylation of acyclic enones (S/L = 100:1)[295]

2% CuH, 1% **173**

toluene, PMHS, -78 °C

OTBS **175** → OTBS (95%) 98% ee

To a 5-mL, round-bottom flask, flame dried and purged with Ar, was added $(Ph_3P)CuH$ (2 mg, 0.006 mmol) and JOSIPHOS ligand **173** (5.4 mg, 0.003 mmol). Toluene (0.3 mL) was added, and the solution was cooled to –78 °C. PMHS (78 μL, 1.21 mmol) was introduced dropwise *via* syringe followed by the addition of the enone (neat; 82 mg, 0.304 mmol). The reaction was allowed to stir for 6 h at –78 °C until complete by TLC (2% Et_2O/ ligroin). It was then quenched at –78 °C with 3-M NaOH, warmed to RT, and diluted with Et_2O/H_2O with stirring for 2 h. The aqueous layer was extracted two times with Et_2O, and the combined organic layers were washed with brine, dried (anhydrous $MgSO_4$), filtered, and concentrated *via* rotary evaporation. The residue was purified by flash chromatography (2% Et_2O/ligroin) to afford the product ketone (78 mg, 95%) as a clear oil. The product (R_f = 0.27, 10% Et_2O/ligroin) was analyzed by GC (BDM 110): ee = 98%. HREIMS calcd. for $C_{15}H_{32}O_2Si$-C_4H_9: 215.1467; found 215.1470.

Representative procedure for asymmetric hydrosilylation of an enoate. (S)-Ethyl 3-phenylbutanoate (S/L = 7700:1)[289]

cat. CuH, ligand **172**

$PhCH_3$, PMHS, *t*-BuOH
0 °C, 12 h
(S/L ≅ 7700:1)

Ph OEt → Ph OEt (92%) 98% ee

To a 25-mL, round-bottom flask, flame dried and purged with Ar, was added CuCl (10 mg, 0.101 mmol), NaO-*t*-Bu (12 mg, 0.125 mmol), and *(R)*-DTBM-SEGPHOS (2 mg, 0.002 mmol). Toluene (10 mL) was added, and the solution was stirred at RT for 20 min, then cooled to 0 °C, and the enoate (neat, 2.47 g, 13.020 mmol) was added. PMHS (1.7 mL, 26 mmol) was introduced *via* syringe followed by *t*-BuOH (1.45 mL, 14.30 mmol). The mixture was stirred at 0 °C until complete by TLC (12 h; 5% EtOAc/hexanes). The reaction was quenched by pouring into saturated $NaHCO_3$, diluted with Et_2O/H_2O, and then stirred for 2 h at RT. The aqueous layer was extracted twice with Et_2O, and the combined organic layers were washed with brine, dried over anhydrous $MgSO_4$, filtered, and concentrated by rotary evaporation. The residue was purified by flash chromatography (2% EtOAc/hexanes) to afford the title compound (2.29 g, 92%) as a clear oil. The product (R_f= 0.15, 5% EtOAc/hexanes) was analyzed by GC (G-TA 100), which indicated an ee of 98%.

Notwithstanding the tendancy of nonracemically ligated CuH to add in a 1,2-sense to aryl ketones in high ee's, and the natural inclination of soft copper reagents in general to undergo conjugate additions, there remained until recently a third mode of reaction by copper reagents involving unsaturated ketones: asymmetric 1,2-addition.[297] With respect to CuH, enones bearing an α-substituent react with either DTBM-SEGPHOS- or 3,4,5-Me_3-BIPHEP-complexed CuH in Et_2O at –25 °C to afford predominantly the 1,2-aduct in good ee's and high islated yields.[298] The former ligand gives rise to the opposite absolute sense of stereoinduction. Diethoxymethylsilane (DEMS) is the stoichiometric source of hydride of choice, used together with catalytic amounts of $Cu(OAc)_2{\bullet}H_2O$. Alkyl substituents in the α-site relative to, *e.g.*, an α-bromoenone, tend to give better ee's. Acyclic educts bearing both aromatic and alkyl substituents at the β-position, as well as cyclic enones, undergo this shift toward regiocontrolled 1,2-hydrosilylation. In those cases where the enone is not substituted in the α-position, only β,β–disubstituted, β-aryl- enones are amenable to enantioselective 1,2-reduction by the same reagents, although the regioselectivities are typically in the 75–80 : 20–25 1,2:1,4 range.[299]

$Cu(OAc)_2{\bullet}H_2O$
ligand (3 mol %)
DEMS (3 equiv)
Et_2O (0.5 M), -25 °C

(96%) 93% ee [with (Me_3-BIPHEP)CuH]

(82%) 90% ee

(90%) 95% ee

82:18 1,2- : 1,4-; 94% ee

[with (DTBM-SEGPHOS)CuH]

General procedure for the enantioselective CuH-catalyzed 1,2-reduction of β,β-disubstituted enones[298]

A conical 3-mL microwave vial containing a conical stir bar was charged with fine-powdered $Cu(OAc)_2{\bullet}H_2O$ (0.5 mg, 3 mol%, 3 mmol) and ligand (3 mol%, 3 mmol). The vial was capped with a rubber septum and placed under an Ar atmosphere, followed by the addition of 200 mL of Et_2O added *via* syringe. At RT, the silane (0.3 mmol) was introduced, and the mixture was stirred for 30 min. The vial was then placed into a precooled acetone bath set to –25 °C and stirred for an additional 5 min. The enone (0.1 mmol) was subsequently introduced *via* syringe. The side of the reaction vial was rinsed with Et_2O (2 × 50 mL). The extent of conversion was monitored by TLC. All reactions were run to completion in *ca.* 24 h. The reaction was quenched at –25 °C after 24 h by the addition of satd. NH_4F/MeOH (0.5 mL). The reaction vial was removed from the cooling bath and warmed to RT. After filtration through SiO_2, the solvent was evaporated *in vacuo* and the crude reaction mixture was analyzed by NMR and GC/free induction decay (FID). This was followed by purification by column chromatography on silica gel. The purified product was analyzed by chiral HPLC or GC for the determination of ee.

Asymmetric copper-catalyzed 1,2-additions of a variety of alkyl Grignard reagents to enones has also been recently described, providing entry to nonracemic tertiary allylic alcohols.[300] α,β-Disubstituted cases work well, run in *t*-BuOMe at low temperatures (–78 °C or –60 °C over 5–10 hours. The best source of chirality was found to be the JOSIPHOS ligand (ent-**136**), while CuBr•DMS (5 mol%) was the favored source of copper(I); interestingly, $Cu(OAc)_2$ led to racemic products. Yields are typically high, and ee's are modest to high depending on the enone substitution pattern.

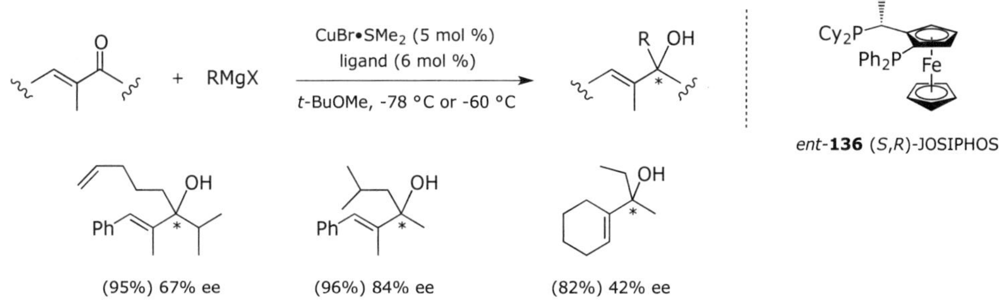

$CuBr \cdot SMe_2$ (5 mol %)
ligand (6 mol %)
+ RMgX
t-BuOMe, -78 °C or -60 °C

ent-**136** (*S*,*R*)-JOSIPHOS

(95%) 67% ee (96%) 84% ee (82%) 42% ee

Conjugate reductions of both nitro olefins and unsaturated nitriles rely on the parent JOSIPHOS-ligated CuH (Scheme 1-18). The former class was initially studied using preformed CuO-*t*-Bu in the presence of a mixture of silanes in toluene containing a slight excess of H_2O relative to substrate.[275] An updated version that avoids air-sensitive CuO-*t*-Bu using CuF_2 in a mixture of toluene/CH_3NO_2 affords similar results,[276] and it avoids the presence of salts (in particular, NaCl generated from CuCl + NaO-*t*-Bu), which can significantly inhibit these reactions. The latter relies on catalytic $Cu(OAc)_2$/PMHS in toluene containing *t*-BuOH. Such proton sources (H_2O, *t*-BuOH, etc.) can make a *huge* difference in reaction rates (*vide supra*). Good ee's and isolated yields are to be expected, as with the examples and procedures below. Even with β,β-diaryl unsaturated nitriles,[272] the JOSIPHOS-complexed reagent is remarkably capable of distinguishing subtle differences between rings and gives excellent ee's. These products are precursors to many functional groups with remote central chirality.

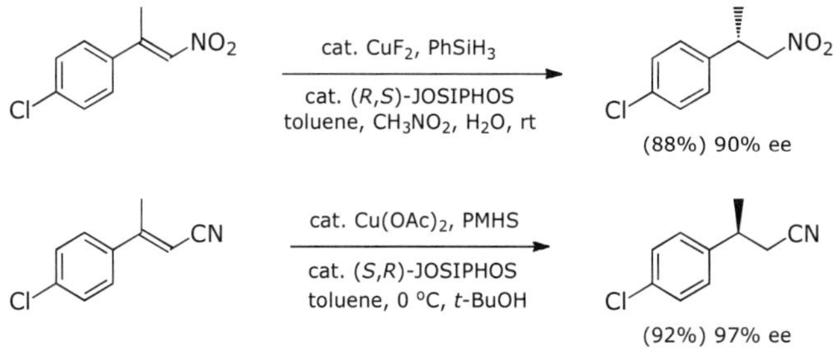

cat. CuF_2, $PhSiH_3$
cat. (*R*,*S*)-JOSIPHOS
toluene, CH_3NO_2, H_2O, rt
(88%) 90% ee

cat. $Cu(OAc)_2$, PMHS
cat. (*S*,*R*)-JOSIPHOS
toluene, 0 °C, *t*-BuOH
(92%) 97% ee

Scheme 1-18

General procedure for the reduction of nitroalkenes using copper(II)-fluoride–derived CuH and nitromethane[276]

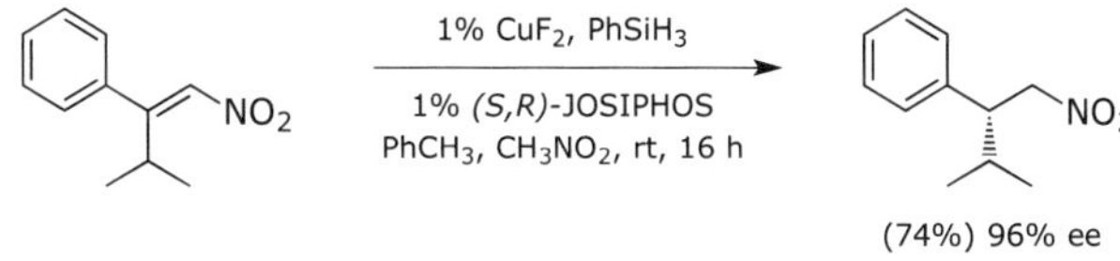

1% CuF_2, $PhSiH_3$
1% (*S,R*)-JOSIPHOS
$PhCH_3$, CH_3NO_2, rt, 16 h
(74%) 96% ee

In a 10-mL Schlenk flask, commercially available (Reidel-de-Haen), anhydrous copper(II)-fluoride (5.1 mg, 50 μmol) and (*R*,*S*)-JOSIPHOS (33 mg, 55 μmol) were dissolved in toluene (5 mL). After stirring for 60 min, PMHS (6.0 μL, 0.10 mmol) was added followed by phenylsilane (32 mg, 37 μL, 0.30 mmol) and water (18.0 μL, 1.00 mmol). After stirring for 5 min, nitromethane (6.1 mg, 5.4 μL, 0.10 mmol) was added and stirring was continued for 1 h. Phenylsilane (108 mg, 124 μL, 1.00 mmol) and the nitroolefin (1.00 mmol) were added with vigorous stirring. After stirring for 12 h, phenylsilane (54 mg, 62 μL, 0.50 mmol) was added and stirring continued for 4 h. TBAF solution (4 mL, 1.0 M in THF, 4.0 mmol) was added, and stirring was continued for 1 h. Water (20 mL) was added, and the mixture was extracted with ether (2 × 30 mL). After drying over anhydrous sodium sulfate, the solvent was evaporated. Flash chromatography (hexane/EtOAc) provided the product as a colorless oil. Anal. calcd. for $C_{11}H_{15}NO_2$: C, 68.37; H, 7.82; N, 7.25. Found C, 68.40; H, 7.72; N, 7.23.

General procedure for the conjugate reduction of α,β-unsaturated nitriles[272]

R' CN → 3% $Cu(OAc)_2$ / 3% JOSIPHOS, PMHS / *t*-BuOH, $PhCH_3$, 0 °C → R' * CN

R = H, *p*-Cl, *p*-CH_3, *o*-CH_3
R' = CH_3, Et, *i*-Pr

(81-98%) ≥94% ee

$Cu(OAc)_2$ (2.72 mg, 0.015 mmol) and nonracemic JOSIPHOS (0.015 mmol) were placed in an oven-dried Schlenk tube, and PMHS (0.12 mL, 2 mmol) and toluene (0.5 mL) were added under nitrogen. The reaction mixture was stirred for 5 min at 0 °C, and then an α,β-unsaturated nitrile (0.5 mmol) was added, followed by *t*-BuOH (0.19 mL, 2.0 mmol). The reaction vessel was washed in with toluene (0.5 mL) and sealed, and the reaction mixture was stirred until no starting material was detected by TLC analysis. The reaction mixture was quenched with water and transferred into a round-bottom flask with the aid of Et_2O (10 mL), and then NaOH (2.5 M, 1.2 mL) was added. The biphasic mixture was stirred vigorously for 0.5 h. The layers were separated, and the aqueous layer was extracted with Et_2O (3 × 20 mL). The combined organic layers were washed with brine, dried over anhydrous $MgSO_4$, and concentrated. The product was purified by chromatography on silica gel or by Kugelrohr distillation.

In the case of β,β-disubstituted-α,β-unsaturated sulfones, BINAP appears to be the ligand of choice for asymmetric hydrosilylations, affording higher ee's relative to those observed using SEGPHOS, DTBM-SEGPHOS (**170**), JOSIPHOS (**136**), or Taniaphos (**137**).[301] Key to success is the 2-pyridylsulfone moiety, as had been noted in earlier related studies, as the simple phenylsulfone analog fails to react. Several precursors to CuH (acetate, chloride, iodide, and fluoride) worked equally well in terms of yields, with $Cu(OAc)_2$ and CuCl/NaO-*t*-Bu in toluene at room temperature being preferred. Excess phenylsilane ($PhSiH_3$) was employed in all cases, along with 5% ligand.

THPO SO2 N → 5% CuCl/NaO-*t*-Bu / x's $PhSiH_3$, 5% (*R*)-BINAP / toluene, rt, 24 h → H THPO SO_2(2-py)

(91%) 91% ee

A solution to the nonreactive nature of phenylsulfones has been found; the hemilabile bidentate ligand methyl DuPHOS monoxide [Me-DuPHOS(O)], together with catalytic amounts (5 mol%) of inexpensive $CuF_2{\bullet}H_2O$ and stoichiometric $PhSiH_3$ in benzene at room temperature, give excellent results (Scheme 1-19).[302] Other solvents such as ethers DME, THF, and MTBE led to depressed yields. Importantly, the presence of aqueous hydroxide (20 mol% in a 5.5 M solution) is crucial for obtaining reproducible results, the concentrated base assumed to be "mopping up" traces of water present that likely react with $PhSiH_3$. Attempts to use Me-DuPHOS in place of the monooxide failed for the model case of the unsaturated sulfone derived from acetophenone (very low yield). And while JOSIPHOS afforded a high ee, a low yield (32%) was obtained as well for this substrate under optimized conditions.

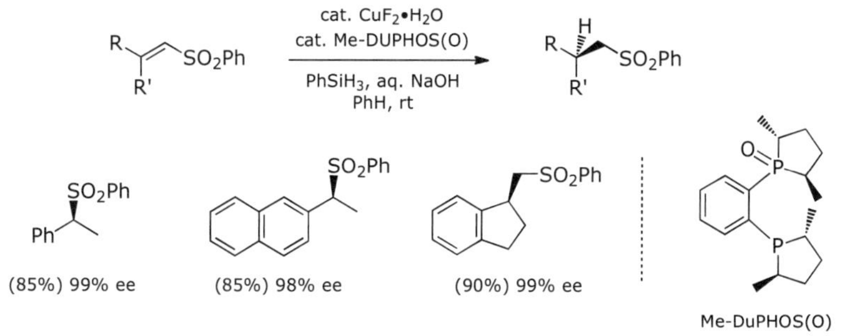

cat. $CuF_2{\bullet}H_2O$
cat. Me-DUPHOS(O)
$PhSiH_3$, aq. NaOH
PhH, rt

(85%) 99% ee (85%) 98% ee (90%) 99% ee

Me-DuPHOS(O)

Scheme 1-19

General procedure for the asymmetric reduction of unsaturated sulfones[302]

A flame-dried, 10-mL, round-bottomed flask equipped with an egg-shaped magnetic stirring bar was charged with $CuF_2{\bullet}H_2O$ (3 mg, 0.025 mmol, 5 mol%) and Me-DuPHOS(O) (8.8 mg, 0.027 mmol, 5.5 mol%) in a glove box. Benzene (1.5 mL) was added to the mixture, and the resulting suspension was stirred under argon at RT for 1 h. $PhSiH_3$ (92 μL, 0.75 mmol, 1.5 equiv) was then added, and the resulting mixture was stirred for exactly 1 min. A 5.5-M aqueous solution of NaOH (18 μL, 0.10 mmol, 20 mol% of NaOH, and equal to 2 equiv water) was then added, and immediately afterward, a solution of the vinylsulfone (0.5 mmoles, 1 equiv) dissolved in a minimum amount of benzene (1.5 to 4 mL) was added through a syringe under argon. The heterogeneous mixture was stirred for 12 h at RT. The mixture was filtered over Celite, and the reaction flask was washed twice with benzene (2 × 2 mL). The filtrate was evaporated under reduced pressure and the crude product purified by flash chromatography on silica gel (10–20% EtOAc/hexane) to afford the desired enantioenriched sulfone as a white powder.

A similar approach, a representative case being shown below, has led to high enantioselectivities in conjugate reductions of vinylic benzoxazoles, benzothiazoles, and related heteroaromatics (*e.g.*, vinylpyridines).[303] JOSIPHOS analog **173**, together with $PhSiH_3$ in toluene assisted by *t*-BuOH, afforded high yields of β-alkylated adducts.

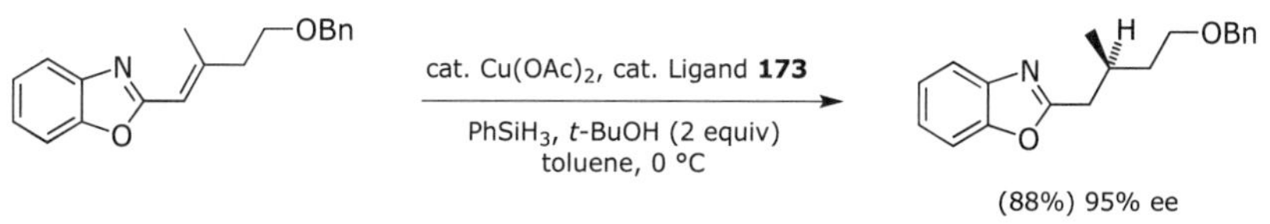

cat. $Cu(OAc)_2$, cat. Ligand **173**
$PhSiH_3$, *t*-BuOH (2 equiv)
toluene, 0 °C

(88%) 95% ee

Several hydrosilylations have been examined using a heterogeneous version of asymmetrically ligated CuH, generated from Cu/C by exposure to PMHS (from Lancaster) and catalytic NaOPh in toluene (Eqn. 1-11).[304] The NaOPh is necessary to presumably form Cu-OPh, thus, breaking the likely polymeric -(O-Cu-O)- that may

$$\text{Cu/C} \xrightarrow[\text{cat. DTBM-SEGPHOS, toluene, rt}]{\text{cat. NaOPh, PMHS}} \text{"(DTBM-SEGPHOS)CuH/C"} \quad \text{(Eqn. 1-11)}$$

be the major species present within the charcoal pores and is a poor precursor to CuH. Aside from the generality associated with this reagent, another distinguishing feature is the observation that the phosphine remains embedded within the support, on copper, and is of similar activity and effectiveness. Hence, the catalyst can be filtered after reaction and recycled without adding any additional salts (*e.g.*, NaOPh) or ligand (Scheme 1-20, L* = DTBM-SEGPHOS). Ratios of S/L up to 10,000:1 using CuH/C have been reported. The CuH/C catalyst appears to be applicable to the same substrate types used with asymmetrically ligated CuH in solution, although in some cases (*e.g.*, enoates), reactions are surprisingly sluggish. The "trick" to achieve high conversion is to use ultrasonication.

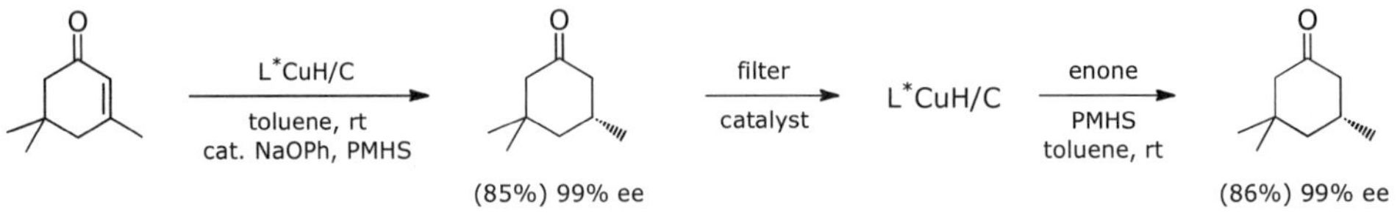

L*CuH/C; toluene, rt; cat. NaOPh, PMHS → (85%) 99% ee; filter catalyst → L*CuH/C; enone; PMHS; toluene, rt → (86%) 99% ee

Scheme 1-20

Heterogeneous asymmetric hydrosilylation of isophorone with Cu/C. (R)-3,3,5-Trimethylcyclohexanone (wet recycle)[304]

To a 25 mL, round-bottom filter flask, flame dried and purged with argon, was added Cu/C (134.4 mg, 0.046 mmol), NaOPh (24 mg, 0.2 mmol), and DTBM-SEGPHOS (1.2 mg, 0.001 mmol). Toluene (10 mL) was added and allowed to stir for 90 min. PMHS (1.2 mL, 20 mmol H^-) was added dropwise and stirred for 30 min. Isophorone (3 mL, 10 mmol) was added neat, and the reaction was placed in a sonication bath until complete as judged by TLC analysis (3 h; 4:1 hexanes/EtOAc). The reaction vessel was inverted and filtered *in vacuo* into a 100-mL, round-bottom flask. The filter cake was washed with toluene (2 × 10 mL) and filtered *in vacuo*. The 100-mL, round-bottom flask was quenched with aqueous NaOH (50 mL, 3 M) and allowed to stir at RT for 3 h. The residue was purified by flash chromatography (4:1 hexanes/EtOAc) to afford the title product (1.195 g, 85% yield) as a clear oil. Analysis of the residue by GC showed 98.5% ee [Chiraldex-BDM column with 75 °C isotherm, R_t = 41.47 (minor) and 44.94 (major)].

Recycling. Toluene (10 mL) was added to the 25 mL, round-bottom filter flask and used to break up the filter cake. PMHS (1.2 mL, 20 mmol H^-) was added dropwise and allowed to stir for 30 min. Isophorone (3 mL, 10 mmol) was added neat, and the reaction was placed in a sonication bath until complete as judged by TLC (3.5 h; 4:1 hexanes/EtOAc). The reaction vessel was inverted and filtered *in vacuo* into a 100 mL, round-bottom flask. The filter cake was washed with Et_2O (2 × 10 mL). The contents of the 100-mL, round-bottom flask were quenched with aqueous NaOH (50 mL, 3 M) and allowed to stir at RT for 3 h. The residue was purified by flash chromatography (4:1 hexanes/EtOAc) to afford the title product (1.210 g, 86% yield) as a clear oil. Analysis of the residue by GC showed 99.0% ee [Chiraldex-BDM column with 75 °C isotherm, R_t = 41.79 (minor) and 44.98 (major) min].

While recyclable nonracemically ligated CuH in charcoal could be viewed as "green" methodology, an alternative to charcoal in toluene as the reaction medium (as above) has recently been reported. By virtue of self-aggregation of the newly introduced, commercially available "designer" surfactant TPGS-750-M (Aldrich #733857) into nanomicelles,[305] CuH-catalyzed asymmetric hydrosilylation of several types of substrates can be carried out in water at room temperature.[306] Only 2 wt% of the surfactant is needed. The source of copper is inexpensive $Cu(OAc)_2 \cdot H_2O$ (3 mol%), while PMHS serves as the stoichiometric source of hydride. The same ligands that have been shown to work very well in organic solvents, such as 3,5-Xyl-MeO-BIPHEP (**169**), and JOSIPHOS (**136**) and its di-*t*-butyl analog PPF-P(*t*-Bu)$_2$ (**173**), serve nicely in these hydrosilylations as well. β,β-Disubstituted enones, enoates, and unsaturated nitriles and lactones are all amenable to this technology.

conditions
$Cu(OAc)_2 \cdot H_2O$ (3 mol %)
Ligand (3 mol %)
PMHS (0.31 equiv)
TPGS-750-M (2 wt. %)
0.25 M in H_2O

n-BuO ... 83%; 99% ee (with **173**)

Ph ... 80%; 93% ee (with **173**)

Ph ... 89%; 98% ee (with **169**)

General procedure for a CuH-catalyzed asymmetric 1,4-reduction of isophorone in water[306]

(isophorone) → cat. L*CuH, PMHS; TPGS-750-M (2 wt. %); H_2O, rt → 88%; 94% ee (with **169**)

To a 5-mL glass vial was added $Cu(OAc)_2 \cdot H_2O$ (3 mg, 0.015 mmol) and (*R*)-3,5-Xyl-MeOBIPHEP (10.4 mg, 0.015 mmol). After an atmosphere exchange with Ar, 2 wt% TPGS-750-M/H_2O (2 mL) was added and the mixture was vigorously stirred for 2 h. Isophorone (75 μL, 0.5 mmol) was added, and the reaction was continued with stirring for 10 min. PMHS (0.3 mL) was added dropwise to the reaction over 6 h. The reaction was stirred vigorously overnight (12 h), then quenched with saturated aqueous NH_4F (0.3 mL), and stirred for an additional 2 h. The resulting solution was filtered through a short plug of silica gel and concentrated under vacuum. The product was purified *via* silica gel flash chromatography (10% Et_2O/hexanes) to afford the desired 3,3,5-trimethylcyclohexanone as a colorless oil (62 mg, 88%). The ee was determined by chiral GC on a β-DEX SM column, 100 °C isotherm, 1 mL/min; racemic $\mathbf{t}_R$: 13.67 and 14.81 min; nonracemic $\mathbf{t}_R$: 13.65 (minor) and 14.85 (major); 94% ee.

Conjugate reductions using *in situ*–formed L*CuH (L* = nonracemic ligand) can be exploited further when run in the presence of an electrophilic trap, which may be of an inter- or intramolecular nature. Thus, aryl ketones react with the copper enolate generated following 1,4-hydride addition to methyl acrylate to favor *anti* products with good-to-excellent ee's (Scheme 1-21).[307] The nonracemic ligand of choice is *(R,S)*-Cy_2-Taniaphos (**137**). Other ligands examined that gave lower dr's and/or ee's include MeO-BIPHEP, BINAP, JOSIPHOS, MANDYPHOS, and WALPHOS. Since chirality transfer must derive from an asymmetrically ligated copper enolate, reactions are best carried out in toluene at –50 °C wherein the aldol event must occur faster than either protioquenching or metathesis with the silane ($PhSiH_3$), otherwise affording the unalkylated ester or an achiral silyl enol ether, respectively. The implication of a strong preference for a *E*-copper enolate is yet to be rationalized. Related trappings with dialkyl ketones have also been reported with modest success.[308]

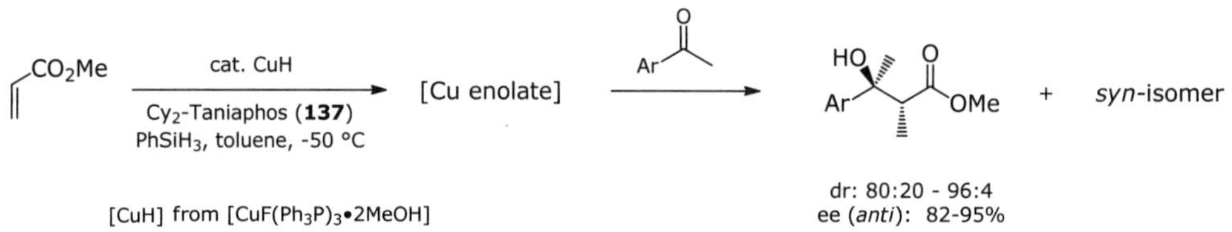

CO_2Me — cat. CuH, Cy_2-Taniaphos (**137**), $PhSiH_3$, toluene, -50 °C → [Cu enolate] — ArC(O)CH₃ → anti-aldol (HO, Ar, OMe) + *syn*-isomer

[CuH] from [CuF(Ph_3P)$_3$•2MeOH]

dr: 80:20 - 96:4
ee (*anti*): 82-95%

Scheme 1-21

Allenic esters (**177**) are also subject to conjugate reduction with catalytic amounts of *(R)*-DTBM-SEGPHOS-complexed CuH, where CuOAc is the precursor copper salt.[309] These initial 1,4-additions can then, depending on conditions, lead to either α- or γ-aldol products. In the presence of PCy_3 and pinacolborane (PinBH), in THF at 0 °C, the subsequent γ-selective aldol reaction with ketones gives good regioselectivity and excellent yields and ee's of *Z*-enoate products. Switching to a CuH catalyst derived from CuF•3PPh_3 and Taniaphos as ligand gives α-selective aldols in high yields and good ee's.

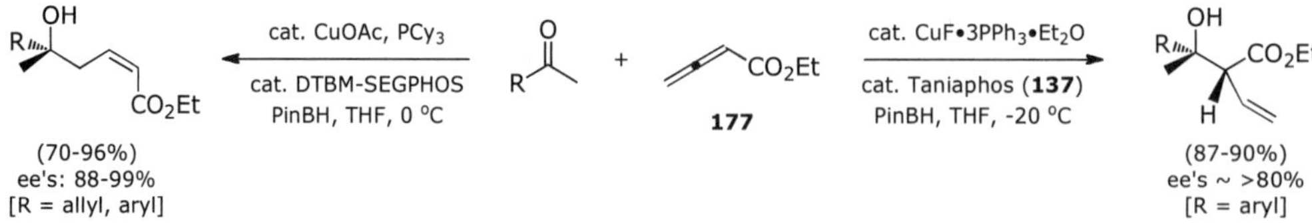

RC(O)CH₃ + **177** (CO_2Et allene):
← cat. CuOAc, PCy_3, cat. DTBM-SEGPHOS, PinBH, THF, 0 °C: γ-product (CO_2Et) (70-96%), ee's: 88-99% [R = allyl, aryl]
→ cat. CuF•3PPh_3•Et_2O, cat. Taniaphos (**137**), PinBH, THF, -20 °C: α-product (CO_2Et) (87-90%), ee's ~ >80% [R = aryl]

The corresponding reductive Mannich reaction involving ketimines can be initiated by CuH, complexed by *(R)*-DIFLUORPHOS (10 mol%).[310] In this case, the silane diethoxymethylsilane (DEMS) is used at low temperatures. PinBH in place of R_3SiH afforded considerably lower ee's. This is interesting in that metathesis leading to regeneration of CuH with a silane or PinBH is supposedly occurring after the stereo-defining 1,2-addition, suggesting that the hydride source may be involved at an earlier stage of the catalysis, perhaps with the reagent itself.[286] The dr's are typically 3:1 to 30:1, with ee's greater than 82%.

cat. CuOAc
cat. (*R*)-DIFLUORPHOS
DEMS, THF, -50 °C
(90%) 91% ee (dr = 6:1)

A procedure has recently been disclosed wherein asymmetric 1,4-reduction takes place on an unsymmetrically β,β-disubstituted Michael acceptor (enones **178**), followed by intramolecular electrophilic enolate trapping.[311, 312] This sequence presumably goes through a transient copper enolate with subsequent rapid aldol addition prior to metathesis with DEMS, which would otherwise afford a silyl enol ether. The catalyst of choice is a JOSIPHOS-complexed, *in situ*–formed CuH that leads to three new contiguous asymmetric centers of defined relative and absolute stereochemistry. Of special note is that essentially identical results are obtained when using *heterogeneous* conditions: ligated CuH-in-charcoal (L*CuH/C). Likewise, the corresponding 1,4-reduction-intramolecular aldol has been carried out in water under micellar conditions[313]; hence, several options exist for this ligand-accelerated catalysis.

5% $Cu(OAc)_2$
1% Ligand, DEMS
toluene, -10 °C

178 (R = alkyl, aryl) (88%) 96% ee (*S*,*R*)-PPF-P(*t*-Bu)$_2$ DEMS

Related intramolecular versions leading to newly formed 5- and 6-membered ring lactones are also known, these being smoothly catalyzed by $Cu(OAc)_2$/TMDS in DME or THF (Eqn. 1-12).[314] Several biaryl *bis*-phosphines were surveyed; *e.g.*, (*R*)-Xyl-MeO-BIPHEP (**169**) and the parent ligand (*S*)-SEGPHOS (**170**, Ar = Ph), giving the best ee's (≤83%). Silane TMDS gave somewhat cleaner reactions relative to PMHS. Solvents CH_3CN, toluene, and DCM led to inferior results. Similar reactions with unsaturated amides, in the achiral series (using DPPF), also show promise.[315]

cat. $Cu(OAc)_2$•H_2O
cat. Ligand **171**
THF, TMDS, rt

n = 1,2 (51-79%) 49-83% ee (Eqn. 1-12)

(3R,4R)-3-(4-Chlorobenzyl)-4-hydroxy-4-methyltetrahydropyran-2-one[314]

cat. $Cu(OAc)_2$•H_2O
cat. (*R*)-Xyl-MeO-BIPHEP
TMDS, THF, rt
(71%) 83% ee

A solution of $Cu(OAc)_2$•H_2O (2.0 mg, 0.01 mmol) and *(R)*-3,5-Xyl-MeO-BIPHEP (6.9 mg, 0.01 mmol) in THF (1 mL) was stirred for 15 min prior to the addition of TMDS (35 μL, 0.20 mmol). The initially green solution was stirred until it became yellow (*ca.* 5 min); after which a solution of the substrate (51 mg, 0.20 mmol) in THF (0.5 mL + 0.5 mL rinse) was then added rapidly *via* cannula. The reaction was stirred at RT for 24 h. The reaction was quenched by the addition of 1-M HCl (1 mL), and the mixture was stirred for 1 h before being diluted with saturated aqueous NH_4Cl solution (20 mL). The mixture was extracted with DCM (3 × 15 mL), and the combined organic layers were dried (anhydrous $MgSO_4$) and concentrated *in vacuo*. Purification of the residue by column chromatography (40% EtOAc/petrol) gave a white solid (36 mg, 71%); 83% ee; $[\alpha]_D$ -46.9 (c = 0.96, $CHCl_3$). The structure of the product was further confirmed by X-ray analysis.

N-Heterocyclic carbene-complexes of copper hydride: (NHC)CuH

N-Heterocyclic carbenes (NHCs) as ligands offer considerable promise as alternatives to phosphine complexes of CuH.[263] They are quite different from tertiary phosphines both electronically (as strong σ-donors and weak π-acceptors, with greater interaction with the metal) and structurally (with a likely linear and planer array between the carbene carbon, copper, and hydrogen). Usually, the CuH catalyst is prepared from the starting imidazolium salt **179** by initial treatment with base and CuCl, followed by *in situ* conversion in the presence of a silane (Scheme 1-22). Achiral complexes such as **180**, Ar = 2,6-diisopropylphenyl, and mesityl can be used in place of phosphines for conjugated reductions of cyclic enones,[316] hydrosilylations of hindered and functionalized ketones,[317] and in 3-component couplings initiated by the addition of hydride to acrylate followed by an *anti*-selective aldol.[318] Substrate-to-catalyst loadings up to 1000:1 have been documented. Asymmetric versions in terms of the catalyst (*i.e.*, a nonraceminc (NHC)CuH) have yet to be described, although applications are likely to be forthcoming.

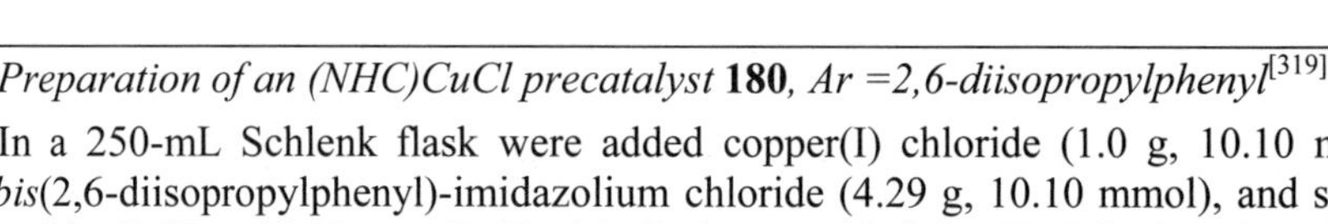

Scheme 1-22. Route to (NHC)CuH derivatives.

Preparation of an (NHC)CuCl precatalyst **180**, *Ar =2,6-diisopropylphenyl*[319]

In a 250-mL Schlenk flask were added copper(I) chloride (1.0 g, 10.10 mmol), 1,3-*bis*(2,6-diisopropylphenyl)-imidazolium chloride (4.29 g, 10.10 mmol), and sodium *t*-butoxide (0.97 g, 10.10 mmol). To this flask was added dry THF (100 mL) under Ar, and the mixture was magnetically stirred for 20 h at RT. The mixture was then filtered through a plug of Celite and the solvent evaporated *in vacuo* to give a white solid (4.59 g, 9.40 mmol, 93%).

The new NHC-containing precatalyst (*R,R,R,R*)-CuPhEt[320] can be readily converted into its copper hydride derivative upon treatment with KO-*t*-Bu and either (volatile) diethyl- or (nonvolatile) diphenylsilane in THF.[321] Triethylsilane leads to a less reactive reagent, and with triphenylsilane, no reduction is observed. The resulting hydrido species is especially reactive toward hydride delivery, presumably due to its steric bulk that lowers its aggregation state, effecting hydrosilylation of ketones in less than one hour at room temperature. More remarkable is that this nonracemic complex is discriminating toward *diakyl* ketones, in addition to the more commonly seen aryl alkyl ketones. As shown in the examples below, this reagent can distinguish methyl from ethyl, and ethyl from propyl groups in the educt.

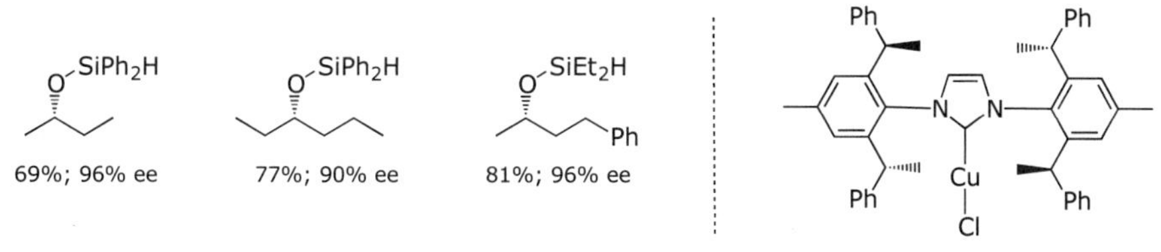

$SiPh_2H$ 69%; 96% ee — $SiPh_2H$ 77%; 90% ee — $SiEt_2H$ Ph 81%; 96% ee

(*R,R,R,R*)-CuPhEt

General procedure for hydrosilylation of ketones with (R,R,R,R)-CuPhEt[321]

To an oven-dried vial fitted with a septum screw cap and purged with Ar was added copper carbenoid CuPhEt (0.020 mmol) and freshly sublimed potassium *t*-butoxide (0.012 mmol) in freshly distilled THF (2 mL). The cloudy white mixture was stirred for 10 min at RT before the addition of the silane (3.0 mmol). An immediate color change to bright yellow was observed, and the reaction was stirred for an additional 20 min before the addition of the ketone (1.0 mmol). Solid ketones were added as concentrated solutions in THF. The progress of the reaction was monitored by GC or TLC. Upon completion (or observation of the cessation of product formation), the solution was treated with charcoal and filtered through a plug of Celite. The Celite plug was rinsed with EtOAc, and the combined organic layers were concentrated under reduced pressure. The residue was analyzed on a β-DEXcst (14% cyanopropyl-phenyl/86% dimethylpolysiloxane) by CSP-GC.

One example of a tandem, group 10, metal-catalyzed sequence that converts a *trans*-proparagylic epoxide (**182**) into a stereodefined (*cis*-) disubstituted dihydrofuran (**184**) proceeds through the intermediacy of an α-hydroxyallene (**183**).[322] Of several NHC precursor salts, Arduengo's **185**[323] and Glorius's **186**[324] led to the highest yields. Copper salts other than CuCl ($CuCl_2$, CuF_2, $Cu(OAc)_2 \cdot H_2O$) and silanes other than PMHS (*e.g.*, Et_3SiH) gave less reactive complexes and/or lower stereoselectivities. The (NHC)CuH results in a predictable S_N2'-*anti* sense of addition to give allene **183**, which can then be cyclized using a gold(III) catalyst to product **184**.[325]

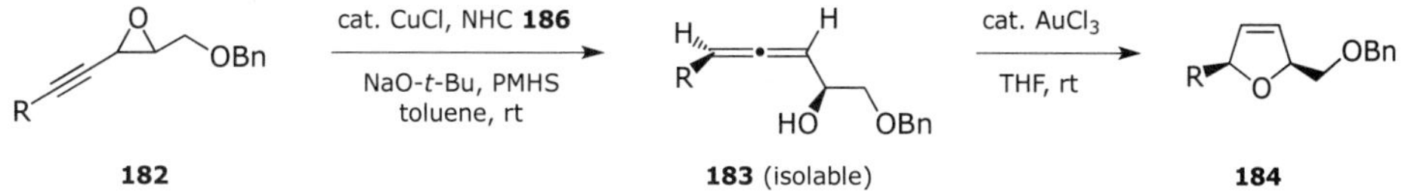

182 —(cat. CuCl, NHC **186**; NaO-*t*-Bu, PMHS, toluene, rt)→ **183** (isolable) —(cat. $AuCl_3$; THF, rt)→ **184**

Mes–N⁺ N–Mes Cl⁻ **185**

TfO⁻ **186**

Representative procedure; CuH addition to a propargylic epoxide[322]

2.5% CuCl, 2.5% **185**; PMHS, $PHCH_3$, 0 °C, 30 min → (86%)

In a Schlenk flask, CuCl (4 mg, 0.039 mmol), NaO-*t*-Bu (11 mg, 0.12 mmol), and SiMes*HCl (**185**; 13 mg, 0.039 mmol) were suspended under Ar in dry, degassed toluene (2 mL). The mixture was heated to 40 °C for 1 h and then allowed to cool to RT over 1 h. The reaction mixture turned from a white suspension to a grayish-yellow mixture. PHMS (0.21 mL, 3.26 mmol) was added, and the mixture was stirred for 5 min at RT and then cooled to 0 °C. The solution turned dark yellow with slight foaming. After the addition of *trans*-2,3-epoxy-6,6-dimethylhept-5-yn-1-ol (250 mg, 1.63 mmol), the mixture was stirred at 0 °C for 30 min (complete consumption of the substrate determined by TLC). It was then poured into a cold (0 °C) solution of n-$Bu_4NF \cdot 3H_2O$ (1.03 g, 3.26 mmol) in THF (CAUTION: FOAMING!), and the stirred mixture was warmed to RT over 2 h. After the addition of aqueous NH_4Cl and extraction with Et_2O, the combined organic layers were filtered through a short column of silica gel, charcoal, and Celite. The solvent was removed under reduced pressure, and the crude product was purified by flash column chromatography (SiO_2, cyclohexane/EtOAc, 4:1 to 2:1) to yield 215 mg (86%) of 6,6-dimethylhepta-3,4-diene-1,2-diol as a pale yellow oil.

In addition to epoxides **182** as educts, related studies from two independent laboratories have recently shown that propargylic methyl carbonates (**187**) are excellent substrates for *in situ*–generated CuH chelated by either an NHC ligand (*e.g.*, the homolog of **186**, in toluene)[326] or Xantphos (used in THF).[327] Cyclic carbonates are also readily available and react accordingly. Importantly, in addition to the mild conditions involved (PMHS, 0 °C up to 50 °C in THF), complete transfer of central to axial chirality (**188**) is to be expected.

cat. $Cu(OAc)_2$, cat. Ligand; PMHS, THF, 0 °C

(*S*)-**187**; 99% ee → (*S*)-**188**; (77%) 99% ee

Xantphos

Bottom line comments. With the advent of new and remarkably innately biased biaryl, ferrocenyl-based, and NHC ligands, nonracemically complexed CuH-catalyzed hydrosilylations have begun to compete with asymmetric hydrogenation on several levels. Thus, while economics have always been on the side of base metals like copper, TONs and ee's have begun to look "Noyori-like." Interestingly, Takasago has recently acknowledged that copper-based asymmetric hydrogenation has industrial potential.[274, 294] More CuH chemistry using newly engineered ligands and nonracemic NHCs is likely.

10. Miscellaneous Reactions Catalyzed by Cu(I)

10.1 C-H Activation

Reviews. Ornellas, S. D.; Storr, T. E.; Williams, T.J.; Baumann, C. G.; Fairlamb, I. J. S. *Curr. Org. Syn.* **2011**, *8*, 79; Kakiuchi, F.; Kochi, T. *Synthesis* **2008**, 3013; Herrerias, C. I.; Yao, X.; Li, Z.; Li, C-J. *Chem. Rev.* **2007**, *107*, 2546; Alberico, D.; Scott, M. E.; Lautens, M. *Chem. Rev.* **2007**, *107*, 174; Goj, L. A.; Gunnoe, T. B. *Curr. Org. Chem.* **2005**, *9*, 671.

The arguments for replacing through substitution an aromatic C-H bond rather than an aryl C-halide bond are convincing, and hence, a tremendous amount of attention has rightly been directed to such transformations involving catalysis. By far,

cat. CuI, DMF; 1,10-phenanthroline; LiO-*t*-Bu, 100-125 °C

(89%)

CN N NO_2 (51%) [with K_3PO_4]

most is driven by palladium, and only recently has copper entered this arena notwithstanding its early participation in such processes (albeit in stoichiometric quantities).[328] One means of generating arylcopper intermediates is through proton abstraction in educts that offer such opportunities (*i.e.*, have protons with $pK_a < 35$ in DMSO). Alkoxide bases at elevated temperatures that avoid etherification and benzyne formation, such as lithium *t*-butoxide and Et_3COLi, can be used for arylation (with aryl iodides) of electron-rich heteroaromatics (as illustrated below).[329] More acidic protons (pK_a <27) in certain cases allow for K_3PO_4 to serve as base; 1,10-phenanthroline is the preferred ligand in both these and electron-deficient cases.

General procedure for C-H activation cross-coupling reactions[329]

Outside the glove box, a 1-dram vial equipped with a magnetic stir bar was charged with the haloarene, phenanthroline (10 mol%), substrate, and solvent (DMF or a 1/1 mixture of DMF and xylenes). If anhydrous DMPU was used, the reaction was set up inside the glove box. The vial was flushed with argon, capped, and placed inside a glove box. To this mixture was added CuI (10 mol%) and base (1.7–4.0 equiv). The sealed vial was then taken out of the glove box, stirred at RT for 5 min, and then placed in a preheated oil bath. After completion of the reaction, the mixture was cooled to RT and diluted with EtOAc (50 mL). The resulting solution was washed with brine (15 mL), dried over anhydrous $MgSO_4$, and concentrated under vacuum to a volume of *ca.* 1 mL.

The mixture containing the product was subjected to column chromatography on silica gel (hexanes followed by appropriate solvent to elute the products). After concentrating the fractions containing the product, the residue was dried under reduced pressure to yield pure product.

2,4-Difluoro-3-(pyridin-2-yl)benzophenone[329]

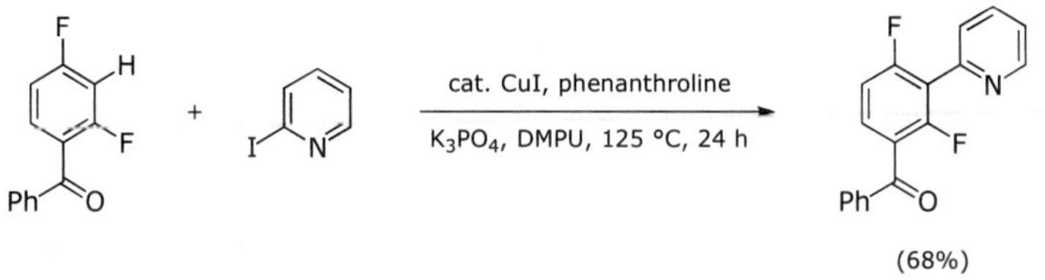

cat. CuI, phenanthroline

K_3PO_4, DMPU, 125 °C, 24 h

(68%)

Copper(I) iodide (19 mg, 0.1 mmol), 2-iodopyridine (205 mg, 1.0 mmol), 1,10-phenanthroline (18 mg, 0.1 mmol), 2,4-difluorobenzophenone (436 mg, 2.0 mmol), K_3PO_4 (424 mg, 2.0 mmol), and anhydrous DMPU (0.6 mL) were used according to the procedure above, at 125 °C for 24 h. After column chromatography (hexanes, then 1:1 EtOAc/hexanes), 200 mg (68%) of a light tan oil was obtained; R_f = 0.45 (1:1 EtOAc/hexanes). Anal. calcd. for $C_{18}H_{11}F_2NO$: C, 73.22; H, 3.75; N, 4.74; found C, 72.55; H, 3.88; N, 4.74.

While most methods for C-H activation direct an electrophilic partner to the *ortho*-position of the stabilizing substituent, a procedure has been developed that accomplishes *meta*-selective arylation.[330] Pivalamide derivatives of anilines (as in the case below), where electron-donating substituents on the aryl ring are favored, direct incoming aromatic groups to the *meta*-site. Electron-withdrawing substituents on the anilide tend to give lower yields. The effect has been attributed to an "oxy-cupration," where *ortho*-stabilization by oxygen from the amide following an initially formed Cu(III) species ultimately results *via* rearomatization and reductive elimination to afford the observed *meta*-arylated adduct (Scheme 1-23). Related technology has appeared that now accomplishes *para*-selective arylation of aniline and phenol derivatives.[331]

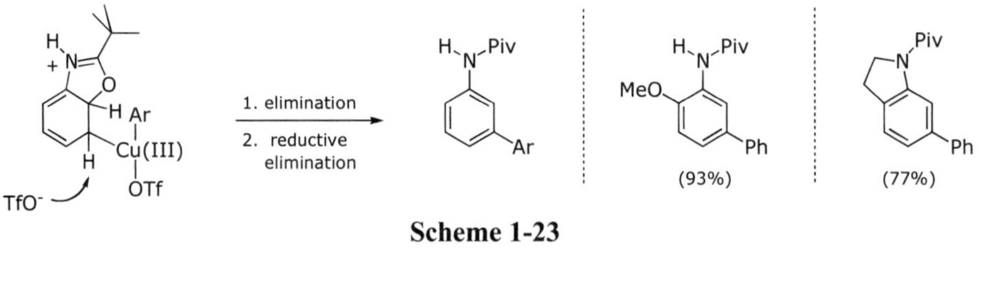

1. elimination
2. reductive elimination

(93%)

(77%)

Scheme 1-23

General procedure for the meta-arylation of anilides[330]

10 mol % $Cu(OTf)_2$

$ClCH_2CH_2Cl$, 70 °C

TfO⁻

(70%)

To a solution of the anilide (0.50 mmol) in 1,2-dichloroethane (2.5 mL) was added the iodonium salt (0.75 – 3.00 mmol) and $Cu(OTf)_2$ (18 mg, 0.050 mmol). The reaction was stirred for the specified time and at the specified temperature before dilution with DCM (25 mL) and washing with satd. aqueous sodium bicarbonate solution (25 mL). The aqueous phase was extracted further with DCM (25 mL), and the combined organics were dried over anhydrous magnesium sulfate and evaporated *in vacuo*. The crude reside was purified by flash column chromatography to yield the pure 3-arylanilide.

A method for carboxylation of both C-H as well as N-H bonds relies on the bulky NHC-ligated species Cu(IPr)OH (shown below), which in the presence of CsOH in refluxing THF, results in the corresponding carboxylic acids upon acidic workup. Substrates with pK_a's below *ca.* 28 react readily, including benzoxazoles, benzthiazoles, oxazoles, and polyfluorinated benzenes. Yields reported are in the 77–93% range.[332]

(90%) (82%) (77%) (93%)

Cu(IPr)OH

10.2 Protodecarboxylation

A convenient microwave-assisted protocol for decarboxylation of benzoic acids catalyzed by Cu_2O (10 mol%) along with 1,10-phenanthroline has been reported.[333] A mixture of two polar solvents, NMP and quinoline (3:1), led to rapid absorption of microwave energy and, hence, to a quick rise in temperature. Reactions, depending on substrate acid, take place between 160 °C and 190 °C, affording product arenes in only between 5 and 15 min in good yields. Neither CuBr nor CuOAc was a satisfactory alternative source of copper.

cat. Cu_2O
1,10-phenanthroline
NMP/quinoline (3:1)

	(84%)	(64%)	(85%)	(80%)
Method:	B	A	B	B

Protodecarboxylation of aromatic carboxylic acids[333]

Method A. An oven-dried, 10-mL microwave vial was charged with the carboxylic acid (1.0 mmol), Cu_2O (7.2 mg, 0.05 mmol), and 1,10-phenanthroline (18 mg, 0.10 mmol). After the reaction mixture was placed under an inert atmosphere, a mixture of NMP (1.5 mL) and quinoline (0.5 mL) was added *via* syringe. The resulting mixture was submitted to microwave irradiation at 190 °C for 15 min at a maximum power of 150 W and subsequently air-jet cooled to RT. The maximum pressure detected during the reaction was 5.5 bar. The mixture was then diluted with aqueous HCl (5 N, 10 mL) and extracted repeatedly with diethyl ether (2 mL portions). The combined organic layers were washed with water and brine, dried over anhydrous $MgSO_4$, and filtered. The corresponding arene was obtained in pure form after removal of the solvents by distillation over a Vigreux column.

Method B. Method B is analogous to Method A but with a lower loading of the copper/1,10-phenanthroline catalyst and microwave irradiation at 190 °C for 5 min at a maximum power of 150 W. The following amounts were used: carboxylic acid (1.0 mmol), Cu_2O (1.5 mg, 0.01 mmol), and 1,10-phenanthroline (3.6 mg, 0.02 mmol).

10.3 Copper-Catalyzed Carbometalation

Review. Fox, J. M.; Yan, N. *Curr. Org. Chem.* **2005**, *9*, 719.

The *cis*-directing ability of a 3-carboalkoxy group in cyclopropenes can be used to great advantage in net carbozincation reactions catalyzed by either CuI or CuCN.[334] Toluene is the solvent in which best diastereoselectivities are realized; both THF and Et_2O led to poorer results.

CO_2Et, R; cat. CuI, Me_2Zn; toluene, 0 °C

CO_2Et, R: (70%) [dr >95:5] R = n-C_6H_{13}

H, CO_2Et, R, Ph: (73%) [dr 94:6] R = n-C_6H_{13}

Ph, CO_2Me, H: (82%) [dr >95:5] R = H

Rel-(1S,2R) Ethyl 1-(dimethylphenylsilyl)-2-phenylcyclopropane carboxylate[334]

$PhMe_2Si$, CO_2Et; 20 mol % CuCN, Ph_2Zn; toluene, 0 °C to rt; $PhMe_2Si$, CO_2Et, Ph

(85%) dr >95:5

A 10 mL, round-bottom flask was charged with Ph_2Zn (55 mg, 0.25 mmol) and CuCN (4.5 mg, 0.050 mmol) in toluene (3 mL). The flask was cooled by an ice bath and a solution of ethyl 1-(dimethylphenylsilyl)-cycloprop-2-ene carboxylate (62 mg, 0.25 mmol) in toluene (1 mL) was added dropwise *via* syringe. After 5 min, the ice bath was removed and the resulting mixture was allowed to stir for 3 h while warming to RT.

The mixture was again cooled by an ice bath, and the reaction was quenched by the addition of aqueous HCl (0.1 M). The mixture was extracted with DCM (3 × 25 mL). The combined organic phase was then dried over anhydrous $MgSO_4$, filtered, and concentrated under reduced pressure. Analysis of the crude product by ^1H NMR showed >95% diastereomeric purity. Flash column chromatography (with a gradient of hexanes to 3–5% ether in hexanes) provided 69 mg (0.21 mmol, 85%) of the product as an oil; purity by ^1H NMR was measured to be >95%.

In addition to zinc reagents R_2Zn that are commercially available (*e.g.*, Me_2Zn, Et_2Zn, and Ph_2Zn), those prepared from Grignard reagents and $ZnCl_2$ also gave good yields and high dr's. Educts bearing a nonracemic oxazolidinone as auxiliary, in the presence of $MgBr_2•OEt_2$, afforded adducts reflecting high levels of diastereoselectivities (>95:5) along with predictable regiochemistry. Trapping the intermediate cyclopropylzinc species with reactive electrophiles other than a proton (*e.g.*, iodine and allyl bromide) leads to products of 3-component coupling.

1. cat. $CuBr•SMe_2$, Et_2Zn, $MgBr_2•OEt_2$, -78 °C
2. I_2
(66%)

The addition of Grignard reagents to nonracemic cyclopropenylcarbinols **189** in the presence of CuI (20 mol%) takes place in ether with excellent chirality transfer from educt to the resulting alkylidenecyclopropene (**190**).[335] Both alkyl and aryl Grignards (2 equiv) work equally well. A mechanism in which the initially deprotonated alcohol **191** delivers the group on copper in a *syn*-1,2-fashion, followed by transmetalation to magnesium and *syn*-elimination from **192**, account for the observed transfer of chirality. Yields reported are in the 66–88% range, while ee's are very high (95–99%).

cat. CuI, RMgX
Et_2O, -50 °C to rt
189 **191** **192** **190**

More recent studies describe treatment of alcohols **189** with a CuI/LAH mixture, ostensibly providing a CuH reagent to effect a net S_N2' delivery, thus, placing hydrogen in a stereodefined allylic position (Eqn. 1-13).[336] Secondary alcohols invariably gave products of *E*-configuration. CuI/LAD mixtures led to >95% incorporation of deuterium. Hydroalumination of the strained olefin is not a competing side reaction in this chemistry.

OH Ph cat. CuI, LAH Et_2O, -50 °C to rt H H Ph (Eqn. 1-13)

(76%) 93:7 *E:Z*

11. Conclusions and Outlook

Copper occupies a special place in the Periodic Table. It has several fundamental features to its credit that simultaneously attract and yet challenge the creativity of chemists. As the base substance among time-honored group 11 coinage metals, its role in the history of alchemy is secure. And today, much of the rich history of even modern organocopper chemistry has been discussed in other, more lengthy works. Hence, the idea behind this chapter mainly on copper(I)-*catalyzed* processes has been succinctly phrased by Prof. Reinhard Hoffmann,[337] who said, "Think of it this way: when your car breaks down, you want to open the *Manual* and see which screw to turn, not the history of the parts." And so, with the practitioner in mind and dozens of working procedures provided, as well as associated references to the original literature, a sense of "how to" hopefully prevails.

Looking forward, it seems safe to conclude that copper chemistry will be done mainly, if not almost exclusively, in a catalytic mode; that copper's reactivity will be tailored and its selectivity tuned with refinements mainly in ligands, among other reaction variables. Thus, in a sense, organocopper chemistry has become "greener"; less transition metal in any reaction is usually welcomed from an environmental perspective, and that is unlikely to change, at least for copper. Copper chemistry in the absence of organic solvents, *i.e.*, in pure water, is rare, but it is here.[306, 338] Its cross-coupling chemistry resulting in a variety of carbon-carbon bonds, especially in the asymmetric mode, appears to be thriving and on a path of increasing importance. The increasing popularity of copper catalysis is surely driven today, in part, by favorable economics relative to that by Pd; indeed, there is even a call by the U.S. National Science Foundation to develop replacement technologies that rely on base, rather than on precious, metal catalysis.[339] And although still very early, the recent realization that even trace amounts of copper can have a beneficial impact on cross-couplings ostensibly mediated by the "cross-coupling metal," Pd, has not gone unnoticed. Thus, the renaissance of copper chemistry, along with its associated applications to cross-couplings in particular, is likely to continue. Of course, as Beletskaya skillfully and cogently reminds us,[340] the "ancestor" of current Pd-catalyzed cross-couplings is … copper.

12. Acknowledgments

It was literally impossible for this author to organize, write, draw structures, get and check references, find procedures, and proofread this opus entirely on his own.

For these invaluable contributions by group members working behind the scenes; to those who did so much of this work, I warmly say, "thank you." Initially, graduate students John Unger, Danielle Nihan, David Chung, Grant Aguinaldo, Ben Baker, Ching-Tien Lee, Ricardo Corral, Tom Butler, Ben Taft, Sean Riznikove, and postdocs Malati Raghunath and Ben Amoreli all assisted with literature searches and drawings, all overseen by graduate student Tue Petersen. The final, updated manuscript, completed in late 2009, was brought to fruition by the efforts, in particular, of Žarko Bošković, along with fellow graduate students Alex Abela, Ralph Moser, and Shenlin Huang, and postdocs Takashi Nishikata, Christophe Duplais, Helena Leuser, and Subir Ghorai. This has truly been a collaborative effort; a contribution from all of us, many being pictured below. And, for sure, at a time when financial support is extremely tough to come by, we are indebted to the U.S. National Science Foundation for continuous funding of our work in the organocopper area.

The team of group collaborators (left to right): Dr. Takashi Nishikata, Ralph Moser, Alex Abela, Dr. Subir Ghorai, Dr. Christophe Duplais, BHL, Dr. Arkady Krasovskiy, Žarko Bošković, Shenlin Huang, and Dr. Helena Leuser

13. References

[1] Gilman, H.; Jones, R.G.; Woods, L.A.; *J. Org. Chem.* **1952**, *17*, 1630.

[2] Rostovtsev, V.V.; Green, L.G.; Fokin, V.V.; Sharpless, K.B.; *Angew. Chem., Int. Ed.* **2002**, *41*, 2596.

[3] Tornøe, C.W.; Christensen, C.; Meldal, M.; *J. Org. Chem.* **2002**, *67*, 3057.

[4] Bräse, S.; Gil, C.; Knepper, K.; Zimmermann, V.; *Angew. Chem., Int. Ed.* **2005**, *44*, 5188.

[5] (a) Huisgen, R.; *Pure Appl. Chem.* **1989**, *61*, 613; (b) Huisgen, R.; Szeimies, G.; Mobius, L.; *Chem. Ber.* **1967**, *100*, 2494.

[6] Bock, V.D.; Hiemstra, H.; van Maarseveen, J.H.; *Eur. J. Org. Chem.* **2006**, 51.

[7] Boren, C.B.; Narayan, S.; Rasmussen, L.K.; Zhang, L.; Zhao, H.; Lin, Z.; Jia, G.; Fokin, V.V.; *J. Am. Chem. Soc.* **2008**, *130*, 8923.

[8] (a) Rodionov, V.O.; Fokin, V.V.; Finn, M.G.; *Angew. Chem., Int. Ed.* **2005**, *44*, 2210; (b) Himo, F.; Lovell, T.; Hilgraf, R.; Rostovtesev, V.V.; Noodleman, L.; Sharpless, K.B.; Fokin, V.V.; *J. Am. Chem. Soc.* **2005,** *127*, 210.

[9] Cantillo, D.; Avalos, M.; Babiano, R.; Cintas, P.; Jimenez, J.L.; Palacios, J.C.; *Org. Biomol. Chem.* **2011**, *9*, 2952.

[10] Tanaka, K.; Kageyama, C.; Fukase, K.; *Tetrahedron Lett.* **2007**, *48*, 6475.

[11] Fazio, F.; Bryan, M.C.; Blixt, O.; Paulson, J.C.; Wong, C.-H.; *J. Am. Chem. Soc.* **2002**, *124,* 14397.

[12] (a) Díez-Gonzáles, S.; Nolan, S.P.; *Angew. Chem., Int. Ed.* **2008**, *47*, 8881; (b) Díez-Gonzáles, S.; Correa, A.; Cavallo, L.; Nolan, S.P.; *Chem. Eur. J.* **2006**, *12*, 7558.

[13] Lal, S.; Díez-González, S.; *J. Org. Chem.* **2011**, *76*, 2367.

[14] Shao, C.; Wang, X.; Zhang, Q.; Luo, S.; Zhao, J.; Hu, Y.; *J. Org. Chem.* **2011**, *76*, 6832.

[15] Beckmann, H.S.G.; Wittmann, V.; *Org. Lett.* **2007**, *9*, 1.

[16] (a) Chan, T.R.; Hilgraf, R.; Sharpless, K.B.; Fokin, V.V.; *Org. Lett.* **2007**, *6*, 2853; (b) Detz, R.J.; Heras, S.A.; de Gelder, R.; van Leeuwen, P.W.N.M.; Hiemstra, H.; Reek, J.N.H.; van Maarseveen, J.H.; *Org. Lett.* **2006**, *8*, 3227.

[17] (a) Yoo, E.J.; Ahlquist, M.; Kim, S.H.; Bae, I.; Fokin, V.V.; Sharpless, K.B.; Chang, S.; *Angew. Chem., Int. Ed.* **2007**, *46*, 1730; (b) Cho, S.H.; Chang, S.; *Angew. Chem., Int. Ed.* **2007**, *46*, 1897.

[18] Kumar, D.; Reddy, V.B.; Varma, R.S.; *Tetrahedron Lett.* **2009**, *50*, 2065.

[19] Appukkuttan, P.; Dehaen, W.; Fokin, V.V.; Van der Eycken, E.; *Org. Lett.* **2004**, *6*, 4223.

[20] Wang, F.; Fu, H.; Jiang, Y.; Zhao, Y.; *Green Chem.* **2008**, *10*, 452.

[21] Ackerman, L.; Potukuchi, H. K.; Landsberg, D.; Vicente, R.; *Org. Lett.* **2008**, *10*, 3081.

[22] Wang, Y.; Liu, J.; Xia, C.; *Adv. Synth. Catal.* **2011**, *353*, 1534.

[23] Wegner, J.; Ceylan, S.; Kirschning, A.; *Chem. Commun.* **2011**, *47*, 4583.

[24] Kupracz, L.; Hartwig J.; Wegner, J.; Ceylan, S.; Kirschning, A.; *Beilstein J. Org. Chem.* **2011**, *7*, 1441.

[25] Kuijpers, B.H.M.; Dijkmans, G.C.T.; Groothuys, S.; Quaedflieg, P.J.L.M.; Blaauw, R.H.; van Delft, F.L.; Rutjes, F.P.J.T.; *Synlett* **2005**, 3059.

[26] Jin, T.; Kamijo, S.; Yamamoto, Y.; *Eur. J. Org. Chem.* **2004**, 3789.

[27] Kamijo, S.; Jin, T.; Huo, Z.; Yamamoto, Y.; *J. Org. Chem.* **2004**, *69*, 2386.

[28] Moni, L.; Ciogli, A.; D'Acquarica, I.; Dondoni, A.; Gasparrini, F.; Marra, A.; *Chem. Eur. J.* **2010**, 16, 5712.

[29] Hafrén, J.; Zou, W.; Córdova, A.; *Macromol. Rapid Commun.* **2006**, *27*, 1362.

[30] Chassaing, S.; Sido, A.S.S.; Alix, A.; Kumarraja, M.; Pale, P.; Sommer, J.; *Chem. Eur. J.* **2008**, *14*, 6713.

[31] Orgueira, H.A.; Fokas, D.; Isome, Y.; Chan, P.C.-M.; Baldino, C.M.; *Tetrahedron Lett.* **2005**, *46*, 2911.

[32] Pachón, L.D.; van Maarseveen, J.H.; Rothenberg, G.; *Adv. Synth. Catal.* **2005**, *347*, 811.

[33] Alonso, F.; Moglie, Y.; Radivoy, G.; Yus, M.; *Tetrahedron Lett.* **2009**, *50*, 2358.

[34] Lipshutz, B.H.; Taft, B.R.; *Angew. Chem., Int. Ed.* **2006**, *45*, 8235.

[35] Thorwirth, R.; Stolle, A.; Ondruschka, B.; Wild, A.; Schubert, U.S.; *Chem. Commun.* **2011**, *47*, 4370.

[36] Hartwig, J.F.; *Acc. Chem. Res.* **2008**, *41*, 1534.

[37] Tye, J.W.; Weng, Z.; Johns, A.M.; Incarvito, C.D.; Hartwig, J.F.; *J. Am. Chem. Soc.* **2008**, *130*, 9971.

[38] Xu, H.; Wolf, C.; *Chem. Commun.* **2009**, 1715.

[39] Son, S.U.; Park, I.K.; Park, J.; Hyeon, T.; *Chem. Commun.* **2004**, 778.

[40] Franc, G.; Jutand, A.; *Dalton. Trans.* **2010**, *39*, 7873.

[41] Ma, D.; Cai, Q.; Zhang, H.; *Org. Lett.* **2003**, *5*, 2453.

[42] Rao, H.; Fu, H.; Jiang, Y.; Zhao, Y.; *J. Org. Chem.* **2005**, *70*, 8107.

[43] Gujadhur, R.; Venkataraman, D.; Kintigh, J.T.; *Tetrahedron Lett.* **2001**, *42*, 4791.

[44] Patil, N.M.; Kelkar, A.A.; Chaudhari, R.V.; *J. Mol. Catal. A: Chem.* **2004**, *223*, 45.

[45] Kelkar, A.A.; Patil, N.M.; Chaudhari, R.V.; *Tetrahedron Lett.* **2002**, *43*, 7143.

[46] Hu, Z.; Ye, W.; Zou, H.; Yu, Y.; *Synth. Commun.* **2010**, *40*, 222.

[47] Yang, K.; Qiu, Y.; Li, Z.; Wang, Z.; Jiang, S.; *J. Org. Chem.* **2011**, *76*, 3151.

[48] Tao, C.-Z.; Liu, W.-W.; Sun, J.-Y.; Cao, Z.-L.; Li, H.; Zhang, Y. F.; *Synthesis* **2010**, 1280.

[49] Sperotto, E.; van Klink, G.P.M., de Vries, J.G.; van Koten, G.; *Tetrahedron* **2010**, *66*, 3478.

[50] Shafir, A.; Buchwald, S.L.; *J. Am. Chem. Soc.* **2006**, *128*, 8742.

[51] Lu, Z.; Twieg, R.J.; Huang, S.D.; *Tetrahedron Lett.* **2003**, *44*, 6289.

[52] Zhu, D.; Wang, R.; Mao, J.; Xu, L.; Wu, F.; Wan, B.; *J. Mol. Catal. A: Chem.* **2006**, *256*, 256.

[53] Yeh, V.S.C.; Wiedeman, P.E.; *Tetrahedron Lett.* **2006**, *47*, 6011.

[54] Gajare, A.S.; Toyota, K.; Yoshifuji, M.; Ozawa, F.; *Chem. Commun.* **2004**, 1994.

[55] Kwong, F.Y.; Klapars, A.; Buchwald, S.L.; *Org. Lett.* **2002**, *4*, 581.

[56] Yadav, L.D.S.; Yadav, B.S.; Rai, V.K.; *Synthesis* **2006**, 1868.

[57] Klapars, A.; Antilla, J.C.; Huang, X.; Buchwald, S.L; *J. Am. Chem. Soc.* **2001**, *123*, 7727.

[58] Klapars, A.; Huang, X.; Buchwald, S.L.; *J. Am. Chem. Soc.* **2002**, *124*, 7421.

[59] Strieter, E.R.; Blackmond, D.G.; Buchwald, S.L.; *J. Am. Chem. Soc.* **2005**, *127*, 4120.

[60] Hosseinzadeh, R.; Tajbakhsh, M.; Mohadjerani, M.; Mehdinejad, H.; *Synlett* **2004**, 1517.

[61] Cristau, H.-J.; Cellier, P.P.; Spindler, J.-F.; Taillefer, M.; *Chem. Eur. J.* **2004**, *10*, 5607.

[62] Moriwaki, K.; Satoh, K.; Takada, M.; Ishino, Y.; Ohno, T.; *Tetrahedron Lett.* **2005**, *46*, 7559.

[63] Zhang, H.; Cai, Q.; Ma, D.; *J. Org. Chem.* **2005**, *70*, 5164.

[64] Lu, Z.; Twieg, R.J.; *Tetrahedron Lett.* **2005**, *46*, 2997.

[65] Job, G.E.; Buchwald, S.L.; *Org. Lett.* **2002**, *4*, 3703.

[66] Kim, K.-Y.; Shin, J.-T.; Lee, K.-S.; Cho, C.-G.; *Tetrahedron Lett.* **2004**, *39*, 117.

[67] de Lange, B.; Lambers-Verstappen, M.H.; Schmieder-van de Vondervoort, L.; Sereinig, N.; de Rijk, R.; de Vries, A.H.M.; de Vries, J.G.; *Synlett* **2006**, 3105.

[68] Boduszek, B.; Oleksyszyn, J.; Kam, C.-M.; Selzler, J.; Smith, R.E.; Powers, J.C.; *J. Med. Chem.* **1994**, *37*, 3969.

[69] Kwong, F.Y.; Buchwald, S.L.; *Org. Lett.* **2003**, *5*, 793.

[70] Yang, C.-T.; Fu, Y.; Huang, Y.-B.; Guo, Q.-X.; Liu, L.; *Angew. Chem., Int. Ed.* **2009**, *48*, 7398.

[71] Shafir, A.; Lichtor, P.A.; Buchwald, S.L.; *J. Am. Chem. Soc.* **2007**, *129*, 3490.

[72] Liu, Z-J.; Vors, J-P.; Gesin, E.R.F.; Bolm, C.; *Adv. Synth. Catal.* **2010**, *352*, 3158.

[73] Xia, N.; Taillefer, M.; *Angew. Chem., Int. Ed.* **2009**, *48*, 337.

[74] Zhu, Y.; Wei, Y.; *Can. J. Chem.* **2011**, *89*, 645.

[75] Thakur, K.G.; Ganapathy, D.; Sekar, G.; *Chem. Commun.* **2011**, *47*, 5076.

[76] Zeng, X.; Huang, W.; Qiu, Y.; Jiang, S.; *Org. Biomol. Chem.* **2011**, *9*, 8224.

[77] Rao, H.; Fu, H.; Jiang, Y.; Zhao, Y.; *Angew. Chem., Int. Ed.* **2009**, *48*, 1114.

[78] Tao, C.-Z.; Li, J.; Fu, Y.; Liu, L.; Guo, Q.-X.; *Tetrahedron Lett.* **2008**, *49*, 70.

[79] Gao, X.; Fu, H.; Qiao, R.; Jiang, Y.; Zhao, Y.; *J. Org. Chem.* **2008**, *73*, 6864.

[80] Lan, J.-B.; Chen, L.; Yu, X.-Q.; You, J.-S.; Xie, R.-G.; *Chem. Commun.* **2004**, 188.

[81] Altman, R.A.; Buchwald, S.L.; *Org. Lett.* **2006**, *8*, 2779.

[82] Xie, Y.-X.; Pi, S.-F.; Wang, J.; Yin, D.-L.; Li, J.-H.; *J. Org. Chem.* **2006**, *71*, 8324.

[83] Kaddouri, H.; Vicente, V.; Ouali, A.; Ouazzani, F.; Taillefer, M.; *Angew. Chem., Int. Ed.* **2009**, *48*, 333.

[84] Taillefer, M.; Xia, N.; Ouali, A.; *Angew. Chem., Int. Ed.* **2007**, *46*, 934.

[85] Guo, D.; Huang, H.; Zhou, Y.; Xu, J.; Jiang, H.; Chen, K.; Liu, H.; *Green Chem.* **2010**, *12*, 276.

[86] Teo, Y.-C; Yong, F.-F.; Lim, G.S.; *Tetrahedron Lett.* **2011**, *52*, 7171.

[87] Zhu, L.; Li, G.; Luo, L.; Guo, P.; Lan, J.; You, J.; *J. Org. Chem.* **2009**, *74*, 2200.

[88] Chen, H.; Wang, D.; Wang, X., Huang, W.; Chai, Q.; Ding, K.; *Synthesis* **2010**, 1505.

[89] Ma, D.; Xia, C.; *Org. Lett.* **2001**, *3*, 2583.

[90] Okano, K; Tokuyama, H.; Fukuyama, T.; *J. Am. Chem. Soc.* **2006**, *128*, 7136.

[91] Yamada, K.; Kurokawa, T.; Tokuyama, H.; Fukuyama, T.; *J. Am. Chem. Soc.* **2003**, *125*, 6630.

[92] Tsutsui, H.; Hayashi, Y.; Narasaka, K.; *Chem. Lett.* **1997**, 317.

[93] Tsutsui, H.; Ichikawa, T.; Narasaka, K.; *Bull. Chem. Soc. Jpn.* **1999**, *72*, 1869.

[94] He, C.; Chen, C.; Cheng, J.; Liu, C.; Liu, W.; Li, Q.; Lei, A.; *Angew. Chem., Int. Ed.* **2008**, *47*, 6414.

[95] Daşkapan, T.; *Tetrahedron Lett.* **2006**, *47*, 2879.

[96] Berman, A.M.; Johnson, J.S.; *J. Am. Chem. Soc.* **2004**, *126*, 5680.

[97] Erdik, E.; Daşkapan, T.; *Synth. Comm.* **1999**, *29*, 3989.

[98] Berman, A.M.; Johnson, J.S.; *J. Org. Chem.* **2006**, *71*, 219.

[99] Berman, A.M.; Johnson, J.S.; *Org. Synth.* **2006**, *83*, 31.

[100] Berman, A.M.; Johnson, J.S.; *J. Org. Chem.* **2005**, *70*, 364.

[101] Antilla, J.C.; Buchwald, S.L.; *Org. Lett.* **2001**, *3*, 2077.

[102] Collman, J.P.; Zhong, M.; Zeng, L.; Costanzo, S.; *J. Org. Chem.* **2001**, *66*, 1528.

[103] Choudary, B.M.; Sridhar, C.; Kantam, M.L.; Venkanna, G.T.; Sreedhar, B.; *J. Am. Chem. Soc.* **2005**, *127*, 9948.

[104] Kantam, M.L.; Venkanna, G.T.; Sridhar, C.; Kumar, K.B.S.; *Tetrahedron Lett.* **2006**, *47,* 3897.

[105] Kantam, M.L.; Venkanna, G.T.; Sridhar, C.; Sreedhar, B.; Choudary, B.M.; *J. Org. Chem.* **2006**, *71*, 9522.

[106] (a) Islam, S.M.; Mondal, S.; Mondal, P.; Singha Roy, A.; Tuhina, K., Mobarak, M.; *Inorg. Chem. Commun.* **2011**, *14*, 1352.

[107] Islam, S.M.; Mondal, S.; Mondal, P.; Singha Roy, A.; Tuhina, K., Salam, N.; Paul, S.; Hossain, D.; Mobarak, M.; *Transition Met. Chem.* **2011**, *36*, 447.

[108] (a) Pan, K.; Ming, H.; Yu, H.; Liu, Y.; Kang, Z.; Zhang, H.; Lee, S-T.; *Cryst. Res. Technol.* **2011**, *46*, 1167. (b) Wang, Y.; Biradar, A.V.; Wang, G.; Sharma, K.K. Duncan, C.T.; Rangan, S.; Asefa, T.; *Chem. Eur. J.* **2010**, *16*, 10735.

[109] Babu, S.G.; Karvembu, R.; *Ind. Eng. Chem. Res.* **2011**, *50*, 9594.

[110] Ahmadi, S.J.; Sadjadi, S.; Hosseinpour, M.; Abdollahi, M.; *Monatsh Chem.* **2011**, *142*, 801.

[111] Goldberg, I.; *Chem. Ber.* **1906**, *39*, 1691.

[112] Lv, X.; Boa, W.; *J. Org. Chem.* **2007**, *72*, 3863.

[113] Shen, R.; Porco, J.A.; *Org. Lett.* **2000**, *2*, 1333.

[114] Jiang, L.; Job, G.E.; Klapars, A.; Buchwald, S.L.; *Org. Lett.* **2003**, *5*, 3667.

[115] Pan, X.; Cai, Q.; Ma, D.; *Org. Lett.* **2004**, *6*, 1809.

[116] Martín, R.; Rivero, M.R.; Buchwald, S.L.; *Angew. Chem., Int. Ed.* **2006**, *45*, 7079.

[117] Han, C.; Shen, R.; Su, S.; Porco, J.A.; *Org. Lett.* **2004**, *6*, 27.

[118] Toumi, M.; Couty, F.; Evano, G.; *Angew. Chem., Int. Ed.* **2007**, *46*, 572.

[119] Bolshan, Y.; Batey, R.A.; *Angew. Chem., Int. Ed.* **2008**, *47*, 2109.

[120] Larsson, P-F.; Correa, A.; Carril, M.; Norrby, P-O.; Bolm, C.; *Angew. Chem., Int. Ed.* **2009**, *48*, 5691.

[121] Ullmann, F.; *Chem. Ber.* **1904**, *37*, 853.

[122] Xing, X.; Padmanaban, D.; Yeh, L.-A.; Cuny, G.D.; *Tetrahedron* **2002**, *58*, 7903.

[123] Boger, D.L.; Patane, M.A.; Zhou, J.; *J. Am. Chem. Soc.* **1994**, *116*, 8544.

[124] Theil, F.; *Angew. Chem., Int. Ed.* **1999**, *38*, 2345.

[125] Marcoux, J.-F.; Doye, S.; Buchwald, S.L.; *J. Am. Chem. Soc.* **1997**, *119*, 10539.

[126] Chen, Y.-J.; Chen, H.-H.; *Org. Lett.* **2006**, *8*, 5609.

[127] Buck, E.; Song, Z.J.; *Org. Synth.* **2005**, *82*, 69.

[128] Fagan, P.J.; Hauptman, E.; Shapiro, R.; Casalnuovo, A.; *J. Am. Chem. Soc.* **2000**, *122*, 5043.

[129] Palomo, C.; Oiarbide, M.; Lopez, R.; Gomez-Bengoa, E.; *Chem. Commun.* **1998**, 2091.

[130] Gujadhur, R.K.; Bates, C.G.; Venkataraman, D.; *Org. Lett.* **2001,** *3,* 4315.

[131] Ma, D.; Cai, Q.; *Org. Lett.* **2003**, *5*, 3799.

[132] Cristau, H.-J.; Cellier, P.P.; Hamada, S.; Spindler, J.-F.; Taillefer, M.; *Org. Lett.* **2004**, *6*, 913.

[133] Cai, Q.; Zou, B.; Ma, D.; *Angew. Chem., Int. Ed.* **2006**, *45*, 1276.

[134] Buck, E.; Song, Z.J.; Tschaen, D.; Dormer, P.G.; Volante, R.P.; Reider, P.J.; *Org. Lett.* **2002**, *4*, 1623.

[135] Sigma-Aldrich, catalog #345083.

[136] Mehmood, A.; Devine, W.G.; Leadbeater, N.E.; *Top. Catal.* **2010**, *53*, 1073.

[137] Niu, J.; Zhou, H.; Li, Z.; Xu, J.; Hu, S.; *J. Org. Chem.* **2008**, *73*, 7814.

[138] Naidu, A.B.; Jaseer, E.A.; Sekar, G.; *J. Org. Chem.* **2009**, *74*, 3675.

[139] Naidu, A.B.; Raghunath, O.R.; Prasad, D.J.C.; Sekar, G.; *Tetrahedron Lett.* **2008**, *49*, 1057.

[140] Naidu, A.B.; Sekar, G.; *Tetrahedron Lett.* **2008**, *49*, 3147.

[141] Niu, J.; Guo, P.; Kang, J.; Li, Z.; Xu, J.; Hu, S.; *J. Org. Chem.* **2009**, *74*, 5075.

[142] Liu, X.; Fu, H.; Jiang, Y.; Zhao, Y.; *Synlett* **2008**, 221.

[143] Xia, N.; Taillefer, M.; *Chem. Eur. J.* **2008**, *14*, 6037.

[144] Chang, J.W.W.; Chee, S.; Mak, S.; Buranaprasertsuk, P.; Chavasiri, W.; Chan, P.W.H.; *Tetrahedron Lett.* **2008**, *49*, 2018.

[145] Schareina, T.; Zapf, A.; Cotté, A.; Müller, N.; Beller, M.; *Tetrahedron Lett.* **2008**, *49*, 1851.

[146] Miao, T.; Wang, L.; *Tetrahedron Lett.* **2007**, *48,* 95.

[147] Benyahya, S.; Monnier, F.; Taillefer, M.; Man, M.W.C.; Bied, C.; Ouazzani, F.; *Adv. Synth. Catal.* **2008**, *350*, 2205.

[148] Lipshutz, B.H.; Unger, J.B.; Taft, B.R.; *Org. Lett.* **2007**, *9*, 1089.

[149] Irfan, M.; Fuchs, M.; Glasnov, T.N.; Kappe, C.O.; *Chem. Eur. J.* **2009**, *38*, 11608.

[150] Zhang, J.; Zhang, Z.; Wang, Y.; Zheng, X.; Wang, Z.; *Eur. J. Org. Chem.* **2008**, 5112.

[151] Lohmann, S.; Andrews, S.P.; Burke, B.J.; Smith, M.D.; Attfield, J.P.; Tanaka, H.; Kaneko, K.; Ley, S.V.; *Synlett* **2005**, 1291.

[152] Jung, N.; Bräse, S.; *J. Comb. Chem.* **2009**, *11*, 47.

[153] Palomo, C.; Oiarbide, M.; Lopez, R.; Gomez-Bengoa, E.; *Tetrahedron Lett.* **2000**, *41*, 1283.

[154] Bates, C.G.; Gujadhur, R.K.; Venkataraman, D.; *Org. Lett.* **2002,** *4,* 2803.

[155] Savarin, C.; Srogl, J.; Liebeskind, L.S.; *Org. Lett.* **2002,** *4,* 4309.

[156] Sim, H.-S.; Feng, X.; Yun, J.; *Chem. Eur. J.* **2009**, *15*, 1939.

[157] Chen, I-H.; Yin, L.; Itano, W.; Kanai, M.; Shibasaki, M.; *J. Am. Chem. Soc.* **2009**, *131*, 11664.

[158] Feng, X.; Yun, J.; *Chem. Commun.* **2009**, 6577. Chen, I-H.; Kanai, M.; Shibasaki, M.; *Org. Lett.* **2010**, *12*, 4098.

[159] Fleming, W.J.; Müller-Bunz, H.; Lillo, V.; Fernández, E.; Guiry, P.J.; *Org. Biomol. Chem.* **2009**, *7*, 2520.

[160] Park, J.K.; Lackey, H.H.; Rexford, M.D.; Kovnir, K.; Shatruk, M.; McQuade, D.Y.; *Org. Lett.* **2010**, *12*, 5008.

[161] O'Brien, J.M.; Lee, K.; Hoveyda, A.H.; *J. Am. Chem. Soc.* **2010**, *132*, 10630.

[162] Noh, D.; Chea, H.; Ju, J.; Yun, J.; *Angew. Chem., Int. Ed.* **2009**, *48*, 6062.

[163] Lee, Y.; Hoveyda, A.H.; *J. Am. Chem. Soc.* **2009**, *131*, 3160.

[164] Corberán, R.; Mszar, N.W.; Hoveyda, A.H.; *Angew. Chem., Int. Ed.* **2011**, *50*, 7079.

[165] Jang, H.; Zhugralin, A.R.; Lee, Y.; Hoveyda, A.H.; *J. Am. Chem. Soc.* **2011**, *133*, 7859.

[166] Carosi, L.; Hall, D.G.; *Angew. Chem., Int. Ed.* **2007**, *46*, 5913.

[167] Ito, H.; Ito, S.; Sasaki, Y.; Matsuura, K.; Sawamura, M.; *J. Am. Chem. Soc.* **2007**, *129*, 14856.

[168] (a) Guzman-Martinez, A.; Hoveyda, A.H.; *J. Am. Chem. Soc.* **2010**, *132*, 10634. (b) Shimizu, M.; *Angew. Chem., Int. Ed.* **2011**, *50*, 5998.

[169] Park, J.K.; Lackey, H.H.; Ondrusek, B.A.; McQuade, D.Y.; *J. Am. Chem. Soc.* **2011**, *133*, 2410.

[170] Beenen, M.A.; An, C.; Ellman, J.A.; *J. Am. Chem. Soc.* **2008**, *130*, 6910. Laitar, D.S.; Tsui, E.Y.; Sadighi, J.P.; *J. Am. Chem. Soc.* **2006**, *128*, 11036.

[171] Adams J.; Kauffman, M.; *Cancer Invest.* **2004**, *22*, 304.

[172] Lipshutz, B.H.; Siegmann, K.; Garcia, E.; Kayser, F.; *J. Am. Chem. Soc.* **1993**, *115*, 9276.

[173] Whitesides, G.M.; San Filippo, J.; Casey, C.P.; Panek, E.J.; *J. Am. Chem. Soc.* **1967**, *89*, 5302.

[174] (a) Kauffmann, T.; *Angew. Chem., Int. Ed. Engl.* **1974**, *13*, 291; (b) Surry, D.S.; Spring, D.R.; *Chem. Soc. Rev.* **2006**, *35*, 218.

[175] Bertz, S.H.; Gibson, C.P.; *J. Am. Chem. Soc.* **1986**, *108*, 8286.

[176] Surry, D.S.; Su, X.; Fox, D.J.; Franckevicius, V.; Macdonald, S.J.F.; Spring, D.R.; *Angew. Chem., Int. Ed.* **2005**, *44*, 1870.

[177] Iyoda, M.; Rahman, M.J.; Matsumoto, A.; Wu, M.; Kuwatani, Y.; Nakao, K.; Miyake, Y.; *Chem. Lett.* **2005**, *34,* 1474.

[178] Lipshutz, B.H., in *Organometallics in Synthesis. A Manual*, (ed. Schlosser, M.), Wiley, Chichester, 2002, pp. 665–815.

[179] (a) Lipshutz, B.H.; James, B.; *J. Org. Chem.* **1994**, *59*, 7585; (b) Kronenburg, C.M.P.; Jastrzebski, J.T.B.H.; Spek, A.L.; van Koten, G.; *J. Am. Chem. Soc.* **1998**, *120*, 9688; (c) Krause, N.; *Angew. Chem., Int. Ed.* **1999**, *38*, 79.

[180] (a) Spring, D.R.; Krishnan, S.; Schreiber, S.L.; *J. Am. Chem. Soc.* **2000**, *122,* 5656; (b) Spring, D.R.; Krishnan, S.; Blackwell, H.E.; Schreiber, S.L.; *J. Am. Chem. Soc.* **2002**, *124,* 1354.

[181] Coleman, R.S.; Gurrala, S.R.; *Org. Lett.* **2005**, *7*, 1849.

[182] Coleman, R.S.; Gurrala, S.R.; Mitra, S.; Raao, A.; *J. Org. Chem.* **2005**, *70,* 8932.

[183] (a) Coleman, R.S.; Grant, E.B.; *Tetrahedron Lett.* **1993**, *34*, 2225; (b) Coleman, R.S.; Grant, E.B.; *J. Am. Chem. Soc.* **1995**, *117*, 10889.

[184] Miyake, Y.; Wu, M.; Rahman, M.J.; Iyoda, M.; *Chem. Commun.* **2005**, 411.

[185] Miyake, Y.; Wu, M.; Rahman, M.J.; Kuwatani, Y.; Iyoda, M.; *J. Org. Chem.* **2006**, *71,* 6110.

[186] (a) Mulrooney, C.A.; Li, X.; DiVirgilio, E.S.; Kozlowski, M.C.; *J. Am. Chem. Soc.* **2003**, *125*, 6856; (b) Xie, X.; Phuan, P.-W.; Kozlowski, M.C.; *Angew. Chem., Int. Ed.* **2003**, *42*, 2168; (c) Kozlowski, M.C.; Li, X.; Carroll, P.J.; Wu, Z.; *Organometallics* **2002**, *21*, 4513; (d) Li, X.; Hewgley, J.B.; Mulrooney, C.A.; Yang, J.; Kozlowski, M.C.; *J. Org. Chem.* **2003**, *68*, 5500.

[187] (a) Temma, T.; Habaue, S.; *J. Polym. Sci., Part A: Polym. Chem.* **2005**, *43*, 6287; (b) Temma, T.; Habaue, S.; *Tetrahedron Lett.* **2005**, *46*, 5655.

[188] DiVirgilio, E.S.; Dugan, E.C.; Mulrooney, C.A.; Kozlowski, M.C.; *Org. Lett.* **2007**, *9*, 385.

[189] Sugimura, T.; Yamada, H.; Inoue, S.; Tai, A.; *Tetrahedron: Asymmetry* **1997**, *8*, 649.

[190] (a) Qiu, L.; Kwong, F.Y.; Wu, J.; Lam, W.H.; Chan, S.; Yu, W.-Y.; Li, Y.-M.; Guo, R.; Zhou, Z.; Chan, A.S. C.; *J. Am. Chem. Soc.* **2006**, *128,* 5955; (b) Zhang, Z.; Qian, H.; Longmire, J.; Zhang, X.; *J. Org. Chem.* **2000**, *65*, 6223.

[191] Shimizu, H.; Nagasaki, I.; Saito, T.; *Tetrahedron* **2005**, *61*, 5405.

[192] Do, H-Q.; Daugulis, O.; *J. Am. Chem. Soc.* **2011**, *133*, 13577.

[193] (a) del Amo, V.; Dubbaka, S.R.; Krasovskiy, A.; Knochel, P.; *Angew. Chem., Int. Ed.* **2006**, *45*, 7838; (b) Kienle, M.; Dubbaka, S.R.; del Amo, V.; Knochel, P.; *Synthesis* **2007**, 1272; (c) Boudet, N.; Dubbaka, S.R.; Knochel, P.; *Org. Lett.* **2008**, *10*, 1715.

[194] Oisaki, K.; Zhao, D.; Kanai, M.; Shibasaki, M.; *J. Am. Chem. Soc.* **2006**, *128*, 7164; Suto, Y.; Kanai, M.; Shibasaki, M. *J. Am. Chem. Soc.* **2007**, *129*, 500; (allylboronates) Wada, R.; Oisaki, K.; Kanai, M; Shibasaki, M.; *J. Am. Chem. Soc.* **2004**, *126*, 8910. Villalobos, J.M.; Srogl, J.; Liebeskind, L.S.; *J. Am. Chem. Soc.* **2007**, *129*, 15734.

[195] Wada, R.; Shibuguchi, T.; Makino, S.; Oisaki, K.; Kanai, M; Shibasaki, M.; *J. Am. Chem. Soc.* **2006**, *128*, 7687.

[196] (a) Hatano, M; Asai, T.; Ishihara, K.; *Tetrahedron Lett.* **2008**, *49*, 379; (b) Colombo, F.; Benaglia, M.; Orlandi, S.; Usuelli, F.; Celentano, G.; *J. Org. Chem.* **2006**, *71*, 2064.

[197] (a) Gommermann, N.; Knochel, P.; *Chem. Eur. J.* **2006**, *12*, 4380; (b) Knöpfel, T.F.; Aschwanden, P.; Ichikawa, T.; Watanabe, T.; Carreira, E.M.; *Angew. Chem., Int. Ed.* **2004**, *43*, 5971; (c) Detz, R.J.; Delville, M.M.E.; Hiemstra, H.; van Maarseveen, J.H.; *Angew. Chem., Int. Ed.* **2008**, *47*, 3777.

[198] Bisai, A.; Singh, V.K.; *Org. Lett.* **2006**, *8*, 2405.

[199] Yan, X.-X.; Peng, Q.; Li, Q.; Zhang, K.; Yao, J.; Hou, X.-L.; Wu, Y.-D.; *J. Am. Chem. Soc.* **2008**, *130*, 14362.

[200] Tan, C.; Liu, X.; Wang, L.; Wang, J.; Feng, X.; *Org. Lett.* **2008**, *10*, 5305.

[201] González, A.S.; Arrayás, R.G.; Rivero, M.R.; Carretero, J.C.; *Org. Lett.* **2008**, *10*, 4335.

[202] Hernández-Toribio, J.; Arrayás, R.G.; Carretero, J.C.; *J. Am. Chem. Soc.* **2008**, *130*, 16150.

[203] (a) Suzuki, Y.; Yazaki, R.; Kumagai, N.; Shibasaki, M.; *Angew. Chem., Int. Ed.* **2009**, *48*, 5026; (b) Yin, L.; Kanai, M.; Shibasaki, M.; *J. Am. Chem. Soc.* **2009**, *131*, 9610.

[204] (a) Bonnaventure, I.; Charette, A.B.; *J. Org. Chem.* **2008**, *73*, 6330; (b) Bonnaventure, I.; Charette, A.B.; *Tetrahedron* **2009**, *65*, 4968.

[205] Pizzuti, M.G.; Minnaard, A.J.; Feringa, B.L.; *J. Org. Chem.* **2008**, *73*, 940.

[206] Yang, T.; Ferrali, A.; Sladojevich, F.; Campbell, L.; Dixon, D.J.; *J. Am. Chem. Soc.* **2009**, *131*, 9140.

[207] Chen, I-H.; Oisaki, K.; Kanal, M.; Shibasaki, M.; *Org. Lett.* **2008**, *10*, 5151.

[208] Jerphagnon, T.; Pizzuti, M.G.; Minnaard, A.J.; Feringa, B.L.; *Chem. Soc. Rev.* **2009**, *38*, 1039.

[209] Hawner, C.; Li, K.; Cirriez, V.; Alexakis, A.; *Angew. Chem., Int. Ed.* **2008**, *47*, 8211.

[210] Shizuka, M.; Snapper, M.; *Angew. Chem., Int. Ed.* **2008**, *47*, 5049.

[211] Bos, P.H.; Minnaard, A.J.; Feringa, B.L.; *Org. Lett.* **2008**, *10*, 4219.

[212] den Hartog, T.; Harutyunyan, S.R.; Font, D.; Minnaard, A.J.; Feringa, B.L.; *Angew. Chem., Int. Ed.* **2008**, *47*, 398.

[213] Hénon, H.; Mauduit, M.; Alexakis, A.; *Angew. Chem., Int. Ed.* **2008**, *47*, 9122.

[214] Robert, T.; Velder, J.; Schmalz, H.-G.; *Angew. Chem., Int. Ed.* **2008**, *47*, 7718.

[215] López, F.; Harutyunyan, S.R.; Minnaard, A.J.; Feringa, B.L.; *J. Am. Chem. Soc.* **2004**, *126*, 12784.

[216] Feringa, B.L.; Badorrey, R.; Peña, D.; Harutyunyan, S.R.; Minnaard, A.J.; *Proc. Nat. Acad. Sci. U.S.A.* **2004**, *101*, 5834.

[217] Martin, D.; Kehrli, S.; d'Augustin, M.; Clavier, H.; Mauduit, M.; Alexakis, A.; *J. Am. Chem. Soc.* **2006**, *128*, 8416.

[218] Kawamura, K.; Fukuzawa, H.; Hayashi, M.; *Org. Lett.* **2008**, *10*, 3509.

[219] García, J.M.; González, A.; Kardak, B.G.; Odriozola, J.M.; Oiarbide, M.; Razkin, J.; Palomo, C.; *Chem. Eur. J.* **2008**, *14*, 8768.

[220] Brown, M.K.; Degrado, S.J.; Hoveyda, A.H.; *Angew. Chem., Int. Ed.* **2005**, *44*, 5306. Wu, J.; Mampreian, D.M.; Hoveyda, A.H.; *J. Am. Chem. Soc.* **2005**, *127*, 4584.

[221] Oisaki, K.; Zhao, D.; Kanai, M.; Shibasaki, M.; *J. Am. Chem. Soc.* **2007**, *129*, 7439.

[222] Müller, D.; Hawner, C.; Tissot, M.; Palais, L.; Alexakis, A. *Synlett* **2010**, *11*, 1694.

[223] Ireland, T.; Grossheimann, G.; Wieser-Jeunesse, C.; Knochel, P.; *Angew. Chem., Int. Ed.* **1999**, *38*, 3212.

[224] López, F.; Harutyunyan, S.R.; Meetsma, A.; Minnaard, A.J.; Feringa, B.L.; *Angew. Chem., Int. Ed.* **2005**, *44*, 2752.

[225] Wang, S.-Y.; Ji, S.-J.; Loh, T.-P.; *J. Am. Chem. Soc.* **2007**, *129*, 276.

[226] Wang, S.-Y.; Lum, T.-K.; Ji, S.-J.; Loh, T.-P.; *Adv. Synth. Catal.* **2008**, *350*, 673.

[227] Howell, G.P.; Fletcher, S.P.; Geurts, K.; ter Horst, B.; Feringa, B.L.; *J. Am. Chem. Soc.* **2006**, *128*, 14977.

[228] Des Mazery, R.; Pullez, M.; López, F.; Harutyunyan, S.R.; Minnaard, A.J.; Feringa, B.L.; *J. Am. Chem. Soc.* **2005**, *127*, 9966.

[229] (a) Palais, L.; Babel, L.; Quintard, A.; Belot, S.; Alexakis, A. *Org. Lett.* **2010**, *12*, 1988. (b) Quintard, A.; Alexakis, A. *Adv. Synth. Catal.* **2010**, *352*, 1856.

[230] Hajra, A.; Yoshikai, N.; Nakamura, E.; *Org. Lett.* **2006**, *8*, 4153.

[231] Fillion, E.; Wilsily, A.; *J. Am. Chem. Soc.* **2006**, *128*, 2774.

[232] Müller, D.; Hawner, C.; Tissot, M.; Palais, L.; Alexakis, A.; *Synlett* **2010**, *11*, 1694.

[233] May, T.L.; Dabrowski, J.A.; Hoveyda, A.H.; *J. Am. Chem. Soc.* **2011**, *133*, 736.

[234] Lee, K.; Hoveyda, A.H. *J. Am. Chem. Soc.* **2010**, *132*, 2898.

[235] (a) Trost. B.M.; *Chem. Pharm. Bull.* **2002**, *50*, 1; (b) Negishi, E., in *Handbook of Organopalladium Chemistry for Organic Synthesis* (ed. Negishi, E.), Wiley, New York, 2002.

[236] (a) Trost, B.M.; Hachiya, I.; *J. Am. Chem. Soc.* **1998**, *120*, 1104; (b) Trost, B.M.; Hildbrand, S.; Dogra, K.; *J. Am. Chem. Soc.* **1999**, *121*, 10416; (c) Trost, B.M.; Dogra, K.; *J. Am. Chem. Soc.* **2002**, *125*, 7256.

[237] Lloyd-Jones, G.C.; Pfaltz, A.; *Angew. Chem., Int. Ed.* **1995**, *34*, 462.

[238] (a) Lautens, M.; Dockendorff, C.; Fagnou, K.; Malicki, A.; *Org. Lett.* **2002**, *4*, 1311; (b) Tseng, N.-W.; Mancuso, J.; Lautens, M.; *J. Am. Chem. Soc.* **2006**, *128*, 5338; (c) Cho, Y.; Zunic, V.; Senboku, H.; Olsen, M.; Lautens, M.; *J. Am. Chem. Soc.* **2006**, *128*, 6837; (d) McManus, H.A.; Fleming, M.J.; Lautens, M.; *Angew. Chem., Int. Ed.* **2007**, *46*, 433; (e) Menard, F.; Lautens, M.; *Angew. Chem., Int. Ed.* **2008**, *47*, 2085.

[239] (a) Matsushima, Y.; Onitsuka, K.; Kondo, T.; Mitsudo, T.; Takahashi, S.; *J. Am. Chem. Soc.* **2001**, *123*, 10405; (b) Onitsuka, K.; Okuda, H.; Sasai, H.; *Angew. Chem., Int. Ed.* **2008**, *47*, 1454.

[240] Helmchen, G.; Dahnz, A.; Dübon, P.; Schelwies, M.; Weihofen, R.; *Chem. Commun.* **2007**, 675.

[241] Lee, Y.; Akiyama, K.; Gillingham, D.G.; Brown, K.; Hoveyda, A.H.; *J. Am. Chem. Soc.* **2008**, *130*, 446.

[242] Van Zijl, A.W.; Arnold, L.A.; Minnaard, A.J.; Feringa, B.L.; *Adv. Synth. Catal.* **2004**, *346*, 413.

[243] Tissot-Croset, K.; Alexakis, A.; *Tetrahedron Lett.* **2004**, *45*, 7375.

[244] Falciola, C.A.; Tissot-Croset, K.; Alexakis, A.; *Angew. Chem., Int. Ed.* **2006**, *45*, 5995.

[245] Falciola, C.A.; Alexakis, A.; *Angew. Chem., Int. Ed.* **2007**, *46*, 2619.

[246] (a) López, F.; van Zijl, A.W.; Minnaard, A.J.; Feringa, B.L.; *Chem. Commun.* **2006**, 409; (b) Van Zijl, A.W.; López, F.; Minnaard, A.J.; Feringa, B.L.; *J. Org. Chem.* **2007**, *72*, 2558.

[247] Geurts, K.; Fletcher, S.P.; Feringa, B.L.; *J. Am. Chem. Soc.* **2006**, *128*, 15572.

[248] Hartog, T.D.; Maci, B.; Minnaard, A.J.; Feringa, B.L.; *Adv. Synth. Catal.* **2010**, *352*, 999.

[249] Langlois, J.-B.; Alexakis, A.; *Chem. Commun.* **2009**, 3868.

[250] Ohmiya, H.; Yokobori, U.; Makida, Y.; Sawamura, M.; *J. Am. Chem. Soc.* **2010**, *132*, 2895.

[251] Shintani, R.; Takatsu, K.; Takeda, M.; Hayashi, T.; *Angew. Chem., Int. Ed.* **2011**, *50*, 8656.

[252] (a) Jung, B.; Hoveyda, A.H.; *J. Am. Chem. Soc.* **2012**, *134*, 1490; (b) Meng, F.; Jung, B.; Haeffner, F.; Hoveyda, A.H.; *Org. Lett.* **2013**, *15*, 1414.

[253] Gao, F.; Lee Y.; Mandai, K.; Hoveyda, A.H.; *Angew. Chem., Int. Ed.* **2010**, *49*, 8370.

[254] Gao, F.; McGrath K.P.; Lee Y.; Hoveyda, A.H.; *J. Am. Chem. Soc.* **2010**, *132*, 14315.

[255] Dabrowski, J.A.; Gao, F.; Hoveyda, A.H.; *J. Am. Chem. Soc.* **2011**, *133*, 4778.

[256] Wurtz, A.; *Ann. Chem. Phys.* **1844**, *11*, 250.

[257] Brestensky, D.M.; Huseland, D.E.; McGettigan, C.; Stryker, J.M.; *Tetrahedron Lett.* **1988**, *29*, 3749.

[258] Bezman, S.A.; Churchil, M.R.; Osborn, J.A.; Wormald, J.; *J. Am. Chem. Soc.* **1971**, *93*, 2063.

[259] Chandler, C.A.; Phillips A.J.; *Org. Lett.* **2005**, *7*, 3493.

[260] Sakamoto, S.; Sakazaki, H.; Hagiwara, K.; Kamada, K.; Ishii, K.; Noda, T.; Inoue, M.; Hirama, M.; *Angew. Chem., Int. Ed.* **2004**, *43*, 6505.

[261] Crimmins, M.T.; McDougall, P.J.; Emmitte, K.A.; *Org. Lett.* **2005**, *7*, 4033.

[262] Mahoney, W.S.; Brestensky, D.M.; Stryker, J.M.; *J. Am. Chem. Soc.* **1988**, *110*, 291.

[263] Diez-Gonzalez, S.; Nolan, S.P.; *Acc. Chem. Res.* **2008**, *41*, 349.

[264] Lawrence, N.J.; Drew, M.D.; Bushell, S.M.; *J. Chem. Soc., Perkin Trans. 1* **1999**, 3381.

[265] Chen, J.-X.; Daeuble, J.F.; Stryker, J.M.; *Tetrahedron* **2000**, *56*, 2789.

[266] Lipshutz, B.H.; Keith, J.; Papa, P.; Vivian, R.; *Tetrahedron Lett.* **1998**, *39*, 4627.

[267] Chiu, P.; *Synthesis* **2004**, 2210.

[268] Ito, H.; Ishizuka, T.; Arimoto, K.; Miura, K.; Hosomi, A.; *Tetrahedron Lett.* **1997**, *38*, 8887.

[269] Lipshutz, B.H.; Servesko, J.M.; Petersen, T.B.; Papa, P.P.; Lover, A.A.; *Org. Lett.* **2004**, *6*, 1273.

[270] Lipshutz, B.H.; Chrisman, W.; Noson, K.; Papa, P.; Sclafani, J.A.; Vivian, R.W.; Keith, J.M.; *Tetrahedron* **2000**, *56*, 2779.

[271] Rainka, M.P.; Aye, Y.; Buchwald, S.L.; *Proc. Nat. Acad. Sci. U.S.A.* **2004**, *101*, 5821.

[272] (a) Lee, D.; Yang, Y.M.; Yun, J.; *Synthesis* **2007**, 2233; (b) Lee, D.; Kim, D.M.; Yun, J.; *Angew. Chem., Int. Ed.* **2006**, *45*, 2785.

[273] Hughes, G.; Kimura, M.; Buchwald, S.L.; *J. Am. Chem. Soc.* **2003**, *125*, 11253.

[274] Shimizu, H.; Igarashi, D.; Kuriyama, W.; Yusa, Y.; Sayo, N.; Saito, T.; *Org. Lett.* **2007**, *9*, 1655.

[275] Czekelius, C.; Carreira, E.M.; *Angew. Chem., Int. Ed.* **2003**, *42*, 4793.

[276] Czekelius, C.; Carreira, E.M.; *Org. Lett.* **2004**, *6*, 4575.

[277] Courmarcel, J.; Mostefaï, N.; Sirol, S.; Choppin, S.; Riant, O.; *Isr. J. Chem.* **2001**, *41*, 231.

[278] Baker, B.A.; Bošković, Ž.V.; Lipshutz, B.H.; *Org. Lett.* **2008**, *10*, 289; Pelšs, A.; Kumpulainen, E.T.T.; Koskinen, A.M.P.; *J. Org. Chem.* **2009**, *74*, 7598.

[279] (a) Gathy, T.; Riant, O.; Peeters, D.; Leyssens, T. *J. Org. Chem.* **2011**, *696*, 3425. (b) Zhang, W.; Li, W.; Qin, S.; *Org. Biomol. Chem.* **2012**, *10*, 597.

[280] Schmid, R.; Broger, E.A.; Cereghetti, M.; Crameri, Y.; Foricher, J.; Lalonde, M.; Muller, R.K.; Scalone, M.; Schoettel, G.; Zutter, U., in *New Developments in Enantioselective Hydrogenation*, 1996, pp. 131 – 138.

[281] Saito, T.; Yokozawa, T.; Ishizaki, T.; Moroi, T.; Sayo, N.; Miura, T.; Kumobayashi, H.; *Adv. Synth. Catal.* **2001**, *343*, 264.

[282] Blaser, H.U.; Brieden, W.; Pugin, B.; Spindler, F.; Studer, M.; Togni, A.; *Top. Catal.* **2002**, *19*, 3.

[283] Zhang, X.C.; Wu, Y.; Yu, F.; Wu, F.F.; Wu, J.; Chan, A.S.C.; *Chem. Eur. J.* **2009**, *15*, 5888.

[284] Mostefaï, N.; Sirol, S.; Courmarcel, J.; Riant. O.; *Synthesis* **2007**, *8*, 1265.

[285] Yu, F.; Zhou, J.-N.; Zhang, X.-C.; Sui, Y.-Z.; Wu, F.-F.; Xie, L.-J.; Chan, A.S.C.; Wu, J.; *Chem. Eur. J.* **2011**, *17*, 14234.

[286] Lipshutz, B.H.; Frieman, B.A.; *Angew. Chem., Int. Ed.* **2005**, *44*, 6345.

[287] (a) Lipshutz, B.H.; Noson, K.; Chrisman, W.; Lower, A.; *J. Am. Chem. Soc.* **2003**, *125*, 8779. (b) Lipshutz, B.H.; Noson, K.; Chrisman, W.; *J. Am. Chem. Soc.* **2001**, *123*, 12917.

[288] Lipshutz, B.H.; Shimizu, H.; *Angew. Chem., Int. Ed.* **2004**, *43*, 2228.

[289] Lipshutz, B.H.; Servesko, J.M.; Taft, B.R.; *J. Am. Chem. Soc.* **2004**, *126*, 8352.

[290] Lipshutz, B.H.; Tanaka, N.; Taft, B.R.; Lee, C.T.; *Org. Lett.* **2006**, *8*, 1963.

[291] Goeden, G.V.; Caulton, K.G.; *J. Am. Chem. Soc.* **1981**, *103*, 7354.

[292] Lipshutz, B.H.; Caires, C.C.; Kuipers, P.; Chrisman, W.; *Org. Lett.* **2003**, *5*, 3085.

[293] Lipshutz, B.H.; Lower, A.; Noson, K.; *Org. Lett.* **2002**, *4*, 4045.

[294] Shimizu, H.; Nagano. T.; Sayo, N.; Saito, T.; Ohshima, T.; Mashima, K.; *Synlett* **2009**, 3143.

[295] Lipshutz, B.H.; Servesko, J.M.; *Angew. Chem., Int. Ed.* **2003**, *42*, 4789.

[296] Hou, C.-J.; Guo, W.-L.; Hu, X.-P.; Deng, J.; Zheng, Z.; *Tetrahedron: Asymmetry* **2011**, *22*, 195.

[297] Malkov, A.V.; *Angew. Chem., Int. Ed.* **2010**, *49*, 9814.

[298] Moser, R.; Bošković, Ž.V.; Crowe, C.S.; Lipshutz, B.H.; *J. Am. Chem. Soc.* **2010**, *132*, 7852.

[299] Voigtritter, K.R.; Isley, N.A.; Moser, R.; Aue, D.H.; Lipshutz, B.H.; *Tetrahedron* **2012,** *68*, 3414.

[300] Madduri, A.V.R.; Minnaard, A.J.; Harutyunyan, S.R.; *Chem. Commun.* **2012**, *48*, 1478.

[301] Llamas, T.; Arrayas, R.G.; Carretero, J.C.; *Angew. Chem., Int. Ed.* **2007**, *46*, 3329.

[302] Desrosiers, J.N.; Charette, A.B.; *Angew. Chem., Int. Ed.* **2007**, *46*, 5955.

[303] Rupnicki, L.; Saxena, A.; Lam, H.W.; *J. Am. Chem. Soc.* **2009**, *131*, 10386.

[304] Lipshutz, B.H.; Frieman, B.A.; Tomaso, A.E.; *Angew. Chem., Int. Ed.* **2006**, *45*, 1259.

[305] Lipshutz, B.H.; Ghorai, S.; Abela, A.R.; Moser, R.; Nishikata, T.; Duplais, C.; Krasovskiy, A.; *J. Org. Chem.* **2011**, *76*, 4379.

[306] Huang, S.; Voigtritter, K.R.; Unger, J.B.; Lipshutz, B.H.; *Synlett* **2010**, *13*, 2041.

[307] Deschamp, J.; Chuzel, O.; Hannedouche, J.; Riant, O.; *Angew. Chem., Int. Ed.* **2006**, *45*, 1292.

[308] Zhao, D.B.; Oisaki, K.; Kanai, M.; Shibasaki, M.; *Tetrahedron Lett.* **2006**, *47*, 1403.

[309] Zhao, D. B.; Oisaki, K.; Kanai, M.; Shibasaki, M.; *J. Am. Chem. Soc.* **2006**, *128*, 14440.

[310] Du, Y.; Xu, L.W.; Shimizu, Y.; Oisaki, K.; Kanai, M.; Shibasaki, M.; *J. Am. Chem. Soc.* **2008**, *130*, 16146.

[311] Lipshutz, B.H.; Amorelli, B.; Unger, J.B.; *J. Am. Chem. Soc.* **2008**, *130*, 14378.

[312] Chung, W.K.; Chiu, P.; *Synlett* **2005**, 55.

[313] (a) Lipshutz, B.H.; Ghorai, S.; *Aldrichimica Acta* **2008**, *41*, 59; (b) Lipshutz, B.H., Ghorai, S.; *Aldrichimica Acta* **2012**, *45*, 3.

[314] Lam, H.W.; Joensuu, P.M.; *Org. Lett.* **2005**, *7*, 4225.

[315] Lam, H.W.; Murray, G.J.; Firth, J.D.; *Org. Lett.* **2005**, *7*, 5743.

[316] Jurkauskas, V.; Sadighi, J.P.; Buchwald, S.L.; *Org. Lett.* **2003**, *5*, 2417.

[317] Díez-Gonzáles, S.; Kaur, H.; Zinn, F.K.; Stevens, E.D.; Nolan, S.P.; *J. Org. Chem.* **2005**, *70*, 4784.

[318] Welle, A.; Díez-Gonzáles, S.; Tinant, B.; Nolan, S.P.; Riant, O.; *Org. Lett.* **2006**, *8*, 6059.

[319] Kaur, H.; Zinn, F.K.; Stevens, E.D.; Nolan, S.P.; *Organometallics* **2004**, *23*, 1157.

[320] Albright, A.; Eddings, D.; Black, R.; Welch, C.J.; Gerasimchuk, N.N.; Gawley, R.E.; *J. Org. Chem.* **2011**, *76*, 7341.

[321] Albright, A.; Gawley, R.E.; *J. Am. Chem. Soc.* **2011**, *133*, 19680.

[322] Deutsch, C.; Lipshutz, B.H.; Krause, N.; *Angew. Chem., Int. Ed.* **2007**, *46*, 1650.

[323] Arduengo, A.J.; Krafczyk, R.; Schmutzler, R.; Craig, H.A.; Goerlich, J.R.; Marshall, W.J.; Unverzagt, M.; *Tetrahedron* **1999**, *55*, 14523.

[324] Altenhoff, G.; Goddard, R.; Lehmann, C.W.; Glorius, F.; *J. Am. Chem. Soc.* **2004**, *126*, 15195.

[325] Krause, N.; Hoffmann-Röder, A.; Canisius, J.; *Synthesis* **2002**, 1759.

[326] Deutsch, C.; Lipshutz, B.H.; Krause, N.; *Org. Lett.* **2009**, *11*, 5010.

[327] Zhong, C.M.; Sasaki, Y.; Ito, H.; Sawamura, M.; *Chem. Commun.* **2009**, 5850.

[328] Daugalis, O.; Do, H.-Q.; Shabashov, D.; *Acc. Chem. Res.* **2009**, *42*, 1074.

[329] Do, H.-Q.; Kashif, Khan, R.M.; Daugulis, O.; *J. Am. Chem. Soc.* **2008**, *130*, 15185.

[330] Phipps, R.J.; Gaunt, M.J.; *Science* **2009**, *323*, 1593.

[331] Ciana, C-L.; Phipps, R.J.; Brandt, J.R.; Meyer, F-M.; Guant, M.J.; *Angew. Chem., Int. Ed.* **2011**, *50*, 458.

[332] Boogaerts, I.I.F.; Fortman, G.C.; Furst, M.R.L.; Cazin, C.S.J.; Nolan, S.P.; *Angew. Chem., Int. Ed.* **2010**, *49*, 8674; Boogaerts, I.I.F.; Nolan, S.P.; *Chem. Comun.* **2011**, *47*, 3021.

[333] Goossen, L.J.; Manjolinho, F.; Khan, B.A.; Rodríguez, N.; *J. Org. Chem.* **2009**, *74*, 2620.

[334] Tarwade, V.; Liu, X.; Yan, N.; Fox, J.M.; *J. Am. Chem. Soc.* **2009**, *131*, 5382.

[335] Simaan, S.; Masarwa, A.; Bertus, P.; Marek, I.; *Angew. Chem., Int. Ed.* **2006**, *45*, 3963.

[336] Simaan, S.; Marek, I.; *Chem. Commun.* **2009**, 292.

[337] January 11, 2007, at UCSB, Santa Barbara, California

[338] Lipshutz, B.H.; Huang, S.; Leong, W.W.Y.; Zhong, G.; Isley, N.A.; *J. Am. Chem. Soc.,* **2012**, *134*, 19985.

[339] http://www.nsf.gov/funding/pgm_summ.jsp?pims_id=13635; accessed January 1, 2012.

[340] Beletskaya, I.; Cheprakov, A.V.; *Coord. Chem. Rev.* **2004**, *248*, 2337.

Chapter Two

Organorhodium Chemistry

Iwao Ojima, Alexandra A. Athan, Stephen J. Chaterpaul, Joseph J. Kaloko, and Yu-Han Gary Teng

Department of Chemistry
Stony Brook University
Stony Brook, NY, USA

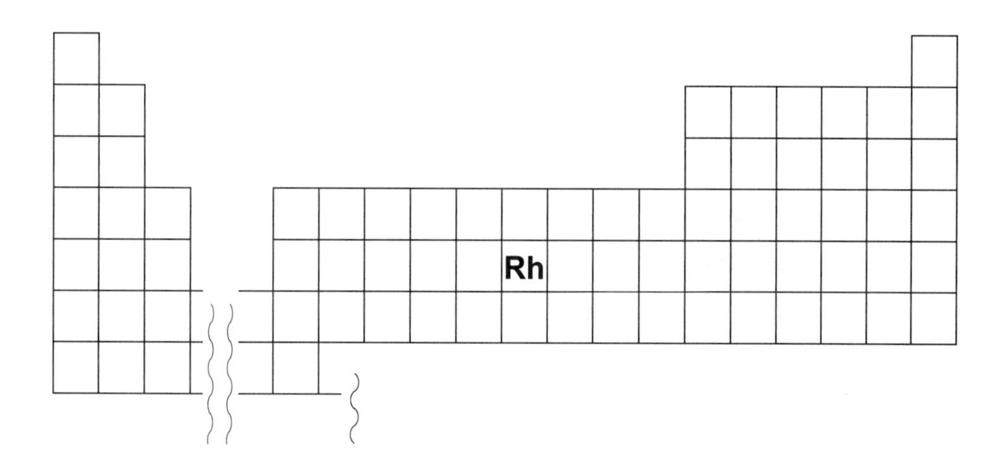

Contents

	List of Abbreviations	**138**
1.	**Introduction**	**139**
2.	**Rhodium Complexes**	**139**
	2.1 Carbonyl Rhodium Complexes	139
	2.2 Phosphine Rhodium Complexes	140
	2.3 Neutral Rhodium Complexes	141
	2.4 Cationic Rhodium Complexes	142
3.	**Hydroformylation**	**142**
	3.1 Highly Linear Selective Hydroformylation of Alkenes	145
	3.2 Asymmetric Hydroformylation	152
	3.3 Tandem Hydroformylation Reactions	155
4.	**Silylformylation**	**160**
	4.1 Tandem Silylformylation/Olefination	161
	4.2 Intramolecular Silylformylation of Alkynes	162
5.	**Cyclopropanation**	**164**
6.	**Cyclohydrocarbonylation**	**168**
7.	**Asymmetric Hydrogenation**	**173**
	7.1 Synthesis of α-Amino Acids by Hydrogenation of 2-Acetamido Acrylic Acid Derivatives	176
	7.2 Synthesis of β-Amino Acids by Hydrogenation of β-(Acylamino)acryl-ates	178
	7.3 Asymmetric Hydrogenation of β,β-Disubstituted α-Dehydroamino Acid Derivatives	181
	7.4 Asymmetric Hydrogenation of Enamides	182
	7.5 Hydrogenation of Itaconic Acid and Dimethyl Itaconate	184
	7.6 Asymmetric Hydrogenation of Enol Esters	186
	7.7 Hydrogenation of Unsaturated Acids and Esters	187
8.	**Hydroacylation of Alkenes and Alkynes**	**193**
9.	**C-H Activation by Metal-Carbenoids and Nitrenoid-Induced C-H Insertion**	**195**
10.	**Rh-Catalyzed C-H Activation**	**204**
11.	**Rh-Catalyzed 1,4-Conjugate Additions of Boronic Acids**	**208**
12.	**Rh-Catalyzed Hydroboration**	**217**
13.	**Rh-Catalyzed Reductive Couplings**	**220**

14. Miscellaneous Reactions Catalyzed by Rh-Complexes 226

15. Rhodium-Catalyzed Cycloadditions and Cyclizations 230
15.1 Cycloaddition and Cyclization Mechanisms 230
15.2 Rh(I)-Catalyzed [2+2+1] Cycloaddition (Pauson-Khand Reaction) 232
15.3 Asymmetric Rh(I)-Catalyzed [2+2+1] Cycloaddition 234
15.4 Alternative Sources of CO 235
15.5 Other Pauson-Khand Type Reactions (PKTR) 236
15.5.1 PK-Type Reactions of Diynes 236
15.5.2 PK-Type Reactions of Dienynes 238
15.5.3 PK-Type Reactions of Allenynes 240
15.5.4 Rh(I)-Catalyzed [4+1] and [3+2] Cycloadditions 245
15.5.5 Rh(I)-Catalyzed [2+2+2] Cycloadditions 248

16. Rh(I)-Catalyzed Cycloisomerization and Alder-ene Reactions 262
16.1 Enyne Cycloisomerizations 262
16.2 Allenyne Cycloisomerizations 268
16.3 Rh(I)-Catalyzed [4+2] Cycloadditions 273
16.4 Rh(I)-Catalyzed [5+2] Cycloadditions 280
16.5 Rh(I)-Catalyzed [2+2+2+1] Cycloadditions 282
16.6 Rh(I)-Catalyzed Reductive Cyclizations 285
16.7 Tandem 1,4-Addition-Aldol Cyclizations 294
16.8 Rh(I)-Catalyzed Domino Reactions in Carbocyclizations 298

17. Closing Remarks 301
18. Acknowledgments 301

19. References 302

List of Abbreviations

acac	acetylacetone
BPE	1,2-bis(phospholano)ethane
CNF	central nervous system
cod	1,5-cyclooctadiene
coe	cyclooctene
cy-hex	cyclohexane
DCE	dichloroethane
DEAD	diethyl azodicarboxylate
DME	1,2-dimethoxyethane
DPPB	diphenyl-phosphinobenzoate
dppe	1,2-bis(diphenylphosphino)ethane
dppp	1,3-bis(diphenylphosphino)propane
MPPIM	methyl 2-hydroxy-1-(2-phenylacetyl)imidazolidine-4-carboxylate anion
MTBE	methyl *tert*-butyl ether
nbd	norbornadiene
NTTL	*N*-naphthoyl-*tert*-leucine
oct	octanoate
PCC	pyridinium chlorochromate
pfd	perfluorodecanoate
p-TSA	*p*-toluenesulfonic acid
t-AmOH	*t*-amyl alcohol
TFA	trifluoroacetic acid
TFE	2,2,2-trifluoroethanol
THF	tetrahydrofuran
TsOH	tosylic acid

1. Introduction

Rhodium is a rare white-silvery metal classified as a member of the platinum group metals. As a result, rhodium is commonly used as a catalyst in chemical reactions. It is also used in several chemical feedstock processes, including hydroformylation. Furthermore, rhodium has been used to catalyze more complex processes, such as higher order cycloaddition reactions. In addition, rhodium has been used in simpler reaction types such as hydrogenation and cycloisomerization. The fact that rhodium is effective for a wide range of chemical processes makes it an attractive metal for catalysis.

2. Rhodium Complexes

Organorhodium compounds are found in oxidation states ranging from +6 to –3. However, the most common oxidation states are +1 and +3. Rh(I) species exist as both a tetra-coordinated, square planar complex as well as a penta-coordinated trigonal bipyramidal complex. Rhodium, like platinum group metals, exhibits reversible oxidation states making a broad range of organic-catalytic transformations possible.

The commercially available rhodium(III) chloride trihydrate ($RhCl_3{\cdot}3H_2O$) complex has been found to be the best starting material for the preparation of various Rh compounds. Additionally, this complex can also be prepared using known methods.[1] An alternative starting material is anhydrous rhodium(III) chloride, but it is insoluble in many conventional solvents, which often leads to low yields.

In many cases, rather than synthesizing the active Rh complexes for reactions, Rh pre-catalyst systems can be treated with various additives to generate active catalyst species *in situ*. Chiral phosphorous ligands are used extensively in asymmetric transformations, while silver hexafluoroantimonate ($AgSbF_6$), silver triflate (AgOTf), and silver tetrafluoroborate ($AgBF_4$) are common additives to neutral Rh complexes to generate cationic species *in situ*.

2.1 Carbonyl Rhodium Complex

The most common homoleptic rhodium complexes are $Rh_4(CO)_{12}$ and $Rh_6(CO)_{16}$. $Rh_4(CO)_{12}$ can be used in catalysis directly or can serve as a catalyst precursor. It can also be used to synthesize other rhodium carbonyl clusters.

Preparation of Tri-μ-carbonyl-nonacarbonyltetrarhodium [$Rh_4(CO)_{12}$][2]

$$RhCl_3.3H_2O + 2\,Cu + 4\,CO + NaCl \longrightarrow Na[RhCl_2(CO)_2] + 2\,Cu(CO)Cl$$

$$4\,Na[RhCl_2(CO)_2] + 6\,CO + 2\,H_2O \longrightarrow Rh_4(CO)_{12} + 2\,CO_2 + 4\,NaCl + 4\,HCl$$

**Caution*: Carbon monoxide is highly toxic; the reaction must be carried out in a well-ventilated hood.

A two-necked, 1-L, round-bottomed flask, equipped with CO inlet, magnetic stir bar, and a dropping funnel was charged with activated copper powder (1.5 g, 24 mmol), sodium chloride 90.6 g, 10 mmol), and water (200 mL). The water was well degassed and saturated with CO. Under a stream of CO, the vigorously stirred mixture was treated with a degassed solution of rhodium(III) chloride trihydrate (2.6 g, 9.9 mmol) in water (50 mL) from the dropping funnel at a rate of about 1 mL/min. The mixture was allowed to stir for an additional 2 h. During this time, some $Rh_4(CO)_{12}$ began to separate as an orange powder. The funnel was charged with a degassed solution of disodium citrate in water (0.4 M, 50 mL), and the solution was added to the reaction mixture over a period of 50 min. The reaction mixture was allowed to stir for an additional 20 h. The resulting orange suspension was filtered under CO atmosphere. The solid was washed with water and then dried *in vacuo* at room temperature. The resulting solid was exracted with the minimum amount of CH_2Cl_2 (5×5 mL) under CO. The solution was quickly evaporated to dryness *in vacuo* on a water bath at room temperature, and the resulting crystals were dried under vacuum to give $Rh_4(CO)_{12}$ (1.55–1.75 g, 80–90%).

2.2 Phosphine Rhodium Complex

Since the discovery of Wilkinson's catalyst $RhCl(PPh_3)_3$ in 1966,[3] this Rh complex has been used in various catalytic reactions. It is a useful catalyst for hydrogenation of numerous olefins, and other analogous chiral complexes have been developed extensively for asymmetric hydrogenation. Wilkinson's catalyst is also a good catalyst for decarbonylation.[4] The stoichiometric reaction of $RhCl(PPh_3)_3$ with aldehydes affords the corresponding hydrocarbon and $RhCl(CO)(PPh_3)_2$.

Preparation of Tris(triphenylphosphine)chlororhodium(I), $RhCl(PPh_3)_3$[3]

$$RhCl_3.3H_2O + 4\ PPh_3 \longrightarrow RhCl(PPh_3)_3 + 2\ H_2O + OPPh_3 + 2\ HCl$$

A hot solution of triphenylphosphine (12 g, 46 mmol) in ethanol (350 mL) was added to a solution of rhodium(III) chloride trihydrate (2.0 g, 7.6 mmol) in 95% ethanol (70 mL) in a 500-mL, round-bottomed flask fitted with nitrogen inlet, reflux condenser, and a gas exit bubbler. After heating the solution at reflux for 2 h, the hot solution was filtered to give the crystalline product, which was washed with anhydrous ether and dried under vacuum to yield $RhCl(PPh_3)_3$ (6.25 g, 88% based on Rh) as an orange-red solid.

Preparation of Chlorocarbonylbis(triphenylphosphine)rhodium(1), $RhCl(CO)(PPh_3)_2$[5]

$$RhCl_3.3H_2O + 3\ PPh_3 + HCHO \longrightarrow$$

$$RhCl(CO)(PPh_3)_2 + 2\ H_2O + OPPh_3 + 2\ HCl + H_2$$

**Caution*: Carbon monoxide is highly toxic; the reaction must be carried out in a well-ventilated hood.

Rhodium (III) chloride 3-hydrate (2 g, 0.0076 mol) in 70 mL of absolute ethanol was slowly added to 300 mL of boiling absolute ethanol containing a twofold excess of triphenylphosphine (7.2 g, 0.0275 mol). The solution, which was turbid, became clear in about 2 min. Then, sufficient (10–20 mL) 37% formaldehyde solution was added to cause the red solution to become pale yellow in about 1 min, and yellow microcrystals precipitated. After cooling, the collected crystals were washed with ethanol and ethyl ether and dried in air or vacuum. They can be recrystallized from the minimum of hot benzene. Yield: 4.5 g (85% based on $RhCl_3 \cdot H_2O$).

2.3 Neutral Rhodium Complex

Preparation of Dichlorotetracarbonyldirhodium(I), $Rh_2(CO)_4Cl_2$[6]

$$2\ RhCl_3.3H_2O + 6\ CO \longrightarrow Rh_2(CO)_4Cl_2 + 6\ H_2O + 2\ COCl_2$$

**Caution*: Carbon monoxide is highly toxic; the reaction must be carried out in a well-ventilated hood.

A gas washing bottle was charged with pulverized rhodium(III) chloride trihydrate (11.0 g, 0.042 mol).The apparatus was then flushed with CO and lowered into a pre-heated oil bath at 100 °C.

The condensed water at the top was removed with absorbent cotton. The reaction was heated for 3–5 h to completion. The crude product yield was 8.3 g (96%) as an orange solid, which is pure enough for most purposes, but it may be further purified by recrystallization from hexane or sublimation at 80 °C (0.1 mmHg).

Preparation of Di-μ-chloro-bis(η^4-1,5-cyclooctadiene)-dirhodium(I), $[Rh(C_8H_{12})Cl]_2$[7]

$$2\ RhCl_3.3H_2O + 2\ C_8H_{12} + 2\ CH_3CH_2OH + 2\ Na_2CO_3$$
$$\longrightarrow [Rh(C_8H_{12})Cl]_2 + 2\ CH_3CHO + 4\ NaCl + 2\ CO_2 + 8\ H_2O$$

A 100-mL, two-necked, round-bottomed flask was fitted with a reflux condenser connected to a nitrogen bubbler. The flask was charged with rhodium(III) chloride trihydrate (2.0 g, 7.6 mmol) and sodium carbonate decahydrate (2.2 g, 7.7 mmol) under nitrogen. The degassed ethanol-water (5:1, 3 mL) and 1,5-cyclooctadiene (cod) (3 mL) were added and the mixture was then heated at reflux with stirring for 18 h. The mixture was cooled and immediately filtered and the product was washed with pentane and then with methanol-water (1:5). The product was dried *in vacuo* to afford $[Rh(C_8H_{12})Cl]_2$ (1.67 g, 94%) as a yellow-orange solid.

*Off-color (olive-green) product sometime would be obtained in the presence of sodium carbonate. To purify, dissolve the off-color solid in dichloromethane-ether and treat the solution with norite, and then heat the mixture to boiling for 15 min. Filter the hot solution through a bed of Celite and concentrate the solution *in vacuo* to afford the expected product.

2.4 Cationic Rhodium Complex

Synthesis of Bis(1,5-cyclooctadiene)rhodium Tetrafluoroborate, [Rh(cod)$_2$]BF$_4$[8]

Triphenylmethyl tetrafluoroborate (2.98 g, 9.1 mmol) in dichloromethane (40 mL) was added dropwise to a solution of acetylacetonato(1,5-cyclooctadiene)rhodium (2.8 g, 9.1 mmol) and 1,5-cyclooctadiene (5 mL excess) in dichloromethane (20 mL) at 25 °C. The yellow solution turned deep red, and was then stirred for 5 min. The solution was added to diethyl ether (400 mL) and the resulting orange-red crystals of bis(cyclo-octa-1,5-diene)rhodium tetrafluoroborate (3.44 g, 94%) were collected.

Synthesis of Bis(acetonitrile)(1,5-cyclooctadiene)rhodium Tetrafluoroborate, [Rh(CN)$_2$(cod)]BF$_4$[8]

Bis(1,5-cyclooctadiene)rhodium tetrafluoroborate (1.4 g, 3.45 mmol) was dissolved in methylene chloride (20 mL), and acetonitrile (5 mL) was added dropwise to the stirred solution. The resultant yellow solution was added to diethyl ether (100 mL) to give bright yellow crystals of *bis*(acetonitrile)(1,5-cyclooctadiene)rhodium tetrafluoroborate (1.15 g, 88%).

Synthesis of 1,2-bis(diphenylphosphino)ethane(1,5-cyclooctadiene)rhodium Tetrafluoroborate, [Rh(dppe)(cod)]BF$_4$[8]

A solution of 1,2-bis(diphenylphosphino)ethane (dppe) (0.064 g, 0.16 mmol) in methylene chloride (20 mL) was added to a stirred (25 °C) solution of [Rh(CN)$_2$(cod)]BF$_4$ (0.06 g, 0.158 mmol) in methylene chloride (10 mL). The solution was added to diethyl ether (80 mL) and the precipitate was recrystallized from methylene chloride-hexane to give orange-yellow crystals of 1,2-*bis*(diphenylphosphino)hane(1,5-cyclooctadiene) rhodium tetrafluoro-borate (0.07 g, 64%).

3. Hydroformylation

Since its discovery by Roelen in 1938,[9] hydroformylation or the "oxo process" has become one of the most important methods for the synthesis of aldehydes. Aldehydes produced *via* hydroformylation are critical feedstock and synthetic precursors in both basic research and industrial applications. The significance of this reaction in chemical industry has drawn much attention in both basic and applied research, leading to the development of more efficient and versatile reaction processes.

Hydroformylation is a reaction process wherein both hydrogen and a formyl group are introduced to an unsaturated bond, especially an olefinic bond. A typical example of this process is the reaction of 1-alkenes with carbon monoxide and hydrogen catalyzed by cobalt or rhodium complexes to give the corresponding homologous aldehydes. It should be noted that both linear and branched aldehyde products can be formed. Linear or branched product selectivity can be controlled by reaction conditions as well as the catalyst species (Scheme 2-1).

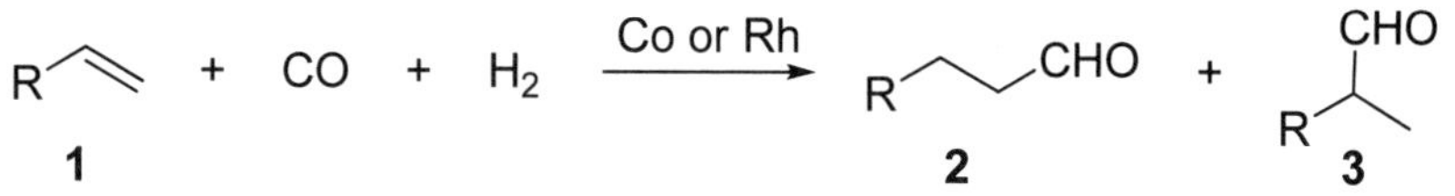

$$\underset{\mathbf{1}}{\text{R}\diagup\!\!\!\diagdown\!\!=} + CO + H_2 \xrightarrow{\text{Co or Rh}} \underset{\mathbf{2}}{\text{R}\diagup\!\diagdown\!\diagup\text{CHO}} + \underset{\mathbf{3}}{\text{R-CH(CHO)-CH}_3}$$

Scheme 2-1. The hydroformylation reaction.

Roelen found that ethylene, H_2, and CO were converted into propanal in the presence of $Co_2(CO)_8$ as catalyst, but at higher pressures, diethyl ketone was formed. He ran the reaction using high temperature (120–170 °C) and high pressure of carbon monoxide/hydrogen ("syn gas") (200–300 bar).[9] Hydroformylation later became a widely used industrial process to produce aldehydes, which can be hydrogenated to give alcohols or oxidized to the corresponding carboxylic acids. The most noteworthy commercial process is the conversion of propene to butanal, which can be subsequently hydrogenated to 1-butanol or converted to 2-ethylhexanol by self-aldol condensation.[10] 2-Ethylhexanol, a crucial intermediate for the production of ester-type plasticizers, is the most important bulk chemical produced by the "oxo process."[10b] Another important commercial application of hydroformylation is the production of long-chain alcohols from C_5 to C_{17} isomeric linear alkenes.[11] These long-chain alcohols are precursors for lubricants, plasticizers, and detergents.[11c] Cobalt was used initially to catalyze industrial hydroformylation reactions until rhodium catalysts were commercialized in the 1970s. Various rhodium complexes demonstrated much higher catalytic activity (10^3–10^4 times) than cobalt complexes.[12] Although the price of cobalt is much lower than that of rhodium, rhodium-catalyzed hydroformylation reactions require lower temperatures (50–80 °C) and pressure (10–50 atm) because of the high activity of rhodium catalysts, which are technologically as well as economically advantageous.[12]

Osborn, Wilkinson, and Young reported the use of Rh(I)/PPh_3 complexes to catalyze hydroformylation reactions.[13] Wilkinson's catalyst, $RhCl(PPh_3)_3$, was the initial Rh/PPh_3 catalyst system used to promote hydroformylation reactions, which served as the historical milestone for further catalyst design and development. Wilkinson *et al.* later developed a halide-free rhodium complex, $HRh(CO)(PPh_3)_3$, which was highly active, giving linear-selective hydroformylation products under mild conditions.[14] The hydroformylation reaction can be performed under mild conditions using a variety of functionalized alkenes.[15] Reactions in aqueous biphase[16] super critical carbon dioxide,[17] or fluorous biphase[18] emerged in response to separation and environmental issues. In fact, a highly efficient "oxo process" using a water soluble rhodium catalyst, $HRh(CO)(TPPTS)_3$ [TPPTS = $P(C_6H_4SO_3Na$-*m*$])_3$, in aqueous biphasic conditions was commercialized by Ruhrchemie/Rhône-Poulenc for the production of butanal.[19]

The mechanism of the Rh(I)-complex–catalyzed hydroformylation is analogous to the mechanism reported by Heck and Breslow[20] for Co-catalyzed hydroformylation. A general mechanism for hydroformylation is illustrated in

Figure 2-1.[21] First, an active Rh(I) catalyst (**A**) (16 electron species) is generated *via* a loss of one CO molecule from coordinatively saturated $HRh(CO)_2(PPh_3)_2$ (**B**). Then, coordination of an alkene substrate **4** to the active catalyst **A** occurs to form π-olefin-Rh complex **C** (18 electron species). This leads to the formation of alkyl-Rh complex **D** (16 electron species) followed by CO coordination to give saturated alkyl-Rh complex **E**. Migratory insertion of CO into the alkyl-Rh bond gives unsaturated acyl-Rh species **F**. Next, oxidative addition of molecular hydrogen to **F** gives acyl-Rh dihydride complex **G** (18 electron species). The final product, aldehyde **5**, is formed *via* reductive elimination and the active catalyst species **A** is regenerated.

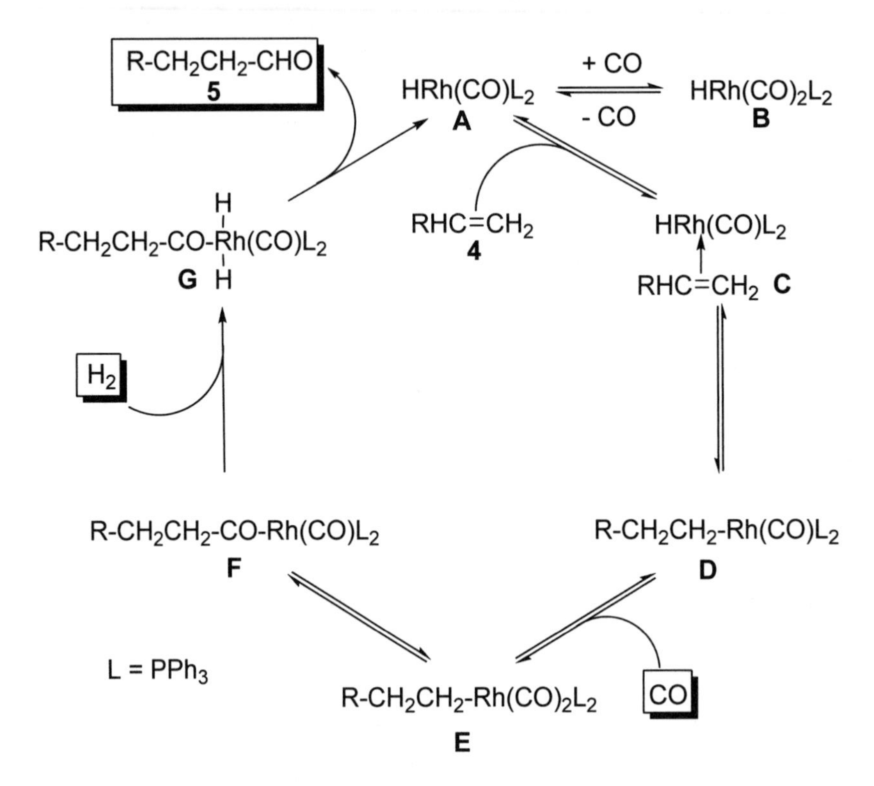

Figure 2-1. Mechanism for the hydroformylation reaction.
Reproduced with permissions from *Organic Reactions*. Ref. 12

Although Figure 2-1 only shows the linear aldehyde product formed from the linear alkyl-Rh and acyl-intermediates for simplicity, the mechanism for the formation of the branched aldehyde product follows the same mechanism. The generation of the branched alkyl-Rh intermediate, $R(CH_3)CH\text{-}Rh(CO)(PPh_3)_2$, leads to the formation of the branched aldehyde product. Studies have shown that excess triphenylphosphine decreases the reaction rate but increases the linear/branched aldehyde product ratio. In addition, excess triphenylphosphine suppresses both hydrogenation and isomerization of the olefin substrate. These results indicate that the regioselectivity of the product is determined during the formation of the alkyl-Rh complex **D** from the π-olefin-complex **C**.[21]

3.1 Highly Linear Selective Hydroformylation of Alkenes

Various catalyst systems with phosphine or phosphite ligands have been used to increase the linear to branched hydroformylation product ratio.[22] The phosphine and phosphate ligands are added in large excess to increase linear selectivity in the presence of rhodium catalysts such as $HRh(CO)(PPh_3)_3$ and $Rh(acac)(CO)_2$.[11a] In the 1960s, Pruett and Smith (Union Carbide) reported the regioselective hydroformylation of α-olefins such as 1-octene **6** catalyzed by Rh/C with various phosphorus ligands.[23] The data showed that using rhodium with excess phosphite ligand afforded a stable and active catalyst system at 6–8 atm and 90 °C. Later, van Leeuwen *et al.* reported the hydroformylation of 1-octene **6** catalyzed by a *tris*(2-*tert*-butyl-4-methylphenyl)phosphite -modified rhodium catalyst (Scheme 2-2).[24]

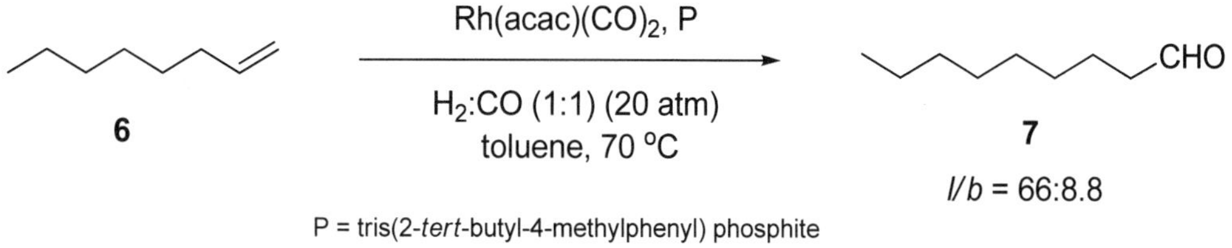

Rh(acac)(CO)$_2$, P

6 → **7** (CHO)

H_2:CO (1:1) (20 atm)
toluene, 70 °C

l/*b* = 66:8.8

P = tris(2-*tert*-butyl-4-methylphenyl) phosphite

Scheme 2-2. Hydroformylation of 1-octene.

Rhodium complexes of chelating phosphorus ligands, BISBI **8** and BIPHEPHOS **9**, can achieve linear to branched ratios of ≥66.5[25] and ≥40,[26] respectively, in the hydroformylation of 1-alkenes. The BISBI ligand, developed by Devon and co-workers, forms a 9-membered ring with the rhodium center.[25a] Furthermore, phosphine and phosphite ligands give higher linear-to-branched product ratios for the hydroformylation of propene ($l/b > 30.1$) compared with the commercial Rh/PPh_3 catalysts ($l/b = 8{:}1$). Other successful bisphosphine ligands that have been developed include NAPHOS **10**, developed by Herrmann and co-workers[27] and XANTPHOS **11**, developed by Kranenburg *et al.*[28] (Figure 2-2).

The linear-to-branched product ratios for the hydroformylation of 1-hexene **12** for various ligands was reported by van Leeuwen (Table 2-2). Both Casey and van Leeuwen have proposed and given good evidence that the regioselectivity in rhodium-diphosphine catalysts is partly related to the ability of the chelating diphosphine to maintain a chelate bite angle of ~120°. This is also referred to as the "bite angle hypothesis."[25b, 29]

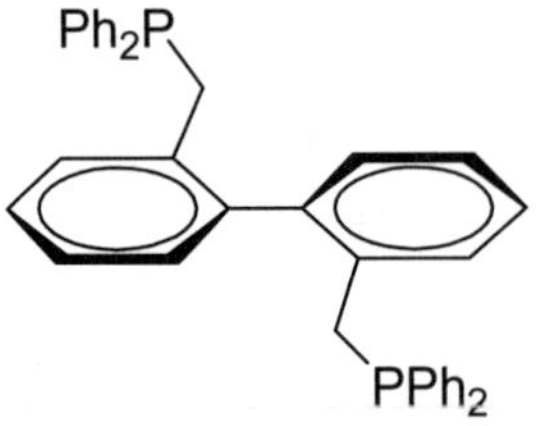

8, BISBI

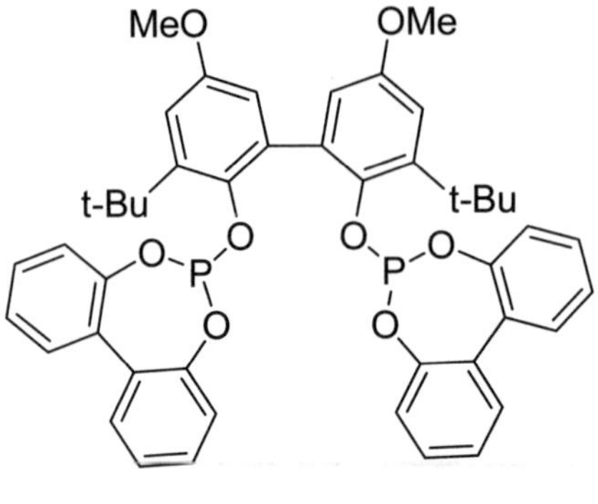

9, BIPHEPHOS

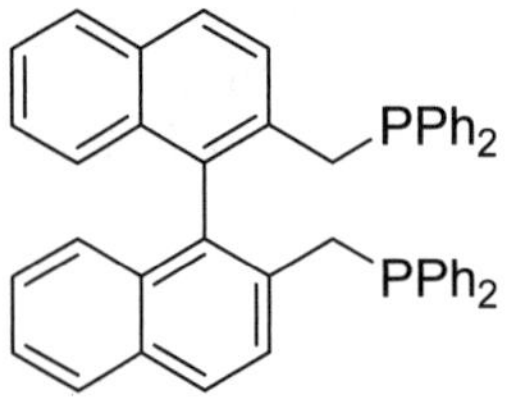

10, NAPHOS

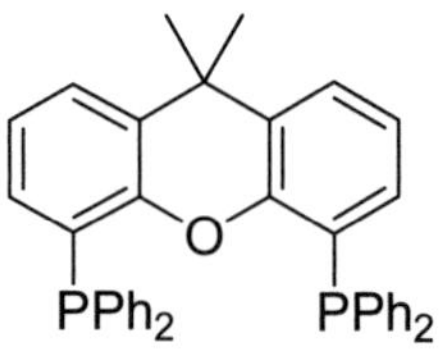

11, XANTPHOS

Figure 2-2. Diphosphine and diphosphite ligands for highly linear selective hydroformylation.

Table 2-1. Hydroformylation of 1-hexene in the presence of Rh-diphosphine catalysts.

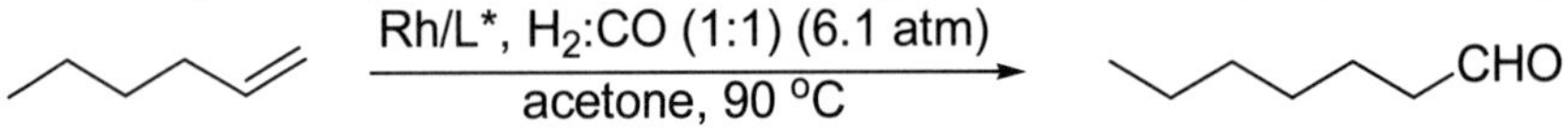

12 → 13: Rh/L*, H_2:CO (1:1) (6.1 atm), acetone, 90 °C

Entry	Catalyst	Init TOF (min^{-1})	Aldehyde linear:branch	% isomerization
1	Rh/PPh_3	13(1)	9:1	>0.5
2	Rh/BISBI	25(2)	70:1	>0.5
3	Rh/NAPHOS	27(1)	120:1	1.5
4	Rh/ XANTPHOS	13(2)	80:1	5.0

General procedure for the hydroformylation of 1-hexene[25b]

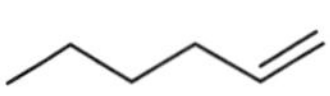

12 → 13: $Rh(acac)(CO)_2$, L, H_2:CO (1:1) (6 atm), benzene, 90 °C

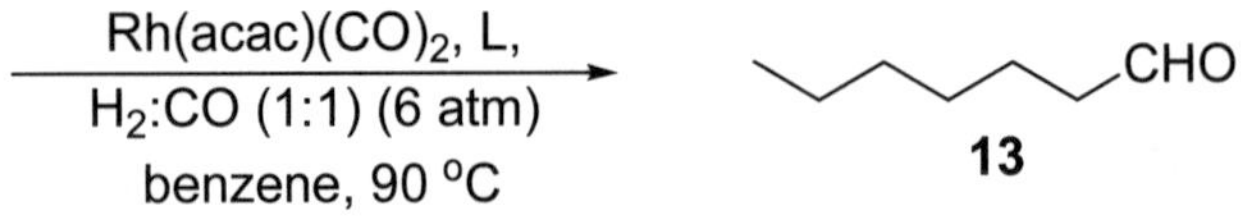

Hydroformylation reactions were performed in a 90-mL Fischer-Porter bottle equipped with a gas inlet valve, liquid sampling valve, and star-head magnetic stir bar. The pressure apparatus was immersed in a constant-temperature bath maintained at 33.6 ± 0.5 °C in a well-ventilated fume hood. A magnetic stirrer placed below the bath provided efficient stirring.

$Rh(acac)(CO)_2$ (7.9 mg, 0.031 mmol) and a chelating diphosphine (0.031 mmol) were placed in the pressure apparatus under nitrogen. The system was flushed with 6 atm of CO/H_2 three times and then pressurized to 6 atm with analyzed CO/H_2 (50.02% CO, 49.98% H_2). Benzene (6.0 mL) and toluene (internal GC standard, 0.20 mL, 1.9 mmol) were added by gastight syringe to the pressurized system. After 1 h of stirring, 1-hexene **12** (2.50 mL, 0.020 mmol) was added. The pressure of the system was maintained throughout the reaction by adding additional CO/H_2 periodically. Samples were removed *via* the liquid sampling valve for analysis. Heptanal and 2-methylhexanal were analyzed by temperature-programmed gas chromatography on an HP5890A chromatograph interfaced to a HP3390A integrator using a 10 m × 0.53 mm methyl silicone capillary column.

The regioselective hydroformylation of conjugated and nonconjugated dienes has been reported.[15b, 30] One example is the low-pressure *bis*-hydroformylation of 1,3-butadiene **14** to adipic aldehyde **16**.[30b] The reaction yielded adipic aldehyde with 50% selectivity over 11 other possible products. The structure of the bisphosphite ligand had the largest effect on the selectivity of this reaction.

Representative procedure for the catalytic hydroformylation of 1,3-butadiene[30b]

14 + 2 H_2 + 2 CO —[Rh], **15**→ **16**

15

$Rh(acac)(CO)_2$ (10.1 mg, 0.0391 mmol) and **15** (32.9 mg, 0.0460 mmol) were added to an autoclave followed by 15 mL of toluene in the glove box. The autoclave was sealed, removed from the glove box, and charged to 30 bar of syngas ($1/1H_2/CO$). The solution was stirred at 1000 rpm while being heated to 90 °C. Once the solution reached 90 °C, the autoclave was pressurized to 40 bar of syngas and the mixture stirred at this temperature and pressure for 1 h prior to the addition of 1,3-butadiene. After catalyst pre-formation, the temperature and pressure were adjusted to the desired values. In this experiment, the temperature was increased to 110 °C and the pressure vented to 40 bar of syngas. 1,3-Butadiene **14** (1.00 M in toluene, 5.0 mL, 5.0 mmol) was added to a Swagelok 304 L SS/DOT-3E 1800, Swagelock Co., Solon, OH, USA (ordering number 304 L-HDF2-40) double-ended cylinder in the glove box, and the cylinder was sealed under Ar. The cylinder was charged to 69 bar of Ar, attached to the autoclave, and added with this overpressure. An overpressure of 20–30 bar of Ar was typically used to add 1,3-butadiene to the reaction mixture. After the mixture was stirred for 1 h, an aliquot was removed for immediate GC analysis using the method described below. For reactions run at 90 °C, the reaction mixtures were typically stirred for 5 h.

In the case of nonconjugated dienes, the terminal olefin moiety reacts preferentially with the Rh catalyst system.[30a, 31] The hydroformylation of 4-vinylcyclohexene **17** results in a single aldehyde product **18** produced in quantitative yield.[30a]

General procedure for the hydroformylation of linear and cyclic mono- and diolefins[30a]

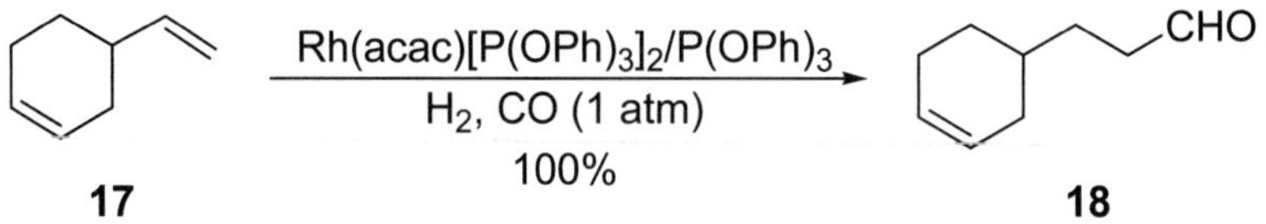

Rh(acac)[P(OPh)$_3$]$_2$/P(OPh)$_3$
H$_2$, CO (1 atm)
100%

17 **18**

CHO

Toluene (benzene) and olefin were carefully distilled before use. Hydroformylation reactions at 10 atm were carried out in a thermostated autoclave of 40 mL volume with a magnetic stirrer. Reactions at 1 atm were performed in a thermostated glass reactor. For each experiment, the amounts used were 2.1×10^{-5} mol of catalyst [Rh(acac){P(OPh)$_3$}$_2$ or Rh(acac)(CO)(PPh$_3$)], 3.2–3.3 × 10^{-5} mol of free ligand [P(OPh)$_3$ or PPh$_3$], and 3.0–4.5 × 10^{-3} mol of olefin in 0.6 mL of toluene or benzene. The reagents were introduced to the autoclave in nitrogen atmosphere and next filled with CO/H$_2$ mixture to required pressure. The reaction products were analyzed using ^{1}H nuclear magnetic resonance (NMR) () and gas chromatography-mass spectrometry (GC-MS) (Hewlett-Packard) instruments.

Highly regioselective hydroformylation of functionalized alkenes has been studied. The rhodium complex with BIPHEPHOS efficiently catalyzes the regioselective hydroformylation of a variety of functionalized terminal alkenes, giving the corresponding aldehydes (Scheme 2-3).[26b]

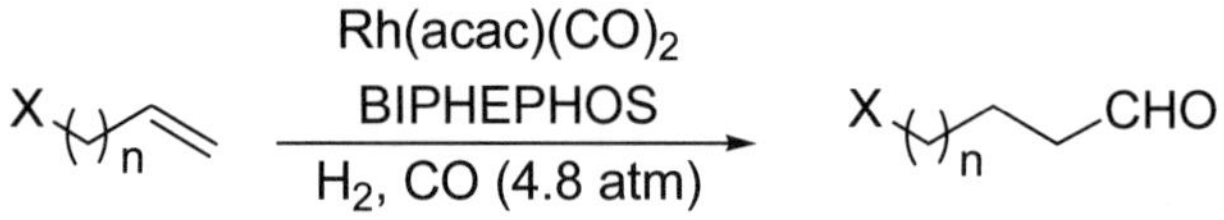

Rh(acac)(CO)$_2$
BIPHEPHOS
H$_2$, CO (4.8 atm)

X n X n CHO

19 **20**

l/b = 10 to >40

X = MeC(O), MeOC(O), BzOC(O), Et$_2$NC(O), (EtO)$_2$CH, (CH$_2$CO)$_2$N; n = 0 to 8

Scheme 2-3. Highly regioselective hydroformylation of functionalized alkenes.

6-Oxoheptanal[26b]

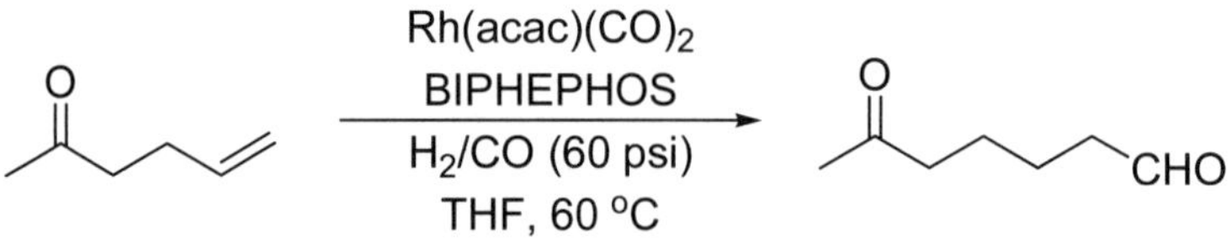

O O

Rh(acac)(CO)$_2$
BIPHEPHOS
H$_2$/CO (60 psi)
THF, 60 °C

CHO

21 **22**

Into a Fisher-Porter bottle, properly fitted with a pressure coupling closure complete with gas inlet, pressure gauge, and pressure release valve, was added (acetylacetonato)dicarbonylrhodium (28 mg, 0.109 mmol, 0.54 mol%), BIPHEPHOS (320 mg, 0.408 mmol), tetrahydrofuran (THF) (36 mL), and 5-hexen-2-one 21 (2.32 mL, 20 mmol). The reaction vessel was degassed three times and purged with CO/H_2 (1:1 mixture) and then heated at 60 °C for 18 h. After cooling to room temperature, the pressure was released and the solution concentrated. The resulting oil was purified by Kugelrohr distillation to give a colorless oil, 6-oxoheptanal 22 (2.21 g, 86% yield).

Phosphine and phosphite moieties exert strong directing effects on the regioselectivity of the hydroformylation of olefinic substrates. The phosphite moiety in cyclic alkenylphosphites allow for 100% regioselectivity and stereoselectivity after reduction of the intermediate aldehyde.[32] Phosphite directing groups are particularly useful for organic syntheses because they can be easily introduced and readily removed after regioselective hydroformylation.

General conditions for hydroformylation[32a]

23 → 24: $Rh_2(OAc)_4$; H_2, CO (34 atm); 80%

24 → 25: $LiAlH_4$; 90%

Benzene (6-17 mL), tetrakis(acetato)dirhodium(II), $[Rh(OAc)_2]_2$, and the substrate in the ratio 1:100 were placed in a 100-mL stainless steel Parr autoclave which was then pressurized with a 1:1 molar mixture of carbon monoxide and hydrogen to 500 psi and heated to temperatures between 30 and 100 °C for 5–90 h. The crude reaction mixture were analyzed the ^1H NMR and ^{13}P immediately after the reaction reached completion, and isomer ratios were determined where appropriate from ^1H NMR spectral data. An alternative method for some reactions involved using benzene (6–10 mL), dodecacarbonyltetrarhodium, $Rh_4(CO)_{12}$, and the substrate in the ratio 1:200 in the autoclave which was then pressurized with a 1:1 molar mixture of carbon monoxide and hydrogen to 500 psi and heated to temperatures between 40 and 70 °C for 22 h.

Reduction to diols

A modified procedure by Sander[33] was used for the reactions of the crude hydroformylation products with lithium aluminum hydride in ether, and a minimum amount of water was used during isolation to maximize recovery of diols. The overall yields which are reported are for the two steps, hydroformylation and reduction.

The rhodium-catalyzed hydroformylation of unsaturated alcohols is useful for the synthesis of important intermediates in organic synthesis.[34] The hydroformylation of homoallylic alcohols has been shown to give isolable lactols, which can be further oxidized to the corresponding lactones.[35]

General procedure[35]

BnO OH [Rh(OAc)2]2, PPh3 / H2/CO (24 atm) → BnO H O OH

26 **27**

PCC / 86% → BnO H O O

28

A steel bomb was charged with alcohol **26** (310 g, 12.8 mmol), triphenylphosphine (2.0 g, 7.55 mmol), $Rh_2(OAc)_4$ (40 mg), and 10 mL of ethyl acetate. The bomb was pressurized with a 1:1 mixture of CO/H_2 to 350 psi and heated to 100 °C for 6 h. After cooling, the contents were removed from the bomb, the ethyl acetate was removed under reduced pressure and the crude hemiacetal **27** was dissolved in CH_2Cl_2 and treated with PCC (5.40 g, 25.0 mmol). The mixture was stirred for 3 h and the product was isolated by filtration through silica gel with ether. Flash chromatography on silica with 20% EtOAc/hexane gave the pure lactone **28** in 86% yield.

The hydroformylation of alkenyl esters gives the corresponding aldehydes in moderate yields with high regioselectivity. An example is the hydroformylation of vinyl acetate **29,** which gives the branched aldehyde product **30** in high yield and high regioselectivity.[36]

General procedure for hydroformylation[36b]

OAc Rh/L / H2/CO (40 atm) / toluene, 60 °C / 95% → CHO OAc L = O P OPh

29 **30** **31**

A Schlenk flask (25 mL) containing a magnetic stirring bar was charged with ligand **31** (0.01 mmol), $Rh(acac)(CO)_2$ (1.00 mg, 0.0038 mmol), and solvent (5 mL). The reaction mixture was stirred for 60 min and then the substrate **29** (10.8 mmol) was added. The yellow reaction mixture was transferred under argon to a Parr autoclave (25 mL). The autoclave was purged three times with syngas and subsequently charged with syngas (CO/H_2 1:1, 50 atm). The autoclave was warmed up to 80 °C and the reaction was started by stirring. After 3 h, the autoclave was cooled and the gases were carefully released in a well-ventilated hood. The reaction mixture was analyzed by GC to determine level of conversion and regioselectivity of the product **30**.

Various α-formyl esters can be generated from the hydroformylation of α,β-unsaturated esters and diesters.[37] Ethyl 2-formylpropanoate **33** can be obtained from the hydroformylation of ethyl acrylate **32**. The reaction was catalyzed by the $Rh_2Cl_2(CO)_2$/phosphine/NEt_3 system at 20–40 °C in the presence of H_2/CO (1/1, 20 atm), and yielded the product with excellent regioselectivity (98–100%).[37a] Particularly effective ligands for this reaction include 1,4-bis(diphenylphosphino)butane (DPPB) **34**, phosphole (*o*-TDPP) **35**, and phosphanorbornadiene (DMTPPN) **36**.

Hydroformylation of ethyl acrylate[37a]

$[RhCl(CO)_2]_2$/L, 56-71%: **32** → **33**

DPPB **34**   *o*-TDPP **35**   DMTTPPN **36**

In a typical run, $[Rh(CO)_2Cl]_2$ and phosphane were weighed and introduced in the autoclave, which was then closed. The atmosphere was replaced by Ar and a toluene solution containing the required quantity of ethyl acrylate **32,** and triethylamine was introduced through the ball valve. This valve was closed and the autoclave was pressurized with CO and H_2 and heated to the required temperature in about 5 min. After the reaction time quoted in the Table,[37a] the autoclave was cooled to –40 °C in *ca.* 1 h in a well-ventilated hood. The reaction products were collected under argon into a Schlenk tube and analyzed by gas chromatography [3 m × 1/8" column 10% carbowax 20M on chromosorb 80-100 mesh] working at 150 °C and N_2 as carrier gas (flow rate 10 Lh^{-1}) with 1,3,5-trimethylbenzene as internal standard for aldehydes analysis, working at 65 °C [0.4 Lh^{-1} flow-rate of N_2] with 2,2-dimethoxypropane as internal standard for ethyl propanoate analysis.

The hydroformylation of glucal derivatives has been reported using $Rh_2[\mu\text{-}S(CH_2)_3NMe_2]_2(cod)_2$ or $Rh_2[\mu\text{-}OMe)_2(cod)_2]$ as the catalyst with $P(O\text{-}2\text{-}t\text{-}BuC_6H_4)_3$. This catalyst system yields products with high regioselectivity and stereoselectivity.[38]

General procedure for the hydroformylation of glucal derivatives[38b]

37 → 38: $Rh_2[\mu\text{-}S(CH_2)_3NMe_2]_2(COD)_2$, $P(O\text{-}2\text{-}t\text{-}BuC_6H_4)_3$, CO/H_2 (1/1, 75 atm), DCE, 120 °C, 64%

In a standard experiment, a solution of the substrate (5 mmol), the catalyst (0.05 mmol), and the phophorous co-catalyst (0.5 mmol) in 15 mL of the solvent was placed in the evacuated autoclave and heated while stirring. Once the system reached thermal equilibrium, the gas mixture was introduced to reach the working pressure. Small samples of the catalytic solution were taken at intervals for analysis. After each run, the system was allowed to cool and the gas vented. The reaction mixture was removed from the autoclave and immediately analyzed byFourier transform–infrared (FT-IR) spectroscopy to determine the metal carbonyl species and by gas chromatography and ^{1}H NMR spectroscopy to determine the conversion and the selectivity of the reaction.

3.2 Asymmetric Hydroformylation

Chiral nonracemic aldehydes can be easily converted into a variety of optically active compounds of chemical and biological interests. As such, asymmetric hydroformylation (AHF) is potentially useful for the preparation of pharmaceuticals.[39] Furthermore, chiral aldehydes can be readily transformed into a variety of value-added chiral chemicals, such as amines, alcohols, and imines, which are important industrial as well as laboratory chemicals.[40] Although AHF has many applications in fine-chemical and pharmaceutical industries, this reactionstill remains underdeveloped.[41] The lack of pursuit can be attributed to several problems: 1) low reaction rates at low temperature; 2) difficulties controlling both regioselectivities and enantioselectivities simultaneously; 3) chiral ligands are limited to specific substrate sets; 4) chiral ligand and catalyst loading must be low, because of their high costs; and 5) chiral aldehyde products may undergo racemization depending on the reaction conditions.[39–40, 42]

Only a few chiral ligands have been successfully applied to AHF. The most popular AHF substrates explored in the literature are simple olefins such as styrene, vinyl acetate and allyl cyanide.[41–43] There has also been some limited success with other substrates such as dihydrofurans using furanoside, as well as phosphine and phosphite based ligands.[43h, 44] Shown here are several examples of the most successful ligand systems for three fundamental substrates (Figure 2-3). The first ligands investigated in Rh-catalyzed-AHF were chiral phosphines, but these systems had little success.[45] Thus far, the BIPHEMPHOS (**43**), and the closely related analog BINAPHOS (**45**) have been the most versatile ligands affording high enatioselectivities for styrene and vinyl acetate (94 and 92% ee, respectively) and moderate enantioselectivity for allyl cyanide (69% ee), with modest branched:linear ratio (7.3–2.2:1).[43f, 43i, 43j]Ojima *et al.*developed a library of chiral monodentate phosphoramidite ligands (**53**) thatwasapplied to the asymmetric hydroformylation of allyl cyanide.[46] Ligand **53** gave an excellent branched to linear product ratio of 96:4 with good enantioselectivity.

More recently, it was demonstrated that the bisphosphine ligands diazaphospholane and Ph-BPE ligands are versatile hydroformylation ligands for all three popular substrates operating at typical hydroformylation temperatures of 50–60 °C (Figure 2-3).[42a, 43i] The diazaphospholane (**49**) ligand afforded enatioselectivities and regioselectivities as high as 82% ee with a 6.6 branched to linear ratio for styrene, 87% ee with 4.1 b:l ratio for allyl cyanide, and 96% ee with 37:1 b:l ratio for vinyl acetate.[40, 43i] The Ph-BPE (**41**) ligand afforded comparable enantioselectivities and regioselectivities of 94% ee with a 45:1 b:l ratio for styrene, 90% ee with a 7.1:1 b:l ration for allyl cyanide, and 82% ee with 340:1 b:l ratio for vinyl acetate.[40] Axtell *et al.*explored two other *bis*-phosphine ligands including TangPhos (**40**) and BINAPINE (**44**); all demonstrated higher enantioselectivities and regioselectivities for all substrates than with other biphosphine-based ligands.[42a] The diphosphite ligands Chiraphite (**46**) and the Furanoside-based ligand (**50**) have shown particular utility in the hydroformylation of styrene at moderate temperatures (20–35°C), affording enantioselectivites of greater than 75% and high regioselectivities (near 50:1).[40,43d,43i,j] Kelliphite, (**42**) a close relative of Chiraphite (**46**), has demonstrated better selectivity for hydroformylation of both allyl cyanide (up 75% ee, b:l 16:1) and vinyl acetate (up 87%, ee, b:l 56:1).[40, 42c, 43d, 43i] The ESPHOS (**39**) ligand system is highly selective for vinyl acetate but has shown little selectivity for styrene.[43j]

Of late, the use of the NOBIN-based (**47**) mixed phosphonite-phosphoramidite ligand system for hydroformylation for styrene and vinyl acetate afforded a 88:12 linear to branched ratio with up to 99% ee and a 93:7 branching ratio with up 96% up ee for styrene and vinyl acetate, respectively. Zhang demonstrated the use of the BINOL-based phosphonite ligand (**48**) system in the hydroformylation of vinyl acetate affording a 98:1 linear to branched ratio with up 80% ee. Wang and Buchwald reported the asymmetric hydroformylation of functionalized 1,1-disubstituted olefins using P-chirogenic ligands BenzP* (**51**) and QuinoxP* (**52**).[47] They prepared linear aldehydes with beta-chirality in up to 91% yield with high enantioselectivity (up to 94% ee) and good chemoselectivities and regioselectivities. Thus far, the bisphosphine-based ligand systems seem to be the most versatile and successful ligand systems for AHF (Figure 2-3).

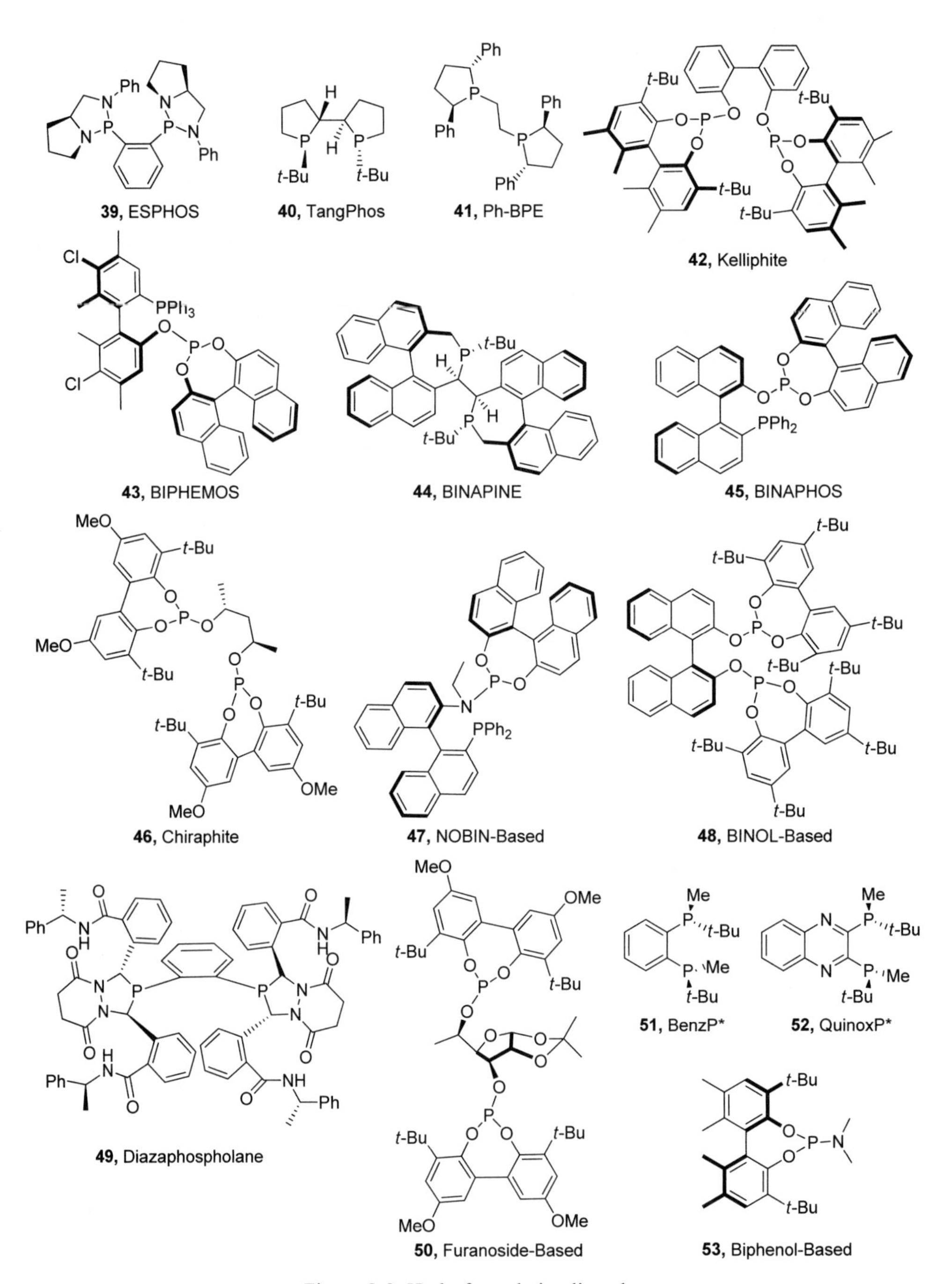

Figure 2-3. Hydroformylation ligands.

General procedure for the asymmetric hydroformylation of olefins catalyzed by rhodium(I) complexes of phosphine-phosphite ligands[43h]

A solution of styrene (2.08 g, 20.0 mmol), dicarbonyl(2,4-pentanedionato)rhodium (2.6 mg, 0.010 mmol), and (*R,S*)-BINAPHOS (31 mg, 0.040 mmol) in benzene (1 mL) was degassed by freeze-pump-thaw cycles and transferred into a 50-mL stainless steel autoclave. Carbon monoxide (50 atm) and dihydrogen (50 atm) were charged, and the solution was stirred at 60 °C for 43 h. Conversion to aldehydes (>99%) and the regioselectivity of the reaction (2-phenylpropanal/3-phenylpropanal = 88/12) were determined by ^{1}H NMR spectroscopy of the crude reaction mixture without evaporation of the solvent. Chromatography on silica gel followed by short pass distillation of the reaction mixture gave a pure sample of (*R*)-(+)-2-phenylpropanal. Optical rotation was used to determine the absolute configuration. The enantiomeric excess of the product was determined to be 94% ee by oxidation of (*R*)-(+)-2-phenylpropanalto the correspondingcarboxylic acid by Jones oxidation followed by gas–liquid chromatography (GLC) analysis using a chiral capillary column (Chrompack β-236M, 0.25 mm × 25 m, 135 °C, He 2 kg.cm^{-2}).

3.3 Tandem Hydroformylation Reactions

Tandem or one-pot syntheses are useful and time-saving transformations. In particular, tandem carbon–carbon bond forming reactions are useful, as they maximize atomic-economy and reduce waste.

Spiropyrans (**55a** and **55b**) and other related systems bearing quaternary centers are important synthons in a large class of natural products with both biological and pharmaceutical importance.[48] These natural products include several important antibiotics and pheromones. Eilbracht and co-workers developed a tandem hydroformylation/cyclization sequence under relatively mild conditions, which led selectively to several spiropyrans **55a** and **55b** (Table 2-2) from relatively easy to synthesize homoallylic alcohols **54**. Furthermore, little or no purification was required for this tandem series of reactions.

Table 2-2. Tandem hydroformylation cyclization of homoallyl alcohols.

OH
R2
R1
[RhCl(cod)]2
CO/H2
dioxane, 110 °C
39-95%
R2
O
R1
+
HO
O
R2
R1
54
55a
55b

Entry	R_1-R_2	Solvent	CO/H_2	Time (h)	Product **55a** (% yield)	R^3	Product **55b** (% yield)
1	-$(CH_2)_4$-	dioxane	90/20	65	39	-OH	15
2	-$(CH_2)_4$-	dioxane	30/20	65	43	-OH	—
3	-$(CH_2)_5$-	dioxane	30/20	65	39	-OH	48
4	-$(CH_2)_5$-	dioxane	30/20	20	64	-OH	—
5	-$(CH_2)_5$-	MeOH	30/20	20	-	-OCH_3	50
6	-$(CH_2)_6$-	dioxane	30/20	65	95	-OH	—
7	-$(CH_2)_7$-	dioxane	30/20	20	70	-OH	9
8	-$CH_2(CMe_2)$-CH_2-CH(Me)-	dioxane	30/20	65	69	-OH	—
9	-$(CH_2)_3$-$CH(CO_2Et)$-	dioxane	30/20	20	21	-OH	26
10	-(CH_2)-O-$(CH_2)_2$-	dioxane	30/20	20	61	-OH	—

Tandem hydroformylation cyclization of homoallyl alcohols[48b]

Homoallyic alcohol **54** (4.8 mmol) and $[Rh(cod)Cl]_2$ (1 mol%) was dissolved in 10 mL of anhydrous dioxane and placed into an autoclave. The reaction autoclave was then charged with 30 bars of carbon monoxide and 20 bars of hydrogen. The autoclave was then heated to 110 °C for 65 h. The crude product obtained after rotary evaporation of the solvent, after which it was filtered through neutral alumina with diethyl ether and further with ethanol as eluent. After evaporation of the solvent, the pure products **55a** and **55b** were obtained in 39–95% yield.

Constructing quaternary carbon centers with adjacent functional groups is a constant goal in the total synthesis arena, especially in the synthesis of complex natural products. Furthermore, development of reactions that generate quaternary centers both stereoselectively and regioselectively are even more in demand. Keränen and Eilbracht developed a sequential enolboration/hydroformylation/aldol addition one-pot casacade sequence resulting in the formation of a cyclic ketone **57** with a new quaternary center.

The Wittig olefination reaction is a key transformation among synthetic reactions as it allows for coupling of two complex fragments. Formation of the aldehyde *in situ* followed by direct one pot Wittig reaction would be an atom economical transformation. It was demonstrated that this process could be accomplished using starting olefin **58** affording product enoate **59**.

Sequential enolboration/hydroformylation/adol addition[49]

BCl(cy-hex)$_2$
Rh(acac)(CO)$_2$
XANTPHOS
CO/H$_2$ (60 bar)
16h

56 → **57**

Triethylamine (1.05 equiv) was pre-complexed under an Ar atmosphere with (*cy*-hex)$_2$BCl (1.05 equiv) in dry dichloromethane (5 mL) at 0 °C for 15 min. The unsaturated carbonyl compound **56** in approximately 1 mL of solvent was added slowly *via* syringe and the enolborane was allowed to stir for an additional 30 min. Before being transferred into the autoclave containing 0.9 mol% Rh(CO)$_2$(acac), 10–15 mL of solvent and 1.8 mol% of XANTPHOS were added. The autoclave was then pressurized to 60 bars with equal pressure of CO and H$_2$ and heated overnight to 80 °C. Upon cooling the autoclave to room temperature, the reaction mixture was removed and concentrated under reduced pressure. Enough MeOH was added to dissolve the solid residue (~25 mL) along with 2 mL of concentrated pH 7 phosphate buffer and 1 mL of 30% H$_2$O$_2$ and the reaction mixture was allowed to stir overnight before being extracted with ether (100 mL), washed with saturated aq NaHCO$_3$ (1 × 75 mL), dried and concentrated prior to further purification when necessary *via* flash choromatography or Kugelrohr distillation. The final product **57** was obtained in 89% yield.

Domino hydroformylation-Wittig olefination-hydrogenation[50]

Ph$_3$P=CMeC(O)OEt
[RhH(CO)(PPh$_3$)$_3$]
20 bar CO/H$_2$ (1:1)
toluene, 90 °C, 48 h

58 → **59**

A solution of [RhH(CO)(PPh$_3$)$_3$] (6.4 mg, 0.007 mmol) in toluene (3 mL) was added at room temperature to alkene **58** (1.0 mmol, 1.0 equiv) and the reaction mixture was stirred further for 5 min. Subsequently, the corresponding ylide (1.0–1.5 mmol, 1.0–1.5 equiv) was added and the reaction mixture was transferred with additional toluene (2 mL) into a stainless steel autoclave. The pressure was adjusted to 20 bars of synthesis gas (CO/H$_2$ 1:1) and heated to 90 °C until complete consumption of starting material was shown by TLC (1–3 days). After cooling to room temperature, the autoclave was depressurized and an appropriate amount of silica gel was added. The solvent was removed *in vacuo* to dryness. The residue was added onto a silica gel column (3 × 1.5 cm) and eluted with *tert*-butyl methyl ether (50–100 mL). After removal of solvent *in vacuo*, the crude product was analyzed by NMR to determine the diastereoselectivity of the reaction. The crude product was purified by flash chromatography to give the product **59** in 75% yield.

Low-molecular-weight aldehydes are important building blocks in the synthesis of chiral secondary amines. Hydroformlyation of simple alkenes can be used to generate chiral branched and linear aldehydes. Clarke and Roff used a phenylphosphatrioxa-adamantane phosphane ligand in the hydroformylation of dimethyl itaconate **60** affording excellent regioselectivity and chemoselectivity in product **62a**. This is the first hydroformylation reaction of 1,2-substituted unsaturated esters affording alpha-hydroformylated products in high selectivity. The use of the phosphane ligand dramatically reduced the formation of the hydrogenated side product **62b** and further increased the regioselectivity of the reaction (**62a**:**62c** = 92:1).[51]

Hydroformylation of unsaturated esters[51]

61, $[Rh(acac)(CO)_2]$, toluene, 50 bar CO/H_2, 50 °C, 20 h

60 → **62a** + **62b** (3%)

61

62c

62a:**62c** = 92/1

$Rh(acac)(CO)_2$ (0.2 mol%) and phosphane **61** (1 mol%) were placed in a clean, dry Schlenk tube. The air was displaced with nitrogen and toluene was added. The alkene **59** was added and the reaction mixture was transferred to an autoclave and stirred under syngas (1:1). The conversion was measured from NMR spectroscopy and products were purified *via* flash chromatography (SiO_2, hexane/EtOAc 4:1). The final product **60a** was obtained in 93% yield as a pale yellow oil.

Amino acids are important building blocks in the synthesis of natural products as well as pharmacologically active compounds. The use of one-pot reactions has been widely explored to reduce the environmental impact of industrial scale synthesis. Robinson and co-workers have reported a one-pot tandem hydroformylation/cyclization of eneamide **63** affording cyclic alpha amino-acids **64a** and **64b** with excellent regioselectivity and enantioselectivity. The enamide starting materials can be readily obtained by asymmetric hydrogen of dienamides with excellent regioselectivity and enantioselectivity using a rhodium Du-PHOS ligand system.[52] The R groups were typically either methyl or hydrogen with methyl giving generally higher selectivity ratios.[52] Similarly, Bates *et al.* reported the synthesis of pseudoconhydrine *via* a tandem hydroformylation/condensation of a tosyl-protected enamide to form the six-membered ring followed by stereoselective dihydroxylation.[53]

Hydroformylation of enamides[52]

Rh-BIPHEPHOS or $RhPPh_3$
CO/H_2 (1:1) (80-400 psi)
benzene, 80-100 °C
20-72 h

63 → **64a** + **64b**

up to 75% yield
conditions to afford exclusively
64a and **64b**

A mixture of substrate **63** (40 mg 0.22 mmol), rhodium(II) acetate dimer (1.0 mg, 2.2 mmol), and BIPHEPHOS (3.5 mg, 4.4 mmol) in deoxygenated benzene (5 mL) were placed in a autoclave that was then pressurized to 400 psi of CO/H_2 gas (1:1 molar ratio). After heating at 100 °C for 72 h, the autoclave was allowed to cool to room temperature and the solvent was removed under reduced pressure. Purification of the crude mixture on silica gel using ethyl acetate and light petroleum gave **64a** and **64b**.

A tandem hydroformylation/acetalization reaction of *p*-menthenic terpenes in the absence of an acid co-catalyst was reported by Vieira *et al.*[54] The reaction took place in ethanol solution in the presence of PPh_3 or P(O-*o*-^{t}BuPh)$_3$.

Typical procedure forthe tandem hydroformylation-acetalization reaction of terpenes[54]

[Rh]
CO/H_2
ethanol

65 → **66**

In a typical run, the solvent (15.0 mL) containing $[Rh(cod)(OMe)]_2$ (3.75×10^{-3} mmol), P(O-*o*-^{t}BuPh)$_2$ (7.5×10^{-3}–1.9×10^{-1} mmol), substrate **65** (3.0 mmol), and dodecane (1.5 mmol, internal standard) were transferred under argon into a stainless steel autoclave, which was pressurized to 40–80 atm (CO/H_2 = 1/2 to 2/1), placed in an oil bath (80–100 °C), and magnetically stirred. After the reaction was carried out and cooled to room temperature, the excess CO and H_2 were slowly vented. The products were separated by column chromatography (silica gel 60) using mixtures of hexane and CH_2Cl_2 as eluents and identified by GC-MS, ^{1}H, and ^{13}C-NMR. The product **66** was obtained in 90% yield.

Chercheja *et al.* reported a metal-catalyzed hydroformylation combined with an organocatalyzed stereoselective Mannich reaction in one pot.[55] *L*-Proline was used as the organocatalyst, and **69** was obtained in 52% yield and 71% ee.

Sequential hydroformylation and enantioselective Mannich reaction[55]

67 + 68 → 69

$[Rh(acac)CO_2]$, CO/H_2 (20/20 bar), L-proline, $P(OPh)_3$, CH_2Cl_2/acetone (4:1), 40 °C, 72 h

To a solution of $[Rh(acac)(CO)_2]$ (1.3 mg, 0.005 mmol, 0.005 equiv) in 10 mL of CH_2Cl_2/acetone (4:1) in a vial was added triphenylphosphite (6.2 mg, 0.02 mmol, 0.02 equiv). The solution was stirred with a magnetic stirrer for 5 min and then charged with alkene **67** (1.0 mmol, 1 equiv), aromatic amine **68** (1.1 mmol, 1.1 equiv), and L-proline (35 mg, 0.3 mmol, 0.3 equiv). The vial was transferred to the autoclave, pressurized with CO and H_2, and heated. After the reaction was completed, the autoclave was cooled down to room temperature, depressurized, flushed with Ar, and opened to obtain a sample for GC analysis. Then, the reaction mixture was filtered through a column filled with silica gel. Additionally, the column was washed with 50 mL of diethyl ether. The filtrate was concentrated under vacuum and the crude product was purified by column chromatography. The final product **69** was obtained in 52% yield.

4. Silylformylation

Silylformylation of olefins and alkynes can be regarded as the silicon version of hydroformylation. The reaction involves the concomitant introduction of a silyl group, derived from a hydrosilane, and a formyl group derived from insertion of carbon monoxide, thus producing functionalized olefins and dienes, which are useful synthons.[56]

The silylformylation of 1-alkynes has been studied extensively affording (*Z*)-1-silyl-2-formyl-1-alkenes high regioselectivity.[57] However, internal alkynes have not demonstrated high regioselectivity. Biffis *et al.* have reported using a cationic dirhodium(II) complex to achieve improved chemoselectivity for the silylformylation of internal alkynes.[58] In addition, intramolecular silylformylation reactions were developed to aid in the regioselectivity of silylformylation of internal alkynes.[56]

Silylformylation of terminal alkynes generally occurs under relatively mild conditions to afford a silylolefin. The major side product of this reaction is hydrosilylation.[57a, 57d] The use of more reactive organosilanes or an increase in the carbon monoxide pressure has been shown to enhance silylformylation.[57a, 57d] Furthermore, Doyle and Shankin demonstrated that a dirhodium complex can be used to overcome this synthetic problem. Dirhodium(II) perfluorobutyrate has been shown to give exclusively silylcarbonylation when used with phenylacetylene **70** favoring the *cis* product with *Z:E* ratios as high as 40:1 (**71a** and **71b**). Additional success with internal alkynes was afforded under these conditions. Aromatic substituted alkynes afforded better results than alkyl alkynes.[57a]

Silylformylation of terminal alkynes[57a]

$$\underset{\mathbf{70}}{Ph{-}C{\equiv}C{-}H} \xrightarrow[\substack{CH_2Cl_2 \\ 0\ ^{\circ}C,\ 5\ h}]{\substack{Rh_2(pfd)_4 \\ Et_3SiH,\ CO}} \underset{\mathbf{71a}}{(OHC)(Ph)C{=}C(H)(SiEt_3)} + \underset{\mathbf{71b}}{(OHC)(Ph)C{=}C(SiEt_3)(H)}$$

82%; *Z/E* = 10:1

Reactions were performed on 2.5-mmol scale by controlled addition (4–5 h) of the alkyne (**70**) in 5 mL of dichloromethane to the same amount of silane and 0.3 mol% of $Rh_2(pfd)_4$ in 25 mL of dicholormethane at 0 °C under an atmosphere of CO (ambient). The final products **71a** and **71b** were obtained in 82% combined yield (*Z/E* = 10:1) after chromatographic separation of catalyst and distillation of product under reduced pressure.

4.1 Tandem Silylformylation/Olefination

Highly functionalized olefins are easily accessible *via* silyformylation of alkynes. When coupled to other reaction systems, silylformylation reactions become even more powerful. Eilbracht *et al.* demonstrated that stabilized phosphorous ylides **74** could be trapped by β-silylated α-β-unsaturated aldehydes formed *via* silylformylation of alkynes in a one-pot synthesis. The tandem reaction proceeds with high yields for both alipathic and aromatic terminal alkynes. Protected propargyl alcohols also showed good reactivity; however, propargyl amines react with low selectivity. The reaction shown below is the best-afforded result for this process.[59]

Another example of a single-pot-type reaction involved replacement of the carbon nucleophile with aniline **76**, thus affording the α,β-unsaturated imine **77**. Both aromatic and alkyl amines afford the desired imine in good-to-excellent yields. The use of triethylsilanes afforded greater *Z:E* ratios in **77**.[60]

Tandem silylformylation/olefination and tandem silylformylation/imination[59, 60]

Ph_3P + CO_2Et; $Rh(COD)BPh_4$, CO (50 bar), DCM, 40 h, 91%; Bu H, CO_2Et, Si

72 **73** **74** **75**

A mixture of alkyne **72** (4.0 mmol), silane **73** (4.3 mmol), ylide **74** (6.0 mmol), and Rh catalyst (1 mol%, based on alkyne) in 10 mL of dry dichloromethane was placed in an autoclave. After flushing the autoclave with Ar, the reactor was pressurized to 50 bar with CO. The reaction mixture was magnetically stirred at 75 °C for 40 h. Then, the autoclave was allowed to cool to room temperature and the solvent was removed by rotary evaporation. The residue was dissolved in *n*-hexane/diethyl ether (2:1) and filtered through alumina. The product **75** was isolated in 91% yield.

NH_2, OMe; $Rh(COD)BH_4$, $SiEt_3H$, CO (50 bar), toluene, 60 °C, 22 h, 94%; Bu, H, N, $SiEt_3$, OMe

72 **76** **77**

Z/E >20:1

A mixture of alkyne **72** (5.0 mmol), silane (5.2 mmol), primary amine **76** (5.2 mmol), and Rh catalyst (1% based on the amount of silane) in 10 mL of dry toluene was placed in an autoclave. After flushing the autoclave with Ar, the reaction was pressurized with 50 bars with CO and the reaction mixture was magnetically stirred at 60 °C for 22 h. Then, the autoclave was allowed to cool to room temperature and the solvent was removed by rotary evaporation. The residue was dissolved in methyl *tert*-butyl ether and filtered through neutral alumina. The isolation of the product **76** was carried out *via* Kugelrohr distillation. The final product **77** was obtained in 94% yield.

4.2 Intramolecular Silylformylation of Alkynes

Intramolecular silylformylation reactions were developed to address the regioselectivity problems associated with intermolecular silylformylation of internal alkynes. Ojima *et al.* demonstrated that rhodium-catalyzed silylformylation of the ω-(dimethylsiloxy)-*i*-alkyne **78** afforded 3-*exo*-(formylmethylene)oxasilacylcoalkane **79** with complete regioselectivity. Both terminal and internal alkynes were employed in this process both achieving complete regioselectivity with moderate isolated yields ranging from 42% to 73%. Varying the chain length of the alkyne yielded different products including five- and six-member silicon heterocycles. In most cases, $Rh(acac)(CO)_2$ catalyst afforded the best results. Additionally, the mixed-metal catalyst $(^tBuNC)_4Rh_2Co_2(CO)_4$ also gave similar results.[61]

Synthesis of 3-exo-(formylmethylene)oxasilacylcoalkane[61]

$Rh(acac)(CO)_2$
CO (10 atm)
toluene, 65 °C, 3 h
96%

78 → **79**

To a 5-mL, round-bottomed flask containing catalyst (2.5×10^{-3} mmol) and a stirring bar, 1.0 mL of toluene was added under carbon monoxide pressure. A solution of 1-(dimethylsiloxy)alkyne **78** (0.50 mmol) in 2.0 mL of toluene was added to the reaction flask, which was then placed in a 300-mL stainless steel autoclave. Carbon monoxide was introduced to substitute the remaining air. After the carbon monoxide pressure was adjusted to 10 atm at room temperature, the autoclave was immersed into an oil bath and allowed to stir magnetically at 65–70 °C for 3–24 h. The reaction time required for completion was dependent on the nature of the catalyst and the substrate used. Then, carbon monoxide was carefully released and the reaction mixture was submitted to GC analysis. The solvent was removed under reduced pressure and the product **79** was purified by bulb-to-bulb distillation.

Silylformylation is a powerful synthetic tool for the synthesis of substituted-formylvinylsilanes. In 2001, the desymmetrization of dimethylsiloxyalkadiynes **80** to afford **81** as well as a novel sequential double silylformylation affording an unsaturated cyclooxasilanes (**82**) was reported.[62] The product afforded by single desymmetrization reaction can subsequently be converted to 1,3,5-triols by way of Tamao oxidation, with subsequent reduction of the resultant keto-aldehyde.[63] Additionally, it was found that use of phenyldimethylhydrosilane decreased the reaction time, whereas use of bulkier hydrosilanes, *e.g.*, *t*-butyldimethylsilane required both increased time and temperature.[62]

General procedure for the silylformylation and sequential double silylformylation of 4-dimethlsiloxy-1,6-heptadiyne[62]

$Rh(acac)(CO)_2$
CO (10 atm)
toluene, rt, 16 h

80 → **81** R = H, 98%
R = Me, 82%

$Rh(acac)(CO)_2$, Et_3SiH, CO (10 atm), toluene, rt, 16 h

80 → **82** R=H, 100%

In a 10-mL, round-bottomed flask under CO atmosphere, $Rh(acac)(CO)_2$ (0.48 mg, 0.0018 mmol) was dissolved in 2 mL of toluene, and a solution 4-dimethysiloxy-1-6-heptdiyne **80** (0.06 g, 0.36 mmol) in toluene (3 mL) was added. The flask was placed in a 300-mL stainless steel autoclave, pressurized with 10 atm of CO, and the reaction mixture stirred at room temperature for 16 h. The solvent was removed under vacuum to afford the desired product **81**.

In a 10-mL, round-bottomed flask under CO atmosphere, $Rh(acac)(CO)_2$ (0.48 mg, 0.0018 mmol) was dissolved in 2 mL of toluene, and a solution 4-dimethysiloxy-1-6-heptdiyne **80** (0.06 g, 0.36 mmol) in toluene (2 mL) and a silane (0.39 mmol, 1.08 equiv) were added. The flask was placed in a 300-mL stainless steel autoclave, pressurized with 10 atm of CO, and the reaction mixture stirred at room temperature for 16 h. The solvent was removed under vacuum to afford the desired product **82**.

5. Cyclopropanation

The cyclopropane ring is a common structural motif found in pyrethriods,[64] the antidepressant tranylcyclopromine,[65] papin and cystein protease inhibitors,[66] potential antipsychotic agents,[67] and anti-HIV compounds.[68] Thus, considerable attention has been paid to the stereoselective synthesis of cyclopropane-containing compounds.[57a, 57e, 69] The use of rhodium in cyclopropanation was made popular by contributions from Doyle and Davies. They have focused, over the years, on using chiral Rh(II) catalysts for asymmetric cyclopropanation *via* the decomposition of diazo compounds in the presence of olefins.[64, 69f–h, 70] Examples of the Rh(II)-catalyzed cyclopropanation of alkynes and allenes have also been reported in the literature.[71]

One example of the rare rhodium-catalyzed 1,6-addition affording a vinylcyclopropane was detailed by Doyle *et al.*[72] The strained vinylcyclopropane adduct **85** was the byproduct of a rhodium-catalyzed tandem carbocyclization of aryl boronate ester **83** bearing a Michael acceptor and strained alkene **84**. Modification and optimization of the conditions afforded the rare rhodium 1,6-addition adduct in excellent yield. Bidentate ligands such as dppf and 1,2-bis(diphenylphosphano)ethane (dppe) gave poor yield, whereas mondentate ligands performed much better under the conditions with the tri-*tert*-butylfluoroborate ligand, affording the best results. The most successful substrates were simple, strained bicyclic compounds. Nitrogen-substituted heterocycles did not fare well under these conditions, but several oxygen analogs afforded the vinylcyclopropane in good-to-excellent yield (79–97%).[72]

Typical Procedure[72]

CO_2t-Bu B(pin) **83** + **84** $[Rh(cod)Cl]_2$, t-$Bu_3PH^+BF_4^-$, KF, H_2O/dioxane, 84% → t-BuO_2C H H H **85**

A solution of 0.3 mL of water and 3 mL of dioxane in a 5-mL, two-necked, round-bottomed flask was purged with nitrogen and stirred for 10 min at 25 °C. $[Rh(cod)Cl]_2$ (3.0 mg, 0.006 mmol of Rh), tri-*tert*-butylphosphonium tetrafluoroborate (3.5 mg, 0.0132 mmol), and potassium fluoride (23.3 mg, 0.40 mmol) were added to the solution and stirred at 25 °C for 10 min. To the bright yellow solution was added the alkene **84**, followed by addition of the boronate ester **83** (56.0 mg, 0.20 mmol), and the reaction mixture was stirred at 80 °C for 3 h. The reaction was quenched with brine, and the aqueous layer was extracted with Et_2O (three times). The combined organic layers were dried with anhydrous $MgSO_4$, filtrated, and concentrated *in vacuo*. The final product **85** was obtained in 84% yield.

Several attempts at asymmetric cyclopropanation were made using a rhodium(II)-based catalyst. Nonetheless, little headway has been made in this area.[72, 73] However, Davies developed a rhodium(II) carboxlyate ligand system **88** that is better able to decompose vinyldiazo compounds such as **87** into vinylcarbenoids.[69c] Davies *et al.* demonstrated the rhodium(II) carboxylates are excellent systems for vinylcyclo-propanation. The proline-based ligand systems afforded the best results in most cases, giving both high enatioselectivity and good chemical yields. Several other alkene species were investigated, including *para*-chlorobenzene, allylacetate, allylethylether, butene, and isopentene. All the examined substrates gave comparable results to styrene at a low temperature.[69a] Doyle has also reported extensive work on the asymmetric enantioselective cyclopropanation reaction[36a, 70a, 74] and has applied this methodology to the synthesis of antidepressant milnacipran.[75]

Cyclopropanation procedure[69a]

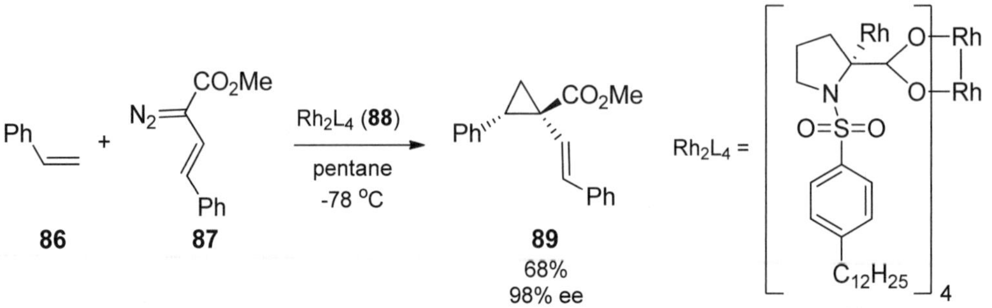

A mixture of the alkene **86** (1.2–20 equiv) and Rh(II) catalyst (0.01 equiv) in CH_2Cl_2 or pentane was stirred at room temperatureunder an Ar atmosphere. To this solution was added the vinyl diazo compound **87** (1 equiv, 0.12 M) in CH_2Cl_2 or pentane over 10 min, and the mixture was stirred for 1–8 h. The mixture was then concentrated *in vacuo*, and the residue was purified on silica using ether/petroleum ether as the eluent. In reactions carried out at –78 °C, the diazo compound was added over 30 min and the reaction was the maintained at –78°C for 24–36 h. Other compounds were prepared from alkenes obtained as gases by condensing a large excess of alkene with a dry ice/acetone cup condenser into a chilled (0 °C) solvent/catalyst solution, followed by addition of the diazo compound and warming to room temperature, and the reaction was worked up as above. Enantiomeric excesses (% ee) were determined by ^{1}H NMR at 200 or 500 MHz using *tris*[3-[(heptafluoropropyl)hydroxymethylene]-(-)-camphorato]praseodymium(III) derivative (0.10–0.35 equiv) and integration of the split signals because of the methoxy or the vinyl group, or by high-profile liquid chromatography (HPLC) using a Diacel Chiralcel OJ (Chiral Technologies, Illkirch – Cedex, France) analytical column.

The role of aziridine rings has grown over the years primarily because they allow easy access to simple chiral amines. However, their synthesis, particularly by catalytic asymmetric methods, has been somewhat more problematic than other members of the same family, *e.g.*, epoxides andcyclopropanes.[76] Krumper*et al.* demonstrated that the aziridine ring can be synthesized by monomeric *bis*-oxazoline rhodium(II) catalyst system **92** *via* carbene transfer to imines.The results demonstrated that this catalyst system can afford aziridines **93a** and **93b** in good yield, with reasonable diastereoselectivity, although the enantioselectivity leaves much to be desired. Both alkyl and aryl R-groups demonstrated good efficacy in this reaction system, with phenyl as either R group affording high yields and good selectivities. The high selectivity was seen in the R_1 = *p*-nitrophenyl (R_2 = phenyl), but the yield was much lower than the alkyl and aryl systems.

Aziridination procedure[76d]

In a glove box, the catalyst **92** (79 mg, 0.12 mmol), imine **90** (430 mg, 2.3 mmol), and ethyl diazoacetate **91** (0.300 mL, 2.8 mmol) were dissolved in THF (5 mL) in a glass vial equipped with a Teflon-coated stirbar. The vial was capped and the reaction mixture was stirred at room temperature for 8 h. The reaction vessel was uncapped occasionally (roughly once every 2 h) to prevent the buildup of a significant overpressure of nitrogen. After 8 h of stirring, the vial was brought out of the box and the reaction mixture was immediatedly chromatographed on silica gel (gradient elution: hexane/2% ethyl acetate in hexane/4% ethyl acetate in hexane).

The first intramolecular cyclopropanation was reported in 1961 by Stork using catalytic cyclization of a diazoketone.[77] Since the discovery of this reaction, all examples of intramolecular cyclopropanation have formed bicylco[3.1.0] and bicylo[4.1.0] ring systems with preference over higher and lower homologs. In the process of examining a synthetic route to optically pure presqualene, Doyle and co-workers serendipitously discovered the formation of macrocycle **95** *via* an intramolecular metal carbene reaction. The 13-membered-marcolide cyclopropane-fused lacatone **95** was obtained in good yield using dirhodium(II) tetraacetate. The use of other dirhodium catalysts such as dirhodium carboxiamides, dirhodium caprolactamate gave the expected [3.1.0]bicylo compound **96** in excellent yield. The dirhodium(II) tetracetate was applied to several other derivatives of the farnesyl diazoacetate **94**, affording the corresponding marcolides in fair yields (35–48%).[78]

General procedure[78]

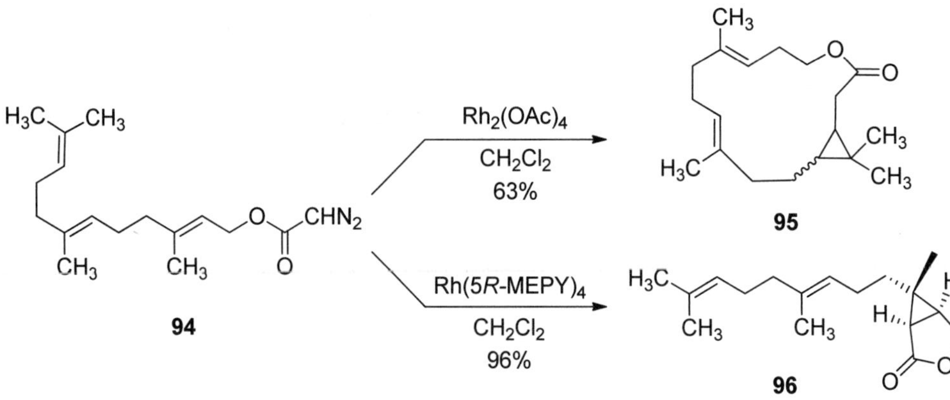

To a solution of Rh catalyst (11 mg, 10 μmol) at room temperature in 20 mL of anhydrous CH_2Cl_2 was added by syringe pump to a solution of *trans, trans*-farnesyl diazoacetete **94** (0.294 g, 1.00 mmol) in 5.0 mL of CH_2Cl_2 over a period of 12 h. After addition was complete, the reaction mixture was filtered through a 3-cm plug of silica to separate the catalyst, the plug was washed three times with 5-mL portions of CH_2Cl_2, and then GC analysis was performed. Chromatographic purification of the reaction mixture on silica (49:1 hexanes:ethyl acetate) afforded the product **95** in 63% yield.

6. Cyclohydrocarbonlyation

Cyclohydrocarbonylation (CHC) is the hydroformylation of a functionalized olefin followed by concomitant intramolecular nucleophillic attack to the newly formed aldehyde moiety leading to a cyclized product.[56] As a variant, the CHC reaction also includes an intramolecular cascade process involving the hydrocarbonylation of a functional alkene, generating an acyl-metal intermediate, which undergoes an intramolecular nucleophilic attack to give the corresponding cyclic compound.[79] CHC reactions have been developed into sophisticated cascade reactions forming bicylic and polycyclic compounds.[79, 80]

Rh-catalyzed cyclohydrocarbonylation of functionalized homoallyic amine has been successfully applied to the synthesis of (+)-prosopinine (**97**) and (-)-deoxyprosophyline (**98**) (Figure 2-4). Both (+)-prosopine and (-)-deoxyprosophyline have been shown to possess both antibiotic and anesthetic properties.[81] This synthesis uses amino acid derivatives that are subjected to CHC conditions to generate the desired substituted piperidine intermediates. The piperidine intermediates are then further converted to the desired alkaloid by subsequent diastereoselective introduction of the alkyl chains at the C-6 position.

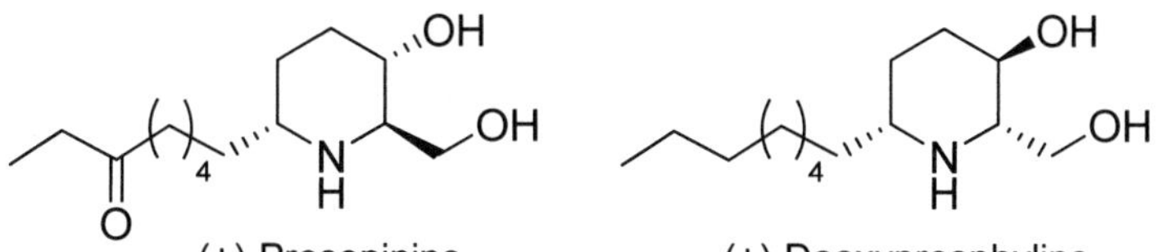

(+)-Prosopinine **97** (+)-Deoxyprosphyline **98**

Figure 2-4. Cyclohydrocarbonylation toward prosopinine and deoxyprosphyline.

Cyclohydrocarbonylation toward prosopinine and deoxyprosphyline[81]

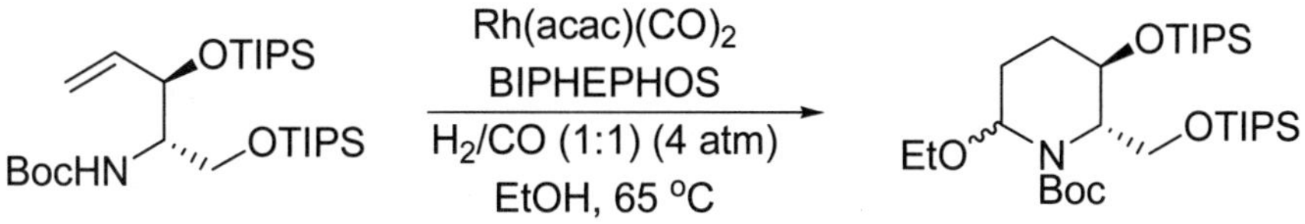

99 → Rh(acac)$(CO)_2$, BIPHEPHOS, H_2/CO (1:1) (4 atm), EtOH, 65 °C, 96% → **100**

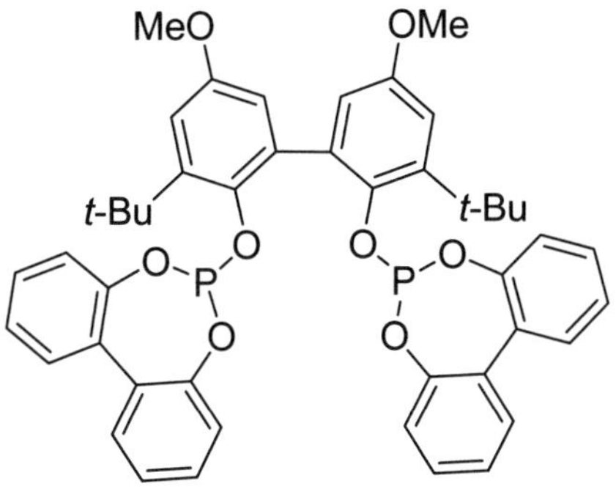

9, BIPHEPHOS

A solution of (3*S*,4*R*)-4-(*tert*-butoxycarbonylamino)-3,5-*bis*-(triisopropylsiloxy)pent-1-ene **99** (142 mg, 0.32 mmol) in ethanol (1 mL) under nitrogen was transferred *via* syringe to a 10-mL, round-bottomed flask equipped with a magnetic stirring bar, containing a solution of Rh(acac)$(CO)_2$ (0.82 mg, 0.0032 mmol) and BIPHEPHOS (**9**) (4.92 mg, 0.0064 mmol) in THF (1 mL) under nitrogen. The reaction flask was placed in a 300-mL stainless steel autoclave and pressurized with 2 atm of carbon monoxide and 2 atm of hydrogen. The autoclave was heated to 65 °C with stirring in an oil bath for 14 h. Then, the autoclave was cooled to room temperature and the gases were carefully released. The solvent was removed on a rotary evaporator and the residue was chromatographed on a silica gel column using hexane/EtOAc (8/1) as the eluent affording the product **100** (96% yield) as a colorless oil.

Ojima *et al.* employed the CHC reaction in the synthesis of enantiopure homokainoids.[80b] Homokainoids are homologs of kainic acid, a naturally occurring potent glutamate receptor agonist. These receptors have attracted considerable attention because of their potential clinical applications in the treatment of many CNS related disorders, both chronic and acute.[82] The synthesis of the intermediate, didehydro-piperidine, was accomplished by CHC reaction in the presence of BIPHEPHOS under relatively mild conditions in high yield with excellent enantiopurity. Both of the β-substituted derivatives were synthesized in good yield with the phenyl derivative requiring longer reaction time under similar conditions.[80b]

Cyclohydrocarbonylation toward homokainoids[80b]

$$\mathbf{101} \xrightarrow[\text{toluene, 75 °C; 99\%}]{\text{Rh(acac)(CO)}_2\text{, BIPHEPHOS, H}_2\text{/CO (1:1) (4 atm)}} \mathbf{102}$$

In a 5-mL, round-bottomed flask, $Rh(acac)(CO)_2$ (1.9 mg, 7.3 μmol, 0.25 mol%), and BIPHEPHOS (11.6 mg 0.0148 mmol 0.5 mol%) were dissolved in toluene (1 mL) under nitrogen. The resulting catalyst solution was degassed by the freeze-thaw procedure at least three times. Methyl (3*S*,4*S*)-5-acetyloxy-4-*t*-butoxycarbonylamino-3-ethenylpentanoate (**101**) (938 mg, 2.95 mmol) was placed in a 100-mL flask. The catalyst solution was transferred to the reaction flask containing (3*S*,4*S*)-5-acetyloxy-4-*t*-butoxycarbonylamino-3-ethenylpentanoate (**101**) by a pipette, and then the total volume was adjusted to 50 mL. The reaction flask was placed in a 300-mL stainless steel autoclave. The autoclave was pressurized with CO (2 atm) followed by H_2 (2 atm). The reaction mixture was stirred at 75 °C for 20–24 h. The reaction was monitored by TLC using EtOAC/*n*-hexane (1:3) as eluent. Upon completion of the reaction, the reaction mixture was concentrated under reduced pressure to give the crude residue. The residue was purified by flash chromatography on silica gel using EtOAc/*n*-hexaneas eluant to give the desired product **102** in 99% yield.

Ojima *et al.* also applied the cyclohydrocarbonylation reaction to the synthesis of natural products crispine A and harmicine.[83] The cyclohydrocarbonylation-bicyclization of *N*-allylic amides of arylacetic acids yielded the tricyclic indolizine alkaloids in good yields. Taddei *et al.* have also applied the cyclohydrocarbonylation reaction toward the synthesis of poly-substituted piperidines.[84] The cyclohydrocarbonylation reaction of enantiopure homoallylamines afforded the corresponding perhydrooxazolo pyridines bearing and oxazolidine. The oxazolidine ring was opened using various nucleophiles to afford the enantiomerically pure piperidines.

General procedure for the cyclohydrocarbonylation reaction[80a]

$$\mathbf{103} \xrightarrow[\text{H}_2\text{/CO (1:1) (4 atm), } p\text{-TSA, toluene, 65 °C}]{\text{Rh(acac)(CO)}_2\text{, BIPHEPHOS}} \mathbf{104}$$

X = O, S, NBoc, m = 1 or 2 n = 0 or 1
R_2 = CO_2Me, CH_2OBn, Me, H

Into a 10-mL, round-bottomed reaction flask were placed $Rh(acac)(CO)_2$ (1.9 mg, 0.0073 mmol, 2 mol%) and BIPHEPHOS (11.6 mg 0.0148 mmol, 4 mol%). The atmosphere of the reaction flask was replaced with nitrogen followed by the addition of degassed toluene (1.0 mL). The resulting solution was allowed to stir until it became homogeneous. The dipeptide substrate **103** (120 mg, 0.363 mmol) and TsOH (7 mg, 0.036 mmol) were placed into a 25-mL reaction flask. The catalyst solution was transferred to the reaction flask *via* syringe and the total volume was adjusted to 6.0 mL (0.06 M). The reaction flask was placed into a 300*mL stainless steel autoclave, and the autoclave was flushed with CO several times. The autoclave was filled with 2 atm of CO, followed by 2 atm of H_2. The autoclave was heated at 65 °C with stirring for 20–24 h. Then, the autoclave was cooled to room temperature and the gases were slowly and carefully released. The reaction mixture was concentrated under reduced pressure to give a viscous oil. The crude oil was purified by flash chromatography on silica gel using EtOAc/*n*-hexane as the eluent to give **104**.

Pyrrolinones are important synthons in the preparation of many important compounds including amino acids, various alkoids and many natural products. These heterocyclic derivatives have also been shown to express anti-tumor properties. In addition, they are inhibitors of cyclooxygenase-2 (COX-2), and HIV-1 protease. They also possess several important and exploitable medicinal properties. Pyrrolinones can be synthesized from functionalized imino-alkynes *via* cyclohhydrocarbonylation in the presence of a zwitterionic rhodium catalyst (ZW) **106** with both high regioselectivity and chemoselectivity as well as in good overall yield (Table 2-3). This versatile reaction system allows for easy access to important synthons.[80c]

Table 2-3. Cyclohydrocarbonylation of iminoynes.

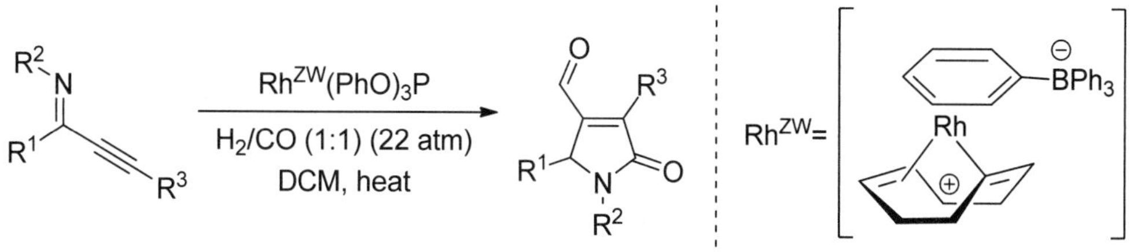

R^2, N, R^1, R^3 — **105** → $Rh^{ZW}(PhO)_3P$; H_2/CO (1:1) (22 atm); DCM, heat → O, R^3, R^1, N, O, R^2 — **107**

Rh^{ZW}= [$BPh_3^{\ominus}$, Rh, ⊕] **106**

Entry	Starting Material	T (°C)	Time (h)	Product	Isolated Yield (%)
a	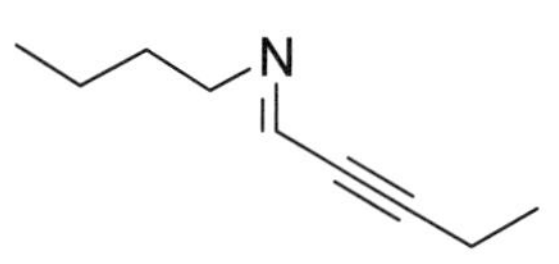 N	100	18	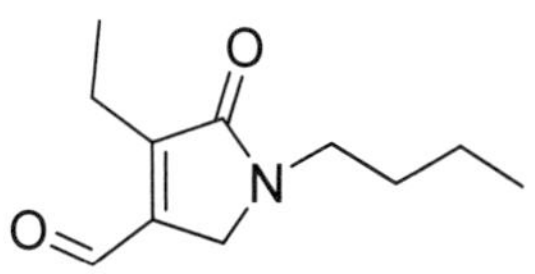 O, N, O	82

b	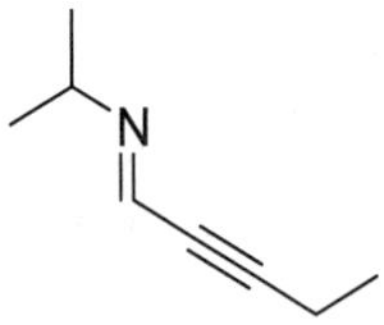	100	24	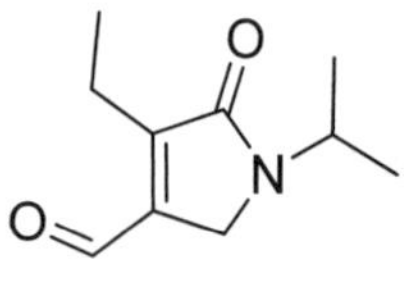	79
c	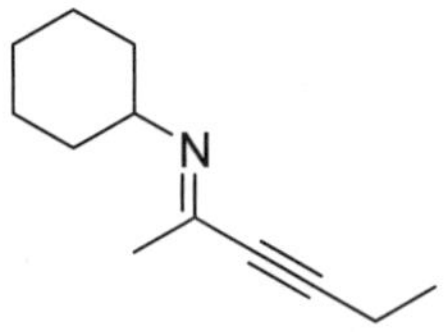	100	36	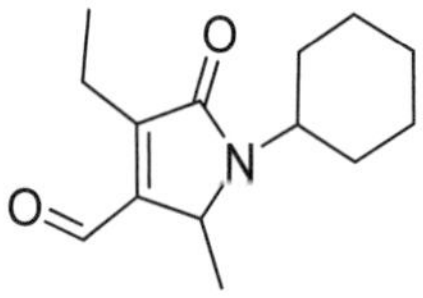	80
d	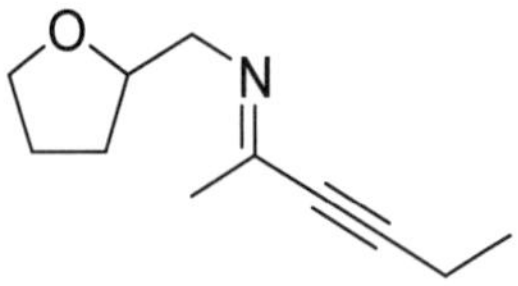	80	36	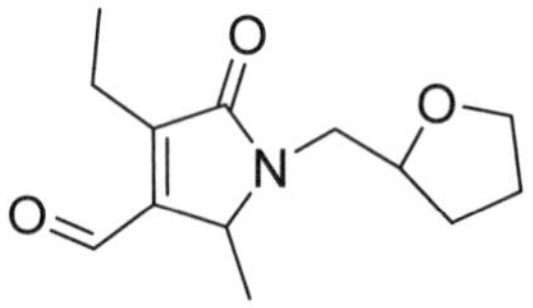	73
e	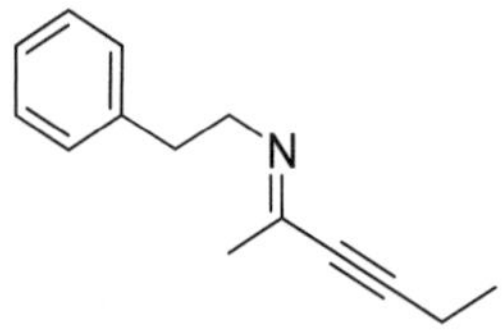	100	18	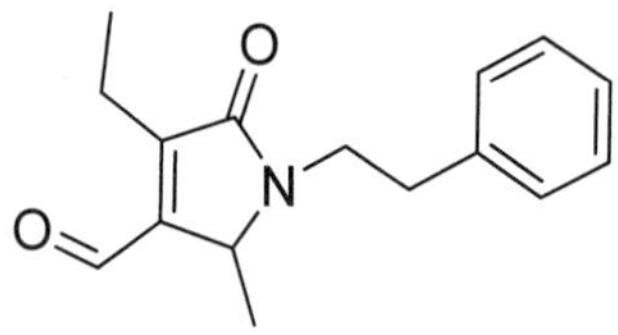	75

In addition, Alper demonstrated that when the keto-analogs were subjected to the same conditions, functionalized furanones were afforded (Scheme 2-4).[80d] The position of the olefin in the furanone can be varied by changing R_2 to an alkyl moiety, thus placing the double bond at the α-position of the furanone ring.

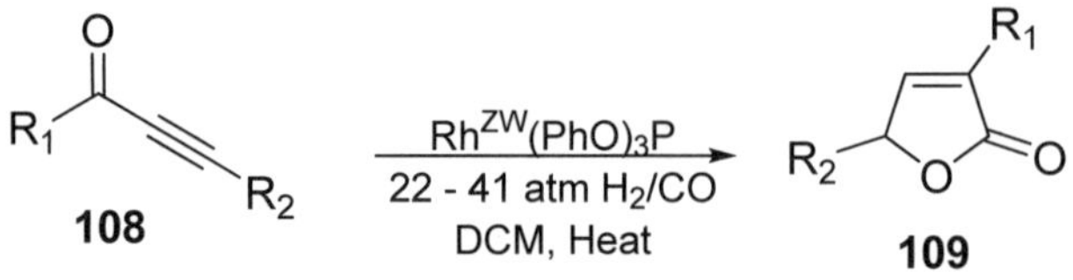

R_1, R_2 **108** → $Rh^{ZW}(PhO)_3P$, 22 - 41 atm H_2/CO, DCM, Heat → R_1, R_2 **109**

Scheme 2-4. Formation of furanones.

General procedure for 4-formylpyrrolin-2-ones and furanones[80c, 80d]

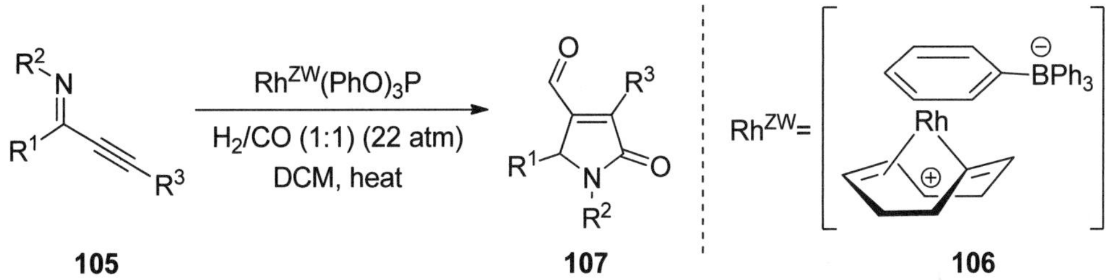

105 → 107 ($Rh^{ZW}(PhO)_3P$; H_2/CO (1:1) (22 atm); DCM, heat); Rh^{ZW} = 106

Into a 45-mL autoclave containing a glass liner and stirring bar was placed zwitterionic rhodium complex (**106**) (0.03 mmol), triphenylphosphite (0.12 mmol), acetylenic imine (**105**) (1.5 mmol), and CH_2Cl_2 (10 mL). The autoclave was flushed three times with carbon monoxide, pressurized from 17.5 to 38.5 atm, followed by the introduction of hydrogen to a total pressure of 21–42 atm. The autoclave was placed into an oil bath at 75–100 °C for 18–36 h and then allowed to cool to room temperature. The autoclave was depressurized, the reaction mixture filtered through Celite, and the solvent removed by rotary evaporation. The resulting residue was purified by chromatography on silica gel using an ethyl acetate/hexanes gradient ranging from 33:67 to 50:50 as the eluent to afford 4-formylpyrolin-2-ones **107**.[80c]

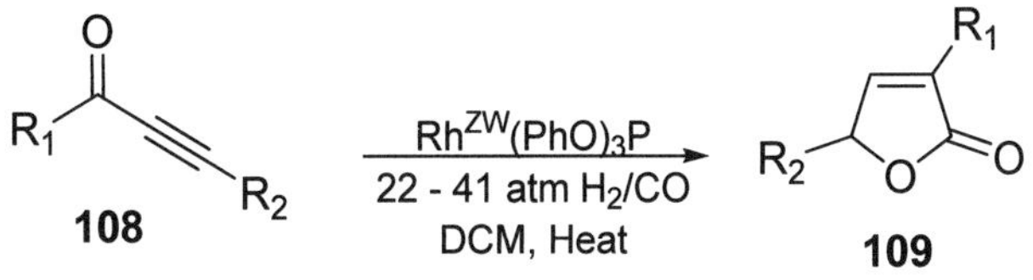

108 → 109 ($Rh^{ZW}(PhO)_3P$; 22 - 41 atm H_2/CO; DCM, Heat)

In a 45-mL autoclave containing a glass liner and a stirring bar was placed zwitterionic rhodium complex $Rh^{ZW}(PhO)_3P$ (0.03 mmol), triphenylphosphite (0.12–0.48 mmol), a keto-alkyne (**108**) (1.5 mmol), and CH_2Cl_2 (10 mL). The autoclave was flushed three times with carbon monoxide, pressurized from 17.5 to 38.5 atm followed by the introduction of hydrogen to a total pressure of 21–42 atm. The autoclave was placed in an oil bath at 75–120 °C for 24–48 h and then allowed to cool to room temperature. The autoclave was depressurized, the reaction mixture filtered through Celite, and the solvent removed by rotary evaporation. The resulting residue was purified by chromatography on silica gel using an ethyl acetate/hexanes gradient ranging from 33:67 to 50:50 as the eluent to afford furanones.[80d]

7. Asymmetric Hydrogenation

Asymmetric hydrogenation is perhaps the most efficient, cost-effective, and environmentally friendly method for constructing chiral compounds. Driven by the demand for enantiomerically pure pharmaceuticals and fine chemicals, new and efficient catalysts have been and continue to be discovered. With the use of transition metal complexes with chiral ligands, both the efficiency and effectiveness of newly synthesized ligands are frequently tested in asymmetric hydrogenation reactions. The importance of asymmetric hydrogenation was recognized in 2001, as both R. Noyori and W.S. Knowles were awarded the Nobel Prize in chemistry for their work on this type of reaction. Currently, a variety of ligands, including chiral bisphosphorus, monophosphorus, phosphoramidite, and ferrocene-based ligands, is used in asymmetric hydrogenations. Representative chiral ligands used for asymmetric hydrogenation are shown in Figure 2-5 and Figure 2-6.

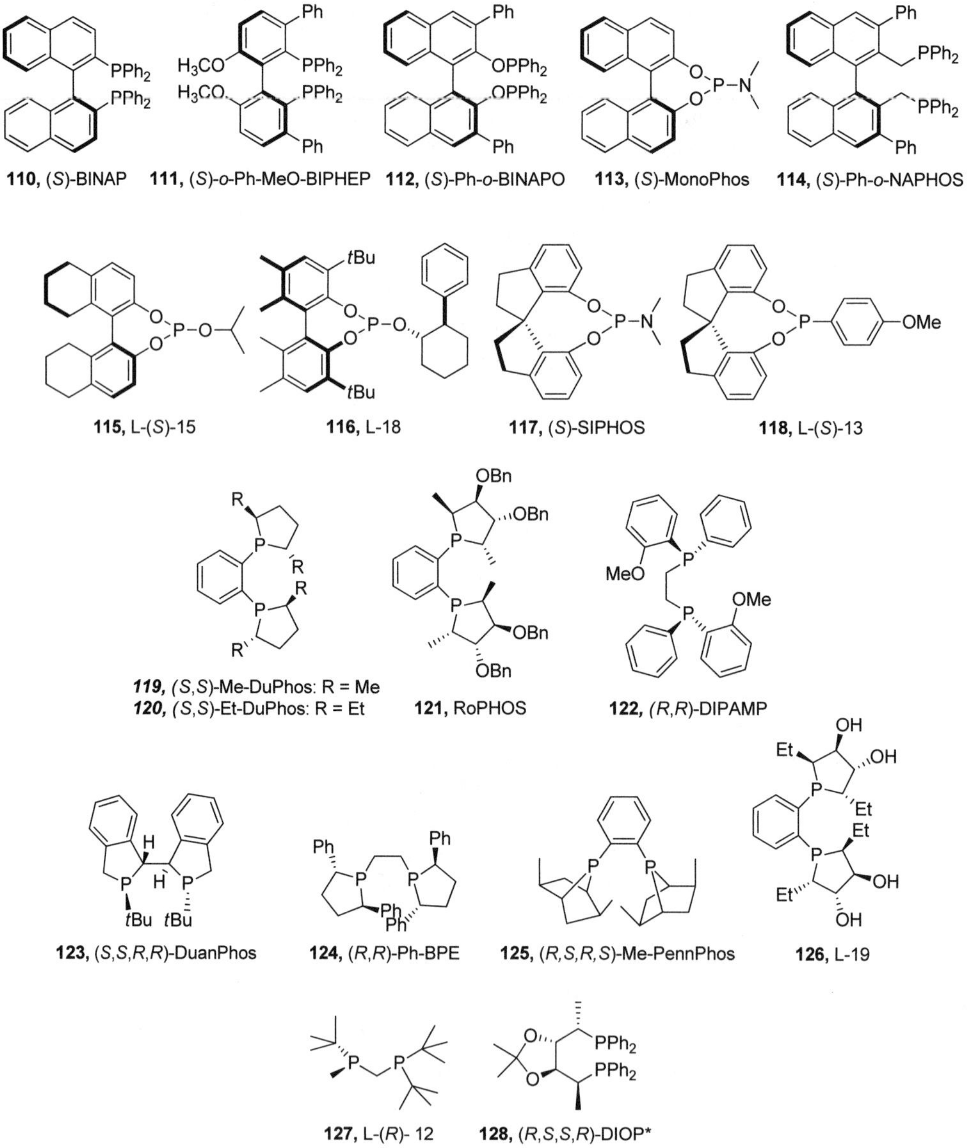

110, (*S*)-BINAP **111**, (*S*)-*o*-Ph-MeO-BIPHEP **112**, (*S*)-Ph-*o*-BINAPO **113**, (*S*)-MonoPhos **114**, (*S*)-Ph-*o*-NAPHOS

115, L-(*S*)-15 **116**, L-18 **117**, (*S*)-SIPHOS **118**, L-(*S*)-13

119, *(S,S)*-Me-DuPhos: R = Me
120, *(S,S)*-Et-DuPhos: R = Et **121**, RoPHOS **122**, *(R,R)*-DIPAMP

123, (*S,S,R,R*)-DuanPhos **124**, (*R,R*)-Ph-BPE **125**, (*R,S,R,S*)-Me-PennPhos **126**, L-19

127, L-(*R*)- 12 **128**, (*R,S,S,R*)-DIOP*

Figure 2-5. Chiral ligands for asymmetric hydrogenation.

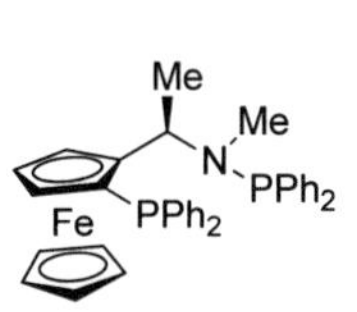

129, (*R,S*)-Me-BoPhoz 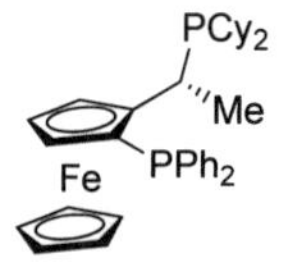 **130,** (*R,S*)-Josiphos 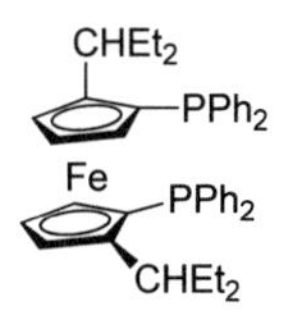 **131,** (S,S)-FerroPHOS 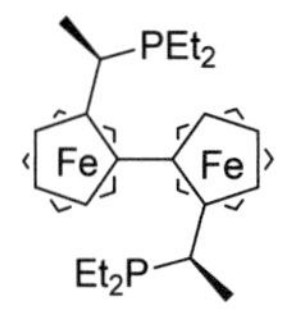 **132,** (*R,R*)-(*S,S*)-EtTRAP 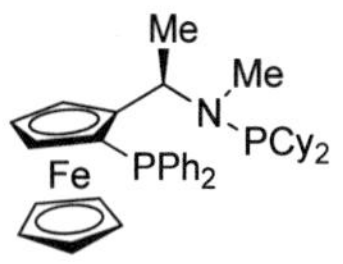 **133,** L-22

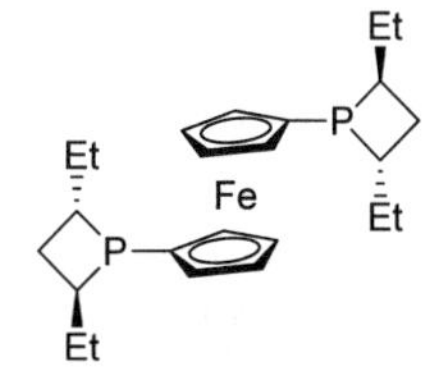

134, (*S,S*)-Et-FerroTANE 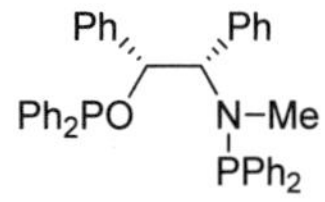 **135,** (1*R*,2*S*)-DPAMP 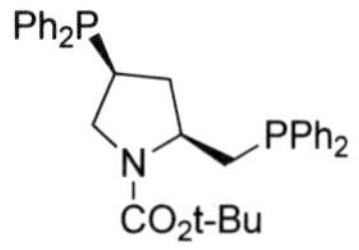 **136,** (*S,S*)-BPPM 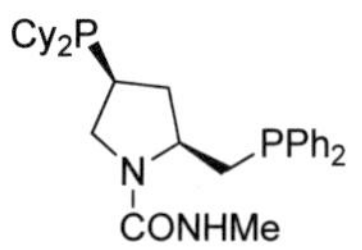 **137,** (*S,S*)-MCCPM

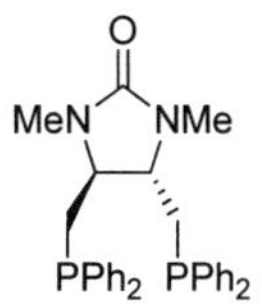

138, L-(*S,S*)-21 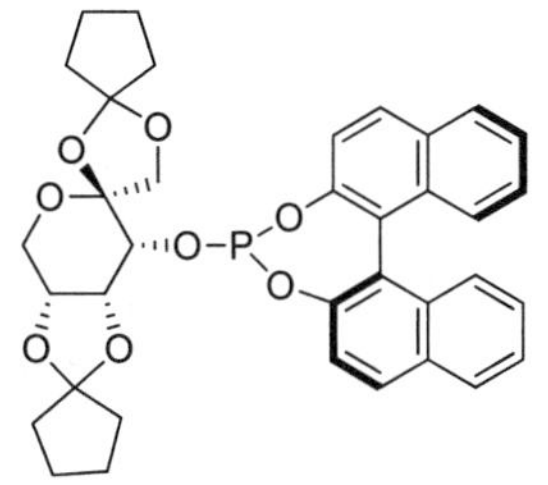 **139,** L-14 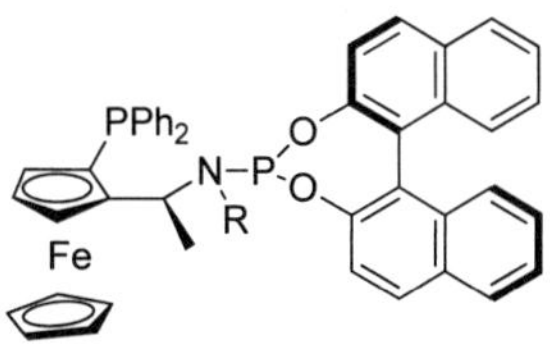 **140,** L-16: R = Me
141, L-17: R = H

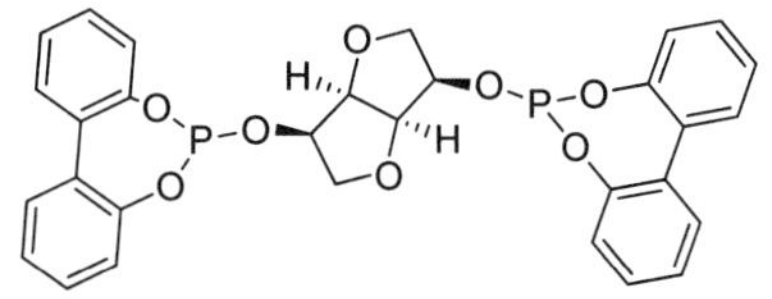

142, L-11 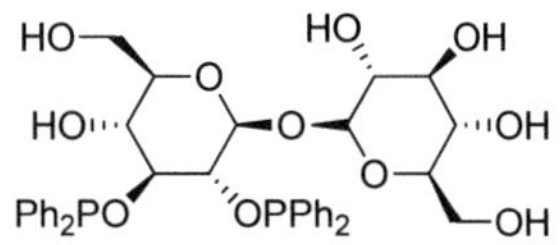 **143,** L-10 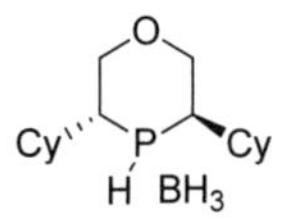 **144,** L-20

Figure 2-6. Chiral ligands for asymmetric hydrogenation.

7.1 Synthesis of α-Amino Acids by Hydrogenation of 2-Acetamido Acrylic Acid Derivatives

Natural and unnatural α-amino acids and their derivatives are important compounds in organic synthesis. Thus, much attention has been given to their synthesis. The asymmetric hydrogenation of α-dehydroamino acid derivatives has become one of the standard methods used to synthesize these compounds. As expected several chiral ligands with great structural diversity have been introduced as effective ligands for Rh-catalyzed asymmetric hydrogenation of α-dehydroamino acid derivatives. A few examples of Rh-catalyzed asymmetric hydrogenation of (*Z*)-2-(acetamido) acrylic acid derivatives **145** with different chiral ligands are shown in Table 2-4. In general, the catalysts displayed high reactivity and high (>95%) enantioselectivities under low or relatively low hydrogen pressure. Chiral ligands such as BINAP,[85] BIPHEP,[86] Et-DuPhos,[87] Josiphos,[88] BINAPO,[89] MonoPhos,[90] DPAMP,[91] and other ligands[92] have been shown to be effective ligands, yielding essentially enantiopure alkyl and aryl substituted α-amino acid derivatives **146**. The commercially available ferrocene based ligand (*R*)-methyl BoPhoz ligand was used in the preparation of enantiomerically pure cyclopropylalanine derivatives in good-to-excellent enantioselectivity.[93]

Table 2-4. Hydrogenation of enamines toward α-amino acid derivatives.

R^2CH=C(NHAc)CO_2R^1 (**145**) —[chiral Rh catalyst, H_2; solvent]→ R^2CH_2C*H(NHAc)CO_2R^1 (**146**)

Entry	R^1	R^2	Conditions	Ligand	% ee (config.)	Ref
1	Me	Ph	EtOH, room temperature, 3 atm H_2	(*S*)-BINAP	100 (*S*)	[85]
2	H	H	CH_2Cl_2, 25 °C, 1.7 atm H_2	(*S*)-*o*-Ph-MeO-BiPHEP	>99 (*S*)	[86]
3	Me	H	MeOH, room temperature, 2 atm, H_2	(*S, S*)-Et-DuPhos	>99 (*S*)	[87]
4	Me	H	MeOH, 25 °C, 10 atm H_2	(*R, R*)-Ph-BPE	99 (*S*)	[92a]
5	Me	Ph	MeOH, 35 °C, 1 atm H_2	(*R*)-(*S*)-JosiPhos	96 (*S*)	[88]
6	H	Ph	EtOH, room temperature, 2 atm H_2	(*S, S*)-Et-FerroPhos	98.9 (*R*)	[92f]
7	Me	Ph	MeOH, 50 °C, 3 atm H_2	(*R, R*)-DIPAMP	96 (*S*)	[92e]

8	Me	H	MeOH, room temperature, 1.4 atm H_2	(*R, R, S, S*)-DuanPhos	>99 (*R*)	[92b]
9	Me	H	MeOH, room temperature, 3 atm H2	(*S*)-Ph-*o*-NAPHOS	98.7 (*S*)	[89]
10	Me	Ph	H_2O, room temperature, 5 atm H_2	L-10	99.9 (*S*)	[92c]
11	Me	Ph	MeOH, room temperature, 50 atm H_2	(1*R, 2S*)-DIPAMP	97 (*R*)	[91]
12	Me	Ph	CH_2Cl_2, room temperature, 1 atm H_2	(*S*)-SIPHOS	96.4 (*S*)	[92d]
13	Me	H	EtOAc, room temperature, 1 atm H_2	(*S*)-MonoPhos	99.6 (*R*)	[90]
14	Bn	cyclopropyl	MeOH, room temperature, 0.68 atm H_2	(*R, S*)-Me-BoPhoz	98	[93]

The DPAMP-Rh(I) catalyst system was applied to the asymmetric hydrogenation of methyl (*Z*)-2-acetamido-3-(3-methoxy-4-acetoxyphenyl)acrylate **147** to furnish the corresponding amino acid **148**, which is a crucial advanced intermediate in the synthesis of L-DOPA. Likewise, a homophenylalanine intermediate, which is a key component of the antihypertensive (*S,S*)-benazepril, was obtained from the hydrogenation of ethyl (*Z*)-2-acetamido-4-phenylcrotonate using the same catalyst and identical reaction conditions.[91]

General procedure for asymmetric hydrogenation of dehydroamino acid derivatives[91]

CO_2Me NHCOMe MeO OAc **147** → [{(1*R*,2*S*)-DPAMP}Rh(COD)]BF_4, H_2 (50 atm), 25 °C, MeOH → CO_2Me NHCOMe MeO OAc **148** 97.4% ee

The cationic Rh catalyst was prepared *in situ* by stirring the mixture of $[Rh(cod)Cl]_2$, $AgBF_4$ and an appropriate chiral DPAMP (**135**) in THF for 2 h. The resulting cationic catalyst was characterized by ^{31}P NMR. In a stainless steel autoclave, dehydroamino acid derivative **147** (0.02 mmol) was dissolved in a degassed solvent (0.4 mL). To the solution was added the prepared catalyst (0.0002 mmol). The reactor was then pressurized with hydrogen and the reaction was run under the chosen conditions. The enantiopurity of the products were determined by GC on a Chrompack Chirasil-L-Val column (Agilent Technologies, Santa

Clara, CA, U.S.A.) or by HPLC on a Chiralcel OD column (Chiral Technologies, Illkirch–Cedex, France).

The DIPAMP-Rh complex was found also to be an effective catalyst system for the asymmetric hydrogenation of the tri-substituted pyridine dehydroamino acid derivative **149**. The product **150** is an advanced intermediate for the synthesis of (*S*)-(-)-acromelobic acid.[94]

General procedure[94]

CO_2Me NHCbz *n*-Pr-O_2C N OMe **149**

(*R*,*R*)-[Rh(DIPAMP)(COD)]BF_4

H_2 (65 psi), MeOH, 48 °C

NHCbz CO_2Me *n*-Pr-O_2C N OMe **150**

89% yield, > 98% ee

The dehydroamino acid derivative **146** (0.117 g, 0.27 mmol) was dissolved in MeOH (10 mL) and nitrogen was bubbled through this solution. (*R*,*R*)-[Rh(DIPAMP)(cod)]BF_4 (13.0 mg, 0.018 mmol, 0.065 equiv) was added under a nitrogen atmosphere and the mixture was degassed under vacuum. The mixture was then hydrogenated (65 psi) at 48 °C for 16 h. The reaction mixture was cooled to room temperature and concentrated on a rotary evaporator, and then the residue was purified by column chromatography on silica gel (25% EtOAc in hexanes) to afford 0.105 g of the desired α-amino acid derivative **147** in 89% yield as viscous oil. Analytical RP HPLC: MeCN 0.05% aqueous acetic acid/60:40, 2.0 mL/min at 225 nm, t_R: 5.44 min, 99% ee.

7.2 Synthesis of β-Amino Acids by Hydrogenation of β-(Acylamino) acrylates

β-Amino acids have been shown to be an important class of compounds both in biomedical research and the pharmaceutical industry. These structural motifs are present in β-lactams and β-peptides and also serve as chiral templates for the synthesis of other pharmaceuticals. Perhaps the most convenient and straightforward method for the synthesis of β-amino acids is the catalytic asymmetric hydrogenation of β-(acylamino)acrylates **151**. Some success has been reported for the Rh-catalyzed asymmetric hydrogenation of (*E*) and (*Z*)- isomers of β-(acylamino)acrylates using chiral phosphorous ligands such as Me-DuPhos,[95] DuanPhos,[92b] TangPhos,[96] and other ligands.[86, 97]

Table 2-5. Asymmetric hydrogenation of β-(acylamino)acrylates.

151 → chiral Rh catalyst, H_2 / solvent → **152**

Entry	R	Geometry	Conditions	Ligand	% ee (config.)	Ref
1	Me	*E*	MeOH, 25 °C, 1 atm H_2	(*S,S*)-Me-DuPhos	98.2 (*S*)	[94]
2	Me	*E*	THF, room temperature, 1.4 atm H_2	(*S,S,R,R*)-TangPhos	99.6 (*R*)	[96]
3	Me	*E*	MeOH, room temperature, 1.4 atm H_2	(*R,R,S,S*)-DuanPhos	>99 (*R*)	[95]
4	*p*-MeO-Ph	*Z*	CH_2Cl_2, 5 °C, 10 atm H_2	**L-17** (**141**)	98 (*R*)	[97a]
5	*p*-F-Ph	*Z*	CH_2Cl_2, 5 °C, 10 atm H_2	**L-17** (**141**)	98 (*R*)	[97a]
6	Me	*Z*	EtOAc, room temperature, 1.4 atm H_2	**L-12** (**127**)	98 (*R*)	[98]
7	Me	*E/Z*	THF, room temperature, 1.4 atm H_2	(*S,S,R,R*)-TangPhos	99.5 (*R*)	[96]
8	Me	*E/Z*	THF, room temperature, 1.4 atm H_2	**L-12** (**127**)	98 (*R*)	[98]
9	*p*-MeO-Ph	*E/Z*	CH_2Cl_2, room temperature, 100 atm H_2	**L-13** (**118**)	98 (*R*)	[99]
10	*p*-MeO-Ph	*E/Z*	CH_2Cl_2, room temperature, 100 atm H_2	**L-13** (**118**)	95 (*R*)	[99]

However, considering the fact that both (*E*) and (*Z*)-isomers of β-aminoacrylic acid derivatives are formed simultaneously during their preparation using most synthetic methods, the development of catalytic systems that can be effective for both isomers is clearly needed. To this end, a phosphite derivative of SIPHOS[100] and TangPhos[101] have been shown to be effective in the hydrogenation of β-aminoacrylic acid derivatives although with limited substrate scope. Wu and Hoge have developed a simple but bulky chiral bisphosphine ligand that proved to be effective in the asymmetric hydrogenation of β-aminoacrylic acids with a broader scope of substrates.[98]

General procedure for asymmetric hydrogenation of β-aminoacrylic acid derivatives[98]

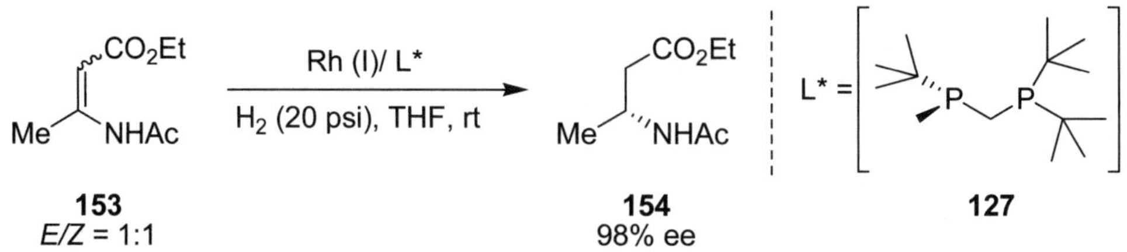

153 E/Z = 1:1 → Rh (I)/ L*, H_2 (20 psi), THF, rt → 154 98% ee; L* = 127

The catalyst (0.005 mmol) and ethyl-3-acetamido-2-butenoate **153** (0.50 mmol) were dissolved in 2.5 mL of THF in a Griffin-Worden pressure vessel, which was then sealed and pressurized to an initial pressure of 20 psi H_2. The mixture was stirred with a Teflon-coated magnet at 25 °C until H_2 uptake was complete (15 min). The H_2 pressure in the bottle was released. The reaction mixture was then analyzed *via* chiral GC to provide the enantiomeric excess (98% ee) of **154**.

Few Rh-catalyst systems have been shown to reduce mixtures of (E) and (Z)-isomers of β-aryl- and β-aminoacrylic acid derivatives effectively.[86] To this end, Fu *et al.* found that a monodentate phosphonite ligand derived from SIPHOS was effective in the asymmetric hydrogenation of β-aryl- and β-aminoacrylic acid derivatives. The authors also found that the electronic properties of the substituents had a significant impact on the reaction rate and enantioselectivities. The reactions of substrates having electron-donating substituents proceeded faster with higher enantioselectivities.[100]

General procedure[100]

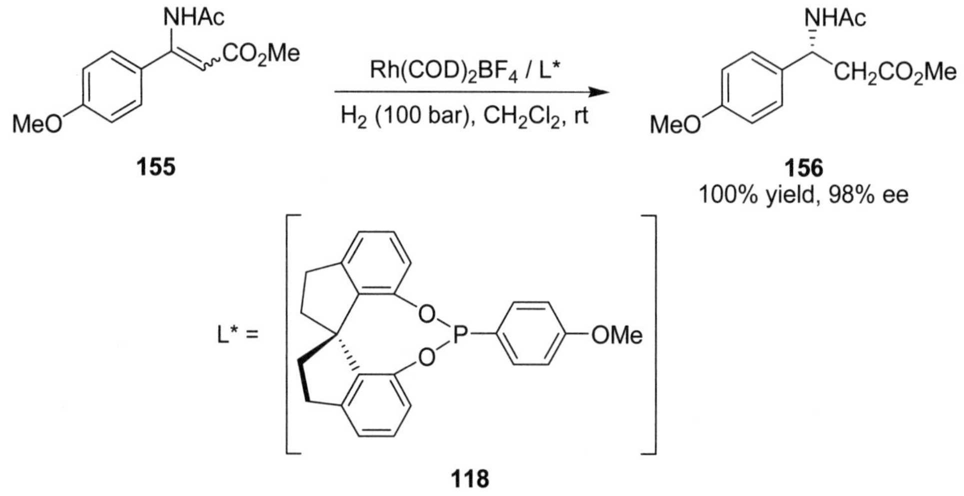

155 → $Rh(COD)_2BF_4$ / L*, H_2 (100 bar), CH_2Cl_2, rt → 156 100% yield, 98% ee; L* = 118

A reaction tube equipped with a stirring bar, in which 5 μmol of $Rh(cod)_2BF_4$, 11 μmol of ligand **118,** and 0.5 mmol of β-dehydroamino ester (**155**) were added under an inert atmosphere into an autoclave. After 5 mL of CH_2Cl_2 were added, the inert atmosphere was replaced by 10 hydrogen/release cycles and the reaction mixture was allowed to stir under

100 bars of H_2 at room temperature for 36 h. The resulting mixture was filtered through a short silica column and **156** was formed in 100% yield; 98% ee, chiral GC Varian Chirasil-L-Val column, 25 m × 0.25 mm × 0.25 μm; N_2 1.8 mL/min, 100 °C then 1 °C/min to 170 °C, t_R = 67.03 min, t_S = 67.82 min.

7.3 Asymmetric Hydrogenation of β,β-Disubstituted α-Dehydro-amino Acid Derivatives

The synthesis of β,β-disubstituted α-amino acids still remains a challenge and there are few examples in which high enantioselectivities (≥ 95%) have been reported in the literature. The Rh-complexes of Me-BPE and Me-DuPhos were found to show excellent enantioselectivies for β,β-disubstituted α-enamide substrates (**157**). Good chemoselectivity was observed in the hydrogenation of substrates that contained other olefin functionalities.[99]

General procedure[99]

157 → (*R*,*R*)-Me-BPE-Rh, 7 atm H_2, benzene, 25 °C → **158**

In a dry box, a Fisher-Porter bottle was charged with β,β-disubstituted α-enamide substrate **157**, anhydrous, degassed benzene and (*R*,*R*)-Me-BPE-Rh catalyst (0.2 mol%). After five vacuum/H_2 cycles, the tube was pressurized to an initial pressure of 7 atm H_2. The mixture was allowed to stir at room temperature for 12 h. The reaction mixture was concentrated on a rotovap, and the residue passed through a short SiO_2 column (EtOAc or EtOAc/hexane: 50/50) to remove catalyst. Without further purification, the enantiomeric excesses were determined directly on the crude products thus obtained, using chiral capillary GC (Chrompack Chirasil-L-Val column, 25 m).

The Me-DuPhos-Rh catalyst system effectively reduced both the (*Z*)- and (*E*)-isomers of β-(acetylamino)-β-methyl-α-dehydroamino acid derivatives **159** with 96% and 98% ee, respectively. With this system, four isomers of *N*,*N*-protected 2,3-diaminobutanoic acid derivatives **160** can be obtained efficiently with high enantioselectivity.[102]

General hydrogenation procedure[102]

159 → 160: (*R*,*R*)-Me-DuPhos-Rh, H_2 (6 atm), PhH, rt

160: 95% yield, > 98% ee

In a dry box, a Fisher-Porter tube was charged with the catalyst (1 mg), deoxygenated benzene (5 mL), and substrate **159** (50–200 mg). Three vacuum/N_2 cycles to purge the gas line of any oxygen followed by three vacuum/N_2 cycles of the vessel were carried out before the tube was pressurized to 6 atm of hydrogen. The mixture was stirred at room temperature for 1 h. The pressure in the vessel was then released, and the contents were evaporated to dryness under reduced pressure. The crude product was passed through a short plug of silica prior to spectroscopic and chromatographic analysis of **160**.

7.4 Asymmetric Hydrogenation of Enamides

Enamides can also be efficiently hydrogenated with chiral Rh-complexes. The Rh complexes of Ph-BPE,[92a] BICP,[103] TangPhos,[101] SIPHOS,[104] and others[105] have been shown to catalyze the hydrogenation of a mixture of (*E*) and (*Z*) β-methyl-α-phenylenamides **161** with excellent enantioselectivities. The hydrogenation of 2- and 3-substituted *N*-acetylindoles with the Ph-TRAP-Rh system was also possible with high ee.[106] (*R,S,S,R*)-DIOP was shown to be an excellent catalyst system for a variety of aromatic enamides.[107]

Table 2-6. Hydrogenation of β-methyl-α-phenylenamides.

161 → 162: chiral Rh catalyst, H_2, solvent

Entry	R	Geometry	Conditions	Ligand	% ee (config.)	Ref
1	H	—	MeOH, 25 °C, 10 atm H_2	(*R,R*)-Ph-BPE	99 (*R*)	[92a]
2	H	—	toluene, 5 °C, 10 atm H_2	(*S*)-SIPHOS	98.7 (*S*)	[104]
3	H	—	CH_2Cl_2, room temperature, 1 atm H_2	**L-21** (**138**)	>98.5 (*R*)	[105a]

4	Me	*E*/*Z*	toluene, room temperature, 2.7 atm H_2	(*R*,*R*)-BICP	95 (*R*)	[103]
5	Me	*E*/*Z*	MeOH, room temperature, 1.4 atm H_2	(*S*,*S*,*R*,*R*)-TangPhos	98 (*R*)	[101]
6	Me	*E*/*Z*	CH_2Cl_2, room temperature, 10 atm H_2	**L-14** (**139**)	96.7 (*S*)	[105b]

General procedure for asymmetric hydrogenation of enamides[107b]

CH3 NHAc (*R*,*S*,*S*,*R*)-DIOP-Rh H_2 (10 atm), MeOH, rt CH3 NHAc

163 **164** > 99% ee

To a solution of $[Rh(cod)_2]SbF_6$ (0.005 mmol) in methanol (3 mL) in a glove box was added (*R*,*S*,*S*,*R*)-DIOP (0.055 mL of 0.1-M solution in toluene, 0.0055 mmol). After the mixture was stirred for 10 min, the enamide **163** (0.25 mmol) was added. The hydrogenation was performed at room temperature under 10 bars of hydrogen for 24–60 h. The hydrogen was released and the reaction mixture was passed through a short silica gel column to remove the catalyst. The enantiomeric excess was measured by GC or HPLC directly without any further purification. The absolute configuration of the product **164** was determined by comparing the observed rotation with the reported value.

The TangPhos system was also shown to be effective in the hydrogenation of an exo-methylene-tetrahydroquinoline substrate **165** in 97% ee.[101] The product **166** is frequently found as a subunit in natural product alkaloids.

Furthermore, the asymmetric hydrogenation reaction has also proven to be useful for industrial scale syntheses. Magano *et al.* reported the asymmetric hydrogenation of enamide precursors to yield the desired β-amino acids that are used to treat general anxiety disorder and insomnia.[108] This method produced these pharmaceuticals on a multi-kilogram scale using (*R*)-mTCFP-Rh(cod)BF_4 and (*R*)-binaphine-Rh(cod)BF_4 as the catalysts. The final products were obtained in good yield with excellent diastereoselectivity (99.8% de) after recrystallization.

General hydrogenation procedure[101]

(1*S*,1′*S*,2*R*,2′*R*)-TangPhos (1, 0.10 mL, 0.05-M solution in methanol, 0.005 mmol) was added to a solution of $[Rh(nbd)_2]SbF_6$ (2.3 mg, 0.0045 mmol) in methanol (3 mL) in a glove box. After the mixture was stirred for 10 min, the substrate **165** (0.5 mmol) was added. The hydrogenation was performed at room temperature under H_2 (1.4 atm) for 12–48 h. After carefully releasing the hydrogen, the reaction mixture was passed through a short silica-gel plug to remove the catalyst. The resulting solution was used directly for chiral GC or HPLC to measure the enantiomeric excess. For the hydrogenation of dehydroamino acids, the enantiomeric excesses were measured after conversion into their corresponding methyl esters by treatment with $TMSCHN_2$ (TMS = trimethylsilyl).

7.5 Hydrogenation of Itaconic Acid and Dimethyl Itaconate

The asymmetric hydrogenation of itaconic acid and its derivatives is often used to test the effectiveness of new ligands. Phosphonite ligands[109] as well as a secondary phosphane ligands[110] were found to be effective in these reactions. *Bis*-phosphorus-chelating ligands such as Et-FerroTANE,[111] Et-DuPhos,[112] Ph-BPE,[92a] and the *trans* chelating *bis*-phosphane ligand Et-TRAP[113] and other bisphosphorus ligands[114] were all found to be effective ligands for the hydrogenation of dimethyl itaconate. In addition, a polyhydroxyl-bisphospholane ligand allowed the hydrogenation of itaconic acid to be carried out in aqueous solution.[115] Ferrocenylaminophosphine,[116] phosphine-phosphoramidite,[117] and chiral ionic phosphite[118] ligands have also been used for the asymmetric hydrogenation of itaconic acid and its derivatives. Ojima and co-workers used a fine-tunable biphenol-based phosphite ligand in the asymmetric hydrogenation of dimethyl itaconate **168a** with essentially perfect (99.6%) enantioselectivity.[109a]

Table 2-7. Hydrogenation of methylenediester **168**.

Entry	R	Conditions	Ligand	% ee (config.)	Ref
1	Me	$ClCH_2CH_2Cl$, 50 °C, 6.8 atm H_2	L-15 (**115**)	99.6 (*R*)	[109a]
2	Me	CH_2Cl_2,-10 °C, 0.3 atm H_2	L-11 (**142**)	98.7 (*R*)	[114]
3	Me	CH_2Cl_2, room temperature, 10 atm H_2	L-16 (**140**)	99.1 (*S*)	[117]
4	Me	CH_2Cl_2, 23 °C, 20 atm H_2	L-18 (**116**)	96.9 (*S*)	[109b]
5	Me	MeOH, 25 °C, 5 atm H_2	(*R,R*)-Et-DuPhos	98 (*R*)	[112]
6	Me	MeOH, room temperature, 5.5 atm H_2	(*S,S*)-Et-FerroTANE	98 (*R*)	[111]
7	Me	MeOH, 25 °C, 10 atm H_2	(*R,R*)-Ph-BPE	99 (*S*)	[92a]
8	Me	CH_2Cl_2, reflux, 1 atm H_2	(*R,R*)-(*S,S*)-Et-TRAP	96 (*S*)	[113]
9	H	MeOH/H_2O (3:97), room temperature, 10 atm H_2	L-19 (**126**)	>99 (*R*)	[115]
10	H	^{i}PrOH, 20 °C, 1.1 atm H_2	L-20 (**144**)	96 (*S*)	[110]
11	H	MeOH, room temperature, 20 atm H_2	(*R,S*)-Me-BoPhoz	97.4 (*R*)	[116]

General procedure[109a]

MeO_2C ... CO_2Me **168a** → $[Rh(cod)_2]SbF_6$ / L*, DCE, 100 psi H_2, 50 °C → MeO_2C ... CO_2Me **169a**

In a 10-mL glass reaction vessel with a magnetic stir bar under nitrogen, a mixture of $[Rh(cod)_2]SbF_6$ (1.4 mg, 0.0025 mmol) and **168a** (3.0 mg, 0.0050 mmol) was dissolved in 5 mL 1,2-dichloroethane, and the solution was stirred at room temperature for 2 min. To this catalyst solution was added dimethyl itaconate (0.08 mL, 0.5 mmol). The reaction vessel was then placed in a 300-mL stainless steel autoclave, and the autoclave was pressurized and purged with hydrogen gas five times. The vessel was subsequently charged with hydrogen gas to 100 psi (6.8 atm). The autoclave was warmed to 50 °C and the reaction was carried out at this temperature for 20 h with sirring. The autoclave was allowed to cool to room temperature, and then the hydrogen gas was carefully released. The reaction mixture was filtered through a short pad of silica gel and the filtrate was subjected to gas chromatographic analysis using a Supelco Beta Dex-225 column for the determination of the conversion and enantiomeric purity of dimethyl methylsuccinate **169a**: The analysis showed that the conversion was 100% with 100% product selectivity and (*R*)-dimethyl methylsuccinate with 99.6% ee was formed.

7.6 Asymmetric Hydrogenation of Enol Esters

Although similar in structure to enamides, the asymmetric hydrogenation of enol esters has proven to be more difficult. Burk *et al.*[112] found the Et-DuPhos ligand **120** to be effective in the asymmetric hydrogenation of a wide range of isomeric mixtures of β-substituted derivatives with high enantioselectivity. High enantioselectivities were also obtained in the hydrogenation of *E*/*Z* isomeric mixtures of enol ester substrates.

Typical procedure for asymmetric hydrogenation[112]

(*S*,*S*)-Et-DuPhos-Rh

CH_2Cl_2, H_2 (60 psi)

170

171
99.8% ee

120, (*S*,*S*)-Et-DuPhos

In a glove box, a Fisher-Porter tube was charged with substrate **170**, deoxygenated MeOH, and catalyst (0.2 mol%). After being removed from the glove box, the vessel was connected to a hydrogen line and subjected to five vacuum/H_2 cycles, and the tube was pressurized to an initial pressure of 5 atm of H_2. The reaction mixture was allowed to stir for 48 h at room temperature. Once the reaction was finished, the solvent was removed *via* rotary evaporation. The oily residue was dissolved in ethyl acetate-hexanes (1:1) and passed through a short plug of silica to remove the catalyst. Without further purification, the enantiomeric excess of **171** was determined with an aliquot of the crude product thus obtained.

The asymmetric hydrogenation of cyclic enol acetate has also proven to be a challenging problem. In this respect, Me-PennPhos **125** was shown to be an effective ligand in the rhodium-catalyzed asymmetric hydrogenation of five- or six-membered cyclic enol acetates such as **172**.[119] In addition, Tang *et al.* found that the TangPhos-Rh catalyst system provided good-to-excellent (92%–99%) enantioselectivities for a diverse set of aryl enol acetates.[120]

General procedure[119]

To a solution of $[Rh(cod)_2]BF_4$ (5.0 mg, 0.012 mmol) in MeOH (10 mL) in a glove box was added (*R*,*S*,*R*,*S*)-Me-PennPhos (**125**) (0.15 mL of a 0.1-M solution in MeOH, 0.015 mmol). After the mixture was stirred for 30 min, the enol acetate **172** (1.2 mmol) was added. The hydrogenation was performed at room temperature under 1.7 atm of hydrogen for 24 h. After the hydrogen was released, the reaction mixture was passed through a short silica gel column to remove the catalyst. The enantiomeric excess of **173** was measured by capillary GC without any further purification. The absolute configurations of the product was determined by comparing the observed optical rotation with that of chiral acetate made from a readily available secondary alcohol.

7.7 Hydrogenation of Unsaturated Acids and Esters

The advances made in the asymmetric hydrogenation of α,β-unsaturated carboxylic acids and esters have been achieved mostly using Ru catalysts.[121] Nevertheless, Hoge and co-workers reported that a three hindered quadrant *bis*-phosphine Rh-catalyst displayed high reactivity and was extremely effective in the hydrogenation of *tert*-butylammonium (3*Z*)-3-cyano-5-methyl-3-hexenoate **174**. Although the transformation could be achieved more quickly, a substrate-to-catalyst ratio of 27,000:1 was achieved with acceptable reaction time and no change in ee. The product of this hydrogenation is an intermediate of pregabalin, a pharmaceutical used to treat epilepsy and pain.[122] Robert *et al.* also had success using Binol-based phosphine–phosphite ligands to catalyze the asymmetric hydrogenation of unsaturated esters and acrylates.[123]

Typical procedure[122]

The substrate **174** (100 g, 442 mmol) was weighed into a hydrogenation bottle in air. The hydrogenation bottle was then transferred to a glove box (O_2 < 5 ppm). To the substrate was added 500 mL degassed MeOH with stirring to dissolve the substrate. Catalyst precursor (9.2 mg) and the ligand **127** was then added to the substrate solution. The hydrogenation vessel was sealed and then pressurized to 50 psi H_2 and stirred vigorously with a Teflon-coated magnet. Pressure of the reaction was maintained at a constant 50 psi H_2. After 40 h, H_2 uptake was complete. GC analysis showed >98% conversion and 98% ee of the product **175** (*S* isomer).

Burk *et al.*[124] used a Me-DuPhos-Rh catalyst in the asymmetric hydrogenation of the unsaturated carboxylic acid **176** to afford a key intermediate **177** in the synthesis of candoxatril, which is a prodrug of candoxatrilat, a potent atrial natriuretic factor potentiator useful in the treatment of hypertension and congestive heart failure.

Typical procedure[124]

MeO, O, t-BuO$_2$C, CO$_2$Na — **176** → [Rh(Me-DuPhos)(COD)]BF$_4$, MeOH, H$_2$ (75 psi), 18 °C → MeO, O, t-BuO$_2$C, CO$_2$Na — **177** >99% ee

A 50-mL Parr pressure vessel was charged with 1.4 g (4 mmol) of carboxylate salt **176** and purged carefully with nitrogen (five vacuum/nitrogen venting cycles). Methanol (6 mL, degassed) was added, and the reactor then was pressurized and vented four times with 5 atm hydrogen under stirring. After the pressure was released, a freshly prepared solution of 0.7 mg (0.12 ímol in MeOH) of [((*R,R*)-Me-DuPHOS)Rh(cod)]BF_4 was added as quickly as possible *via* syringe through an addition port. The vessel was sealed and pressurized with 75 psi hydrogen, and the mixture was allowed to stir at 18 °C for 80 min. After this time, the hydrogen uptake was judged to have ceased. The reactor pressure then was released, the solution was transferred to a round-bottomed flask, and the solvent was evaporated under reduced pressure. The residue was treated with 20 mL of 1-M aqueous hydrochloric acid, and the product was extracted into dichloromethane (3 × 10 mL). The combined organic layers were dried over anhydrous sodium sulfate and filtered, the filter cake was washed three times with 5 mL of dichloromethane, and the solvent was evaporated under reduced pressure. The crude product **177** (free carboxylic acid) was obtained as pale yellow oil: yield 1.28 g (97%). The material thus obtained was subjected directly to full analytical characterization.

Relatively few examples of the asymmetric hydrogenation of β-keto esters using chiral rhodium complexes have been reported compared with their ruthenium analogs. The Josiphos-Rh system was found to be effective in the asymmetric hydrogenation of ethyl 3-oxobutanoate **178** to give **179** with 97% ee.[88]

General procedure[88]

(R)-(S)-Josiphos-Rh; MeOH, H_2 (20 atm)

178 → 179, 97% ee

130, (R,S)-Josiphos

The hydrogenation of ethyl 3-oxobutyrate **178** was carried out in a 50-mL steel autoclave under a hydrogen pressure of 20 bars, using methanol as solvent, and a substrate to catalyst ratio of 100/1. The product **179** was isolated in essentially quantitative yield after a reaction time of 15 h. Determination of optical purities was carried out by GC after derivatization with isopropyl isocyanate (L-Chirasil-Val column).

Boaz *et al.*[116] used a BoPhos derivative **133** in the enantioselective hydrogenation of the cyclic α-keto ester dihydro-4-4-dimethyl-2,3-furandione (**180**) to yield (*R*)-pantolactone (**181**), a key intermediate in the synthesis of vitamin B and coenzyme A.

General procedure[116]

Rh-L*; THF, H_2 (300 psi)

180 → 181, 97.2% ee

L* = 133

Bis(1,5-cyclooctadiene)rhodium trifluoromethanesulfonate (2.3 mg; 5 mol%; 0.01 equiv) and the phosphine-aminophosphine ligand **133** (6 mol; 0.012 equiv) were placed into a reaction vessel and optionally purged with argon (identical results were obtained with or without the purge). THF (2.0 mL) was added and the mixture was stirred under Ar for 15 min. A solution of ketopantolactone (**180**) (0.5 mmol) in THF (3.0 mL) was added to the catalyst solution. The solution was then flushed with H_2 and pressurized to the desired pressure with 300 psi H_2. After 6 h, the mixture was concentrated on a rotary evaporator and analyzed for enantiomeric excess of **181** using standard analytical techniques.

The asymmetric hydrogenation of unfunctionalized ketones still remains a challenging problem. High levels of asymmetric induction have been achieved in the hydrogenation of acetophenone **182** using Me-PennPhos ligand in the presence of 2,6-lutidine and potassium bromide as additives.[125]

General procedure[125]

(*R*,*S*,*R*,*S*)-Me-PennPhos-Rh

MeOH, H_2 (30 atm)

182

183
95% ee

125, (R,*S*,*R*,*S*)-Me-PennPhos

To a solution of $[Rh(cod)Cl]_2$ (2.5 mg, 0.005 mmol) in MeOH (10 mL) was added the ligand **125** (3.7 mg, 0.01 mmol). After the reaction mixture was stirred at room temperature for 10 min, acetophenone (**182**) (1.0 mmol) was added. The orange-yellow solution was stirred for 2 min, and hydrogen was introduced. The hydrogenation was performed in a Parr autoclave at room temperature under 30 atm of hydrogen for 24 h. The residue was passed through a short silica gel column to remove the catalyst and eluted with diethyl ether. The enantiomeric excesses and reaction conversion were measured by gas chromatography on a Supelco β-DEX 120 column. The absolute configuration of the product **183** was determined by comparing the observed rotation with the reported value.

The asymmetric hydrogenation of amino ketones is a simple and economical means to access chiral, nonracemic amino alcohols that serve as important templates for the synthesis of pharmaceuticals and agrochemicals. However, limited success has been realized to date using chiral rhodium complexes. The β-adrenergic blocking agent (*S*)-propranolol (**185**) was obtained from the asymmetric hydrogenation of the amino ketone derivative **184**. The *bis*-phosphine ligand (*S*,*S*)-MCCPM (**137**) was used with low catalyst loading to give the product with 90.8% ee, which was subjected to a single recrystallization to give enantiomerically pure (*S*)-propranolol (**185**).[126a] The same catalyst system was also found to be effective in the asymmetric reduction of other aryloxyamino ketones, yielding key intermediates for the synthesis of centrally acting muscle relaxants, such as (*S*)-mephenoxalone and (*S*)-metaxalone.

(*S*,*S*)-MCCPM-Rh (0.01 mol %)

Et_3N, MeOH, H_2 (20 atm)

184

185
90.8% ee then single recrystallization = 100% ee

137, (*S*,*S*)-MCCPM

Schem 2-5. Asymmetric reduction of aryloxy amino ketones.

Zhang and co-workers[126b] reported that the Rh-DuanPhos system catalyzed highly efficient hydrogenation of a series of β-secondary amino ketones with excellent ee's and high turnover numbers. This method provides direct access to a series of γ-secondary amino alcohols that are key intermediates in the synthesis of an important class of antidepressants, including (*S*)-fluoxetine and (*S*)-duloxetine. A series of aromatic and heteroaromatic substituted β-secondary amino ketone hydrochlorides **186** were hydrogenated to afford the corresponding amino alcohols **187** in high yields (90–93%) and with excellent enantioselectivities. The electronic properties of the substituents on the phenyl ring were well tolerated in terms of reactivity and enantioselectivity.

Table 2-8. Hydrogenation of heteroaromatic substituted β-secondary amino ketone hydrochlorides.

Entry	Ar	R	% ee
a	2-Me-phenyl	Me	99
b	phenyl	Me	98
c	3-Br-phenyl	Me	96
d	4-Br-phenyl	Me	>99
e	2-MeO-phenyl	Me	93
f	4-MeO-phenyl	Me	>99
g	2-naphthyl	Me	99
h	2-thienyl	Me	>99
i	phenyl	Bn	96

The possible amino ketone precursors (**186b** and **186h**) for the synthesis of both (*S*)-fluoxetine and (*S*)-duloxetine, respectively, were subjected to gram-scale asymmetric hydrogenation with a high substrate-to-catalyst ratio (S/C = 6000). Both products were obtained in 75% yield with 98% and >99% ee, respectively. The γ-amino alcohols obtained could be converted to these important antidepressants in a single step.

Some success has been demonstrated using Et-DuPhos-Rh complex in the asymmetric hydrogenation of *N*-acylhydrazone derivatives **188**, which could be converted to the corresponding chiral amines (**189**) in a single step.[127]

Typical procedure[126b]

186b → $[Rh\{(S_c,R_p)\text{-DuanPhos}\}(nbd)]SbF_6$, MeOH, K_2CO_3, H_2 (10 bar), 50 °C (S/C = 6000) → **187b** 75% yield, 98% ee

Rh-complex $[Rh(S_c,R_p)\text{-DuanPhos}(nbd)]SbF_6$ (3 mg) was dissolved in 3 mL of degassed MeOH. To a suspension of phenyl amino ketone derivative **186b** (1.48 g, 7.39 mmol) in 10 mL of degassed MeOH in glove box was added 1 mL of the above complex solution (1 mg, 0.00123 mmol), followed by addition of K_2CO_3 (0.511 g, 3.70 mmol). The resulting mixture was transferred into an autoclave and the hydrogenation was performed under 50 bars of initial hydrogen pressure at 50 °C for 12 h (not optimal). After carefully releasing the hydrogen, the solvent was removed under reduced pressure. The resulting residue was dissolved in 10 mL of NaOH solution (1.0 N) and extracted with CH_2Cl_2 (10 mL × 4). The combined organic layers were dried over anhydrous Na_2SO_4 and concentrated under reduced pressure to give the crude product, which was passed through a short silica gel plug (eluting with 30 mL of EtOAc first, and then a mixture of EtOAc: MeOH: triethylamine = 10:10:1) to afford the amino alcohol **187b** as a pale yellow oil (0.91 g, 75% yield). The enantiomeric excess of **187b** (98%) was determined by chiral HPLC using an OD-H column after being converted to the *N*-acyl derivative as follows: To a solution of the amino alcohol **187b** (45 mg, 0.25 mmol) in CH_2Cl_2 (2 mL) at 0 °C was added acetic anhydride (24 μL, 0.25 mmol). The resulting reaction mixture was stirred at this temperature for 10 min. After removal of the solvent, the residue was purified by flash column chromatography (eluting with EtOAc) to afford the *N*-acyl derivative.

Typical procedure[127]

188 → (*R*,*R*)-Et-DuPhos-Rh, *i*-PrOH, H_2 (4 atm) → **189**

In a nitrogen-filled dry box, a 100-mL Fisher-Porter tube was charged with a stir bar and substrate **188** (0.5 to 1.26 mmol), followed by degassed 2-propanol (10 to 20 mL, *ca.* 0.05 to 0.10 M in substrate) and catalyst (0.1–0.2 mol%). After six vacuum/H_2 cycles to purge the lines of air and two vacuum/H_2 cycles on the reaction mixture, the tube was pressurized to an initial pressure of 60 psi of H_2. The reaction mixture was allowed to stir at 50 °C until no further hydrogen uptake was observed. Complete conversion to product was indicated by GC, TLC, and ^{1}H NMR analyses. The reaction mixture was concentrated, and the residue passed through a short SiO_2 column (EtOAC/hexane or Et_2O/pentane, 50/50) to remove catalyst residues. Without further purification, the enantiomeric excess of **189** was determined using the crude product, thus indicating a 97% ee.

8. Hydroacylation of Alkenes and Alkynes

The Rh(I)-catalyzed intramolecular cyclization of 4-pentenals such as **190** into cyclopentanones **191** was first reported by Sakai and co-workers.[128] The intramolecular hydroacylation of alkynes and alkenes with aldehydes gave either cyclopentenones or cyclopentanones, respectively. These motifs are of importance because of their frequent presence in natural products. High enantioselectivities have been obtained in hydroacylation reactions catalyzed by chiral rhodium complexes. Bosnich and co-workers have reported the hydroacylation of enals in which BINAP was found to be a effective ligand for the reaction. However, the substrate scope was determined to be limited to hindered substituents such as a *t*-butyl group.[129] Sakai *et al.* also found this to be the case using cationic Rh-BINAP complexes.[130] Later, the Me-DUPHOS ligand was observed to be effective for less bulky susbstituents.[131] Hoffman and Carreira also reported successful catalysis of this transformation using rhodium complexes with phosphor-amidite-alkene ligands and achiral phosphine coligands.[132]

General procedure for small-scale hydroacylation reactions[129]

(*S*)-BINAP-Rh

acetone or CH_2Cl_2

^{t}Bu **190** → ^{t}Bu **191**

The catalyst precursor [Rh((*S*)-binap)(nbd)]ClO_4 was converted to[Rh((*S*)-binap)-(solvent)$_2$]ClO_4 prior to the introduction of substrate asfollows. [Rh((*S*)-binap)(nbd)]ClO_4 (2.87 mg, 3.13×10^{-6} mol) was weighed into an NMR tube, which was then capped with a rubber septum. The tube was flushed with argon for at least 60 s, and then solvent (CD_2Cl_2 or acetone-d_6) was added by syringe. Argon was slowly bubbled through the solution for 2 min. The hydrogen gas was then bubbled through the solution for 2 min, causing a color change from orange-yellowto red-orange. Argon was bubbled through the solution for 2 minto purge any hydrogen gas. The 4-pentenal substrate **190** (7.82×10^{-5} mol) was then added by syringe. The reactions were monitored by ^{1} HNMR spectroscopy. Enantiomeric excess was determined by converting the crude cyclopentanones to the diastereomeric hydrazone derivatives using (*S*)-(-)-l-amino-2-(methoxymethyl)pyrrolidine by the method of Enders and Eichenauer.[133] Absolute configuration assignments were made by measuring the optical rotation of the crude cyclopentanones, as isolated above, and comparing the sign of rotation with those of known cyclopentanones. In other cases, assignments were made on the basis of similar behavior of the diastereomeric imine carbon signals.

In addition to simple asymmetric hydroacylation of alkenes, highly efficient desymmetrization of alkenes has also been achieved by Wu and co-workers.[134] Interestingly, the catalyst system (Rh(I)-BINAP) differentiated the enantiotopic faces of the olefins and thus the cyclopentanone products were obtained with excellent ee's. The authors also showed that neutral and cationic Rh(I)-BINAP complexes furnished different products. The neutral catalyst system preferentially furnished the *cis*-3,4-disubstituted cyclopentanone, and the cationic catalyst afforded the *trans* isomer.

Typical procedure[134]

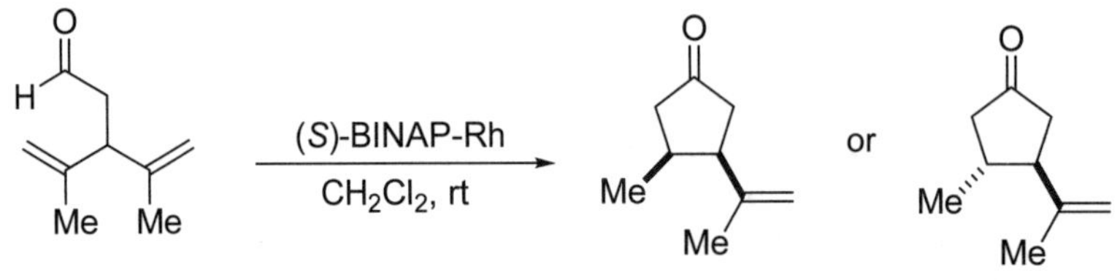

192

(*S*)-BINAP-Rh

CH_2Cl_2, rt

193a >99% ee

or

193b >99% ee

cationic Rh(I) catalyst **193a:193b** = 4:96
neutral Rh(I) catalyst **193a:193b** = 97:3

Aldehyde **192** in CH_2Cl_2 (4 mL) was added dropwise to a solution of the catalyst, Rh(BINAP)Cl (0.5 equiv, for neutral ligand conditions), or $[Rh(BINAP)]ClO_4^-$ (0.05 equiv, for cationic ligand conditions) in CH_2Cl_2 (4 mL). The mixture was stirred for 72 h (neutral ligand system) and 1 h (cationic ligand system). After removal of the solvent, ether (20 mL) was added. The resulting precipitate was filtered off and the filtrate was concentrated *in vacuo* to leave an oily residue, which was subjected to flash chromatography to obtain the products **193a** and **193b**. Optical rotations were measured using a JASCO DIP-360 polarimeter (JASCO Inc., Easton, MD, U.S.A) at 25 °C.

The hydroacylation of alkynals was also reported by Tanaka and Fu, who found that the Rh(I)-Tol-BINAP system was the catalyst of choice for the hydroacylation-desymmetrization of 3-*bis*-alkynals **194** to give 4,4-disubstituted cyclopentenones **195** in excellent yields and high enantioselectivities.[136] Cyclopentenones are important intermediates in the synthesis of natural products such as prostaglandins. This catalyst system was also found to be extremely effective in the kinetic resolution of racemic 3-disubstituted alkynals **196** giving the unreacted aldehyde **197a** with near perfect enantioselection.

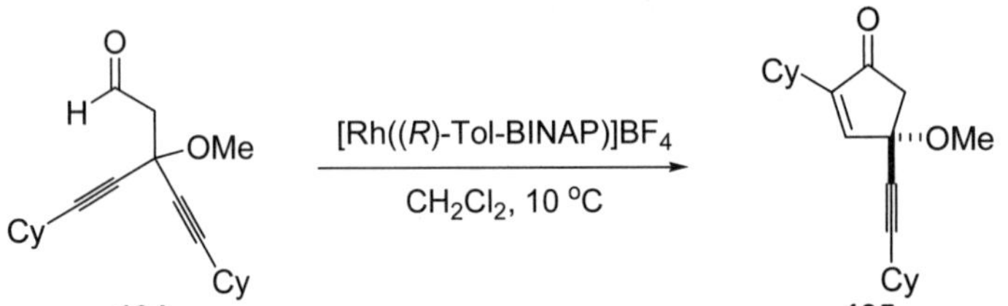

194

[Rh((*R*)-Tol-BINAP)]BF_4

CH_2Cl_2, 10 °C

195
94% yield, 95% ee

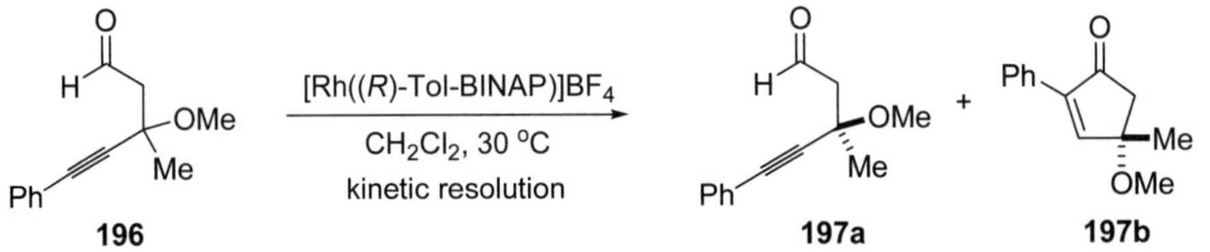

196

[Rh((*R*)-Tol-BINAP)]BF_4

CH_2Cl_2, 30 °C
kinetic resolution

197a
99% ee, 60% conversion

197b

Scheme 2-6. Hydroacylation of alkynals.

The same catalyst system was found to promote an intriguing parallel kinetic resolution in which one enantiomer of the substrate **198** was selectively transformed to cyclobutanone **199a** and the other enantiomer of the substrate was selectively transformed to the cyclopentenone **199b**.[137] Alternatively, the cyclobutanone **199a** or the cyclopentenone **199b** could be selectively obtained by cyclization of an enantiopure 3-methoxy-alkynal using the appropriate enantiomer of the catalyst system.

General procedure for hydroacylation[137]

$[Rh(Tol\text{-}BINAP)]BF_4$, CH_2Cl_2, rt

198 >99% ee → **199a** >99% ee + **199b** >99% ee

(*R*)-Tol-BINAP **199a:199b** = 89 : 11
(*S*)-Tol-BINAP **199a:199b** = 2 : 98

In a N_2-filled glove box, $[Rh(nbd)((R)\text{-Tol-BINAP})]BF_4$ (2.6 mg, 0.0027 mmol) and CH_2Cl_2 (0.5 mL) were added to a Schlenk tube. The Schlenk tube was taken out of the glove box, H_2 was introduced into the tube, and the mixture was stirred at room temperature for 0.5 h. The Schlenk tube was then taken into the glove box, and the CH_2Cl_2 was removed under vacuum. To the residue was added a CH_2Cl_2 (0.2 mL) solution of (*R*)-3-methoxy-5-phenylpent-4-ynal **198** (10.0 mg, 0.053 mmol, >99% ee), prepared by the kinetic resolution of (±)-3-methoxy-5-phenylpent-4-ynal with $[Rh((R,R)\text{-Me-DUPHOS})]BF_4$. The mixture was stirred at room temperature for 24 h, and then the resulting solution was concentrated, and the catalyst was removed by column chromatography (methyl *tert*-butyl ether) to give a residue that was analyzed by GC. The ratio of (*R*)-2- benzylidene-3-methoxycyclobutanone **199a** (>99% ee):(*R*)-4-methoxy-2-phenylcyclopent-2-enone **199b** (>99% ee) was 89:11. The (*R*)-2-benzylidene-3-methoxycyclobutanone (6.2 mg) was isolated by preparative TLC (hexanes:EtOAc = 4:1).

9. C-H Activation by Metal-Carbenoid and Nitrenoid Induced C-H Insertion

Over the past two decades, the activation of unfunctionalized C-H bonds has been extensively studied.[138] In the past few years, significant progress has been made in the area of catalytic metal-carbenoid and nitrenoid C-H insertion chemistry. The transient, metal-carbenoids are mostly derived from easily prepared diazo compounds. The intramolecular version of metal-carbenoid induced C-H insertion was the focus of much of the early research in this area. However, recent progress has been made in the intermolecular version.[139] Earlier work in the area of metal carbenoid-chemistry involved the use of copper complexes as catalysts, but the scope of carbenoid chemistry was greatly expanded with the introduction of dirhodium complexes as catalysts.[138] A variety of dirhodium complexes has been developed for carbenoid C-H activation including dirhodium carboxylate,[73b] dirhodium carboxamidate,[140] and dirhodium phosphate[141] complexes. Doyle's chiral carboxamidate complexes have been effective in catalyzing intramolecular C-H activation of diazoacetates.[142] These catalyst systems have been shown to display high chemoselectivity and regioselectivities along with high enantioselectivity.

Using Doyle's second-generation imidazolidinone catalyst system, $Rh_2(MPPIM)_4$, a series of benzyl substituted γ-butyrolactones was synthesized by intramolecular C-H activation. The strategy was applied to the synthesis of various ligands such as enterolactone (**202**), hinokinin, and isodeoxypodophyllotoxin.[143]

Typical procedure[143b]

$Rh_2(4R\text{-MPPIM})_4$, CH_2Cl_2, 40 °C

200 → **201** (62% yield, 93% ee) ⇉ **202** (-)-enterolactone

A solution of **200** (0.65 g, 2.8 mmol) in 20 mL of rigorously dried $CH_2Cl_2$was added *via* syringe pump at a rate of 2.0 mL/h to a refluxing solution of $Rh_2(MPPIM)_4$(78 mg, 56 mmol, 2.0 mol%) in 40 mL of dry CH_2Cl_2under N_2. The initial blue/purple color of the solution turned to olive green by the end of substrate addition. After addition was complete, the reaction mixture was cooled to room temperature, and the solvent was evaporated under reduced pressure. The residue was purified by flash chromatography on silica gel (hexanes:EtOAc, 4:1), and 0.379 g of pure lactone **201** (1.84 mmol, 66% yield) was isolated as a colorless oil:Enantiomericexcess was determined to be 93% ee by GC analysis on a 30-mChiraldex A-DA column operated at 150 °C for 10min then programmed at 0.2 °C/min to 180 °C.

High levels of enantioselectivities have also been attained in the intermolecular version of this reaction.[139, 144] For instance, Davies and Antoulinakis reported that the $Rh_2(S\text{-DOSP})_4$-catalyzed decomposition of *ortho*, *meta,* and *para* substituted aryldiazoacetates (**204**) in refluxing cyclohexane generated the C-H insertion products (**205**) in generally good yields and in good-to-excellent enantiopurity.[139] Using the same catalyst system, C-H activation of branched alkanes tetrahydrofuran and tetraalkoxysilaneswas also attained. [139, 144]

Table 2-9. Intermolecular C-H activation.

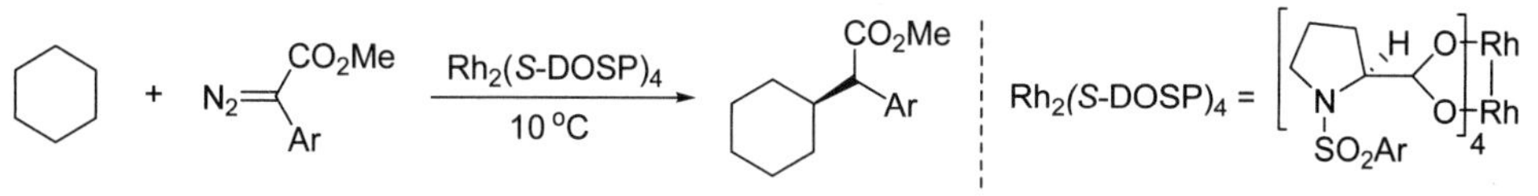

203 + 204 (N_2=C(CO_2Me)Ar) → ($Rh_2(S\text{-DOSP})_4$, 10 °C) → 205

$Rh_2(S\text{-DOSP})_4$ = 206

Ar = p-($C_{11-13}H_{23-27}$)C_6H_4

Entry	Ar	Yield (%)	% ee
a	C_6H_5	80	95
b	*p*-BrC_6H_4	64	95
c	*p*-ClC_6H_4	76	94
d	*p*-(MeO)C_6H_4	23	88
e	*p*-$CF_3C_6H_4$	78	94
f	*p*-MeC_6H_4	63	93
g	*o*-BrC_6H_4	72	90
h	*o*-ClC_6H_4	81	90
i	*m*-BrC_6H_4	62	95
j	*m*-ClC_6H_4	47	94

Typical procedure[139]

A degassed solution of aryldiazoacetate **204** (1 mmol) in anhydrous cyclohexane (10 mL) was added dropwise over 90 min to a stirring degassed solution of $Rh_2(S\text{-DOSP})_4$ (**206**) (0.01 mmol) in anhydrous cyclohexane (5 mL) at 10 °C. The solution was stirred at 10 °C for another 15 min after the addition was over and then warmed to ambient temperature. The solvent was removed under reduced pressure and the residue purified by flash chromatography on silica gel (2% diethyl ether/petroleum ether). Enantiomeric purity of **205** was determined by either ^{1}H-NMR using chiral shift reagents, HPLC using chiral analytical columns, or GC using chiral analytic columns.

C-H insertion has also been shown to be favorable at allylic positions.[139] The high activity of the allylic position allows for the reaction to be carried out at low temperatures, leading to high enantioselectivities.

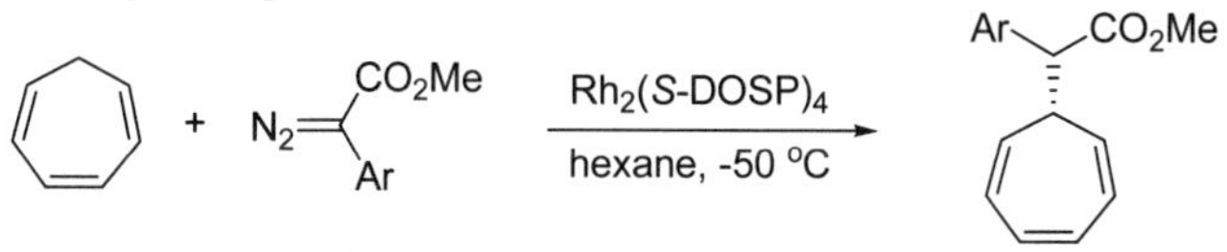

207 + 204 → ($Rh_2(S\text{-DOSP})_4$, hexane, -50 °C) → 208 (91-95% ee)

Scheme 2-7. C-H activation of allylic C-H.

C-H insertion of *N*-Boc-pyrrolidine (**209**) has also been reported. It was found that C2-symmetric amine **210** was formed in a single step using excess diazoacetate **204a**.[145]

Typical procedure[145]

1) $Rh_2(S\text{-DOSP})_4$
2,3-Me_2butane
-50 °C to 58 °C
2) TFA

209 **204a** **210**
78% yield, 97% ee

A solution of methyl phenyldiazoacetate (**204a**) (270 mg, 1.53 mmol), in 2,3-dimethylbutane (12 mL) was added dropwise (3 mL, 1.5 equiv) to a –50 °C solution of $Rh_2(S\text{-DOSP})_4$ (**206**) (28 mg, 0.015 mmol) and *N-tert*-butoxycarbonyl-pyrrolidine carboxylate **209** (43 mg, 0.25 mmol) in 2,3-dimethylbutane (5 mL). The solution was stirred at –50 °C overnight, warmed slowly to room temperature, heated to reflux under argon and the rest of the diazo solution (9 mL, 4.5 equiv) was added dropwise. The solvent was removed *in vacuo*, the crude product dissolved in CH_2Cl_2 and treated with TFA (1 mL) for 4 h at room temperature. The solution was concentrated, dissolved in Et_2O, extracted with 1% HCl (three times), the aqueous phase basified ($NaHCO_3$, 1-M NaOH to pH 10–11) extracted with CH_2Cl_2 (three times), dried (anhydrous $MgSO_4$), concentrated in *vacuo* and purified *via* silica gel chromatography (20% Et_2O/petroleum ether, 2% NEt_3) to give 78% yield (72 mg) of free amine **210**. A small sample of the free amine was dissolved in CH_2Cl_2, treated with trifluoroacetamide (TFAA), purified through a plug of silica gel and subjected to HPLC analysis to give 97% ee.

Davies and co-workers[138] found that C-H donor/acceptor substituted carbenoids were capable of undergoing highly regioselective insertions. Thus, this strategy has been shown to be an aldol reaction equivalent,[144, 146] Claisen condensation equivalent,[147] Mannich reaction equivalent,[145] and other important carbon–carbon bonding forming reaction equivalents.[148]

An asymmetric rhodium-carbenoid insertion at the benzylic position strategy was used as a key step in the synthesis of (+)-imperanene and (-)-α-conidendrin. The C-H activation product with the *S* configuration can be converted to (+)-imperanene *via* a reduction/deprotection. (-)-α-Conidendrin can be synthesized *via* lactonization of the C-H activation product with the *R* configuration. The stereochemistry of these key intermediates is controlled by using opposite enantiomers of the catalyst.[149]

General procedure[149]

MeO TBSO 211 + N_2 O OMe OMe OTBS 212 → $Rh_2(S\text{-DOSP})_4$, 2,2-dimethylbutane, 50 °C → O OMe H MeO OTBS OMe OTBS 213 44% yield, 92% ee

Under an Ar atmosphere, to a refluxing degassed solution of **211** (1.26 g, 5 mmol) and $Rh_2(S\text{-DOSP})_4$ **206** (9.5 mg, 1 mol%) in 2,2-dimethylbutane (2 mL) was added a degassed solution of diazo compound **212** (181 mg, 0.5 mmol) in 2,2-dimethylbutane (5 mL) in 45 min using a syringe pump. The reaction mixture was refluxed for additional 15 min then cooled to room temperature. The solvent was removed *in vacuo* and the residue was purified by flash chromatography on silica gel (10:1–5:1 pentane/ether) to afford **213** as a yellow oil.

Oxazolidinones are frequently used as chiral auxiliaries in asymmetric transformations, as ligands for metal catalysts, and as precursors to vicinal amino alcohols. These common structural motifs in natural products and pharmaceuticals have spurred interest in the development of methods for their preparation. Espino and Du Bois reported the synthesis of indane-appended oxazolidinone **215** *via* Rh-catalyzed C-H insertion of carbamate ester **214**.[150] A variety of other carbamates was found to be competent substrates for this reaction. The role of magnesium oxide as additive was presumed to be as scavenger of acetic acid generated from the iodinane oxidant.

2–Oxazolidinone preparation[150]

H_2N O O 214 → $PhI(OAc)_2$, $Rh_2(OAc)_4$, MgO, CH_2Cl_2, 40 °C → H N O O 215

To a solution of carbamate **214** (1.26 mmol) in 8 mL of CH_2Cl_2 were added successively MgO (117 mg, 2.89 mmol, 2.3 equiv), $PhI(OAc)_2$ (486 mg, 1.51 mmol, 1.4 equiv), and Rh(II) catalyst (63 μmol, 0.05 equiv). The mixture was stirred vigorously and heated at 40 °C for 12 h. After cooling to 25 °C, the reaction mixture was diluted with 10 mL of CH_2Cl_2 and filtered through a pad of Celite (30 × 20 mm). The filter cake was rinsed with 2 × 10 mL of CH_2Cl_2. The combined filtrates were evaporated under reduced pressure and the isolated residue purified by chromatography on silica gel to give **215** in 86% yield.

Wehn *et al.* have also reported the intramolecular aziridination of chiral homoallyl sulfamate esters, such as **216**, with moderate selectivity.[151] The bicyclic products, such as **217**, could be converted to polyfunctionalized amines **219** *via* **218** in a few steps (Scheme 2-8).

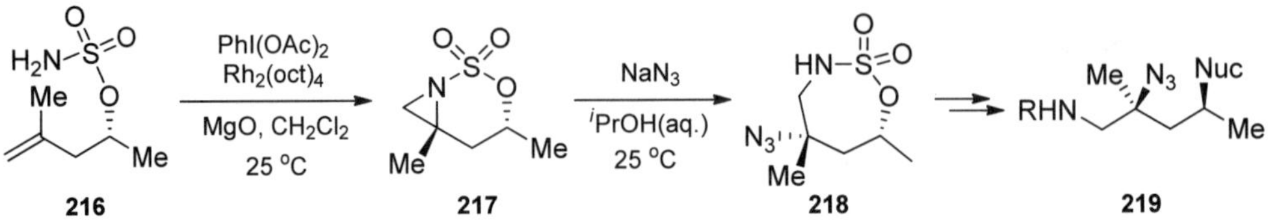

Scheme 2-8. Synthesis of polyfunctionalized amines *via* aziridine intermediates.

Typical procedure[151]

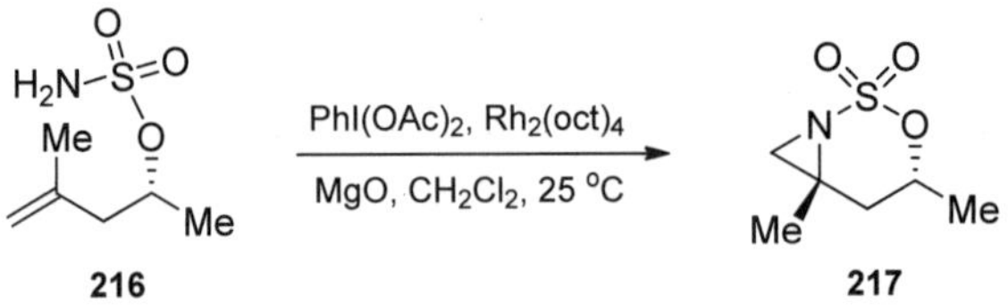

To a solution of sulfamate ester **216** (1.25 mmol) in 8 mL of CH_2Cl_2 were added successively MgO (116 mg, 3.0 mmol, 2.3 equiv), $PhI(OAc)_2$ (443 mg, 1.4 mmol, 1.1 equiv), and 25 μmol (0.02 equiv) of $Rh_2(oct)_4$ as indicated. The mixture was stirred vigorously and heated at 40 °C until complete consumption of starting material was indicated by thin layer chromatography (1–2 h). The reaction mixture was cooled to 25 °C, diluted with 20 mL of CH_2Cl_2, and filtered through a pad of Celite (30 × 20 mm). The filter cake was rinsed with 2 × 15 mL of CH_2Cl_2 and the combined filtrates were evaporated under reduced pressure. The residue was purified by chromatography on silica gel to afford the product **217** in 92% yield.

The Rh(II)-catalyzed intermolecular oxidative sulfamidation of aldehydes was reported by Chan and co-workers.[152] Sulfonamide **220** reacted with **221** to afford the coupling product **222** in excellent yield. Overall, electron-deficient sulfonamides reacted more efficiently than electron-donating sulfonamides under the coupling conditions.

Typical procedure[152]

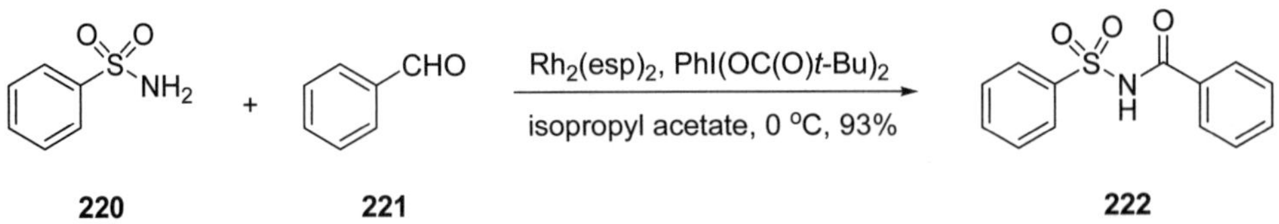

To a mixture of benzenesulfonamide (**220**) (314 mg, 2 mmol), bis(*tert*-butylcarbonyloxy)iodobenzene (1.62 g, 4 mmol) in isopropyl acetate (10 mL) was added benzaldehyde (**221**) (300 μL, 3 mmol). The mixture was cooled to 0 °C and $Rh_2(esp)_2$ (esp = α,α,α',α'-tetramethyl-1,3-benzenedipropionate) (30 mg, 0.04 mmol) was added in a single portion. After stirring for 1 h, the solution was warmed to room temperature and allowed to stir for 2 h. The mixture was concentrated *in vacuo* and charged with 9:1 hexane:acetone (5 mL) and slurried for 2 h. The solids were filtered, washed with 9:1 hexanes:acetone (5 mL), and dried in a vacuum oven to afford the product **222** (492 mg, 93%).

The enantioselective C-H functionalization by metal-nitrenes has been more problematic than the parallel reaction with metal-carbenes because fewer options are available for modulating the reactivity of the metal nitrenes.[153] A highly efficient rhodium-catalyzed intermolecular C-H amination reaction with a sulfonimidamide as the nitrene precursor was reported by Liang *et al.*[154] The optimized procedure was found to be efficient for electron-rich C-H bonds, whereas the selectivity dropped slightly in the presence of an electron-withdrawing group. The high efficiency and selectivity of the transformation was correlated to a pronounced matched effect between the chiral rhodium catalyst and the chiral sulfonimidamide. High enantioselectivity was obtained only when both catalyst and sulfonimidamide configurations matched. Also worth mentioning is the fact that the trapping agent is used as the limiting reagent in this transformation. Good-to-excellent diasteroselectivities (80–99% de) were obtained on a variety of aromatic substrates with benzylic sites, whereas the asymmetric induction was considerably lower when substrates with allylic sites were used.

Typical procedure[154]

223 + **224** ($[Rh_2((S)\text{-NTTL})_4]$, $PhI(CO_2t\text{-}Bu)_2$, DCE, MeOH (3:1)) → **225**

88% yield, 99% de

In an oven-dried tube was introduced activated 4 Å molecular sieves (100 mg), $[Rh_2((S)\text{-NTTL})_4]$[69g] (8.7 mg, 0.006 mmol), and (-)-*N*-(*p*-toluenesulfonyl)-*p* toluenesulfonimidamide (**224**) (78 mg, 0.24 mmol). The tube was capped with a rubber septum and purged with Ar. 1,1,2,2-Tetrachloroethane (0.75 mL) and methanol (0.25 mL) were added under argon, and the mixture was stirred for 5 min before addition of the substrate **223** (0.2 mmol). The tube was cooled to –35 °C, and *bis*(*tert*-butylcarbonyloxy)iodobenzene (115 mg, 0.28 mmol) was added. The mixture was stored in the freezer (–35 °C) for 3 days. After dilution with dichloromethane (3 mL), the molecular sieves were removed by filtration and the filtrate was evaporated to dryness under reduced pressure. The oily residue was purified by flash chromatography on silica gel (dichloromethane/ethyl acetate: 20/1), affording the C-H insertion product **225** as a white solid in 88% yield and >99% de.

The use of Rh(II) catalysts for the formation of 1,3-dipoles from diazo compounds *via* rhodium-carbenoids has facilitated the use of the dipolar cycloaddition reaction in key steps in the preparation of natural products.[155] The synthesis of aspidosperma alkaloids **227** and their derivatives is important because these alkaloids contain the highly functionalized vindoline nucleus which is found in the clinically useful antineoplastic agents vincristine and vinblastine. Padwa and Price used a Rh(II)-catalyzed 1,3-dipole formation in an intramolecular 1,3-dipolar cycloaddition reaction as a key step in the synthesis of the pentacyclic skeleton of dehydrovindorosine alkaloid in excellent yield.[155] A similar route was used by Kataoka *et al.* in their second-generation approach to zaragozic acids.[156]

General procedure[155]

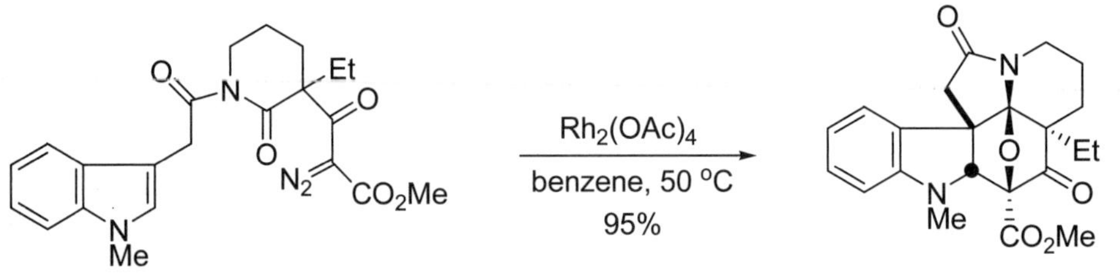

226 → 227 ($Rh_2(OAc)_4$, benzene, 50 °C, 95%)

To a solution of 450 mg (1.1 mmol) of diazoacetate **226** in 5 mL of benzene under nitrogen was added 2 mg of rhodium(II) acetate. The mixture was heated in an oil bath at 50 °C for 4 h and concentrated under reduced pressure, and the residue was subjected to flash silica gel chromatography to give 410 mg (95%) of 3-carbomethoxy-4,10-dioxo-3,19-epoxy-1-methylaspidospermadine **227** as a white solid.

An intramolecular 1,3-dipole formation followed by cycloaddition was used by Dauben *et al.*[157] in their approach to a tigliane natural product, phorbol. The reaction afforded compound **229**, which contains the oxo-bridged BCD ring system of phorbol.

Srikrishna *et al.* used the rhodium-catalyzed carbenoid mediated C-H activation of a tertiary methyl group to form the angular triqhinane system of sesquiterpenes.[158] The highly regioselective transformation of the α-diazo-β-keto ester to the corresponding triquinane was catalyzed by rhodium acetate. The use of Rh(II) catalysts for the formation of 1,3-dipoles from diazo compounds *via* rhodium carbenoids followed by 1,3-dipolar cycloadditions was also applied as a key step in the second-generation synthesis of zaragozic acid by Kataoka and co-workers (Scheme 2-9).[156]

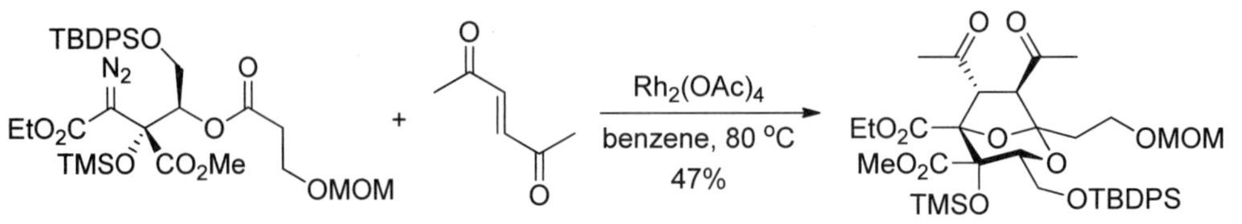

232 + 233 → 234 ($Rh_2(OAc)_4$, benzene, 80 °C, 47%)

Scheme 2-9. Formation of 1,3-dipoles from diazo compounds.

Typical procedure[157]

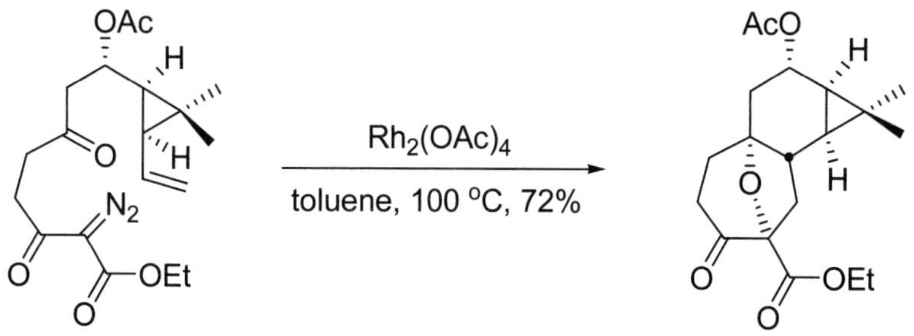

$$\mathbf{228} \xrightarrow[\text{toluene, 100 °C, 72\%}]{Rh_2(OAc)_4} \mathbf{229}$$

A 10-mL flask containing 4 mL of toluene and 37 mg of cyclopropyl-diazo compound **228** was heated in an oil bath at 100 °C. A catalytic amount (~1–2 mg) of rhodium acetate was added and nitrogen evolution instantly became evident. The solution was stirred for 1 h, cooled, and concentrated under reduced pressure. The reaction mixture was purified by column chromatography on silica gel (15% EtOAc/hexane) to provide **229** (only one isomer detected by ^{1}H NMR) as a thick oil (21 mg, 62%). The material was taken up in 2 mL of ethyl acetate, filtered through a pipette column (1 cm of silica gel) into a vial, and evaporated by flowing nitrogen over the solution until a white foamy solid formed. This was redissolved in a minimum of hot hexane and allowed to sit at room temperature for 3 days. Crystals were isolated and the supernatant was refrigerated at 5 °C. X-ray quality crystals were obtained from this bath after 3 days, mp = 115 °C.

Formation of triquinane esters[158]

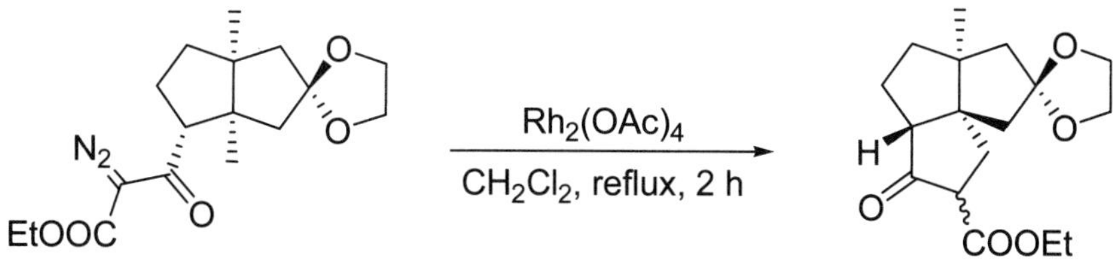

$$\mathbf{230} \xrightarrow[CH_2Cl_2\text{, reflux, 2 h}]{Rh_2(OAc)_4} \mathbf{231}$$

To a magnetically stirred refluxing suspension of $Rh_2(OAc)_4$ (3 mg) in dry CH_2Cl_2 (4 mL) was added a solution of the α-diazo-β-keto ester **230** (60 mg, 0.18 mmol) in dry CH_2Cl_2 (14 mL) over a period of 30 min and refluxed for 2 h. Evaporation of the solvent and purification of the residue on a silica gel column using ethyl acetate-hexane (1:19) as eluent furnished an epimeric mixture of the triquinane ester **231** (48 mg, 87%) along with its enol form as an oil.

10. Rh-Catalyzed C-H Activation

The activation of unreactive C-H bonds remains a challenge for synthetic organic chemists. Activation of such bonds provides the opportunity to functionalize relatively cheap and abundant hydrocarbons. The clear advantages of directly forming carbon–carbon bonds from carbon–hydrogen bonds have driven the development of a variety of reactions in this area.[159] Chatani *et al.* have reported carbonylation of sp^3 C-H bonds of secondary amines.[160] In this reaction, various secondary amines were employed as substrates, and it was found that the presence of a pyridine ring adjacent to the amine group was essential for the carbonylation to proceed.

Typical procedure for carbonylation at sp^3 C-H bonds adjacent to nitrogen atom in alkylamines[160]

$[Rh(cod)Cl]_2$

CO, $H_2C{=}CH_2$

i-PrOH, 160 °C

235 **236**

A 50-mL stainless steel autoclave was charged with 2-(1-pyrrolidinyl) pyridine (**235**) (1 mmol, 148 mg), 2-propanol (3 mL), and $[RhCl(cod)]_2$ (0.04 mmol, 20 mg). After the system was flushed with 10 atm of ethylene three times, it was pressurized with ethylene to 5 atm and then with carbon monoxide to additional 10 atm. The autoclave was immersed in an oil bath at 160 °C. After 40 h had elapsed, it was removed from the oil bath and allowed to cool for 1 h. The gases were then released. The contents were transferred to a round-bottomed flask with toluene, and the volatiles were removed *in vacuo*. The residue was subjected to column chromatography on silica gel (eluent: hexane/Et_2O = 5/1) to give a mixture of 2-(1-pyrrolidinyl)pyridine and 1-[1-(2-pyridinyl)2-pyrrolidinyl]-1-propanone (179 mg). The yield of propanone was determined by comparing the integrations of the 6-H signals on the pyridine for the starting material and product in the ^1H NMR spectrum of the mixture. Purification by HPLC afforded the analytically pure product **236**.

The functionalization of benzylic or arene C-H bonds with boranes leads to synthetically useful boranates. Shimada and co-workers showed that $[RhCl(P\textit{i}Pr_3)_2(N_2)]$ is an effective catalyst precursor for the borylation of aromatic and benzylic C-H bonds with the use of pinacolborane, resulting in high selectivity for benzylic C-H functionalization.[161]

Procedure for hydroboration[161]

$[RhCl(P\textit{i}\text{-}Pr_3)_2N_2]$

pinacolborane

237 **238**

In an N_2-filled glove box, a 15-mL tube with a Young's tap was charged with $[RhCl(P\textit{i}Pr_3)_2N_2]$ (2.0 mg, 0.0041 mmol, 1.0 mol%), pinacolborane (58 mL, 0.40 mmol), *n*-dodecane (standard for GC analysis, 0.065 mmol), and *p*-xylene **237** (2.0 mL). The mixture was heated at 140 °C for 80 h. Small samples were removed periodically for GC(FID) and GC-MS analysis. The yield and product ratio were determined by GC and were based on pinacolborane. The product **238** can be isolated by bulb-to-bulb distillation or silica gel chromatography and was obtained in 41% yield with a product distribution of 98:2 (4-$MeC_6H_4CH_2Bpin$: 2,5-$Me_2C_6H_3Bpin$).

The *ortho*-alkylation of aromatic ketimines with functionalized olefins has been reported by Lim and co-workers.[162] Among many olefin substrates bearing functional groups, acrylates and acrylamides showed good reactivity. The reaction could also be extended to other olefins, including phenyl vinyl sulfone and acrylonitrile, albeit with lower reactivity. The corresponding ketones were obtained after hydrolysis of the ketimines.

General procedure for ortho-alkylation of aromatic ketimine with functionalized alkene by a Rh(I) catalyst[162]

NBn, Me + OMe, O → 1) $Rh(PPh_3)_3Cl$, toluene, 150 °C; 2) H_3O^+ → O, Me, OMe, O

239 **240** **241**

A screw-capped pressure vial (1 mL) equipped with a magnetic stirring bar was charged with aromatic ketimine **239** (0.324 mmol), alkene **240** (0.389 mmol), $(PPh_3)_3RhCl$ (16.2 μmol), and toluene. The vial was closed and heated at 150 °C with vigorous stirring for 2 h. After cooling to room temperature, the reaction mixture was hydrolyzed with 1 *N* HCl. The organic layer was extracted with Et_2O, and dried over anhydrous $MgSO_4$, filtered, and concentrated using a rotatory evaporator. The product was purified by column chromatography (*n*-hexane: EtOAc = 5:2) on silica gel to give **241**.

Although there is a growing interest in the applications of C-H activation to organic synthesis, enantioselective versions of this transformation are currently not abundant. Thalji *et al.* reported a highly enantioselective catalytic reaction involving aromatic C-H bond activation of an aryl vinyl ether.[163] A BINOL-based phosphoramidite ligand was exceptionally active, allowing the reaction to proceed at room temperature, which is a dramatic improvement over achiral catalyst systems that require relatively high temperatures.

The dihydropyrroloindole core is a frequently recurring motif in natural products and drug candidates.[164] As an extension of the asymmetric variant of catalytic C-H bond activation, Wilson *et al.* used a Rh-catalyzed enantioselective C-H bond activation reaction to form the key dihydropyrroloindole core of a known protein kinase C (PKC) inhibitor with 90% ee.[165] The authors found that using the *bis*-trifluoromethyl groups on the imine moiety was critical in obtaining high yields and enantioselectivity in this transformation.

Typical procedure[163]

BnN 242 → [$RhCl(coe)_2]_2$ / L*, toluene, rt → BnN 243, 95% yield, 96% ee | L* = Ph, O, P–N, Ph 244

In a dry box, to a medium-walled glass reaction vessel was added a solution of $[RhCl(coe)_2]_2$ (5 mol%), chiral ligand **244** (10 or 15 mol%), and aryl vinyl ether **242** (60.0 mg, 0.226 mmol) in toluene (0.1 M). Critical to reaction conversion (but not enantioselectivity) is pre-mixing $[RhCl(coe)_2]_2$ and the chiral ligand before addition of **242**. Additionally, pure starting material must be used. The mixture was then stirred at room temperature for 23 h. The mixture was concentrated and the product was hydrolyzed by adding 1 *N* HCl (aq) and vigorously stirring for 3 h. After extraction with EtOAc (three times), the combined extracts were dried, filtered, and concentrated. The residue was dry-loaded onto a silica gel column for purification (10% EtOAc/hexanes) and **243** was obtained as a white solid. mp: 52–54 °C.

Typical procedure[165]

CF_3, CF_3, N, H, N, OMe 245 → 1) $[RhCl(coe)_2]_2$ / L*, toluene, 90 °C; 2) 10% AcOH/THF → O, H, N, OMe 246 | L* = Ph, O, P–N, Ph 244

To a stirred solution of fresh $[RhCl(coe)_2]_2$ (48.9 mg, 0.0682 mmol) and phosphoramidite ligand **244** (73.6 mg, 0.136 mmol) in toluene (3.4 mL) was added imine-alkene substrate **245** (310 mg, 0.682 mmol) dissolved in toluene (6.8 mL, 0.1 M overall).

The flask was sealed and heated to 90 °C for 21 h, after which time the reaction mixture was concentrated, redissolved in 10% AcOH/THF (30 mL), and stirred for 6 h at room temperature. The mixture was concentrated, redissolved in CH_2Cl_2, and washed with water (30 mL), saturated aqueous sodium bicarbonate solution (30 mL), and water (30 mL). The organic layer was then dried over anhydrous magnesium sulfate, concentrated, and chromatographed (5% EtOAc/CH_2Cl_2) to provide **246** in 61% yield (95.2 mg) and 90% ee.

The microwave-promoted arylation of heterocyclic compounds by C-H activation was also reported.[159] It was found that the use of 9-cyclohexylbicyclo[4.2.1]-9-phosphanonane as a ligand was critical in obtaining high yields. With this ligand, aryl bromides were efficiently coupled to a variety of heterocyclic compounds as compared with other catalyst systems that are prone to cause hydrodehalogenation.

Typical procedure[159]

$[RhCl(coe)_2]_2$ / L
1,2-dichlorobenzene
i-Pr_2N-*i*-Bu
microwave (250 °C)
90%

247 + **248** → **249**

L = **250a** + **250b**

250a:**250b** = 2:1

To a 5-mL glass microwave vial was added benzimidazole **247** (0.106 g, 0.897 mmol), a magnetic stir bar, and 2 mL of 1,2-dichlorobenzene. The catalyst solution [1.0 mL of a 1,2-dichlorobenzene (3.1 mL) solution of $[RhCl(coe)_2]_2$ (0.0997 g, 0.139 mmol) and a mixture of **250a** and **250b** (0.188 g, 0.837 mmol)] was then added. Then, 4-cyano-1-bromobenzene (**248**) (0.20 mL, 1.8 mmol) and diisopropylbutylamine (0.55 mL, 2.7 mmol) were added to the reaction vessel. The vial was sealed and heated for 40 min at 250 °C. The reaction mixture was then cooled, quenched with excess Et_3N (1 mL), and concentrated under reduced pressure to remove 1,2-dichlorobenzene. The residue was dissolved in a minimal amount of methanol and loaded onto a silica gel samplet (Biotage No. SAM-1107-16016, Biotage AB, Uppsala, Sweden) and purified using flash chromatography eluting with 30% EtOAc/hexanes to provide 0.0792 g, 90% yield (0.4 mmol scale) of **249** as an off-white solid.

The synthesis of phthalimides has also been achieved using the Rh(III)-catalyzed oxidative addition of aromatic amides *via* C-H/N-H activation.[166] The reaction was carried out using $RhCp^*(MeCN)_3(ClO_4)_2$, KH_2PO_4 as an additive under 1 atmosphere of carbon monoxide in the presence of an oxidizing agent. Various phthalimides were obtained in up to 94% yield.

General procedure for phthalimide synthesis[166]

$[RhCp^*(MeCN)_3](ClO_4)_2$, Ag_2CO_3, KH_2PO_4, CO (1 atm), *t*-AmOH, 100 °C

251 → **252**

All reactions were carried out in 16×150 mm test tubes sealed with CO balloons with magnetic stirring at 600 rpm in oil bath. To the test tubes, $RhCp^*(MeCN)_3(ClO_4)_2$ (6.5 mg, 0.01mmol) followed by substrate **251** (27.0 mg, 0.2 mmol), Ag_2CO_3 (115.8 mg, 0.42 mmol), KH_2PO_4 (57.2 mg, 0.42 mmol), and *t*-AmOH solvent (0.37 mL) was added. The reaction mixture bubbled with CO and then was sealed with CO balloons and heated to 100 °C for 24 h under vigorous stirring. Upon completion, the reaction mixture was cooled to room temperature and diluted with ethyl acetate and then filtered through a small pad of Celite. The filtrate was concentrated *in vacuo*. The NMR yield of desired product **252** was determined by integration using an internal standard (1,3,5-trimethoxybenzene).

11. Rh-Catalyzed 1,4-Conjugate Addition of Boronic Acids

The first Rh-catalyzed nonasymmetric 1,4-conjugate addition of aryl- and alkenyl-boronic acids to α,β-unsaturated ketones was reported by the Miyaura laboratory.[167] Since this seminal report, extensive studies have been performed in modifying the original conditions to achieve high enantioselectivity.[168] There are several advantages of this reaction over other 1,4-conjugate addition reactions.[168b] The relative stability of organoboronic acids toward oxygen, compared with other organometallic reagents, allows the reaction to be carried out in aqueous media.[169] In addition, organoboronic acids are less reactive to enones in the absence of a Rh-catalyst, and no 1,2-addition products are observed. Thus, aryl and alkenyl groups can be selectively introduced to the β-position. In 1998, the first examples of the asymmetric variant of this reaction were reported by Takaya *et al.*[170] using BINAP as the chiral ligand. The scope of the reaction was found to be broad and excellent enantioselectivities (91–99%) were obtained in the conjugate addition of a variety of aryl and alkenyl boronic acids to various cyclic and acyclic α,β-unsaturated ketones. The reaction conditions reported are often used as the standard set of conditions in other asymmetric conjugate additions.[171]

A detailed NMR study of the catalytic cycle of the Rh-catalyzed 1,4-conjugate addition was performed by Hayashi and co-workers.[172] An important finding was that the acetylacetonato (acac) ligand used as a catalyst precursor in the original asymmetric transformation reported by Takaya *et al.*[173] retarded the transmetalation step because of the high stability of the Rh-acac species. Thus, the use of $[Rh(OH)(BINAP)]_2$ as the catalyst enabled the reaction to proceed at lower temperature (35 °C compared with 100 °C with $[Rh(acac)(C_2H_4)_2]$ as the catalyst precursor). Because of the lower reaction temperature used for the new catalyst system, the enantioselectivity and chemical yield were higher and less of the boron reagents were needed compared with the original procedure.[168b]

Alkenylcatecholboranes **258**, derived from the hydroboration of alkynes, *e.g.*, **256**, with catecholborane, are also useful alkenylating reagents.[173] Using the original procedure reported by Takaya *et al.*,[173] the desired products (**259**) were obtained with excellent enantioselectivity but low yield. The low yield was caused by the hydrolysis of the alkenylcatecholborane **258**, rendering the reaction media acidic. Addition of triethylamine to the reaction medium greatly improved the chemical yield without erosion of enantioselectivity. Several other alkenylboranes derived from terminal alkynes were applied and products were obtained with high enantioselectivity. A one-pot synthetic procedure of enantio-enriched β-alkenyl ketones from alkynes and catecholborane without isolation of alkenylcatecholboranes was also reported.

Typical procedure[170]

Rh(acac)(C_2H_4)$_2$, (*S*)-BINAP; dioxane/H_2O (10:1), 100 °C

253 + **254** → **255**, > 99% yield, 97% ee

1,4-Dioxane (1.0 mL) was added to a flask charged with Rh(acac)(C_2H_4)$_2$ (3.1 mg, 12 μmol), (*S*)-BINAP (7.5 mg, 12 μmol), and PhB(OH)$_2$ (**254**) (244 mg, 2.00 mmol), and then it was flushed with nitrogen, followed by addition of water (0.1 mL) and 2-cyclohexenone (**253**) (39 mg, 0.40 mmol). The resulting mixture was then stirred at 100 °C for 5 h. After evaporation of the solvent, the residue was dissolved in ethyl acetate. The solution was washed with saturated sodium bicarbonate and dried over anhydrous Na_2SO_4. Chromatography on silica gel (hexane/AcOEt = 5/1) gave 3-phenylcyclohexanone **255** (70 mg, > 99% yield) as a colorless oil.

Typical procedure[173]

n-C_5H_{11} (**256**) + HB (**257**) → 0 to 70 °C → *n*-C_5H_{11} BCat (**258**) → Rh(acac)(C_2H_4)$_2$, (*S*)-BINAP; Et_3N, 100 °C, dioxane/H_2O (10:1) → C_5H_{11} (**259**), 85% yield, 95% ee

Catecholborane (**257**) (245 mg, 2.00 mmol) was added to 1-heptyne (**256**) (216 mg, 2.20 mmol) at 0 °C. The mixture was stirred at room temperature for 30 min and at 70 °C for 3 h. Unreacted materials were removed under reduced pressure. To the residue were added a solution of 2-cyclohexenone (**253**) (39 mg, 0.40 mmol), Rh(acac)(C_2H_4)$_2$ (3.1 mg, 12 μmol), and (*S*)-BINAP (7.5 mg, 12 μmol) in dioxane (1.0 mL), triethylamine (409 mg, 4.00 mmol), and H_2O (0.1 mL). The whole mixture was heated at 100 °C for 3 h. Addition of 20% aqueous sodium hydroxide followed by ether extraction and silica gel chromatography (hexane/ethyl acetate = 5/1) gave 66 mg (85% yield) of (*S*)-3-((*E*)-l-heptenyl)cyclohexanone **259**, whose enantiomeric purity was determined to be 95% ee by HPLC analysis with a chiral stationary phase column.

Another one-pot reaction for the conjugate addition of aryl groups to enones has also been reported.[174] Lithium trimethylarylborates were generated *in situ* by treatment of the corresponding aryl bromides with butyllithium and trimethoxyborane. It was found that the amount of water present in the reaction medium had an effect on the chemical yields without adversely affecting enantioselectivities.

Pucheault *et al.*[175] used potassium organotrifluoroborates, which are generally more stable than organoboronic acids, in Rh-catalyzed 1,4-conjugate additions. With organotrifluoroborates, cationic Rh-catalyst precursors were necessary to achieve high conversions and high enantioselectivity. In addition, enantioselectivity and chemical yield were strongly dependent on the solvent and amount of water. Thus, the use of toluene and excess water was critical for high yield and enantioselectivity. Using the optimized reaction conditions, a wide range of aryl and alkenyl trifluoroborates was coupled to a variety of cyclic and acyclic ketones, yielding β-vinyl substituted products in high yields and high enantioselectivities. Several other chiral diphosphine ligands that are frequently used in asymmetric transformations were examined. Apart from the standard BINAP ligand, (*R*,*S*)-Josiphos (99% ee) and (*R*)-MeO-biphep (98% ee) gave almost perfect enantioselection in the reaction of 2-cyclohexenone **253** with phenyl trifluoroborate.

Typical procedure[175]

253 + **260** (BF_3K) → [Rh(cod)$_2$]PF_6, (*R*)-MeO-BIPHEP; toluene/H_2O (4:1), 105 °C → **261**

83% yield, 89% ee

A mixture of potassium organotrifluoroborate **260** (2.0 equiv), [Rh(cod)$_2$]PF_6 (3 mol%), and (*R*)- MeO-biphep (3.3 mol%) were placed in a flask and then a degassed toluene/water mixture (2 mL/0.5 mL) was added at room temperature, followed by enone **253** (0.5 mmol). The flask was heated in a pre-heated oil bath at 105–110 °C and the mixture was stirred until completion of the reaction (monitored by GC analysis). After filtration through Celite (eluting with CH_2Cl_2), the solvent was removed under reduced pressure. Purification by chromatography through silica gel afforded analytically pure **261**. HPLC analysis was performed using a Daicel Chiralcel chiral stationary phase column.

Cyclic and acyclic α,β-unsaturated esters were also found to be effective substrates for conjugate addition.[174] Interestingly, a one-pot method in which lithium arylborates, *e.g.*, **263**, were generated *in situ* from the corresponding aryl bromides was found to be effective in the coupling with acyclic esters, *e.g.*, **262**. The enantioselectivity was also found to increase with increasing steric bulk of the ester moiety. Under the standard reaction conditions,[173] five- and six-membered cyclic esters reacted smoothly with a variety of substituted arylboronic acids in excellent yields and enantioselectivity.

Typical procedure[174]

O-*i*-Pr + $[PhB(OMe)_3]^- Li^+$ → ($Rh(acac)(C_2H_4)_2$, (*S*)-BINAP; dioxane/H_2O (10:1), 100 °C) Ph, O-*i*-Pr

262 **263** **264**

Under a nitrogen atmosphere, a hexane solution of butyllithium (650 μL, 1.00 mmol) was added to bromobenzene (157 mg, 1.00 mmol) in Et_2O (0.5 mL) at 0 °C. The mixture was stirred at room temperature for 1 h and then cooled to −78 °C. Trimethoxyborane (104 mg, 1.00 mmol) was added to the reaction mixture. The mixture was stirred at −78 °C for 30 min and then at room temperature for 1 h. To the mixture were added H_2O (18 mg, 1.00 mmol), isopropyl *trans*-2-hexenoate (**262**) (62 mg, 0.40 mmol), and a solution of $Rh(acac)(C_2H_4)_2$ (3.1 mg, 12 μmol) and (*S*)-BINAP (9.0 mg, 14 μmol) in dioxane (2.0 mL). The whole mixture was heated at 100 °C for 3 h. Addition of saturated aqueous sodium bicarbonate followed by ethyl acetate extraction and chromatography on silica gel (hexane:ethyl acetate = 10:1) gave 90 mg (96% yield) of isopropyl 3-phenylhexanoate **264** as a colorless oil. HPLC analysis was performed on a Shimadzu LC-9A (Shimadzu Corp. Nakagyo-ku, Kyoto, Japan) and a JASCO PU- 980, with a JASCO UV-970 UV detector (Jasco Inc., Easton, MD, U.S.A), liquid chromatographic system with chiral stationary phase columns; Chiralcel OD-H, OJ and OG, (95% ee).

The asymmetric conjugate addition reaction has been applied to α,β-unsaturated amides,[176] alkenylphosphonates,[177] and ethenesulfonamides.[178] Hayashi and co-workers[179] have also applied the standard reaction conditions to nitroalkenes. The reaction was found to be applicable to a variety of arylboronic acids. The reaction was also extended to acyclic nitroalkenes without loss of selectivity, but not to nitrocycloheptene or nitrocyclopentene. Interestingly, the thermodynamically less stable *cis* isomers, *e.g.*, **266**, were formed in the reaction of nitrocyclohexene (**265**) with a wide range of arylboronic acids, *e.g.*, **254**. The initial products could be readily converted to the more stable *trans* isomers by treatment with base.

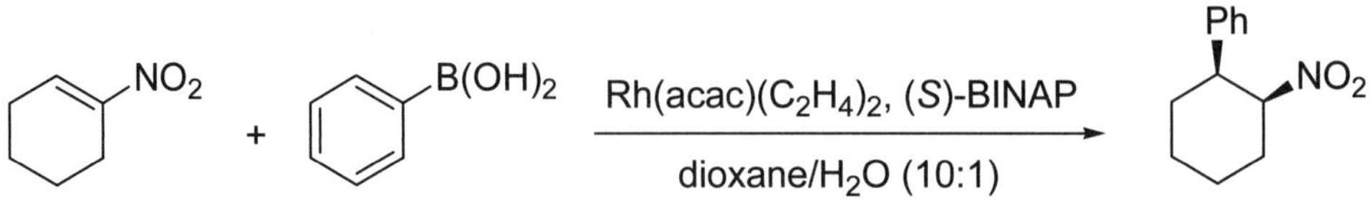

$$\text{265} + \text{254} \xrightarrow[\text{dioxane/H}_2\text{O (10:1)}]{\text{Rh(acac)(C}_2\text{H}_4)_2\text{, (}S\text{)-BINAP}} \text{266 (98.5\% ee)}$$

Scheme 2-10. Asymmetric conjugate addition reaction.

Conjugate addition of organoboronic acids to nitroalkenes[179]

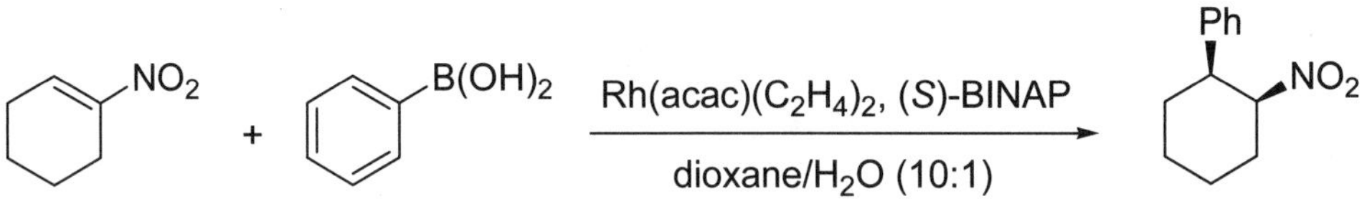

$$\text{265} + \text{254} \xrightarrow[\text{dioxane/H}_2\text{O (10:1)}]{\text{Rh(acac)(C}_2\text{H}_4)_2\text{, (}S\text{)-BINAP}} \text{266 (98.5\% ee)}$$

To a mixture of Rh(acac)(C_2H_4)$_2$ (3.1 mg, 12.0 μmol), (*S*)-BINAP (8.2 mg, 13.2 μmol), and phenylboronic acid **254** (244 mg, 200 mmol) was added 1,4-dioxane (1.0 mL), and the mixture was stirred at room temperature for 3 min. 1-Nitrocyclohexene **265** (50.9 mg, 0.40 mmol) and water (0.1 mL) were added and the whole mixture was stirred at 100 °C for 3 h. After evaporation of the solvent, the residue was dissolved in ethyl acetate. The solution was washed with saturated sodium bicarbonate and dried over anhydrous magnesium sulfate. Chromatography on silica gel (hexane/ethyl acetate = 20/1) gave 2-phenyl-1-nitrocyclohexane **266** (64.9 mg, 79% yield).

An intramolecular tandem 1,4-conjugate addition-aldol cyclization reaction of keto-enones, *e.g.*, **267**, with naphthyl- and phenyl-boronic acids was reported by Cauble and co-workers.[180] BINAP was found to be the ligand of choice, whereas other chiral diphosphine ligands such as Me-Duphos gave racemic products. The products were obtained in good to high yields with essentially perfect diasteroselectivity, although the highest enantiomeric excess obtained was 95%. It is noteworthy that three contiguous stereogenic centers were created in a single manipulation in this process.

Rhodium-catalyzed conjugate additions have also been achieved using oraganosilane,[181] organotin,[182] and organotitanium[183] reagents as nucleophiles. An interesting "*cine*" substitution reaction occurred when alkenyl sulfones were subjected to the reaction with aryltitanium reagents.[184] The addition of aryltitanium reagents to linear alkenyl sulfones gave products in which the sulfone group was substituted with the aryl group at the β-carbon regioselectively. Deuterium labeling experiments provided insights into the reaction mechanism and established that a key step involved the *anti* elimination of rhodium and the sulfonyl group from an alkyl-rhodium intermediate. The reaction of cyclohexenylsulfone **269** with aryltitanium triisopropoxides gave the *cine* substituted products, *e.g.*, **270**, in excellent yields and practically perfect enantioselectivity.

Typical procedure[180]

267 → 268: $PhB(OH)_2$, $[Rh(COD)Cl]_2$, (*R*)-BINAP, dioxane/H_2O

To an oven-dried, Ar-purged flask was added $[Rh(cod)Cl]_2$ (2.5 mol%) and (*R*)-BINAP (7.5 mol%), followed by 5 mL of 1,4-dioxane. The solution was allowed to stir at ambient temperature for 30 min, at which point $PhB(OH)_2$ (200 mol%) was added followed by aqueous KOH (10 mol%), H_2O (500 mol%), and finally substrate **267** (0.5 mmol). The flask was immediately placed in a 95 °C oil bath and allowed to stir. After complete consumption of the substrate, the reaction mixture was partitioned between H_2O and Et_2O, and the aqueous layer was washed several times with Et_2O. The organic extracts were combined, washed with brine, dried over anhydrous Na_2SO_4, concentrated *in vacuo*, and finally subjected to chromatography on silica (SiO_2: EtOAc/hexanes) to yield **268** as a white solid (69%): mp 85–86 °C.

Typical procedure[184]

269 → 270: 4-$MeOC_6H_4Ti(OPr$-*i*$)_3$, $[Rh(OH)((S)$-BINAP$)]_2$, THF, 40 °C; **270**: 99% yield, 99.9% ee

$[Rh(OH)((S)$-binap$)]_2$ (3.0 μmol) and alkenyl sulfone **269** (0.200 mmol) were placed under argon in a Schlenk tube and treated with a solution of 4-$MeOC_6H_4Ti(OPr$-*i*$)_3$ (0.400 mmol) in THF (0.5 mL) at room temperature. The Schlenk tube was immersed in a bath maintained at 40 °C for 12 h. The resulting mixture was cooled to room temperature, treated with hexane (*ca.* 3 mL) and H_2O (*ca.* 50 μL), and filtered through Celite (eluent: hexane). After concentration of the filtrate, the residue was purified by Preparative Thin Layer Chormatography (PTLC) on silica gel (hexane/EtOAc = 10/1) to give the corresponding product **270** (99% yield). The enantiomeric purity was determined by HPLC analysis with chiral stationary phase column, Chiralcel OB-H (hexane, 99.9% ee).

Investigations aimed at finding new and easily accessible ligands for asymmetric transformations are an active area of research. It is widely known that chelating dienes such as 1-5-cyclooctadiene (cod) and norbornadiene (nbd) are stable ligands for late transition metal complexes, but use of their chiral versions has not been widely explored. Hayashi *et al.*[185a] and Defieber *et al.*[186] have independently shown that new chiral dienes based on norbornadiene and (*S*)-carvone, respectively, were effective ligands in the conjugate addition of boronic acids, *e.g.*, **271**, to a wide range of acceptors with good-to-excellent enantioselectivity. The acceptors included cyclohexenone (**253**), cyclopentenone, cycloheptanone, and acyclic α,β-unsaturated ketones and esters. Hayashi also found that the bicyclo[2.2.2]octa-2,5-diene-derived ligand **274** was effective in the conjugate addition of arylboroxines to a,b-unsaturated ketones and other conjugated enones.[185b]

Rhodium-catalyzed asymmetric 1,4-addition of boronic acid or boroxine to enone[185]

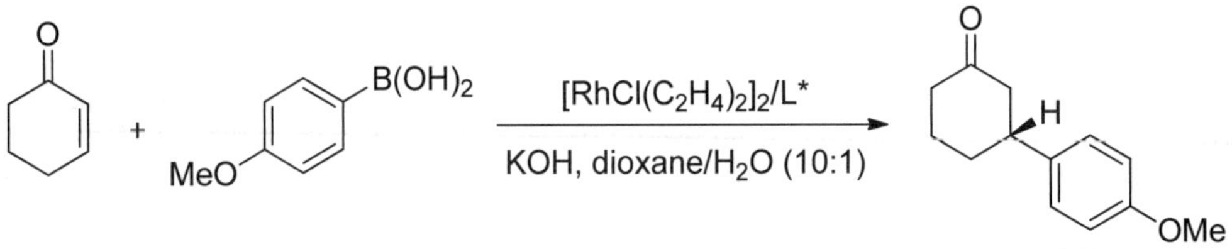

$[RhCl(C_2H_4)_2]_2$/L*; KOH, dioxane/H_2O (10:1)

253 + **271** → **272**

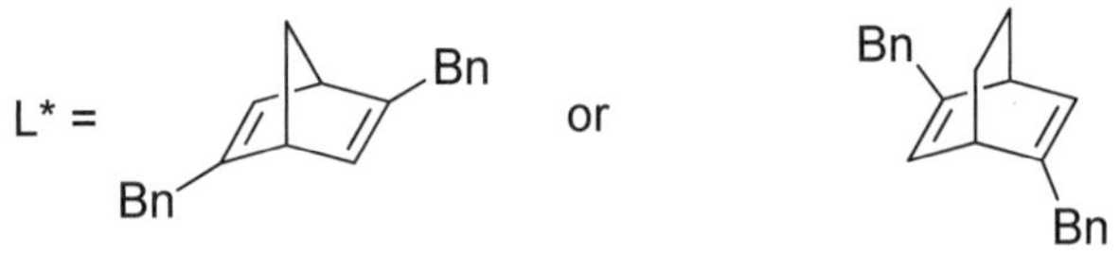

L* = **273** or **274**

A solution of $[RhCl(C_2H_4)_2]_2$ (1.8 mg, 9.0 μmol Rh) and (1*R*,4*R*)-**273** (2.7 mg, 9.9 μmol) in 1 mL of dioxane was stirred at room temperature for 15 min. To this reaction mixture was added KOH (0.1 mL, 1.5M, 0.15 mmol) in water, and the solution was stirred for 15 min. Subsequently, phenylboronic acid **271** (73.1 mg, 0.60 mmol) and 2-cyclohexenone (**253**) (28.8 mg, 0.30 mmol) were added to this solution. After stirring at 30 °C for 1 h, the reaction mixture was quenched with saturated $NaHCO_3$ in water and extracted with Et_2O five times. The combined organic layers were dried over anhydrous magnesium sulfate and concentrated under reduced pressure. The residue was purified by preparative TLC (silica gel, hexane/EtOAc = 3/1) to give 49.5 mg (92% yield) of the product **272** which was 97% ee. The products obtained by the rhodium-catalyzed asymmetric 1,4-addition were fully characterized by comparison of their spectral and analytical data with those reported in the literature.

Shintani *et al.* have also applied similar reaction conditions to those used for 1,4-conjugate additions of boronic acids to α-β-unsaturated ketones to the arylative cyclization of alkynals, *e.g.*, **275**. Using a variety of boronic acids, the enantioselectivity was found to be generally high (75–96% ee) with good-to-excellent yields.[187]

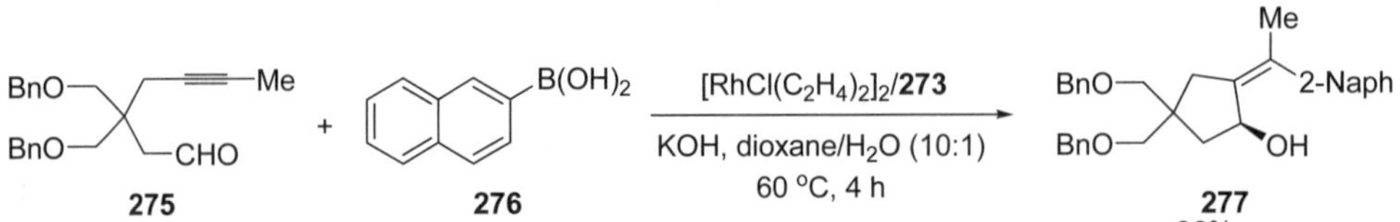

$[RhCl(C_2H_4)_2]_2$/**273**; KOH, dioxane/H_2O (10:1), 60 °C, 4 h

275 + **276** → **277** (96% ee)

Scheme 2-11. 1,4-Conjugate addition of boronic acids to α,β-unsaturated ketones.

Tedrow *et al.* reported the use of electronically differentiated 4-oxobutenamides, *e.g.*, **278**, as substrates in the Rh-catalyzed conjugate addition reaction.[188] The sterically demanding *p*-chiral phosphine ligands, Tangphos and DuanPhos, afforded excellent regioselectivity and enantioselectivity. A variety of commercially available arylboronic acids was found to be good coupling partners. Both aryl and alkyl oxobutamides afforded high regioselectivities and enantioselectivities.

Typical procedure[188]

4-MeOC$_6$H$_4$B(OH)$_2$
[{(*S*,*S*,*R*,*R*)-Duanphos}Rh(nbd)]BF$_4$
Et$_3$N, THF/H$_2$O (19:1), 65 °C

278 → **279**
93% yield, >99:1 dr, 98% ee

A 20-mL scintillation vial was charged with (*E*)-1-morpholino-4-phenylbut-2-ene-1,4-dione (**278**) (0.5 g, 2.0 mmol), 4-methoxyphenylboronic acid (456 mg, 3.0 mmol), [((*S*,*S*,*R*,*R*)-Duanphos)Rh(nbd)]BF$_4$ (27 mg, 0.04 mmol), triethylamine (0.42 mL, 3.0 mmol), and tetrahydrofuran:water (19:1, 5 mL). The reaction mixture was warmed to 65 °C and stirred for 16 h. After completion, the reaction mixture was poured into 20 mL of ethyl acetate and washed with 10 mL of saturated sodium bicarbonate. The organics were dried over anhydrous magnesium sulfate, filtered and concentrated *in vacuo*. The crude material was analyzed by HLPC (Method A: T_{ret} = 4.3 min; T_{ret} = 4.1 min; >99:1; Method B: T_{ret} *R* = 1.08 min; T_{ret} *S* = 1.47 min = 98% ee) and then purified by flash column chromatography on silica gel (30% to 100% ethyl acetate in hexanes) to afford the product **279** as a colorless solid (701 mg, 96%): mp (133–135 °C). $[\alpha]^{25}$D -157.4° (c 1.2 in CH$_2$Cl$_2$).

Rh-catalyzed conjugate addition to enones has largely been limited to aryl and alkenyl groups. The addition of terminal alkynes to enones catalyzed by rhodium complexes is rare and with limited substrate scope.[189] Nishimura and co-workers reported the Rh-catalyzed asymmetric 1,3-rearrangement of the alkynyl group from alkynyl-alkenyl carbinols, *e.g.*, **280**, *via* a β-alkynyl elimination-conjugate addition pathway.[190] This novel reaction serves as a synthetic equivalent to the asymmetric conjugate alkynylation of enones. A variety of racemic aryl- and alkyl-substituted carbinols gave the conjugate addition products in high yields and high enantioselectivity. The *tert*-butyldimethylsilyl substituted carbinols, *e.g.*, **280**, gave product **281** in high yield, whereas smaller silyl groups such as trimethylsilyl gave low yields.

Typical procedure[190]

SiMe$_2$-*t*-Bu
HO
Ph
[Rh(OH)(cod)]$_2$, (*R*)-BINAP
toluene, 60 °C
SiMe$_2$-*t*-Bu
O
Ph

280

281
91% yield, 98% ee

A mixture of [Rh(OH)(cod)]$_2$ (0.005 mmol) and (*R*)-BINAP (0.012 mmol) in toluene (0.5 mL) was heated at 60 °C for 5 min. To the mixture was added the alcohol substrate **280** (0.20 mmol) and toluene (0.5 mL), and it was stirred at 60 °C for 3 h. The mixture was passed through a short column of silica gel with ethyl acetate as eluent. After evaporation of the solvent, the residue was subjected to preparative TLC (SiO_2, hexane/ethyl acetate) to give product **281** as a colorless oil (91% yield). The enantiomeric excess of the product was measured by HPLC Chiralcel OJ-H column × 2, 0.2 mL/min, hexane/2-propanol = 500/1, 224 nm, t_1 = 41.7 min (*S*), t_2 = 43.9 min (*R*); 98% ee (*S*), $[\alpha]^{20}_D$ -9 (C 0.75, $CHCl_3$).

Application of the same conditions to the cyclic allylic alcohol **282** derived from indenone gave moderate (71% ee) enantioselectivity. However, the use of chiral ligand (*S,S*)-Ph-bod* (**284**) in place of BINAP, and cesium carbonate as an additive, gave the desired product **283** with excellent enantioselectivity.

Typical procedure[190]

Si(*i*-Pr)$_3$
HO
[RhCl(C$_2$H$_4$)$_2$]$_4$, (*S,S*)-Ph-bod*
Cs$_2$CO$_3$, toluene, 80 °C
O
H
Si(*i*-Pr)$_3$
(*S,S*)-Ph-bod* =
Ph
Ph

282

283
91% yield, 97% ee

284

A mixture of (*S,S*)-Ph-bod* **284** (3.1 mg, 0.012 mmol) and $[RhCl(C_2H_4)_2]_2$ (1.9 mg, 0.005 mmol) in toluene (0.5 mL) was stirred at room temperature for 10 min. To this solution were added Cs_2CO_3 (6.5 mg, 0.020 mmol), alcohol **282** (62.4 mg, 0.20 mmol), and toluene (0.5 mL), and the mixture was heated at 80 °C for 15 h. The reaction mixture was passed through a short column of silica gel with ethyl acetate as eluent. After evaporation of the solvent, the residue was subjected to preparative TLC (eluent, hexane/ ethyl acetate = 10/1) to give product **283** (57.0 mg, 0.18 mmol, 91% yield) as a white solid. The enantiopurity was measured by HPLC (Chiralpak AD-H column, 0.2 mL/min, hexane/2-propanol = 200/1, 224 nm, t_1 = 28.1 min (*R*), t_2 = 30.8 min (*S*); 97% ee (*R*), $[\alpha]^{20}_D$ -46 (C 1.11, $CHCl_3$).

12. Rh-Catalyzed Hydroboration

In 1985, Männing and Nöth first reported the hydroboration of alkenes catalyzed by Wilkinson's catalyst.[191] Since this pioneering work, the development of transition-metal–catalyzed hydroboration has been investigated extensively. Burgess and Ohlmeyer demonstrated asymmetric catalysis with the use of BINAP and Diop-derived Rh-catalysts.[192] Hayashi *et al.* later reported improvement of the enantioselectivity for the hydroboration of styrenes using Rh-BINAP complexes (up to 96% ee at –78 °C).[193] Other catalyst systems have also been shown to be effective in this reaction.[194] Schnyder *et al.* used a ferrocene-based pyrazole-containing ligand for the hydroboration of styrene, which achieved 98% ee with moderate regioselectivity.[194a] Brown *et al.*[194b] reported the hydroboration of *para*-methoxystyrene with 94% enantioselectivty using the axially chiral Quinap ligand. Maxwell *et al.*[195] described the asymmetric hydroboration of vinylarenes using a Quinazolinap ligand. Smith and Takacs[196] reported an efficient amide-directed asymmetric hydroboration of trisubstituted alkenes using TADDOL-derived phenyl monophosphite ligands. Demay *et al.*[197] reported the hydroboration reaction using C_2-symmetric 1,2-diphosphane ligands. Evans and Fu[198] suggested a catalytic cycle for catalytic hydroboration, which was supported by computational studies.[199]

Asymmetric hydroboration and oxidation of boranes provides a simple and attractive route to a variety of benzylic alcohols with excellent enantiopurity. These benzylic alcohols are useful compounds in organic synthesis. The asymmetric hydroboration of vinylarene derivatives was accomplished by Guiry *et al.* using axially chiral Quinazolinap ligands with excellent enantioselectivity. A variety of Quinazolinap-derived ligands were found to be effective for a wide range of styrene derivatives including electron-donating, electron-deficient, cyclic, and acyclic derivatives.[200] For styrene derivatives, higher enantioselectivity (95% ee) was obtained for the *p*-methoxy derivative compared with the *p*-chloro or unsubstituted analogs. However, in all cases, the regioselectivities were moderate. Excellent enantioselectivities and regioselectivities were obtained for both *cis* and *trans* isomers of β-methylstyrene (97% ee and 95% ee, respectively) using 2-methyl quinazolinap ligand **287**. Using the same ligand, essentially perfect regioselectivity and enantioselectivity was observed for the hydroboration of indene **285**.

General procedure for asymmetric hydroboration[200]

(*R*)- or (*S*)-Diphenyl[1-(2-substituted-quinazolin-4-yl)(2-naphthyl)]phosphine(1,5-cyclo-octadiene)rhodium trifluoromethanesulfonate (5 µmol) in THF (2 mL) was placed under nitrogen in a Schlenk tube. Freshly distilled catecholborane (53 µL, 0.5 mmol) was added *via* microliter syringe, and the light-brown solution was stirred for 5 min at 20 °C. Indene **285** (0.5 mmol) was injected, and the reaction mixture was stirred for 2 h at room temperature (in some cases, extended reaction times were necessary). The reaction was then cooled to 0 °C, and ethanol (1 mL) was added followed by 1-M NaOH (3 mL) and H_2O_2 (3 mL). The ice bath was removed, and the solution was stirred for 1 h at room temperature. The reaction mixture was transferred to a separatory funnel and diethyl ether (10 mL) was added. The organic layer was washed with 1 M NaOH (10 mL), brine (10 mL) and dried over anhydrous $MgSO_4$. The solution was filtered, and the solvent was removed *in vacuo* to give hydroborated product **286** as an oil. Percent conversion and regioselectivity were determined by ^{1}H NMR. The enantiopurity was determined by chiral GC or HPLC analysis.

The use of dicatecholdiborane gives the corresponding diboronate ester, which can be oxidized to the corresponding diol, thereby complementing the Sharpless asymmetric dihydroxylation reaction. On the other hand, the intermediate diboronate ester may be subjected to a cross-coupling reaction followed by an oxidation.[201] Morgan *et al.*[202] reported a Rh-quinap–catalyzed enantioselective diboration of *trans* alkenes using dicatecholdiborane. Both alkyl- and aryl-substituted *trans* alkenes, *e.g.*, **288**, were effective substrates, and the oxidation of the intermediate 1,2-*bis*-catechol esters gave 1,2-diols such as **289** in good yields and excellent enantioselectivity. These reaction conditions were not effective for *cis* alkenes or monosubstituted or 1,1-disubstituted alkenes.

Typical procedure[202]

An oven-dried, 20-mL vial equipped with a stir bar was charged with 7.4 mg (0.025 mmol) of (bicyclo[2.2.1]hepta-2,5-diene)(2,4-pentanedionato)rhodium(I), 11.1 mg (0.025 mmol) of (*S*)-Quinap, and 1.0 mL of THF under an inert atmosphere of Ar in a dry box. The resultant yellow solution was stirred for 5 min. After this time, 132 mg (0.55 mmol) of dicathecholdiborane was added to the solution under Ar. The solution turned immediately from yellow to dark brownish-red. The solution was allowed to stir for 5 min. After this time, (0.50 mmol) of *trans*-alkene **288** was added to the solution under Ar. The vial was sealed with a screw cap and removed from the dry box, where the solution was allowed to stir for 15 h at ambient temperature. After this time, 1 mL of THF was added to the solution, followed by dropwise addition of 0.800 mL of 3-M NaOH and then 0.800 mL of 30% H_2O_2 dropwise. The solution was allowed to stir at ambient temperature for 3 h. The solution was then quenched with 1 mL of saturated aqueous $Na_2S_2O_3$ and 10 mL of 1-M NaOH. The mixture was extracted with ethyl acetate (3 × 25 mL) and the combined organic layers were washed with brine (1 × 10 mL). The organic layers were then dried over anhydrous $MgSO_4$, filtered, and the solvent removed by rotary evaporation. The crude material was purified by silica gel chromatography (2:1 hexanes:ethyl acetate) to provide the product **289**. The enantiomeric excess was determined by chiral GLC (β-dex, Supelco, 160 °C, Sigma–Aldrich Corporation, Bellefonte, Pennsylvania, U.S.A.). Stereochemical ratios were determined in with authentic racemic materials prepared by osmium tetraoxide dihydroxylation. Relative stereochemistry was determined in comparison with ^{1}H NMR reported for the syn diol.[203] Absolute stereochemistry established in comparison with authentic 1*R*,2*R* isomer prepared *via* Sharpless asymmetric dihydroxylation (analogous to β-methylstryene).[204]

Enantiopure benzylamines are important intermediates in the synthesis of pharmaceutically active compounds and chiral ligands for asymmetric transformations. Fernandez and co-workers[205] reported that primary and secondary benzylamines could be obtained in moderate yields by converting the initially formed catecholboronate ester into a trialkylborane by reaction with either diethylzinc or methylmagnesium chloride, followed by treatment of the trialkylborane thus obtained with hydroxylamine-*O*-sulfonic acid, which yielded primary benzylamine **291.** When the trialkylborane was treated with *N*-substituted chloramines, a secondary benzylamine, *e.g.*, **292**, was formed. A variety of vinylarenes gave the corresponding benzylamines in moderate to good yields and good-to-excellent enantioselectivity (78–98% ee).

Typical procedures for the synthesis of primary and secondary amines via asymmetric hydroboration[205b]

1) (*S*)-Quinap-Rh
catecholborane, THF, 25 °C
2) MeMgCl, THF, 25 °C
then H_2NOSO_3H
NH_2
Me
MeO
291
56%, 95% ee
MeO
290
1) (*R*)-Quinap-Rh
catecholborane, THF, 25 °C
2) $ZnEt_2$, toluene
then $PhCH_2NHCl$, 0 °C
NHBn
Me
MeO
292
75%, 91% ee

Typical procedure for the synthesis of primary amines

The freshly prepared (*S*)-[1-(2-diphenylphosphino-1-napthyl)isoquinoine](cyclooctadiene)rhodium(I) trifluoromethanesulfonate (1 mol%), THF (0.5 mL) and alkene **290** (0.5 mmol) were placed in a vial under Ar. Freshly distilled catecholborane (0.5 mmol) was added with stirring and left for 1 h. MeMgCl (1.0 mmol) was added and the solution stirred for 30 min. The resultant solution was added to pre-dried hydroxylamine-*O*-sulfonic acid (1.5 mmol) in THF (0.7 mL) and stirred under Ar overnight. Hydrochloric acid (1 M, 2 mL) was added and the mixture poured into water (4 mL). The aqueous layer was extracted with ether (3 × 20 mL), and then made strongly alkaline with NaOH (1 M, 3 mL). The mixture was extracted with diethyl ether (3 × 20 mL). The ether extracts were combined, dried (anhydrous $MgSO_4$), and the solvent removed in *vacuo*. In those cases in which GC analysis was employed, the amines were converted into their acetamides: Primary amines (0.5 mmol) were dissolved in toluene (2 mL) and acetic acid (0.6 mmol) was added by syringe along with one equivalent of 1,1'-carbonyldiimidazole. The mixture was stirred vigorously overnight and then extracted with toluene. The organic extracts were washed with NaOH (1 M, 3 × 20 mL), dried (anhydrous $MgSO_4$) and the solvent removed in *vacuo*.

Procedure for the synthesis of secondary amines

To a solution of [Rh(cod)(acac)] (0.01 mmol, 0.2 mol%) and (*R*)-Quinap (0.01 mmol, 0.2 mol%) in dry THF (2 mL) under Ar was added trimethylsilyltriflate (0.03 mmol) with stirring. The volume was reduced to 0.5 mL in *vacuo*, and pentane (2 × 10 mL) was added to precipitate the catalyst. The pentane was removed with a syringe and the catalyst was placed in *vacuo*, then under Ar. Toluene (1 mL) was added, followed by alkene **290** (1.0 mmol), and then catecholborane (1.0 mmol) was added slowly *via* syringe. The mixture was stirred for 2 h and $ZnEt_2$ (1.0 mmol) was then added slowly *via* syringe. The mixture was stirred for 2 h and a white precipitate formed. The alkylchloramine was generated by the addition of sodium perchlorite (1.2 mmol) to the amine solution (1.2 mmol) in diethyl ether (6 mL) at 0 °C. The alkylborane solution pre-cooled to 0 °C was added to the chloramine solution. The borane residue was dissolved further with ether (2 mL), which was also added to the chloramine solution. The reaction mixture was stirred at 0 °C for 20 min, allowed to warm to room temperature, and then stirred for 1 h. Aqueous HCl (1.0 M, 5 mL) was added and the mixture was stirred for 10 min and then washed with diethyl ether (3 × 10 mL). Aqueous NaOH (2.0 M, 5 mL) was added to the aqueous layer, which was then extracted with diethyl ether (3 × 10 mL), dried (anhydrous $MgSO_4$), filtered, and evaporated to give the product **292**, which was further purified by column chromatography (silica gel, Et_2O:pentane = 1:5). The enantiopurity of the product amine **292** was determined directly by ^1H NMR spectroscopy (d_6-acetone) with (+) or (-)-mandelic acid as the chiral NMR shift reagent.

13. Rh-Catalyzed Reductive Coupling

Catalytic enantioselective aldol reactions provide access to a variety of β-hydroxy carbonyl compounds, which are useful intermediates for the synthesis of natural products as well as commodity chemicals. A drawback, however, involves the use of latent enolates that must be prepared by stoichiometric methods. Enolate formation is most often accomplished through deprotonation of carbonyl compounds or activation of related enol derivatives. Transition-metal–catalyzed reductive methods for the generation of enolates and enol derivatives from α,β-unsaturated carbonyl compounds have been reported.[206] Rh-catalyzed 1,4-hydrosilylation of α,β-unsaturated carbonyl compounds yields enolates that can be trapped with electrophiles. A series of Rh-catalyzed reductive couplings of hydrosilane, α,β-unsaturated carbonyl compounds, and electrophilic acceptors has been reported by Matsuda and co-workers.[207] The reductive coupling of α,β-unsaturated carbonyl compounds with allylic carbonates as acceptors gave esters in high yields with moderate regioselectivity.[207a] Mannich-type products were obtained in good yields from the reductive coupling with aldimines.[207b] The synthesis of aryl-substituted amide **295** was also possible using aryl isocyanates as acceptors.[207c] Isocyanates **293** afforded the corresponding aryl-substituted amide **295** in excellent yield. It is worth mentioning that these reductive couplings required cationic Rh-catalyst precursors to achieve synthetically useful yields.

Typical procedure[207c]

$$\text{PhNCO } (\mathbf{293}) + \text{CH}_2\text{=CHCO}_2\text{Me } (\mathbf{294}) \xrightarrow[\text{CH}_2\text{Cl}_2,\ 45\ ^\circ\text{C},\ 90\%]{\text{Et}_2\text{MeSiH},\ [\text{Rh(cod)}\{\text{P(OPh)}_3\}_2]\text{OTf}} \text{PhNHC(O)CH(CH}_3\text{)CO}_2\text{Me } (\mathbf{295})$$

To a solution of $[Rh(cod)P(OPh)_3)_2]OTf$ (9.9 mg, 0.010 mmol) in CH_2Cl_2 (4 mL) was added a mixture of phenyl isocyanate (**293**) (121 mg, 1.0 mmol), methyl acrylate (see structure; **294**) (182 mg, 2.1 mmol), and diethylmethylsilane (205 mg, 2.0 mmol) in CH_2Cl_2 (2 mL). The resulting mixture was then refluxed for 3 h under a N_2 atmosphere. The solvent was removed under reduced pressure, and the residue was purified by flash column chromatography on silica gel (hexane/ethyl acetate 4:1 as an eluent) to afford the product **295** (188 mg, 0.91 mmol) as a white solid (90% yield).

Taylor *et al.*[208] applied an asymmetric variant of this reaction using a Rh-BINAP catalyst and aldehydes, *e.g.*, **296**, as acceptors to yield aldol products such as **298**. Although a variety of aromatic, cyclic, and acyclic aldehydes was effective acceptors, the scope of the α,β-unsaturated ester was limited to phenyl acrylate **297,** and the enantioselectivity and *syn/anti* selectivities still need to be improved. Shiomi *et al.*[209] reported an asymmetric reductive coupling reaction of enones and aromatic aldehydes using chiral Rh(Phebox) catalysts. Diphenylmethylsilane was used as a hydride donor in this case, and the desired β-hydroxyketones were produced with up to 93% ee.

Representative procedure for catalytic reductive aldol reaction[208]

296 + **297** → [RhCl(cod)]$_2$, (*R*)-BINAP; Et_2MeSiH; $ClCH_2CH_2Cl$, 25 °C → **298**

syn : *anti* = 3.4:1
87% ee (*syn*)

A 10-mL, flamed-dried, round-bottomed flask was charged with 10.0 mg of chloro(1,5-cyclooctadiene)rhodium(I) dimer (0.02 mmol), 33.0 mg (*R*)-BINAP (0.053 mmol), and 500 μL of 1,2-dichloroethane. The resulting solution was stirred at room temperature for 1 h. After 1 h, 481 μL of 1,2-dichloroethane and 174 μL of diethylmethylsilane (0.97 mmol) were added to the mixture, and the reaction vessel was stirred for 30 min. Next, 1.15 mL of stock benzaldehyde/phenyl acrylate solution (0.07 M in aldehyde and 0.84 M in acrylate, 0.81-mmol aldehyde, 0.97-mmol acrylate) was added dropwise to the solution. The vessel was then sealed and allowed to stir for 24 h. Solvent was then evaporated from the reaction mixture and 1 mL each of THF, MeOH, and 4 N HCl were added. This mixture was stirred at room temperature for an additional 30 min. Ethyl acetate was then used to extract the product (3 × 7 mL). The combined organic layers were washed with a saturated aqueous sodium bicarbonate solution (2 × 20 mL), dried over anhydrous $MgSO_4$, and filtered. The solvent was removed by rotary evaporation to yield crude product, which was purified *via* flash chromatography (9:1 then 5:1 hexanes:ethyl acetate) to yield the product **298**.

Jang and co-workers[206b] reported a Rh-catalyzed protocol for the reductive generation of metal enolates from α,β-unsaturated compounds using hydrogen as the terminal reductant. The metal enolates generated were subjected to electrophilic trapping by either appendant or exogenous aldehydes. The optimized reaction conditions proved general for the intramolecular reductive coupling of aromatic, heteroamatic, and aliphatic enones to form five- and six-membered aldol products **300**, which were obtained in good-to-excellent yields and selectivities. A slight excess of the α,β-unsaturated partner **299** proved beneficial for the intermolecular variant of the reductive aldol condensation of aromatic and heteroaromatic aldehydes.

Typical procedure[206b]

299 → Rh(cod)$_2$OTf, (*p*-$CF_3C_6H_4)_3P$; H_2 (1 atm), KOAc, DCE, 25 °C → **300**

71% yield, *syn:anti*=24:1

To a 50-mL, round-bottomed flask charged with Rh(cod)$_2$OTf (0.148 mmol, 10 mol%) and (*p*-$CF_3C_6H_4)_3P$ (0.356 mmol, 24 mol%) was added 1,2-dichloroethane (DCE) (0.1 M, 14.8 mL). The mixture was stirred for 10 min, at which point substrate **299**

(1.48 mmol, 100 mol%) and KOAc (0.44 mmol, 30 mol%) were added. The system was purged with hydrogen gas for 1 min and the reaction was allowed to stir at room temperature under 1 atm of hydrogen until complete consumption of substrate. The product **300** was purified by flash column chromatography, and was obtained in 71% yield.

The cationic Rh-catalyzed reductive cyclization of 1,6-diynes and 1,6-enynes mediated by hydrogen afforded the corresponding 1,2-dialkylidenecyclopentane and monoalkylidenecyclopentane, respectively.[210] A highly enantioselective reductive cyclization of 1,6-enynes using BINAP, PHANEPHOS, and Cl,MeO-BIPHEP ligands was later reported (Table 2-10).[211] The reactions of structurally diverse 1,6-enynes afforded the corresponding alkylidene-substituted carbocyclic and heterocyclic products in good yields. The chemical yields and enantioselectivity were largely influenced by the structural features of the chiral phosphine ligands employed.

Rhee and Krische[212] have also reported that reductive hydrogenation of acetylenic aldehydes using a cationic (*R*)-Cl,OMe-BIPHEP-Rh catalyst enabled the formation of cyclic allylic alcohols **304a–304d** in good yields and exceptional enantioselectivity (Table 2-11). Acetylenic aldehydes **303a–303d** with pre-existing stereogenic centers were found to engage in highly diastereoselective reductive cyclization. The authors found that addition of catalytic amounts of Brønsted acid dramatically improved the chemical yields of the products. Both electron-deficient alkynes and simple unactivated alkynes were effective substrates for the reductive cyclization. In addition, terminal and internal alkynes were equally effective. High levels of enantioselectivity were obtained for acetylenic aldehydes possessing geminal substituents either adjacent to the aldehyde or at the propargylic position.

Table 2-10. Rh-catalyzed reductive cyclization of 1,6 enynes mediated by hydrogen.

Entry	Enyne (**301**)	Chiral phosphine	Product (**302**)	Yield (%)	% ee
a	OBn, O	(*R*)-Cl,MeO-BIPHEP	OBn, O, CH_3, H	82	98
b	TsN	(*R*)-BINAP	TsN, CH_3, H	79	93
c	O, BnN	(*R*)-PHANEPHOS	O, BnN, CH_3, H	73	91

d	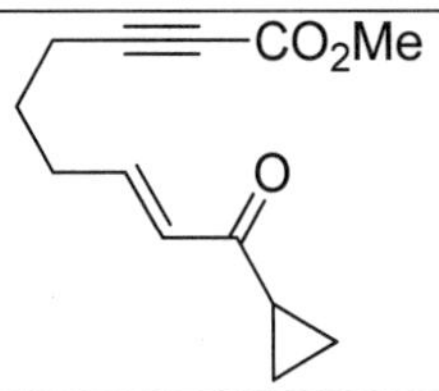 [structure: enone–alkyne substrate with CO_2Me and cyclopropyl ketone]	(*R*)-Cl,MeO-BIPHEP	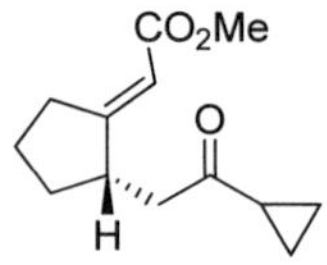 [structure: cyclopentane product with CO_2Me alkylidene and cyclopropyl ketone]	68	98

Typical procedure[211]

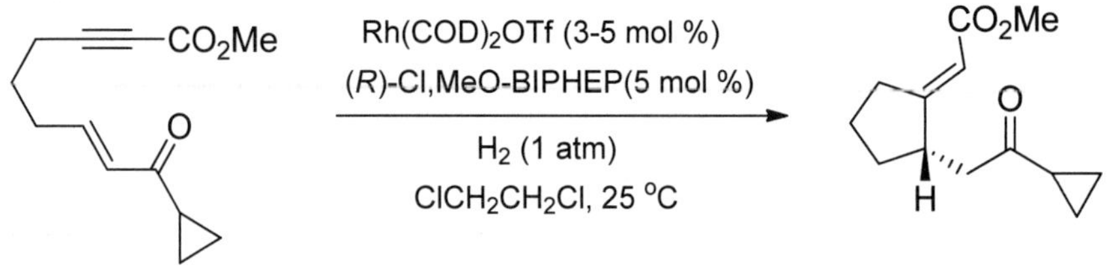

$$\mathbf{301d} \xrightarrow[\text{H}_2\text{ (1 atm)},\ \text{ClCH}_2\text{CH}_2\text{Cl},\ 25\ ^\circ\text{C}]{\text{Rh(COD)}_2\text{OTf (3-5 mol \%)},\ (R)\text{-Cl,MeO-BIPHEP(5 mol \%)}} \mathbf{302d}$$

302d
68% yield, 98% e.e.

To a solution of enyne **301d** in dichloroethane (0.1 M) at ambient temperature was added $Rh(cod)_2OTf$ (5 mol%) and (*R*)-Cl,MeO-BIPHEP (5 mol%). The system was purged with hydrogen gas and the reaction mixture was allowed to stir under an atmosphere of hydrogen until complete consumption of **301d** was observed, at which point the reaction mixture was evaporated onto silica gel and the product purified by column chromatography on silica gel. The final product **302d** was obtained in 68% yield and 98% ee.

Table 2-11. Reductive cyclization of acetylenic aldehydes.

Entry	Acetylenic aldehydes (**303**)	Product (**304**)	Yield (%)	% ee
a	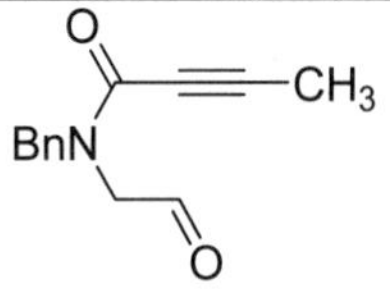 [structure: BnN, O, CH_3]	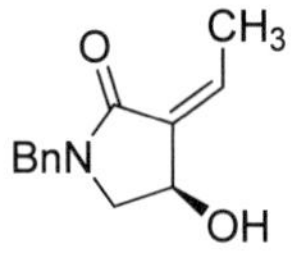 [structure: BnN, O, CH_3, OH]	83	99
b	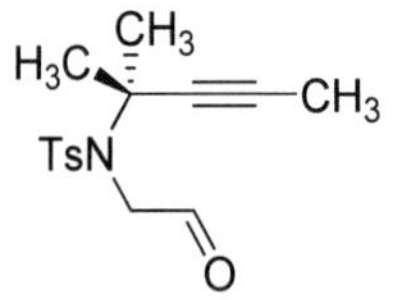 [structure: H_3C, CH_3, TsN, CH_3, O]	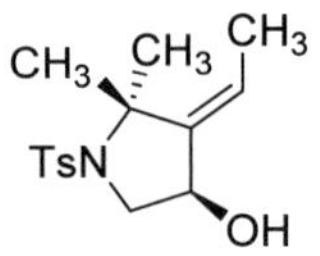 [structure: CH_3, CH_3, CH_3, TsN, OH]	63	98
c	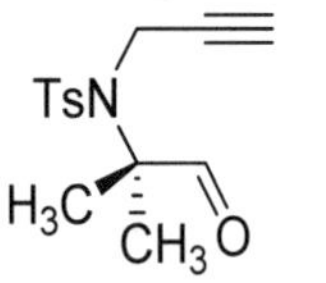 [structure: TsN, H_3C, CH_3, O]	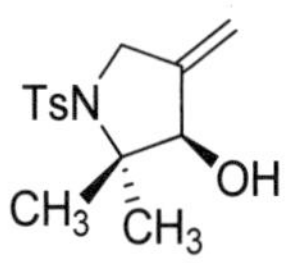 [structure: TsN, CH_3, CH_3, OH]	76	96
d	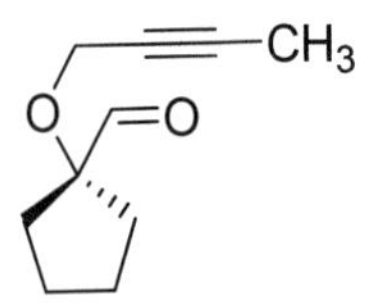 [structure: O, CH_3, =O]	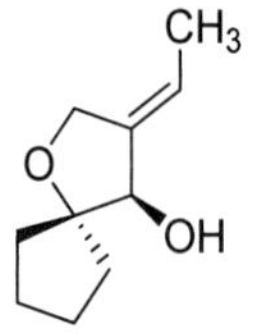 [structure: O, CH_3, OH]	73	99

Representative procedure for the reductive cyclization of acetylenic aldehydes[212]

Rh(COD)$_2$OTf (5 mol %)
(*R*)-Cl,MeO-BIPHEP (5 mol %)
2-naphthoic acid (5 mol %)
H$_2$ (1 atm), DCE, 45 °C

303a → **304a** 83%, 99% ee

305, (*R*)-Cl,MeO-BIPHEP

To a solution of but-2-ynoic acid benzyl-(2-oxo-ethyl)-amide (**303a**) (0.15 mmol) in dichloroethane (0.1 M) at ambient temperature was added Rh(cod)$_2$OTf (5 mol%), 2- naphthoic acid (5 mol%), and (*R*)-Cl,MeO-BIPHEP (6 mol%). The reaction mixture was purged with hydrogen gas and allowed to stir at 45 °C under an atmosphere of hydrogen until complete consumption of the substratewas observed. After concentration on a rotary evaporator, the desired product**304a** was isolated by column chromatography on silica gel eluting with a mixture of hexane:EtOAc = 1:1 in 83% yield and 99% ee.

Kimura *et al.*[213] reported the use of triethylborane to promote the Rh-catalzed reductive coupling of aldehydes with conjugated dienes. The corresponding homoallyl alcohols were produced in excellent yields when [Rh(OH)(COD)]$_2$ was used as the catalyst (Figure 2-7). Triethylborane reacts with the RhOH species (**A**) to produce [Rh]Et (**D**). [Rh]Et then undergoes a β-hydride elimination to produce the RhH active species **F** which then adds to the conjugated diene **306**. The resulting π-allylrhodium intermediate **G** readily reacts with the aldehyde **307** to form the homoallyl alcohol. At this time, the [Rh]Et species **D** is regenerated upon an ethyl transfer from another molecule of triethylborane to the alkoxyrhodium intermediate **J**. A subsequent β-hydride elimination of [Rh]Et regenerates the [Rh]H species.

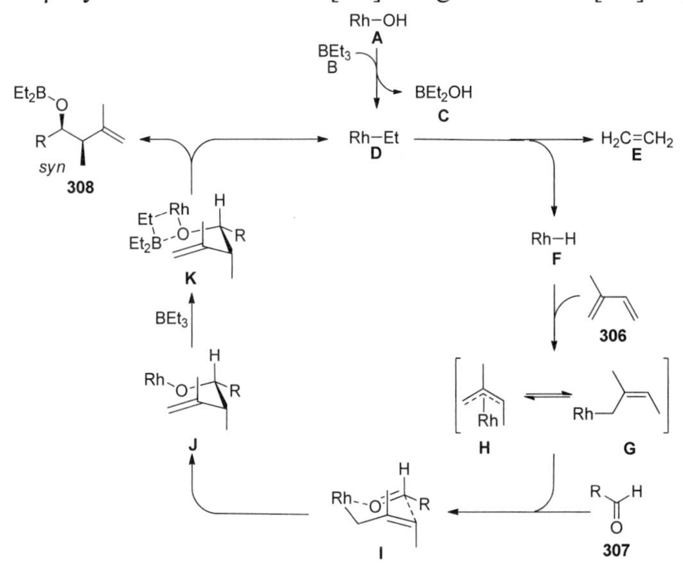

Figure 2-7. Rh-catalyzed reductive coupling of aldehydes with conjugated dienes promoted by triethylborane.[213]

General procedure for the Rh-catalyzed reductive coupling reaction of benzaldehyde with 2,3-dimethyl-1,3-butadidene[213]

cat. Rh
Et_3B, 50 °C

307a **306a** **309a**

To a solution of $[Rh(OH)(cod)]_2$ (22.8 mg, 0.05 mmol) in dry THF (5 mL) were successively added benzaldehyde **307a** (106 mg, 1 mmol), 2,3-dimethyl-1,3-butadiene (**306a**) (400 μL, 4 mmol), and triethylborane (3 mmol, 1.0-M hexane solution) *via* syringe at ambient temperature under nitrogen. The mixture was stirred at 50 °C for 24 h. The mixture was diluted with EtOAc and washed with 2 N HCl, sat. $NaHCO_3$, and then brine. The extract was dried (anhydrous $MgSO_4$), concentrated *in vacuo*, and the residual oil was subjected to column chromatography on silica gel (hexane/EtOAc = 12/1 v/v) to give the product **309a** (188 mg, 99% yield, R_f = 0.57; hexane/ethyl acetate = 4/1 v/v).

14. Miscellaneous Reactions Catalyzed by Rh-Complexes

Wilkinson's catalyst, $Rh(PPh_3)_3Cl$, has been shown to be an effective catalyst in the isomerization of propargyl alcohols such as **310** to *trans* **311** in a stereoselective manner.[214] The reaction is highly *trans* selective and the products such as **311** are intermediates in the synthesis of dipeptide isosters.

Typical procedure[214]

BocHN, OH, CO_2Et
$Rh(PPh_3)_3Cl$, $^nBu_3P$
toluene, 110 °C
BocHN, O, CO_2Et

310 **311**

A solution of the propargyl alcohol **310** (299 mg; 1 mmol), *tris*-triphenylphosphinerhodium(I) chloride (0.03 mmol, 27.75 mg), and tri-*n*-butylphosphine (0.05 mmol, 10.12 mg) in 5 mL of anhydrous toluene was heated under Ar at reflux for 12 h. The reaction mixture was then cooled to room temperature, concentrated *in vacuo*, and purified by column chromatography on silica gel using petroleum ether-AcOEt (85:15) as eluent to give *trans* enone **311**.

Takaya and co-workers[215] found that the low-valent $Rh_4(CO)_{12}$ complex was an effective catalyst for the addition of isocyanoacetates, *e.g.*, **313**, to 1,3-dicarbonyl compounds, *e.g.*, **312**, to give polysubstituted pyrroles such as **314**. The authors proposed the formation of an isocyanoalkylrhodium intermediate **A** *via* α-C-H activation of isocyanoacetate **313** followed by insertion of the 1,3-dicarbonyl **312.** This is followed by a decarbonylation and cyclocondensation of the resulting enamino intermediate (Figure 2-8). The reaction of isocyanoacetates with ketones gave the corresponding α,β-unsaturated formamides, which are important precursors to *N*-formylamino acid derivatives.

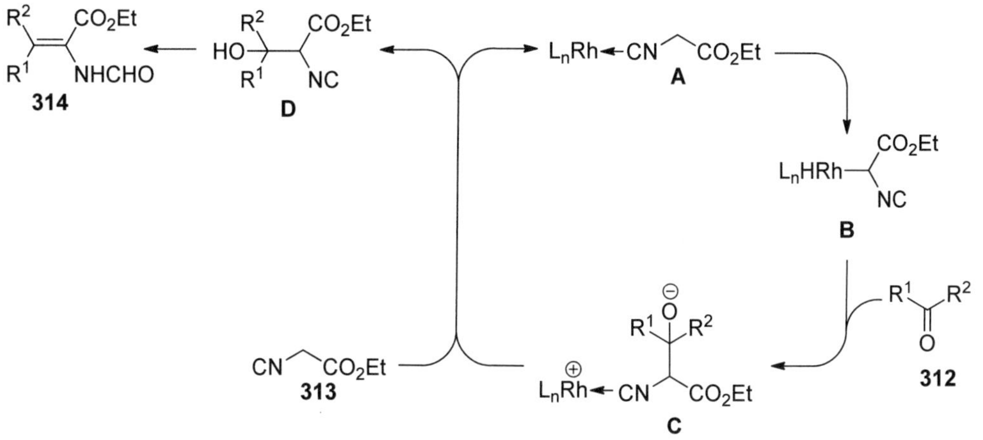

Figure 2-8. Mechanism for the formation of α,β-unsaturated formamides.

Rhodium-catalyzed synthesis of pyrroles[215]

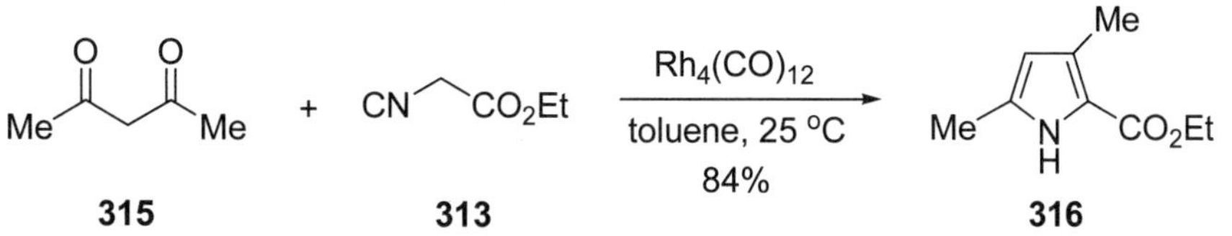

A mixture of $Rh_4(CO)_{12}$ (5.6 mg, 0.0075 mmol), ethyl isocyanoacetate (**313**) (113.1 mg, 1.0 mmol), acetylacetone (**315**) (200.3 mg, 2.0 mmol), and dry toluene (0.5 mL) was placed in a 15-mL Teflon-cocked test tube equipped with a magnetic stir bar. After stirring the mixture for 4 h under Ar at room temperature, the solvent was removed under reduced pressure. The resulting dark orange mixture was purified by column chromatography (SiO_2, hexane/ethyl acetate = 10/1) to afford pyrrole **316** as a colorless solid (151.5 mg, 84%).

Krug and Hartwig[216] recently reported that arylrhodium(I) complex **317** underwent insertion of benzaldehyde into the aryl-Rh bond to give a rhodium alkoxide intermediate, which led to the formation of either ketone **318** or diarylmethanol **319**, depending on the reaction conditions. Ketone **318** was formed under nonaqueous conditions, whereas diarylmethanol **319** was formed under aqueous conditions. This type of insertion reaction is limited to aryl aldehydes.

Representative procedure for the reactions of Rh(I) complexes with aldehydes in C_6D_6[216]

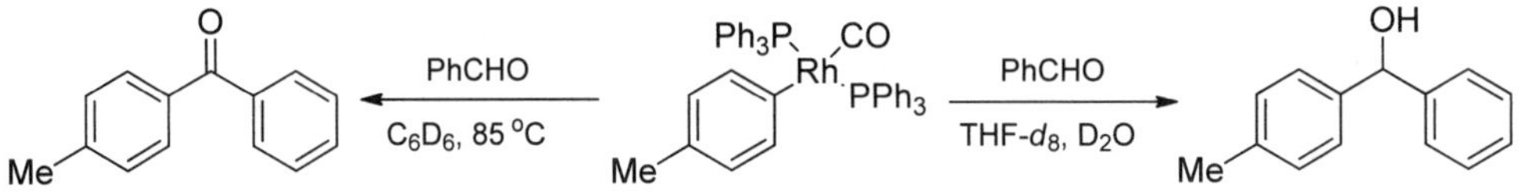

318 **317** **319**

To a small vial was added 9.0 mg (0.011mmol) of **317,** and *ca.* 1 mg of Cp_2Fe. C_6D_6 (0.7 mL) *via* syringe, and the yellow suspension was then transferred to an NMR tube, which was subsequently warmed in a 55 °C oil bath for 5 min to ensure complete dissolution of Rh-complex. A ^{1}H NMR spectrum was acquired, and the solution was transferred to a small vial equipped with a magnetic stirbar. Aldehyde (1.5 equiv) was added *via* syringe (or as a solid) to the solution, and the vial was capped and placed in an 85 °C oil bath for a specified period of time. A ^{1}H NMR spectrum was acquired aftercompletion of the reaction, and a yield was calculated based on the amount of ketone **318** formed relative to the amount of **317** used in the reaction.

Typical procedure for the reactions of Rh(I) complexes with aldehydes in THF-d_8 and D_2O[216]

To a small vial was added 9.0 mg (0.011 mmol) of **317,** and *ca.* 1 mg of Cp_2Fe or 1,3,5-trimethoxybenzene. Addition of THF-d_8 (0.7 mL) *via* syringe generated a clear, yellow solution. A ^{1}H NMR spectrum was acquired, and the solution was transferred to a small vial equipped with a magnetic stir bar. Aldehyde (20 equiv) was added v*ia* syringe (or as a solid) to the solution, and the vial was capped with a piercable septum. Degassed D_2O (0.1 mL) was added *via* syringe, and the resulting solution was stirred at the specified temperature and period of time. A ^{1}H NMR spectrum was acquired following completion of the reaction, and a yield was calculated on the basis of the amount of carbinol formed relative to the amount of **317** used in the reaction.

Matsuda and co-workers[217] reported the catalytic enantioselective Rh-catalyzed reaction of 3-(2-hydroxyphenyl)cyclobutanones such as **320** to give 4-mono- and di-substituted 3,4-dihydrocoumarins **321** and **322** in high yields and excellent enantioselectivities (92–97% ee) by means of asymmetric carbon–carbon bond cleavage. Deuterium-labeling experiments were crucial in developing this new cascade reaction. Whereas Rh-SEGPHOS, Rh-BINAP and Rh-Tol-BINAP complexes prepared *in situ* were found to be effective for constructing chrial tertiary carbon center from 3-mono-substituted cyclobutanones, Tol-BINAP proved to be superior to SEGPHOS in constructing chiral quaternary carbon centers from 3-disubstituted cyclobutanones. The proposed mechanism (Figure 2-9) suggests that ring-opening of the cyclobutanone skeleton by β-carbon elimination forming intermediate **C** is the enantiodifferentiating step. Interestingly, the enantioselective Rh-catalyzed reaction of 3,3-disubstituted cyclobutanones with electron-deficient olefins such as acrylonitrile provided products such as **321** through 1,4-addition of the olefins to the aromatic ring. These products arose from the generation of arylrhodium species such as **D** *via* a 1,4-rhodium shift.

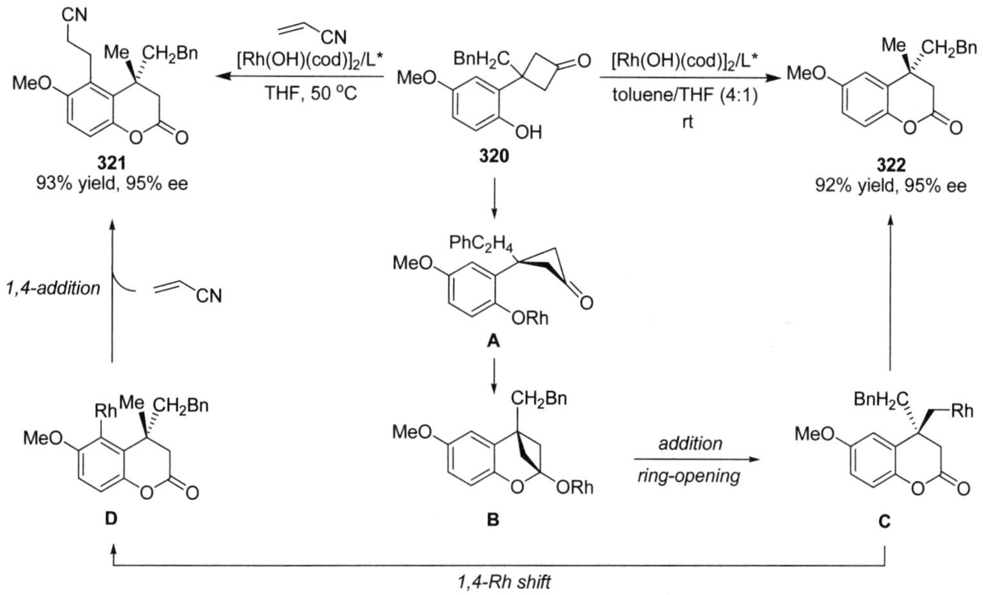

Figure 2-9. Mechanism for the addition/ring-opening of cyclobutanone.

General procedure for rhodium-catalyzed enantioselective addition/ring opening of cyclobutanones[217]

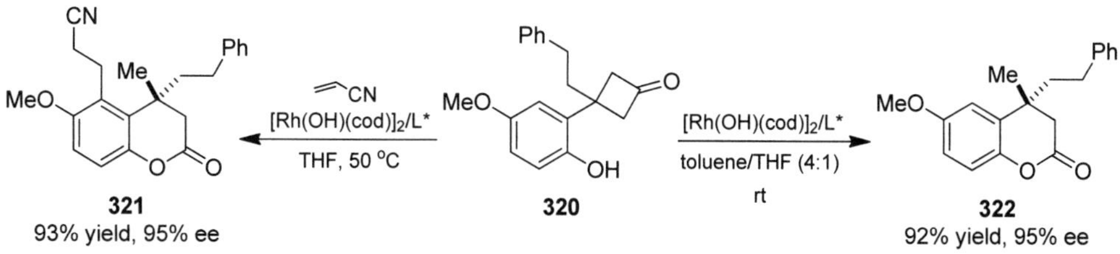

To a mixture of $[Rh(OH)(cod)]_2$ (3.5 mol%) and (*R*)-Tol-BINAP (8 mol%) in toluene/THF (4:1) (1.6 mL) was added **320** (59.3 mg, 0.200 mmol) in toluene/THF (0.4 mL), and the reaction mixture was stirred for 19 h at room temperature. The reaction mixture was filtered through a pad of FLORISIL (U.S. Silica, Frederick, MD, U.S.A.) (Et_2O–AcOEt) and concentrated *in vacuo*. The residue was purified by preparative thin-layer chromatography on silica gel (hexane:AcOEt = 8:1) to afford **322** (54.9 mg, 0.185 mmol, 92% yield). The enantiopurity was determined to be 95% ee by chiral HPLC [Daicel CHIRALCEL OD-H column, Chiral Technologies, Inc, West Chester, PA, U.S.A, hexane:*i*-PrOH = 95:5, 0.6 mL/min, retention times: t_1 = 41.2 min (minor); t_2 =48.5 min (major)].

A rhodium-catalyzed cascade reaction of a cyclobutanone with an electron-deficient alkene[217]

To a mixture of $[RhOH(cod)]_2$ (3.20 mg, 7.0 μmol, 3.5 mol%) and (*R*)-Tol-BINAP (10.9 mg, 16.1 μmol, 8.1 mol%) in THF (1.4 mL) was added acrylonitrile (108 mg, 2.04 mmol). Cyclobutanone **320** (59.3 mg, 0.200 mmol) in THF (0.6 mL) was then added dropwise to the reaction mixture at 50 °C over 1 h using a syringe pump. After being stirred for 10 h at

50 °C, the reaction mixture was filtered through a pad of FLORISIL (U.S. Silica, Frederick, MD, U.S.A.) (Et_2O–AcOEt). The filtrate was concentrated, and the residue was purified by preparative thin-layer chromatography on silica gel (hexane:AcOEt = 3:1) to afford **321** (65.0 mg, 0.186 mmol, 93% yield). The enantiopurity was determined to be 95% ee by chiral HPLC [Daicel CHIRALCEL OD-H column, , Chiral Technologies, Inc, West Chester, PA, U.S.A, hexane:*i*-PrOH = 80:20, 0.6 mL/min, retention times: t_1 = 32.1 min (minor); t_2 = 38.1 min (major)].

15. Rhodium-Catalyzed Cycloadditions and Cyclizations

Transition metal-mediated cycloaddition and cyclization reactions have played a vital role in the advancement and applications of modern synthetic organic chemistry. Rhodium-catalyzed cycloadditions/cyclizations have attracted significant attention because of their versatility in the transformations of activated and unactivated acetylenes, olefins, allenes, etc. These reactions are particularly valuable because of their ability to increase molecular complexity through a convergent and highly selective combination of acyclic components. In addition, these reactions allow for the preparation of molecules with chemical, biological, and medicinal importance with greater atom economy. Recent developments in rhodium-catalyzed cycloaddition and cyclization reactions are described in this section.

15.1 Cycloaddition and Cyclization Mechanisms

The [2+2+1],[218] [4+1],[219] [2+2+2],[220] [5+2],[221] [4+2],[222] and [2+2+2+1][223] cycloadditions share a similar mechanism involving oxidative metallacycle formation in the first step. For simplicity, the mechanism of [2+2+1] cycloaddition is used as a representative example (Figure 2-10). The first step is the complexation of the rhodium catalyst with enyne **323**, and subsequent isomerization to the rhodacyclopentene **A**. The following step involves insertion of other carbon component(s), followed by reductive elimination of the catalyst. In the case of the [2+2+1] process, carbon monoxide inserts into the rhodacyclopentene complex to form intermediate **B**, with subsequent reductive elimination of the Rh catalyst to yield product **324**.

The characteristic feature in the mechanism of cyclization reactions is the reductive cascade ring closure. The mechanisms of silylcarbocyclizations (SiCaC,[224] SiCaT,[225] CO-SiCaC,[224c] CO-SiCaB,[226] and CO-SiCaT[223b, 227]), reductive cyclization,[210, 211] and aldol cycloreduction[180, 206b, 228] can also be categorized into this class. The SiCaC mechanism is illustrated as a representative process (Figure 2-11). The first step is the oxidative addition of a hydrosilane to the Rh catalyst, generating silyl-[Rh]H complex **A**. The acetylene moiety of enyne **323** then inserts into the silyl-Rh bond to form β-silylalkenyl-Rh intermediate **B**, followed by cyclization to intermediate **C**. The product **325** is generated *via* hydrosilane-mediated reductive elimination, which regenerates the silyl-[Rh]H complex **A**. Other related silylcarbocyclizations follow a similar mechanistic pathway with variations in the carbon components.[223b, 224c, 227, 229] For the reductive cyclization/aldol cycloreduction, hydrogen gas is used instead of a hydrosilane.[206b, 228]

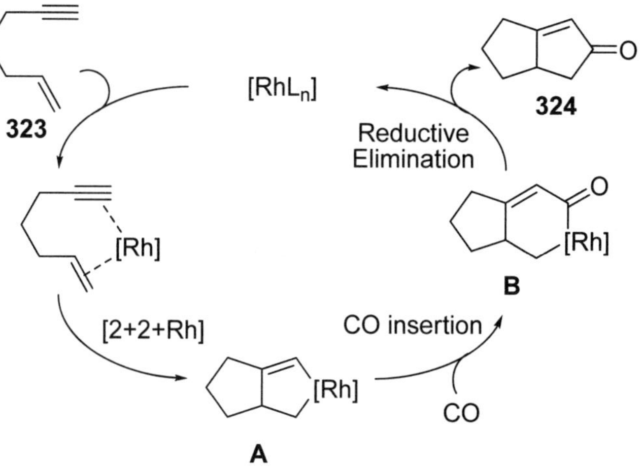

Figure 2-10. Proposed catalytic cycle of Rh-catalyzed [2+2+1] cycloaddition.[230]

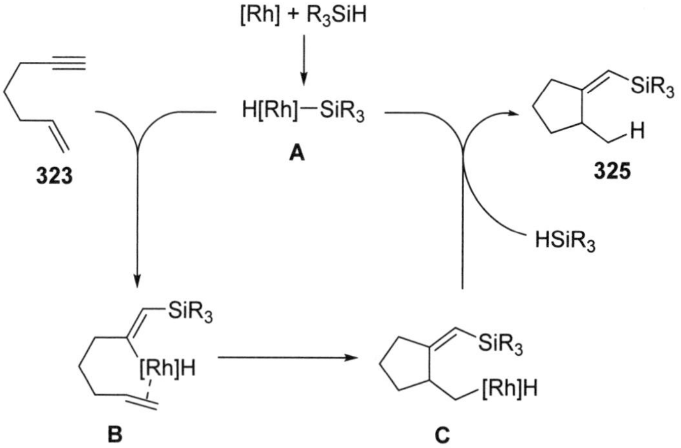

Figure 2-11. SiCaC mechanism.[224c]

15.2 Rh(I)-Catalyzed [2+2+1] Cycloaddition

The preparation of cyclopentenone from the three-component transition-metal–catalyzed [2+2+1] cycloaddition of an alkyne, alkene, and carbon monoxide in the presence of a stoichiometric amount of $Co_2(CO)_8$ was first reported by Khand *et al.* in 1973 (Scheme 2-12).[231] This reaction is attractive for organic synthesis as it tolerates a wide variety of functionalities including esters, ethers, tertiary amines, amides, and alcohols.

Schore introduced the first examples of intramolecular Pauson-Khand reaction (Scheme 2-13),[232] resolving the regioselectivity issue observed in the intermolecular Pauson-Khand reaction. The intramolecular version employs a carbon tether, linking the alkene and alkyne moieties, leading to the formation of a bicyclic product. Since the debut of this reaction, it has been extensively used as the key step in natural product synthesis.[233]

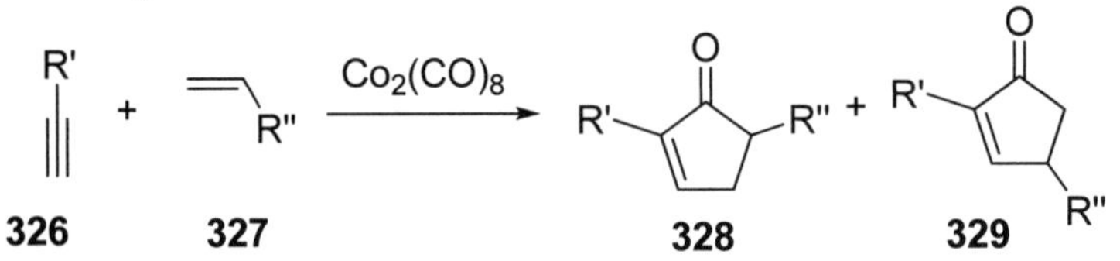

Scheme 2-12. Pauson-Khand reaction.

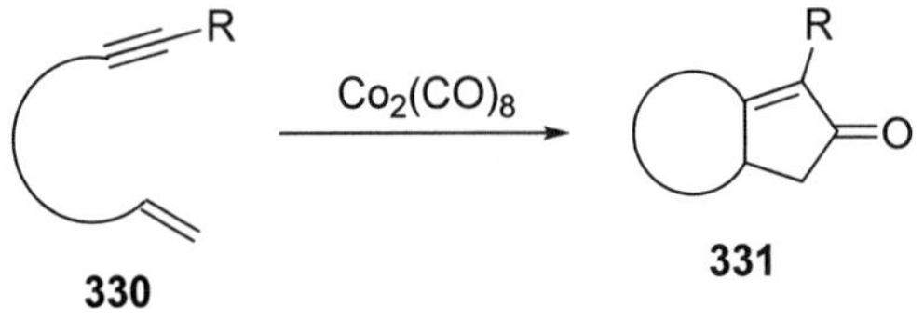

Scheme 2-13. Intramolecular Pauson-Khand reaction.

As follow-ups of the original reaction procedure, numerous improvements have been made, including the lowering of the reaction temperature, use of other transition metals, and the catalytic use of various metal species.[218c, 234] In the 1990s, cobalt, titanium, and ruthenium complexes were found to serve as catalysts for the Pauson-Khand reaction, but these catalytic systems often need to employ medium to high pressures of CO.[235]

Despite significant progress toward catalytic processes, more efficient catalysts were necessary. Cobalt and ruthenium complexes used for this reaction possessed nontunable ligands, which posed serious limitations to expand the reaction scope.[218a] Titanium catalysts were shown to be more versatile and also used in the catalytic enantioselective version of the reaction. Nevertheless, the applications of the titanium-based catalyst were hampered by its high moisture sensitivity.[218a, 235c]

In 1998, the first rhodium-catalyzed Pauson-Khand reaction was reported by Koga *et al.* using $[Rh(CO)_2Cl]_2$ under an atmospheric pressure of CO.[218b] In the same year, Jeong *et al.* reported the use of a variety of rhodium complexes bearing phosphine ligands for the catalytic Pauson-Khand reaction of enyne **332** (Scheme 2-14).[218a] The mononuclear neutral rhodium catalysts **334** and **335** bearing electron-donating ligands required the presence of AgOTf to generate a cationic Rh(I) species *in situ*. However, when dinuclear rhodium catalyst *trans*-$[Rh(CO)Cl(dppp)]_2$ (**337**) was employed, the reaction gave **333** in high yield without the additive. This work paved a way to develop the enantioselective Pauson-Khand reaction using chiral phosphine ligands.

Further catalyst screening identified $[Rh(CO)_2Cl]_2$ and $[Rh(cod)_2Cl]_2$ as good catalysts for this reaction in xylene at 130 °C and ambient pressure of CO.[236] However, $Rh_4(CO)_{12}$, $Rh(acac)(CO)_2$, and $[Rh(CO)_2Cl]_2$-PPh_3 were found to be ineffective.[236] As for the scope of the enyne substrates, enynes **338** with alkyl and aryl substituents at the acetylene terminus afforded bicyclic products **339** in excellent yields (Scheme 2-15).[236] However, **338d** (R = H, X = $CH(CO_2Et)_2$) with a terminal acetylene moiety gave **339d** in only 55% yield. 1-TMS-enyne **338e** (R = TMS, X = $CH(CO_2Et)_2$) was found to be highly sensitive to the catalyst used. Thus, the reaction catalyzed by $[Rh(CO)_2Cl]_2$ in dibutyl ether at 130–150 °C (condition B) gave **339e** in 76% yield, whereas use of $[RhCl(CO)dppp]_2$ in toluene at 110 °C (condition A) did not produce **339e** at all (Scheme 2-15).[236]

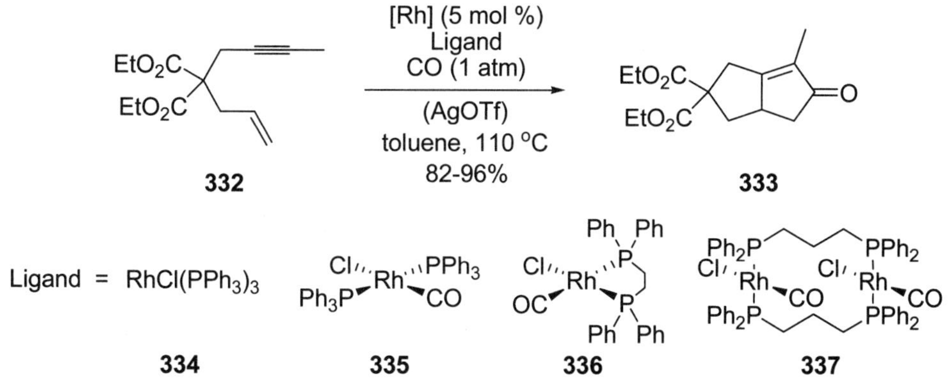

[Rh] (5 mol %), Ligand, CO (1 atm), (AgOTf), toluene, 110 °C, 82-96%: **332** → **333**

Ligand = $RhCl(PPh_3)_3$ (**334**), **335**, **336**, **337**

Scheme 2-14. Rhodium-phosphine catalysts effective for the Pauson-Khand reactions.

[Rh], CO (1 atm), Condition A or B: **338** → **339**

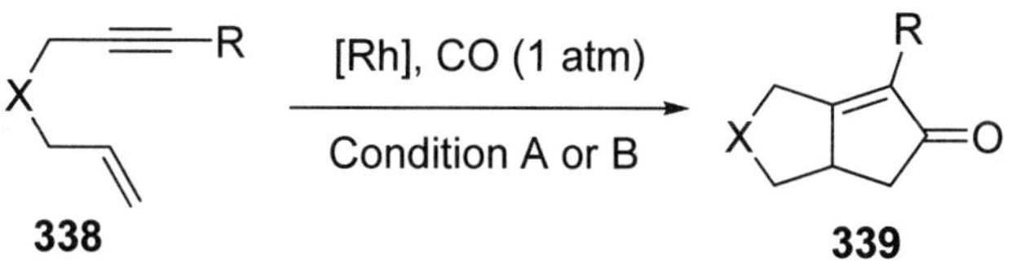

R = Me, Et, Ph, H, TMS; X = $C(CO_2Et)_2$, NTs, O

(A) $[RhCl(CO)dppp]_2$, toluene, 110 °C: 0–99% (B) $[Rh(CO)_2Cl]_2$, Bu_2O, 130–150 °C: 55–94%

Scheme 2-15. Rh-catalyzed Pauson-Khand reactions.

General procedure for the Rh-catalyzed intramolecular Pauson-Khand reaction[236]

EtO_2C, EtO_2C, Et — **338b** → $[Rh(CO)_2Cl]_2$, CO (1 atm), dibutyl ether, 130 °C, 18 h → EtO_2C, EtO_2C, Et, O — **339b**

To a solution of $[Rh(CO)_2Cl]_2$ (3.9 mg, 0.01 mmol) in dibutyl ether (4 mL) was added a solution of diethyl 6-nonen-3-yn-6,6-dicarboxylate (**338b**, 266.7 mg, 1.00 mmol) in dibutyl ether (6 mL) under an atmospheric pressure of CO, and the mixture was heated at 130 °C (the oil bath temperature) for 18 h. After evaporation of the solvent *in vacuo*, the crude products were purified by preparative thin-layer chromatography (silica gel, hexane: ethyl acetate = 4:1, three times) to afford diethyl 8-ethyl-7-oxobicyclo[3.3.0]oct-1-(8)-ene-3,3-dicarboxylate (**339b**, 268.7 mg; 91% yield).

15.3 Asymmetric Rh(I)-Catalyzed [2+2+1] Cycloaddition

Although tremendous advances in the catalytic Pauson-Khand reaction have been made, the development of an asymmetric version did not share the same degree of success. Several asymmetric Pauson-Khand reactions were reported using chiral auxiliaries.[218c, 237] However, those systems required stoichiometric amounts of cobalt as well as the chiral source. Attempts at using a catalytic amount of cobalt did not give satisfactory results.[238] By contrast, the use of titanium chiral catalyst (*S,S*)-(EBTHI)Ti$(CO)_2$ (EBTHI = ethylene-1,2-bis(η^5-4,5,6,7-tetrahydro-1-indenyl) afforded enantiomerically enriched products in excellent yields.[239] After the success of chiral Ti catalysts, rhodium catalysts bearing chiral phosphine ligands were explored. These Rh-phosphine catalyst systems found greater success than their titanium counterparts and further opened an avenue for the development of new asymmetric catalytic reactions.[218a] The search for optimum conditions for the Rh-catalyzed Pauson-Khand reactions commenced with the use of chiral monodentate phosphoramidite ligand SIPHOS[240] and BINAP-based bidentate phosphine ligands (Scheme 2-16).[230, 241] The high enantioselectivity, up to 84% ee, was realized with SIPHOS in the presence of $AgSbF_6$ as an additive,[240] but the best result (99% ee) for the formation of **339** was achieved using $[Rh(CO)Cl_2]_2$, 3,5-diMeC_6H_4-BINAP and AgOTf.[241b]

Typical procedure for the catalytic enantioselective Pauson-Khand reaction at ambient temperature[241b]

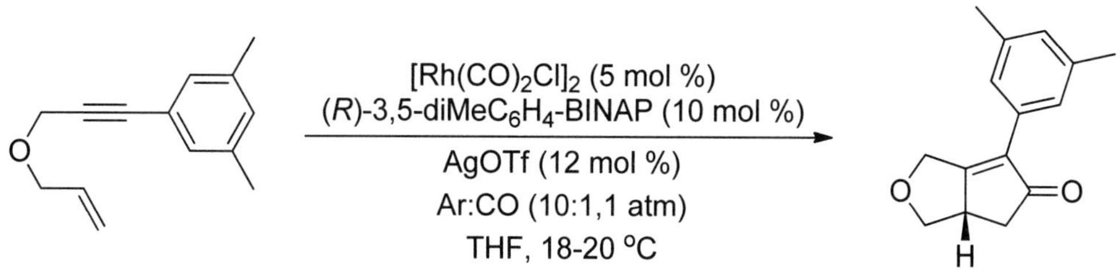

338j → 339j: $[Rh(CO)_2Cl]_2$ (5 mol %), (*R*)-3,5-diMeC_6H_4-BINAP (10 mol %), AgOTf (12 mol %), Ar:CO (10:1,1 atm), THF, 18-20 °C

$[Rh(CO)_2Cl]_2$ (3.4 mg, 0.009 mmol, 5 mol%) and (*R*)-3,5-diMeC_6H_4-BINAP (12.8 mg, 0.017 mmol, 10 mol%) were placed in THF (2 mL), and the mixture was stirred for 30 min at 20 °C under an atmospheric pressure of Ar. A solution of AgOTf (4.5 mg, 0.017 mmol, 10 mol%) in THF (1 mL) was added, and the resultant reaction mixture was stirred for another 30 min at 20 °C. The Ar atmosphere was replaced with CO in Ar (1:10, 1 atm), and then a solution of **338j** (30 mg, 0.174 mmol) in THF (1 mL) was introduced. The reaction mixture was stirred at 20 °C. After completion of the reaction, a gaseous mixture was released in the hood. The crude reaction mixture was concentrated *in vacuo*, and the residue was purified by column chromatography on silica gel using *n*-hexane/ethyl acetate mixture as eluent to afford product **339j**. The enantiomeric excess of the product was determined to be 92% ee by chiral HPLC analysis using Daicel columns.

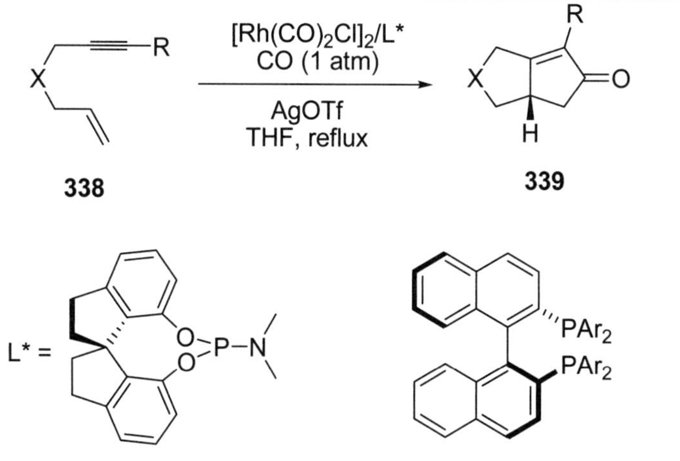

338 → 339: $[Rh(CO)_2Cl]_2$/L*, CO (1 atm), AgOTf, THF, reflux

L* = **340**, (*R*)-SIPHOS; **341**, (*R*)-3,5-$Me_2C_6H_3$-BINAP, Ar = 3,5-$Me_2C_6H_3$

Scheme 2-16. Asymmetric Rh-catalyzed Pauson-Khand reaction.

15.4 Alternative Sources of CO

Rhodium catalysts are known to be effective in the hydroformylation of unsaturated bonds and also in the decarbonylation of aldehydes.[242] This efficacy arises from facile formation of rhodium carbonyl complexes from either carbon monoxide or an aldehyde. The use of carbon monoxide in carbonylation reactions is considered a drawback, especially in academic laboratories. As such, alternative CO sources have been explored.

In 2002, Morimoto and Shibata independently reported the use of a rhodium carbonyl complex obtained *via* aldehyde decarbonylation for a Pauson-Khand type reaction (Scheme 2-17).[243] This success prompted further investigation into CO gas-free carbonylation reactions. Lee *et al.* reported the use of a formate ester in the CO gas-free asymmetric Pauson-Khand type reaction.[244] Park *et al.*[245] reported the use of alcohol as a CO source, and Ikeda and co-workers[246] used aldoses as a source of CO.

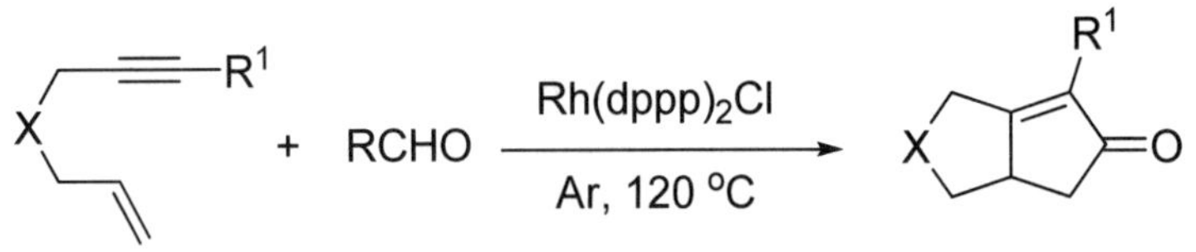

R^1 X + RCHO — $Rh(dppp)_2Cl$, Ar, 120 °C → X R^1 =O

338 **339**

Scheme 2-17. Pauson-Khand reaction catalyzed by a Rh-CO complex generated by aldehyde decarbonylation.

Experimental procedure for the Pauson-Khand-type reaction with cinnamaldehyde under solvent-free conditions[243b]

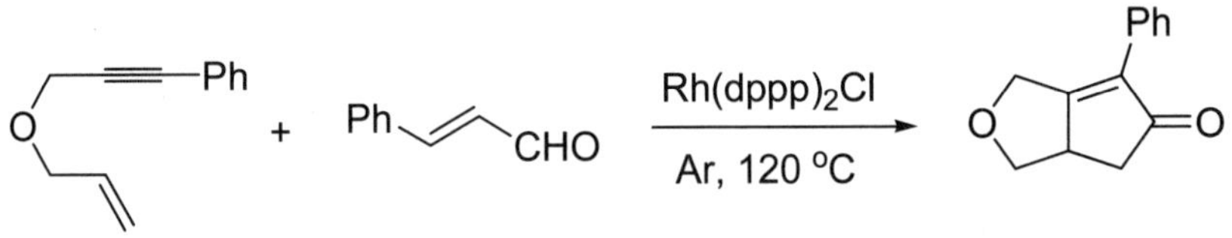

Ph O + Ph CHO — $Rh(dppp)_2Cl$, Ar, 120 °C → O Ph =O

338i **342** **339i**

$Rh(dppp)_2Cl$ (101 mg, 0.105 mmol), enyne **338i** (362 mg, 2.10 mmol), and cinnamaldehyde (333 mg, 2.52 mmol) were placed in a reaction flask, and the mixture was stirred at 120 °C for 3 h under Ar. The flask was then attached to a bulb-to-bulb distillation apparatus. After removal of styrene and a small amount of cinnamaldehyde, the pure product **339i** was distilled (334 mg, 1.67 mmol, 80% yield).

15.5 Other Pauson-Khand (PK) Type Reactions

15.5.1 PK-Type Reactions of Diynes

Cyclopentadienones are reactive and versatile diene systems that can be used to construct polycyclic compounds[247] and polymeric materials.[248] The early preparation of cyclopentadienones *via* Rh-catalyzed [2+2+1] cycloaddition of two alkynes and CO often required a stoichiometric amount of a metal complex and a stepwise process, as shown in Scheme 2-18.[249] The rhodium-pentadiene complex

344 was prepared in excellent yield. However, the corresponding carbonylated product **345** was obtained only in fair to good yields.

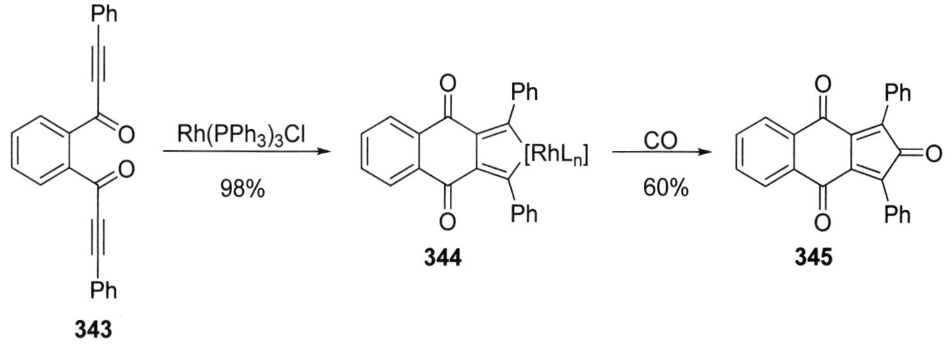

Scheme 2-18. Formation of cyclopentadienones *via*a Rh complex.

Under catalytic conditions, Rh complexeswere only modestly active in the preparation of cyclopentadienones by means of [2+2+1] cycloaddition.[250] In contrast, diyne **346** was readily transformed to cylcopentadienone **347A** by iridium catalysts in 65–99% yields along with minor isomerized product **347B** (Scheme 2-19). Under the same reaction conditions, $[Rh(cod)Cl]_2$ and $[Rh(cod)Cl]_2/PPh_3$ afforded **347A** and **347B** in 29% and 9% yields, respectively.[250]

When isocyanide was used as a carbon monoxide equivalent, Rh-catalyst systems were able to transform diyne **346** into bicyclco-iminocyclopentadienes **349A** and **349B** in high yields (Scheme 2-20).[251] The first example of isocyanide-diyne cycloaddition forming iminocyclopentadienes was reported by Tamao*et al.* using a stoichiometric amount of Ni(0) catalyst in 1989.[252] In 2002, Shibata *et al.* reported the formation of the iminocyclopentadienes using a catalytic amount of a Rh complex.[251]

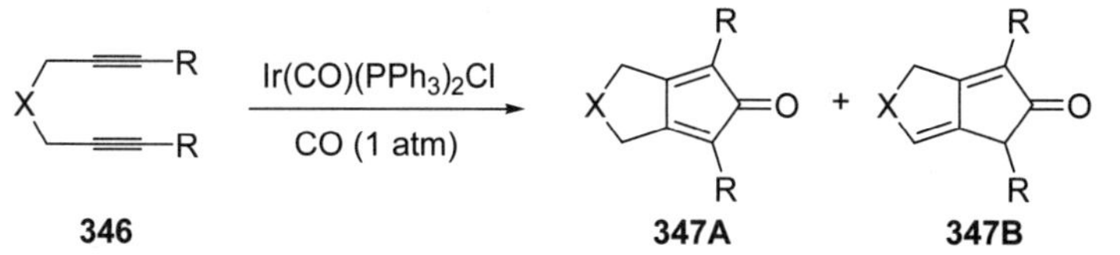

Scheme 2-19. Formation of cyclopentadienones by an Ir complex.

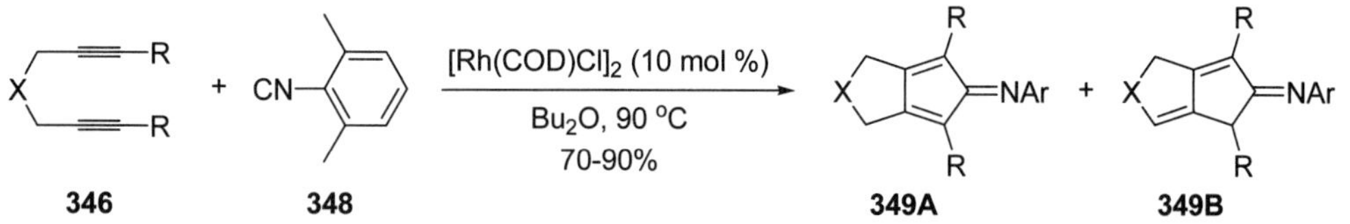

Scheme 2-20. Formation of iminocyclopentadienes catalyzed by aRh complex.

General procedure[251]

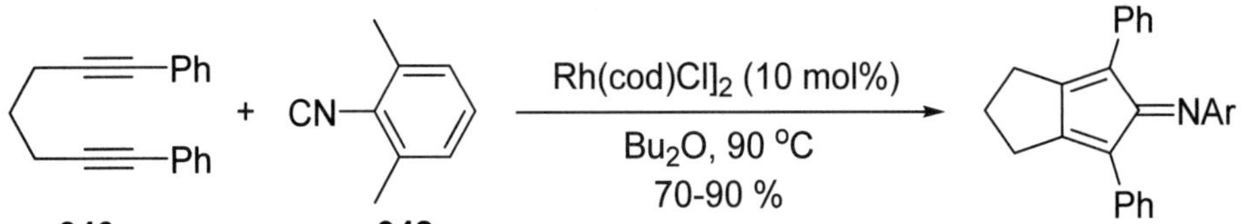

Ph + CN Rh(cod)Cl]$_2$ (10 mol%) Bu$_2$O, 90 °C 70-90 % Ph =NAr Ph Ph

346a 348 349Aa

$[Rh(cod)Cl]_2$ (4.9 mg, 0.01 mmol) was placed in a flask and a dibutyl ether solution (2 mL) of 1,7-dcphenylhepta-1,6-diyne **346a** (24.7 mg, 0.21 mmol) was added. The resulting mixture was stirred at 90 °C. A dibutyl ether solution (0.2 mL) of 2,6-dimethylphenylisocyanide **348** (2.7 mg, 0.021 mmol) was added five times at intervals of 15 min. After the last addition of the isocyanide, the resulting solution was stirred for 30 min. The solvent was removed under reduced pressure and the crude products were purified by thin layer chromatography to give iminocyclopentadiene **349Aa** (31.3 mg, 0.17 mmol, 83%).

15.5.2 PK-Type Reactions of Dienynes

The scope of Pauson-Khand type reactions has been expanded by exploring the use of various carbon components. Dienyne **350** is a versatile substrate, which can be subjected to two cycloaddition pathways (Scheme 2-21).[253] In 2003, Wender *et al.* reported the Rh-catalyzed PK-type reaction of **350**. Under unoptimized conditions, **350** underwent three competing cycloadditions, *i.e.*, 1) an intramolecular [4+2] cycloaddition to afford product **351**, 2) a new version of a [2+2+1] cycloaddition to afford product **352**, and 3) an unprecedented [4+2+1] cycloaddition to afford product **353**. After further refinement, the [2+2+1] product **352** was obtained in excellent yield.[253]

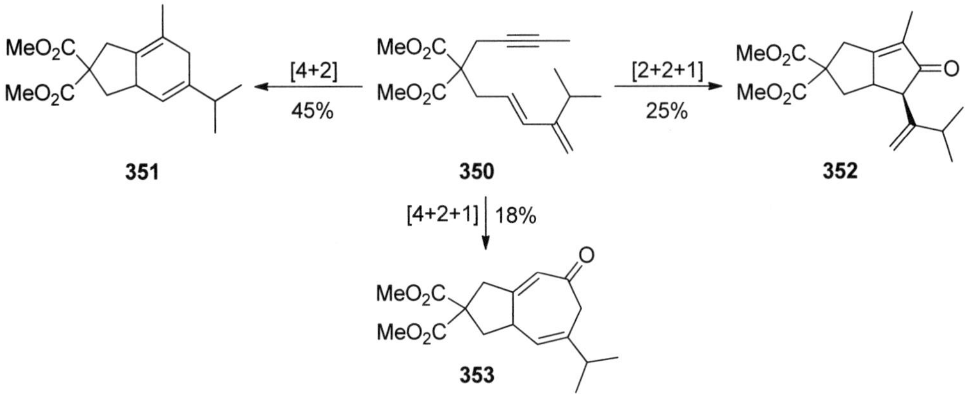

MeO_2C MeO_2C [4+2] 45% MeO_2C MeO_2C [2+2+1] 25% MeO_2C MeO_2C =O

351 350 352

[4+2+1] 18%

MeO_2C MeO_2C O

353

Scheme 2-21. Rh-catalyzed transformation of dienyne **350**.

Procedure for PK-type reaction of dienyne **350**[253]

$RhCl(CO)(PPh_3)_2$
$AgSbF_6$
CO
$ClCH_2CH_2Cl$

350 → **352**

$RhCl(CO)(PPh_3)_2$ (0.039 g, 0.056 mmol), $AgSbF_6$ (0.020 g, 0.058 mmol), and dichloroethane (5.6 mL, 0.01 M) were added into an oven-dried borosilicate glass test tube equipped with a stirring bar and capped with a rubber septum. The resulting suspension was stirred for 1 h under a balloon of CO and vented through a bubbler. Dienyne **350** (0.035 g, 0.1197 mmol) was dissolved in dichloroethane (1.08 mL, 0.11 M) in a separate oven-dried borosilicate glass test tube equipped with a stirring bar and capped with a rubber septum. This solution was stirred under a balloon of CO and vented through a bubbler for 15 min. The catalyst solution (0.12 mL) was added to the solution of **350** and the resulting mixture was stirred vigorously under CO (1 atm) for 14 h. The mixture was purified by flash-column chromatography on silica gel (gradient elution: 0→10% ethyl acetate in pentane). The pure product **352** was obtained in 89% yield as colorless oil.

From all the PK-type reactions discussed thus far, an alkyne group is always one of the two carbon components. However, Wender *et al.* reported a PK-type reaction of diene-alkenes (Scheme 2-22).[254] Although triene-system **356** had been studied in great detail for the intramolecular Diels-Alder reactions,[222b, 255] the PK-type reaction of 1,6-dienes was not known at that time. 1,6-Diene **354** was employed in an attempted reaction using conditions similar to those for PK-type reactions, but no reaction took place (Scheme 2-22).

The reaction of 1,3,8-triene **356** catalyzed by a Rh-complex in the presence of $AgSbF_6$ under a CO atmosphere gave PK-type reaction products **357** and **358** in a combined yield of 82% (Scheme 2-22).[254] A control experiment confirmed that product **358** was formed exclusively through isomerization of **357**. It should be noted that the substitution at the 2-position of the 1,3-diene moiety is well tolerated; however, substitution at the 1- or 3-position is detrimental to the reaction.[254]

General procedure[254]

$[Rh(CO)_2Cl]_2$ (5 mol %)
CO (1 atm), DCE

356a → **357a** + **358a**

In an oven-dried borosilicate glass test tube, equipped with a stir bar and capped with a rubber septum, 1,3,8-triene **356a** (0.019 g, 0.075 mmol) was dissolved in dichloroethane (0.75 mL, 0.10 M). $[RhCl(CO)_2]_2$ (0.0015 g, 0.0038 mmol) was added and the solution was stirred under a balloon of CO and vented to a bubbler for 15 min. The reaction mixture was stirred vigorously under 1 atm of CO at 80 °C for 3 h. The mixture was purified by flash column chromatography on silica gel (gradient elution: 0%→20% Et_2O in pentane). Products **357a** and **358a** were obtained in 99% combined yield as a colorless oil.

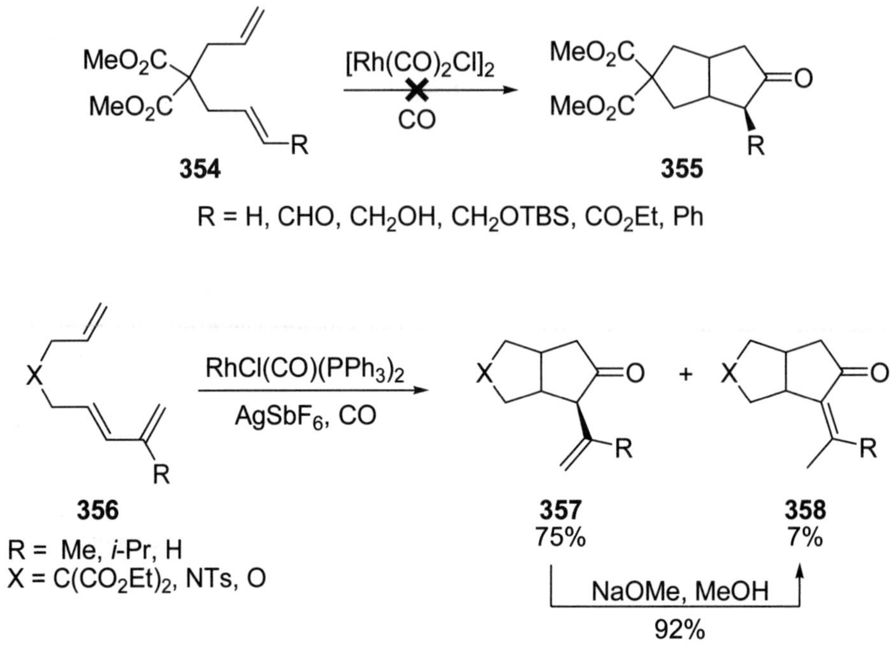

MeO_2C / MeO_2C — $[Rh(CO)_2Cl]_2$, CO (✗) → MeO_2C / MeO_2C … =O, R

354 **355**

R = H, CHO, CH_2OH, CH_2OTBS, CO_2Et, Ph

X — R — $RhCl(CO)(PPh_3)_2$, $AgSbF_6$, CO → X …=O, R + X …=O, R

356 **357** 75% **358** 7%

R = Me, *i*-Pr, H
X = $C(CO_2Et)_2$, NTs, O

NaOMe, MeOH 92%

Scheme 2-22. Rh-catalyzed PK-type reaction of 1,3,8-triene **356**.

15.5.3 PK-Type Reaction of Allenynes

Allenes are important and versatile building blocks in transition-metal–catalyzed, carbon–carbon, bond-forming reactions.[256] However, control of regioselectivity for the two reactive orthogonal double bonds poses a major challenge.[257] The most common strategy to address this issue is to introduce a substituent at the carbon atom adjacent to the allene moiety for imposing geometrical restriction and electronic differentiation to the two orthogonal olefins.[257] However, recent advances in transition-metal–catalyzed transformations of allenes have made it possible to control selectivity solely by the selection of the metal catalyst species used in the transformation.[222a, 258]

In the presence of different transition metal catalysts, allenyne **359** can be diversified into different carbocycles and heterocycles depending on reaction conditions (Scheme 2-23).[258, 259] In the presence of $Mo(CO)_6$, allenyne **359** gave bicyclo[3.3.0] product **360**. By contrast, in the Rh-catalyzed reaction, the terminal olefin of **359** reacted to give the bicyclo[4.3.0] product **361.** In the absence of carbon monoxide, allenyne **359** underwent an Alder-ene reaction to form the cross-conjugated triene **362**.

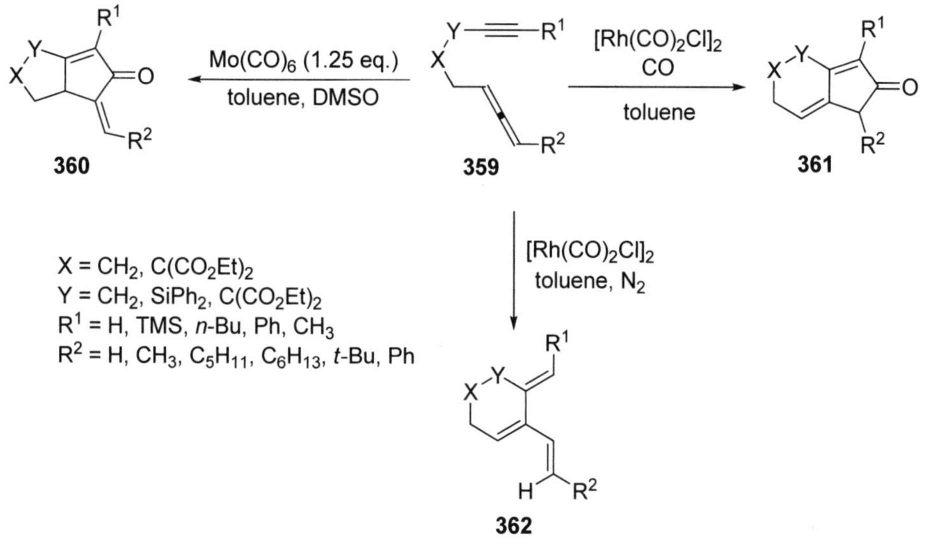

R1
Mo(CO)6 (1.25 eq.)
toluene, DMSO
Y—R1
X
R2
[Rh(CO)2Cl]2
CO
toluene
R1
O
R2
360
359
361
X = CH2, C(CO2Et)2
Y = CH2, SiPh2, C(CO2Et)2
R1 = H, TMS, n-Bu, Ph, CH3
R2 = H, CH3, C5H11, C6H13, t-Bu, Ph
[Rh(CO)2Cl]2
toluene, N2
R1
H
R2
362

Scheme 2-23. Transition-metal–catalyzed transformation of allenyne **359**.

The PK-type reaction of allenes provides powerful means to assemble larger bicyclo[m.3.0] ring systems ($m > 4$). The bicyclo[5.3.0] ring system is of particular interest as it is the core structure of various natural products. Despite successful construction of the bicyclo[3.3.0] and [4.3.0] systems, the construction of bicyclo[5.3.0] **364** from enyne **363** was found to be challenging (Scheme 2-24).[260] The major breakthrough in the synthesis of these bicyclic systems was brought about independently by Brummond *et al.*[258] and Mukai *et al.*[260] using the Rh-catalyzed reaction of allenynes **365** (Scheme 2-24).

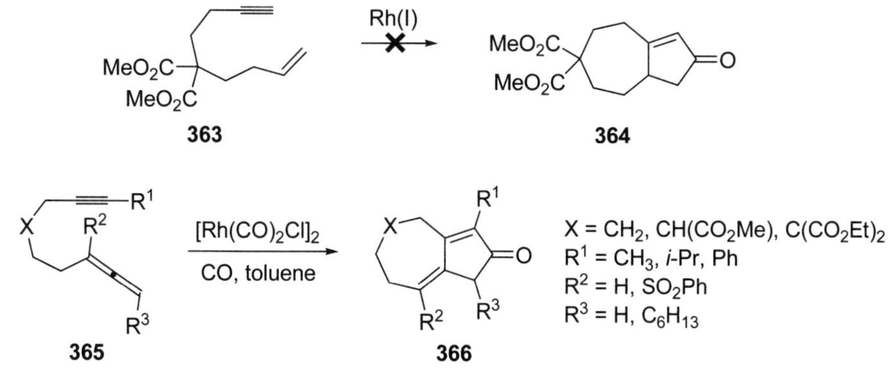

Rh(I)
MeO2C
MeO2C
363
364
R1
R2
[Rh(CO)2Cl]2
CO, toluene
R3
365
366
X = CH2, CH(CO2Me), C(CO2Et)2
R1 = CH3, i-Pr, Ph
R2 = H, SO2Ph
R3 = H, C6H13

Scheme 2-24. Formation of bicyclo[5.3.0]ring system.

During the investigation of substrates for the PK-type reaction of allenynes, Brummond *et al.* developed a series of *N*-alkynyl allenic α-amino acids **367** for diversity-oriented synthesis (DOS) (Scheme 2-25). It was reported that the Alder-ene reaction of **366** was substantially faster (<10 min) than the previous substrate series. This rate acceleration was in part attributed to the Thorpe-Ingold effect arising from the *ortho*-disubstituted methyl and carbomethoxy groups. An earlier substrate series readily underwent PK-type reaction under CO atmosphere without formation of the cross-conjugated product. However, *N*-alkynyl allenic amino acids **367** subjected to standard PK-type reaction conditions afforded the triene product **368** exclusively. It was later found that the addition of PPh_3 and $AgBF_4$ was necessary for the Alder-ene reaction to proceed to give the PK-reaction product **369**. It should be noted that this reaction type is not compatible with terminal alkynes.

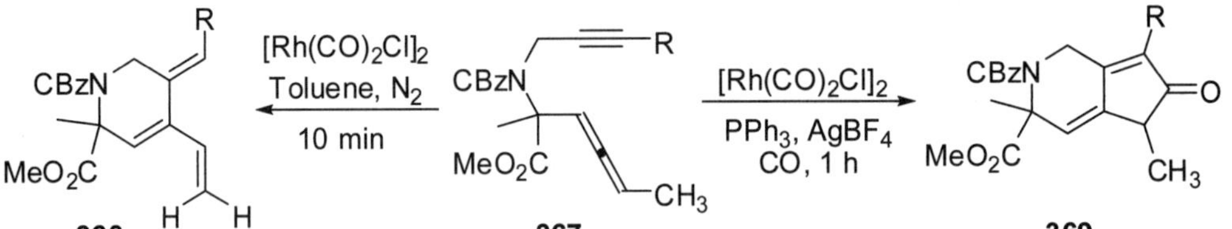

Scheme 2-25. Alder-ene and PK-type reaction of allenyne **367**.

Procedure for the Rh-catalyzed Pauson-Khand type reaction of allenynes[258]

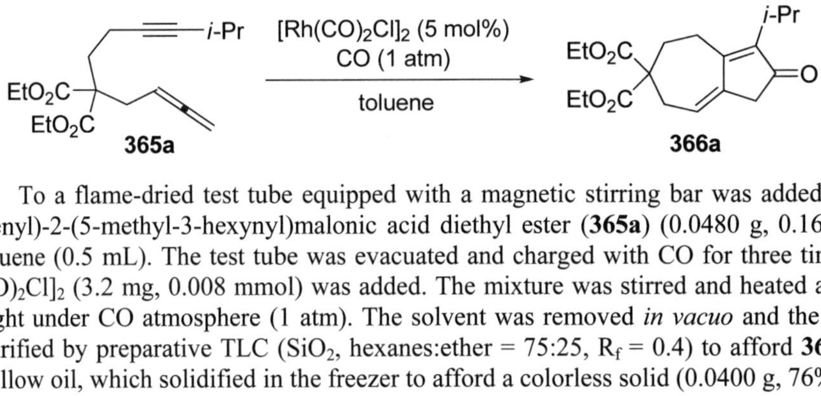

To a flame-dried test tube equipped with a magnetic stirring bar was added 2-(2,3-butadienyl)-2-(5-methyl-3-hexynyl)malonic acid diethyl ester (**365a**) (0.0480 g, 0.16 mmol) and toluene (0.5 mL). The test tube was evacuated and charged with CO for three times and $[Rh(CO)_2Cl]_2$ (3.2 mg, 0.008 mmol) was added. The mixture was stirred and heated at 90 °C overnight under CO atmosphere (1 atm). The solvent was removed *in vacuo* and the residue was purified by preparative TLC (SiO_2, hexanes:ether = 75:25, R_f = 0.4) to afford **366a** as a pale yellow oil, which solidified in the freezer to afford a colorless solid (0.0400 g, 76%).

*General procedure for rhodium-catalyzed Pauson-Khand type reaction of **367***[259a]

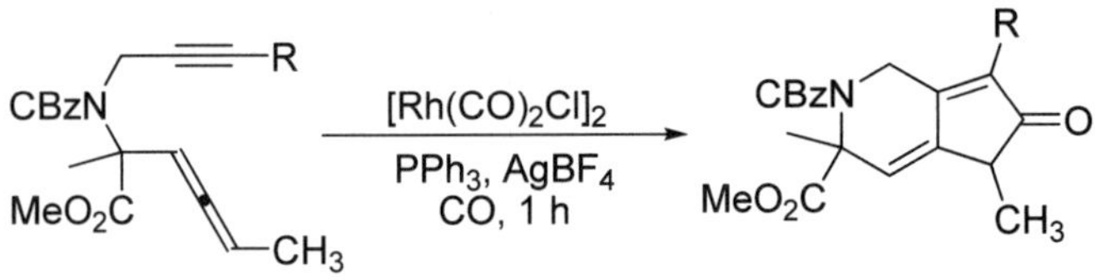

To a flame-dried test tube equipped with a magnetic stir bar was added $[Rh(CO)_2Cl]_2$ (0.05 mmol) and dichloroethane (1 mL). To this solution, PPh_3 (0.15 mmol) in dichloroethane (1 mL) was added dropwise *via* a syringe at room temperature. The test tube was evacuated under vacuum by inserting a needle and charged with CO from a balloon three times. After 5 min, $AgBF_4$ (0.1 mmol of a 0.05-M or 0.1-M solution in dichloroethane) was added dropwise *via* a syringe. The mixture was stirred for an additional 5–10 min at room temperature, and then the allenyne **367** (0.5 mmol) in dichloroethane (2 mL) was added *via* a syringe. The reaction was monitored by TLC and after completion (1–2 h at room temperature or 40 °C in some cases), it was directly purified by column chromatography on silica gel. Gradient elution (hexanes-EtOAc, 9:1 to 3:1, v/v) afforded azabicyclo[4.3.0]octadienes **369**.

In the continued expansion of the substrate type for the allene-alkene type PK-type reaction Brummond *et al.*[259b] and Makino and Itoh[261] independently reported the use of allene-alkene **370**. However, under these conditions, the reaction does not form any bicyclic adduct. Instead it underwent Rh-catalyzed cycloisomerization to the seven-membered ring product **371** (Scheme 2-26).

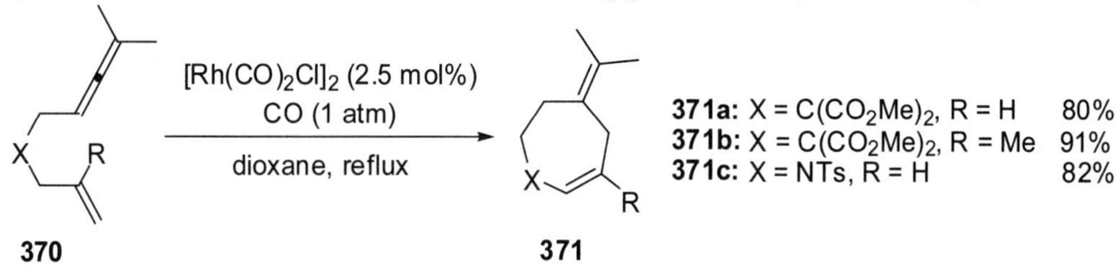

$[Rh(CO)_2Cl]_2$ (2.5 mol%), CO (1 atm), dioxane, reflux

370 → **371**

371a: X = $C(CO_2Me)_2$, R = H 80%
371b: X = $C(CO_2Me)_2$, R = Me 91%
371c: X = NTs, R = H 82%

Scheme 2-26. Allene-ene cycloisomerization.

Inagaki *et al.*[262] reported that allenenes bearing a terminal allene moiety underwent the [2+2+1] cycloaddition to form 6-5 bicyclic products. The best results were achieved using $[RhCl(CO)dppp]_2$ as the catalyst and $AgBF_4$ as the additive under subambient pressures of CO (Scheme 2-27).

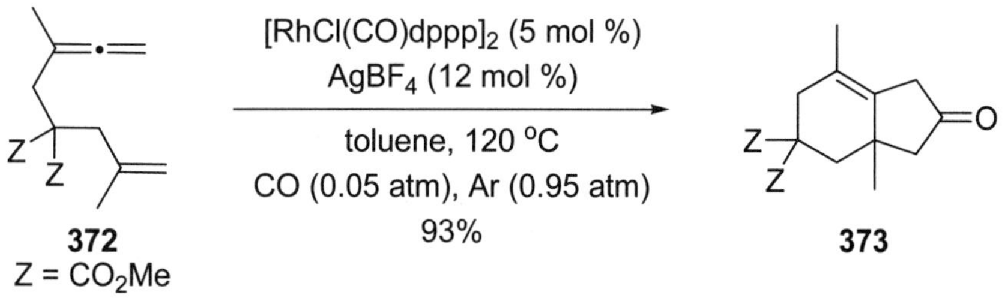

$[RhCl(CO)dppp]_2$ (5 mol %), $AgBF_4$ (12 mol %), toluene, 120 °C, CO (0.05 atm), Ar (0.95 atm), 93%

372 (Z = CO_2Me) → **373**

Scheme 2-27. Rh-catalyzed PK-type reaction of allenene **372**.

Wender *et al.* applied his Rh-catalyzed alkene-diene PK-type reaction described above[254] to allene-dienes **374** and obtained similar results, wherein the proximal double bond of the allene moiety reacted exclusively to give bicyclo[3.3.0]octanones **375** (Scheme 2-28). It should be noted that the substitution pattern at the terminal carbon of the allene moiety has a significant effect on the reaction.

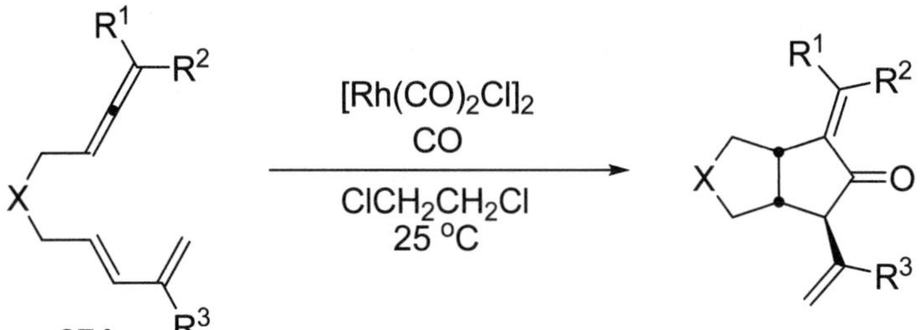

Scheme 2-28. Rh- catalyzed allene-diene PK-type reaction.

The reactions of allene-dienes **374** ($R^1 = R^2 =$ Me) bearing 1,1-dimethyallene moiety afforded [2+2+1] products **375** in 81–97% yield, whereas those of **374** ($R^1 =$ H, $R^2 =$ Me, *t*-Bu) bearing monosubstituted allene moiety gave [2+2+1] products **375** in 39-94% yield as a 2:1 to 7.5:1 mixture of the double bond isomers. The reaction of **374** ($R^1 = R^2 =$ H, $R^3 =$ *i*-Pr, X = TsN) with a terminal allene moiety required a higher reaction temperature (60 °C *vs.* room temperature) and a higher CO pressure (4 atm *vs.* 1 atm) for the reaction to proceed, giving a 2.3:1 mixture of the corresponding **375** and [4+2] product in 62% combined yield.[263]

Typical procedure for allene-diene PKR[263]

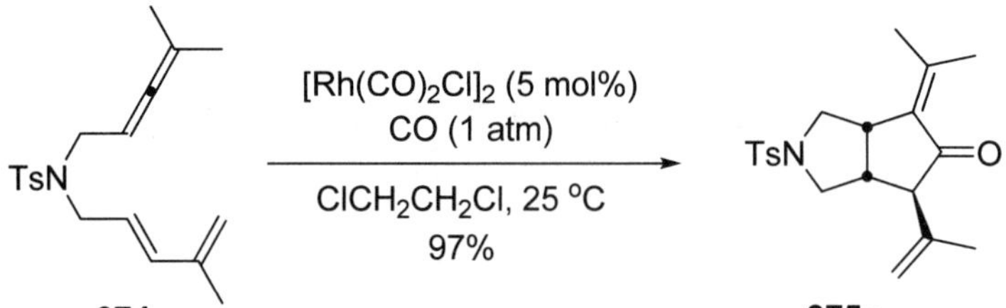

$[Rh(CO)_2Cl]_2$ was weighed into an oven-dried test tube equipped with a magnetic stir bar, and dichloroethane was added by syringe to make a 0.005-M solution. The test tube was capped with a septum, and the solution was stirred under a balloon of CO vented to a bubbler for 45 min. Diene–allene **374a** (21 mg, 0.063 mmol) was weighed into an oven-dried test tube equipped with a magnetic stir bar, dichloroethane (0.114 mL) was added by syringe, and the test tube was capped with a rubber septum. The solution was stirred under a balloon of CO vented to a bubbler for 30 min. The catalyst solution $[Rh(CO)_2Cl]_2$ (0.0127 mL, 0.000063 mmol) was added by syringe, and the reaction mixture was stirred under a balloon of CO for 8 h. The solution was concentrated by rotary evaporation, and the residue was purified by column chromatography on silica gel ($EtOAc/CH_2Cl_2 = 2/98$). The product **375a** was obtained (20.6 mg, 97%) as a white solid.

Inagaki *et al.*[264] expanded further the scope of allenic PK-type reaction by investigating the [2+2+1] cycloaddition of *bis*-(allene)s. The construction of a bicyclo[6.3.0]undecadienone framework was achieved in high yield. Double-bond isomerization in the final bicyclic products was observed in some cases.

General procedure for carbonylative [2+2+1] cycloaddition under an atmosphere of CO[264]

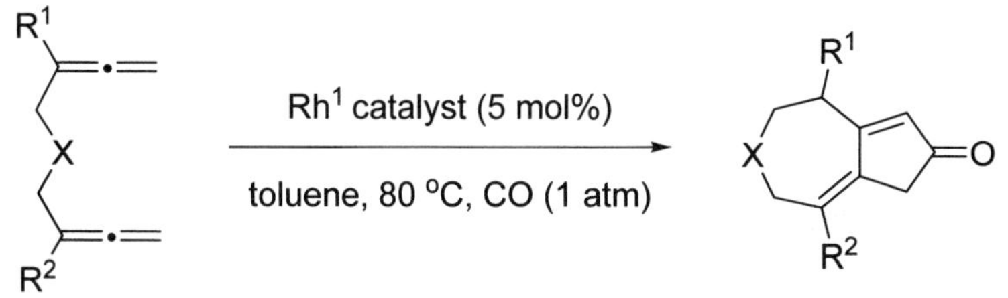

To a solution of *bis*-allene **376** (0.10 mmol) in toluene (1.0 mL) was added 5–10 mol% Rh(I) catalyst. The reaction mixture was heated at 80 °C under an atmosphere of CO until the complete disappearance of the starting material (monitored by TLC). Toluene was evaporated off, and the residual oil was chromatographed with hexane-AcOEt or CH_2Cl_2-AcOEt to afford cyclized product **377**.

15.5.4 Rh(I)-Catalyzed [4+1] and [3+2] Cycloaddition

The Rh-catalyzed [4+1] cycloaddition of vinylallene **378** was first reported using a stoichiometric amount of Wilkinson's catalyst to generate the (vinylallene)rhodium complex **379** in 92% isolated yield. Under the CO atmosphere, (vinylallene)rhodium complex **379** can be readily transformed into cyclopentenone **381** in 96% yield. The formation of the product is presumed to occur *via* isomerization of cyclopentenone **380** (Scheme 2-29).[265]

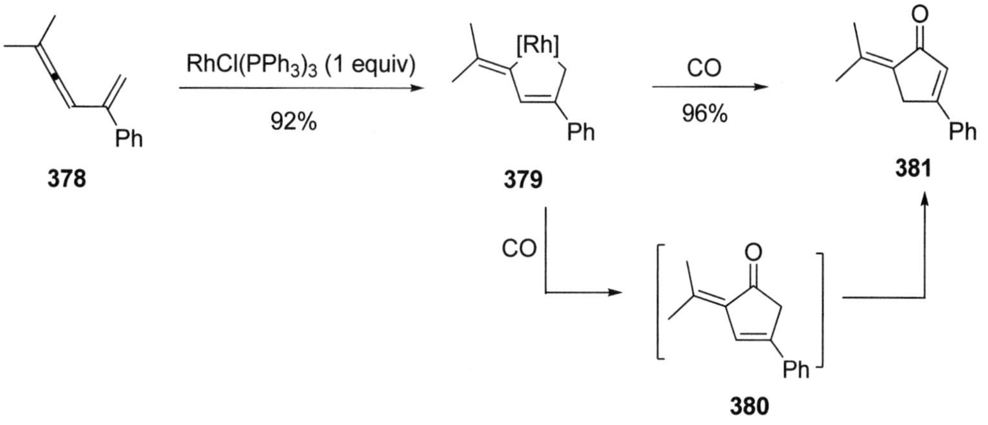

Scheme 2-29. Rh-catalyzed [4+1] cycloaddition of vinylallene.

Soon after the transformation of the (vinylallene)rhodium complex **379** into cyclopentenone **381** was discovered,[265] the catalytic version of the Rh-catalyzed [4+1] cycloaddition was developed using $[Rh(cod)_2]OTf$ with along with the 1,2-*bis*-(diphenylphosphino)-benzene (dppbe) ligand.[219c] It was found that a cationic Rh complex was necessary to promote the reaction efficiently, and in combination with the dppbe ligand, the isomerization was suppressed. The use of high CO pressure (10 atm) provided optimum results, as shown in Scheme 2-30.

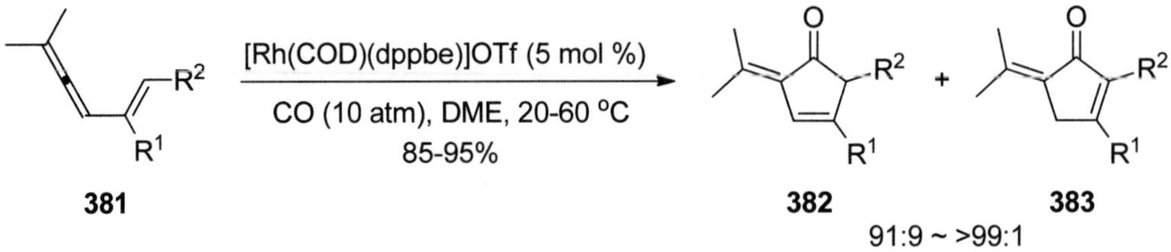

Scheme 2-30. Rh-catalyzed [4+1] cycloaddition.

An asymmetric version of [4+1] cycloaddition was later reported using $[Rh(cod)_2]PF_6$ catalyst with nonracemic diphosphine ligand (*R,R*)-Me-DuPHOS (Scheme 2-31, eq. 1).[219a, 219b] A higher pressure of CO was required to suppress the formation of conjugated triene **386**, which was formed *via* β-hydride elimination of the Rh metallacycle **385** (Scheme 2-31, eq. 2).

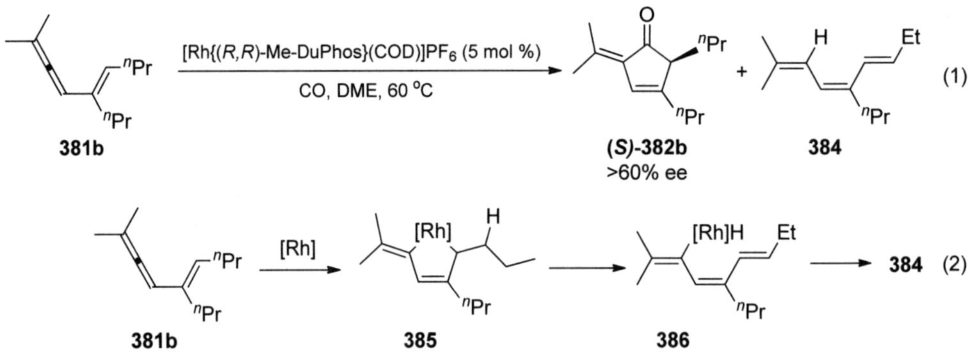

Scheme 2-31. Asymmetric Rh-catalyzed [4+1] reaction.

Preparation and use of [Rh(dppbe)(cod)]OTf[219c]

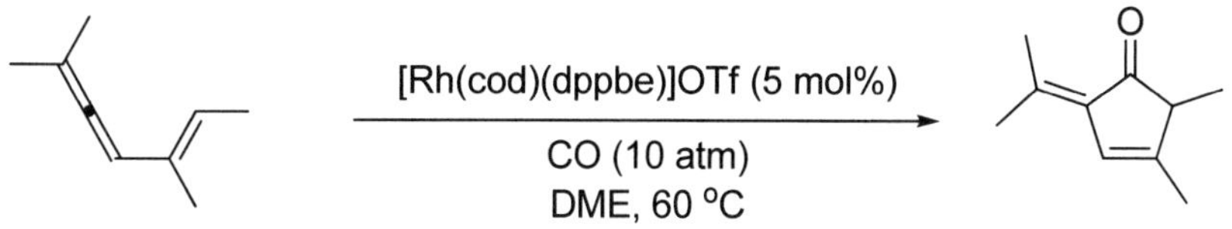

[Rh(cod)(dppbe)]OTf (5 mol%)

CO (10 atm)

DME, 60 °C

381a **382a**

To a solution of $[Rh(cod)_2]OTf$ (144 mg, 0.307 mmol) in CH_2Cl_2 (5 mL) at room temperature was added dropwise a solution of dppbe (146 mg, 0.326 mmol) in CH_2Cl_2 (4 mL). The reaction mixture was stirred for 30 min and then concentrated to 1 mL under vacuum. Et_2O (15 mL) was added, and the resulting precipitates were washed with Et_2O to afford [Rh(dppbe)(cod)]OTf (239 mg, 97%) as an orange solid.

Synthesis of 4,5-dimethyl-2-isopropylidene-3-cyclopentenone ***382a*** *by the Rh-catalyzed [4 + 1] cycloaddition*[219b]

A mixture of [Rh(cod)(dppbe)]OTf (16.5 mg, 20.5 μmol) and **381a** (50.0 mg, 409 mol) in 1,2-dimethoxyethane (2 mL) under 10 atm of CO in an autoclave was stirred in an oil bath at 60 °C for 15 h. After the mixture was cooled, the solvent was removed under vacuum. The residue was subjected to preparative thin layer chromatography (silica gel, ether:hexane = 1:7) to afford **382a** (52.2 mg, 85%).

A novel route to cyclopentadienones was developed based on the Rh-catalyzed [3+2] cycloaddition of cyclopropenone **387** with alkynes (Scheme 2-32).[263] The [3+2] cycloaddition of **387** with a variety of alkynes catalyzed by $[Rh(CO)_2Cl]_2$ (1 mol%) in toluene gave cyclopentadieneones **389** in 17-99% yield. As alkyne components of this process, aryl alkynes, heteroaryl alkynes, cyclohexenyl, and even benzyne were employed. This process is regiospecific, giving a single product in all cases examined. As catalyst species, $[Rh(CO)_2Cl]_2$ appears to be the catalyst of choice because other catalysts, such as $Pd(OAc)_2$, $IrCl(CO)(PPh_3)_2$, Lewis acids, and Brønsted acids, afforded at most a trace amount of **389**.

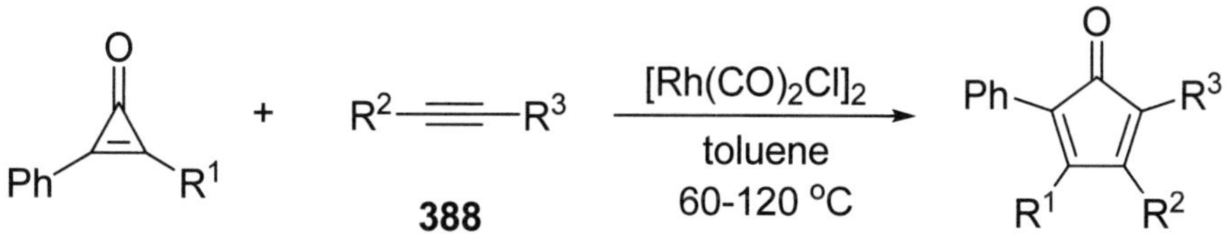

$[Rh(CO)_2Cl]_2$

toluene

60-120 °C

387 **388** **389**

Scheme 2-32. [3+2] cycloaddition route to cyclopentadienone **389**.

Typical procedure for the Rh-catalyzed [3+2] cycloaddition[263]

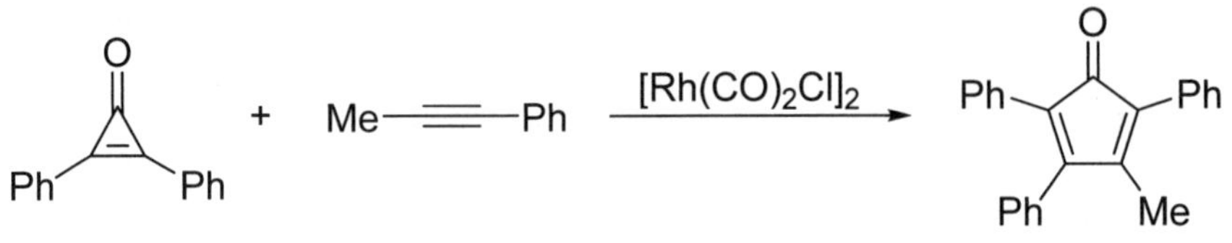

387a + 388a → 389a ($[Rh(CO)_2Cl]_2$)

Diphenylcyclopropenone (**387a**, 375 μmol, 77.3 mg) and 1-phenyl-1-propyne (**388a**, 250 μmol, 25.1 mg, 27 μL) were weighed out into a test tube with a stir bar under nitrogen. To this mixture was added toluene (750 μL) and $[RhCl(CO)_2]_2$ (1.0 mg, 2.5 μmol). The tube was capped with a septum and heated at 80 °C for 2 h under nitrogen. The resulting solution was cooled to room temperature and then purified by flash column chromatography eluting with 100 mL petroleum ether followed by 150 mL 5% diethyl ether/petroleum ether. After purification, cyclopentadienone **389a** was obtained as a deep purple solid (76 mg, 94%).

15.5.5 Rh(I)-Catalyzed [2+2+2] Cycloaddition

Transition-metal–catalyzed intermolecular [2+2+2] cyclotrimerization of alkynes to benzenes has been extensively studied with several catalyst systems involving palladium, cobalt, nickel, rhodium, and other transition metals.[266] This methodology can be applied to the preparation of polysubstituted benzenes. The major challenge of this transformation is control of regioselectivity of unsymmetrical alkynes, particularly in the cross-cyclotrimerization of two or three alkynes.

In 1996, Garcia *et al.* reported cyclotrimerization of (trifluoromethyl)acetylene **390a** using a bimetallic rhodium complex at room temperature to afford 1,3,5-*tris* (trifluoromethyl) benzene **391** in high yield (Scheme 2-33).[267] Recently, Tanaka *et al.* reported a chemoselective and regioselective intermolecular cross-cyclotrimerization of two different alkynes using a cationic Rh complex as catalyst (Scheme 2-34).[220e]

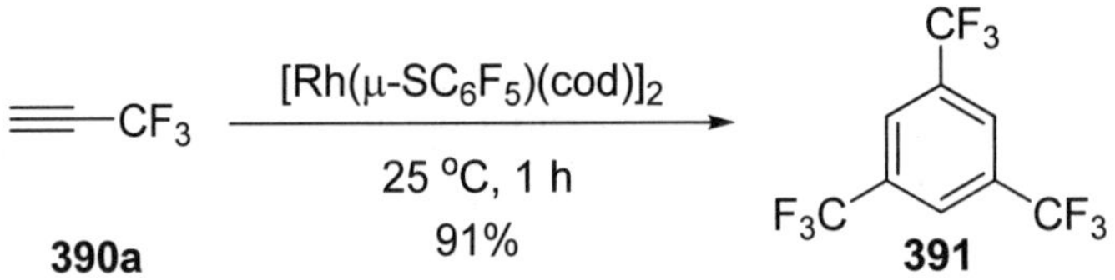

390a → 391 ($[Rh(\mu\text{-}SC_6F_5)(cod)]_2$, 25 °C, 1 h, 91%)

Scheme 2-33. Cyclotrimerization of monosubstituted alkynes.

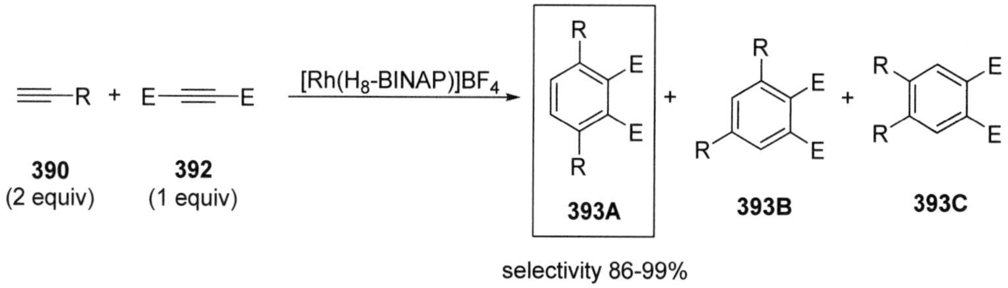

Scheme 2-34. Intermolecular cross-cyclotrimerization.

Typical procedure[267]

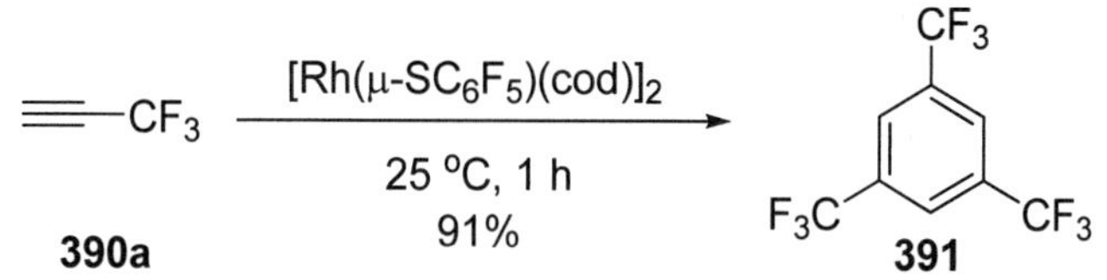

The rhodium catalyst (0.018 mmol) was dissolved in 20 mL of benzene. 1,1,1-Trifluoropropyne (**390a**) was introduced continuously to the stirred reaction mixture for 5 min at room temperature and atmospheric pressure. After evaporation of solvent, the catalyst was separated from the white trimer either by sublimation (at 100 °C, 0.1 mmHg) or by chromatography on silica gel, eluting with hexane-ethyl acetate (5/1 v/v). The product **391** was obtained in 91% yield.

Typical procedure for the intermolecular cross-cyclotrimerization of dialkyl acetylenedicarboxylates and terminal monoynes[220e]

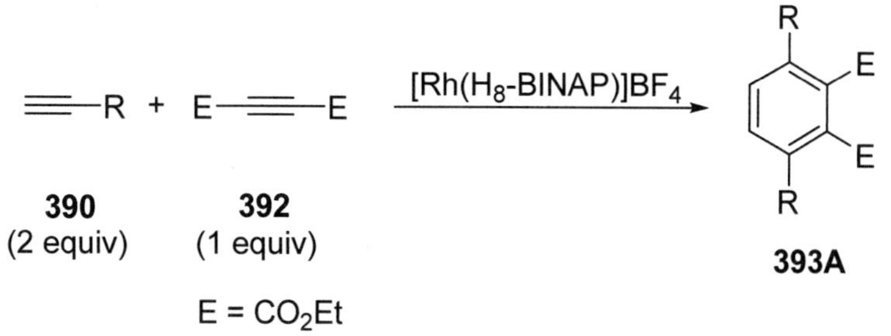

Under an Ar atmosphere, H_8-BINAP (5.7 mg, 0.009 mmol) and $[Rh(cod)_2]BF_4$ (3.7 mg, 0.009 mmol) were dissolved in CH_2Cl_2 (1.0 mL) and the mixture was stirred for 5 min. Hydrogen gas was then introduced into the resulting solution in a Schlenk tube. After stirring for 0.5 h at room temperature, the resulting solution was concentrated to dryness and the residue was redissolved in CH_2Cl_2 (2.0 mL). A solution of 1-dodecyne (**390**: R = n-$C_{10}H_{21}$) (99.8 mg, 0.60 mmol) and diethyl acetylenedicarboxylate (**392**) (51.0 mg, 0.30 mmol) in CH_2Cl_2 (0.5 mL) was then added dropwise to this solution over 1 min, and any substrates remaining in the syringe were rinsed into the reaction mixture further with CH_2Cl_2 (0.5 mL). The mixture was stirred at room temperature (20–25 °C) for 1 h. The resulting solution was then concentrated and the residue was purified by preparative TLC (hexane/ethyl acetate, 10/1), which furnished a mixture of diethyl 3,6-didecylphthalate (**393A**), diethyl 3,5-didecylphthalate (**393B**), and diethyl 4,5-didecylphthalate (**393C**) (133 mg, 0.264 mmol, 88%, **393A**:**393B**:**393C** = 92:6:2). This mixture was purified by preparative TLC (hexane/ethyl acetate, 10:1), which furnished pure 3,6-didecylphthalic acid diethyl ester (**393A**) (120 mg, 0.238 mmol, 79%).

Axially chiral biaryls are an important class of molecules primarily because of their biological activity, as well as their use as chiral ligands.[46, 227, 268] Although enantioselective cyclotrimerization of 1,6-diynes with alkynes[220e, 269] or nitriles[270] catalyzed by transition metals has been developed, it was difficult to realize the intermolecular cross-cyclotrimerization process with three alkynes. However, a cationic Rh-complex with (*S*)-H_8-BINAP was found to catalyze effectively the regioselective and enantioselective intermolecular cross-cyclotrimerization of two alkynes **392** and **394** to give chiral biaryls **395** with 89–96% ee in good yields (Scheme 2-35).[220a]

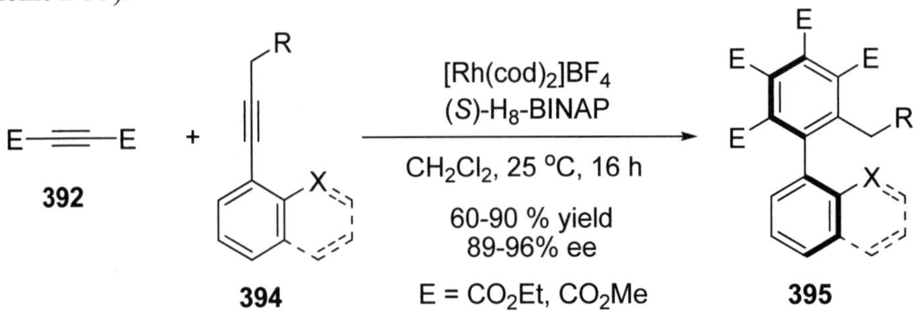

E—≡—E + **394** → **395**

392

$[Rh(cod)_2]BF_4$
(*S*)-H_8-BINAP
CH_2Cl_2, 25 °C, 16 h
60-90 % yield
89-96% ee
E = CO_2Et, CO_2Me

Scheme 2-35. Asymmetric intermolecular cross-cyclotrimerization.

Typical procedure[220a]

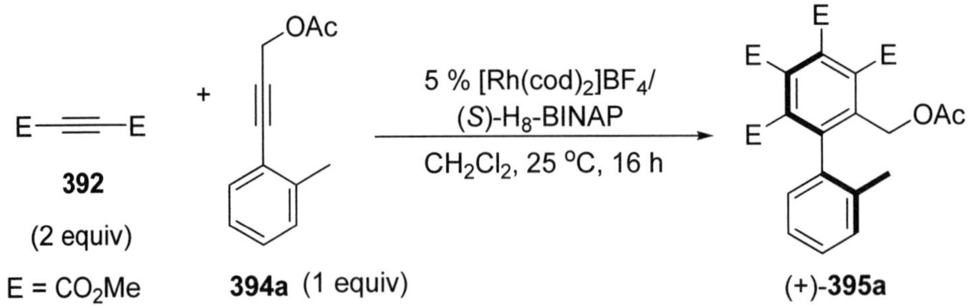

392 (2 equiv) + **394a** (1 equiv) → (+)-**395a**

E = CO_2Me

5 % $[Rh(cod)_2]BF_4$/
(*S*)-H_8-BINAP
CH_2Cl_2, 25 °C, 16 h

Under an Ar atmosphere, a CH_2Cl_2 (1.0 mL) solution of (*S*)-H_8-BINAP (9.5 mg, 0.015 mmol) was added to a CH_2Cl_2 (1.0 mL) solution of $[Rh(cod)_2]BF_4$ (6.1 mg, 0.015 mmol) at room temperature. The mixture was stirred at room temperature for 5 min. Hydrogen gas was introduced to the resulting solution in a Schlenk tube. After stirring at room temperature for 0.5 h, the resulting solution was concentrated to dryness and dissolved in CH_2Cl_2 (0.5 mL). This solution was added to a CH_2Cl_2 (0.5 mL) solution of 3-tolyl-2-propynyl acetate (**394a**) (56.4 mg, 0.300 mmol) and washed remaining catalyst away by using CH_2Cl_2 (0.5 mL). To this solution was added a CH_2Cl_2 (0.5 mL) solution of dimethyl acetylenedicarboxylate (**392**) (85.3 mg, 0.600 mmol), and remaining substrate was washed by using CH_2Cl_2 (0.5 mL). The solution was stirred at room temperature for 16 h. The resulting solution was concentrated and purified by preparative TLC (hexane:ethyl acetate = 2:1), which furnished (+)-(*R*)-5-acetoxymethyl-6-tolylphthalic acid tetramethyl ester ((+)-**395a**, 114.8 mg, 0.243 mmol, 81% yield, 89% ee) as a colorless solid.

Another regioselective approach to [2+2+2] cyclotrimerization is to use a diyne-alkyne combination. Diyne-alkyne cyclotrimerizations have been used in many natural product analog syntheses. Substituted carbazoles are a growing class of natural products with a variety of biological activities. Modification of the carbazole core[271] is well established. However, those processes often result in poor regioselectivity and low yields.

In 2002, Witulski and Alayrac reported a highly efficient synthesis of a variety of substituted carbazoles. The marine carbazole hyellazole (**399**) was synthesized through the regioselective trimerization of diyne **396** and alkyne **397**, followed by deprotection in excellent yield (Scheme 2-36).[272]

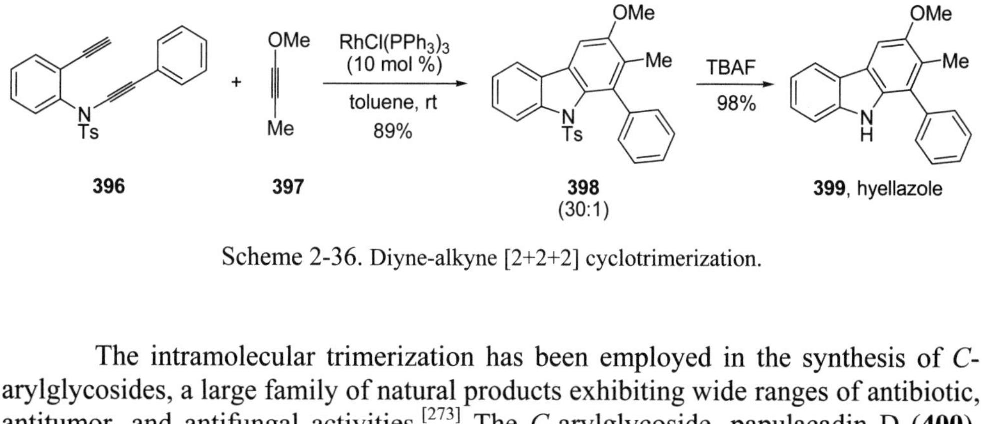

396 + 397 (OMe, Me) → RhCl(PPh$_3$)$_3$ (10 mol %), toluene, rt, 89% → 398 (30:1) → TBAF, 98% → 399, hyellazole

Scheme 2-36. Diyne-alkyne [2+2+2] cyclotrimerization.

The intramolecular trimerization has been employed in the synthesis of *C*-arylglycosides, a large family of natural products exhibiting wide ranges of antibiotic, antitumor, and antifungal activities.[273] The *C*-arylglycoside, papulacadin D (**400**), and the *C*-anthracyclinone glycoside, vineomycinone B (**401**), have been synthesized using this protocol in the key step (Scheme 2-37).

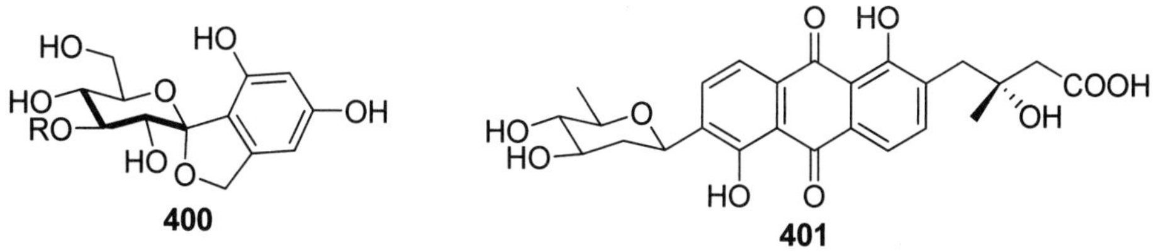

400 **401**

Figure 2-12. Naturally occurring *C*-arylglycosides.

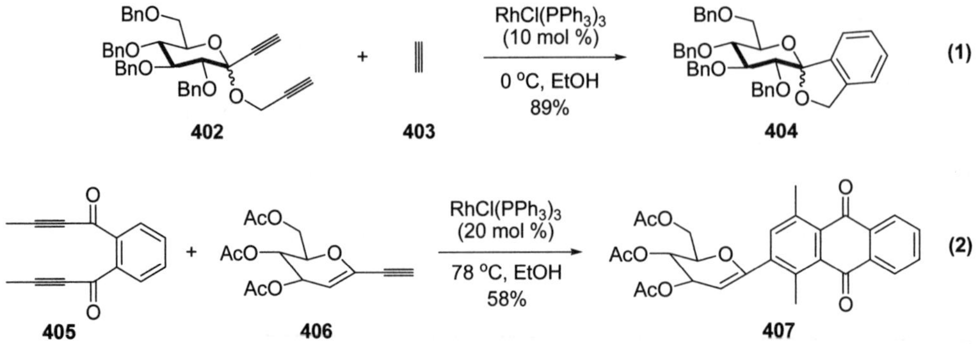

Scheme 2-37. Diene-alkyne cyclotrimerization in the total synthesis of natural products.

Typical procedure for the diyne-alkyne cyclotrimerization[273]

A flame-dried, 5-mL, round-bottomed flask equipped with a stir bar, bubbler, and nitrogen inlet was charged with a solution of the major anomer **402** (0.030 g, 0.05 mmol) in dry ethanol (1.5 mL). The solution was degassed under a stream of nitrogen for 10 min. The system was then cooled to 0 °C with an ice/water bath. Acetylene gas was passed through the solution for 15 min, and then a suspension of Wilkinson's catalyst (0.005 g, 0.005 mmol) in dry, degassed ethanol (1 mL) was added. The reaction was allowed to warm to room temperature and stir for 16 h under a slow stream of acetylene, then diluted with dichloromethane (25 mL). The solution was filtered and concentrated to a volume of ~0.5 mL. This solution was eluted through a plug of silica gel with 10:1 pentane/EtOAc. The silica gel was flushed with dichloromethane and the combined fractions concentrated *in vacuo* to give the cyclization product **404** (28 mg, 89%) as a colorless oil.

Chiral 3-substituted phthalides are central structures in a number of biologically active compounds.[274] The synthesis of chiral phthalides has been achieved by Witulski and Zimmerman, starting with enantiopure substituted propargylic alcohols **409** (Scheme 2-38).[274c] The chiral ester-linked diyne **410** was cyclotrimerized with acetylene by using $RhCl(PPh_3)_3$ as the catalyst to afford enantiopure phthalides **411** in good yields.

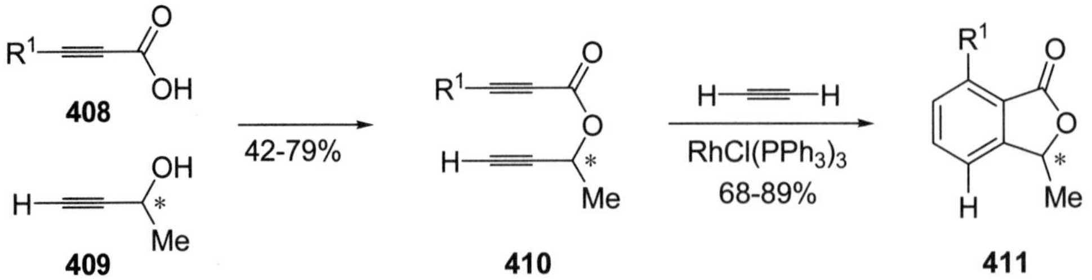

Scheme 2-38. Synthesis of 3-substituted phthalides.

Tanaka *et al.* developed a Rh-catalyzed asymmetric one-pot transesterification and [2+2+2] cyclotrimerization using nonracemic ligand **415** in the synthesis of enantio-enriched 3,3-disubstituted phthalides ($R^3 \neq R^4$) (Scheme 2-39).[220j] The chiral Rh complex with **415** efficiently desymmetrized dipropargyl alcohols **413** (R^3 = R^2-C≡C-) in the reaction with **412** to give phthalides **414** (R^3 = R^2-C≡C-) in up to 87% yield and 93% ee. Also, the kinetic resolution of racemic tertiary propargylic alcohols **413** ($R^3 \neq R^4$) proceeded well by using the same chiral catalyst to afford phthalides **414** ($R^3 \neq R^4$) in up to 89% yield and up to 93% ee. The same methodology was successfully applied to the synthesis of axially chiral biaryls **418** (Scheme 2-40).[275]

The rhodium-catalyzed intramolecular cyclotrimerization of 1,6,11-triyne **419**, forming 5-6-5 fused-ring system **420**, has been extensively studied (Scheme 2-41, eq. 1).[220b, 276] This reaction has also been used as the key step in the synthesis of a marine illudalane sesquiterpenoid, alcyopterosin E (**423**) (Scheme 2-41, eq. 2),[277] as well as in the asymmetric synthesis of chiral diphosphine ligands **425** (Scheme 2-41, eq. 3).[278]

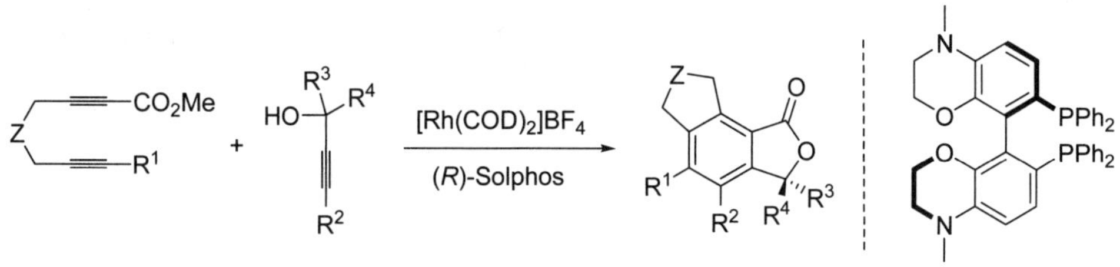

412 **413** achiral or racemic **414** **415**, (*R*)-Solphos

Scheme 2-39. Synthesis of enantio-enriched 3,3-disubstituted phtalides.

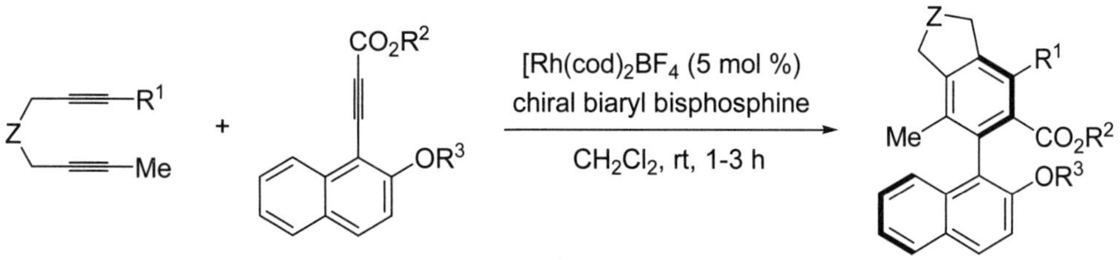

416 (1.2 equiv) **417**

R^1 = Me, Ar
R^2 = *i*-Pr, Et, Me
R^3 = Me, Bn, CH_2OMe

418 up to >99% yield up 97% ee

Scheme 2-40. Synthesis of axially chiral biaryls *via* asymmetric cyclotrimerization.

Typical procedure[220j]

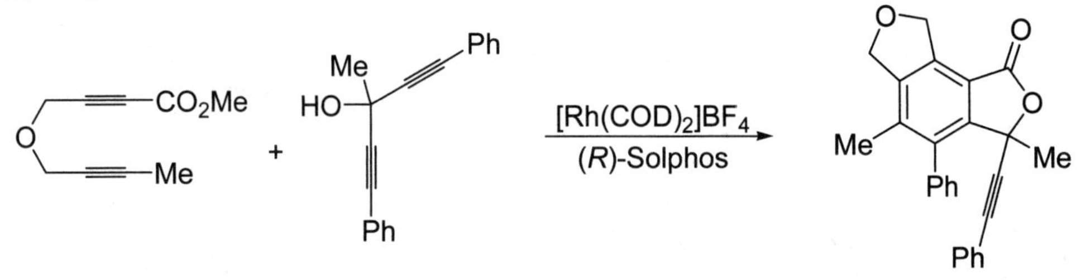

412a 413a 414a

Under an Ar atmosphere, (*R*)-Solphos (**415**) (6.6 mg, 0.010 mmol) and $[Rh(cod)_2]BF_4$ (4.1 mg, 0.010 mmol) were dissolved in CH_2Cl_2 (1.0 mL), and the mixture was stirred at room temperature for 5 min. Hydrogen gas was introduced to the resulting solution in a Schlenk tube. After stirring at room temperature for 0.5 h, the resulting solution was concentrated to dryness and dissolved in CH_2Cl_2 (0.4 mL). To this solution was added a CH_2Cl_2 (0.2 mL) solution of **412a** (147.7 mg, 0.60 mmol), and the remaining substrate was washed away by using CH_2Cl_2 (0.2 mL). To this solution was added dropwise over 10 min a CH_2Cl_2 (1.0 mL) solution of **413a** (33.2 mg, 0.20 mmol) at room temperature, and the remaining substrate was washed away by using CH_2Cl_2 (0.2 mL). The mixture was stirred at room temperature for 1 h. The resulting solution was concentrated and purified by preparative TLC (hexane:EtOAc = 2:1), which furnished 3,5-dimethyl-4-phenyl-3-phenylethynyl-6,8-dihydro-3*H*-2,7-dioxa-as-indacen-1-one (**414a**, 62.1 mg, 0.163 mmol, 82% yield) as a colorless solid.

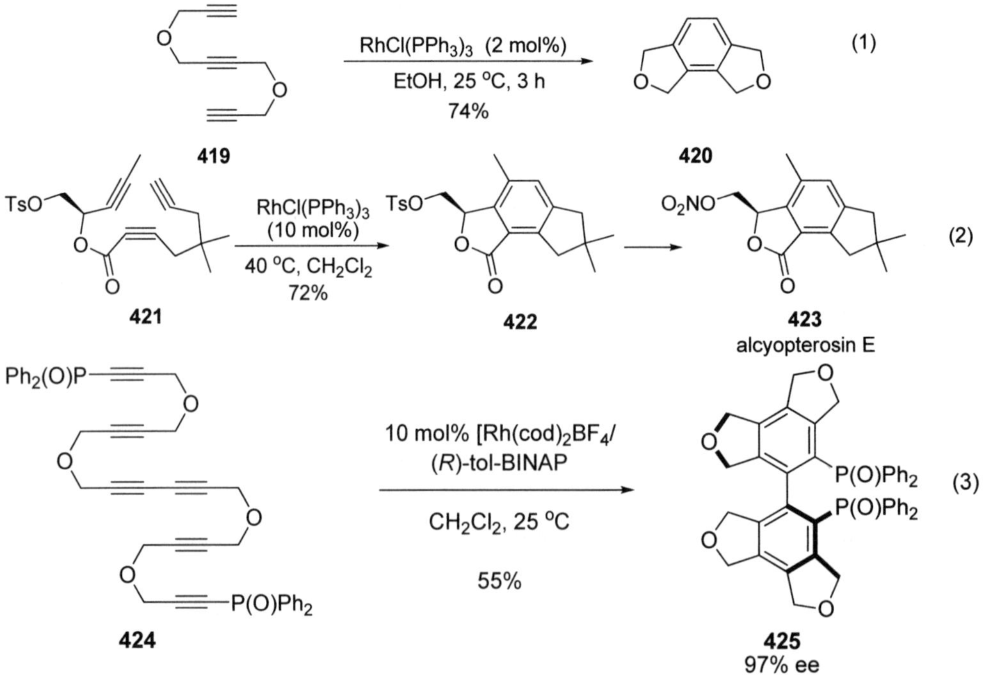

Scheme 2-41. Intramolecular cyclotrimerization of poly-ynes.

The [2+2+2] cycloadditions are not limited to the formation of benzene derivatives. For example, the asymmetric [2+2+2] cycloaddition of 1,6-enyne **426** (R^2 = H) with alkyne **427** catalyzed by $[Rh(cod)_2Cl]_2$/(*S*)-xyl-P-PHOS/$AgBF_4$ gave **428A** in high yield with excellent regioselectivity and enantioselectivity, accompanied by a small amount of **428B** (Scheme 2-42, eq. 1).[220c, 279] When enynes **426a** (R^2 = Me) were used as 1,6-enyne components in this reaction with $[Rh(cod)_2]BF_4$/(*R*)-tol-BINAP as the catalyst, the reaction afforded **428Aa** with chiral quaternary carbon in high yield and excellent enantioselectivity (Scheme 2-42, eq. 2).[220d]

The enantioselective intramolecular [2+2+2] cycloaddition of 1,4-dienynes **429** gave the bicyclo[2.2.1]heptene skeleton **430**, accompanied by a small amount of **431** (Scheme 2-43).[280] For this reaction, dienynes connecting the alkyne and the other alkene components with 1,1-disubstituted alkene moieties were employed. For this reaction, dienynes connecting the alkyne and the other alkene components with 1,1-disubstituted alkene moieties were employed. Each product of this reaction represents a unique class of cycloadducts that possess two quaternary carbon stereocenters.

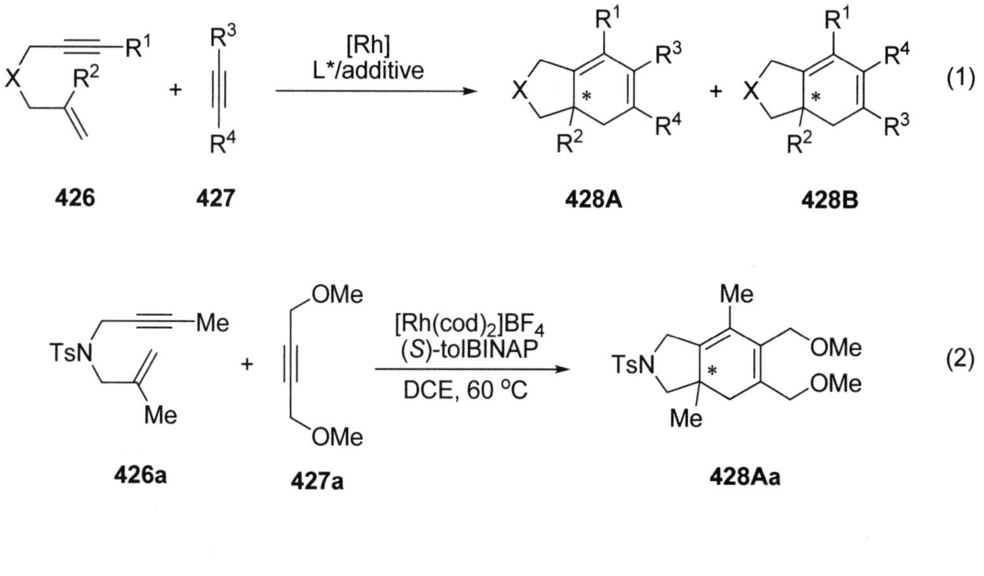

Scheme 2-42. Asymmetric [2+2+2] cycloaddition of 1,6-enynes with alkynes.

Typical procedure[220d]

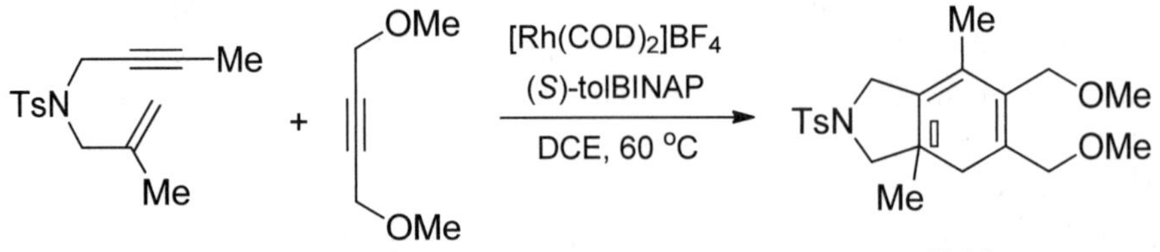

426a **427a** **428Aa**

Under an atmosphere of Ar, (*S*)-tol-BINAP (6.8 mg, 0.010 mmol) and $[Rh(cod)_2]BF_4$ (4.1 mg, 0.010 mmol) were stirred in 1,2-dichloroethane (0.25 mL) at room temperature to give a yellow solution. Then, 1,4-dimethoxybut-2-yne **427a** (22.8 mg, 0.20 mmol or 114.1 mg, 1.00 mmol) and enyne **426a** (0.10 mmol) in 1,2-dichloroethane (0.75 mL) were added to the solution and the mixture was stirred at 60 °C. After completion of the reaction, the solvent was removed under reduced pressure, and the crude products were purified by thin layer chromatography to give chiral cycloadduct **428Aa** (81%) as a colorless oil. The enantiopurity was determined to be 97% ee by HPLC analysis using a chiral column.

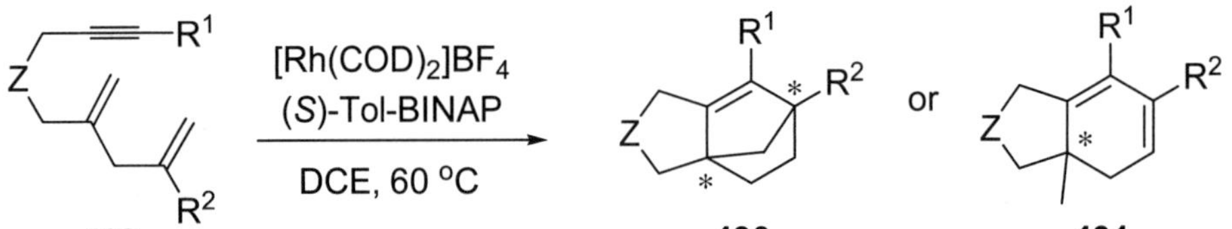

429 **430** R^2 = Me, Ph **431** R^2 = H

Scheme 2-43. Asymmetric intramolecular [2+2+2] cycloaddition of dienynes.

The product selectivity of the reaction depended on the substituent R^2 in the alkene moiety. The formation of tricyclic product **430** was anticipated from the reaction mechanism. However, unexpected product **431** was obtained exclusively when 1,4-dien-ynes **429** (R^2 = H) which lacked a substituent at the 2-position of the 1,4-diene moiety, were used under the same reaction conditions. The proposed mechanism for the formation of **431** is illustrated in Figure 2-13.

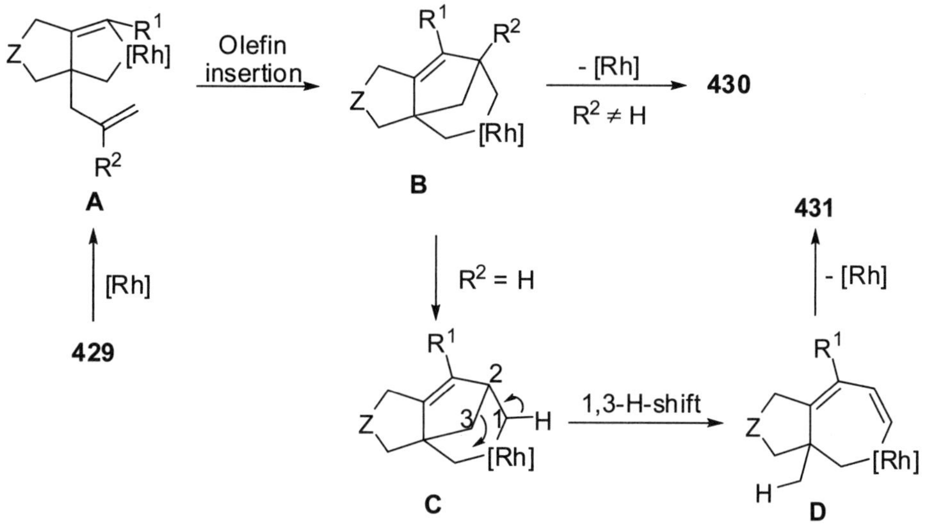

Figure 2-13. Proposed mechanism for the formation **430** and **431**.

Initial oxidative coupling of the ligated Rh complex with both the alkyne and alkene gave the metallacyclopentene **A**, followed by olefin insertion to form metallacycloheptene **B**. Tricyclic compound **430** was obtained by reductive elimination of Rh from **B** when R^2 is not hydrogen. In contrast, when R^2 is hydrogen, a 1,3-hydride shift with concomitant ring opening takes place to afford metallacycle **D**. Subsequent reductive elimination of Rh resulted in the formation of bicyclic compound **431**.

A control reaction using 1,3-dienyne **432**, which was obtained by double bond isomerization of 1,4-dienyne **429**, did not proceed at all even under reflux conditions (Scheme 2-44). This result suggested that bicyclic compound **431** was, indeed, formed through metallcycle **C** and not through the intramolecular [4+2] cycloaddition pathway.

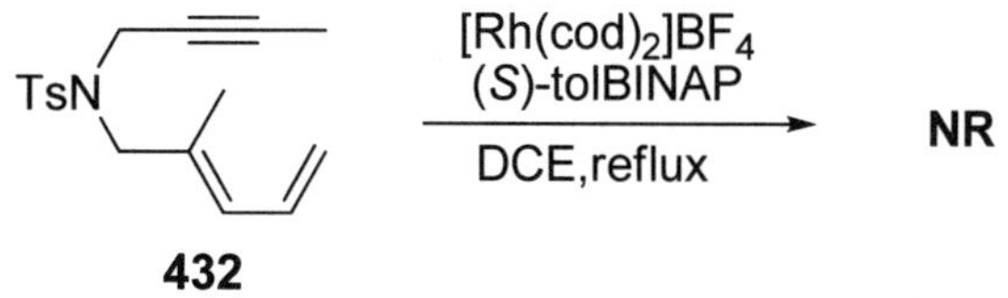

Scheme 2-44. Attempted intramolecular Diels-Alder reaction of 1,3-dienyne **432**.

Shibata further expanded the scope of the [2+2+2] cycloaddition by varying the substitution pattern of dienyne **433** to form 5-6-5 tricyclic and 5-6-5-6 tetracyclic systems **434** with excellent enantioselectivity (up to 99% ee) (Scheme 2-45).[281]

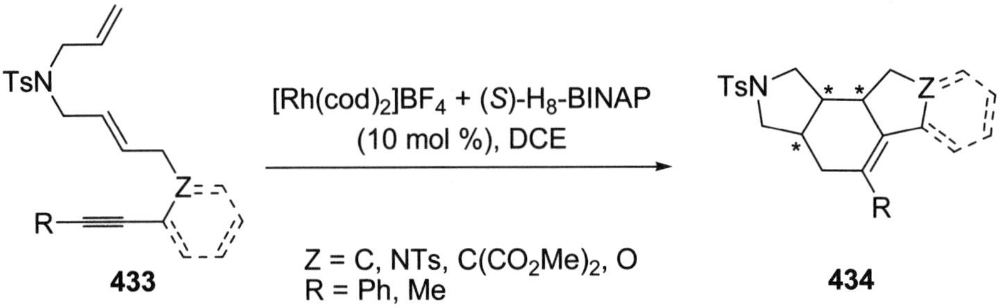

TsN [Rh(cod)$_2$]BF$_4$ + (*S*)-H$_8$-BINAP (10 mol %), DCE

433 Z = C, NTs, C(CO$_2$Me)$_2$, O; R = Ph, Me **434**

Scheme 2-45. Asymmetric [2+2+2] cycloaddition of dienynes.

Typical experimental procedure[280]

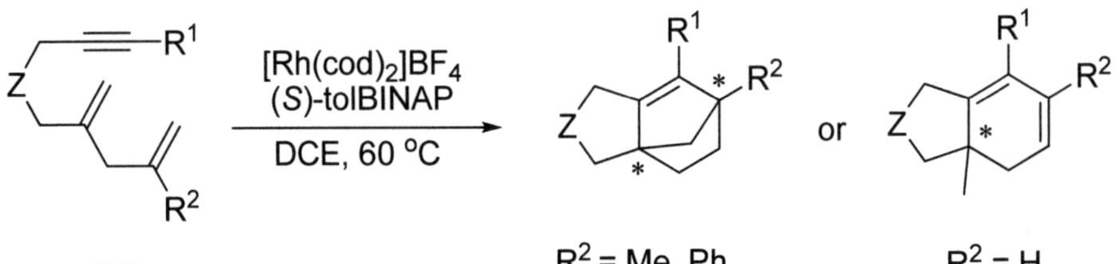

[Rh(cod)$_2$]BF$_4$ (*S*)-tolBINAP DCE, 60 °C

429 → R^2 = Me, Ph **430** or R^2 = H **431**

Under an atmosphere of Ar, (*S*)-tol-BINAP (6.8 mg, 0.010 mmol) and [Rh(cod)$_2$]BF$_4$ (4.1 mg, 0.010 mmol) were stirred in 1,2-dichloroethane (0.25 mL) at room temperature to give a yellow solution. Then, dienyne **429** (0.10 mmol) in 1,2-dichloroethane (0.75 mL) were added to the solution and the mixture was stirred at the appropriate temperature. After completion of the reaction, the solvent was removed under reduced pressure, and the crude product was purified by thin layer chromatography to give a chiral cycloadduct. The enantioselectivity was determined by HPLC analysis using a chiral column.

The formation of pyridones *via* [2+2+2] cycloaddition has been shown to be effectively catalyzed by metal complexes of cobalt,[282] ruthenium,[283] and zirconium/nickel.[284] Recently, Rh complexes were found to catalyze [2+2+2] cycloaddition of 1,6-diynes **435** with various heterocumulenes **436** such as isocyanates (Y = O, Z = R'-N), isothiocyanates (Y = R'-N, Z = S), and carbon disulfide (Y = Z = S) to yield the corresponding bicyclic pyridones **437** (Y = O, Z = R'-N) and thiopyrans **437** (Y = R'-N or S, Z = S) (Scheme 2-46).[220f, 285]

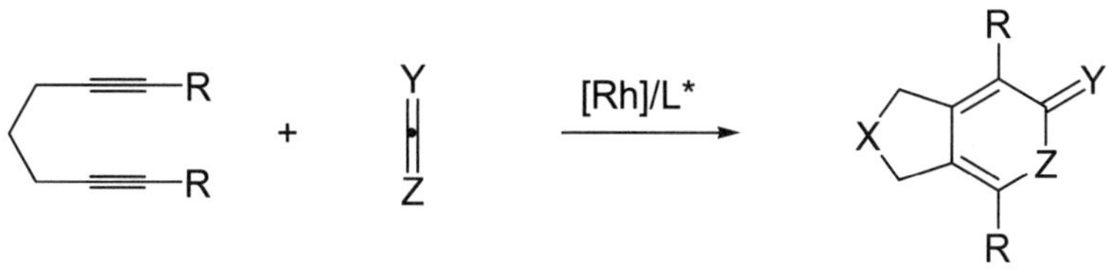

[Rh]/L*

435 436a-c 437

a: Y = O, Z = NR'
b: Y = NR', Z = S
c: Y = Z = S

Scheme 2-46. Rh-catalyzed [2+2+2] cycloaddition of 1,6-diyne with heterocumulenes.

A cationic Rh complex with (*R*)-DTBM-SEGPHOS (**440**) was found to be a highly effective catalyst for the formation of arylpyridones with high regioselectivities and enantioselectivities. As examples, the [2+2+2] reaction of unsymmetrical 1,6-diynes **438**, bearing an *ortho*-substituted phenyl at the 1-position, with isocyanates **436a** catalyzed by the Rh^+-complex with **440** afforded axially chiral arylpyridone **439** as the exclusive product with up to 92% ee (Scheme 2-47).[285]

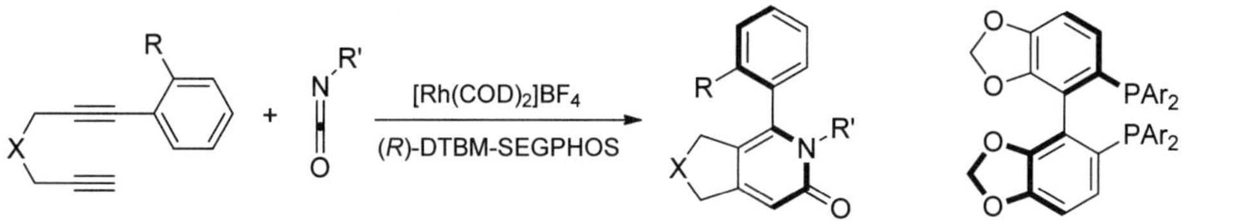

$[Rh(COD)_2]BF_4$
(*R*)-DTBM-SEGPHOS

438 **436a** **439** 85-92% ee

440, (*R*)-DTBM-SEGPHOS
Ar = 4-MeO-3,5-(t-Bu)$_2C_6H_2$

Scheme 2-47. Asymmetric[2+2+2] cycloaddition of diynes with isocyanates.

Typical procedure[285]

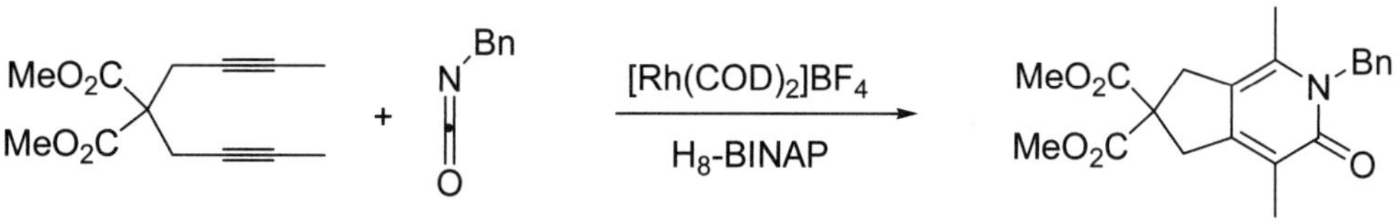

$[Rh(COD)_2]BF_4$
H_8-BINAP

435a **436a-1** **437a-1**

Under an Ar atmosphere, H_8-BINAP (15.8 mg, 0.0250 mmol) and $[Rh(cod)_2]BF_4$ (10.2 mg, 0.0250 mmol) were dissolved in CH_2Cl_2 (2.0 mL), and the reaction mixture was stirred at room temperature for 5 min. Hydrogen gas was introduced to the resulting solution in a Schlenk tube. After stirring at room temperature for 0.5 h, the resulting mixture was concentrated to dryness. To the CH_2Cl_2 (3.5 mL) solution of the residue was added a CH_2Cl_2 (0.5 mL) solution of 2,2-dibut-2-ynylmalonic acid dimethyl ester (**435a**, 118.1 mg, 0.500 mmol) and benzyl isocyanate (**436a-1**, 73.2 mg, 0.550 mmol) at room temperature, and the remaining substrate was washed away by using CH_2Cl_2 (1.0 mL). The mixture was stirred at room temperature for 18 h. The resulting mixture was concentrated and purified by preparative TLC (hexane:ethyl acetate = 1:1), which furnished 2-benzyl-1,4-dimethyl-3-oxo-2,3,5,7-tetrahydro-[2]pyrindine-6,6-dicarboxylic acid dimethyl ester (**437a-1**, 183.6 mg, 0.497 mmol, 99% yield) as a colorless oil.

Another application of the [2+2+2] cycloaddition used alkenyl isocyanates **441** and terminal alkynes **390** for the formation of izidine alkaloid skeletons **442** and **443** with a lactam and a vinylogous amide moieties, respectively (Scheme 2-48).[220g, 220h, 286]

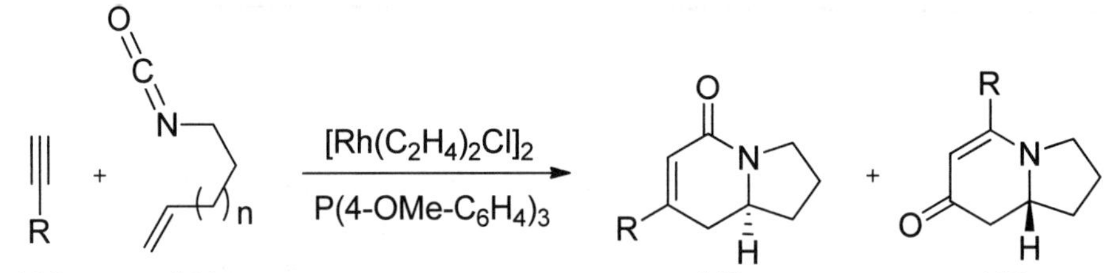

Scheme 2-48. [2+2+2] Cycloaddition of alkenyl isocyanates and terminal alkynes.

The regioselectivity of this cycloaddition process depends on the nature of the acetylene moieties. Thus, when alkyl acetylenes were used, lactam **442** was the major product. However, when aryl acetylenes were used, the process involved a CO migration to afford vinylogous amide **443** as the major product (Figure 2-14).[220g] It appears that the product selectivity is dependent on the electronic properties as well as the steric bulk of the substituents of the alkynes used. When R is aryl, the catalytic sequence predominantly goes through pathway **A**, with some exceptions when the *para*-substituent of phenyl is an electron-withdrawing group. The initial oxidative cyclization of alkyne and isocyanates takes place in an orientation where the C-N bond is formed. Subsequent CO migration, alkene insertion, and reductive elimination furnishes vinylogous amide **443**. When R is alkyl, the reaction proceeds through pathway **B**. Initial oxidative cyclization occurs in an orientation where a C-C bond was formed. Subsequent alkene insertion and reductive elimination gives lactam **442**. However, obviously a more detailed mechanistic study is necessary to rationalize the observed regioselectivity for the formation of two metallacycles.

This process was successfully applied to the enantioselective total synthesis of (+)-lasubine II (**448**) in only three steps from isocyanate **441b** and alkyne **444** (Scheme 2-49). The asymmetric [2+2+2] cycloaddition of **441b** with **44** catalyzed by $[Rh(C_2H_4)_2Cl]_2$ with chiral phosphoramidite ligands proceeded in toluene at 110 °C to give quinolizinone **445** with 98% ee in 62% yield, accompanied by 20–23% of pyridone **446** as the side product. The formation of **446** suggests that the alkene moiety is the last 2π carbon component incorporated, which is in agreement with the proposed mechanism (Figure 2-14). Quinolizinone **445** was subjected to diastereoselective hydrogenation and subsequent Mitsunobu reaction to afford (+)-lasubine II **448** in good overall yield (Scheme 2-49).

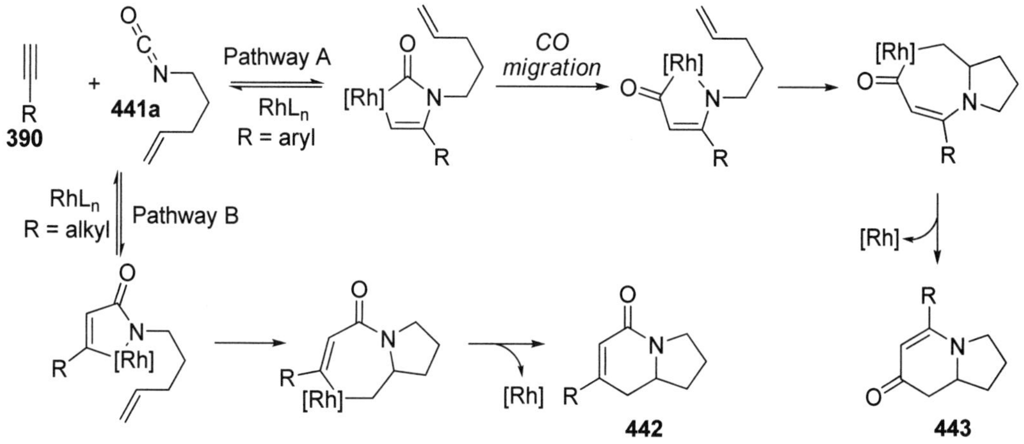

Figure 2-14. Proposed mechanism for the [2+2+2] cycloaddition of alkenyl isocyanates with terminal alkynes.

General procedure for the Rh-catalyzed [2+2+2] cycloaddition of alkenyl isocyanates and terminal alkynes[220g]

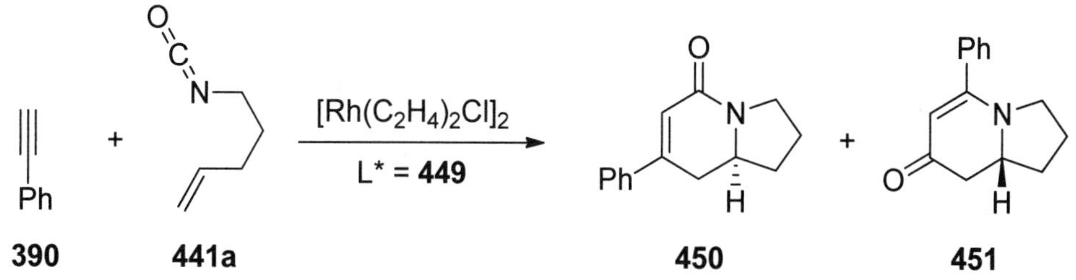

A flame-dried, round-bottomed flask was charged with $[Rh(ethylene)_2Cl]_2$ (0.05 equiv) and the phosphoramidite ligand **449** (0.1 equiv), and it was fitted with a flame-dried reflux condenser in an inert atmosphere (N_2) glove box. After removal from the glove box, 1.0 mL toluene was added *via* syringe and the resulting yellow solution was stirred at ambient temperature under Ar flow for 15 min. To this solution was added a solution of alkyne **390** (2.0 equiv) and isocyanate **441a** (0.270 mmol) in 2 mL of toluene *via* syringe or cannula. After an additional 1 mL of toluene was used to wash down the remaining residue, the resulting solution was heated to 110 °C in an oil bath and maintained at reflux for *ca.* 16 h. The reaction mixture was cooled to ambient temperature, concentrated *in vacuo*, and purified by flash column chromatography on silica gel (elution typically 100% ethyl acetate). Evaporation of solvent afforded the analytically pure product.

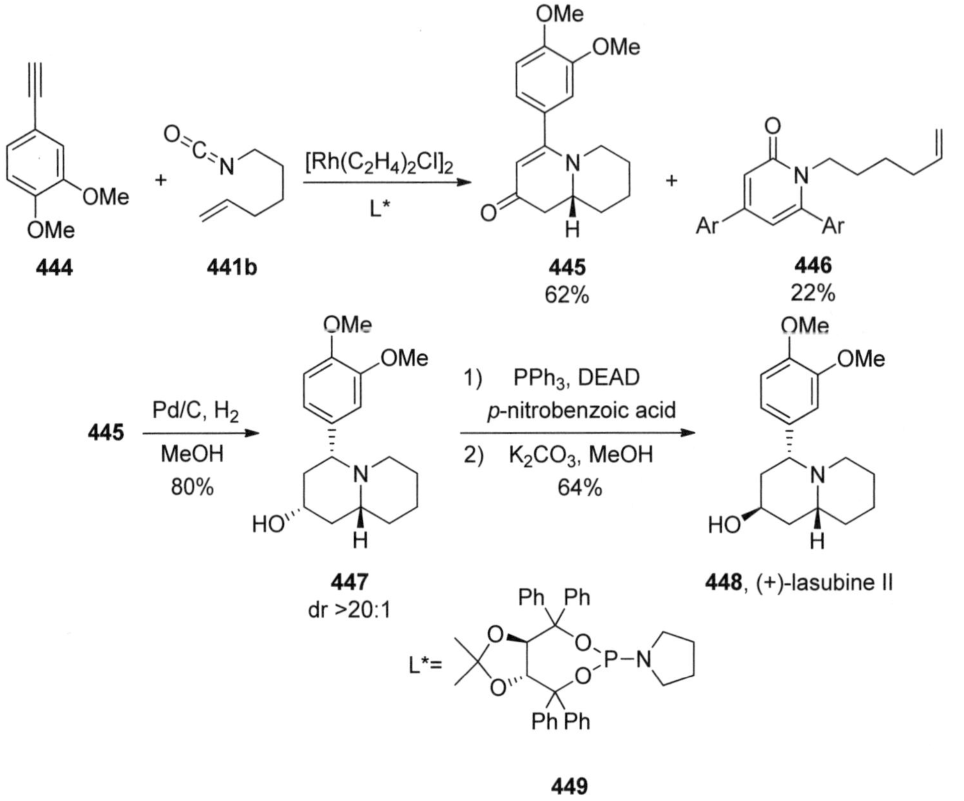

Scheme 2-49. Synthesis of lasubine-II.

16. Rh(I)-Catalyzed Cycloisomerization and Alder-Ene Reactions

16.1 Enyne Cycloisomerization

Transition-metal–catalyzed intramolecular cycloisomerization is one of the most useful carbocyclization reactions, and specifically, the Rh-catalyzed cycloisomerization of 1,6-enynes provides a powerful tool in organic synthesis. Cyclization of 1,6-enyne **452** catalyzed by Wilkinson's catalyst gave 1-*exo*-methylene-2-cyclohexene **453** *via* a 6-*exo*-trig mode in 83% yield (Scheme 2-50).[276b] Terminal substitution on the alkene moiety dramatically suppressed the cyclization, and substitution of the terminal alkyne moiety was detrimental to the reaction as well.

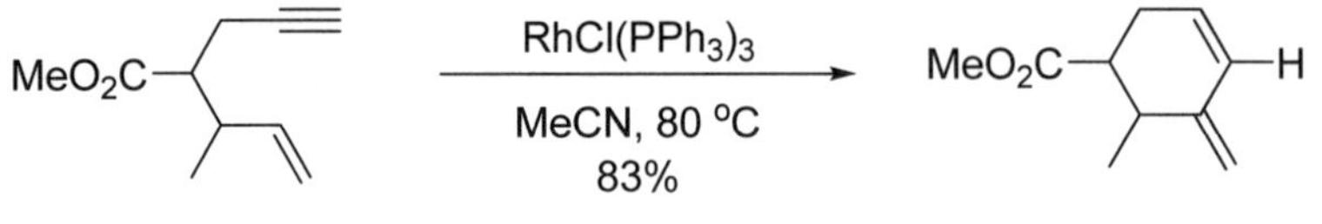

Scheme 2-50. Cylcoisomerization of 1,6-enynes.

The early proposed mechanism involved an intermolecular hydrogen transfer from one alkyne to another.[276b] Recently, Kim and Lee proposed a mechanism involving a vinylidene-Rh species (Figure 2-15).[287] Insertion of Rh(I) into the C-H bond of terminal alkyne is a well-known process. Rearrangement of the acetylene-Rh-hydride moiety leads to the vinylidene-Rh species. Subsequent [2+2] cycloaddition, ring-opening of the rhodacyclobutane *via* β-hydride elimination, and reductive elimination of the Rh catalyst gives **454**. A deuterium-labeling study supported the proposed mechanism.[287]

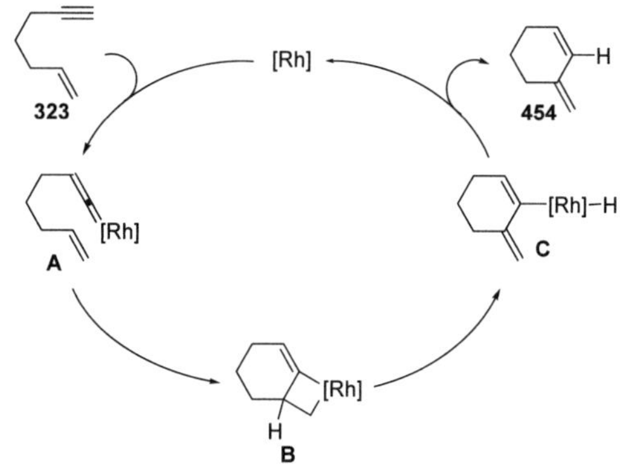

Figure 2-15. Proposed mechanism for the cycloisomerization of enyne **323**.

The first use of vinylidene-metal species in C-C bond formation with a ruthenium catalyst was reported by Trost in 1990.[288] Since then, the use of vinylidene-Rh species in organic synthesis has been extended to the formation of heterocycles through cycloisomerization of alkynols and alkynyl anilines.[289] For example, the cycloisomerization of alkynol **455**, leading to the formation of pyran **456** was used for the synthesis of amino sugar **457** (Scheme 2-51).[289a]

Benzofuran is the core of many complex natural products such as psoralens, a family of natural products that exhibits a wide range of bioactivities.[290] Indoles are also found in many natural and organic compounds such as tryptophan and indole alkaloids. These core structures, **459** and **461**, have been synthesized through the Rh-catalyzed cycloisomerizations of alkynylphenols **458** and alkynylanilines **460**, respectively (Scheme 2-52).

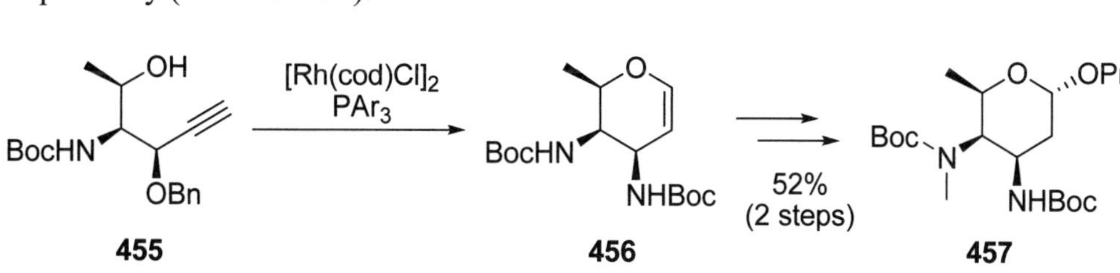

Scheme 2-51. Cycloisomerization of alkynol **455.**

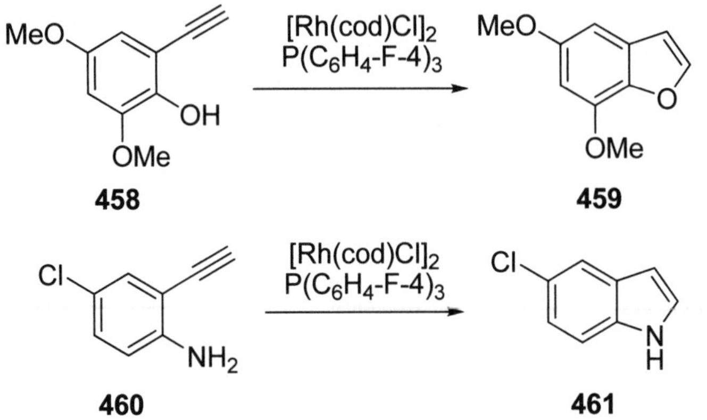

Scheme 2-52. Cycloisomerization of ethynylphenol and ethynylaniline.

The intramolecular Alder-ene reaction of 1,6-enynes **464** was found to be effected by cationic Rh-catalysts generated *in situ* from $[RhL_2Cl]_2$ (L = bidentate phosphorus ligands) and $AgSbF_6$ to afford 3-*exo*-alkylidene-4-vinyltetrahydrofurans **465** (X = O), pyrrolidines **465** (X = $PhSO_2N$) or cyclopentane **465** (X = $C(CO_2Et)_2$) in 50–100% yield (Scheme 2-53).[291] It was found that the Rh^+-complex with dppb was effective only for the reactions of enynes **464** containing a *cis*-olefin moiety; *i.e.*, enynes with a *trans*-olefin moiety simply did not react. In sharp contrast, the Rh^+ complex with bis(diphenylphosphinoxy)–dicyclopentane (BICPO) (**466**) was highly effective for the reactions of enynes **464** bearing either a *cis*- or *trans*-alkene moiety.[291] This reaction was naturally applied to enantioselective cycloisomerization using enantiopure (*R,R,R,R*)-BICPO (**466**) as well as (*R,R*)-Me-DuPhos and (*R,R,R,R*)-BICP using enynes **464** with a *cis*-olefin moiety to give **465** with 65–98% ee in 39–99% yield (Scheme 2-53).[292]

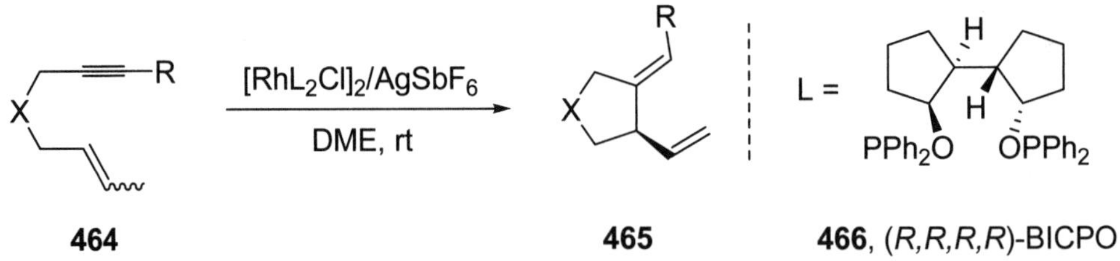

Scheme 2-53. Intramolecular Alder-ene cyclization.

Typical experimental procedure[289b]

Cl, NH_2 **460** → $[Rh(cod)Cl]_2$, PPh_3; DMF, 45 °C; 77 % → Cl, N H **461**

N,N-Dimethylformamide (2.50 mL) was degassed (Ar balloon) for 20 min and then added through a cannula to a mixture of **460** (76 mg, 0.500 mmol), $[Rh(cod)Cl]_2$ (2.5 mg, 0.005 mmol), and triphenylphosphine (5.2 mg, 0.020 mmol). The solution was stirred for 2 h at 85 °C under Ar. The reaction mixture was then cooled to 25 °C and extracted from saturated aq. $NaHCO_3$ (50 mL) with diethyl ether (3 × 10 mL). The combined organic extracts were washed with saturated aq. $NaHCO_3$ (10 mL), dried over anhydrous $MgSO_4$, filtered, and evaporated *in vacuo*. Purification by flash column chromatography on silica gel (petroleum ether/diethyl ether/triethylamine 90:10:3→50:50:3→25:75:3) gave the product **461** (58 mg, 77%) as pale tan crystals.

Typical experimental procedure[289a]

C_9H_{19}, OH **462** → $[Rh(cod)Cl]_2$, $P(C_6H_3\text{-}F_2\text{-}3,5)_3$; DMF, 85 °C; 77% → C_9H_{19}, O **463**; Ar = (F, F)

A mixture of 1-tridecyn-4-ol (**462**) (100 mg, 0.509 mmol), $[Rh(cod)Cl]_2$ (6.3 mg, 0.013 mmol, 2.5 mol%), and $P(C_6H_3\text{-}F_2\text{-}3,5)_3$ (104 mg, 0.280 mmol, 55 mol%) in DMF (2.6 mL, 0.2 M) was placed in a preheated oil bath at 85 °C and stirred for 2 h. The reaction mixture was cooled to room temperature, diluted with ether (30 mL), and washed with water (2 × 10 mL). The aqueous layer was extracted with ether (2 × 20 mL). The organic phases were combined, dried over anhydrous $MgSO_4$, and concentrated. The residual oil was purified by column chromatography (deactivated silica gel with 1% triethylamine, petroleum ether:EtOAc = 20:1) to give **463** (69 mg, 0.352 mmol) as a clear oil (77% yield).

An isomerization/enantioselective intramolecular Alder-ene cyclization of enynes **467** was reported by Okamoto and co-workers (Scheme 2-54).[293] The reaction was catalyzed by a cationic Rh(I)-(*R*)-BINAP complex to afford various bicyclic and trycyclic dihydrobenzofurans and dihydronaphthofurans **469** in good yields and excellent enantioselectivity. Another example of an Alder-ene cyclization is the reaction of allylic ester **470**, which yielded α-methylene-γ-butyrolactone **471** in 50% yield. It was found that other transition metals, such as Pd, Ru, and Ti, were unable to catalyze this reaction (Scheme 2-55).[291]

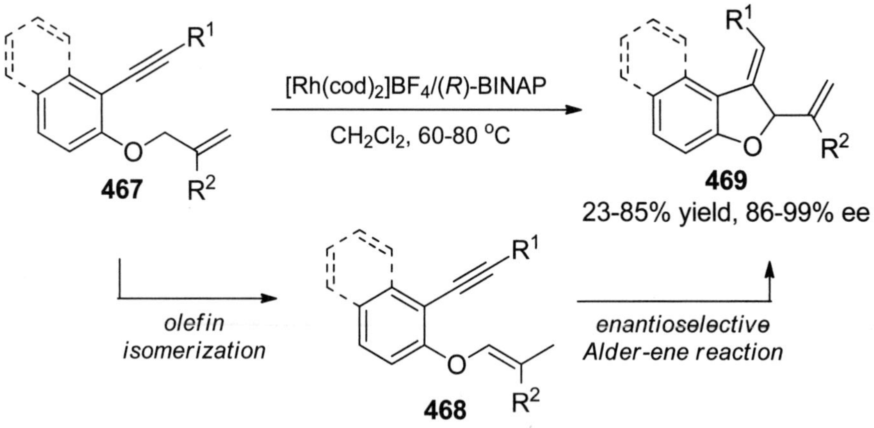

Scheme 2-54. Isomerization/enantioselective Alder-ene cyclization of enynes.

Scheme 2-55. Rh-catalyzed Alder-ene cyclization of enyne **483**.

When (*R,R*)-Me-DuPhos (*R,R*-**119**), (*R,R,R,R*)-BICPO (**466**),[291] or (*R*)-BINAP (*R*-**110**)[294] were used, the Rh-catalyzed asymmetric cycloisomerization of **464a** afforded **465a** with an ee >99.5% (Scheme 2-56). This asymmetric process was applied to the synthesis of key intermediate (*R*)-(+)-**473** (99% yield, 99% ee), useful for the total synthesis of (+)-pilocarpine (**474**) (Scheme 2-57).[295]

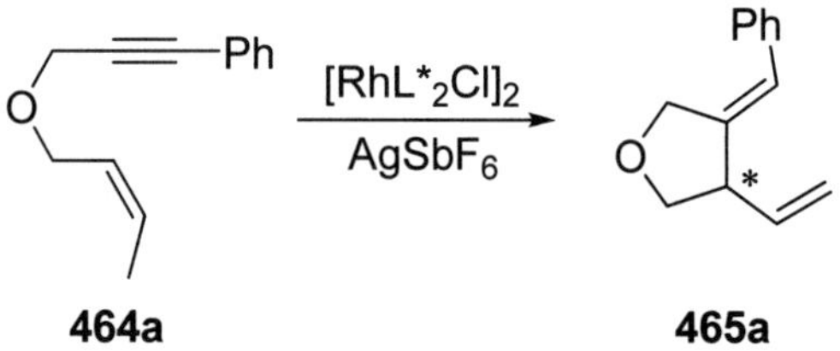

L^*_2	Yield (%)	% ee
(*R,R*)-Me-DuPhos (*R,R*-**119**)	62	96
(*R,R,R,R*)-BICPO (**466**)	81	65
(*R*)-BINAP (*R*-**110**)	96	>99.5

Scheme 2-56. Rh-catalyzed asymmetric cycloisomerization of enyne **477a**.

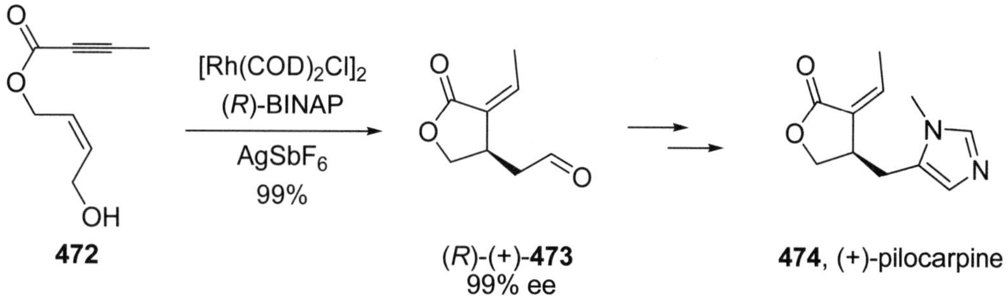

[$Rh(COD)_2Cl]_2$
(*R*)-BINAP
$AgSbF_6$
99%

472 (*R*)-(+)-**473** 99% ee **474**, (+)-pilocarpine

Scheme 2-57. Synthesis of (+)-pilocarpine *via* asymmetric cycloismerization of enyne **472**.

*Procedure for the asymmetric Alder-ene reaction of **464a***[294]

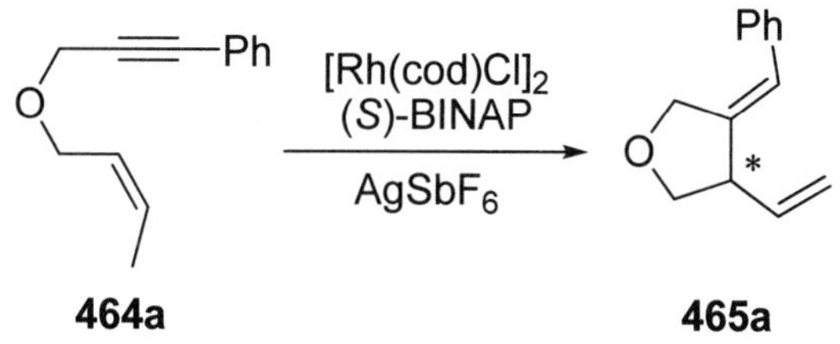

$[Rh(cod)Cl]_2$
(*S*)-BINAP
$AgSbF_6$

464a **465a**

In a dried Schlenk tube, $[Rh(cod)Cl]_2$ (4.9 mg, 0.01 mmol) and (*S*)-BINAP (*S*-110) (13.8 mg, 0.022 mmol) were dissolved in freshly distilled 1,2-dichloroethane (2 mL). Then, freshly prepared enyne **464a** (37.2 mg, 0.2 mmol) was added to the solution at room temperature under nitrogen. After the mixture had been stirred for 1 min, $AgSbF_6$ (0.04 mmol) was added, and the reaction was complete within 5 min. The reaction mixture was directly subjected to column chromatography on silica gel to give compound **465a** (35.8 mg, 96% yield, >99.9% ee).

Another reaction that has been extensively studied is the γ-lactam formation.[295, 296] The initial cyclization study using a secondary amide tether did not afford any desired product. It was believed that the predominant *trans* conformation of the secondary amide prevented the cyclization (Figure 2-16).[295] In fact, the protected enynes with a tertiary amide linked were readily transformed into a variety of γ-lactams. Under the optimized conditions, enyne-amide **475** afforded functionalized γ-lactams **476** in high yield and good-excellent enantioselectivities (Scheme 2-58).

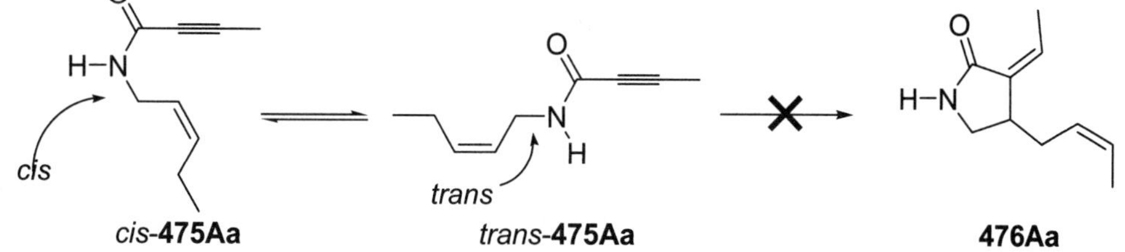

cis-**475Aa** *trans*-**475Aa** **476Aa**

Figure 2-16. Blocked cyclization of the γ-lactams *trans*-form of enyne-amide.

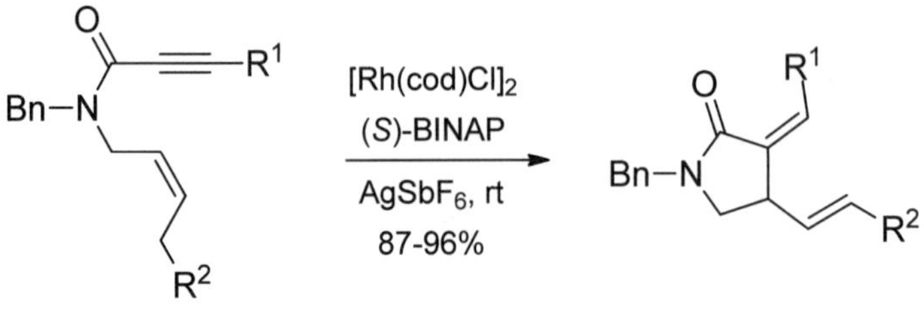

$[Rh(cod)Cl]_2$, (*S*)-BINAP, $AgSbF_6$, rt, 87-96%

475 → **476** >99% ee

R_1 = Ph, Me, n-C_5H_{11}, CH_2OBn, CH_2OMOM
R_2 = H, Et, OMe

Scheme 2-58. Rh-catalyzed Alder-ene reaction of enynes with a tertiary amide tether.

Typical procedure[295]

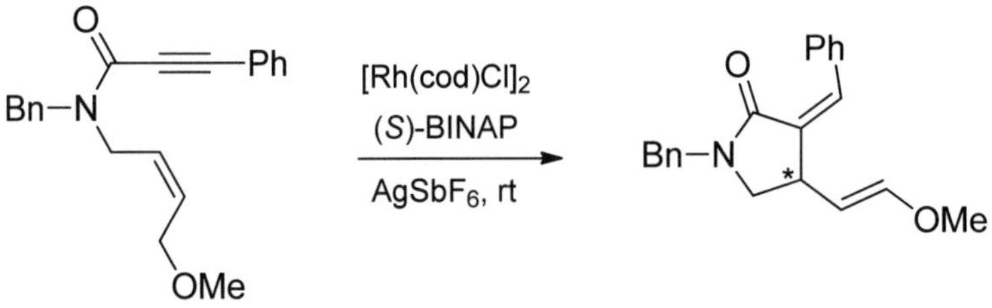

$[Rh(cod)Cl]_2$, (*S*)-BINAP, $AgSbF_6$, rt

475a → **476a**

In a flame-dried Schlenk tube, the $[Rh(cod)Cl]_2$ (2.5 mg, 0.005 mmol) and (*S*)-BINAP (6.9 mg, 0.012 mmol) were dissolved in freshly distilled 1,2-dichloroethane (1 mL), and then freshly prepared **475a** (28.7 mg, 0.1 mmol) was added to the solution at room temperature under nitrogen. After stirring for 1 min, $AgSbF_6$ (0.02 mmol in $ClCH_2CH_2Cl$) was added to the mixture. The reaction was completed within 5 min. The reaction mixture was directly subjected to column chromatography, from which (+)-**476a** was obtained (24.9 mg, 87%, over 99% *ee*).

16.2 Allenyne Cycloisomerization

Brummond demonstrated that cross-conjugated trienes **478** could be prepared by the Rh-catalyzed Alder-ene reaction of allenynes **477** (Scheme 2-59).[297] The Alder-ene reaction of allenynes tolerates various functionalities. The reactions of **477** (R^2 = H) with a methylallene moiety gave a single product in good yield in every case. The reactions of **477** (R^2 = C_5H_{11}) with a pentylallene moiety also afforded the corresponding products in good yields but as a 3:1–6:1 mixture of *E*- and *Z*-isomers. Use of $[Rh(cod)Cl]_2$ with $AgSbF_6$ substantially accelerated the reaction. Several cationic Rh-complexes were examined and it was found that only $[Rh(cod)Cl]_2/AgSbF_6$ was able to increase the *E*/*Z*-selectivity moderately. It was also found that the reactions of allenynes **477** catalyzed by $[Ir(cod)Cl]_2/AgBF_4$ yielded the *E*-isomers of **478** with high selectivity (*E*:*Z* = >20:1).

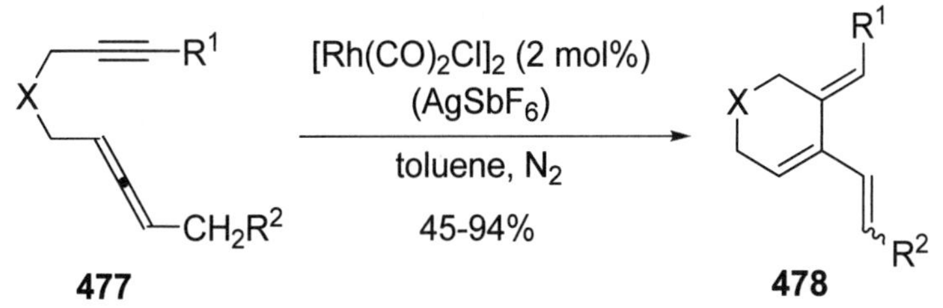

X = $C(CO_2Me)_2$, NTs, O; R^1 = H, Me, TMS; R^2 = H, C_5H_{11}

Scheme 2-59. Rh-catalyzed Alder-ene reactions of allenynes.

Typical procedure[297]

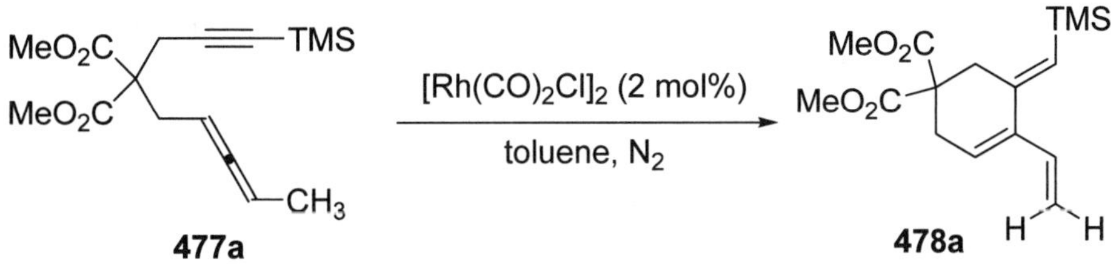

To a flame-dried test tube equipped with a magnetic stirring bar was added 2-(2,3-pentadienyl)-2-(3-trimethylsilanyl-2-propynyl)malonic acid diethyl ester (**477a**, 41 mg, 0.13 mmol) and toluene (0.5 mL). The test tube was evacuated and charged with N_2 (this step was repeated 3 times) and $[Rh(CO)_2Cl]_2$ (2.5 mg, 6×10^{-3} mmol) was added. The mixture was stirred at room temperature for 1 h under N_2 atmosphere (1 atm). The solvent was removed *in vacuo* and the residue was purified by flash chromatography (SiO_2, hexanes:ethyl acetate=9:1) to afford **478a** as a colorless oil (18.3 mg, 45%).

The reaction of α-amino ester-linked alleneynes **479** afforded cross-conjugated trienes **480** in high yields (Scheme 2-60). Alder-ene reactions of these *N*-alkynyl allenic amino esters were substantially faster (<10 min) than the previously used substrates. The rate acceleration was attributed in part to the Thorpe-Ingold effect of the *gem*-disubstituted methyl and carbomethoxy groups.

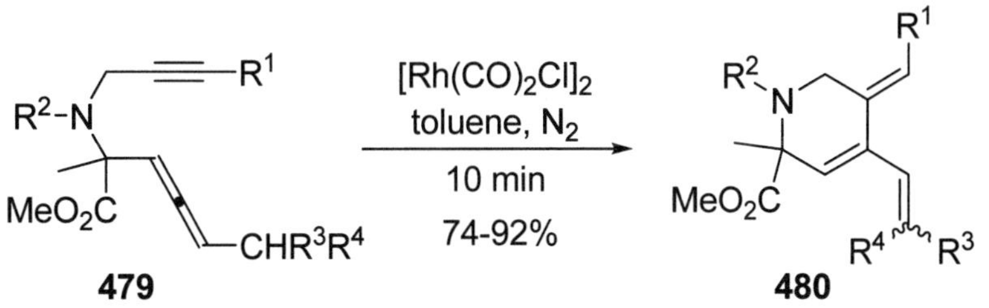

Scheme 2-60. Alder-ene of allenyne with an α-amino ester linker.

*General procedure for the Rh-catalyzed formal Alder-ene reaction of N-alkynyl allenic α-amino acid esters **479***[259a]

To a 10-mL test tube equipped with a magnetic stirring bar was added the alkynyl allene **478** (0.1 mmol) as a solution in benzene. The solvent was removed under vacuum and the atmosphere replaced with N_2. Toluene (1 mL) was added followed by addition of $[Rh(CO)_2Cl]_2$ (0.005 mmol) at room temperature. After completion of the reaction based on TLC (10 min), the solution was applied to a plug of silica gel. Elution with (hexanes:EtOAc, 9:1, v/v) afforded the product **480**.

Rh-catalyzed Alder-ene reactions of allenynes with an amide-linker **481** and **483** afforded δ- and ε-lactams **482** and **484**, respectively (Scheme 2-61).[259c] Higher temperature was required for the reaction to proceed efficiently, which could be attributed to the presence of the secondary amide, preferring *trans* conformation.[294] Yields were found to be lower when the substrates bearing a terminal acetylyne moiety (R^1 = H) were employed.

The Rh-catalyzed Alder-ene reaction of **483** afforded ε-lactams **484** in 56–65% yield (Scheme 2-61, eq. 2).[259c] This reaction only took place with tertiary amides, and no cyclized product was obtained with secondary amides. These are the first examples of δ- and ε-lactam formation under the Alder-ene reaction conditions.

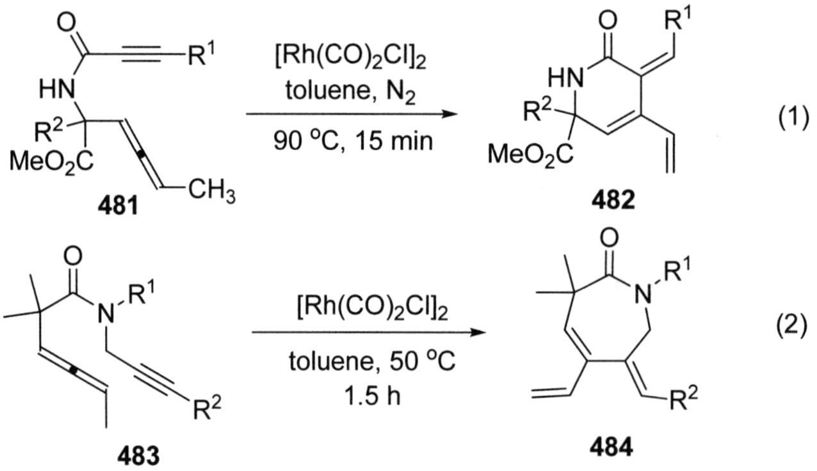

O, HN, R^1, R^2, MeO_2C, CH_3, **481**; $[Rh(CO)_2Cl]_2$, toluene, N_2, 90 °C, 15 min; O, HN, R^1, R^2, MeO_2C, **482** (1)

O, N, R^1, R^2, **483**; $[Rh(CO)_2Cl]_2$, toluene, 50 °C, 1.5 h; O, N, R^1, R^2, **484** (2)

Scheme 2-61. Rh-catalyzed Alder-ene reactions, forming δ- and ε-lactams.

Typical procedure[259c]

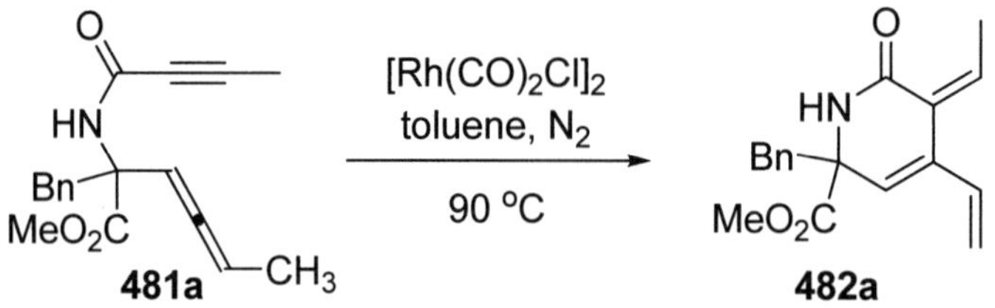

O, HN, Bn, MeO_2C, CH_3, **481a**; $[Rh(CO)_2Cl]_2$, toluene, N_2, 90 °C; O, HN, Bn, MeO_2C, **482a**

Allenyne **481a** (0.996g, 3.35 mmol) was placed in an oven-dried test tube equipped with a magnetic stirring bar, a septum, and an Ar balloon inlet, and dry toluene (112 mL, 0.03 M) was added. The solution was degassed by purging with argon and $[Rh(CO)_2Cl]_2$ (0.13 g, 0.33 mmol) was added. The reaction vessel was then submerged in an oil bath pre-heated to 90 °C. The progress of the reaction was followed by TLC, and after completion (30 min), it was cooled to room temperature The light yellow-brown solution was directly applied to a silica gel column and eluted (0 to 25% EtOAc/hexanes) to give cross-conjugated triene **482a** (0.916 g, 92%) as viscous oil.

Cross-conjugated trienes are useful building blocks for organic synthesis. It has been shown that cross-conjugated trienes can be used in tandem Diels-Alder reactions, forming the unsaturated decalin ring (Figure 2-17).[298]

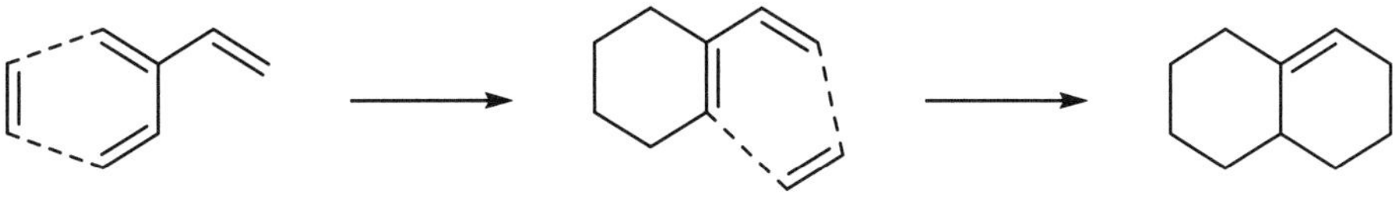

Figure 2-17. Cross-conjugated triene in tandem Diels-Alder reaction.

Based on this synthetic concept, the cross-conjugated trienes obtained by allenic Alder-ene reaction were utilized in the synthesis of polycyclic compounds by Rh(I)-catalyzed tandem Alder-ene/Diels-Alder reactions. Incorporation of a free hydroxyl group into the tether moiety allows the attachment of additional functional groups required for the reactions subsequent to the Alder-ene reaction (Figure 2-18).[299]

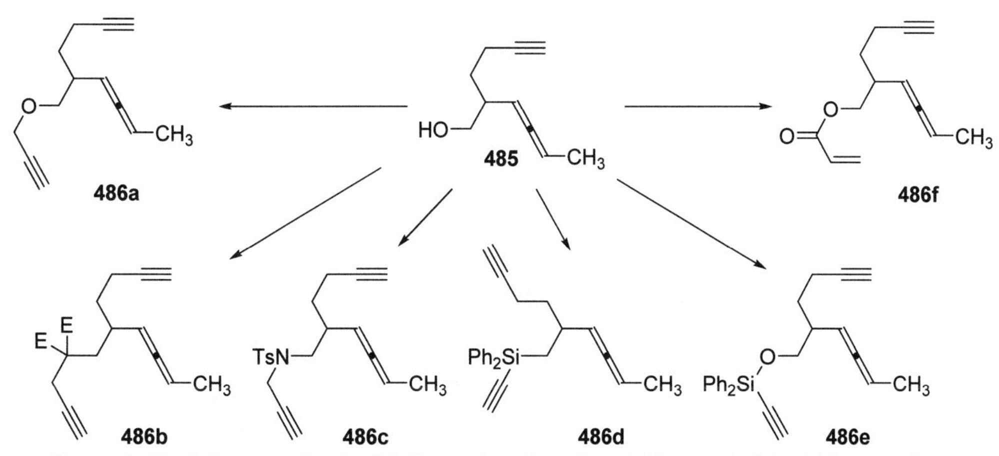

Figure 2-18. Substrates for the Rh(I)-catalyzed tandem Alder-ene/Diels-Alder reaction.

Under optimized conditions, Alder-ene reactions of these substrates **487** proceeded with no interference from the additional functional groups appended to the tether moiety, giving the desired products **588** in good-to-excellent yields (Scheme 2-62).

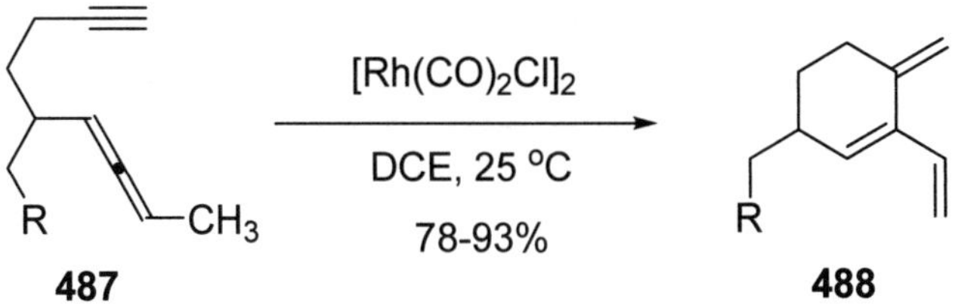

[Rh(CO)$_2$Cl]$_2$
DCE, 25 °C
78-93%

487 488

Scheme 2-62. Rh(I)-catalyzed tandem Alder-ene/Diels-Alder reactions.

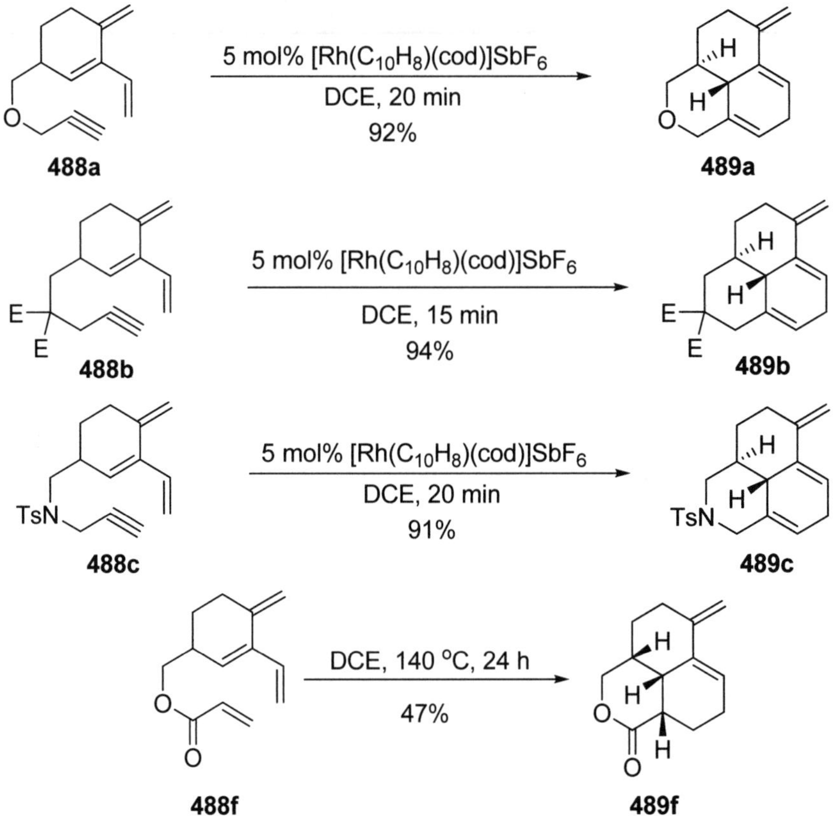

5 mol% [Rh(C$_{10}$H$_8$)(cod)]SbF$_6$
DCE, 20 min
92%
488a 489a

5 mol% [Rh(C$_{10}$H$_8$)(cod)]SbF$_6$
DCE, 15 min
94%
488b 489b

5 mol% [Rh(C$_{10}$H$_8$)(cod)]SbF$_6$
DCE, 20 min
91%
488c 489c

DCE, 140 °C, 24 h
47%
488f 489f

Scheme 2-63. Intramolecular Diels-Alder reaction of **488**.

Compounds **488a–f**, thus obtained, were all subjected to the Diels-Alder reaction conditions.[299] The initial investigation using compound **488f** in the thermal intramolecular [4+2] cycloaddition afforded tricyclic diene **489f** in moderate yield (Scheme 2-63). The attempted use of Lewis acids to promote the reaction only led to decomposition or recovery of starting material. Under Rh-catalyzed intramolecular [4+2] cycloaddition conditions, trienes **488a–c** gave cyclized products **489a–c** in excellent yields (Scheme 2-63). However, the corresponding reactions of silicon-tethered triene-alkynes did not proceed, which might be attributed to the steric bulk of the silyl substituents.[299]

A one-pot Alder-ene/Diels-Alder reaction of **487a** was carried out to test the viability of the tandem Alder-ene/Diels-Alder reaction, which successfully achieved the rapid construction of polycyclic **490** in 87% yield (Scheme 2-64).

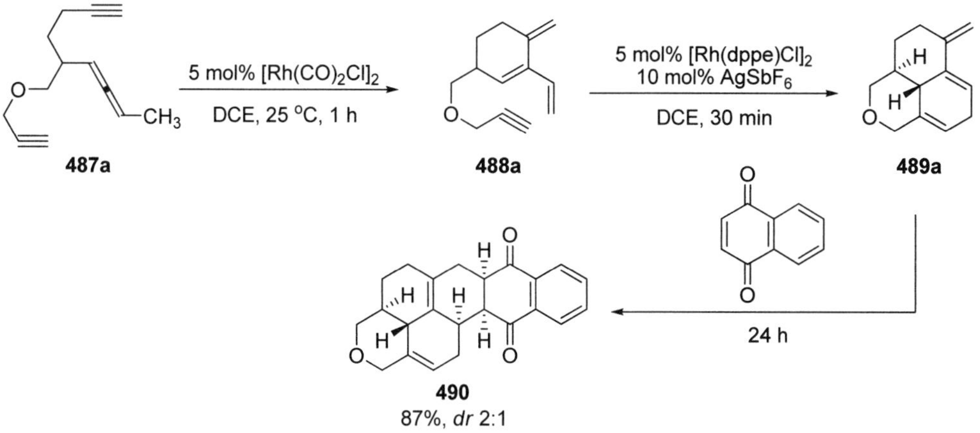

Scheme 2-64. Tandem Alder-ene/Diels-Alder reaction of **487a** in one pot.

*Tandem Rh(I)-catalyzed Alder-ene/Diels-Alder/Diels-Alder reaction of **487a***[299]

Allenediyne **487a** (30 mg, 0.16 mmol) was dissolved in 1.6 mL of 1,2-dichloroethane. To this solution was added $[Rh(CO)_2Cl]_2$ (3 mg, 0.008 mmol). The reaction mixture was stirred for 30 min at room temperature under N_2. TLC showed complete conversion to cross-conjugated triene **488a**. Then, $[Rh(dppe)Cl]_2$ (8 mg, 0.0075 mmol) and $AgSbF_6$ (300 μL of a 0.05-M solution in 1,2-dichloroethane) were added sequentially. The reaction mixture was stirred at room temperature for 45 min. TLC showed the intramolecular Diels-Alder reaction was complete. 1,4-Benzoquinone (38 mg, 0.24 mmol) was added and the reaction mixture was stirred at room temperature for 24 h. The solvent was removed under reduced pressure and the resulting residue was purified *via* flash chromatography on silica gel to give 48 mg (87% yield) of yellow solid. The product **490** consisted of a 2:1 mixture of two cycloadducts according to ^1H NMR analysis. The two diastereomers were then separated using HPLC.

16.3 Rh(I)-Catalyzed [4+2] Cycloadditions

The Diels-Alder reaction is one of the most useful synthetic methodologies in organic synthesis, and it has been employed extensively in the total synthesis of natural products.[301] A conjugated diene reacts with alkenes, alkynes, and other components to form a six-membered ring in a regioselective and stereoselective manner. The electronic properties of dienes and the dienophiles are critical factors for the Diels-Alder reaction to be effective. Dienes or dienophiles that lack electron-donating groups or electron-withdrawing groups often fail to undergo the cycloaddition except under extreme reaction conditions. Complexation of transition metals to activate π-bonds of dienes and dienophiles to promote the [4+2] cycloaddition makes the transition-metal–catalyzed [4+2] cycloadditions a valuable solution to such limitations.

In 1987, Matsuda *et al.* reported the first Rh-catalyzed intermolecular [4+2] cycloaddition (Scheme 2-65).[302] The cationic rhodium-complex–catalyzed co-dimerization of isoprene (**491**) with alkynes **390**, affording the corresponding cyclohexadienes **492** in 44–85%, yields and excellent regioselectivity.

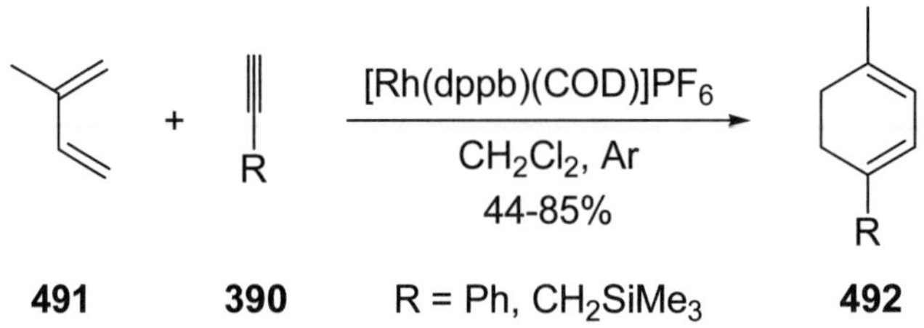

[Rh(dppb)(COD)]PF_6

CH_2Cl_2, Ar

44-85%

491 **390** R = Ph, CH_2SiMe_3 **492**

Scheme 2-65. Rh-catalyzed [4+2] intermolecular cycloaddition of isoprene with 1-alkynes.

The first intramolecular Rh-catalyzed [4+2] cycloaddition was reported by Livinghouse and co-workers in 1990 (Scheme 2-66).[300b] Ether-tethered dienyne **493a** was treated with Wilkinson's catalyst under mild thermal conditions to give the bicyclic product **494a** in excellent yield and high diastereoselectivity.

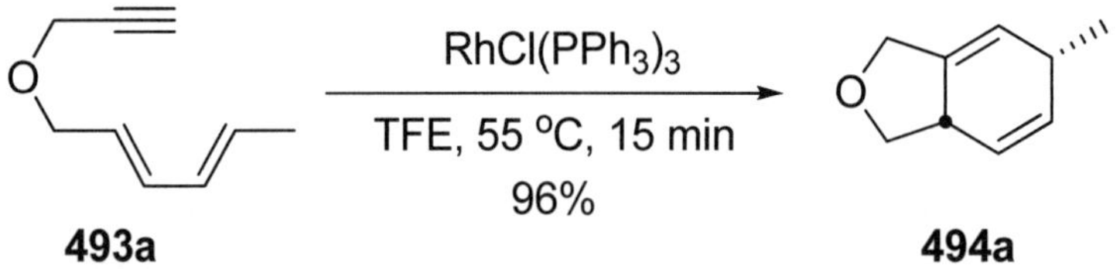

$RhCl(PPh_3)_3$

TFE, 55 °C, 15 min

96%

493a **494a**

Scheme 2-66. First intramolecular Rh-catalyzed [4+2] cycloaddition of dieneyne **493a**.

Van Leeuwen and Roobeek reported that use of electron-deficient phosphite ligands in Rh(I)-catalyzed hydroformylation reactions could dramatically increase reaction efficiency.[303] Based on this literature precedent, Livinghouse and co-workers screened a variety of phosphite ligands, and significant enhancement was observed with a *tris*-(hexafluoro-2-propyl)phosphite-modified rhodium complex.[300b, 304] This catalytic system was applicable to both terminal alkyne and alkene cyclizations. Improvements were made for this process by Gilbertson and Hoge employing $[Rh(DIPHOS)(CH_2Cl_2)_2]SbF_6$ (DIPHOS = diphenylphosphinoethane) as the catalyst. The [4+2] cycloaddition of **493** and **495** proceeded at room temperature to afford bicyclic products **494** and **496**, respectively, in good yields (Scheme 2-67).[300c] This process was successfully applied to the synthesis of the core of (+/–)-ptilocaulin.[305]

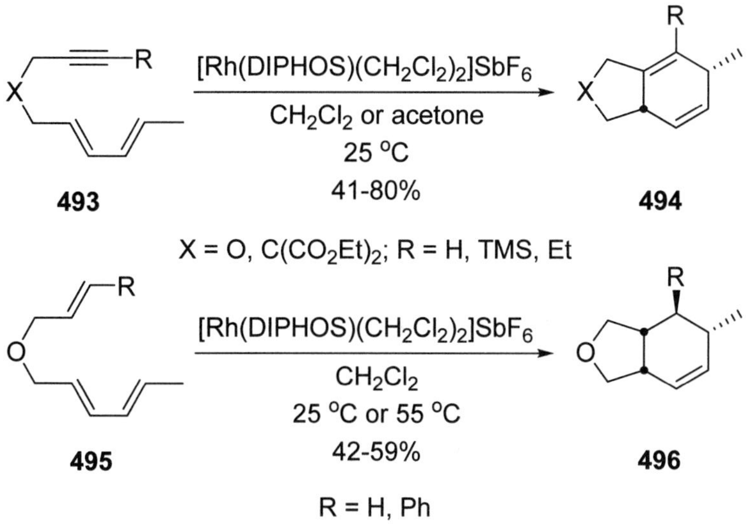

493 → 494

$[Rh(DIPHOS)(CH_2Cl_2)_2]SbF_6$

CH_2Cl_2 or acetone

25 °C

41-80%

$X = O, C(CO_2Et)_2$; R = H, TMS, Et

495 → 496

$[Rh(DIPHOS)(CH_2Cl_2)_2]SbF_6$

CH_2Cl_2

25 °C or 55 °C

42-59%

R = H, Ph

Scheme 2-67. Rh-DIPHOS-catalyzed [4+2] cycloadditions.

A cationic $[Rh(naphthalene)(cod)]^+$ complex was found to be an excellent catalyst for the [4+2] reaction of dienynes **493A** to give **494A** in 84–98% yield (Scheme 2-68, eq. 1).[306] Also, a cationic $[Rh(dppb)Cl]^+$ complex was shown to be a highly efficient catalyst for the reaction of dienynes **493**, affording **494** in 75–99% yield (Scheme 2-68, eq. 2).[300a]

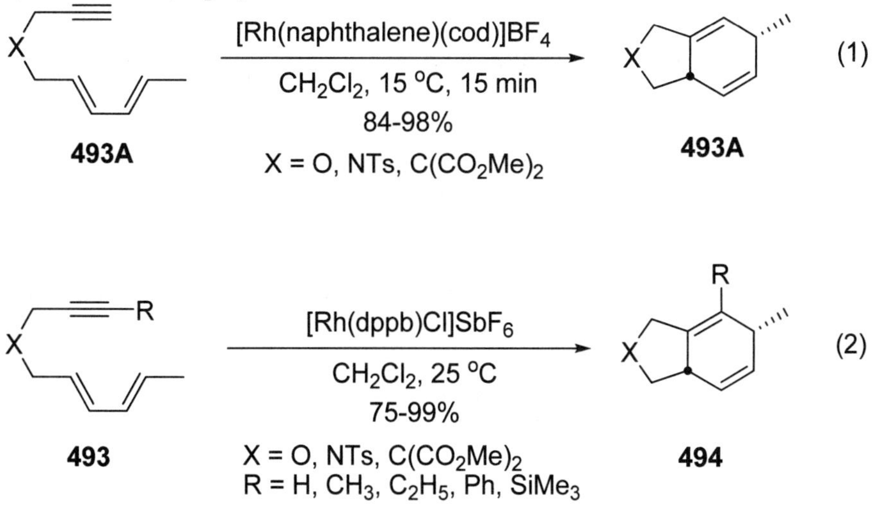

493A → 493A (1)

$[Rh(naphthalene)(cod)]BF_4$

CH_2Cl_2, 15 °C, 15 min

84-98%

$X = O, NTs, C(CO_2Me)_2$

493 → 494 (2)

$[Rh(dppb)Cl]SbF_6$

CH_2Cl_2, 25 °C

75-99%

$X = O, NTs, C(CO_2Me)_2$

$R = H, CH_3, C_2H_5, Ph, SiMe_3$

Scheme 2-68. [4+2] cycloaddition of dienynes catalyzed by cationic Rh complexes.

*Procedure for the [4+2] cycloaddition reaction of **493b***[300a]

$[Rh(dppb)Cl]SbF_6$

CH_2Cl_2, 25 °C

493b **494b**

In a glove box, a dry 25-mL Schlenk tube was charged with **493b** (106 mg, 0.5 mmol), $[Rh(dppb)Cl]_2$ (7 mg, 0.0062 mmol), and DCE (5 ml). The reaction mixture was stirred for 1 min, followed by addition of $AgSbF_6$ (4.3 mg, 0.0124 mmol) at room temperature. The resulting orange solution was stirred at room temperature under nitrogen for 10 min. The reaction mixture was diluted with ether (5 mL) and filtered through Celite to remove silver chloride, then concentrated and purified by flash chromatography (silica gel, 5% ether in hexane) to give a cycloadduct **494b** in 99% yield as a colorless oil.

The asymmetric intramolecular [4+2] cycloaddition of triene **495a** was first realized using a Rh-DIOP catalyst, which afforded cycloadduct **496a** in high yield and good enantioselectivity (Scheme 2-69, eq. 1).[307] For the reaction of dienyne **493c**, Rh-catalyst with a DIOP derivative gave better selectivity (87% ee) (Scheme 2-69, eq. 2).[307] Gilbertson *et al.* reported that a BINAP-Rh complex was the most effective catalyst in the [4+2] cycloaddition of triene **495a** to give **496a** with up to 98% ee, whereas(*S,S*)-Me-DuPhos-Rh complex worked best for the reactions of dienynes **493**, giving **494** in high yield and 88–95% ee (Scheme 2-69, eq. 3).[308]

It has been shown that the diene ligands such as cod and nbd used for the generation of nonracemic diphosphine-Rh^+ catalysts have critical influence on the resulting enantioselectivity.[309] Based on this observation, chiral diene ligands **497** were used to investigate their effects on enantioselectivity in combination with chrial diphosphine ligands, BINAP, Me-DuPhos, and Et-DuPhos. In the absence of chiral diphosphine, Rh-complex with chiral diene **497a** catalyzed the reaction of dienyne **493b** to give **494b** in 91% and 26% ee.[309] Clear results ascribed to the matching and mismatching pairs of chiral diene **497a** with (*R,R*)- and (*S,S*)-DuPhos were obtained; *i.e.*, the combination of **497a** with (*R,R*)-Duphos gave **494b** with 91% ee, whereas that with (*S,S*)-DuPhos afforded the opposite enantiomer with only 9% ee.[309]

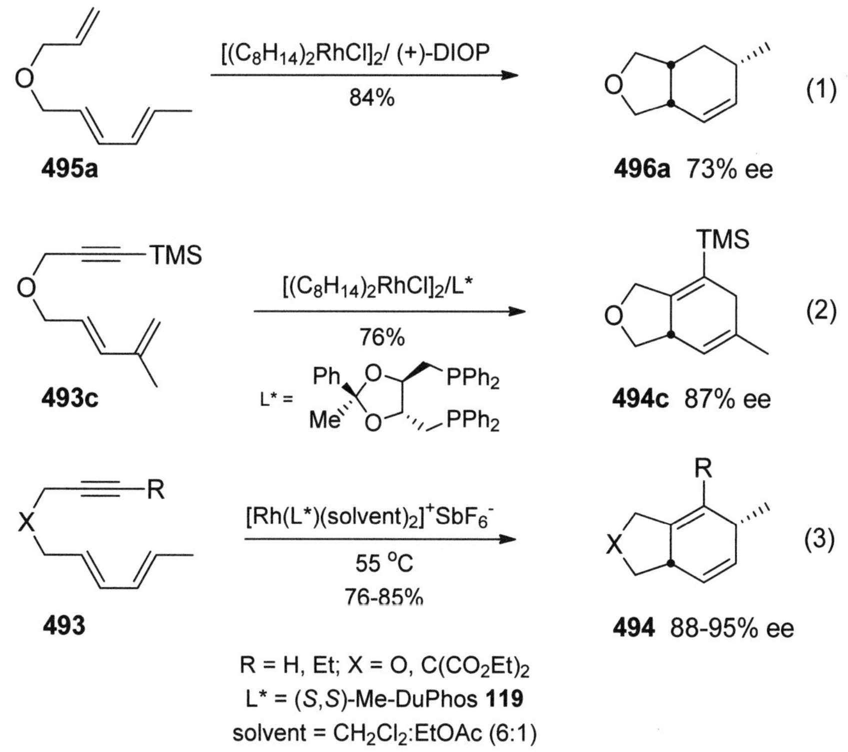

$[(C_8H_{14})_2RhCl]_2$/ (+)-DIOP

84%

(1)

495a

496a 73% ee

TMS

$[(C_8H_{14})_2RhCl]_2$/L*

76%

(2)

L* = Ph, Me, O, O, PPh_2, PPh_2

493c

494c 87% ee

R

$[Rh(L^*)(solvent)_2]^+SbF_6^-$

55 °C

76-85%

(3)

493

494 88-95% ee

R = H, Et; X = O, $C(CO_2Et)_2$

L* = (*S*,*S*)-Me-DuPhos **119**

solvent = CH_2Cl_2:EtOAc (6:1)

Scheme 2-69. Asymmetric [4+2] cycloadditions of trienes and dienynes.

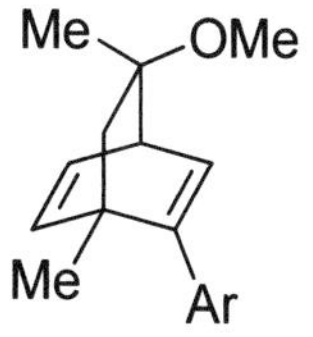

Me, OMe, Me, Ar

497

a: Ar = 3,5-Me_2-C_6H_3
b: Ar = Ph
c: Ar = 4-*t*-Bu-C_6H_4

Figure 2-19. Chiral diene ligands.

General experimental procedure[308]

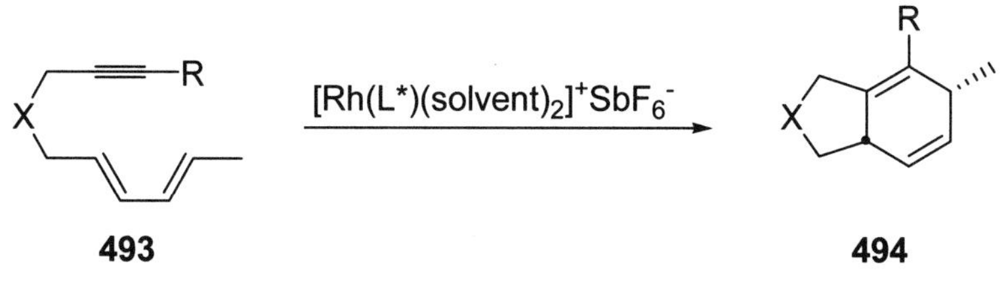

$[Rh(L^*)(solvent)_2]^+SbF_6^-$

493

494

[Rh(nbd)(diphosphine)]$^+$(SbF$_6$)$^-$ (37 mg, 0.044 mmol) was placed in a Schlenk tube and dissolved in 6 mL of freshly distilled, deoxygenated solvent. Hydrogen gas was then bubbled through the solution for 2 min (solution color changed to dark red) followed by N_2 gas for 2 min. The substrate **493** (0.735 mmol) was added to the catalyst solution along with 1 mL of solvent. The Schlenk tube was then freeze-pump-thaw degassed (three cycles) and stirred under N_2 at ambient temperature. The course of the reaction was followed by TLC (5% ethyl acetate/hexane). After the reaction was deemed complete, the solvent was removed under reduced pressure. The products were purified by flash chromatography. The selectivity for the reactions was determined by GC analysis of the crude reaction mixture, using a chiral capillary column (Chiraldex B-TA 10 m × 0.25 mm). The enantiomeric excesses are reported to (0.5% with baseline resolution of enantiomers.

Wender *et al.* reported the first Rh-catalyzed [4+2] cycloaddition of tethered allene-diene **498** using P(O-biphenyl-*o*)$_3$ as the ligand to construct *exo*-methylene-bicyclo[4.3.0]octane **499** in high yield (Scheme 2-70, eq. 1).[222a] It was also demonstrated that the chemoselective binding of Rh-catalyst to the allene π-system could be altered by changing the electronic properties of the catalyst used (Scheme 2-70, eq. 2) with *tris*-(hexafluoroprop-2-yl)phosphite as the ligand. For example, the Rh catalyst selectively reacted with the terminal olefin of the allene moiety of **500** to give 6-6-fused ring product **501**.[222a]

Furthermore, the stereoselectivity at the bridged carbon could be controlled by using appropriate catalysts. For example, the reaction of allene-diene **502** catalyzed by a Rh-complex with ligand P[OCH(CF$_3$)-(C$_6$H$_4$OMe-2)]$_3$ gave *trans*-**503** as the prominent product, whereas that with P[OCH(CF$_3$)-{C$_6$H$_3$(OMe)$_2$-2,6}]$_3$/AgOTf afforded *cis*-**503** with excellent selectivity (Scheme 2-71).[222a]

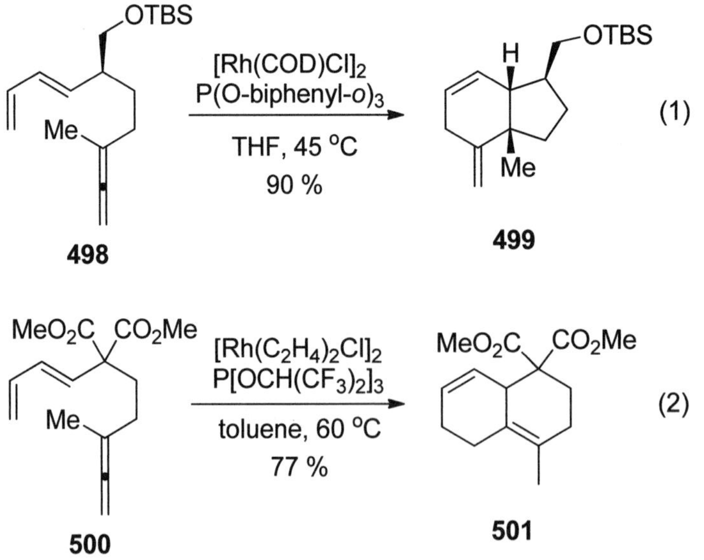

OTBS
Me
498
[Rh(COD)Cl]$_2$
P(O-biphenyl-*o*)$_3$
THF, 45 °C
90 %
H
OTBS
Me
499
(1)

MeO$_2$C CO$_2$Me
Me
500
[Rh(C$_2$H$_4$)$_2$Cl]$_2$
P[OCH(CF$_3$)$_2$]$_3$
toluene, 60 °C
77 %
MeO$_2$C CO$_2$Me
501
(2)

Scheme 2-70. Intramolecular [4+2] cycloadditions of allenic dienes.

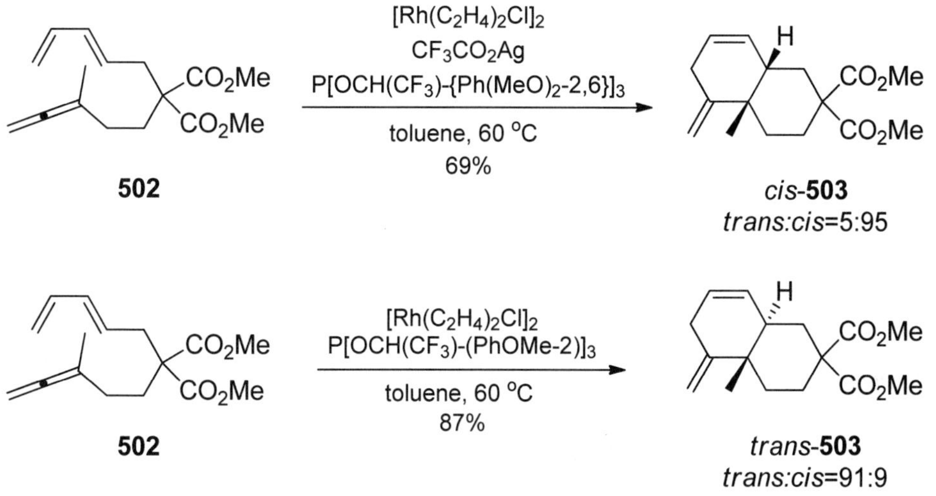

$[Rh(C_2H_4)_2Cl]_2$, CF_3CO_2Ag, $P[OCH(CF_3)\text{-}\{Ph(MeO)_2\text{-}2,6\}]_3$, toluene, 60 °C, 69%

502 → *cis*-**503**, *trans:cis*=5:95

$[Rh(C_2H_4)_2Cl]_2$, $P[OCH(CF_3)\text{-}(PhOMe\text{-}2)]_3$, toluene, 60 °C, 87%

502 → *trans*-**503**, *trans:cis*=91:9

Scheme 2-71. Control of stereoselectivity in the intramolecular [4+2] cycloaddition of allene-diene **516**.

Typical procedure[222a]

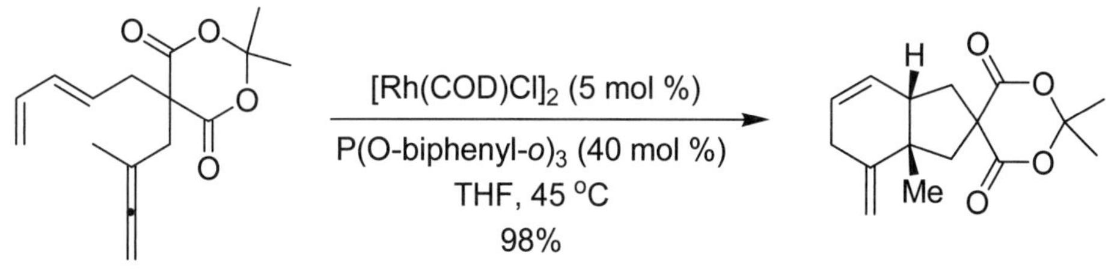

$[Rh(COD)Cl]_2$ (5 mol %), P(O-biphenyl-*o*)$_3$ (40 mol %), THF, 45 °C, 98%

504 → **505**

A 250-mL Schlenck flask was flushed with nitrogen and charged with **504** (1.08 mmol) and freshly distilled anhydrous THF (100 mL). *Tris-o*-biphenylphosphite (0.432 mmol) and $[Rh(cod)Cl]_2$ (0.054 mmol) were then added. The pale yellow solution was stirred at 45 °C for 10.5 h and then filtered through a plug of neutral alumina and concentrated *in vacuo*. Flash chromatography of the residue on silica gel afforded **505** (1.06 mmol) in 98 % yield.

Murakami *et al.* developed new methodology for the preparation of substituted benzenes **508** wherein a vinylallene moiety was used as a part of the diene component **506** and a monosubstituted alkyne **507** as the dienophile (Scheme 2-72).[310]

R^1 = TMS, Ph; R^2 = H, Et, *n*-Bu, Ph, $HO(CH_2)_2$, $Cl(CH_2)_2$

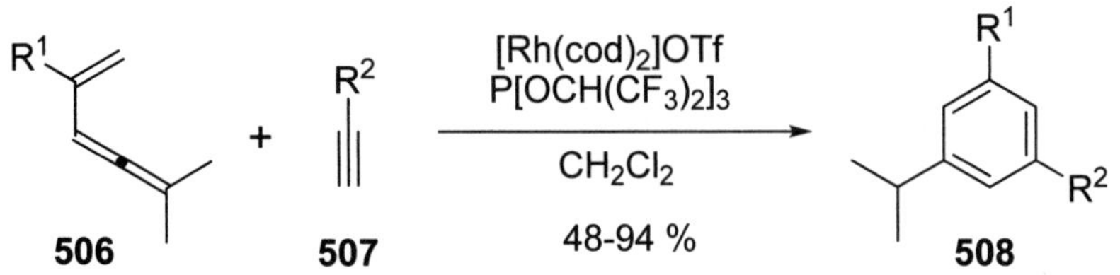

Scheme 2-72. Rh-catalyzed intermolecular [4+2] cycloaddition of allene-ene with alkynes.

General procedure[310]

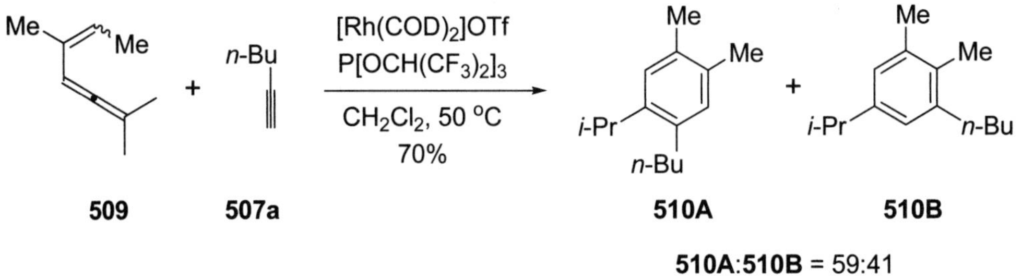

A mixture of $[Rh(cod)_2]OTf$ (9.6 mg, 20.5 mmol), $P[OCH(CF_3)_2]_3$ (10.9 mg, 20.5 mmol), **509** (50.0 mg, 409 mmol), and 1-hexyne (67.2 mg, 818 mmol) in 1,2-dimethoxyethane (2 mL) was stirred at 50 °C for 24 h. After the mixture was cooled, the solvent was removed under vacuum. The residue was subjected to gel permeation chromatography to afford a 59:41 mixture of **510A** and **510B** (58.2 mg, 70%).

16.4 Rh(I)-Catalyzed [5+2] Cycloadditions

The first example of the transition-metal–catalyzed [5+2] cycloaddition of vinylcyclopropane (VCP) with alkyne was reported in 1995 by Wender *et al.* (Scheme 2-73).[222a] Various tethered alkyne-VCPs underwent Rh-catalyzed [5+2] cycloaddition to give bicyclo[5.3.0] products **512** in excellent yields, irrespective of the steric and electronic properties of the R^1 group.

In addition to the [5+2] cycloaddition of tethered alkyne-VCPs **511**, the reactions of tethered alkene-VCPs **513**, yielding **514**[311] and tethered allene-VCPs **515**, affording **516**[221a] have also been developed (Scheme 2-74).

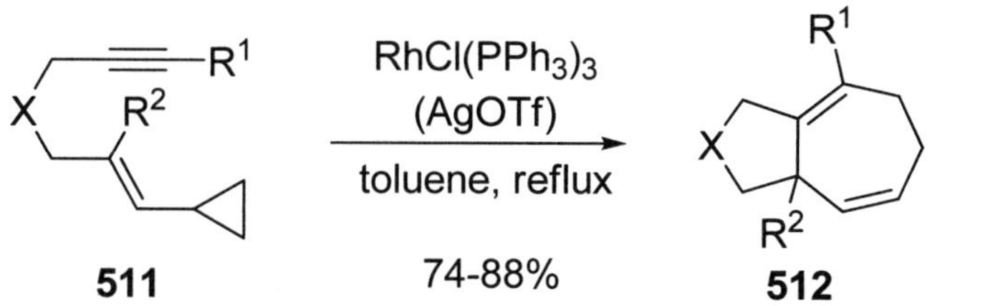

RhCl(PPh3)3 (AgOTf) toluene, reflux 74-88%

511 512

X = C(CO2Me)2, O; R1 = Me, TMS, Ph, CO2Me; R2 = H, Me

Scheme 2-73. Rh-catalyzed [5+2] cycloadditions.

The first example of the asymmetric [5+2] cycloaddition of alkene-VCP **517** was reported using a Chiraphos-Rh catalyst, which gave *cis*-fused bicyclo[5.3.0]decene **518** in 80% yield and 63% ee (Scheme 2-75, eq. 1). The reaction of alkyne-VCP **519** catalyzed by $[\{RhCl(C_2H_4)_2\}_2]$ complex in the presence of $NaBAr^F_4$ gave **520** with 99% ee (Scheme 2-75, eq. 2).[312]

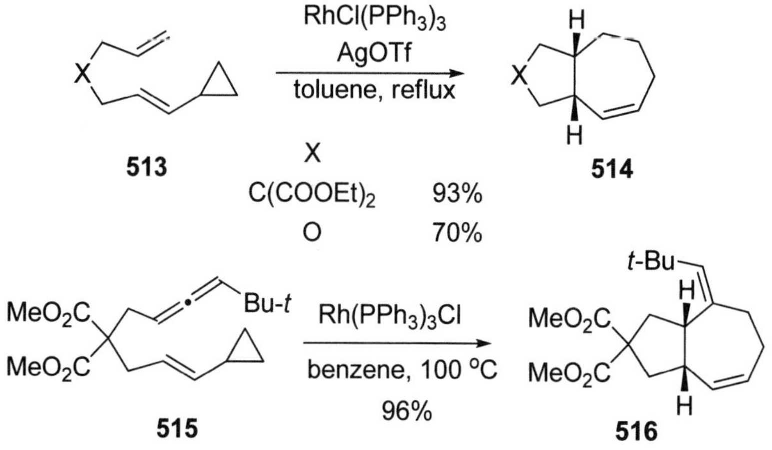

RhCl(PPh3)3 AgOTf toluene, reflux

513 514

X	
C(COOEt)2	93%
O	70%

Rh(PPh3)3Cl benzene, 100 °C 96%

515 516

Scheme 2-74. [5+2] Cycloaddition of vinylcyclopropanes.

Typical procedure for the Rh-catalyzed [5+2] cycloaddition[311]

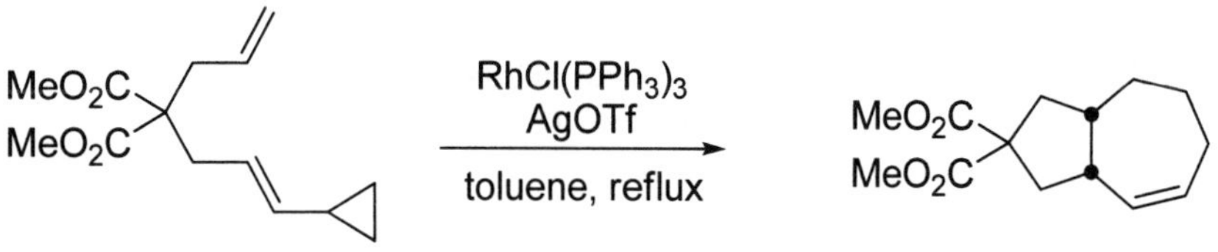

RhCl(PPh3)3 AgOTf toluene, reflux

517 518

Tris-(triphenylphsphine)rhodium(I) chloride (3.65 mg, 0.004 mmol, 0.001 equiv) and silver triflate (1.02 mg, 0.004 mmol, 0.001 equiv) were added sequentially, each in one bath, to a base-washed, oven-dried Schlenk flask containing freshly distilled, oxygen-free toluene

(2 mL) under an Ar atmosphere. The solution was stirred for 5 min at room temperature, after which ene-vinylcyclopropane **517** (1.00 g, 3.95 mmol, 1.0 equiv) in toluene (2 mL) was added over 10 s. The resulting solution was heated at 110 °C for 17 h. After cooling to room temperature, the reaction mixture was filtered through a plug of alumina and concentrated *in vacuo*. Purification by flash chromatography on silica gel (5% ethyl acetate in hexane) gave cycloadduct **518** (0.857 g, 86%) as a clear, colorless oil.

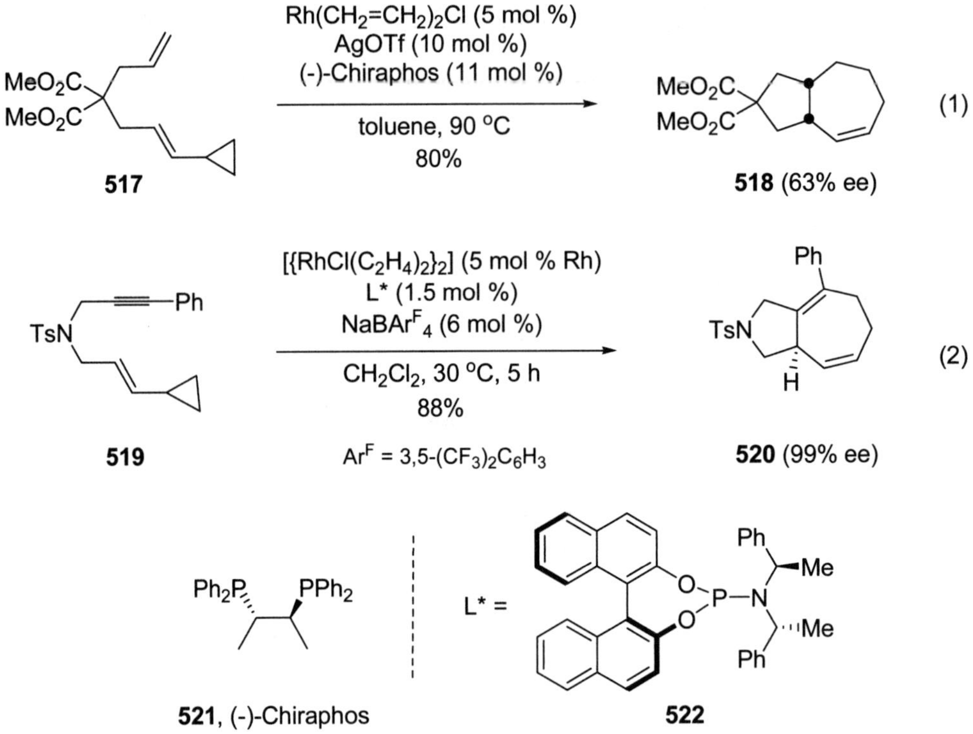

$Rh(CH_2{=}CH_2)_2Cl$ (5 mol %)
AgOTf (10 mol %)
(-)-Chiraphos (11 mol %)
toluene, 90 °C
80%

517 → **518** (63% ee) (1)

$[\{RhCl(C_2H_4)_2\}_2]$ (5 mol % Rh)
L* (1.5 mol %)
$NaBAr^F_4$ (6 mol %)
CH_2Cl_2, 30 °C, 5 h
88%

$Ar^F = 3,5\text{-}(CF_3)_2C_6H_3$

519 → **520** (99% ee) (2)

521, (-)-Chiraphos

L* = **522**

Scheme 2-75. Asymmetric [5+2] cycloadditions.

16.5 Rh(I)-Catalyzed [2+2+2+1] Cycloaddition

In 2000, Ojima and Lee reported a novel Rh-catalyzed carbonylative silicon-initiated cascade tricyclization (CO-SiCaT) of dodec-11-ene-1,6-diyne **523**, which proceeded at room temperature to give tricyclic 5-7-5 product **524** in good-to-excellent yields (Scheme 2-76).[227]

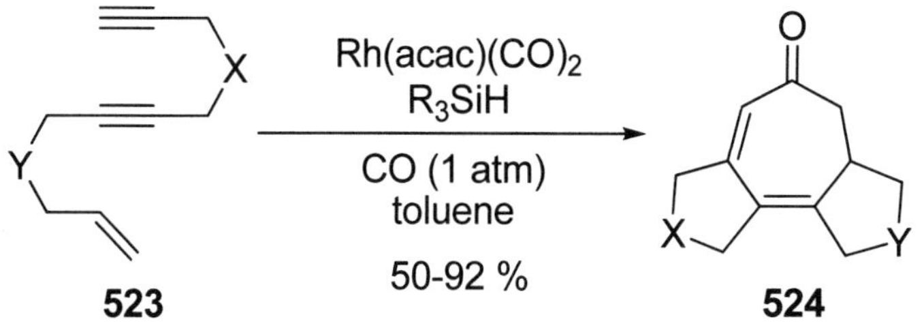

Rh(acac)(CO)$_2$
R$_3$SiH
CO (1 atm)
toluene
50-92 %

523 524

X, Y = esters, ethers, O, NR'

Scheme 2-76. Rh-catalyzed carbonylative silicon-initiated cascade tricyclizations (CO-SiCaT).

During the study on the scope and limitations of CO-SiCaT reactions, it was found that in the absence of hydrosilane, the reaction of 1-substituted endiynes **525** catalyzed by [Rh(cod)Cl]$_2$ afforded carbonylative tricyclization products **526** in good yield *via* a novel [2+2+2+1] cycloaddition, accompanied by a small amount of noncarbonylative product (**527**) *via* a [2+2+2] cycloaddition (Scheme 2-77).[223] The reaction of **525** under the CO-SiCaT conditions gave noncarbonylated product **527** exclusively. Also, it should be noted that the attempted reaction of enediynes **525** with a terminal acetylene moiety under the standard [2+2+2+1] conditions did not take place.

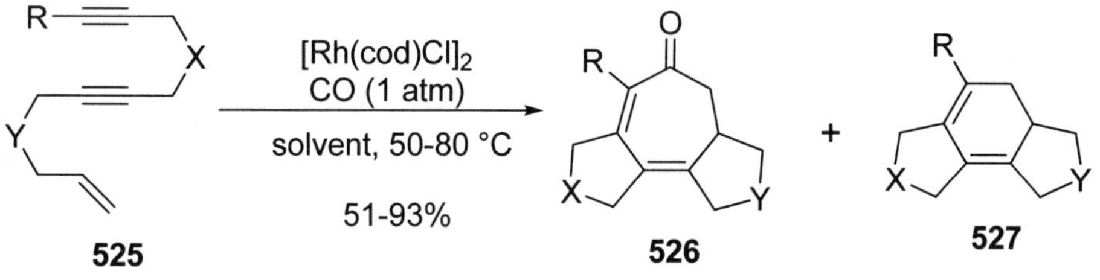

[Rh(cod)Cl]$_2$
CO (1 atm)
solvent, 50-80 °C
51-93%

525 526 527

R= TMS. alkyl; X, Y=esters, ethers, O, NR'

Scheme 2-77. [2+2+2+1] cycloadditions of endiynes **525**.

Although these novel Rh-catalyzed [2+2+2+1] cycloadditions of endiynes **525** with CO gave similar products to those from CO-SiCaT reactions, the mechanism of the reaction is totally different. The CO-SiCaT reaction is a stepwise process of carbocyclizations,[227] whereas the Rh-catalyzed [2+2+2+1] cycloaddition proceeds *via* a series of metallacycles (Figure 2-20).[223]

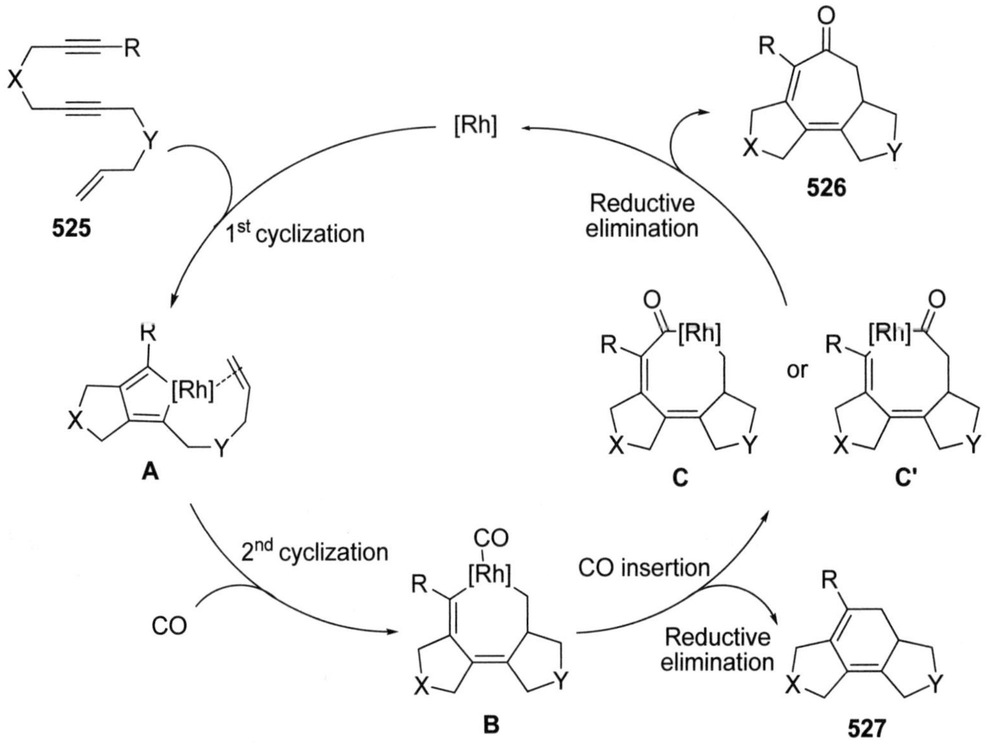

Figure 2-20. Proposed mechanism for the Rh-catalyzed [2+2+2+1] cycloaddition of enediynes.

In the proposed mechanism, the Rh catalyst selectively reacts with the diyne moiety of **525** to form metallacycle **A.** Alkene insertion to the Rh-C bond and CO coordination gives metallacycle **B**. From metallacycle **B**, CO insertion to the Rh-C bond gives metallacycle **C** or **C'**, and subsequent reductive elimination affords cycloadduct **526**. Reductive elimination of Rh from metallacycle **B** prior to CO insertion gives the [2+2+2] cycloadduct **527**. Kaloko *et al.* further expanded the scope of this novel [2+2+2+1] cycloaddition process to cyclohexene-diyne substrates **528**,[313] which gave the corresponding tetracyclic products **529** as single diastereomers in high yields (Scheme 2-78).

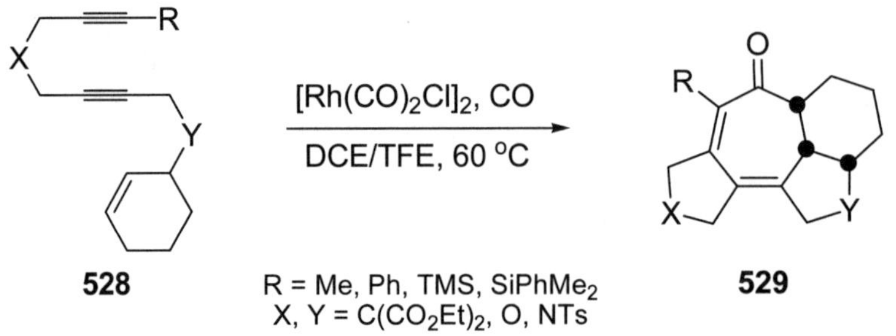

Scheme 2-78. Rh-catalyzed [2+2+2+1] cycloadditions of cyclohexene-diynes.

Typical procedure for the [2+2+2+1] cycloaddition of enediyne ***503a***[223b]

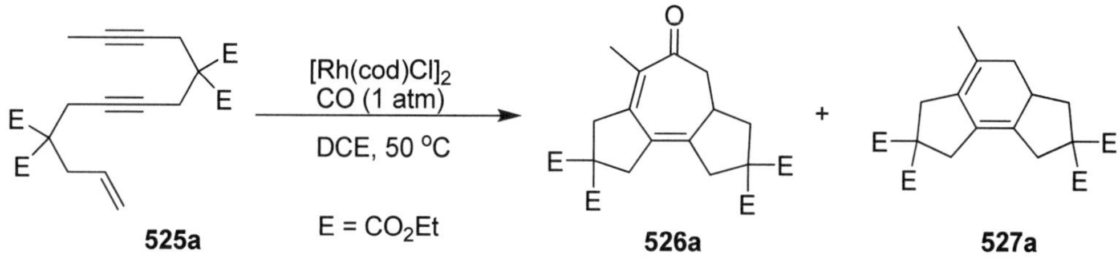

$[Rh(cod)Cl]_2$
CO (1 atm)
DCE, 50 °C
E = CO_2Et
525a **526a** + **527a**

Enediyne **525a** (100 mg, 0.216 mmol) was introduced to a Schlenck tube, followed by 1,2-dichloroethane (2.16 mL) under Ar, and then CO was bubbled into the solution at room temperature. After 15 min, $[Rh(cod)Cl]_2$ (5.3 mg, 0.0108 mmol, 5 mol%) was added under CO and the resulting mixture was stirred at room temperature for an additional 5 min. Then, the reaction mixture was heated to 50 °C with stirring and kept for 16 h under CO (ambient pressure, bubbled into the solution). All volatiles were removed from the reaction mixture under reduced pressure, and the crude product was purified by flash chromatography on silica gel (EtOAc/hexanes = 1:9-3:7) to give **526a** as light yellow oil (92.5% yield by ^{1}H NMR; 93 mg, 88% isolated yield) and **527a** as colorless oil (5.5% yield by ^{1}H NMR; <4 mg, <4% isolated yield).

16.6 Rh(I)-Catalyzed Reductive Cyclizations

The first Rh-catalyzed carbonylative silylcarbocyclization (CO-SiCaC) was reported in 1992[224a, 314] in which silylcyclopentenone **531** was isolated as a minor product in the silylformylation of 1-hexyne **530** (Scheme 2-79).

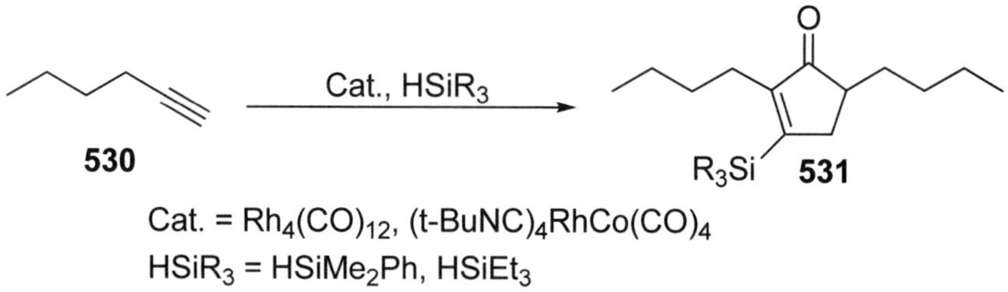

530 Cat., $HSiR_3$ → **531** (R_3Si)

Cat. = $Rh_4(CO)_{12}$, $(t\text{-BuNC})_4RhCo(CO)_4$
$HSiR_3$ = $HSiMe_2Ph$, $HSiEt_3$

Scheme 2-79. Rh-catalyzed carbonylative silylcarbocyclization (CO-SiCaC) of 1-hexyne **530**.

Under optimized conditions using Et_3SiH and $(t\text{-BuNC})_4RhCo(CO)_4$ as the catalyst at 60 °C, **531** was formed in 54% yield.[314] A possible mechanism proposed for this intermolecular CO-SiCaC is shown in Figure 2-21.[314] According to this mechanism, the reactions proceed *via* intermolecular trapping of the β-silylacryloyl-[Rh] complex **A** by another molecule of 1-hexyne to give intermediate **B**. Subsequent carbocyclization gives intermediate **C** that undergoes β-hydride elimination to form cyclopentadienone-[Rh]H complex **D**. Highly regioselective hydrometalation of the olefin moiety at the sterically less congested site forms intermediate **E** and subsequent reductive elimination affords **531**.

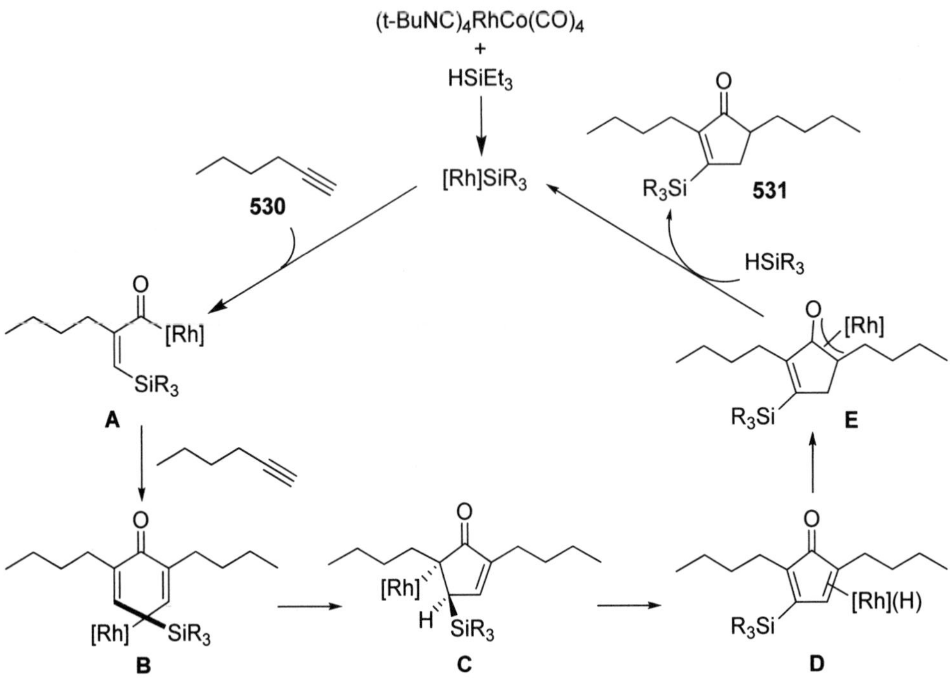

Figure 2-21. Proposed mechanism of intermolecular CO-SiCaC of 1-hexyne **530**.

Immediately after the discovery of the intermolecular CO-SiCaC reaction of 1-alkynes, the intramolecular version of the SiCaC reaction was investigated using 1,6-enynes. The first SiCaC and CO-SiCaC reaction of 1,6-enynes was reported in 1992.[224a] The reaction of allyl propargyl ether **532** with $PhMe_2SiH$ catalyzed by $Rh_2Co_2(CO)_{12}$ gave 3-(silylmethylene)-4-methylhydrofuran (**533**) in 62% yield. The result clearly indicates that the β-silylethenyl-[Rh] species can efficiently be trapped by the olefin moiety of a 1,6-enyne (Scheme 2-80).[224a]

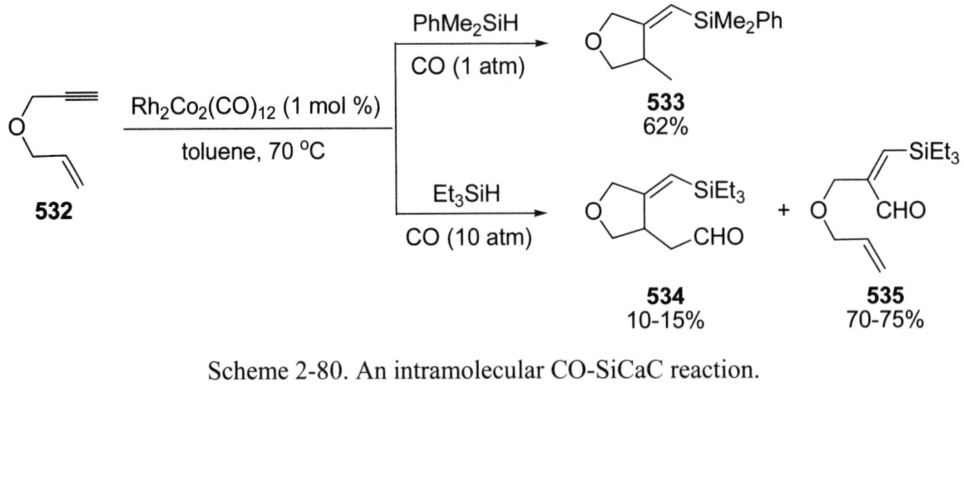

Scheme 2-80. An intramolecular CO-SiCaC reaction.

When the reaction of **532** was carried out under 10 atm of CO using Et_3SiH, the CO-SiCaC reaction took place to give the corresponding aldehyde **534** as a minor product (15020%) together with the silylformylation product **535** (70–75%) (Scheme 2-80).[224a]

The SiCaC reaction of enynes has been investigated in detail and has been applied to a wide range of substrates (Scheme 2-81).[224c] Rhodium carbonyl clusters, such as $Rh_4(CO)_{12}$ and $Rh_2Co_2(CO)_{12}$, were found to be effective catalysts to promote this transformation. The reaction of 1,6-enyne **536** (X = $C(CO_2Et)_2$) with $PhMe_2SiH$ catalyzed by $Rh_4(CO)_{12}$ completed within 1 min at room temperature, giving the corresponding silylmethylene-2-methylcyclopentane **537** (X = $C(CO_2Et)_2$) in 99% yield (Scheme 2-81, eq. 1). When the reaction of **536** catalyzed by $Rh_4(CO)_{12}/P(OEt)_3$[224c] or $Rh_4(CO)_{12}$[315] was carried out under 20 atmospheres of CO, the CO-SiCaC reaction took place almost exclusively to give **538** in good-to-excellent yield (Scheme 2-81, eqs. 2 and 3). Rhodium-atalyzed SiCaC and CO-SiCaC reactions show good functional group tolerance, thus providing efficient methods for the construction of synthetically useful exo-methylene-cyclopentane, -tetrahydrofuran, and -pyrrolidine[316] skeletons. In 2004, Shibata *et al.* reported the silylcarbocyclization of allenynes (Scheme 2-82).[229]

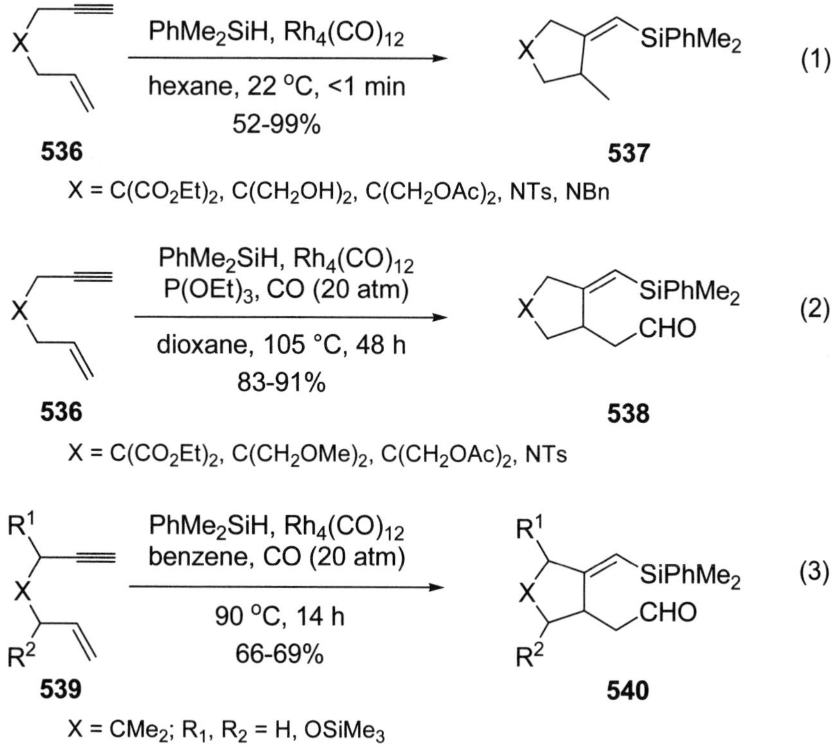

$PhMe_2SiH$, $Rh_4(CO)_{12}$; hexane, 22 °C, <1 min; 52-99% (1)

536 → **537** ($SiPhMe_2$)

X = $C(CO_2Et)_2$, $C(CH_2OH)_2$, $C(CH_2OAc)_2$, NTs, NBn

$PhMe_2SiH$, $Rh_4(CO)_{12}$, $P(OEt)_3$, CO (20 atm); dioxane, 105 °C, 48 h; 83-91% (2)

536 → **538** ($SiPhMe_2$, CHO)

X = $C(CO_2Et)_2$, $C(CH_2OMe)_2$, $C(CH_2OAc)_2$, NTs

$PhMe_2SiH$, $Rh_4(CO)_{12}$, benzene, CO (20 atm); 90 °C, 14 h; 66-69% (3)

539 (R^1, R^2) → **540** (R^1, R^2, $SiPhMe_2$, CHO)

X = CMe_2; R_1, R_2 = H, $OSiMe_3$

Scheme 2-81. SiCaC and CO-SiCaC reactions of enynes.

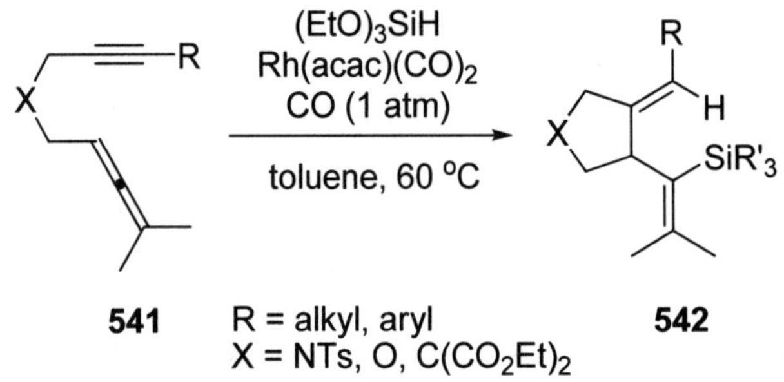
R
X
(EtO)3SiH
Rh(acac)(CO)2
CO (1 atm)
toluene, 60 °C
H
SiR'3
541
R = alkyl, aryl
X = NTs, O, C(CO2Et)2
542

Scheme 2-82. Silylcarbocyclization of allenynes.

The proposed mechanism is shown in Figure 2-22. Regioselective insertion of the proximal double bond of allene **541** into the Si-[Rh] bond of the hydrosilane-[Rh] oxidative adduct, H[Rh]SiR$_3$, yields intermediate **A**. Subsequent alkyne insertion forms the cyclic vinyl-Rh complex **B** and the subsequent silane-promoted reductive elimination gives the SiCaC product **542** and regenerates the active Rh-Si complex. When allenes without substituents at the acetylene and allene turmini were employed, silylated products were obtained. The result suggests that the substituents on both alkyne and allene termini play a critical role in the chemoselectivity between the alkyne and allene moieties, as well as the regioselectivity for the two double bonds of the allene.

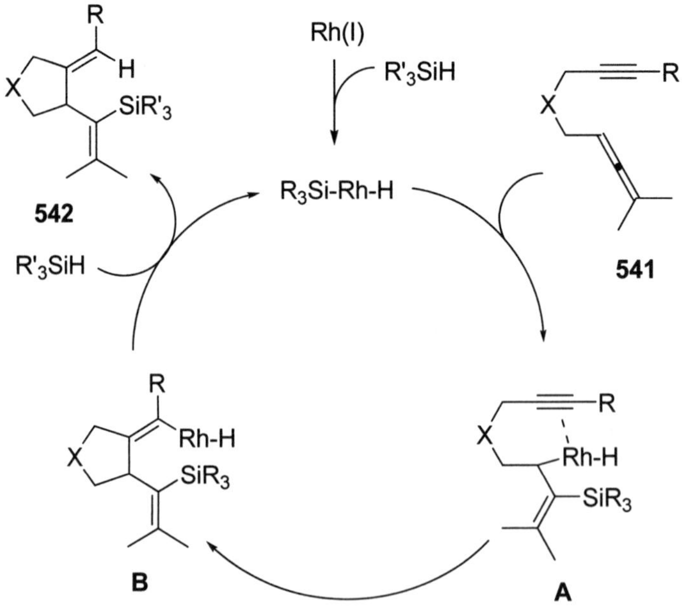
R
H
X
SiR'3
542
R'3SiH
Rh(I)
R'3SiH
R3Si-Rh-H
R
X
541
R
Rh-H
X
SiR3
B
R
X
Rh-H
SiR3
A

Figure 2-22. Proposed mechanism for the silylcarbocyclization of allenynes.

Typical procedure for the Rh-catalyzed SiCaC reaction[224c]

EtO2C EtO2C [Rh] HSiMe2Ph EtO2C EtO2C SiMe2Ph and/or EtO2C EtO2C SiMe2Ph CHO

536a **537a** **538a**

A reaction vessel equipped with a stirring bar and a CO inlet was charged with $Rh_4(CO)_{12}$ (3.8 mg, 0.005 mol, 0.5 mol%). After purging the vessel with CO, hexane (1.0 mL) was added to dissolve the catalyst. Me_2PhSiH (68 mg, 0.5 mmol) was added *via* syringe. After stirring for 5 min at ambient temperature, the reaction mixture was then cannulated into a 5-mL, round-bottomed flask containing a solution of **538a** (238 mg, 1 mmol), Me_2PhSiH (138 mg, 1 mmol) in hexane (1.5 mL) under CO atmosphere without stirring. The resulting mixture was stirred for less than 1 min, and the reaction mixture was submitted to GC analysis. After the reaction was complete, all volatiles were removed under reduced pressure, and the crude product was purified by column chromatography on silica gel to give SiCaC product **537a** in 99% yield.

General procedure for the Rh-catalyzed CO-SiCaC reaction[224c]

A typical procedure is described for the reaction of **536a**. A reaction vessel equipped with a stirring bar and a CO inlet was charged with $Rh_4(CO)_{12}$ (3.8 mg, 0.005 mmol, 0.5 mol%). After purging the vessel with CO, 1,4-dioxane (2 mL) was added to dissolve the catalyst. After stirring for 5 min at ambient temperature, $P(OEt)_3$ (17 mg, 0.10 mmol; 5 equiv to Rh) in 1,4-dioxane (2 mL) was added *via* syringe, and the mixture was stirred for an additional 10 min. The color of the solution turned from bright red to dark red during this period. The resulting catalyst solution was then cannulated into a 100-mL, round-bottomed flask containing a solution of **536a** (238 mg, 1 mmol), Me_2PhSiH (145 mg, 1.05 mmol) in 1,4-dioxane (50 mL) under CO atmosphere without stirring. The resulting mixture was placed into a 300-mL stainless steel autoclave, pressurized with CO gas (20 atm), and then heated to 105 °C with stirring for 48 h. After releasing CO, the reaction mixture was submitted to GC analysis. All volatiles were removed under reduced pressure and the crude product was purified by column chromatography on silica gel using hexanes/EtOAc (15/1) as eluent to give CO-SiCaC product **538a** in 91% yield.

Typical procedure for the Rh-catalyzed SiCaC of an allenyne[229]

TsN HSi(OEt)3 Rh(acac)(CO)2 CO (1 atm) toluene, 60 °C TsN H SiR'3

541a **542a**

(Acetylacetonato)dicarbonylrhodium(I) (2.2 mg, 8.46×10^{-3} mmol, 2 mol%) and triethoxysilane (209 mg, 1.27 mmol) in dry toluene (3.6 mL) were stirred under an atmosphere of carbon monoxide at room temperature. After the mixture became homogeneous, a solution of *N*-(but-2-ynyl)-*N*-(4-methylpenta-2,3-dienyl)tosylamine (**541a**)

(128 mg, 0.423 mmol) in dry toluene (5.1 mL) was added to the mixture and stirred for 30 min at 60 °C. The solvent was removed under reduced pressure to give the crude product, which was further purified by preparative thin layer chromatography to give the pure product **542a** (166.0 mg, 84%).

In 1992, Tamao *et al.* reported the first example of the Ni-catalyzed silylcarbo-cyclization of diynes.[317] Thereafter, Ojima *et al.* extensively developed rhodium-catalyzed silylcarbocyclizations of diynes,[224b] enynes,[224a, 224c] triynes,[225] and endiynes.[227, 318]

The reaction of 4,4-*bis*-(carbethoxy)hepta-1,6-diyne (**543a**) with *t*-$BuMe_2SiH$ at 50 °C and 15 atm of CO proceeded through a novel carbonylative silylcarbobicyclization (SiCaB) process to afford 2-silylbicyclo[3,3,0]octen-3-one **544a** in 98% yield (Scheme 2-83).[224b, 226] Compound **544a** can be quantitatively isomerized to the more stable 2-silylbicyclo[3,3,0]oct-1-en-3-one **545a** using a catalytic amount of $RhCl_3{\cdot}3H_2O$.[224b, 226]

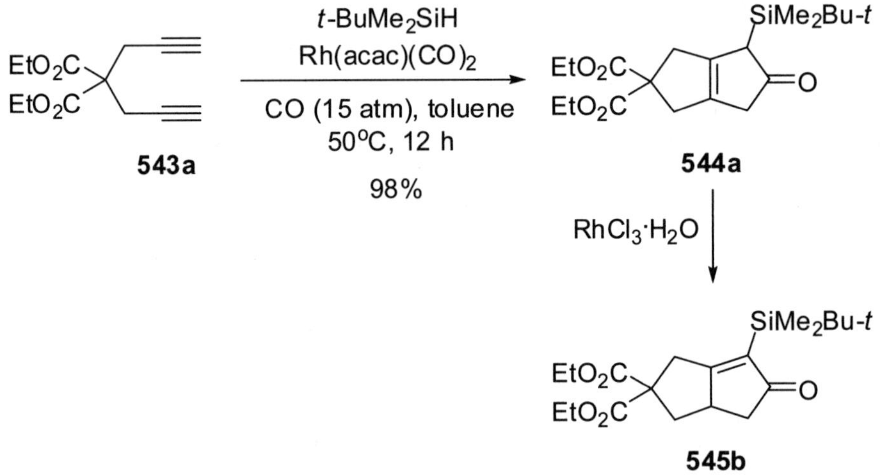

Scheme 2-83. Carbonylative silylcarbobicyclization (SiCaB) of 1,6-diyne **543a**.

A typical procedure for the SiCaC reaction of ***543a***[224b]

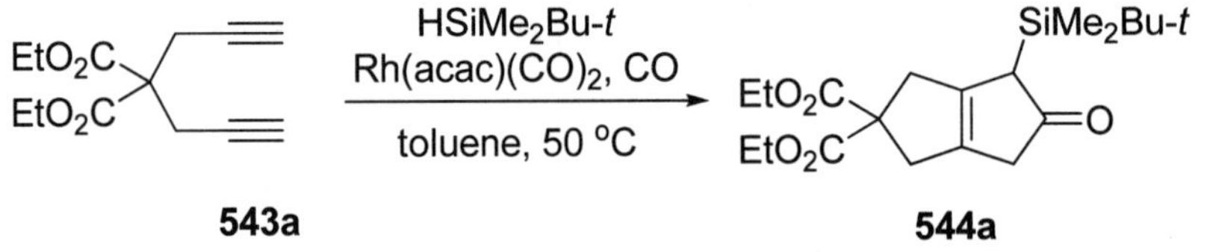

To a 25-mL, Pyrex, round-bottomed flask containing $Rh(acac)(CO)_2$ (1 mol%) and a magnetic stirring bar in toluene (5 mL), was added a solution of $HSiMe_2Bu^t$ (4 mmol) and 4,4-*bis*-(carbethoxy)hepta-1,6-diyne (**543a**) (2 mmol) in toluene (5 mL) *via* a syringe under carbon monoxide atmosphere. The reaction vessel was placed in an autoclave and charged with 10 atm of CO. Carbon monoxide was released slowly, and this process was repeated twice more. The CO pressure was adjusted to 15 atm. The reaction mixture was stirred magnetically at 50 °C and 15 atm of CO for 12 h. The autoclave was cooled in an ice bath, CO was carefully released, and the reaction mixture was submitted to GC and TLC analysis. After evaporation of the solvent under reduced pressure, the crude product was immediately submitted to flash chromatography on silica gel (hexanes/EtOAc) to give **544a** in 98% yield.

Another type of SiCaB reaction takes place when 1,6-diynes **543** bearing a basic amine moiety at the 4-position is employed. For example, the reaction of *N*-benzyl-, *N*-*n*-hexyl- or *N*-allyldipropargylamine with Et_3SiH (1.6 equiv) catalyzed by $Rh(acac)(CO)_2$ at 65 °C and 50 atm of CO gave the corresponding 2-silyl-7-azabicyclo[3.3.0]octa-5,8-dien-3-one **545** as the exclusive product (Scheme 2-84).

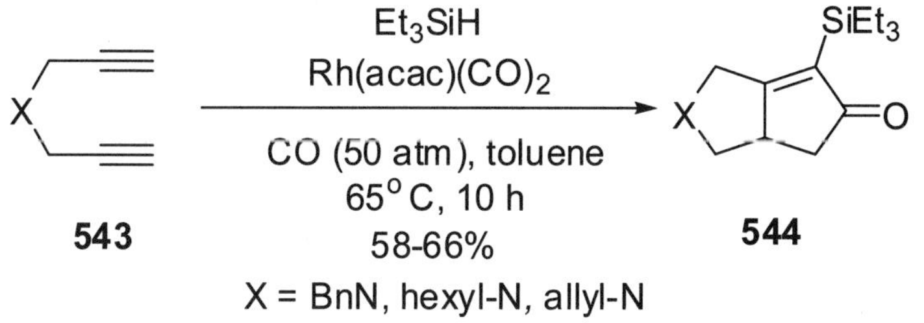

Et_3SiH, $Rh(acac)(CO)_2$; CO (50 atm), toluene, 65 °C, 10 h; 58-66%; X = BnN, hexyl-N, allyl-N

543 → **544** ($SiEt_3$, X, O)

Scheme 2-84. Carbonylative silylcarbobicyclization (SiCaB) of 1,6-diyne **543** bearing a basic amine moiety.

Cascade silylcarbocyclization reactions have been developed based on the fact that it is possible to realize successive intramolecular carbocyclizations as long as the competing reductive elimination is slower than the carbometalation. For example, the reaction of dodec-6-ene-1,11-diyne (**545**) with $PhMe_2SiH$ catalyzed by $Rh(acac)(CO)_2$ at 50 °C and atmospheric pressure of CO gave *bis*-(*exo*-methylenecyclopentyl) **546** in 55% isolated yield (Figure 2-23).[318] The reaction is stereospecific, *i.e.*, (6*E*)- and (6*Z*)-dodec-6-ene-1,11-diynes, (*E*)-**545** and (*Z*)-**545**, afford (*R**,*R**)-**546** and (*S**,*R**)-**546**, respectively. A possible mechanism for this reaction is illustrated in Figure 2-23. It should be noted that no tricyclic product was formed even though the third carbocyclization in the intermediate **C** is conceptually possible.

Replacement of the carbon–carbon double bond in **545** with a triple bond connecting the two alkyne moieties allows the third cyclization to occur in the cascade SiCaC process. Thus, the silylcarbotricyclization (SiCaT) of various triynes (**547**) was effectively catalyzed by Rh complexes such as $Rh(acac)(CO)_2$, $[Rh(cod)Cl]_2$, $[Rh(nbd)Cl]_2$, $Rh_4(CO)_{12}$, and $Rh_2Co_2(CO)_{12}$ to give the corresponding fused tricyclic benzene derivatives **548** and **549** in 57–99% yields (Scheme 2-85).[225]

The reactions catalyzed by rhodium carbonyl clusters, $Rh_4(CO)_{12}$ and $Rh_2Co_2(CO)_{12}$, proceeded smoothly at room temperature with high selectivity for the formation of **548** when X is a carbon tether. Heteroatom-tethered triynes are highly selective for the formation of **549**. The SiCaT reaction was also applied to 1,7,12- and 1,7,13-triynes, which gave 6-6-5 and 6-6-6 fused tricyclic benzene derivatives, respectively, in high yields.[225]

A proposed mechanism for the SiCaT reaction, using the 1,6,11-triyne system as an example, is illustrated in Figure 2-24. The reaction proceeds through insertion of one of the terminal alkynes into the Si-[Rh] complex, generating a β-silylethenyl-[Rh] intermediate which undergoes addition to the second and third alkyne moieties to form intermediate **A**. Subsequent carbocyclization forms **B**, which is followed by β-hydride elimination to give the tricyclic silylbenzene derivative **548**. Alternatively, β-silylethenyl -[Rh] intermediate **A** isomerizes to the thermodynamically more favorable intermediate **C** *via* a zwitterionic carbene species, *i.e.*, the "Ojima-Crabtree mechanism."[319] Subsequent carbocyclization gives **D**, which undergoes β-silyl elimination to afford nonsilylated product **549**.

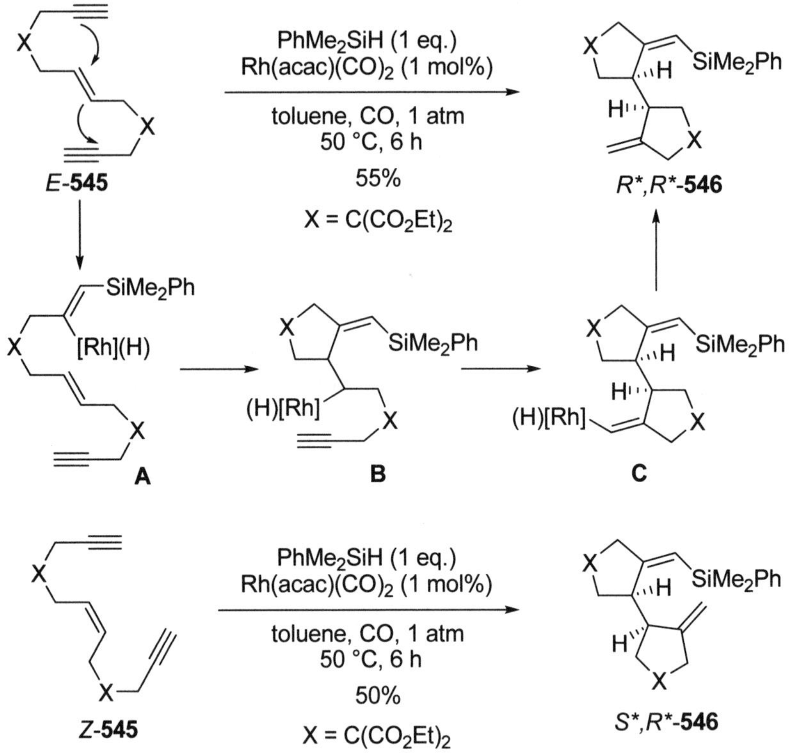

Figure 2-23. Cascade SiCaC reactions of endiynes.

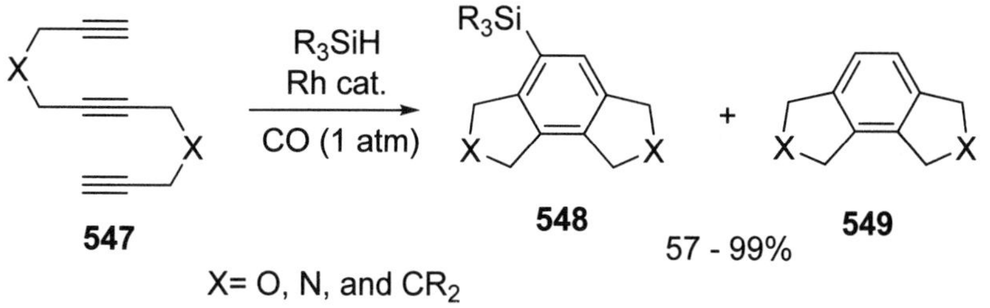

R_3SiH
Rh cat.
CO (1 atm)

547 548 549

X= O, N, and CR_2

57 - 99%

Scheme 2-85. SiCaT of triynes.

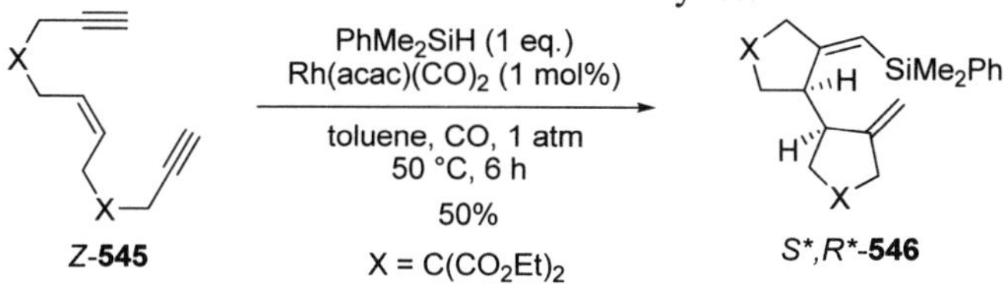

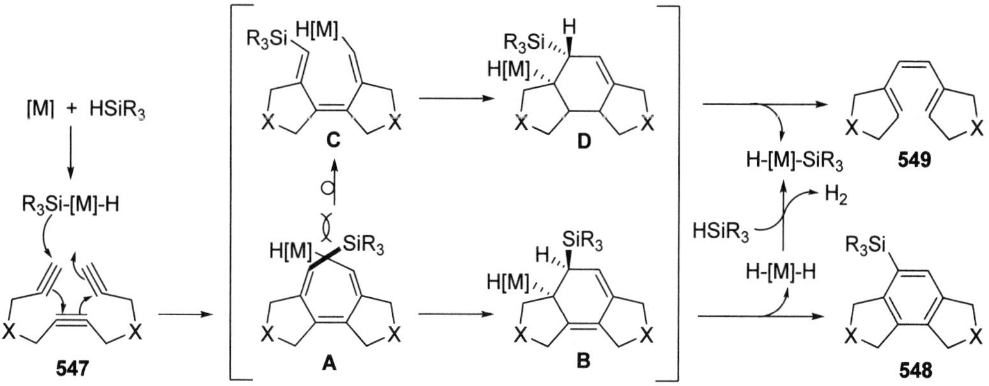

$PhMe_2SiH$ (1 eq.)
$Rh(acac)(CO)_2$ (1 mol%)
toluene, CO, 1 atm
50 °C, 6 h
50%
X = $C(CO_2Et)_2$

Z-545 S*,R*-546

$[M] + HSiR_3$
R_3Si-[M]-H
H-[M]-SiR_3
$HSiR_3$
H_2
H-[M]-H
547 A B C D 548 549

Figure 2-24. A proposed mechanism for the SiCaT of tryines.

General procedure for the Rh-catalyzed SiCaT reactions of 1,6,11-tryines[225]

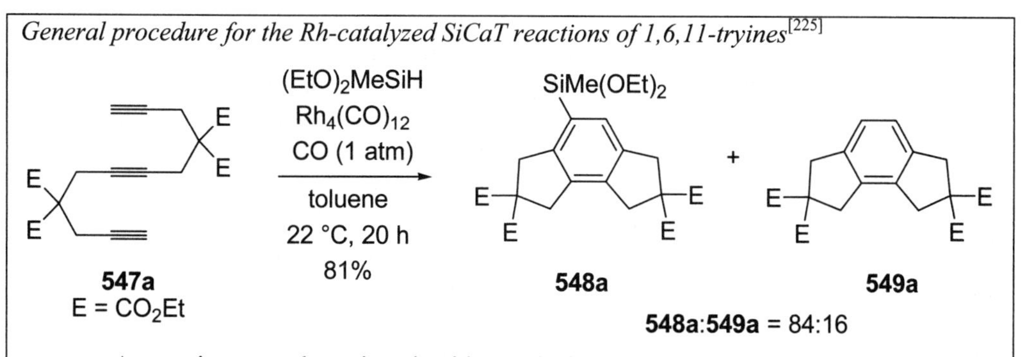

A reaction vessel equipped with a stir bar and a CO inlet was charged with $Rh_4(CO)_{12}$ (3.5 mg, 0.0135 mmol, 1 mol%). After purging the vessel with CO, toluene (6 mL) was added to dissolve the catalyst. $(EtO)_2MeSiH$ (415 μL, 2.70 mmol, 2 equiv) was then added and the mixture was stirred at 22°C. After 1 min, substrate **547a** (602 mg, 1.35 mmol) in toluene (10 mL) was added, and the reaction mixture was stirred under CO (1 atm) for 22 h.

All volatiles were then removed under reduced pressure to give a mixture of **548a** and **549a** (**548a**:**549a** = 84:16), in 97% total yield (^{1}H NMR analysis). The crude products were purified by column chromatography on silica gel using hexanes/EtOAc (13/1) as eluent. Compounds **548a** and **549a** were isolated in 81% yield (isolated ratio = 84:16).

16.7 Tandem 1,4-Addition-Aldol Cyclization

In the early 2000s, Krische and co-workers reported a series of reductive aldol cyclization using molecular hydrogen as the reductant (Scheme 2-86).[206b, 228] This method represents a novel approach to reductive aldol cyclization, which differs from traditional literature methods that use hydrosilane and diethylzinc as terminal reductants.[320]

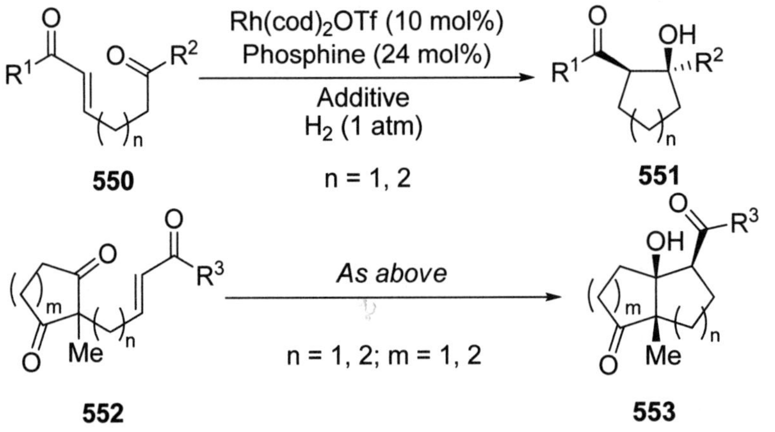

Rh(cod)$_2$OTf (10 mol%)
Phosphine (24 mol%)
Additive
H_2 (1 atm)
n = 1, 2

550 → **551**

As above
n = 1, 2; m = 1, 2

552 → **553**

Scheme 2-86. Reductive aldol cyclizations using molecular hydrogen as reductant.

To probe the feasibility of C-C bond formation under hydrogenation conditions, enone **550a** was subjected to hydrogenation using Rh(cod)$_2$OTf as a pre-catalyst under one atmosphere of hydrogen.[206b] When triphenylphosphine was used as ligand, equal amounts of aldol cyclization product **551a** and conjugate reduction product **554a** were obtained. It was hypothesized that deprotonation of the $Rh^{III}L_n(H)_2$ species to generate Rh-monohydride [$Rh^{III}L_nH$] would disable the conjugate reduction pathway. Indeed, the presence of a weak base (KOAc) increased the amount of the aldol product to 59%. By using P(C_6H_4-CF_3-*p*)$_3$ as the ligand to enhance the Lewis acidity of the Rh metal along with KOAc, the aldol product **551a** was obtained in 89% with only 0.1% of the reduced product **554a** (Scheme 2-87). Under optimized conditions, various aromatic and heteroaromatic enone-aldehyde substrates **550b** underwent reductive aldol cyclization to form five- and six-membered ring products **551b** in 70–90% yields and 5:1–24:1 diastereomeric ratios (Scheme 2-88).

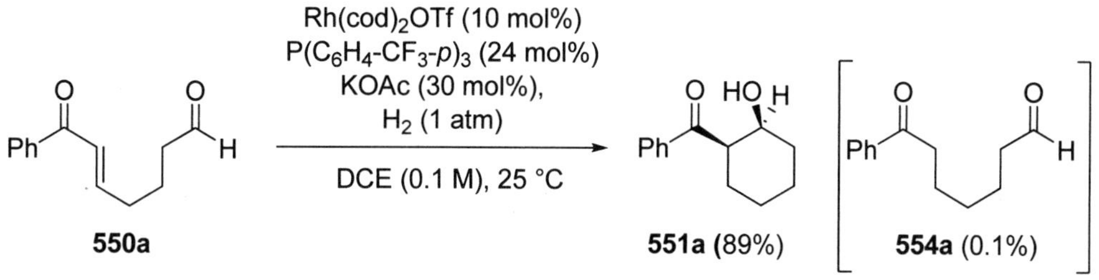

$Rh(cod)_2OTf$ (10 mol%)
$P(C_6H_4\text{-}CF_3\text{-}p)_3$ (24 mol%)
KOAc (30 mol%),
H_2 (1 atm)
DCE (0.1 M), 25 °C

550a → **551a** (89%) [**554a** (0.1%)]

Scheme 2-87. Rh-catalyzed reductive aldol cyclization under hydrogenation conditions.

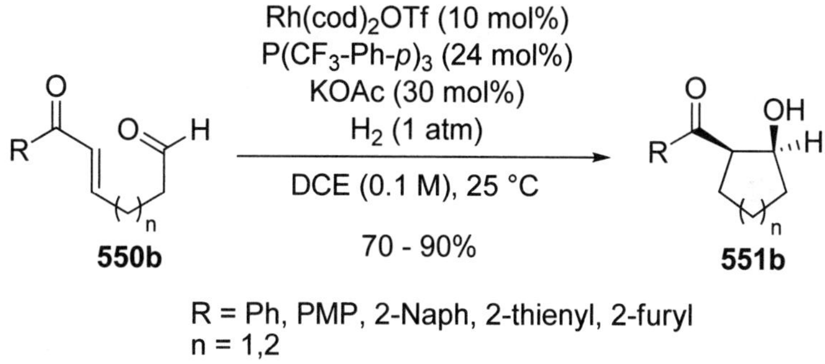

$Rh(cod)_2OTf$ (10 mol%)
$P(CF_3\text{-}Ph\text{-}p)_3$ (24 mol%)
KOAc (30 mol%)
H_2 (1 atm)
DCE (0.1 M), 25 °C
70 - 90%

550b → **551b**

R = Ph, PMP, 2-Naph, 2-thienyl, 2-furyl
n = 1,2

Scheme 2-88. Rh-catalyzed reductive aldol cyclizations of enone-aldehyde.

The reductive aldol cyclization of enone-ketone substrate **550c**, wherein the aromatic enone moiety is tethered to a ketone moiety, afforded five- and six-membered ring products **551c** in good yields and excellent diastereoselectivities (dr > 95:5) (Scheme 2-89).[228a] In the reaction, however, considerable amounts of the reduced products **554c** were formed.[228a]

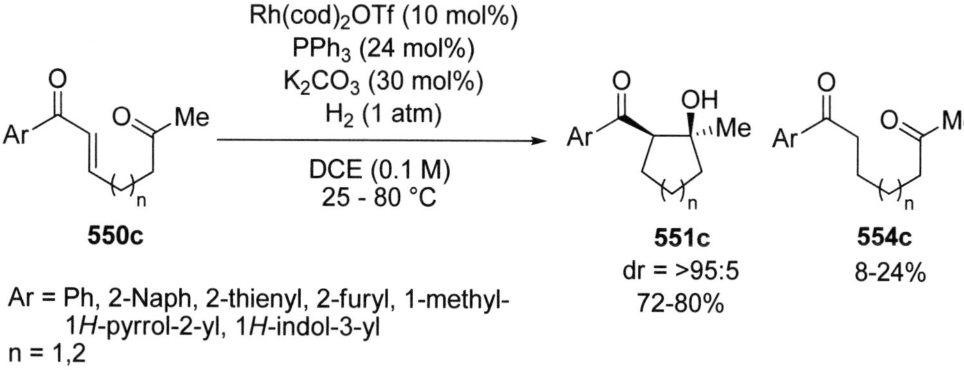

$Rh(cod)_2OTf$ (10 mol%)
PPh_3 (24 mol%)
K_2CO_3 (30 mol%)
H_2 (1 atm)
DCE (0.1 M)
25 - 80 °C

550c → **551c** (dr = >95:5, 72-80%) + **554c** (8-24%)

Ar = Ph, 2-Naph, 2-thienyl, 2-furyl, 1-methyl-1*H*-pyrrol-2-yl, 1*H*-indol-3-yl
n = 1,2

Scheme 2-89. Rh-catalyzed reductive aldol cyclizations of keto-enones.

While the reductive aldol cyclization products of keto-enones were accompanied by significant amounts of 1,4-reduction products, use of a 1,3-dione as the coupling partner proved to be efficient in attenuating the conjugate reduction pathway. The dione moiety is a more reactive electrophile than are simple ketone moieties because of its inductive effects and relief of dipole–dipole interactions of the two carbonyls. The reaction of dione-containing substrates **552** under the standard reaction conditions led to the formation of bicyclic aldol products **553** with excellent diastereoselectivity (>95:5) without formation of the 1,4-reduction product in most cases (Scheme 2-90).[228a] Reactions of enal-ketones **552** were also studied.[228b] Under optimal reaction conditions, five- and six-membered ring products **553** were formed accompanied by a moderate quantity of the conjugate reduction products. Representative results are summarized in Table 2-12.

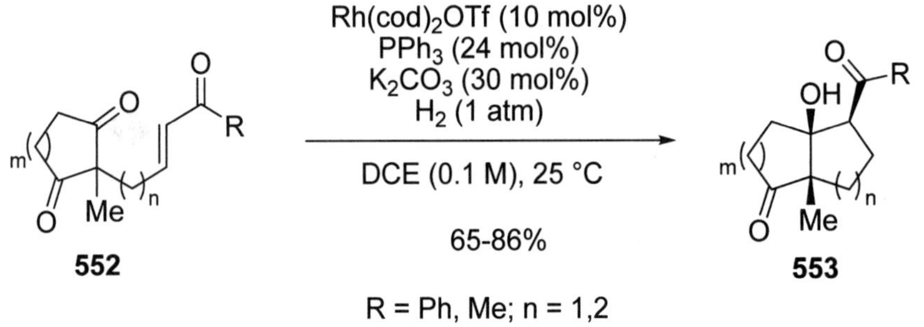

$Rh(cod)_2OTf$ (10 mol%)
PPh_3 (24 mol%)
K_2CO_3 (30 mol%)
H_2 (1 atm)
DCE (0.1 M), 25 °C
65-86%

552 → **553**

R = Ph, Me; n = 1,2

Scheme 2-90. Rh-catalyzed reduction aldol cyclization of enone-dione.

Table 2-12. Rh-catalyzed reductive aldol cyclization of enal-ketones.

Entry	Enal-ketone (552)	Aldol product (553) (syn:anti)	Reduction
a	m = 1, n = 1	72% (2:1)	16%
b	m = 2, n = 1	73% (10:1)	21%
c	m = 1, n = 2	63% (5:1)	30%
d	m = 2, n = 2	59% (4:1)	29%
e	n = 1	63% (1:3)	21%
f	n = 2	61% (2:1)	20%

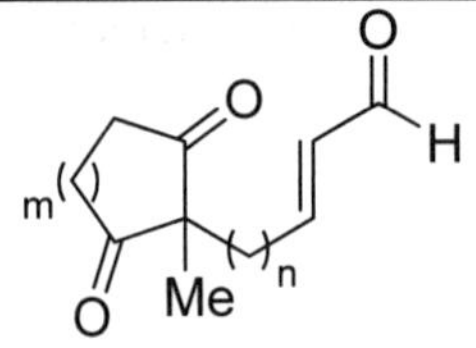

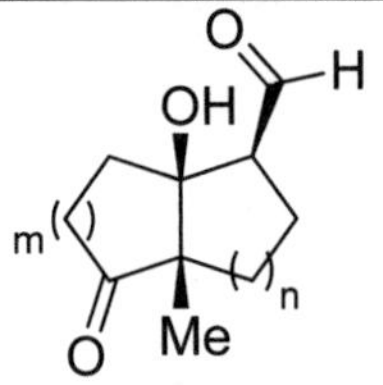

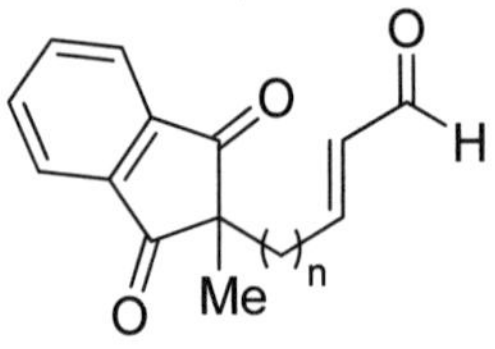

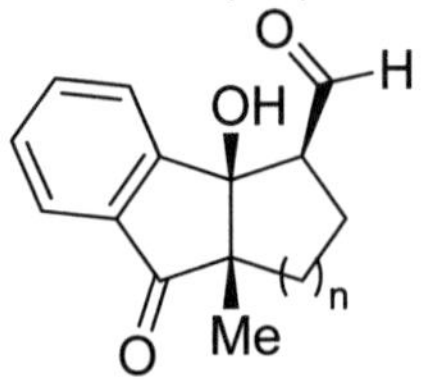

g	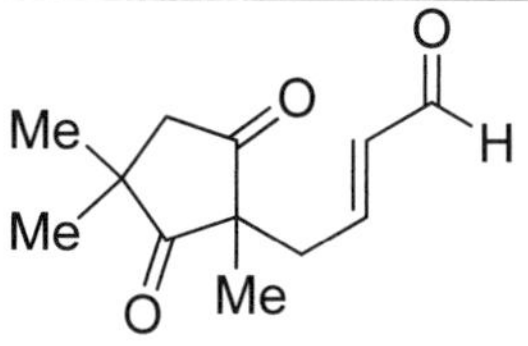	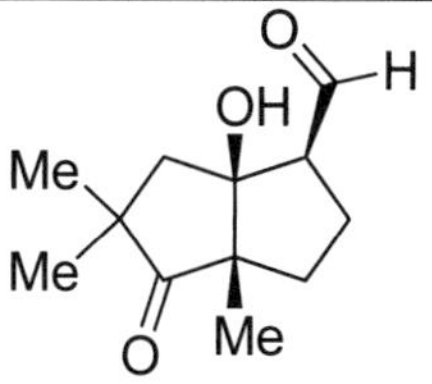 63% (2:1)	21%
h	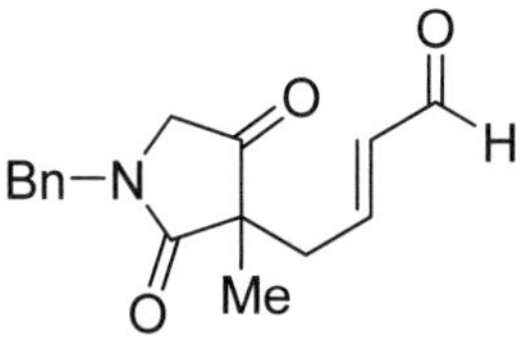	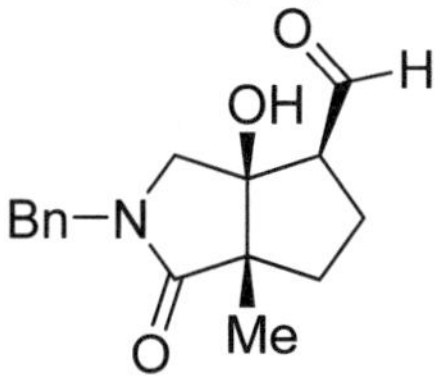 67% (2:1)	
i	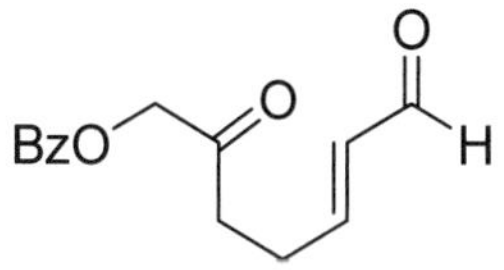	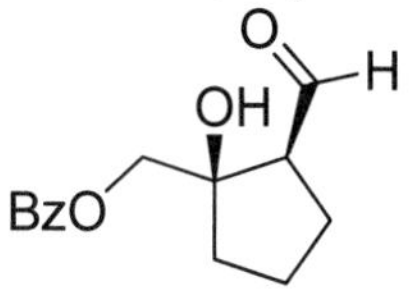 61% (5:1)	20%

$Rh(cod)_2OTf$ (10 mol%), (2-furyl)$_3$P (24 mol%), K_2CO_3 (1 equiv) H_2 (1 atm), THF, 40 °C

Representative procedure for the Rh-catalyzed aldol cycloreduction of a keto-enal[228b]

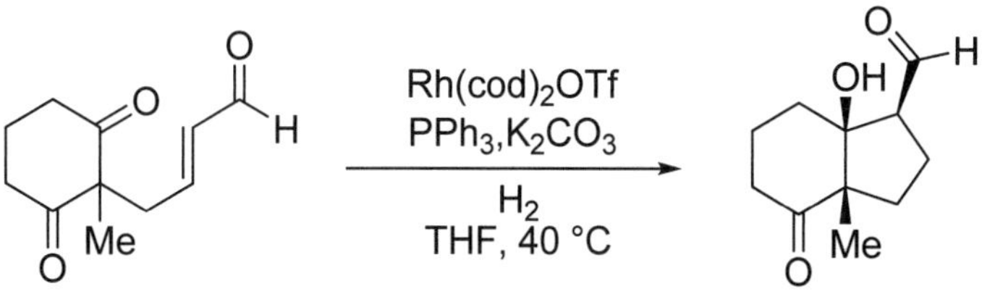

$Rh(cod)_2OTf$, PPh_3, K_2CO_3, H_2, THF, 40 °C

552b → **553b**

To a dried, 25-mL, round-bottomed flask charged with Rh(cod)OTf (24 mg, 0.0515 mmol, 10 mol%) and triphenylphosphine (32 mg, 0.123 mmol, 24 mol%) was added 10 mL of dry THF. The mixture was stirred for 10 min under an Ar atmosphere, at which point **552b** (100 mg, 0.515 mmol, 100 mol%) and potassium carbonate (72 mg, 0.515 mmol, 100 mol%) were added. The system was purged with hydrogen gas and the reaction mixture was allowed to stir at 40 °C under 1 atm of hydrogen until complete consumption of the substrate, at which point the solvent was evaporated and the product purified *via* column chromatography on silica gel to give **553b** in 72% yield (dr 2:1).

A proposed mechanism for the Rh-catalyzed reductive aldol cyclization of enone/enal-ketone/aldehyde **550A** (n = 1) is illustrated in Figure 2-25.[228] Oxidative addition of hydrogen is followed by deprotonation to generate Rh-monohydride intermediate **A**. Enone/enal hydrometallation results in the formation of Rh-enolate **B**, and the subsequent addition to a ketone/aldehyde moiety generates the Rh-aldolate **C**. Oxidative addition of hydrogen to form Rh-aldolate **D**, followed by reductive elimination yields the desired aldol product **551A** and regenerates the active Rh-monohydride complex **A**.

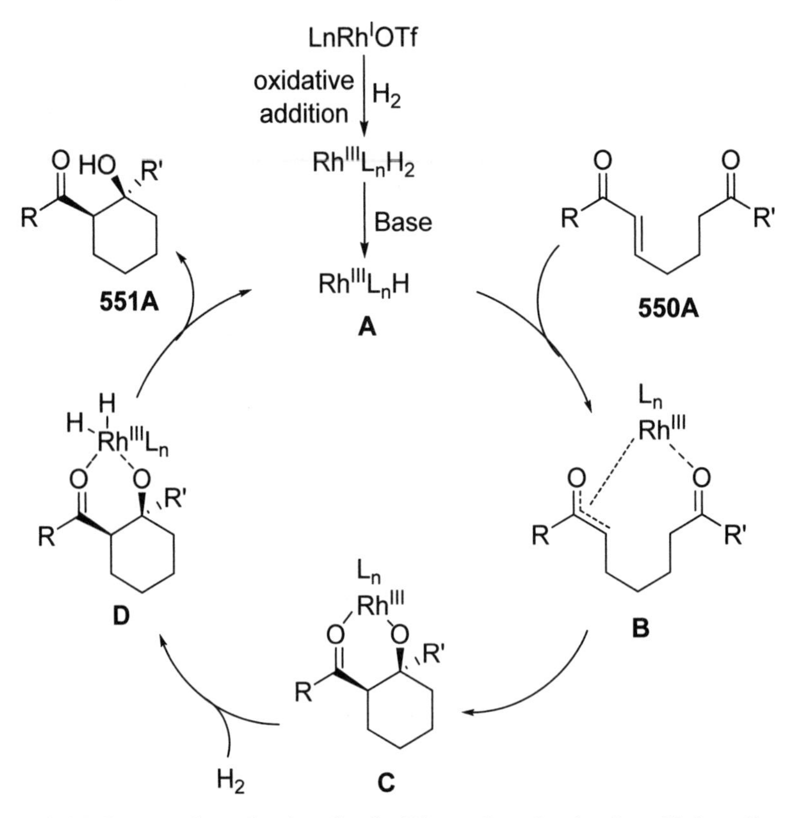

Figure 2-25. Proposed mechanism for the Rh-catalyzed reductive aldol cyclization.

16.8 Rh(I)-Catalyzed Domino Reactions in Carbocyclizations

Domino reactions have attracted significant attention given that this methodology enables multistep synthesis to be completed in one pot. The early work done by Jeong *et al.* used a dual-catalyst system for the preparation of bicyclopentenone **339**. A palladium complex was used to catalyze the allylic substitution of **556** to form the enyne intermediate **338**. Subsequent Rh-catalyzed Pauson-Khand reaction furnished the desired product **339** in high yield (Scheme 2-91).[321]

In the late 1990s, Evans reported a series of allylic substitution reactions using a modified Wilkinson's catalyst. The modified Wilkinson's catalyst was proven to be effective for the allylic substitution reaction but was not an active catalyst for Pauson-Khand reactions.[322] These reports encouraged further search for a single catalyst system that could be used to catalyze both an allylic substitution reaction and a Pauson-Khand reaction in one pot. By vigorous screening of various rhodium catalysts, Evans reported the first regioselective and diastereoselective sequential allylic substitution/Pauson-Khand reaction of allylic carbonates **557** and the Na or Li salt of propargyl malonate, tosylamine, or alcohol **558** using $[RhCl(CO)dppp]_2$ as the optimum catalyst (Scheme 2-92).[323] This process gave a mixture of two diastereomers, *exo*-**560** and *endo*-**560** in 63-84% with 3:1 ~ ≥19:1 dr. When virtually enantiopure **557** (R^2 = 2-Naph, ≥99% ee) was used for the reaction with **558** (R^1 = H, X = TsN, M = Li), **560a** was obtained in 82% yield with 43:1 dr and 98% ee.

Ashfeld *et al.* expanded the scope of this Rh-catalyzed domino reaction to other types of cycloadditions.[324] It was reported that $[Rh(CO)_2Cl]_2$ was an effective catalyst for allylic alkylation with high regioselectivity. When an enantiomerically pure carbonate was employed, the allylic alkylation gave the product with retention of absolute configuration (98% ee). In addition to studying allylic alkylation reactions, a series of novel domino reactions such as cycloisomerizations and [5+2] cycloadditions was explored (Scheme 2-93).[324]

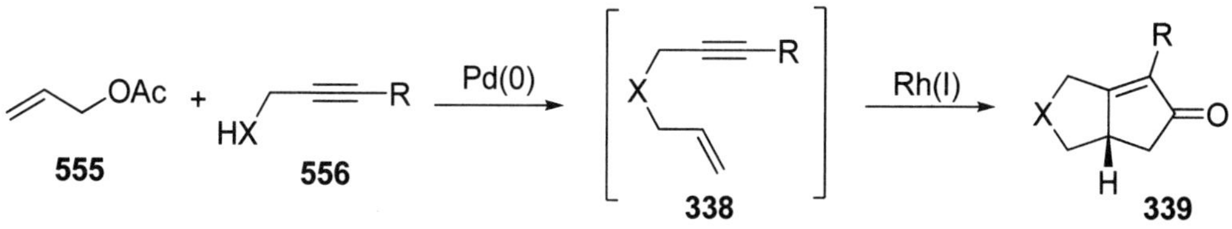

Scheme 2-91. Reductive domino allylic substitution and Pauson-Khand reaction.

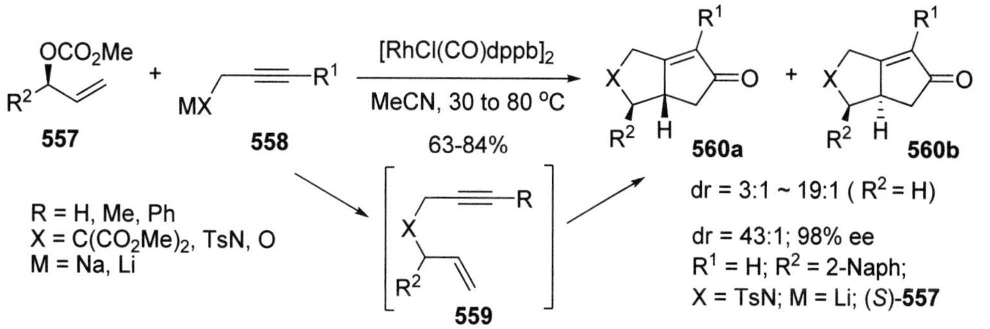

Scheme 2-92. Rh-catalyzed domino allylic substitution and Pauson-Khand reactions.

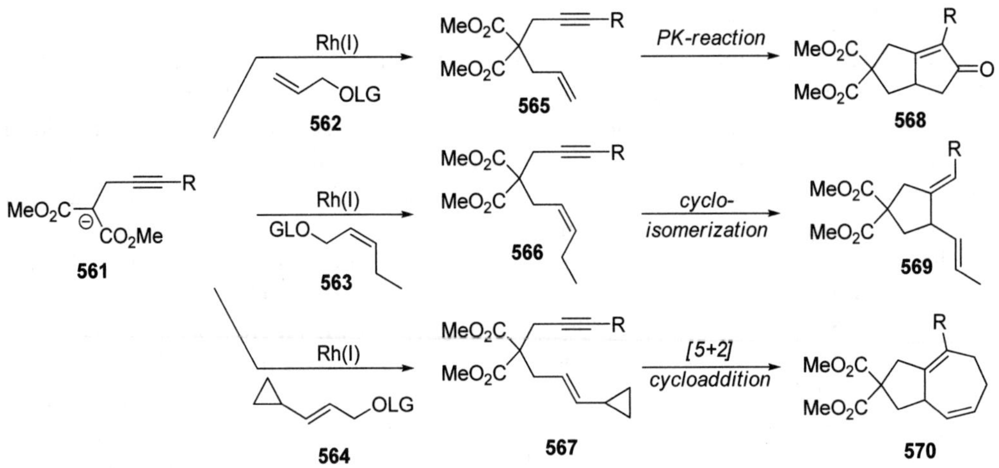

MeO2C CO2Me 561 Rh(I) OLG 562 MeO2C MeO2C R 565 PK-reaction MeO2C MeO2C R O 568
Rh(I) GLO 563 MeO2C MeO2C R 566 cyclo-isomerization MeO2C MeO2C R 569
Rh(I) OLG 564 MeO2C MeO2C R 567 [5+2] cycloaddition MeO2C MeO2C R 570

Scheme 2-93. Rh-catalyzed domino allylic substitution/carbocyclizations.

Typical procedure for the [Rh(CO)$_2$Cl]$_2$-catalyzed tandem allylic alkylation/Pauson-Khand annulation[324b]

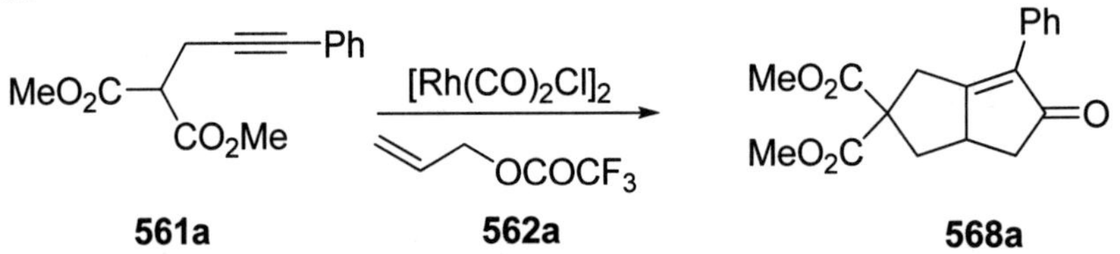

MeO2C CO2Me Ph 561a [Rh(CO)2Cl]2 OCOCF3 562a MeO2C MeO2C Ph O 568a

Malonate **561a** (0.26 mmol) was added to a suspension of NaH (60% dispersion in mineral oil, 0.22 mmol) in dry, degassed THF (1 mL), and the reaction mixture was stirred for 15 min at room temperature. The solvent was removed under reduced pressure and the resulting yellow oil was azeotroped with toluene (2 × 3 mL) under reduced pressure. The residue was dissolved in THF (1 mL) and transferred *via* syringe to a solution of allyl trifluoromethyl carbonate (**562a**, 0.2 mmol) and [Rh(CO)$_2$Cl]$_2$ (10 mol%) in THF (1 mL) under a CO atmosphere. The reaction mixture was stirred at room temperature for 1 h, and then the solution was heated under reflux until the starting material was consumed (12–24 h) (as indicated by TLC analysis). The solvent was removed under reduced pressure and the residue was purified by flash chromatography on silica gel eluting with hexanes/EtOAc (3/1) to furnish **568a** in 68% yield.

17. Concluding Remarks

This chapter compiled representative synthetically useful reactions and processes catalyzed by rhodium complexes. An impressive array of reaction types has been covered together with typical procedures, including hydroformylation, silylformylation, cyclo-propanation, cyclohydrocarbonylations, asymmetric hydrogenation, hydroacylation, C-H activations, conjugate additions, hydroboration, reductive couplings, cycloadditions and cyclizations, cycloisomerizations and Alder-ene reactions, and other miscellaneous reactions. The authors sincerely hope that the readers will find this chapter highly informative and useful for their use of these reactions and processes as well as their design of chemical syntheses by exploiting the highly efficient rhodium-complex–catalyzed reactions. The authors foresee continuous exploration of newer and more efficient rhodium-complex–catalyzed reactions and processes in the future through innovative use of this highly versatile metal, which will play a key role in the synthesis of various pharmaceuticals, agrochemicals, and organic materials, as well as in energy-related research.

18. Acknowledgments

The authors acknowledge the long-term support of the National Science Foundation as well as the National Institute of General Medical Sciences for their research on the discovery and development of a variety of rhodium-complex–catalyzed reactions and processes over the years. Generous support from the Mitsubishi Chemical Corporation is also gratefully acknowledged.

19. References

[1] Johnson, S.A.; Basolo, F.; *Inorg. Syn.* **1963**, *7*, 214.

[2] Martinengo, S.; Giordano, G.; Chini, P.; *Inorg. Syn.* **1990**, *28*, 242.

[3] Osborn, J.A.; Jardine, F.H.; Young, J.F.; Wilkinson, G.; *J. Chem. Soc. A* **1966**, 1711.

[4] Ohno, K.; Tsuji, J.; *J. Am. Chem. Soc.* **1968**, *90*, 99.

[5] Evans, D.; Osborn, J.A.; Wilkinson, G.; *Inorg. Syn.* **1971**, *11*, 99.

[6] McCleverty, J.A.; Wilkinson, G.; *Inorg. Syn.* **1966**, *8*, 211.

[7] Giordano, G.; Crabtree, R.H.; *Inorg. Syn.* **1990**, *28*, 88.

[8] Green, M.; Kuc, T.A.; Taylor, S.H.; *J. Chem. Soc. Sect. A* **1971**, 2334.

[9] Roelen, O.; *Ger. Offen* **1938**, *949*, 548.

[10] (a) Matar, S.; Hatch, L.F., *Chemistry of Petrochemical Processes*, Gulf Professional, Woburn, **2001**; (b) Thurman, L.R.; Harris, J.B.; *U. S. Pat* **1992**, US 5,227,544.

[11] (a) Cornilis, B., in *New Syntheses with Carbon Monoxide*; (ed. Falbe, J.), Springer-Verlag: Berlin, 1980, pp. 1 – 225; (b) Tkatchenko, I., in *Comprehensive Organic Chemistry*; (ed. Wilkonson, G.), Pergamon: Oxford, 1982; Vol. 8, pp. 101 – 223; (c) Botteghi, C.; Ganzerla, R.; Lenarda, M.; Moretti, G.; *J. Mol. Catal.* **1987**, *40*, 129.

[12] Ojima, I.; Tsai, C.Y.; Tzamarioudaki, M.; Bonafoux, D., in *Organic Reactions*; (ed. Overman, L.E.), Wiley: 2000; Vol. 56, p. 1 – 354.

[13] Osborn, J.A.; Wilkinson, G.; Young, J. F.; *Chem. Comm.* **1965**, 17.

[14] Evans, D.; Osborn, J.A.; Wilkinson, G.; *J. Chem. Soc. A* **1968**, 3133.

[15] (a)Melean, L.G.; Rodriguez, M.; Romero, M.; Alvarado, M.L.; Rosales, M.; Baricelli, P.J.; *Appl. Catal., A: General* **2011**, *394*, 117; (b) Watkins, A.L.; Landis, C.R.; *Org. Lett.* **2011**, *13*, 164; (c) Jin, H.; Subramaniam, B.; *Chem. Eng. Sci.* **2004**, *59*, 4887; (d) Sellin, M.F.; Webb, P.B.; Cole-Hamilton, D.J.; *Chem. Commun.* **2001**, *2001*, 781.

[16] (a) Yuan, M.; Chen, H.; Li, R.; Li, Y.; Li, X.; *Appl. Catal., A* **2003**, *251*, 181; (b) Peng, Q.; Yang, Y.; Wang, C.; Liao, X.; Yuan, Y.; *Catal. Lett.* **2003**, *88*, 219; (c) Paganelli, S.; Zanchet, M.; Marchetti, M.; Mangano, G.; *J. Mol. Catal. A: Chem.* **2000**, *157*, 1; (d) Desset, S.L.; Reader, S.W.; Cole-Hamilton, D.J.; *Green Chem.* **2009**, *11*, 630; (e) Legrand, F.-X.; Six, N.; Slomianny, C.; Bricout, H.; Tilloy, S.; Monflier, E.; *Adv. Synth. Catal.* **2011**, *353*, 1325; (f) Desset, S.L.; Hintermair, U.; Gong, Z.X.; Santini, C.C.; Cole-Hamilton, D.J.; *Top. Catal.* **2010**, *53*, 963.

[17] (a) Pedros, M.G.; Masdeu-Bulto, A.M.; Bayardonb, J.; Sinoub, D.; *Catal. Lett.* **2006**, *107*, 205; (b) Rathke, J.W.; Klingler, R.J.; Krause, T.R.; *Organometallics* **1991**, *10*, 1350; (c) Koeken, A.C.J.; van den Broeke, L.J.P.; Deelman, B.J.; Keurentjes, J.T.F.; *J. Mol. Catal. A: Chem.* **2011**, *346*, 1; (d) Kunene, T.E.; Webb, P.B.; Cole-Hamilton,

D.J.; *Green Chem.* **2011**, *13*, 1476; (e) Ren, W.; Rutz, B.; Scurto, A.M.; *J. of Supercritical Fluids* **2009**, *51*, 142.

[18] (a) Aghmiza, A.; Claver, C.; Masdeu-Bultó, A.M.; Maillard, D.; Sinou, D.; *J. Mol. Catal. A: Chem.* **2004**, *208*, 97; (b) Horváth, I.T.; Rábai, J.; *Science* **1994**, *266*, 72.

[19] (a) Cornils, B.; Kuntz, E.G.; *J. Organomet. Chem.* **1995**, *502*, 177; (b) Kohlpaintner, C.W.; Fischer, R.W.; Cornils, B.; *Appl. Catal., A* **2001**, *221*, 219.

[20] (a) Breslow, D.S.; Heck, R.F.; *Chem. Ind. (London)* **1960**, 467; (b) Heck, R.F.; Breslow, D.S.; *J. Am. Chem. Soc.* **1961**, *83*, 4023.

[21] Collman, J.P.; Hegedus, L.S.; Norton, J.R.; Finke, R.G., *Principles and Applications of Organotransition Metal Chemistry*, University Science Books, Mill Valley, CA, **1987**.

[22] (a) Selent, D.; Franke, R.; Kubis, C.; Spannenberg, A.; Baumann, W.; Kreidler, B.; Borner, A.; *Organometallics* **2011**, *30*, 4509; (b) Vogl, C.; Paetzold, E.; Fischer, C.; Kragl, U.; *J. Mol. Catal. A: Chem.* **2005**, *232*, 41; (c) da Silva, A.C.; de Oliveira, K.C. B.; Gusevskaya, E.V.; dos Santos, E.N.; *J. Mol. Catal. A: Chem.* **2002**, *179*, 133.

[23] Pruett, R.L.; Smith, J.A.; *J. Org. Chem.* **1969**, *34*, 327.

[24] Van Rooy, A.; de Bruijin, J.N.H.; Roobeek, K.F.; Kamer, P.C.J.; Van Leeuwen, P.W.N. M.; *J. Organomet. Chem.* **1996**, *507*, 69.

[25] (a) Devon, T.J.; Phillips, G.W.; Puckette, T.A.; Stavinoha, J.L.; Vanderbilt, J.J.; *U. S. Pat* **1987**, US 4; (b) Casey, C.P.; Whiteker, G.T.; Melville, M.G.; Petrovich, L.M.; Gavney, J., J.A.; Powell, D.R.; *J. Am. Chem. Soc.* **1992**, *114*, 5535.

[26] (a) Billig, E.; Abatjoglou, A.G.; Bryant, D.R.; *U. S. Pat* **1987**, US 4,668,651; (b) Cuny, G.D.; Buchwald, S.L.; *J. Am. Chem. Soc.* **1993**, *115*, 2066.

[27] Herrmann, W.A.; Schmid, R.; Kohlpainter, C.W.; Priermeier, T.; *Organometallics* **1995**, *14*, 1961.

[28] Kranenburg, M.; van der Burgt, Y.E.M.; Kamer, P.C.J.; van Leeuwen, P.W.N.M.; *Organometallics* **1995**, *14*, 3081.

[29] van der Veen, L.A.; Boele, M.D. K.; Bregman, F.R.; Kamer, P.C.J.; van Leeuwen, P.W. N.M.; Goubitz, K.; Fraanje, J.; Schenk, H.; Bo, C.; *J. Am. Chem. Soc.* **1998**, *120*, 11616.

[30] (a) Trzeciak, A.M.; Ziółkowski, J.J.; *J. Organomet. Chem.* **1994**, *479*, 213; (b) Smith, S.E.; Rosendahl, T.; Hofmann, P.; *Organometallics* **2011**, *30*, 3643; (c) Behr, A.; Reyer, S.; Tenhumberg, N.; *Dalton Trans.* **2011**, *40*, 11742; (d) Yu, S.C.; Chie, Y.M.; Zhang, X.W.; Dai, L.Y.; Zhang, X.M.; *Tetrahedron Lett.* **2009**, *50*, 5575.

[31] (a) Botteghi, C.; Paganelli, S.; *J. Organomet. Chem.* **1991**, *417*, C41; (b) Bakker, A.G.; Bachasingh, A.K.; *Eur. Pat.* **1991**, BR 9005224.

[32] (a) Jackson, W.R.; Moffat, M.R.; Perlmutter, P.; Tasdelen, E. E.; *Aust. J. Chem.* **1992**, *45*, 823; (b) Jackson, W.R.; Perlmutter, P.; Tasdelen, E.E.; *J. Chem. Soc., Chem. Commun.* **1990**, 763.

[33] Sander, M.; *Chem. Ber.* **1960**, *93*.

[34] (a) Pitchai, R.; Gaffney, A.M.; Nandi, M.K.; Han, Y.Z.; U. S. Pat **1994**, US 5,276,210; (b) Dureanleau, R.G.; Knifton, J.F.; U. S. Pat **1987**, US 4,678,857.

[35] Wuts, P.G.M.; Obrzut, M.L.; Thompson, P.A.; *Tetrahedron Lett.* **1984** *25*, 4051.

[36] (a) Doyle, M.P.; Shanklin, M.S.; Zlokazov, M.V.; *Synlett.* **1994**, 615; (b) Shuklov, I.A.; Dubrovina, N.V.; Jiao, H.; Spannenberg, A.; Börner, A.; *Eur. J. Org. Chem.* **2010**, *2010*, 1669.

[37] (a) Neibecker, D.; Reau, R.; *New J. Chem.* **1991**, *15*, 279; (b) Yu, S.C.; Chie, Y.M.; Zhang, X.M.; *Adv. Synth. Catal.* **2009**, *351*, 537.

[38] (a) Polo, A.; Fernandez, E.; Claver, C.; Castillon, S.; *J. Chem. Soc., Chem. Commun.* **1992**, 639; (b) Fernandez, E.; Ruiz, A.; Claver, C.; Castillo, S.; *Organometallics* **1998**, *17*, 2857.

[39] van Leeuwen, W.N.M.P.; Claver, C., *Rhodium Catalyzed Hydroformylation*, Kulwer Academic Publishers, Dordrecht, The Netherlands, **2000**.

[40] Axtell, A. T.; Cobley, C. J.; Klosin, J.; Whiteker, G. T.; Zanotti-Gerosa, A.; Abboud, K. A.; *Angew. Chem., Int. Ed.* **2005**, *44*, 5834.

[41] Yan, Y.; Zhang, X.; *J. Am. Chem. Soc.* **2006**, *128*, 7198.

[42] (a) Axtell, A.T.; Klosin, J.; Abboud, K.A.; *Organometallics* **2006**, *25*, 5003; (b) Zou, Y.; Yan, Y.; Zhang, X.; *Tetrahedron Lett.* **2007**, *48*, 4781; (c) Dieguez, M.; Pamies, O.; Claver, C.; *Tetrahedron: Asymm.* **2004**, *15*, 2113.

[43] (a) Buisman, G.J.H.; Martin, M.E.; Vos, E.J.; Klootwijk, A.; Kamer, P.C.J.; van Leeuwen, P.W.N.M.; *Tetrahedron: Asymm.* **1995**, *6*, 719; (b) Lambers-Verstappen, M.M.H.; de Vries, J.G.; *Adv. Syn. Cat.* **2003**, *345*, 478; (c) Buisman, G.J.H.; Vos, E.J.; Kamer, P.C.J.; van Leeuwen, P.W.N.M.; *J. Chem. Soc., Dalton Trans.* **1995**, 409; (d) Cobley, C.J.; Gardner, K.; Klosin, J.; Praquin, C.; Hill, C.; Whiteker, G.T.; Zanotti-Gerosa, A.; *J. Org. Chem.* **2004**, *69*, 4031; (e) Cobley, C.J.; Klosin, J.; Qin, C.; Whiteker, G.T.; *Org. Lett.* **2004**, *6*, 3277; (f) Horiuchi, T.; Ohta, T.; Shirakawa, E.; Nozaki, K.; Takaya, H.; *Tetrahedron* **1997**, *53*, 7795; (g) Nozaki, K.; Matsuo, T.; Shibahara, F.; Hiyama, T.; *Organometallics* **2003**, *22*, 594; (h) Nozaki, K.; Sakai, N.; Nanno, T.; Higashijima, T.; Mano, S.; Horiuchi, T.; Hidemasa, T.; *J. Am. Chem. Soc.* **1997**, *119*, 4413; (i) Clark, T.P.; Landis, C.R.; Freed, S.L.; Klosin, J.; Abboud, K.A.; *J. Am. Chem. Soc.* **2005**, *127*, 5040; (j) Breeden, S.; Cole-Hamilton, D.J.; Foster, D.F.; Schwarz, G.J.; Wills, M.; *Angew. Chem., Int. Ed.* **2000**, *39*, 4106.

[44] (a) Nakano, K.; Tanaka, R.; Nozaki, K.; *Helv. Chim. Acta* **2006**, *89*, 1681; (b) Dieguez, M.; Pamies, O.; Claver, C.; *Chem. Comm.* **2005**, 1221.

[45] (a) Botteghi, C.; Paganelli, S.; Schionato, A.; Marchetti, M.; *Chirality* **1991**, *3*, 355; (b) Agbossou, F.; Carpentier, J.; Mortreux, A.; *Chem Rev.* **1995**, *95*, 2485.

[46] Hua, Z.; Vassar, V.C.; Choi, H.; Ojima, I.; *Proc. Natl. Acad. Sci. U.S.A.* **2004**, *101*, 5411.

[47] Wang, X.; Buchwald, S.L.; *J. Am. Chem. Soc.* **2011**, *133*, 19080.

[48] (a) Jaramillo, C.; Knapp, S.; *Synthesis* **1994**, *1994*, 1; (b) Kitsos-Rzychon, B.; Eilbracht, P.; *Tetrahedron* **1998**, *54*, 10721.

[49] Keraenen, M.D.; Eilbracht, P.; *Org. Biomol. Chem.* **2004**, *2*, 1688.

[50] Breit, B.; Zahn, S.K.; *Tetrahedron* **2005**, *61*, 6171.

[51] Clarke, M.L.; Roff, G.J.; *Chem. Eur. J.* **2006**, *12*, 7978.

[52] Teoh, E.; Campi, E.M.; Jackson, W.R.; Robinson, A.J.; *New J. Chem.* **2003**, *27*, 387.

[53] Bates, R.W.; Sivarajan, K.; Straub, B.F.; *J. Org. Chem.* **2011**, *76*, 6844.

[54] Vieira, C.G.; da Silva, J.G.; Penna, C.A.A.; dos Santos, E.N.; Gusevskaya, E.V.; *Appl. Catal., A* **2010**, *380*, 125.

[55] Chercheja, S.; Rothenbucher, T.; Eilbracht, P.; *Adv. Synth. Catal.* **2009**, *351*, 339.

[56] Varchi, G.; Ojima, I.; *Curr. Org. Chem.* **2006**, *10*, 1341.

[57] (a) Doyle, M.P.; Shankiin, M.S.; *Organometallics* **1993**, *12*, 11; (b) Matsuda, I.; Ogiso, A.; Sato, S.; Izumi, Y.; *J. Am. Chem. Soc.* **1990**, *112*, 6120; (c) Matsuda, I.; Sakakibara, J.; Nagashima, H.; *Tetrahedron Lett.* **1991**, *32*, 7431; (d) Ojima, I.; Ingallina, P.; Donovan, R.J.; Clos, N.; *Organometallics* **1991**, *10*, 38; (e) Zhou, J.-Q.; Alper, H.; *Organometallics* **1994**, *13*, 1586.

[58] Biffis, A.; Conte, L.; Tubaro, C.; Basato, M.; Aronica, L.A.; Cuzzola, A.; Caporusso, A. M.; *J. Organomet. Chem.* **2010**, *695*, 792.

[59] Eilbracht, P.; Hollmann, C.; Schmidt, A.M.; Barfacker, L.; *Eur. J. Org. Chem.* **2000**, 1131.

[60] Bärfacker, L.; Hollmann, C.; Eilbracht, P.; *Tetrahedron* **1998**, *54*, 4493.

[61] Ojima, I.; Vidal, E.; Tzamarioudaki, M.; Matsuda, I.; *J. Am. Chem. Soc.* **1995**, *117*, 6797.

[62] Bonafoux, D.; Ojima, I.; *Org. Lett.* **2001**, *3*, 1303.

[63] Tamao, K.; Tohma, T.; Inui, N.; Nakayama, O.; Ito, Y.; *Tetrahedron Lett.* **1990**, *31*, 7733.

[64] Nishii, Y.; Maruyama, N.; Wakasugi, K.; Tanabe, Y.; *Bioorg. Med. Chem* **2001**, *9*, 33.

[65] Csuk, R.; Schabel, M.J.; von Scholz, Y.; *Tetrahedron: Asymm.* **1996**, *7*, 3505.

[66] Kumar, J.S.; Roy, S.; Datta, A.; *Bioorg. Med. Chem. Lett.* **1999**, *9*, 513.

[67] Zhang, X.; Hodgetts, K.; Rachwal, S.; Zhao, H.; Wasley, J.W.F.; Carven, K.; Brodbeck, R.; Kieltyka, A.; Hoffman, D.; Bacolod, M.D.; Girard, B.; Tran, J.; Thurkauf, A.; *J. Med. Chem* **2000**, *43*, 3923.

[68] Csuk, R.; Kern, A.; Mohr, K.Z.; *Z. Naturforsch., B: Chem. Sci.* **1999**, *54*, 1463.

[69] (a) Davies, H.M.L.; Bruzinski, P.; Hutcheson, D.K.; Kong, N.; Fall, M.J.; *J. Am. Chem. Soc.* **1996**, *118*, 6897; (b) Davies, H.M.L.; Bruzinski, P.R.; Fall, M.J.; *Tetrahedron Lett.* **1996**, *37*, 4133; (c) Davies, H.M.L.; Huby, N.J.S.; Cantrell, W.R.,

Jr.; Olive, J.L.; *J. Am. Chem. Soc.* **1993**, *115*, 9468; (d) Davies, H.M.L.; Hutcheson, D.K.; *Tetrahedron Lett.* **1993**, *34*, 7243; (e) Davies, H.M.L.; Peng, Z.-Q.; Houser, J.H.; *Tetrahedron Lett.* **1994**, *35*, 8939; (f) Doyle, M.P.; Austin, R.E.; Bailey, A.S.; Dwyer, M.P.; Dyatkin, A.B.; Kalinin, A.V.; Kwan, M.M.Y.; Liras, S.; Oalmann, C.J.; *J. Am. Chem. Soc.* **1995**, *117*, 5763; (g) Ghanem, A.; Lacrampe, F.; Schurig, V.; *Helv. Chim. Acta* **2005**, *88*, 216; (h) Lui, H.W.; Walsh, C.T.; in *Vol.* (eds. Patai S. and Pappoport Z.), Wiley, Chichester, **1987**, p. 959; (i) Mueller, P.; Ghanem, A.; *Org. Lett.* **2004**, *6*, 4347; (j) Mueller, P.; Lacrampe, F.; *Helv. Chim. Acta* **2004**, *87*, 2848.

[70] (a) Doyle, M.P.; Kalinin, A.V.; *J. Org. Chem.* **1996**, *61*, 2179; (b) Lindsay, V.N.G.; Nicolas, C.; Charette, A.B.; *J. Am. Chem. Soc.* **2011**, *133*, 8972; (c) Verdecchia, M.; Tubaro, C.; Biffis, A.; *Tetrahedron Lett.* **2011**, *52*, 1136; (d) Nishimura, T.; Maeda, Y.; Hayashi, T.; *Angew. Chem. Int. Ed.* **2010**, *49*, 7324.

[71] (a) Briones, J.F.; Davies, H.M.L.; *Tetrahedron* **2011**, *67*, 4313; (b) Goto, T.; Takeda, K.; Shimada, N.; Nambu, H.; Anada, M.; Shiro, M.; Ando, K.; Hashimoto, S.; *Angew. Chem. Int. Ed.* **2011**, 6803; (c) Gregg, T.M.; Algera, R.F.; Frost, J.R.; Hassan, F.; Stewart, R.J.; *Tetrahedron Lett.* **2010**, *51*, 6429.

[72] Tseng, N.-W.; Mancuso, J.; Lautens, M.; *J. Am. Chem. Soc.* **2006**, *128*, 5338.

[73] (a) Brunner, H.; Klushchanzoff; Wutz, K.; *Bull. Chem. Soc. Belg* **1989**, *98*, 63; (b) Kennedy, M.; McKervey, M.A.; Maguire, A.R.; Roos, G.H.P.; *J. Chem. Soc. Chem. Comm.* **1990**, 361.

[74] Doyle, M.P.; Peterson, C.S.; Protopopova, M.N.; Marnett, A.B.; Parker, D.L.; Ene, D.G.; Lynch, V.; *J. Am. Chem. Soc.* **1997**, *119*, 8826.

[75] Doyle, M.P.; Hu, W.H.; *Adv. Synth. Catal.* **2001**, *343*, 299.

[76] (a) Doyle, M.P.; *Chem Rev.* **1986**, *86*, 919; (b) Tanner, D.; *Angew. Chem., Int. Ed.* **1994**, *33*, 599; (c) Osborn, H.M.I.; Sweeny, J.B.; *Tetrahedron: Asymm.* **1997**, *8*, 1693; (d) Krumper, J.R.; Gerisch, M.; Suh, J.M.; Bergman, R.G.; Tilley, T.D.; *J. Org. Chem.* **2003**, *68*, 9705.

[77] Stork, G.; Ficini, J.; *J. Am. Chem. Soc.* **1961**, *83*, 4678.

[78] Doyle, M.P.; Protopopova, M.N.; Poulter, C.D.; Rogers, D.H.; *J. Am. Chem. Soc.* **1995**, *117*, 7281.

[79] Chiou, W.-H.; Lee, S.-Y.; Ojima, I.; *Can. J. Org. Chem.* **2005**, *83*, 681.

[80] (a) Chiou, W.-H.; Mizutani, N.; Ojima, I.; *J. Org. Chem.* **2007**, *72*, 1871; (b) Chiou, W.-H.; Schoenfelder, A.; Sun, L.; Mann, A.; Ojima, I.; *J. Org. Chem.* **2007**, *72*, 9418; (c) van den Hoven, B.G.; Alper, H.; *J. Am. Chem. Soc.* **2001**, *123*, 10214; (d) van den Hoven, B.G.; El Ali, B.; Alper, H.; *J. Org. Chem.* **2000**, *65*, 4131; (e) Park, H.S.; Alberico, E.; Alper, H.; *J. Am. Chem. Soc.* **1999**, *121*, 11697; (f) Quan, L.G.; Lamrani, M.; Yamamoto, Y.; *J. Am. Chem. Soc.* **2000**, *122*, 4827; (g) Mizutani, N.; Chiou, W.-H.; Ojima, I.; *Org. Lett.* **2002**, *4*, 4575.

[81] Ojima, I.; Vidal, E.S.; *J. Org. Chem.* **1998**, *63*, 7999.

[82] (a) Bleakman, D.; Lodge, D.; *Neuropharmacology* **1998**, *37*, 1187; (b) Johansen, T.N.; Greenwood, J.R.; Frydenvang, K.; Maden, U.; Krogsgaard-Larsen, P.; *Chirality* **2003**, *15*, 167.

[83] Chiou, W.-H.; Lin, G.-H.; Hsu, C.-C.; Chaterpaul, S.J.; Ojima, I.; *Org. Lett.* **2009**, *11*, 2659.

[84] Arena, G.; Zill, N.; Salvadori, J.; Girard, N.; Mann, A.; Taddei, M.; *Org. Lett.* **2011**, *13*, 2294.

[85] Miyashita, A.; Yasuda, A.; Takaya, H.; Toriumi, K.; Ito, T.; Souchi, T.; Noyori, R.; *J. Am. Chem. Soc.* **1980**, *102*, 7932.

[86] Wu, S.; He, M.; Zhang, X.; *Tetrahedron: Asymm.* **2004**, *15*, 2177.

[87] Burk, M.J.; Feaster, J.E.; Nugent, W.A.; Harlow, R.L.; *J. Am. Chem. Soc.* **1993**, *115*, 10125.

[88] Togni, A.; Breutel, C.; Schnyder, A.; Spindler, F.; Landert, H.; Tijani, A.; *J. Am. Chem. Soc.* **1994**, *116*, 4062.

[89] Zhou, Y.-G.; Zhang, X.; *Chem. Comm.* **2002**, 1124.

[90] van den Berg, M.; Minnaard, A.J.; Schudde, E.P.; van Esch, J.; de Vries, A.H.M.; de Vries, J.G.; Feringa, B.L.; *J. Am. Chem. Soc.* **2000**, *122*, 11539.

[91] Lou, R.; Mi, A.; Jiang, Y.; Qin, Y.; Li, Z.; Fu, F.; Chan, A.S.C.; *Tetrahedron* **2000**, *56*, 5857.

[92] (a) Pilkington, C. J.; Zanotti-Gerosa, A.; *Org. Lett.* **2003**, *5*, 1273; (b) Liu, D.; Zhang, X.; *Eur. J. Org. Chem.* **2005**, 646; (c) Yonehara, K.; Hashizume, T.; Mori, K.; Ohe, K.; Uemura, S.; *J. Org. Chem.* **1999**, *64*, 5593; (d) Fu, Y.; Xie, J.-H.; Hu, A.-G.; Zhou, H.; Wang, L.X.; Zhou, Q.L.; *Chem. Comm.* **2002**, 480; (e) Vineyard, B.D.; Knowles, W.S.; Sabacky, M.J.; Bachman, G.L.; Weinkauff, D.J.; *J. Am. Chem. Soc.* **1977**, *99*, 5946; (f) Kang, J.; Lee, J.H.; Ahn, S.H.; Choi, J.S.; *Tetrahedron Lett.* **1998**, *39*, 5523.

[93] Boaz, N.W.; Debenham, S.D.; Large, S.E.; Moore, M.K.; *Tetrahedron: Asymm.* **2003**, 3575.

[94] Adamczyk, M.; Akireddy, S.R.; Reddy, R.E.; *Tetrahedron* **2002**, *58*, 6951.

[95] Heller, D.; Holz, J.; Drexler, H.J.; Lang, J.; Drauz, K.; Krimmer, H.P.; Borner, A.; *J. Org. Chem.* **2001**, *66*, 6816.

[96] Tang, W.; Zhang, X.; *Org. Lett.* **2002**, *4*, 4159.

[97] (a) Hu, X.P.; Zheng, Z.; *Org. Lett.* **2005**, *7*, 419; (b) Birch, M.; Challenger, S.; Crochard, J.P.; Fradet, D.; Jackman, H.; Luan, A.; Madigan, E.; McDowall, N.; Meldrum, K.; Gordon, C.M.; Widegren, M.; Yeo, S.; *Org. Process Res. Dev.* **2011**, *15*, 1172.

[98] Wu, H.P.; Hoge, G.; *Org. Lett.* **2004**, *6*, 3645.

[99] Burk, M.J.; Gross, M.F.; Martinez, J.P.; *J. Am. Chem. Soc.* **1995**, *117*, 9375.

[100] Fu, Y.; Hou, G.H.; Xie, J.H.; Xing, L.; Wang, L.X.; Zhou, Q.L.; *J. Org. Chem.* **2004**, *69*, 8157.

[101] Tang, W.; Zhang, X.; *Angew. Chem., Int. Ed.* **2002**, *41*, 1612.

[102] Robinson, A.J.; Stanislawski, P.; Mulholland, D.; He, L.; Li, H.Y.; *J. Org. Chem.* **2001**, *66*, 4148.

[103] Zhu, G.; Zhang, X.; *J. Org. Chem.* **1998**, *63*, 9590.

[104] Hu, A.G.; Fu, Y.; Xie, J.-H.; Zhou, H.; Wang, L.-X.; Zhou, Q.-L.; *Angew. Chem., Int. Ed.* **2002**, *41*, 2348.

[105] (a) Lee, S.G.; Zhang, Y.J.; Song, C.E.; Lee, J.K.; Choi, J.H.; *Angew. Chem., Int. Ed.* **2002**, *41*, 847; (b) Huang, H.; Zheng, Z.; Luo, H.; Bai, C.; Hu, X.; Chen, H.; *Org. Lett.* **2003**, *5*, 4137.

[106] Kuwano, R.; Sato, K.; Kurokawa, T.; Karube, D.; Ito, Y.; *J. Am. Chem. Soc.* **2000**, *122*, 7614.

[107] (a) Liu, D.; Li, W.; Zhang, W.; *Tetrahedron: Asymm.* **2004**, *15*, 2181; (b) Li, W.; Zhang, X.; *J. Org. Chem.* **2000**, *65*, 5871.

[108] Magano, J.; Conway, B.; Bowles, D.; Nelson, J.; Nanninga, T.N.; Winkle, D.D.; Wu, H.; Chen, M.H.; *Tetrahedron Lett.* **2009**, *50*, 6329.

[109] (a) Hua, Z.; Vassar, V.C.; Ojima, I.; *Org. Lett.* **2003**, *5*, 3831; (b) Gergely, I.; Hegedus, C.; Gulyas, H.; Szollosy, A.; Monsees, A.; Riermeier, T.; Bakos, J.; *Tetrahedron: Asymm.* **2003**, *14*, 1087.

[110] Ostermeier, M.; Prieß, J.; Helmchen, G.; *Angew. Chem., Int. Ed.* **2002**, *41*, 612.

[111] Berens, U.; Burk, M.J.; Gerlach, A.; Hems, W.; *Angew. Chem., Int. Ed.* **2000**, *39*, 1981.

[112] Burk, M.J.; Kalberg, C.S.; Pizzano, A.; *J. Am. Chem. Soc.* **1998**, *120*, 4345.

[113] Kuwano, R.; Sawamura, M.; Ito, Y.; *Tetrahedron: Asymm.* **1995**, *6*, 2521.

[114] Reetz, M.T.; Neugebauer, T.; *Angew. Chem., Int. Ed.* **1999**, *38*, 179.

[115] Li, W.; Zhang, Z.; Xiao, D.; Zhang, X.; *J. Org. Chem.* **2000**, *65*, 3489.

[116] Boaz, N.W.; Debenham, S.D.; Mackenzie, E.B.; Large, S.E.; *Org. Lett.* **2002**, *4*, 2421.

[117] Hu, X.P.; Zheng, Z.; *Org. Lett.* **2004**, *6*, 3585.

[118] Zhao, Y.W.; Huang, H.M.; Shao, J.P.; Xia, C.G.; *Tetrahedron: Asymm.* **2011**, *22*.

[119] Jiang, Q.; Xiao, D.; Zhang, Z.; Cao, P.; Zhang, X.; *Angew. Chem., Int. Ed.* **1999**, *38*, 516.

[120] Tang, W.; Liu, D.; Zhang, X.; *Org. Lett.* **2003**, *5*, 205.

[121] (a) Grasa, G.A.; Zanotti-Gerosa, A.; Ghosh, S.; Teleha, C.A.; Kinney, W.A.; Maryanoff, B.E.; *Tetrahedron Lett.* **2008**, *49*, 5328; (b) Yoshimura, M.; Ishibashi, Y.; Miyata, K.; Bessho, Y.; Tsukamoto, M.; Kitamura, M.; *Tetrahedron* **2007**, *63*,

11399; (c) Qiu, L. Q.; Li, Y.-M.; Kwong, F.Y.; Yu, W.-Y.; Fan, Q.-H.; Chan, A.S.C.; *Adv. Synth. Catal.* **2007**, *349*, 517; (d) Tellers, D.M.; McWilliams, J.C.; Humphrey, G.; Journet, M.; DiMichele, L.; Hinksmon, J.; McKeown, A.E.; Rosner, T.; Sun, Y.K.; Tillyer, R.D.; *J. Am. Chem. Soc.* **2006**, *128*, 17063; (e) Yamamoto, N.; Murata, M.; Morimoto, T.; Achiwa, K.; *Chem. Pharm. Bull.* **1991**, *39*, 1085.

[122] Hoge, G.; Wu, H.P.; Kissel, W.S.; Pflum, D.A.; Greene, D.J.; Bao, J.; *J. Am. Chem. Soc.* **2004**, *126*, 5966.

[123] Robert, T.; Abiri, Z.; Sandee, A.J.; Schmalz, H.G.; Reek, J.N.H.; *Tetrahedron: Asymm.* **2010**, *21*, 2671.

[124] Burk, M.J.; Bienewald, F.; Challenger, S.; Derrick, A.; Ramsden, J.A.; *J. Org. Chem.* **1999**, *64*, 3290.

[125] Jiang, Q.; Jiang, Y.; Xiao, D.; Cao, P.; Zhang, X.; *Angew. Chem., Int. Ed.* **1998**, *37*, 1100.

[126] (a) Takahashi, H.; Sakuraba, S.; Takeda, H.; Achiwa, K.; *J. Am. Chem. Soc.* **1990**, *112*, 5876; (b) Liu, D.; Gao, W.; Wang, C.; Zhang, X., *Angew. Chem. Int. Ed.* **2005**, *44*, 1687.

[127] Burk, M.J.; Feaster, J.E.; *J. Am. Chem. Soc.* **1992**, *114*, 6266.

[128] Sakai, K.; Ide, J.; Oda, O.; Nakamura, N.; *Tetrahedron Lett.* **1972**, *13*, 1287.

[129] Barnhart, R.W.; Wang, X.; Noheda, P.; Bergens, S.H.; Whelan, J.; Bosnich, B.; *J. Am. Chem. Soc.* **1994**, *116*, 1821.

[130] Wu, X.-M.; Funakoshi, K.; Sakai, K.; *Tetrahedron Lett.* **1992**, *33*, 6331.

[131] Barnhart, R.W.; McMorran, D.A.; Bosnich, B.; *Chem. Comm.* **1997**, 589.

[132] Hoffman, T.J.; Carreira, E.M.; *Angew. Chem. Int. Ed.* **2011**, *50*, 10670.

[133] Enders, D.; Eichenauer, H.; *Tetrahedron Lett.* **1977**, *18*, 191.

[134] Wu, X.M.; Funakoshi, K.; Sakai, K.; *Tetrahedron Lett.* **1993**, *34*, 5927.

[135] Taura, Y.; Tanaka, M.; Wu, X.-M.; Funakoshi, K.; Sakai, K.; *Tetrahedron* **1991**, *47*, 4879.

[136] Tanaka, K.; Fu, G.C.; *J. Am. Chem. Soc.* **2002**, *124*, 10296.

[137] Tanaka, K.; Fu, G.C.; *J. Am. Chem. Soc.* **2003**, *125*, 8078.

[138] Davies, H.M.L.; Beckwith, R.E.J.; *Chem. Rev.* **2003**, *103*, 2861.

[139] Davies, H.M.L.; Antoulinakis, E.G.; *J.Organomet. Chem.* **2001**, *617*, 47.

[140] Timmons, D.J.; Doyle, M.P.; *J.Organomet. Chem.* **2001**, *617*, 98.

[141] (a) McCarthy, N.; McKervey, M.A.; Ye, T.; McCann, M.; Murphy, E.; Doyle, M.P.; *Tetrahedron Lett.* **1992**, *33*, 5983; (b) Pirrung, M.C.; Zhang, J.; *Tetrahedron Lett.* **1992**, *33*, 5987.

[142] Doyle, M.P.; Forbes, D.C.; *Chem. Rev.* **1998**, *98*, 911.

[143] (a) Doyle, M.P.; Protopopova, M.N.; Zhou, Q.-L.; Bode, J.W.; Simonsen, S.H.; Lynch, V.; *J. Org. Chem.* **1995**, *60*, 6654; (b) Bode, J.W.; Doyle, M.P.; Protopopova, M.N.; Zhou, Q.L.; *J. Org. Chem.* **1996**, *61*, 9146.

[144] Davies, H.M.L.; Antoulinakis, E.G.; *Org. Lett.* **2000**, *2*, 4153.

[145] Davies, H.M.L.; Hansen, T.; Hopper, D.W.; Panaro, S.A.; *J. Am. Chem. Soc.* **1999**, *121*, 6509.

[146] Davies, H.M.L.; Stafford, D.G.; Hansen, T.; Churchill, M.R.; Keil, K.M.; *Tetrahedron Lett.* **2000**, *41*, 2035.

[147] Davies, H.M.L.; Yang, J.; Nikolai, J.; *J. Organomet. Chem.* **2005**, *690*, 6111.

[148] (a) Davies, H.M.L.; Hansen, T.; Churchill, M.R.; *J. Am. Chem. Soc.* **2000**, *122*, 3063; (b) Nadeau, E.; Li, Z.J.; Morton, D.; Davies, H.M.L.; *Synlett* **2009**, *2009*, 151.

[149] Davies, H.M.L.; Jin, Q.; *Tetrahedron: Asymm.* **2003**, *14*, 941.

[150] Espino, C.G.; Du Bois, J.; *Angew. Chem., Int. Ed.* **2001**, *40*, 598.

[151] Wehn, P.M.; Lee, J.; DuBois, J.; *Org. Lett.* **2003**, *5*, 4823.

[152] Chan, J.; Baucom, K.D.; Murry, J.A.; *J. Am. Chem. Soc.* **2007**, *129*, 14106.

[153] Davies, H.M.L.; *Angew. Chem., Int. Ed.* **2006**, *45*, 6422.

[154] Liang, C.; Robert-Peillard, F.; Fruit, C.; Müller, P.; Dodd, R.H.; Dauban, P.; *Angew. Chem., Int. Ed.* **2006**, *45*, 4641.

[155] Padwa, A.; Price, A.T.; *J. Org. Chem.* **1998**, *63*, 556.

[156] Kataoka, O.; Kitagaki, S.; Watanabe, N.; Kobayashi, J.-i.; Nakamura, S.-i.; Shiro, M.; Hashimoto, S.-i.; *Tetrahedron Lett.* **1998**, *39*, 2371.

[157] Dauben, W.G.; Dinges, J.; Smith, T.C.; *J. Org. Chem.* **1993**, *58*, 7635.

[158] Srikrishna, A.; Sheth, V.M.; Nagaraju, G.; *Synlett* **2011**, *2011*, 2343.

[159] Lewis, J.C.; Wu, J.Y.; Bergman, R.G.; Ellman, J.A.; *Angew. Chem., Int. Ed.* **2006**, *45*, 1589.

[160] Chatani, N.; Asaumi, T.; Ikeda, T.; Yorimitsu, S.; Ishii, Y.; Kakiuchi, F.; Murai, S.; *J. Am. Chem. Soc.* **2000**, *122*, 12882.

[161] Shimada, S.; Batsanov, A.S.; Howard, J.A.K.; Marder, T.B.; *Angew. Chem., Int. Ed.* **2001**, *40*, 2168.

[162] Lim, S.G.; Ahn, J.A.; Jun, C.H.; *Org. Lett.* **2004**, *6*, 4687.

[163] Thalji, R.K.; Ellman, J.A.; Bergman, R.G.; *J. Am. Chem. Soc.* **2004**, *126*, 7192.

[164] Danishefsky, S.; Schkeryantz, J.; *Synlett* **1995**, 475.

[165] Wilson, R.M.; Thalji, R.K.; Bergman, R.G.; Ellman, J.A.; *Org. Lett.* **2006**, *8*, 1745.

[166] Du, Y.; Hyster, T.K.; Rovis, T.; *Chem. Commun.* **2011**, *47*, 12074.

[167] Sakai, M.; Hayashi, H.; Miyaura, N.; *Organometallics* **1997**, *16*, 4229.

[168] (a) Fagnou, K.; Lautens, M.; *Chem. Rev.* **2003**, *103*, 169; (b) Hayashi, T.; Yamasaki, K.; *Chem. Rev.* **2003**, *103*, 2829.

[169] Tang, Y.Q.; Lv, H.; He, X.N.; Lu, J.M.; Shao, L.X.; *Catal. Lett.* **2011**, *141*, 705.

[170] Takaya, Y.; Ogasawara, M.; Hayashi, T.; *Tetrahedron Lett.* **1998**, *39*, 8479.

[171] (a) Kuriyama, M.; Nagai, K.; Yamada, K.I.; Miwa, Y.; Taga, T.; Tomioka, K.; *J. Am. Chem. Soc.* **2002**, *124*, 8932; (b) Reetz, M.T.; Moulin, D.; Gosberg, A.; *Org. Lett.* **2001**, *3*, 4083.

[172] Hayashi, T.; Takahashi, M.; Takaya, Y.; Ogasawara, M.; *J. Am. Chem. Soc.* **2002**, *124*, 5052.

[173] Takaya, Y.; Ogasawara, M.; Hayashi, T.; Sakai, M.; Miyaura, N.; *J. Am. Chem. Soc.* **1998**, *120*, 5579.

[174] Takaya, Y.; Senda, T.; Kurushima, H.; Ogasawara, M.; Hayashi, T.; *Tetrahedron: Asymm.* **1999**, *10*, 4047.

[175] Pucheault, M.; Darses, S.; Genêt, J.-P.; *Eur. J. Org. Chem.* **2002**, 3552.

[176] (a) Senda, T.; Ogasawara, M.; Hayashi, T.; *J. Org. Chem.* **2001**, *66*, 6852; (b) Sakuma, S.; Miyaura, N.; *J. Org. Chem.* **2001**, *66*, 8944.

[177] Hayashi, T.; Senda, T.; Takaya, Y.; Ogasawara, M.; *J. Am. Chem. Soc.* **1999**, *121*, 11591.

[178] Zilaout, H.; van den Hoogenband, A.; de Vries, J.; Lange, J.H.M.; Terpstra, J.W.; *Tetrahedron Lett.* **2011**, *52*, 5934.

[179] Hayashi, T.; Senda, T.; Ogasawara, M.; *J. Am. Chem. Soc.* **2000**, *122*, 10716.

[180] Cauble, D.F.; Gipson, J.D.; Krische, M.J.; *J. Am. Chem. Soc.* **2003**, *125*, 1110.

[181] (a) Oi, S.; Taira, A.; Honma, Y.; Inoue, Y.; *Org. Lett.* **2003**, *5*, 97; (b) Mori, A.; Danda, Y.; Fujii, T.; Hirabayashi, K.; Osakada, K.; *J. Am. Chem. Soc.* **2001**, *123*, 10774.

[182] (a) Oi, S.; Moro, M.; Ito, H.; Honma, Y.; Miyano, S.; Inoue, Y.; *Tetrahedron* **2002**, *58*, 91; (b) Oi, S.; Moro, M.; Fukuhara, H.; Kawanishi, T.; Inoue, Y.; *Tetrahedron Lett.* **1999**, *40*, 9259.

[183] Hayashi, T.; Tokunaga, N.; Yoshida, K.; Han, J.W.; *J. Am. Chem. Soc.* **2002**, *124*, 12102.

[184] Yoshida, K.; Hayashi, T.; *J. Am. Chem. Soc.* **2003**, *125*, 2872.

[185] (a) Hayashi, T.; Ueyama, K.; Tokunaga, N.; Yoshida, K.; *J. Am. Chem. Soc.* **2003**, *125*, 11508; (b) Chen, F.-X.; Kina, A.; Hayashi, T.; *Org. Lett.* **2006**, 8, 341.

[186] Defieber, C.; Paquin, J.F.; Serna, S.; Carreira, E.M.; *Org. Lett.* **2004**, *6*, 3873.

[187] Shintani, R.; Okamoto, K.; Otomaru, Y.; Ueyama, K.; Hayashi, T.; *J. Am. Chem. Soc.* **2005**, *127*, 54.

[188] Zigterman, J.L.; Woo, J.C.S.; Walker, S.D.; Tedrow, J.S.; Borths, C.J.; Bunel, E.E.; Faul, M.M.; *J. Org. Chem.* **2007**, *72*, 8870.

[189] Lerum, R.V.; Chisholm, J.D.; *Tetrahedron Lett.* **2004**, *45*, 6591.

[190] Nishimura, T.; Katoh, T.; Takatsu, K.; Shintani, R.; Hayashi, T.; *J. Am. Chem. Soc.* **2007**, *129*, 14158.

[191] Detlef Männig, H.N.; *Angew. Chem., Int. Ed.* **1985**, *24*, 878.

[192] Burgess, K.; Ohlmeyer, M.J.; *J. Org. Chem.* **1988**, *53*, 5178.

[193] Hayashi, T.; Matsumoto, Y.; Ito, Y.; *J. Am. Chem. Soc.* **1989**, *111*, 3426.

[194] (a) Schnyder, A.; Hintermann, L.; Togni, A.; *Angew. Chem., Int. Ed.* **1995**, *34*, 931; (b) Brown, J.M.; Hulmes, D.; Layzell, T.P.; *J. Chem. Soc., Chem. Comm.* **1993**, 1673; (c) McCarthy, M.; Guiry, P.J.; Hooper, M.W.; *Chem. Comm.* **2000**, 1333.

[195] Maxwell, A.C.; Flanagan, S.P.; Goddard, R.; Guiry, P.J.; *Tetrahedron: Asymm.* **2010**, *21*, 1458.

[196] Smith, S.M.; Takacs, J.M.; *J. Am. Chem. Soc.* **2010**, *132*, 1740.

[197] Demay, S.; Volant, F.; Knochel, P.; *Angew. Chem., Int. Ed.* **2001**, *40*, 1235.

[198] Evans, D.A.; Fu, G.C.; *J. Org. Chem.* **1990**, *55*, 2280.

[199] Dorigo, A.E.; Schleyer, P.V.; *Angew. Chem., Int. Ed.* **1995**, *34*, 115.

[200] Connolly, D.J.; Lacey, P.M.; McCarthy, M.; Saunders, C.P.; Carroll, A.M.; Goddard, R.; P.J.G.; *J. Org. Chem.* **2004**, *69*, 6572.

[201] Miller, S.P.; Morgan, J.B.; Nepveux, F.J.; Morken, J.P.; *Org. Lett.* **2004**, *6*, 131.

[202] Morgan, J.B.; Miller, S.P.; Morken, J.P.; *J. Am. Chem. Soc.* **2003**, *125*, 8702.

[203] Mohan, R.S.; Whalen, D.L.; *J. Org. Chem.* **1993**, *58*, 2663.

[204] Norrby, P.-O.; Becker, H.; Sharpless, K.B.; *J. Am. Chem. Soc.* **1996**, *118*, 35.

[205] (a) Fernandez, E.; Hooper, M.W.; Knight, F.I.; Brown, J.M.; *Chem. Comm.* **1997**, 173; (b) Fernandez, E.; Maeda, K.; Hooper, M.W.; Brown, J.M.; *Chem. Eur. J.* **2000**, *6*, 1840.

[206] (a) Taylor, S.J.; Morken, J.P.; *J. Am. Chem. Soc.* **1999**, *121*, 12202; (b) Jang, H.Y.; Huddleston, R.R.; Krische, M.J.; *J. Am. Chem. Soc.* **2002**, *124*, 15156; (c) Matsuda, I.; Takahashi, K.; Sato, S.; *Tetrahedron Lett.* **1990**, *31*, 5331.

[207] (a) Muraoka, T.; Matsuda, I.; Itoh, K.; *J. Am. Chem. Soc.* **2000**, *122*, 9552; (b) Muraoka, T.; Kamiya, S.; Matsuda, I.; Itoh, K.; *Chem. Comm.* **2002**, 1284; (c) Muraoka, T.; Matsuda, I.; Itoh, K.; *Organometallics* **2001**, *20*, 4676.

[208] Taylor, S.J.; Duffey, M.O.; Morken, J.P.; *J. Am. Chem. Soc.* **2000**, *122*, 4528.

[209] Shiomi, T.; Adachi, T.; Ito, J.; Nishiyama, H.; *Org. Lett.* **2009**, *11*, 1011.

[210] Jang, H.Y.; Krische, M.J.; *J. Am. Chem. Soc.* **2004**, *126*, 7875.

[211] Jang, H.Y.; Hughes, F.W.; Gong, H.; Zhang, J.; Brodbelt, J.S.; Krische, M.J.; *J. Am. Chem. Soc.* **2005**, *127*, 6174.

[212] Rhee, J.U.; Krische, M.J.; *J. Am. Chem. Soc.* **2006**, *128*, 10674.

[213] Kimura, M.; Nojiri, D.; Fukushima, M.; Oi, S.; Sonoda, Y.; Inoue, Y.; *Org. Lett.* **2009**, *11*, 3794.

[214] Eddine Saiah, M.K.; Pellicciari, R.; *Tetrahedron Lett.* **1995**, *36*, 4497.

[215] Takaya, H.; Kojima, S.; Murahashi, S.I.; *Org. Lett.* **2001**, *3*, 421.

[216] Krug, C.; Hartwig, J.F.; *J. Am. Chem. Soc.* **2002**, *124*, 1674.

[217] Matsuda, T.; Shigeno, M.; Murakami, M.; *J. Am. Chem. Soc.* **2007**, *129*, 12086.

[218] (a) Jeong, N.; Lee, S.; Sung, B.K.; *Organometallics* **1998**, *17*, 3642; (b) Koga, Y.; Kobayashi, T.; Narasaka, K.; *Chem. Lett.* **1998**, *27*, 249; (c) Brummond, K.M.; Kent, J.L.; *Tetrahedron* **2000**, *56*, 3263.

[219] (a) Murakami, M.; Itami, K.; Ito, Y.; *J. Am. Chem. Soc.* **1997**, *119*, 2950; (b) Murakami, M.; Itami, K.; Ito, Y.; *J. Am. Chem. Soc.* **1999**, *121*, 4130; (c) Murakami, M.; Itami, K.; Ito, Y.; *Organometallics* **1999**, *18*, 1326.

[220] (a) Tanaka, K.; Nishida, G.; Ogino, M.; Hirano, M.; Noguchi, K.; *Org. Lett.* **2005**, *7*, 3119; (b) Kinoshita, H.; Shinokubo, H.; Oshima, K.; *J. Am. Chem. Soc.* **2003**, *125*, 7784; (c) Evans, P.A.; Sawyer, J.R.; Lai, K.W.; Huffman, J.C.; *Chem. Comm.* **2005**, 3971; (d) Shibata, T.; Arai, Y.; Tahara, Y.; *Org. Lett.* **2005**, *7*, 4955; (e) Tanaka, K.; Toyoda, K.; Wada, A.; Shirasaka, K.; Hirano, M.; *Chem. Eur. J.* **2005**, *11*, 1145; (f) Tanaka, K.; Wada, A.; Noguchi, K.; *Org. Lett.* **2006**, *8*, 907; (g) Yu, R.T.; Rovis, T.; *J. Am. Chem. Soc.* **2006**, *128*, 12370; (h) Yu, R.T.; Rovis, T.; *J. Am. Chem. Soc.* **2006**, *128*, 2782; (i) Shibata, T.; Kurokawa, H.; Kanda, K.; *J. Org. Chem.* **2007**, *72*, 6521; (j) Tanaka, K.; Osaka, T.; Noguchi, K.; Hirano, M.; *Org. Lett.* **2007**, *9*, 1307.

[221] (a) Wender, P.A.; Dyckman, A.J.; *Org. Lett.* **1999**, *1*, 2089; (b) Wender, P.A.; Barzilay, C.M.; Dyckman, A.J.; *J. Am. Chem. Soc.* **2001**, *123*, 179; (c) Wegner, H.A.; deMeijere, A.; Wender, P.A.; *J. Am. Chem. Soc.* **2005**, *127*, 6530; (d) Wender, P.A.; Haustedt, L.O.; Lim, J.; Love, J.A.; Williams, T.J.; Yoon, J.Y.; *J. Am. Chem. Soc.* **2006**, *128*, 6302.

[222] (a) Wender, P.A.; Jenkins, T.E.; Suzuki, S.; *J. Am. Chem. Soc.* **1995**, *117*, 1843; (b) Fallis, A.G.; *Acc. Chem. Res.* **1999**, *32*, 464; (c) Tanaka, K.; Fu, G.C.; *Org. Lett.* **2002**, *4*, 933.

[223] (a) Bennacer, B.; Fujiwara, M.; Ojima, I.; *Org. Lett.* **2004**, *6*, 3589; (b) Bennacer, B.; Fujiwara, M.; Lee, S.Y.; Ojima, I.; *J. Am. Chem. Soc.* **2005**, *127*, 17756.

[224] (a) Ojima, I.; Donovan, R.J.; Shay, W.R.; *J. Am. Chem. Soc.* **1992**, *114*, 6580; (b) Ojima, I.; Zhu, J.; Vidal, E.S.; Kass, D.F.; *J. Am. Chem. Soc.* **1998**, *120*, 6690; (c) Ojima, I.; Vu, A.T.; Lee, S.Y.; McCullagh, J.V.; Moralee, A.C.; Fujiwara, M.; Hoang, T.H.; *J. Am. Chem. Soc.* **2002**, *124*, 9164.

[225] Ojima, I.; Vu, A.T.; McCullagh, J.V.; Kinoshita, A.; *J. Am. Chem. Soc.* **1999**, *121*, 3230.

[226] Ojima, I.; Fracchiolla, D.A.; Donovan, R.J.; Banerji, P.; *J. Org. Chem.* **1994**, *59*, 7594.

[227] Ojima, I.; Lee, S.Y.; *J. Am. Chem. Soc.* **2000**, *122*, 2385.

[228] (a) Huddleston, R.R.; Krische, M.J.; *Org. Lett.* **2003**, *5*, 1143; (b) Koech, P.K.; Krische, M.J.; *Org. Lett.* **2004**, *6*, 691.

[229] Shibata, T.; Kadowaki, S.; Takagi, K.; *Organometallics* **2004**, *23*, 4116.

[230] Jeong, N.; Sung, B.K.; Choi, Y.K.; *J. Am. Chem. Soc.* **2000**, *122*, 6771.

[231] Khand, I.U.; Knox, G.R.; Pauson, P.L.; Watts, W.E.; Foreman, M.I.; *J. Chem. Soc., Perkin Trans. 1* **1973**, 977.

[232] Schore, N.E.; Croudace, M.C.; *J. Org. Chem.* **1981**, *46*, 5436.

[233] (a) Gybin, A.S.; Smit, V.A.; Caple, R.; Veretenov, A.L.; Shashkov, A.S.; Vorontsova, L.G.; Kurella, M.G.; Chertkov, V.S.; Karapetyan, A.A.; *J. Am. Chem. Soc.* **1992**, *114*, 5555; (b) Exon, C.; Magnus, P.; *J. Am. Chem. Soc.* **1983**, *105*, 2477.

[234] Geis, O.; Schmalz, H.G.; *Angew. Chem., Int. Ed.* **1998**, *37*, 911.

[235] (a) Lee, B.Y.; Chung, Y.K.; Jeong, N.; Lee, Y.; Hwang, S.H.; *J. Am. Chem. Soc.* **1994**, *116*, 8793; (b) Jeong, N.; Hwang, S.H.; Lee, Y.W.; Lim, J.S.; *J. Am. Chem. Soc.* **1997**, *119*, 10549; (c) Hicks, F.A.; Kablaoui, N.M.; Buchwald, S.L.; *J. Am. Chem. Soc.* **1996**, *118*, 9450; (d) Morimoto, T.; Chatani, N.; Fukumoto, Y.; Murai, S.; *J. Org. Chem.* **1997**, *62*, 3762; (e) Kondo, T.; Suzuki, N.; Okada, T.; Mitsudo, T.; *J. Am. Chem. Soc.* **1997**, *119*, 6187.

[236] Kobayashi, T.; Koga, Y.; Narasaka, K.; *J. Organomet. Chem.* **2001**, *624*, 73.

[237] (a) Adrio, J.; Carretero, J.C.; *J. Am. Chem. Soc.* **1999**, *121*, 7411; (b) Verdaguer, X.; Moyano, A.; Pericas, M.A.; Riera, A.; Bernardes, V.; Greene, A.E.; Alvarez-Larena, A.; Piniella, J.F.; *J. Am. Chem. Soc.* **1994**, *116*, 2153; (c) Fonquerna, S.; Moyano, A.; Pericas, M.A.; Riera, A.; *J. Am. Chem. Soc.* **1997**, *119*, 10225.

[238] (a) Sturla, S.J.; Buchwald, S.L.; *J. Org. Chem.* **2002**, *67*, 3398; (b) Hiroi, K.; Watanabe, T.; Kawagishi, R.; Abe, I.; *Tetrahedron Lett.* **2000**, *41*, 891.

[239] Hicks, F.A.; Buchwald, S.L.; *J. Am. Chem. Soc.* **1999**, *121*, 7026.

[240] Fan, B.-M.; Xie, J.-H.; Li, S.; Tu, Y.-Q.; Zhou, Q.-L.; *Adv. Syn. Cat.* **2005**, *347*, 759.

[241] (a) Kim, D.E.; Choi, C.; Kim, D.E.; Jeulin, S.; Ratovelomanana-Vidal, V.; Genet, J.-P.; Jeong, N.; *Adv. Syn. Cat.* **2007**, *349*, 1999; (b) Kim, D.E.; Kim, I.S.; Ratovelomanana-Vidal, V.; Genet, J.-P.; Jeong, N.; *J. Org. Chem.* **2008**, *73*, 7985.

[242] (a) Breit, B.; Seiche, W.; *Synthesis* **2001**, 0001; (b) Beck, C.M.; Rathmill, S.E.; Park, Y. J.; Chen, J.; Crabtree, R.H.; Liable-Sands, L.M.; Rheingold, A.L.; *Organometallics* **1999**, *18*, 5311.

[243] (a) Morimoto, T.; Fuji, K.; Tsutsumi, K.; Kakiuchi, K.; *J. Am. Chem. Soc.* **2002**, *124*, 3806; (b) Shibata, T.; Toshida, N.; Takagi, K.; *Org. Lett.* **2002**, *4*, 1619; (c) Shibata, T.; Toshida, N.; Takagi, K.; *J. Org. Chem.* **2002**, *67*, 7446.

[244] Lee, H.W.; Chan, A.S.C.; Kwong, F.Y.; *Chem. Comm.* **2007**, 2633.

[245] Park, J.H.; Cho, Y.; Chung, Y.K.; *Angew. Chem. Int. Ed.* **2010**, *49*, 5138.

[246] Ikeda, K.; Morimoto, T.; Kakiuchi, K.; *J. Org. Chem.* **2010**, *75*, 6279.

[247] (a) Jikyo, T.; Eto, M.; Harano, K.; *Tetrahedron* **1999**, *55*, 6051; (b) Jikyo, T.; Eto, M.; Harano, K.; *J. Chem. Soc. Perkin Trans.* **1998**, 3463.

[248] Kumar, U.; Neenan, T.X.; *Macromolecules* **1995**, *28*, 124.

[249] Muller, E.; *Synthesis* **1974**, 761.

[250] Shibata, T.; Yamashita, K.; Ishida, H.; Takagi, K.; *Org. Lett.* **2001**, *3*, 1217.

[251] Shibata, T.; Yamashita, K.; Katayama, E.; Takagi, K.; *Tetrahedron* **2002**, *58*, 8661.

[252] Tamao, K.; Kobayashi, K.; Ito, Y.; *J. Org. Chem.* **1989**, *54*, 3517.

[253] Wender, P.A.; Deschamps, N.M.; Gamber, G.G.; *Angew. Chem., Int. Ed.* **2003**, *42*, 1853.

[254] Wender, P.A.; Croatt, M.P.; Deschamps, N.M.; *J. Am. Chem. Soc.* **2004**, *126*, 5948.

[255] Takao, K.I.; Munakata, R.; Tadano, K.I.; *Chem Rev.* **2005**, *105*, 4779.

[256] Zimmer, R.; Dinesh, C.U.; Nandanan, E.; Khan, F.A.; *Chem. Rev.* **2000**, *100*, 3067.

[257] Hashmi, A.S.K.; *Angew. Chem.* **2000**, *39*, 3590.

[258] Brummond, K.M.; Chen, H.; Fisher, K.D.; Kerekes, A.D.; Rickards, B.; Sill, P.C.; Geib, S.J.; *Org. Lett.* **2002**, *4*, 1931.

[259] (a) Brummond, K.M.; Mitasev, B.; *Org. Lett.* **2004**, *6*, 2245; (b) Brummond, K.M.; Chen, H.; Mitasev, B.; Casarez, A.D.; *Org. Lett.* **2004**, *6*, 2161; (c) Brummond, K.M.; Painter, T.O.; Probst, D.A.; Mitasev, B.; *Org. Lett.* **2007**, *9*, 347.

[260] Mukai, C.; Nomura, I.; Yamanishi, K.; Hanaoka, M.; *Org. Lett.* **2002**, *4*, 1755.

[261] Makino, T.; Itoh, K.; *J. Org. Chem.* **2004**, *69*, 395.

[262] Inagaki, F.; Itoh, N.; Hayashi, Y.; Matsui, Y.; Mukai, C.; *Beilstein J. Org. Chem.* **2011**, *7*, 404.

[263] Wender, P.A.; Croatt, M.P.; Deschamps, N.M.; *Angew. Chem., Int. Ed.* **2006**, *45*, 2459.

[264] Inagaki, F.; Narita, S.; Hasegawa, T.; Kitagaki, S.; Mukai, C.; *Angew. Chem. Int. Ed.* **2009**, *48*, 2007.

[265] Murakami, M.; Itami, K.; Ito, Y.; *J. Am. Chem. Soc.* **1996**, *118*, 11672.

[266] (a) Ojima, I.; Tzamarioudaki, M.; Li, Z.; Donovan, R.J.; *Chem. Rev.* **1996**, *96*, 635; (b) Saito, S.; Yamamoto, Y.; *Chem Rev.* **2000**, *100*, 2901; (c) Lautens, M.; Klute, W.; Tam, W.; *Chem. Rev.* **1996**, *96*, 49.

[267] Garcia, J.J.; Sierra, C.; Torrens, H.; *Tetrahedron Lett.* **1996**, *37*, 6097.

[268] Bringmann, G.; Breuning, M.; Tasler, S.; *Synthesis* **1999**, 525.

[269] Shibata, T.; Fujimoto, T.; Yokota, K.; Takagi, K.; *J. Am. Chem. Soc.* **2004**, *126*, 8382.

[270] Gutnov, A.; Heller, B.; Fischer, C.; Drexler, H.-J.; Spannenberg, A.; Sundermann, B.; Sundermann, C.; *Angew. Chem., Int. Ed.* **2004**, *43*, 3795.

[271] (a) Pindur, U.; Erfanian-Abdoust, H.; *Chem. Rev.* **1989**, *89*, 1681; (b) Grotjahn, D.B.; Vollhardt, K.P.C.; *J. Am. Chem. Soc.* **1986**, *108*, 2091.

[272] Witulski, B.; Alayrac, C.; *Angew. Chem., Int. Ed.* **2002**, *41*, 3281.

[273] McDonald, F.E.; Zhu, H.Y.H.; Holmquist, C.R.; *J. Am. Chem. Soc.* **1995**, *117*, 6605.

[274] (a) Wang, Y.-H.; Zhang, Z.-K.; Yang, F.-M.; Sun, Q.-Y.; He, H.-P.; Di, Y.-T.; Mu, S.-Z.; Lu, Y.; Chang, Y.; Zheng, Q.-T.; Ding, M.; Dong, J.-H.; Hao, X.-J.; *J. Nat. Prod.* **2007**, *70*, 1458; (b) Seibert, S.F.; Eguereva, E.; Krick, A.; Kehraus, S.; Voloshina, E.; Raabe, G.; Fleischhauer, J.; Leistner, E.; Wiese, M.; Prinz, H.; Alexandrov, K.; Janning, P.; Waldmann, H.; Konig, G.M.; *Org. Biomol. Chem.* **2006**, *4*, 2233; (c) Witulski, B.; Zimmermann, A.; *Synlett* **2002**, 1855.

[275] Ogaki, S.; Shibata, Y.; Noguchi, K.; Tanaka, K.; *J. Org. Chem.* **2011**, *76*, 1926.

[276] (a) Grigg, R.; Scott, R.; Stevenson, P.; *Tetrahedron Lett.* **1982**, *23*, 2691; (b) Grigg, R.; Stevenson, P.; Worakun, T.; *Tetrahedron* **1988**, *44*, 4967; (c) Neeson, S.J.; Stevenson, P.J.; *Tetrahedron* **1989**, *45*, 6239.

[277] Witulski, B.; Zimmermann, A.; Gowans, N.D.; *Chem. Comm.* **2002**, 2984.

[278] Mori, F.; Fukawa, N.; Noguchi, K.; Tanaka, K.; *Org. Lett.* **2011**, *13*, 362.

[279] Evans, P.A.; Lai, K.W.; Sawyer, J.R.; *J. Am. Chem. Soc.* **2005**, *127*, 12466.

[280] Shibata, T.; Tahara, Y.; *J. Am. Chem. Soc.* **2006**, *128*, 11766.

[281] Shibata, T.; Otomo, M.; Endo, K.; *Synlett* **2010**, *2010*, 1235.

[282] Earl, R.A.; Vollhardt, K.P.C.; *J. Org. Chem.* **1984**, *49*, 4786.

[283] Yamamoto, Y.; Takagishi, H.; Itoh, K.; *Org. Lett.* **2001**, *3*, 2117.

[284] (a) Hoberg, H.; Oster, B.; *Synthesis* **1982**, 324; (b) Varela, J.A.; Saa, C.; *Chem. Rev.* **2003**, *103*, 3787.

[285] Tanaka, K.; Wada, A.; Noguchi, K.; *Org. Lett.* **2005**, *7*, 4737.

[286] Friedman, R.K.; Oberg, K.M.; Dalton, D.M.; Rovis, T.; *Pure Appl. Chem.* **2010**, *82*, 1353.

[287] Kim, H.; Lee, C.; *J. Am. Chem. Soc.* **2005**, *127*, 10180.

[288] Trost, B.M.; Dyker, G.; Kulawiec, R.J.; *J. Am. Chem. Soc.* **1990**, *112*, 7809.

[289] (a) Trost, B.M.; Rhee, Y.H.; *J. Am. Chem. Soc.* **2003**, *125*, 7482; (b) Trost, B.M.; McClory, A.; *Angew. Chem., Int. Ed.* **2007**, *46*, 2074.

[290] (a) Trumble, J.T.; Millar, J.G.; *J. Agric. Food Chem.* **1996**, *44*, 2859; (b) Kobertz, W.R.; Essigmann, J.M.; *J. Am. Chem. Soc.* **1997**, *119*, 5960; (c) Kim, K.H.; Nielsen, P.E.; Glazer, P.M.; *Biochemistry* **2006**, *45*, 314.

[291] Cao, P.; Wang, B.; Zhang, X.; *J. Am. Chem. Soc.* **2000**, *122*, 6490.

[292] Cao, P.; Zhang, X.; *Angew. Chem. Int. Ed.* **2000**, *39*, 4104.

[293] Okamoto, R.; Okazaki, E.; Noguchi, K.; Tanaka, K.; *Org. Lett.* **2011**, *18,* 4894.

[294] Lei, A.; He, M.; Wu, S.; Zhang, X.; *Angew. Chem. Int. Ed.* **2002**, *41*, 3457.

[295] Lei, A.; He, M.; Zhang, X.; *J. Am. Chem. Soc.* **2002**, *124*, 8198.

[296] Wang, J.P.; Xie, X.M.; Ma, F.F.; Peng, Z.Y.; Zhang, L.; Zhang, Z.G.; *Tetrahedron* **2010**, *66*, 4212.

[297] Brummond, K.M.; Chen, H.; Sill, P.; You, L.; *J. Am. Chem. Soc.* **2002**, *124*, 15186.

[298] (a) Hashimoto, S.; Sakata, S.; Sonegawa, M.; Ikegami, S.; *J. Am. Chem. Soc.* **1988**, *110*, 3670; (b) Kanoh, N.; Ishihara, J.; Murai, A.; *Synlett* **1995**, *1995*, 895.

[299] Brummond, K.M.; You, L.; *Tetrahedron* **2005**, *61*, 6180.

[300] (a) Wang, B.; Cao, P.; Zhang, X.; *Tetrahedron Lett.* **2000**, *41*, 8041; (b) Jolly, R.S.; Luedtke, G.; Sheehan, D.; Livinghouse, T.; *J. Am. Chem. Soc.* **1990**, *112*, 4965; (c) Gilbertson, S.R.; Hoge, G.S.; *Tetrahedron Lett.* **1998**, *39*, 2075.

[301] (a) Kagan, H.B.; Riant, O.; *Chem. Rev.* **1992**, *92*, 1007; (b) Nicolaou, K.C.; Snyder, S.A.; Montagnon, T.; Vassilikogiannakis, G.; *Angew. Chem., Int. Ed.* **2002**, *41*, 1668.

[302] Matsuda, I.; Shibata, M.; Sato, S.; Izumi, Y.; *Tetrahedron Lett.* **1987**, *28*, 3361.

[303] van Leeuwen, P.W.N.M.; Roobeek, C.F.; *J. Organomet. Chem.* **1983**, *258*, 343.

[304] O'Mahony, D.J.R.; Belanger, D.B.; Livinghouse, T.; *Synlett* **1998**, 443.

[305] Shen, K.S.; Livinghouse, T.; *Synlett* **2010**, *2010*, 247.

[306] Paik, S.J.; Son, S.U.; Chung, Y.K.; *Org. Lett.* **1999**, *1*, 2045.

[307] McKinstry, L.; Livinghouse, T.; *Tetrahedron* **1994**, *50*, 6145.

[308] Gilbertson, S.R.; Hoge, G.S.; Genov, D.G.; *J. Org. Chem.* **1998**, *63*, 10077.

[309] Aikawa, K.; Akutagawa, S.; Mikami, K.; *J. Am. Chem. Soc.* **2006**, *128*, 12648.

[310] Murakami, M.; Ubukata, M.; Tami, K.I.; Ito, Y.; *Angew. Chem. Int. Ed.* **1998**, *37*, 2248.

[311] Wender, P.A.; Husfeld, C.O.; Langkopf, E.; Love, J.A.; *J. Am. Chem. Soc.* **1998**, *120*, 1940.

[312] Shintani, R.; Nakatsu, H.; Takatsu, K.; Hayashi, T.; *Chem. Eur. J.* **2009**, *15*, 8692.

[313] Kaloko, J.J.; Gary, Y.H.; Ojima, I.; *Chem. Commun.* **2009**, *30*, 4569.

[314] Ojima, I.; Donovan, R.J.; Eguchi, M.; Shay, W.R.; Ingallina, P.; Korda, A.; Zeng, Q.P.; *Tetrahedron* **1993**, *49*, 5431.

[315] Fukuta, Y.; Matsuda, I.; Itoh, K.; *Tetrahedron Lett.* **1999**, *40*, 4703.

[316] Denmark, S.E.; Liu, J.H.C.; Muhuhi, J.M.; *J. Org. Chem.* **2011**, *76*, 201.

[317] Tamao, K.; Kobayashi, K.; Ito, Y.; *Synlett* **1992**, 539.

[318] Ojima, I.; McCullagh, J.V.; Shay, W.R.; *J. Organomet. Chem.* **1996**, *521*, 421.

[319] (a) Ojima, I.; Clos, N.; Donovan, R.J.; Ingallina, P.; *Organometallics* **1990**, *9*, 3127; (b) Tanke, R.S.; Crabtree, R.H.; *J. Am. Chem. Soc.* **1990**, *112*, 7984.

[320] (a) Baik, T.G.; Luis, A.L.; Wang, L.C.; Krische, M.J.; *J. Am. Chem. Soc.* **2001**, *123*, 5112; (b) Wang, L.C.; Jang, H.Y.; Roh, Y.; Lynch, V.; Schultz, A.J.; Wang, X.; Krische, M.J.; *J. Am. Chem. Soc.* **2002**, *124*, 9448; (c) Lipshutz, B.H.; Amorelli, B.; Unger, J.B.; *J. Am. Chem. Soc.* **2008**, *130*, 14378; (d) Lam, H.W.; Joensuu, P.M.; *Org. Lett.* **2005**, *7*, 4225; (e) Lam, H.W.; Murray, G.J.; Firth, J.D.; *Org. Lett.* **2005**, *7*, 5743; (f) Lam, H.W.; Joensuu, P.M.; Murray, G.J.; Fordyce, E.A.F.; Prieto, O.; Luebbers, T.; *Org. Lett.* **2006**, *8*, 3729; (g) Joensuu, P.M.; Murray, G.J.; Fordyce, E.A.F.; Luebbers, T.; Lam, H.W.; *J. Am. Chem. Soc.* **2008**, *130*, 7328.

[321] Jeong, N.; Seo, S.D.; Shin, J.Y.; *J. Am. Chem. Soc.* **2000**, *122*, 10220.

[322] (a) Evans, P.A.; Leahy, D.K.; *J. Am. Chem. Soc.* **2000**, *122*, 5012; (b) Evans, P.A.; Robinson, J.E.; Nelson, J.D.; *J. Am. Chem. Soc.* **1999**, *121*, 6761; (c) Evans, P.A.; Nelson, J.D.; *J. Am. Chem. Soc.* **1998**, *120*, 5581.

[323] Evans, P.A.; Robinson, J.E.; *J. Am. Chem. Soc.* **2001**, *123*, 4609.

[324] (a) Ashfeld, B.L.; Miller, K.A.; Smith, A.J.; Tran, K.; Martin, S.F.; *Org. Lett.* **2005**, *7*, 1661; (b) Ashfeld, B.L.; Miller, K.A.; Smith, A.J.; Tran, K.; Martin, S.F.; *J. Org. Chem.* **2007**, *72*, 9018.

Chapter Three

Organonickel Chemistry

John Montgomery
Department of Chemistry
University of Michigan
Ann Arbor, MI, USA

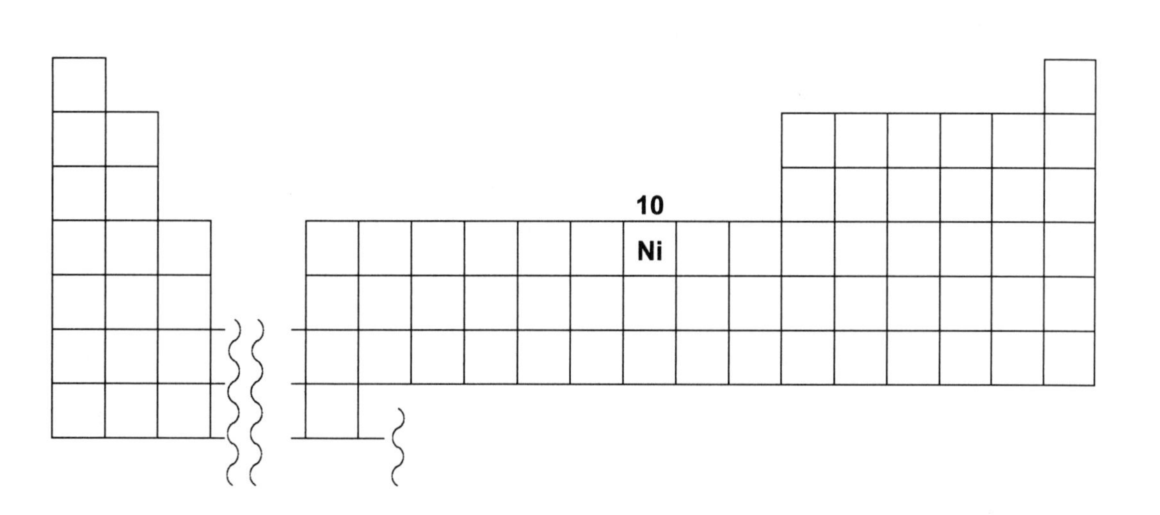

Contents

1	**Introduction**	**322**
2	**Cross-Couplings and Homocouplings**	**325**
2.1	**Early Developments in Nickel-Catalyzed Cross-Couplings**	**325**
2.2	**Synthetic Developments in Nickel-Catalyzed Cross-Couplings**	**327**
2.3	**Mechanistic Aspects**	**334**
2.4	**Participation of Phenol Derivatives and Other C-O Substrates**	**336**
2.5	**Homo- and Hetero-Reductive Couplings of Organic Halides and Related Substrates**	**343**
2.6	**Cross-Couplings Involving C-H Functionalization**	**347**
3	**Cycloaddition Processes**	**348**
3.1	**Early Developments in Nickel-Catalyzed Cycloadditions**	**348**
3.2	**[4+2] Cycloadditions**	**349**
3.3	**[2+2+2] Multicomponent Cycloadditions**	**353**
3.4	**[3+2] and [2+2+1] Cycloadditions**	**357**
3.5	**Cycloadditions in Medium-Ring Construction**	**363**
4	**Reductive Coupling Processes**	**367**
4.1	**Early Developments**	**367**
4.2	**Reductive Couplings of Electron-Deficient Alkenes**	**368**
4.3	**Reductive Couplings of Aldehydes and Dienes**	**373**
4.4	**Reductive Couplings of Aldehydes and Alkynes**	**376**
4.5	**Other Classes of Reductive Couplings**	**381**

5 Carbonyl Additions and Conjugate Additions 384

5.1 Nozaki-Hiyama-Kishi Couplings 384

5.2 Other Additions to Aldehydes or Carbon Dioxide 388

5.3 Conjugate Addition Processes 390

6 1,2- and 1,4- Additions to Alkenes, Alkynes, and Dienes 395

6.1 Hydrometallation 395

6.2 Carbometallation 399

6.3 Hydrovinylation, hydrocyanation, and other hydrocarbations 401

6.4 Carbocyanation 408

6.5 Hydroamination 412

7 Conclusions and Outlook 414

8 Acknowledgments 414

9 References 415

1. Introduction

General review of the historical development of nickel catalysis. Wilke, G.; *Angew. Chem., Int. Ed.* **1988**, *27*, 185-206.

General reviews of the use of nickel in synthetic organic chemistry applications. Jolly, P. W.; Wilke, G. *The Organic Chemistry of Nickel*, Vols. 1 & 2, Academic Press: New York, 1974; Montgomery, J. in *Science of Synthesis (Houben-Weyl Methods of Molecular Transformations)*, ed. Lautens, M., Vol. 1, Thieme: Stuttgart, 2001, p. 11-62; Tamaru, Y. (Ed.) *Modern Organonickel Chemistry*, Wiley-VCH: Weinheim, 2005.

General review of nickel catalysis in industrial processes. Keim, W.; *Angew. Chem., Int. Ed.* **1990**, *29*, 235-244.

General review of nickel catalysis in polymerizations. Ittel, S.D.; Johnson, L.K.; Brookhart, M.; *Chem. Rev.* **2000,** *100*, 1169-1203.

The catalysis of organic reactions by nickel complexes has a long history, dating back to the ground-breaking studies from Reppe in the 1940s describing a series of impressive transformations including the cyclotetramerization of acetylene to cyclooctatetraene. These studies included a description of the involvement of a nickel metallacycle, which is a class of reactive intermediates that plays an essential role in many groups of reactions covered in this chapter. Soon after this important report from Reppe, an extensive program studying the structure and reactivity of organonickel species was initiated in Mülheim in the laboratories of Wilke. Many insights into the oligomerization of small molecules such as ethylene and acetylene emerged from his program. Additionally, a broad array of novel nickel complexes was characterized that provides key insight into catalytic processes being discovered in contemporary programs. The studies from Wilke were among the earliest to illustrate that ligand tuning could alter the course of a catalytic reaction, which, of course, is a finding that is at the heart of the utility of contemporary organometallic chemistry. Another notable advance in the development of organonickel chemistry was the 1972 discovery by Kumada and Corriu that nickel catalysts promote the cross-coupling of Grignard reagents and organic halides. This key advance stimulated the development of related palladium-catalyzed processes, which are among the most central processes in organic synthesis.

The utility of nickel-catalyzed reactions for the derivatization of simple feedstocks has played an incredibly important role in the development of large-scale manufacturing processes. Numerous classes of industrial processes have employed nickel catalysts at one time or another. Examples include the hydrogenation of benzene to cyclohexane, the synthesis of acrylic acid and propionic acid by carbonylative processes of acetylene and ethylene, the synthesis of acetic acid by carbonylation of methanol as a forerunner to the Monsanto process, and olefin oligomerizations in the Shell higher olefin process (SHOP). The interested reader is referred to an excellent review from Keim on industrial processes that involve nickel catalysis.[1] Additionally, nickel catalysis has more recently been recognized as exceptionally useful in the synthesis of functionalized polymers. For example, key work from Brookhart has described nickel diimine catalysts that are more functional group tolerant than early metal alternatives, while suppressing competing β-hydride elimination processes, thus, allowing the synthesis of high-molecular-weight polymers.[2] Processes such as these will not be covered in this review, with the focus here being on contemporary processes amenable to complex, small-molecule synthesis applications. However, the historical importance of these earlier developments must be recognized not only for its industrial significance but also for having defined reactivity and mechanistic considerations that have enabled many developments described in this chapter.

A common feature of the early advances in organonickel chemistry is that they did not gain rapid acccptancc by the synthetic organic community for the synthesis of relatively complex molecules. The oligomerization of simple π-systems was simply not sufficiently selective to be useful when two- or three-component heterocouplings were desired. The cross-coupling reactions failed to gain popularity because of the low functional group tolerance of Grignard reagents, in comparison with the very robust palladium-catalyzed procedures of organozincs, organoboranes, and organostannanes that soon followed. Adding to these complexities that restricted the use of the nickel-catalyzed procedures, the starting Ni(0) catalysts were often unstable and required more careful handling than alternative metal catalysts. Additionally, the ability of nickel to access various oxidation states (0, I, II, and III) often complicated mechanistic analyses and led to the perception that the mechanisms were ill defined, thus, affording complex mixtures of products.

Although a period of relatively slow acceptance followed the early advances in the field of organonickel chemistry, the last decade has observed a dramatic resurgence in the popularity of organonickel chemistry among organic chemists. Whereas features of the chemistry noted in the previous paragraph limited the acceptance of organonickel chemistry in its infancy, many of these restrictions have been overcome, allowing recent solutions to major challenges in organic synthesis. Many of the recent advances described in this review build on the early challenges. For example, selective heterocouplings of structurally different π-components can now be achieved in remarkably challenging settings, in contrast to the challenges of simple oligomerizations of small molecules that were noted originally. Similarly, rapid advances in nickel-catalyzed cross-couplings are being achieved that do not simply provide an alternative to palladium chemistry but provide highly desirable outcomes that are unprecedented with other metals. In these cases, the multiple, easily accessed oxidation states of nickel that were originally viewed as a hindrance have instead enabled these unprecedented modes of reactivity. In addition to the previously discussed advances that build on reactivity trends noted in the early developments in organonickel chemistry, new reactivity trends are still being uncovered, with impressive recent advances in coupling processes initiated by C-H or C-C activation being especially notable examples.

Given the long and illustrious history of organonickel chemistry over the past 65 years, many excellent reviews have appeared. Rather than repeating these descriptions of early developments, the focus of this chapter instead will be to highlight the advances that have most impacted contemporary organic synthesis, with an emphasis on the most recent developments. The organization of the chapter is assembled by the net transformation achieved rather than by the substrates employed or the mechanism involved. While there are remaining mechanistic questions and uncertainties in many (if not most, or perhaps even all) reactions described in this chapter, the current state of mechanistic understanding is provided whenever possible. The descriptions provided are a mixture of proposals from the authors who originally described the work as well as the opinions of the author of this review. Many of the synthetically useful transformations described herein have not been studied in a great level of mechanistic detail, so much of the mechanistic discussion contained in this chapter merits critical analysis with more complete experimentation. The hope is that both experts and nonexperts in organonickel chemistry 1) can benefit from this description of the most recent advances in the field and 2) recognize the most important challenges for the field from this point forward.

To facilitate readability, ligand abbreviations are typically provided in the schemes or figures where the abbreviations are used. However Table 3-1 defines ligand abbreviations that are used at multiple points in the chapter. These structures are only provided in Table 3-1.

Table 3-1. Ligand Abbreviations.

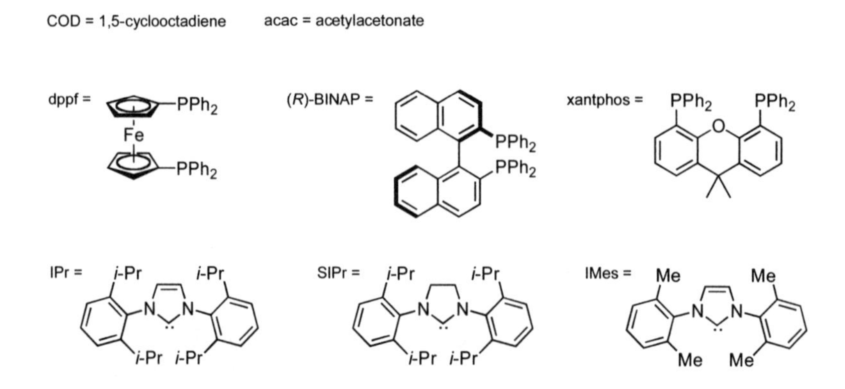

2. Cross-Couplings and Homocouplings

Reviews. Kumada, M.; *Pure Appl. Chem.* **1980,** *52*, 669-679; Hu, X.L.; *Chem. Sci.* **2011,** *2*, 1867-1886.; Phapale, V.B.; Cárdenas, D.J.; *Chem. Soc. Rev.* **2009,** *38*, 1598-1607; Netherton, M.R.; Fu, G.C.; *Adv. Synth. Catal.* **2004,** *346*, 1525-1532; Jana, R.; Pathak, T.P.; Sigman, M.S.; *Chem. Rev.* **2011**, *111*, 1417-1492; Rosen, B.M.; Quasdorf, K.W.; Wilson, D.A.; Zhang, N.; Resmerita, A.M.; Garg, N.K.; Percec, V.; *Chem. Rev.* **2011**, *111*, 1346-1416.

2.1 Early Developments in Nickel-Catalyzed Cross-Couplings

The use of nickel catalysis in cross-coupling reactions played an important role in the historical development of this Nobel Prize–winning field. In 1972, Kumada.[3] and Corriu[4] independently reported on the nickel-catalyzed coupling of sp^2 and sp^3 Grignard reagents with sp^2 organic halides. Significant findings in subsequent papers from Kumada included the origin of alkyl group isomerization during additions of secondary sp^3 Grignard reagents,[5] and the observation of dynamic kinetic resolution during couplings of secondary sp^3 Grignard or organozinc reagents with sp^2 organic halides when chiral ligands were employed (Scheme 3-1).[6–8]

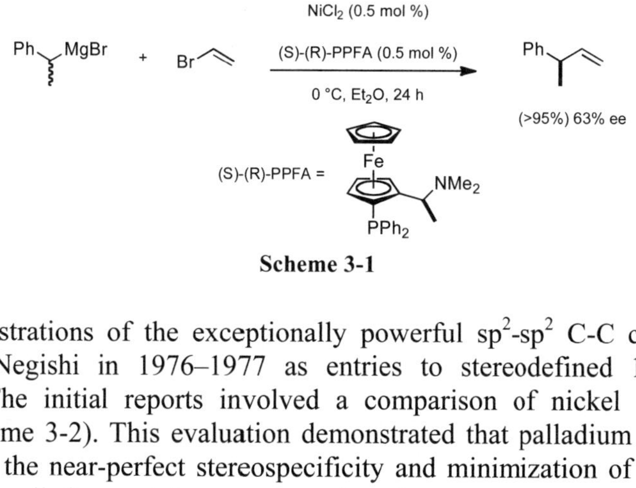

Ph MgBr + Br; $NiCl_2$ (0.5 mol %), (S)-(R)-PPFA (0.5 mol %), 0 °C, Et_2O, 24 h; (>95%) 63% ee

(S)-(R)-PPFA = Fe, NMe_2, PPh_2

Scheme 3-1

Demonstrations of the exceptionally powerful sp^2-sp^2 C-C couplings were provided by Negishi in 1976–1977 as entries to stereodefined 1,3-dienes and biaryls.[9, 10] The initial reports involved a comparison of nickel and palladium catalysts (Scheme 3-2). This evaluation demonstrated that palladium catalysts were superior given the near-perfect stereospecificity and minimization of homocoupling observed when palladium catalysts were employed.

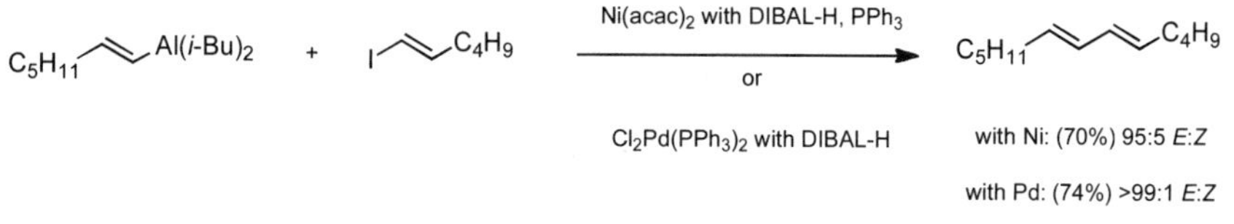

C_5H_{11} $Al(i\text{-}Bu)_2$ + I C_4H_9; $Ni(acac)_2$ with DIBAL-H, PPh_3 or $Cl_2Pd(PPh_3)_2$ with DIBAL-H; C_5H_{11} C_4H_9

with Ni: (70%) 95:5 *E*:*Z*

with Pd: (74%) >99:1 *E*:*Z*

Scheme 3-2

The tremendous popularity of palladium catalysis has overshadowed the utility of nickel catalysis in most processes involving the union of sp^2-sp^2 and sp-sp^2 carbon–carbon bonds. However, the development of nickel-catalyzed cross-couplings emerged as useful in couplings of aryl chloride substrates. The high reactivity of nickel catalysts toward aryl chlorides, coupled with the well-defined mechanistic pathways of aryl chlorides compared with the corresponding aryl iodides and bromides, renders the aryl chloride-based processes especially useful.[11]

Following these early illustrations of the vast potential of nickel-catalyzed cross-couplings, a series of landmark studies by Kochi in the late 1970s formulated a detailed mechanistic analysis that began to elucidate the origins of the unique capabilities of nickel catalysts in cross-coupling processes.[12] A key initial advance involved the demonstration that an isolated Ni(II)(aryl)(methyl) complex undergoes reductive elimination too slowly to be competent in the catalytic cross-coupling of methyl magnesium bromide and bromoarenes (Scheme 3-3).[13] This study illustrated that the rate of reductive elimination is greatly enhanced by the presence of the bromoarene that serves as a starting component of the catalytic process. This observation was attributed to the role of the bromoarene as an oxidant, which reacts with the Ni(II)(aryl)(methyl) species *via* charge transfer to produce a Ni(III)(aryl)(methyl) species. The Ni(III) species then undergoes rapid reductive elimination.

Br + MeMgBr $\xrightarrow{Br_2Ni(PEt_3)_2}$ Me

slow

fast

$(Et_3P)_2Ni$ Me $\xrightarrow[\text{fast}]{\text{PhBr}}$ $[(Et_3P)_2Ni^+$ Me $PhBr^{\cdot -}]$

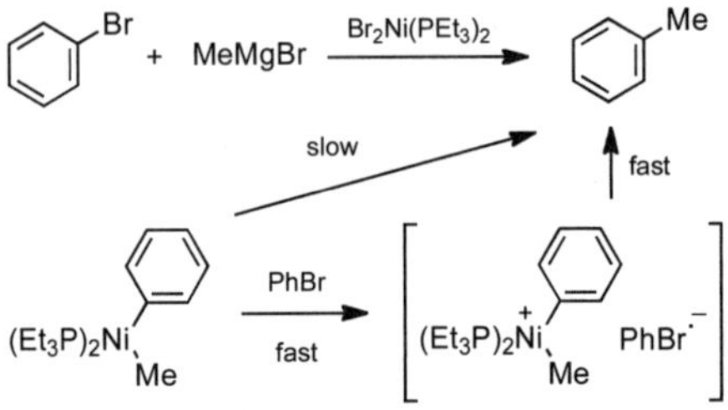

Scheme 3-3

A subsequent study from Kochi focused on the mechanism of oxidative addition of Ni(0) complexes with aromatic halides.[14] In this report, a picture emerged that is much more complex than the corresponding well-defined, two-electron oxidative additions to haloaromatics that are the hallmark of the analogous palladium-catalyzed processes. Although this comprehensive mechanistic study will not be presented here in detail, a key conclusion from this study is that the oxidative addition proceeds by initial electron transfer from Ni(0) to the aryl halide to generate a Ni(I) ion pair intermediate (Scheme 3-4). This species may then fragment in two distinct ways. The common Ni(I) ion pair species may collapse to generate the Ni(II)(aryl)(halide) species, or alternatively, fragmentation to generate an aryl radical and a Ni(I) halide species may occur.

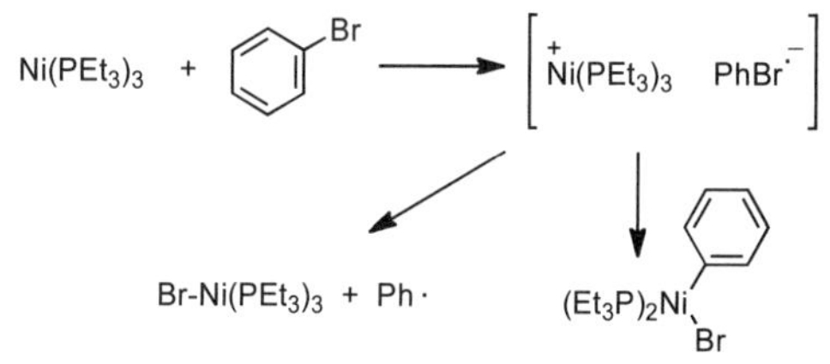

Scheme 3-4

Whereas these early demonstrations of nickel catalysis in cross-couplings established important advances in both synthetic utility and mechanistic understanding, the tremendous utility of the corresponding palladium-catalyzed processes quickly overshadowed the corresponding use of nickel. The popularity of palladium compared with nickel can be attributed to the easier handling of palladium catalysts, the slightly higher yield and stereospecificity observed for palladium in sp^2-sp^2 couplings, and the more straightforward mechanistic analysis of palladium-catalyzed reactions, given the complexities of electron-transfer processes involving nickel. However, the distinct mechanistic pathways involved in nickel-catalyzed processes do present unique advantages in some classes of cross-couplings. As highlighted in the subsequent sections, many recent advances are beginning to demonstrate clearly these advantages in coupling processes that cannot be readily achieved with palladium or other metals.

2.2 Synthetic Developments in Nickel-Catalyzed Cross-Couplings

Much of the extensive and unique utility of nickel-catalyzed cross-couplings has more recently been realized in processes with at least one sp^3 carbon center being involved in the key bond formation.[15, 16] Until recently, cross-couplings involving sp^3 centers were generally viewed as inefficient processes because of the low reactivity of sp^3 electrophiles toward oxidative addition and the potential for competing β-hydride eliminations and alkyl group isomerizations that prevent formation of the desired product. These complexities could be avoided by the use of reactive electrophiles that lack β-hydrogens, as demonstrated in the synthesis of coenzyme Q using a nickel-catalyzed vinyl aluminum coupling with a benzylic chloride (Scheme 3-5).[17]

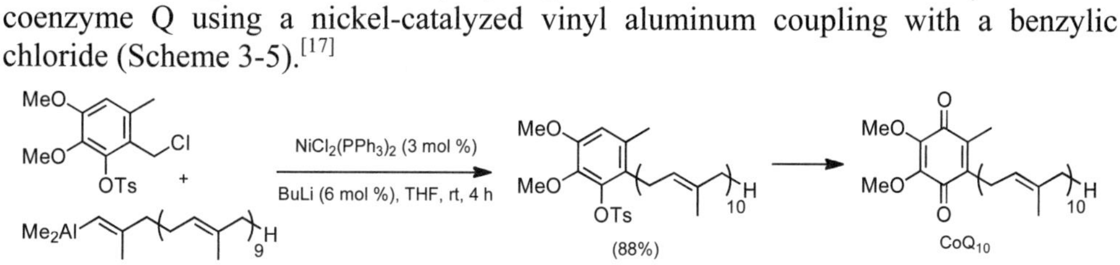

Scheme 3-5

In cases where a simple alkyl functionality (nonbenzylic or allylic) is desired at the site of C-C bond formation, early strategies largely involved the use of an sp^2 electrophile and sp^3 nucleophile (Scheme 3-6). Although these developments more frequently involved palladium catalysis, various nickel-catalyzed strategies involving either homogeneous or heterogeneous catalysts have also been successfully employed.[18, 19]

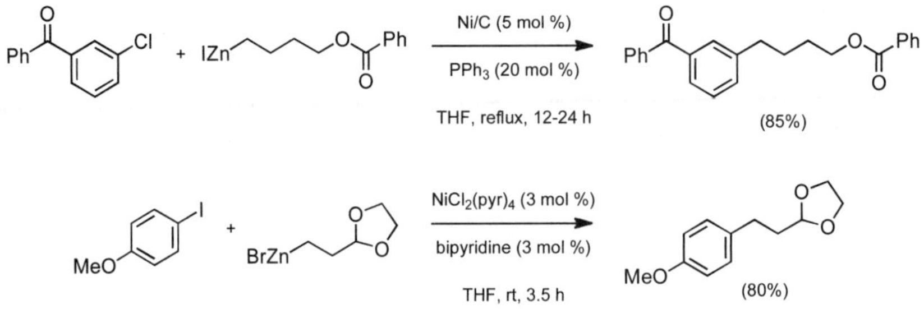

Scheme 3-6

However, the truly unique reactivity of nickel compared with palladium began to be demonstrated most clearly with advances using sp^3 electrophiles in cross-couplings. The perceptions about the limited tolerance of the sp^3-electrophile structure began to change after a report from Knochel that described the efficient nature of nickel-catalyzed cross-couplings between sp^3-carbon centers.[20] Unsaturated alkyl bromides were found to be surprisingly effective in cross-couplings with diethylzinc in the presence of $Ni(acac)_2$ as catalyst (Scheme 3-7). The efficiency of the process was attributed to the coordination of the remote tethered alkene to nickel, with the alkene π-acidity promoting the C-C reductive elimination, while impeding undesired β-hydride elimination processes. Substrates lacking the unsaturation produce organozinc products. Whereas this effect was initially limited to substrates that possess a remote alkene, the process was generalized by finding that trifluoromethylated arene and styrene additives could promote a similar outcome with substrates that lack the tethered alkene.[21, 22]

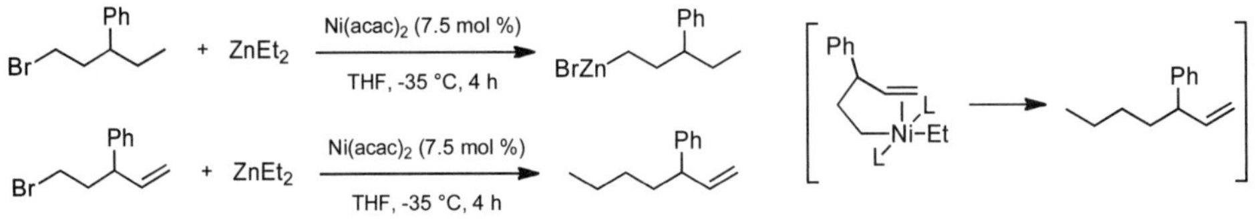

Scheme 3-7

Representative procedure: synthesis of benzyl octanoate[22]

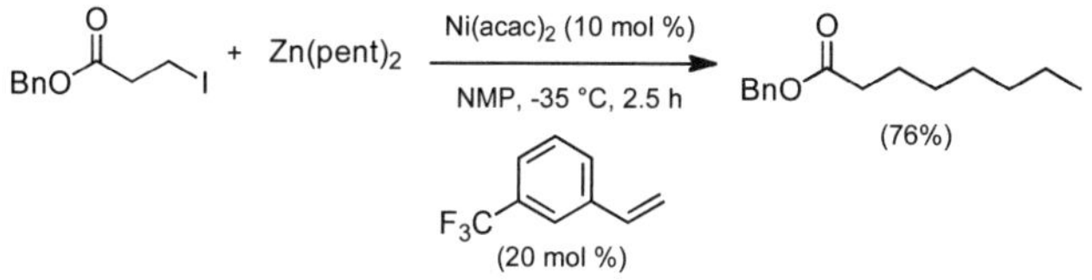

A 25-mL, two-necked flask equipped with an Ar inlet and a rubber septum was charged with [Ni(acac)$_2$] (0.128 g, 0.5 mmol, 10 mol%), tetrahydrofuran (THF) (3.4 mL), *N*-methyl-2-pyrrolidone (NMP) (1.7 mL), *m*-trifluoromethylstyrene (0.15 mL, 1 mmol, 20 mol%), and benzyl 3-iodopropanoate (1.45 g, 5 mmol) at RT. The flask was cooled to −78 °C, and dipentylzinc (2 mL, 10 mmol) was added carefully. The reaction mixture was allowed to warm to −35 °C and stirred for 2.5 h before being poured into an ice-cold saturated aqueous solution of NH_4Cl and extracted with diethyl ether. The organic phase was washed with brine and dried over anhydrous Na_2SO_4. The resulting crude oil, obtained after evaporation of the solvents, was purified by flash chromatography (hexanes/diethyl ether, 10/1) to give the desired product as a colorless oil (0.89 g, 3.8 mmol, 76%).

A related observation from Kambe illustrated that 1,3-butadiene additives effectively promote cross-couplings of sp^3 alkyl halides with sp^3 organomagnesium and organozinc reagents (Scheme 3-8).[23, 24] In these instances, both Ni(acac)$_2$ and $NiCl_2$ were effectively used as a catalyst. This report suggested a nickel (II/IV) cycle involving the formation of an active Ni(II) catalyst by the oxidative cyclization of Ni(0) with two butadienes to form a metallacyclic catalyst. This mechanistic proposal enables transmetallation of the organomagnesium or organozinc component prior to oxidative addition of the organic halide, which reverses the sequence commonly invoked in the more widely employed palladium-catalyzed systems. Although the involvement of a high oxidation state Ni(IV) species has not been documented under the reaction conditions or studied in any detail, the utility of the novel 1,3-diene effect was clearly demonstrated in these important early studies.

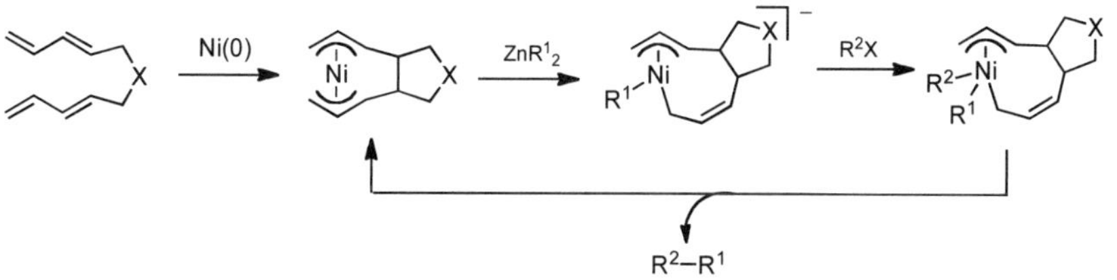

Scheme 3-8

Following these encouraging reports describing the nickel-catalyzed couplings of sp^3 halides and sp^3 transmetallating agents, an extensive program by Fu moved the field forward toward widespread applicability through the discovery of efficient asymmetric catalysts with remarkable scope and efficiency. Through these reports, the unique features of nickel-catalyzed cross-couplings were realized in a broad range of settings that could not be tolerated by competitive technologies including the vast array of palladium-catalyzed methods currently available. An initial important breakthrough was the efficient participation of secondary sp^3 electrophiles in cross-coupling reactions. The Pybox ligand motif was identified as particularly effective in Negishi couplings of functionalized primary sp^3 organozincs with secondary alkyl bromides and iodides.[25] Related advances were then made in Suzuki couplings involving aryl and alkenyl boronic acids using bathophenanthroline as ligand,[26] and in Hiyama couplings involving aryl trifluorosilanes using norephedrinc as ligand.[27]

Representative procedure: synthesis of ethyl-6-ethyloctanoate[25]

Et, Et—I + BrZn~~~C(O)OEt → Ni(COD)$_2$ (4 mol %), DMA rt, 20 h → product OEt (62%)

s-Bu-Pybox (8 mol %)

General procedure: Ni(cod)$_2$ (5.6 mg, 0.020 mmol; 1,5-cyclooctadiene) and *s*-Bu-Pybox (13.2 mg, 0.040 mmol) were added to a vial. The air was removed by evacuating/refilling with Ar (three times), and then *N,N*-dimethylacetamide (DMA) (0.80 mL) was added. The resulting mixture was stirred for 20 min at room temperature (RT), and then the resulting deep-blue solution was treated in turn with a solution of the alkylzinc halide (0.75 M in DMA; 1.06 mL, 0.80 mmol) and the alkyl halide (0.50 mmol). The reaction mixture was stirred for 20 h at RT, and then it was quenched with iodine chips (~100 mg, to iodinate the excess organozinc reagent). After stirring at RT for 10 min, the dark-brown mixture was passed through a short pad of silica gel (to remove DMA, inorganic salts, and iodine). The filtrate was then concentrated, and the residue was purified by flash chromatography. Using the above general procedure, 3-iodopentane (99 mg, 0.50 mmol) and 4-ethoxycarbonylbutylzinc bromide were converted into the desired product (62 mg, 62%) as a colorless oil.

Representative procedure: synthesis of trans-1-t-butyldimethylsiloxy-2-(4-trifluoromethylphenyl)indane[26]

(OTBS)indanyl-Br + (HO)$_2$B–C$_6$H$_4$–CF$_3$ → Ni(COD)$_2$ (4 mol %), KO-*t*-Bu (1.6 equiv), *s*-BuOH, 60 °C, 5 h → (OTBS)indanyl–C$_6$H$_4$–CF$_3$ (63%)

Bathophenanthroline (Ph, Ph) (8 mol %)

General procedure: Ni(cod)$_2$ (5.6 mg, 0.020 mmol), bathophenanthroline (13.2 mg, 0.040 mmol), the arylboronic acid (0.60 mmol), and KO-*t*-Bu (90 mg, 0.80 mmol) were added to a vial equipped with a stir bar. The vial was evacuated/refilled with Ar three times, and then *s*-butanol (3.0 mL) was added. The mixture was stirred at RT for 10 min, and to the resulting deep-purple solution was added the alkyl halide (0.50 mmol). The reaction mixture was stirred at 60 °C for 5 h, and then it was passed through a short pad of silica gel (to remove *s*-butanol and polar compounds). The filtrate was concentrated, and the residue was purified by flash chromatography. Using the above general procedure, trans-2-bromo-1-*t*-butyldimethylsiloxylindane (164 mg, 0.50 mmol) and 4-trifluoromethylphenylboronic acid (114 mg, 0.60 mmol) were converted into the desired product (123 mg, 63%) as a colorless oil.

Representative procedure: synthesis of 1-morpholino-2-phenylpropan-1-one[27]

$NiCl_2$•glyme (10 mol %)
LiHMDS (12 mol %), H_2O (8 mol %)
CsF (3.8 equiv), DMA, 60 °C, 16 h
(15 mol %)
(86%)

General procedure: DMA (0.6 mL) was added to a mixture of lithium bis(trimethylsilyl)amide (LiHMDS) (20.1 mg, 0.12 mmol), dry CsF (576 mg, 3.8 mmol; note: prior to use, this must be dried in a vacuum oven at 80 °C for 48 h), and H_2O (1.4 μL, 0.080 mmol) in a 4-mL glass vial in a glove box. A solution of $NiCl_2$•glyme (22.0 mg, 0.10 mmol) and norephedrine (18.1 mg, 0.12 mmol) in DMA (1.0 mL) was added, and the reaction mixture was stirred at RT for 5 min. Next, the arylsilane (1.50 mmol) was added, and the reaction mixture was stirred for an additional 5 min. Last, the alkyl halide (1.00 mmol) was added. The reaction vial was capped, removed from the glove box, and stirred at 60 °C for 16 h. After 16 h, the reaction mixture was allowed to cool to RT, and then EtOH (0.5 mL) and Et_2O (1.5 mL) were added. This heterogeneous mixture was poured onto a pad of silica gel (Et_2O as the eluent). The product was then purified by flash chromatography. Using the preceding general procedure, 2-chloro-1-morpholinopropan-1-one (1.00 mmol) and trifluorophenylsilane (1.50 mmol) were converted into the desired product (191 mg, 87%) as a colorless oil.

Concurrently with the preceding developments that presented a very broad scope with secondary alkyl bromides, dynamic kinetic resolution processes were demonstrated with various classes of racemic alkyl halides. The mechanistic implications of this finding will be discussed subsequently, but the synthetic implications are substantial in the asymmetric preparation of sp^3-sp^3 and sp^3-sp^2 carbon–carbon-bond linkages. On a strategic level, this class of cross-couplings has enormous potential, as it allows the unfunctionalized aliphatic portion of complex molecules to now be considered as a point of disconnection in retrosynthetic planning. Representative examples of products that may be asymmetrically prepared are provided (Figure 3-1), with the newly formed bond and class of cross-coupling reagent denoted by the arrows.[28–33] Representative procedures for these asymmetric couplings are provided for the γ-alkylation of carbonyl compounds[28] and for the α-arylation of carbonyl compounds *via* Suzuki couplings.[29]

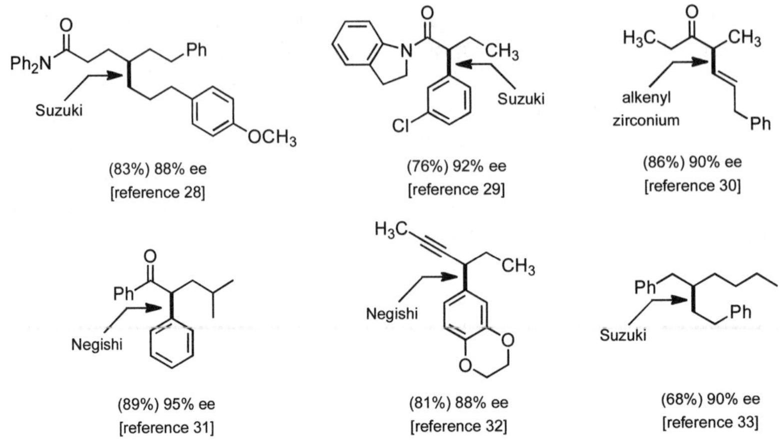

Figure 3-1

Representative procedure: synthesis of 4-methyl-8-(2-methyl-1,3-dioxolan-2-yl)-N,N-diphenyloctanamide[28]

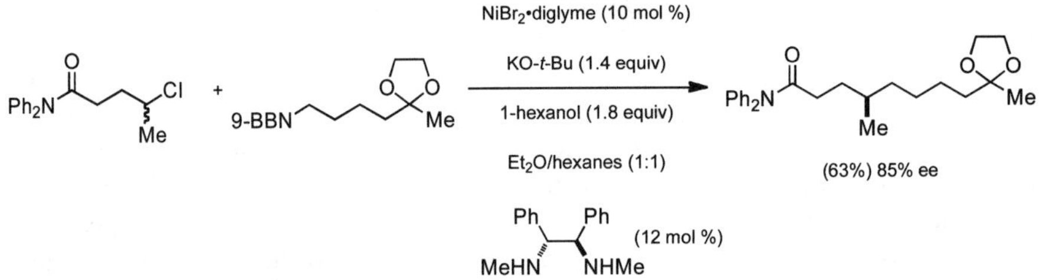

General procedure: In a nitrogen-filled glove box, a solution of the organoboron reagent (670 μL, 1.0 mmol; 1.5 M) was added to a slurry of potassium *tert*-butoxide (78.5 mg, 0.70 mmol) and 1-hexanol (113 μL, 92 mg, 0.90 mmol) in a 4-mL vial equipped with a stir bar. The vial was sealed with a Teflon-lined septum cap, and the mixture was stirred vigorously for 30 min and then used in the next step. In a glove box, $NiBr_2$•diglyme (17.6 mg, 0.050 mmol), (*R,R*)-ligand (14.5 mg, 0.060 mmol), hexanes (3.1 mL), and Et_2O (1.4 mL) were added in turn to a 20-mL vial equipped with a stir bar. The vial was sealed with a Teflon-lined septum cap, and the mixture was stirred vigorously for 45 min (a light-blue slurry forms). The solution of the activated organoboron reagent was then added to the slurry, and the vial was sealed with a Teflon-lined cap and stirred for 30 min (the reaction mixture turns brown). The electrophile (0.50 mmol in 0.5 mL of Et_2O; purified) was added to the slurry *via* syringe and the vial that contained the electrophile was then rinsed with additional Et_2O (0.5 mL), and this solution was added to the slurry. The mixture was sealed with a Teflon-lined cap and stirred vigorously at RT for 24 h (outside of the glove box). Next, the reaction mixture was passed through a short plug of silica gel, eluting with Et_2O. The solution was concentrated to furnish an oil, which was purified by reverse-phase flash chromatography on C-18 silica gel with 10→100% acetonitrile/water. Using the above general procedure, 4-chloro-*N,N*-diphenylpentanamide (144 mg, 0.50 mmol) and a solution of the alkylborane prepared by hydroboration of

2-(but-3-en-1-yl)-2-methyl-1,3-dioxolane with 9-BBN dimer (1.5 M in Et_2O; 670 μL, 1.0 mmol) were converted into the desired product (123 mg, 63%, 85% ee) as a colorless oil.

Representative procedure: synthesis of 4-(tert-butyldimethylsilyloxy)-1-(indolin-1-yl)-2-phenylbutan-1-one[29]

$NiBr_2$•diglyme (8 mol %)
KO-*t*-Bu (1.3 equiv)
i-BuOH (1.5 equiv)
toluene, -5 °C, 24 h
OTBS + 9-BBN
Cl
F_3C CF_3
MeHN NHMe
(12 mol %)
OTBS
(77%) 85% ee

General procedure: In a nitrogen-filled glove box, $NiBr_2$•diglyme (8.8 mg, 0.040 mmol, 8.0%), (*S*,*S*)-ligand (18.8 mg, 0.050 mmol, 10%), the electrophile (0.50 mmol), and toluene (2.5 mL) were added to a 10-mL flask. The following materials were added in turn to a 4-mL vial: KO-*t*-Bu (73 mg, 0.65 mmol, 1.3 equiv), *i*-BuOH (69 μL, 0.75 mmol, 1.5 equiv), the aryl-(9-BBN) reagent (0.75 mmol, 1.5 equiv), and toluene (2.5 mL). The flask and the vial were each capped with a rubber septum, and the two mixtures were stirred for 10 min. Next, the vessels were removed from the glove box and placed in a –5 °C bath, and the mixtures were stirred for 10 min. The solution in the vial was then transferred by syringe to the slurry in the 10-mL flask, which was attached to a nitrogen-filled balloon. The reaction mixture was stirred at –5 °C for 24 h (it turned orange after a few minutes). Next, the mixture was poured into a separatory funnel and washed with a saturated solution of sodium carbonate (5 mL; if the aqueous layer was very viscous, then distilled water [3 mL] was added). The aqueous phase was extracted with EtOAc (5 mL × 2), and the organic layers were combined and washed with brine (5 mL), dried over anhydrous Na_2SO_4, and concentrated. The resulting residue was purified by flash chromatography. Using the above general procedure, 4-(*tert*-butyldimethylsilyloxy)-2-chloro-1-(indolin-1-yl)butan-1-one (179 mg, 0.50 mmol) and 9-phenyl-9-borabicyclo[3.3.1]nonane (149 mg, 0.75 mmol) were converted into the desired product 152 mg (77% yield, 85% ee) as a yellow solid.

Homogeneous catalyst systems have been primarily exploited in C-C and C-heteroatom, bond-forming, cross-coupling processes. These catalysts are typically generated either directly from $Ni(cod)_2$ or *in situ* from stable, inexpensive pre-catalysts such as $Ni(acac)_2$, $NiCl_2$•glyme, or $(R_3P)_2NiCl_2$. Although $NiCl_2$ and $NiBr_2$ are very stable and inexpensive, they are polymeric insoluble solids that typically require prior conversion into a ligated, soluble, monomeric form. Alternatively, effective heterogeneous catalysts have been prepared as nickel(II) embedded on charcoal (Ni/C). This material may be reduced with BuLi or H_2, and efficient C-C and C-N cross-couplings have been realized.[34, 35] A detailed spectroscopic analysis demonstrated that couplings with Ni/C likely proceeded by a homogeneous pathway, although catalyst recovery by filtration was high, leaving only trace metal in solution.[36]

2.3 Mechanistic Aspects

The mechanistic aspects of nickel-catalyzed cross-couplings are complex, and the general understanding of these processes continues to evolve.[19, 37–40] Based on data presented to date, there is a general consensus that simple Ni(0)/Ni(II) mechanisms analogous to those commonly invoked for palladium-catalyzed processes do not adequately explain the experimental observations in many of the most widely studied nickel-catalyzed processes, particularly those of couplings that involve the union of sp^3-hybridized carbon centers. Dating back to the fundamental studies from Kochi described previously, electron-transfer processes and odd electron species are commonly invoked in mechanistic descriptions of nickel-catalyzed cross-couplings. Although the mechanistic details seem to vary substantially depending on electrophile, transmetallating agent, and ligand structure in addition to other factors, several key studies have provided considerable insight into possible mechanistic pathways. A recent review classifies possible mechanisms by ligand type employed and provides an excellent critical analysis of the current understanding of mechanism in the broadly useful case of couplings involving nonactivated alkyl halides.[38] An important take-home message[41, 42] is that the common formalisms of ligand charge and metal oxidation state are often not accurate representations of the true electronic structure of intermediate complexes. Key redox events can be ligand based in the most active catalyst systems, rendering ambiguity on how metal oxidation states are best defined.

A simple, generalized mechanistic scheme, which is consistent with the major synthetic features of the reaction, can be formulated as follows (Scheme 3-9).[28] Generation of an active Ni(I)(halide) catalyst **1** is followed by transmetallation of the transmetallating component to generate a Ni(I)(alkyl) species **2**. The addition of this species to the sp^3 alkyl halide generates a Ni(II)(alkyl)(halide) intermediate **3** along with extrusion of the alkyl free radical. Free radical recombination with the Ni(II)(alkyl)(halide) species **3** generates a Ni(III)(alkyl)(alkyl')halide species **4**. Reductive elimination generates the coupled product and regenerates the Ni(I)(halide) catalyst **1**. This mechanism explains basic features such as the dynamic kinetic resolution feature, given that alkyl halide stereochemistry is lost during free radical generation. With chiral ligands employed, it is the recombination of the alkyl radical with the Ni(II)(alkyl)(halide) species **3** that is the enantioselectivity-determining step. Whereas typically racemic alkyl halides are used, either enantiomer of the starting alkyl halide converges to the same enantiomeric composition of product when a chiral ligand is employed. Evidence has been presented in Suzuki couplings involving a simple diamine catalyst that the initial consumption of the alkyl halide is irreversible.[28, 31] In one case, the unreacted alkyl halide was kinetically resolved,[31] although in other cases, the unreacted alkyl halide is racemic throughout the reaction.

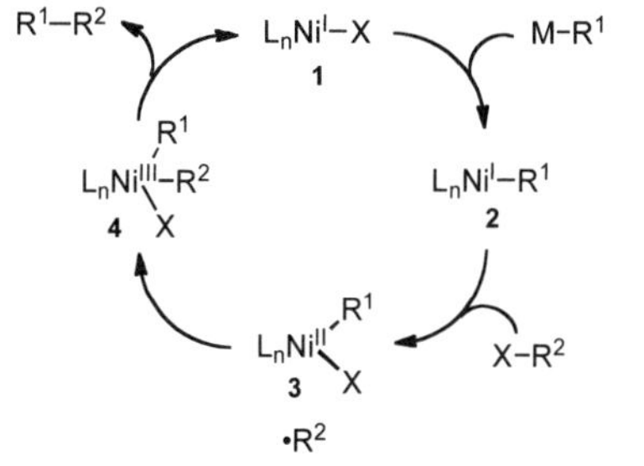

Scheme 3-9

Although detailed mechanisms of the numerous active catalyst systems will not be described here, the well-defined case of a terpyridine ligand serves as an important model for the current understanding of the operative mechanistic pathways.[41, 42] In the proposed mechanism of this well-defined system (Scheme 3-10), a Ni(alkyl) intermediate **5**, analogous to structure **2** in Scheme 3-9, reduces the alkyl halide to generate proposed intermediate **6**, which is analogous to structure **3** in Scheme 3-9. This species is then further converted into intermediate **7**, which is analogous to intermediate **4** in Scheme 3-9. An important outcome of this study is the demonstration that significant spin density resides in the ligand in structure **5** and that the electronic structure of nickel in complex **5** is best described as Ni(II) bound to an anionic terpyridine. Considerable mechanistic insights with other active catalysts for alkyl halide cross-couplings have been described with tridentate pincer ligands and mixed amino-amide ligands[43] and with *N*-heterocyclic carbene ligands.[44] Each of these studies concluded that free radical generation followed by recombination with nickel were integral features of the catalytic cycle.

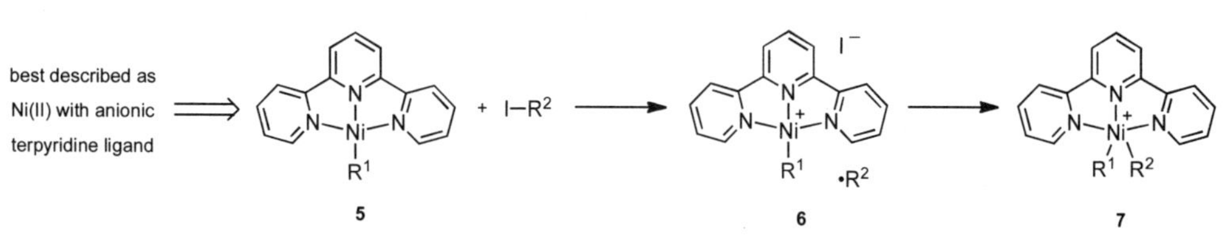

Scheme 3-10

Although most of the recent mechanistic work has focused on the synthetically important couplings of sp^3 alkyl halides, relatively little mechanistic work has appeared in systems involving traditional ligand systems and sp^2-sp^2 couplings of aryl halides since the early work of Kochi. Considerable insights have appeared in the polymer literature in work from McNeil describing mechanistic features of chain growth polymerizations involving Kumada (Grignard) couplings with bromoarenes and thiophenes.[45, 46] Detailed rate studies and characterization of catalyst resting states were consistent with a Ni(0)/Ni(II) catalytic cycle. Mechanistic changes were found as ligand structures were varied. Interestingly, the rate-determining step of cross-couplings was reductive elimination with 1,2-bis(diphenylphosphano)ethane (dppe) as ligand and transmetallation with 1,2-bis(diphenylphosphano)propane (dppp) as ligand.

2.4 Participation of Phenol Derivatives and Other C-O Substrates

Whereas the above discussion in cross-couplings largely focuses on the addition reactions of alkyl, aryl, and vinyl halides, the chemistry of simple phenol derivatives is pervasive in the nickel literature. Functionalizations of aryl and alkenyl triflates are efficient in a manner that patterns the widely used palladium-catalyzed cross-coupling methods. However, compared with palladium-catalyzed methods, a much broader range of phenol derivatives undergoes efficient oxidative additions to nickel(0). The range of phenol-derived substrates and related enol derivatives that participates in nickel-catalyzed processes includes aryl and vinyl triflates, mesylates, tosylates, sulfamates, phosphonates, carbamates, carbonates, pivalates, acetates, simple methyl ethers, and even underivatized phenols themselves (Figure 3-2). Additionally, while employed less frequently than phenol derivatives, aryl trimethylammonium salts are also efficient substrates for nickel-catalyzed cross-couplings.[47–49] An excellent recent review extensively describes the substrates that undergo C-O cleavage during nickel-catalyzed cross-coupling reactions, and the reader is referred to this source for details beyond the brief summary provided in this chapter.[50]

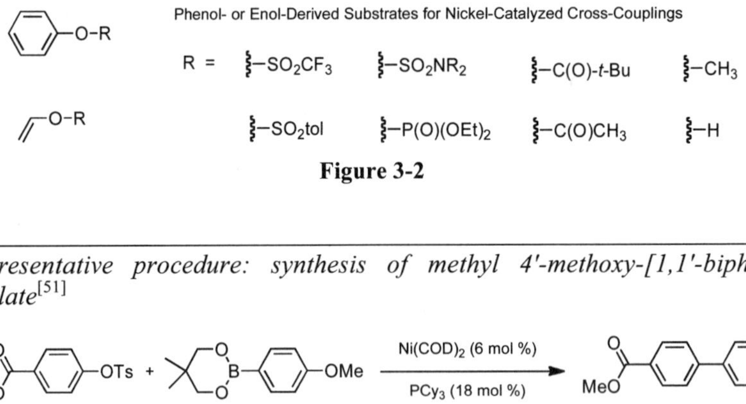

Phenol- or Enol-Derived Substrates for Nickel-Catalyzed Cross-Couplings

Ph–O–R; vinyl–O–R

R = $-SO_2CF_3$, $-SO_2NR_2$, $-C(O)$-*t*-Bu, $-CH_3$, $-SO_2tol$, $-P(O)(OEt)_2$, $-C(O)CH_3$, $-H$

Figure 3-2

Representative procedure: synthesis of methyl 4'-methoxy-[1,1'-biphenyl]-4-carboxylate[51]

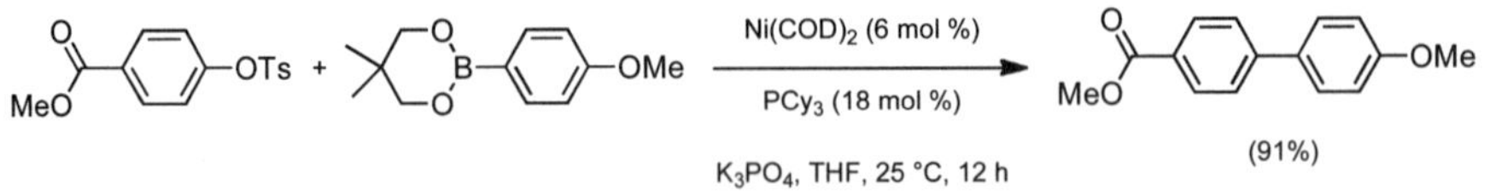

$Ni(COD)_2$ (6 mol %), PCy_3 (18 mol %), K_3PO_4, THF, 25 °C, 12 h; (91%)

General procedure: In a glove box with N_2 atmosphere, a Schlenk tube was charged with aryl halide, tosylate or mesylate (1.0 equiv), aryl neopentylglycolboronate (1.5 equiv), potassium phosphate (0.9 mmol, 3.0 equiv), $Ni(cod)_2$ (0.018 mmol, 0.06 equiv), tricyclohexylphosphine (0.11 mmol, 0.18 equiv), and a Teflon-coated stirbar. Dry THF (2 mL) was added, and the reaction mixture was stirred for 8 h at RT. The reaction mixture was diluted with dichloromethane (DCM)

(10 mL) and filtered. The filtered solid was washed with DCM (100 mL), and the organic solvent was concentrated. The coupling products were precipitated in MeOH, and the white crystals were collected by filtration and washed with cold MeOH. Using the preceding general procedure, methyl 4-(tosyloxy)benzoate and 2-(4-methoxylphenyl)-5,5-dimethyl-1,3,2-dioxaborinane were converted into the desired product (91% yield).

Representative procedure: synthesis of 1-(4-methoxyphenyl)naphthalene[52, 53]

-OSO$_2$NMe$_2$ + (HO)$_2$B–⟨⟩–OMe → NiCl$_2$(PCy$_3$)$_2$ (5 mol %), K$_3$PO$_4$ (4.5 equiv), toluene, 110 °C, 24 h → –⟨⟩–OMe (95%)

Via the aryl sulfamate: A 1-dram vial was charged with anhydrous powdered K_3PO_4 (382 mg, 1.80 mmol, 4.5 equiv, obtained from Acros Organics [Geel, Belgium]) and a magnetic stir bar. The vial and contents were flame dried under reduced pressure and then allowed to cool under N_2. (4-Methoxyphenyl)boronic acid (152 mg, 1.00 mmol, 2.5 equiv), $NiCl_2(PCy_3)_2$ (13.7 mg, 0.02 mmol, 5 mol%), and naphthalen-1-yl dimethylsulfamate (100 mg, 0.40 mmol, 1 equiv) were added. The vial was then evacuated and backfilled with N_2. Toluene (1.5 mL) was added, and the vial was sealed with a Teflon-lined screw cap. The heterogeneous mixture was allowed to stir at 23 °C for 1 h and then heated to 110 °C for 24 h. The reaction vessel was cooled to 23 °C and then transferred to a round-bottom flask containing CH_2Cl_2 (20 mL). Silica gel (3 mL) was added and the solvent was removed under reduced pressure to afford a free-flowing powder. This powder was then dry loaded onto a silica gel column (4.5 × 5 cm) and purified by flash chromatography (2:1 hexanes:benzene) to yield the desired product (89 mg, 95% yield) as a colorless solid.

-OC(O)NEt$_2$ + (HO)$_2$B–⟨⟩–OMe → NiCl$_2$(PCy$_3$)$_2$ (10 mol %), K$_3$PO$_4$ (7.2 equiv), toluene, 130 °C, 24 h → –⟨⟩–OMe (86%)

Via the aryl carbamate: A 1-dram vial was charged with anhydrous powdered K_3PO_4 (670 mg, 3.16 mmol, 7.2 equiv, obtained from Acros) and a magnetic stir bar. The vial and contents were flame dried under reduced pressure and then allowed to cool under N_2. (4-Methoxyphenyl)boronic acid (268 mg, 1.76 mmol, 4 equiv), $NiCl_2(PCy_3)_2$ (30 mg, 0.0439 mmol, 10 mol%), and naphthalen-1-yl diethylcarbamate (107 mg, 0.439 mmol, 1 equiv) were added. The vial was then evacuated and backfilled with N_2. Toluene (1.5 mL) was added, and the vial was sealed with a Teflon-lined screw cap. The heterogeneous mixture was allowed to stir at 23 °C for 1 h and then heated to 130 °C for 24 h. The reaction vessel was cooled to 23 °C and then transferred to a round-bottom flask containing CH_2Cl_2 (20 mL). Silica gel (3 mL) was added and the solvent was removed under reduced pressure to afford a free-flowing powder. This powder was then dry loaded onto a silica gel column (4.5 × 5 cm) and purified by flash chromatography (2:1 hexanes:benzene) to yield the desired product (89 mg, 86% yield).

Representative procedure: synthesis of 2-phenylnaphthalene[54]

OMe + B(O...O) → Ni(COD)$_2$ (10 mol %), PCy$_3$ (40 mol %), CsF (4.5 equiv), toluene, 120 °C, 12 h → (93%)

The reaction was carried out in a 10-mL sample vial with a Teflon-sealed screwcap in a glove box under nitrogen atmosphere. To the orange mixture of Ni(cod)$_2$ (13.8 mg, 0.05 mmol), PCy$_3$ (56.0 mg, 0.2 mmol) and CsF (341.8 mg, 2.25 mmol) in toluene (1.5 mL) was added 2-methoxynaphthalene (79.1 mg, 0.5 mmol) and 2-phenyl-5,5-dimethyl-1,3,2-dioxaborinane (142.5 mg, 0.75 mmol) at RT followed by heating at 120 °C for 12 h. After cooling to RT, the crude mixture was purified by flash column chromatography on silica gel (hexane) to give the desired product (94.7 mg, 93%) as a white powder.

Two recent computational studies have examined the mechanism of cross-couplings involving aryl acetates, carbamates, and sulfamates. In a study of synthetic advances with aryl pivalates,[55, 56] oxidative addition of Ni(0) to the aryl acetate was found to proceed by a three-centered mechanism *via* an η^2 arene complex.[57] Alternatively, in a separate study, oxidative addition involving aryl carbamates and sulfamates was found to proceed by a five-centered transition state, with coordination of the carbonyl oxygen to nickel being a key interaction in the oxidative addition.[52]

In addition to cross-coupling processes, the hydrogenolysis of simple aryl(alkyl) ethers and diaryl ethers has been accomplished. The reduction of simple aryl(methyl) ethers was developed as a strategy to use the directing influence of the methoxy substituent in processes such as electrophilic aromatic substitution, followed by traceless removal of the methoxy substituent.[58] The analogous process using direct hydrogenolysis with H_2 was developed as a means for generating fuels and chemical feedstocks from biomass and the liquefaction of coal.[59]

Representative procedure: synthesis of 2-methylquinoline[58]

Me, N, OMe → Ni(COD)$_2$ (5 mol%), PCy$_3$ (10 mol %), 1,1,3,3-tetramethyldisiloxane (1 equiv), toluene, 110 °C, 14 h → Me, N (64%)

General procedure: An oven-dried screw-cap test tube containing a stir bar was charged with the aryl ether (0.50 mmol), Ni(cod)$_2$ (7.0 mg, 5 mol%), and PCy$_3$ (14.0 mg, 10 mol%) inside a dry box. Then, the flask was removed from the dry box and 1,1,3,3- tetramethyldisiloxane (88 μL, 0.50 mmol) and toluene (1 mL) were added by syringe under a positive Ar atmosphere. The mixture was stirred in a preheated oil bath (110 °C) for 14 h. The mixture was then allowed to warm to RT, diluted with ethyl acetate (5 mL), and filtered through a Celite plug (Sigma-Aldrich, St. Louis, MO), eluting with additional ethyl acetate (10 mL). The filtrate was concentrated and purified by column chromatography on silica gel (eluting with hexanes/ethyl acetate mixtures). Following the general procedure, 6-methoxy-2-methylquinoline (86.6 mg, 0.5 mmol) was used employing Ni(cod)$_2$ (14.0 mg, 10 mol%) and PCy$_3$ (28.0 mg, 20 mol%) to give the desired product (45.8 mg, 64%) as a yellow oil.

Just as the scope of oxidative additions of Ni(0) to phenol derivatives proceeds with a much greater range of oxygen leaving groups than typically occurs with the corresponding palladium-catalyzed processes, a similar trend is observed in oxidative additions to allylic alcohol derivatives. Athough the commonly employed allylic acetates and allylic carbonates are efficient participants in various nickel-catalyzed allylic functionalization processes, simple allylic ethers and underivatized allylic alcohols themselves can efficiently participate in various processes (Scheme 3-11). As a representative example depicted subsequently, simple silylated allylic alcohol derivatives undergo couplings with Grignard reagents by using NiCl$_2$(dppf) as catalyst.[60] As evident from the product stereochemistry, the process proceeds by a net inversion, likely by oxidative addition with inversion, followed by transmetallation and reductive elimination with retention of stereochemistry.

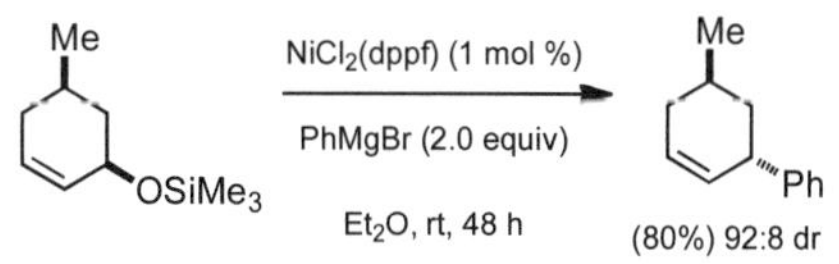

Me OSiMe$_3$ — NiCl$_2$(dppf) (1 mol %), PhMgBr (2.0 equiv), Et$_2$O, rt, 48 h → Me Ph (80%) 92:8 dr

Scheme 3-11

In a further demonstration of the broad scope of nickel(0) oxidative additions to C-O bonds, cyclic anhydrides were illustrated to be versatile substrates in cross-coupling processes.[61] For example, enantioselective desymmetrizations with diethylzinc and cyclic anhydrides provide efficient access to functionalized 1,4-dicarbonyl products. The enantioselectivity-determining step in this process is the oxidative addition of nickel(0) to anhydride C-O bond.[62] Several other functionalization processes derived from oxidative addition of nickel(0) to anhydrides were previously described.[63]

Representative procedure: synthesis of (1R,2S)-2-propionylcyclohexanecarboxylic acid[61]

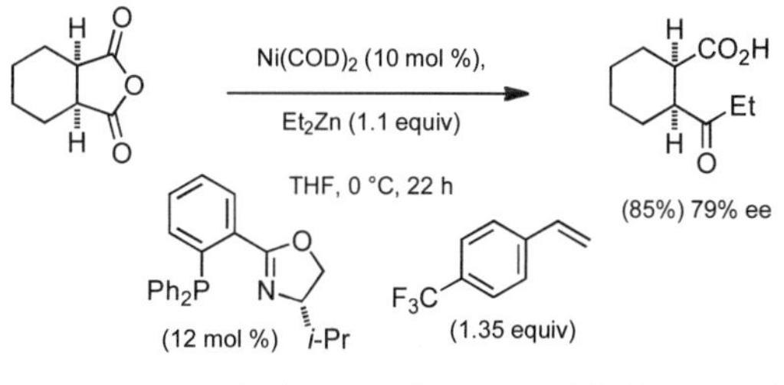

H O H O — Ni(COD)$_2$ (10 mol %), Et$_2$Zn (1.1 equiv), THF, 0 °C, 22 h → H CO$_2$H Et H O (85%) 79% ee

Ph$_2$P N O *i*-Pr (12 mol %); F$_3$C (1.35 equiv)

A flame-dried, round-bottom flask was charged with 9.9 mg (0.0360 mmol) of Ni(cod)$_2$ and 16.6 mg (0.0445 mmol) of the chiral phosphine ligand in an inert atmosphere (N_2) glove box. After removal from the glove box, 1.0 mL of THF was added *via* syringe and the solution was stirred at RT for 15 min. The solution was then cooled to 0 °C in an ice bath and 0.01 mL (0.0675 mmol) of 4-(trifluoromethyl)styrene followed by 0.5 mL (1.0-M solution in hexane, 0.5 mmol) of Et_2Zn were added *via* syringe. Hexahydroisobenzofuran-1,3-dione (55.0 mg, 0.357 mmol) in 1 mL of THF was then added *via* cannula, and the reaction was stirred at 0 °C for 22 h. The reaction mixture was then diluted with 10 mL of ether and quenched with 10 mL of 10% aq. HCl (v/v). The layers were separated, and the aqueous layer was extracted with ether (2 × 10 mL). The organic layers were combined and extracted with sat. aq. Na_2CO_3 (2 × 7 mL); the basic layers were then combined and brought to pH = 1–2 with conc. HCl. The acidified aqueous layer was extracted with Et_2O (3 × 10 mL), and the combined organic extracts were washed with brine, dried over anhydrous $MgSO_4$, filtered, and concentrated *in vacuo* to yield 55.9 mg (85%) of the desired product as a white solid in 79% ee.

The tremendous utility of catalytic C-O and C-N bond formation processes, pioneered by Buchwald and Hartwig, is largely based on the efficiency of palladium-catalyzed systems. However, efficient nickel-catalyzed processes have also been reported for this important class of transformations. The catalytic coupling of aryl chlorides with primary or secondary amines is a particular niche where utility of the nickel-catalyzed process was found.[64] A representative sampling of aniline derivatives prepared by the activity of $Ni(cod)_2$/dppf as the pre-catalyst is depicted below (Figure 3-3).

$Ni(COD)_2$/(dppf)-catalyzed couplings of aryl chlorides with primary and secondary amines.

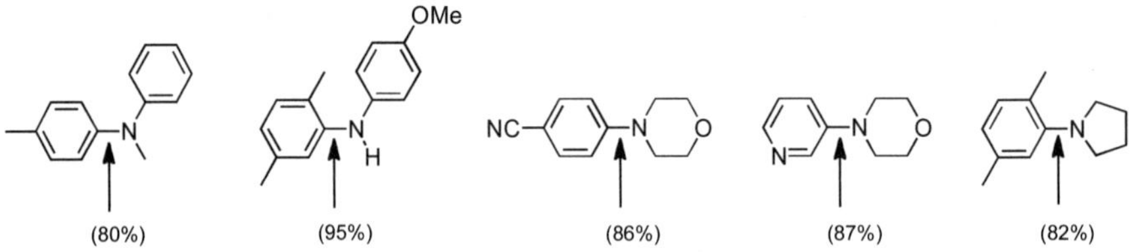

OMe
N
N
H
NC
N
O
N
O
N
N
(80%) (95%) (86%) (87%) (82%)

Figure 3-3

Representative procedure: synthesis of N-(4-methoxyphenyl)-2,5-dimethylaniline[64]

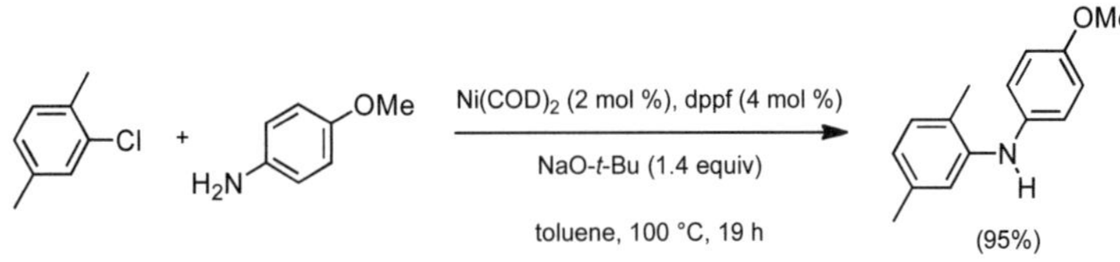

Cl + H_2N OMe
$Ni(COD)_2$ (2 mol %), dppf (4 mol %)
NaO-*t*-Bu (1.4 equiv)
toluene, 100 °C, 19 h
OMe
N
H
(95%)

General procedure: A sealable Schlenk tube was charged with $Ni(cod)_2$ (6 mg, 0.02 mmol, 2 mol%), dppf (22 mg, 0.04 mmol, 4 mol%), and sodium *tert*-butoxide (135 mg, 1.4 mmol) under Ar in a Vacuum Atmospheres Company (Hawthorne, CA) glove box. Toluene (1 mL) was added, followed by the aryl chloride (1.0 mmol), amine (1.2 mmol), and additional toluene (3 mL). The tube was sealed, removed from the glove box, and heated to 100 °C with stirring until the starting halide had been consumed as judged by gas chromatography (GC) analysis. The reaction mixture was cooled to RT, diluted with ether (10 mL), filtered, and concentrated. The crude product was purified by flash chromatography on silica gel. Following the general procedure, 2-chloro-1,4-dimethylbenzene (1.0 mmol) and 4-methoxyaniline (1.2 mmol) were used to give the desired product (216 mg, 95%) as a pale yellow solid.

In addition to extensive studies of phenol derivatives, other classes of C-O electrophiles have been used in nickel-catalyzed cross-couplings. Simple benzylic ethers have been demonstrated to undergo nickel-catalyzed cross-couplings with MeMgI (Scheme 3-12).[65] Processes of this type using racemic (1,1′-binaphthalene-2,2′-diyl)bis(diphenylphosphine) (BINAP) as ligand proceed with net inversion of configuration. This substrate-controlled outcome contrasts with the more extensively studied couplings of secondary alkyl halides, which involve destruction of the original substrate chirality and allow ligand-controlled asymmetric processes. Although the displacement of simple methyl ether required naphthyl-type extended aromatic rings, an alternative procedure employing methoxy ethyl ethers as traceless directing groups increased the scope to include simple monocyclic aromatic substituents.[66]

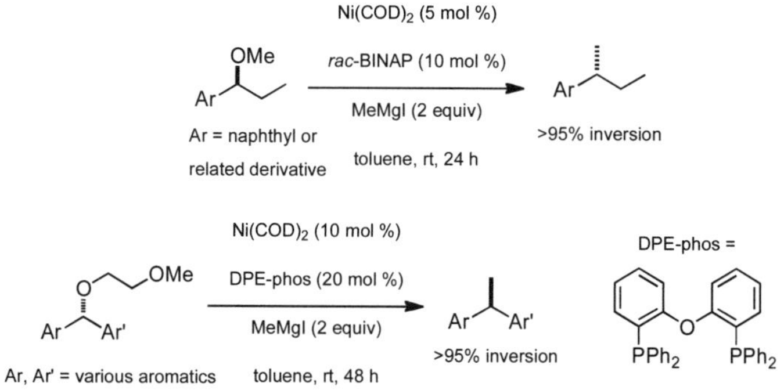

Ni(COD)$_2$ (5 mol %)
OMe
rac-BINAP (10 mol %)
Ar
MeMgI (2 equiv)
toluene, rt, 24 h
Ar = naphthyl or related derivative
Ar
>95% inversion

Ni(COD)$_2$ (10 mol %)
DPE-phos (20 mol %)
MeMgI (2 equiv)
toluene, rt, 48 h
OMe
O
Ar Ar'
Ar, Ar' = various aromatics
Ar Ar'
>95% inversion
DPE-phos =
O
PPh$_2$ PPh$_2$

Scheme 3-12

Representative procedure: synthesis of (S)-2-sec-butylnaphthalene[65]

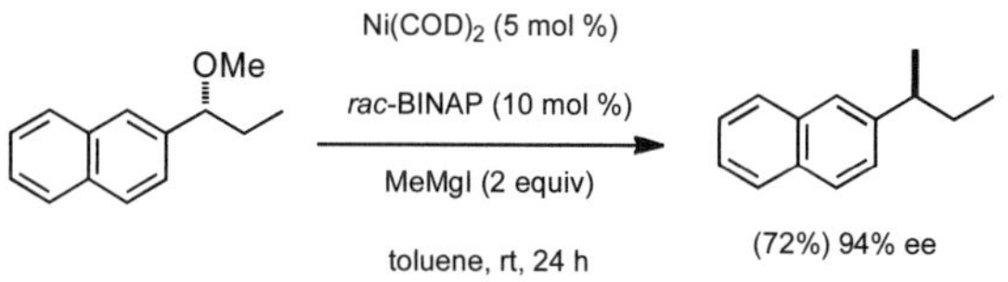

OMe
Ni(COD)$_2$ (5 mol %)
rac-BINAP (10 mol %)
MeMgI (2 equiv)
toluene, rt, 24 h
(72%) 94% ee

To a flame-dried vial in a glove box was added *bis*-(1,5-cyclooctadiene)nickel (6.9 mg, 0.025 mmol) and racemic-2,2'-bis(diphenylphosphino)-1,1'-binaphthyl (31 mg, 0.050 mmol). The vial was removed from the glove box and toluene (4 mL) was added. After stirring for 5 min, (*R*)-2-(1-methoxypropyl)naphthalene (0.100 g, 0.500 mmol) was added, followed by dropwise addition of methylmagnesium iodide (0.35 mL, 1.0 mmol, 2.9 M in Et_2O). The reaction was stirred for 24 h before quenching with saturated aqueous ammonium chloride (3 mL) and extracting with EtOAc (3 × 5 mL). The combined organics were washed with brine (3 mL), dried over anhydrous $MgSO_4$, and concentrated *in vacuo*. The product was purified by flash chromatography with silver-impregnated silica (100% pentane) to afford the title compound as a clear, colorless oil (65.9 mg, 72%).

Recent advances have also demonstrated the use of more unconventional C-O electrophiles for cross-coupling processes (Scheme 3-13). In one advance, epoxides undergo nickel-catalyzed cross-couplings with arylboronic acids concomitant with a 1,2-hydride migration.[67] This outcome is attributed to initial rearrangement of the epoxide to an aldehyde followed by addition of the boronic acid to the aldehyde. Nickel catalysis is involved in both the rearrangement and the aldehyde addition steps. Additionally, aminals and mixed acetals have been used as substrates in nickel-catalyzed cross-coupling processes.[68, 69] These latter processes provide a useful entry to highly functionalized isoindolinone and chromene derivatives.

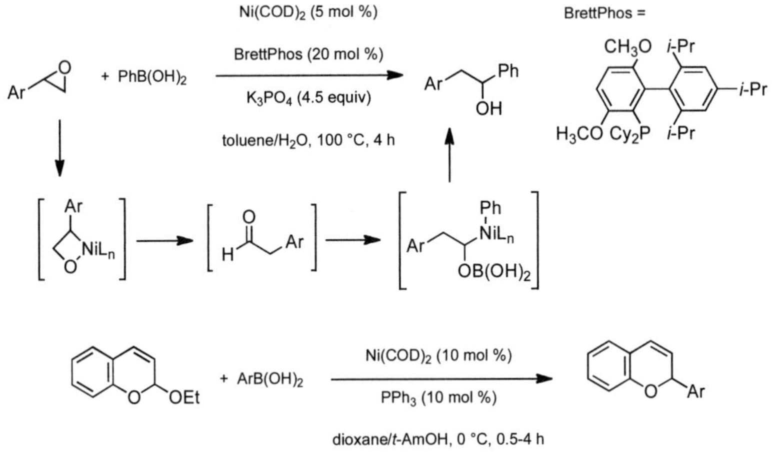

Scheme 3-13

Although the rearrangement of epoxides illustrated above precluded the direct cross-coupling capture of the kinetically produced four-membered oxametallacycle, the analogous process with aziridines does proceed with direct capture of the corresponding four-membered azametallacycle (Scheme 3-14).[70] A simple procedure employing $NiCl_2$•diglyme as pre-catalyst with a fumarate ligand allowed efficient Negishi couplings between organozincs and styrenyl aziridines. Diastereoselectivities were high with 1,2-disubstituted aziridines, proceeding with formal retention of stereochemistry. In contrast, an enantiopure monosubstituted aziridine provided the inversion stereochemistry with low enantioselectivity. Although these intriguing mechanistic complexities remain to be fully resolved, one rationale suggested by the authors is that the net retention illustrated in the subsequent examples may arise from a sulfonamide-directed radical recombination following either an S_N2 or single electron transfer-mediated oxidative addition.

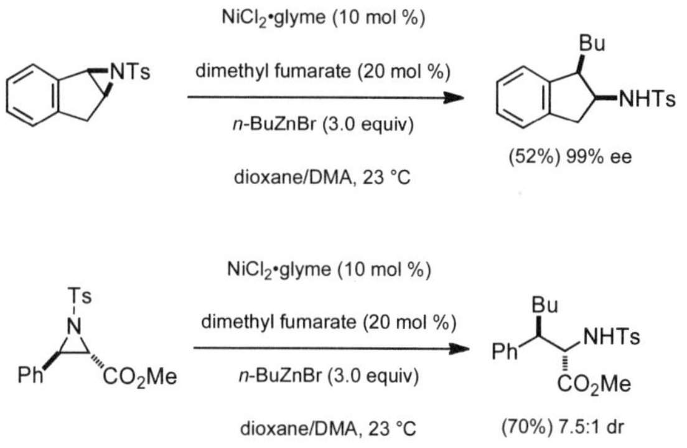

NTs

$NiCl_2$•glyme (10 mol %)
dimethyl fumarate (20 mol %)
n-BuZnBr (3.0 equiv)
dioxane/DMA, 23 °C

Bu
NHTs
(52%) 99% ee

Ts
N
Ph CO_2Me

$NiCl_2$•glyme (10 mol %)
dimethyl fumarate (20 mol %)
n-BuZnBr (3.0 equiv)
dioxane/DMA, 23 °C

Bu
Ph NHTs
CO_2Me
(70%) 7.5:1 dr

Scheme 3-14

2.5 Homo- and Hetero-Reductive Couplings of Organic Halides and Related Substrates

The nickel-catalyzed homocoupling of aryl halides has a long history, and the development of chemoselective heterocouplings of this type constitutes an active and evolving area of research.[50] A key study from Semmelhack in 1971 illustrated the homocoupling of haloarenes after exposure to $Ni(cod)_2$.[71] Kumada then illustrated that the process could be rendered catalytic in nickel by the use of zinc dust as the stoichiometric reductant.[72]

The mechanism of aryl halide homocoupling has been the subject of considerable controversy. The solvent polarity and the presence or concentration of reductant both play an important role, and it is likely that different mechanisms are operative under different conditions. A mechanistic analysis by Kochi provided a detailed description of the complex nature of processes of this type.[73] The previous discussion on early developments in nickel cross-coupling processes (section 2.1) described the competing pathways that enable production of Ni(I) and Ni(II) adducts from the oxidative addition of Ni(0) to aryl halides. Drawing on this discussion, the key features of the proposed mechanism for haloarene homocoupling may be summarized as follows (Scheme 3-15). Oxidative addition proceeds by initial electron transfer from Ni(0) complex **8** to the aryl halide to generate a Ni(I) ion pair intermediate **9**. Species **9** may collapse to generate the Ni(II)(aryl)(halide) species **10**, or alternatively, fragmentation of **9** to generate an aryl radical and Ni(I) halide species **11** may occur. Ni(I) halide species **11** undergoes oxidative addition to the halobenzene to generate species **12**. An aryl transfer process between **10** and **12** then results in the production of complexes **13** and **14**. In this mechanism, the role of the reductant is likely the regeneration of Ni(0) species **8** from Ni(II) species **13**. The conversion of **9** into **11** is solely an initiation step, as **11** is continually regenerated from reductive elimination of **14**.

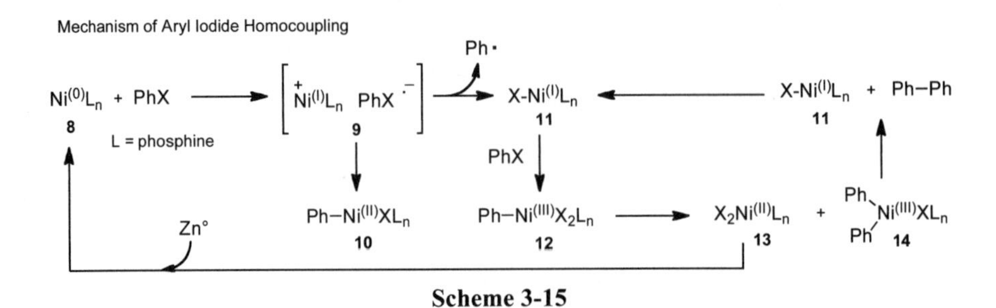

Scheme 3-15

Subsequent studies modified this mechanism, to expand the proposed role of zinc as a reductant in the initial production and the continual regeneration of a Ni(I) aryl species **15**.[74] This proposed catalytic cycle, which better explains the experimental observations in aryl chloride homocoupling, involves generation of **15** initially from reduction of adduct **10** (Scheme 3-16). Species **15** then undergoes oxidative addition with the aryl halide to generate diaryl Ni(III) species **14**. Reductive elimination of **14** forms the product biaryl along with Ni(I) halide **11**. Oxidative addition of aryl halide to **11** generates Ni(III) species **12**, which undergoes reduction with zinc dust to regenerate the key intermediate **15**. For additional discussion of other proposed mechanisms, the reader is referred to an excellent review on the subject.[50]

Mechanism of Aryl Iodide Homocoupling

Ni(0)Ln + PhX (**8**; L = phosphine) → Ph–Ni(II)XLn (**10**) —Zn°→ Ph–Ni(I)Ln (**15**) ←Zn°— Ph–Ni(III)X2Ln (**12**)

15 —PhX→ Ph2Ni(III)XLn (**14**) → X-Ni(I)Ln (**11**) + Ph–Ph

11 —PhX→ **12**

Scheme 3-16

Recent advances have enabled the participation of phenol-derived substrates, including triflates and mesylates, to participate in homocoupling processes to generate functionalized biaryls.[75] In these processes, the addition of a tetraalkylammonium iodide was beneficial in enhancing the rate of homocouplings. A broad range of symmetrical biaryls may be generated efficiently by this procedure.

Representative procedure: synthesis of 4,4'-dicarbomethoxybiphenyl[75]

MeO2C–C6H4–OSO2Me → MeO2C–C6H4–C6H4–CO2Me (99%)

NiCl2(PPh3)2 (10 mol %)
Zn powder (1.7 equiv), Et4NI (1.5 equiv)
THF, reflux, 5-10 h

General procedure: All reactions were carried out under nitrogen using oven-dried (110 °C) glassware. In a typical reaction, a 125-mL Schlenk tube was charged with $NiCl_2(PPh_3)_2$ (0.10 mmol), Zn powder (1.7 mmol), Et_4NI (1.5 mmol), and a magnetic stirring bar. After the tube was sealed with a rubber septum, the contents were dried at 22 °C under reduced pressure (1 mmHg) for 10 h. The contents of the tube were then placed under an Ar atmosphere by filling with Ar followed by three evacuation-filling cycles. Freshly distilled THF (0.50 mL) was added *via* a syringe through the rubber septum. The mixture was stirred at RT for 5 min, and during this time, the color of the mixture gradually became deep red-brown. Aryl sulfonate (1.0 mmol) was dissolved in freshly distilled THF (0.50 mL) and added to the catalyst mixture *via* a syringe through the rubber septum. The reaction mixture was heated at reflux and stirred at this temperature for 5–10 h. The reaction mixture was then cooled, filtered, diluted with water, extracted with $CHCl_3$, dried over anhydrous $MgSO_4$, and the solvent was evaporated *in vacuo*. The corresponding biaryl was obtained after column chromatography (silica gel, *n*-hexane/ethyl acetate) and recrystallization from $CHCl_3$/hexanes. Following the general procedure, methyl 4-((methylsulfonyl)oxy) benzoate (1.0 mmol) was used to give the desired product (99%) as a white crystalline solid.

Recent advances have described the homocoupling of primary and secondary sp^3 alkyl halides to generate functionalized alkanes.[76] Although the strategy holds promise in the synthesis of symmetrical hydrocarbons, the most intriguing advance is in the discovery of selective cross-couplings of aryl halides and alkyl halides under reducing conditions (Scheme 3-17).[77] Early reports illustrated that α-chlorocarbonyls undergo cross-couplings with bromoarenes in modest yield under electrochemical reductive conditions.[78] Recent studies have described a more general protocol wherein a range of activated and nonactivated alkyl bromides and iodides undergoes cross-couplings with aryl bromides and iodides.[79] In the original disclosure, an unusual catalyst composition was utilized involving 10.7 mol% NiI_2 with 5 mol% of a bipyridine, 5 mol% of a *bis*-phosphine, and 10 mol% pyridine, with manganese powder as the stoichiometric reductant. A second-generation procedure of substantially broadened scope involved Zn° as the reductant, sodium iodide as an additive, and modified ligand structures.[80] In these methods, the origin of the observed selectivity for homocoupling has not been fully elucidated. However, several pieces of evidence were presented to illustrate that direct insertion (Mn° or Zn°) to the alkyl or aryl halide was not involved. Instead, it seems likely that different oxidation states of nickel undergo selective oxidative addition to the aryl or alkyl halide at different stages of the catalytic process. Related advances involving selective cross-coupling of two different sp^3 alkyl halides were recently described, using $Ni(cod)_2$ with pybox-type ligands, and with zinc dust as the stoichiometic reductant.[81] In this very challenging problem, significant homocoupling was observed, and an excess of one of the alkyl halides was required to achieve moderate yields.

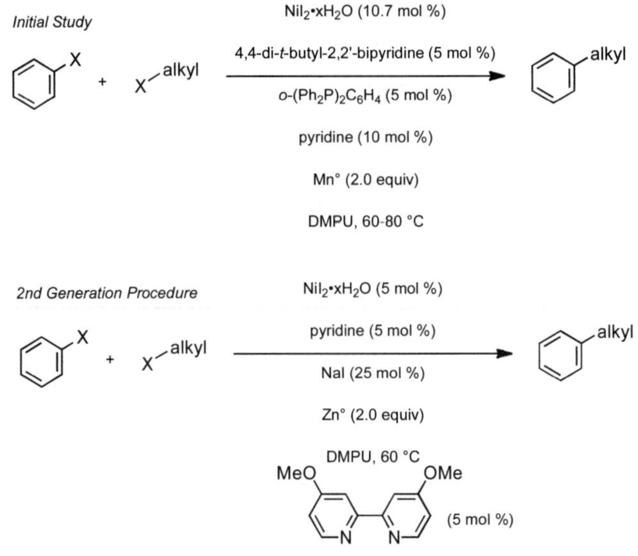

Initial Study

Ph–X + X–alkyl → Ph–alkyl

$NiI_2 \cdot xH_2O$ (10.7 mol %)
4,4-di-*t*-butyl-2,2'-bipyridine (5 mol %)
o-$(Ph_2P)_2C_6H_4$ (5 mol %)
pyridine (10 mol %)
Mn° (2.0 equiv)
DMPU, 60-80 °C

2nd Generation Procedure

Ph–X + X–alkyl → Ph–alkyl

$NiI_2 \cdot xH_2O$ (5 mol %)
pyridine (5 mol %)
NaI (25 mol %)
Zn° (2.0 equiv)
DMPU, 60 °C
4,4'-dimethoxy-2,2'-bipyridine (MeO, OMe) (5 mol %)

Scheme 3-17

Representative procedure: synthesis of t-butyl (3-phenylpropyl)carbamate[80]

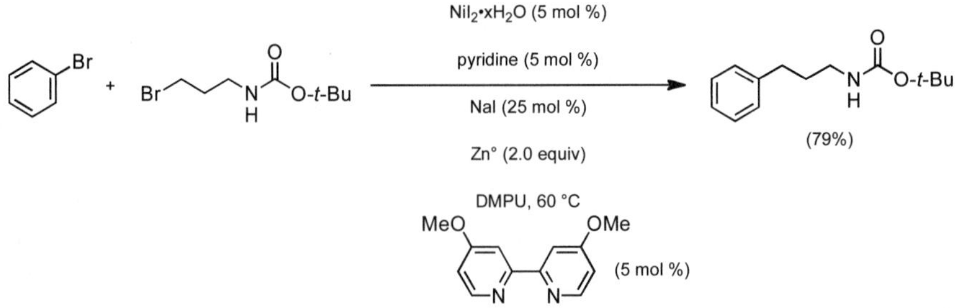

PhBr + Br(CH2)3NHC(O)O-*t*-Bu → Ph(CH2)3NHC(O)O-*t*-Bu (79%)

$NiI_2 \cdot xH_2O$ (5 mol %)
pyridine (5 mol %)
NaI (25 mol %)
Zn° (2.0 equiv)
DMPU, 60 °C
4,4'-dimethoxy-2,2'-bipyridine (MeO, OMe) (5 mol %)

General procedure (on the bench top with no precaution to exclude air or moisture): $NiI_2 \cdot xH_2O$ (x = 3.5, MW = 375.56 determined by elemental analysis, 15.1–30.2 mg, 0.040–0.080 mmol, 0.053–0.107 equiv) 4,4'-dimethoxy-2,2'-bipyridine (for electron rich aryl halides, 8.1–16.2 mg, 0.0375–0.0750 mmol, 0.050–0.100 equiv) or 1,10-phenanthroline (for electron-poor aryl halides, 6.8–13.6 mg, 0.0377–0.0750 mmol, 0.050–0.100 equiv), sodium iodide (28.5 mg, 0.190 mmol, 0.250 equiv), and solid substrates (0.75 mmol, 1.00 equiv) were weighed on weigh paper and transferred to a 1-dram vial equipped with a magnetic stir bar. Solvent, DMPU (1,3-dimethyl-3,4,5,6-tetrahydro-2(1H)-pyrimidinone (3.0 mL), pyridine (3–6 μL, 0.039–0.0775 mmol, 0.052–0.103 equiv), liquid reagents, and unactivated zinc dust (>10 μm, 98 mg, 1.50 mmol, 2.00 equiv) were added. The reaction vials were capped with a polytetrafluoroethylene (PTFE)-faced silicone septum, and the generally green solution was stirred at RT for approximately 5 min before heating to 60 °C or 80 °C in a reaction block on the bench top. After completion (judged as <1 area % of starting materials remaining by GC analysis), the entire reaction mixture was loaded onto a silica gel column for purification. The reaction vial was rinsed with ether and dichloromethane (1 mL each), which were also loaded on the column. After the general procedure, bromobenzene and *t*-butyl (3-bromopropyl)carbamate and 4,4'-dimethoxy-2,2'-bipyridine were used to afford the desired product (79% yield).

Achieving selective heterocouplings of two different aryl or alkyl chlorides under reducing conditions, and thus avoiding the need for preassembly of a stoichiometric transmetallating partner, is an exciting development in cross-coupling technology that will undoubtedly stimulate much future study.

2.6 Cross-Couplings Involving C-H Functionalization

Although primarily examined with palladium catalysis, cross-couplings involving direct functionalization of a C-H bond have also been examined in nickel-catalyzed processes. The vast majority of these nickel-catalyzed C-H functionalizations involves oxazoles, thiazoles, and benzoxazoles, which undergo facile metallation to organonickel intermediates (Scheme 3-18). The processes have been mostly developed between oxazoles and aryl halides or phenol derivatives, which result in thc dircct loss of IICl or HOTf on coupling.[82–84] Alternatively, oxidative couplings between oxazoles and organosilanes have been developed using Cu oxidants.[85]

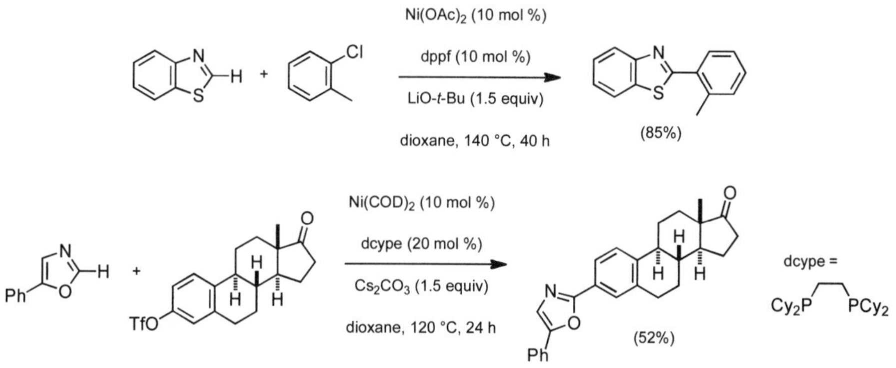

Scheme 3-18

Although electron-rich aromatics are most commonly employed in processes of this type, a nickel-catalyzed coupling of organozincs with electron-deficient heteroaromatics such as pyridines has been developed.[86] This process likely involves initial catalyzed addition of the organozinc to the 2-position of the pyridine, followed by dihydropyridine oxidation to afford product. The authors demonstrated that dihydropyridine oxidation in the presence of organozincs proceeds even in the absence of nickel.

Representative procedure: synthesis of 2-phenylquinoline[86]

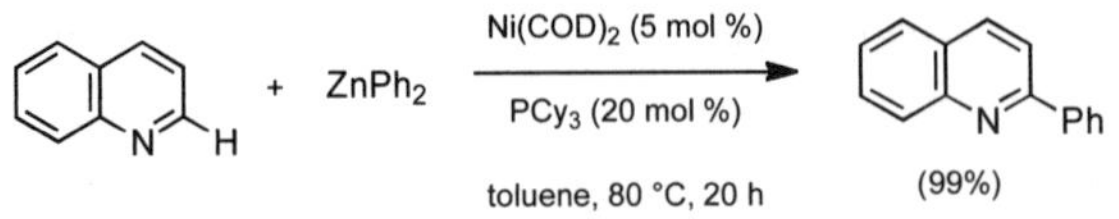

Ni(COD)$_2$ (5 mol %), PCy$_3$ (20 mol %), toluene, 80 °C, 20 h; + ZnPh$_2$; (99%)

To an oven-dried, 10-mL, pressure-screw-capped vial, Ni(cod)$_2$ (3.4 mg, 0.0125 mmol), PCy$_3$ (7.0 mg, 0.025 mmol), Ph$_2$Zn (83.5 mg, 0.38 mmol), quinoline (0.25 mmol), and toluene (1.0 mL) were added in a glove box. The cap was fastened and the mixture was stirred for 20 h at 80 °C. The reaction mixture was subjected to column chromatography on silica gel to afford the desired product (99% yield).

3. Cycloaddition Processes

Note on coverage: This section includes reaction classes in which a new ring is formed *via* two bond formations, irrespective of the mechanism or redox characteristics of the process. Classifications [*i.e.,* X+Y+Z] refer to the number of atoms that each reactant contributes to the new ring created. The reactions covered are always stepwise processes that involve discreet organometallic intermediates; therefore, classifications used in pericyclic processes are not relevant here. Processes that involve nickel-based Lewis acids, although extremely useful, fall outside of the scope of reactions covered in this chapter.

3.1 Early Developments in Nickel-Catalyzed Cycloadditions

Review. Wilke, G.; *Angew. Chem., Int. Ed.* **1988,** *27*, 185 – 206.

The enormous scope and utility of nickel-catalyzed cycloadditions can be traced to pioneering studies of Reppe, who reported in 1948 that nickel catalysts promote the cyclooligomerization of acetylene to cyclooctatetraene.[87] Not only was this unprecedented transformation described, but also a remarkable level of mechanistic insight was articulated in this report. In particular, this report depicted the formation of a 9-membered metallacycle en route to formation of cyclooctatetraene (Scheme 3-19).

(CO)$_2$Ni + 4 HC≡CH → [9-membered nickelacycle] → (CO)$_2$Ni + cyclooctatetraene

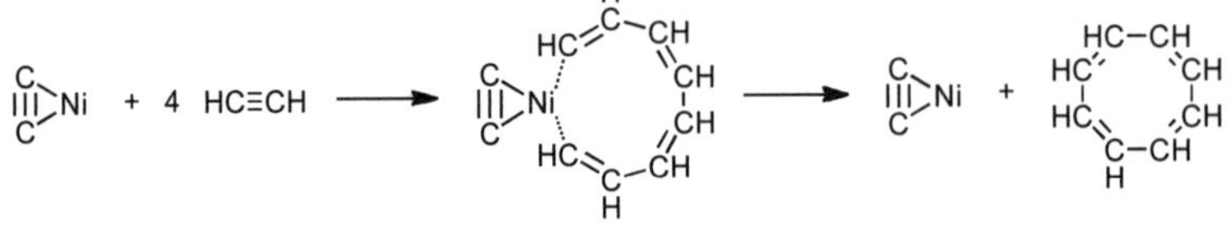

Scheme 3-19

Early reports from Wilke illustrated the remarkably diverse range of structures that are derived from cyclic oligomerizations of butadiene.[88] For example, the two-component cycloaddition of butadiene was illustrated to follow [2+2], [4+2], and [4+4] pathways to generate cyclobutane, cyclohexene, and cyclooctadiene products, in addition to three-component cycloadditions to produce 12-membered ring adducts (Figure 3-4). The involvement of pathways involving oxidative cyclizations to metallacycle intermediates, followed by product-forming reductive elimination, were elucidated during these studies. Much of the understanding of both fundamental reactivity and mechanism in nickel-catalyzed cycloadditions can be attributed to these early studies.

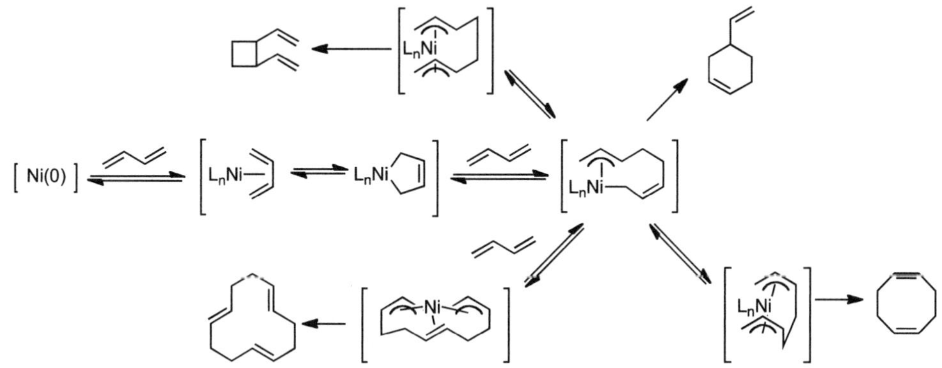

[Ni(0)] L_nNi L_nNi L_nNi L_nNi Ni L_nNi

Figure 3-4

3.2 [4+2] Cycloadditions

Despite the tremendous range of thermal and Lewis acid-promoted [4+2] cycloadditions that can be routinely accomplished, the nickel-catalyzed counterpart to these reactions provides unique opportunities for efficiency and selectivity. The [4+2] cycloaddition of 1,3-dienes with alkynes is a subclass of reactions that enjoys particular benefits from nickel catalysis. In illustrations involving assembly of the all-carbon cyclohexadiene units, reports from Wender [89–92] described the efficient intramolecular cycloaddition of 1,3-dienes with alkynes (Scheme 3-20). Notably, the corresponding thermal processes either failed to proceed or required harsher conditions, providing substantially lower yields than the corresponding nickel-catalyzed processes. The stereochemistry of the diene is conserved in the cycloadditions, as a comparison in Scheme 3-20 of *E,E*- and *E,Z*-dienes illustrates.

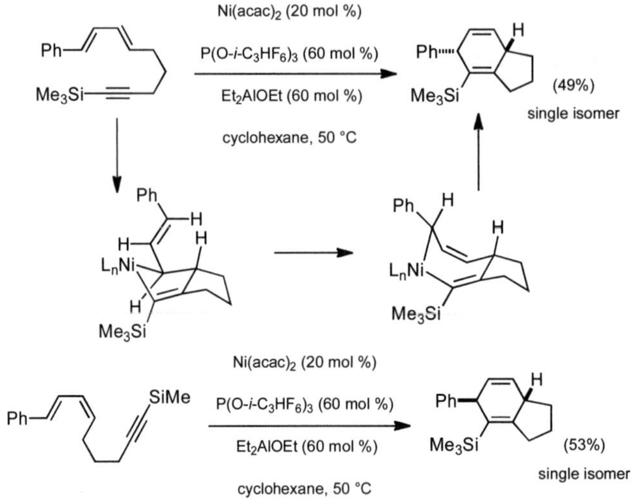

Scheme 3-20

Complex examples of the diene-alkyne [4+2] cycloaddition reaction have been illustrated in syntheses of a yohimban alkaloid and vitamin D analogs (Scheme 3-21). A representative procedure for the synthesis of a vitamin D analog is provided below.

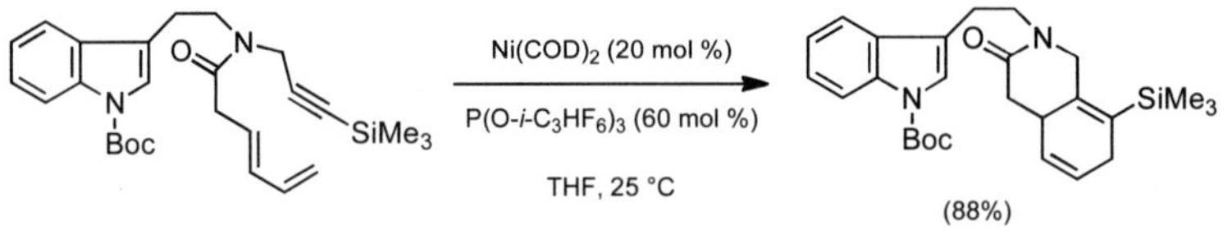

Scheme 3-21

Representative procedure: synthesis of (1S,5R*,7aS*)-[4-(2-methoxymethoxyethyl)-5-(4-methoxyphenyl)-7a-methyl-2,3,5,7a-tetrahydro-1H-inden-l-yloxy]trimethylsilane*[92]

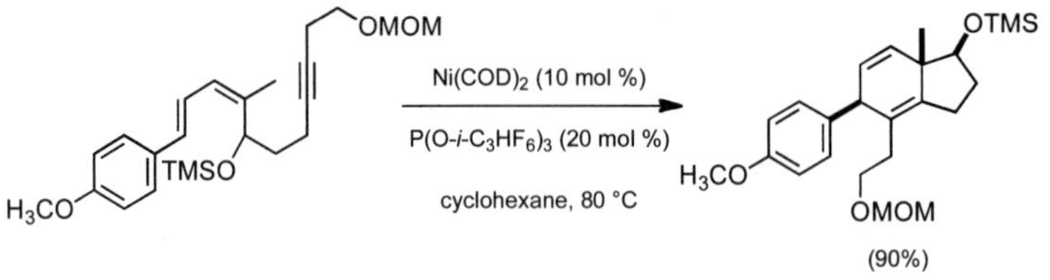

To an acid-washed, base-washed, 200-mL Schlenk flask was added the dienyne (665 mg, 1.60 mmol, 1 equiv). Under a positive nitrogen flow, freshly distilled cyclohexane (160 mL) and *tris*-(hexafluoroisopropyl) phosphite (170 mg, 0.319 mmol, 0.2 equiv) were added. *bis*-1,5-Cyclooctadiene nickel (2.13 μL, 0.075 M in THF, 0.160 mmol, 0.1 equiv) was added, and the reaction was stirred at RT for 1 h. The clear solution slowly changed to a golden yellow solution that was then warmed to

80 °C and stirred for 17.5 h. The reaction was quenched by opening to air and stirring for 30 min. Purification by flash filtration through a 1-inch plug of silica (20% Et_2O/hexanes) followed by flash chromatography (12.5% Et_2O/hexanes) afforded the desired product (597 mg, 90%).

Related [4+2] cycloadditions involving substrates that incorporate heteroatoms into the formed six-membered ring are also known. In representative examples from Matsubara[93] and Ogoshi,[94] the cycloaddition of alkylidene malonates with alkynes provide dihydropyran products, whereas cycloadditions of 1,3-dienes with nitriles provide pyridine products (Scheme 3-22). The formation of pyridines arises from oxidation of initially formed dihydropyridines.

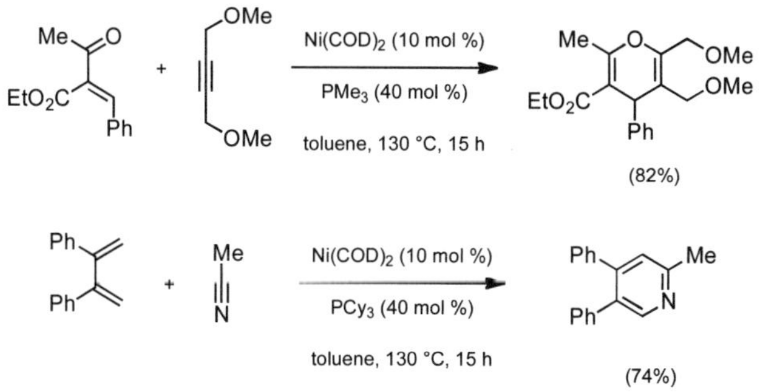

Me O EtO2C Ph + OMe OMe — Ni(COD)2 (10 mol %), PMe3 (40 mol %), toluene, 130 °C, 15 h → Me O OMe OMe EtO2C Ph (82%)

Ph Ph + Me N — Ni(COD)2 (10 mol %), PCy3 (40 mol %), toluene, 130 °C, 15 h → Ph Me Ph N (74%)

Scheme 3-22

In addition to the preceding processes that involve cycloadditions in direct analogy to Diels-Alder-type processes, several formal [4+2] cycloaddition processes have been described that proceed *via* completely different substrate classes and reaction pathways. In one example, a novel two-carbon ring expansion process was reported by Murakami, wherein the addition of cyclobutanones with alkynes provides cyclohexenones by two-carbon ring expansion of the starting material (Scheme 3-23).[95, 96] The mechanism of this process likely involves oxidative cyclization of the ketone and alkyne with Ni(0) to provide a five-membered metallacycle, followed by a ring-expanding β-carbon elimination as key steps of the process.

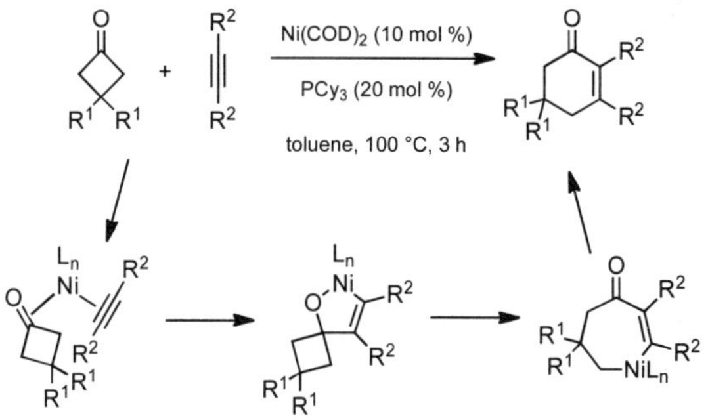

O R1 R1 + R2 R2 — Ni(COD)2 (10 mol %), PCy3 (20 mol %), toluene, 100 °C, 3 h → O R2 R1 R1 R2

Ln Ni R2 O R2 R1 R1 → Ln Ni O R2 R2 R1 R1 → O R2 R1 R1 R2 NiLn

Scheme 3-23

Representative procedure: synthesis of 5-methyl-5-phenyl-2,3-dipropyl-2-cyclohexenone[95]

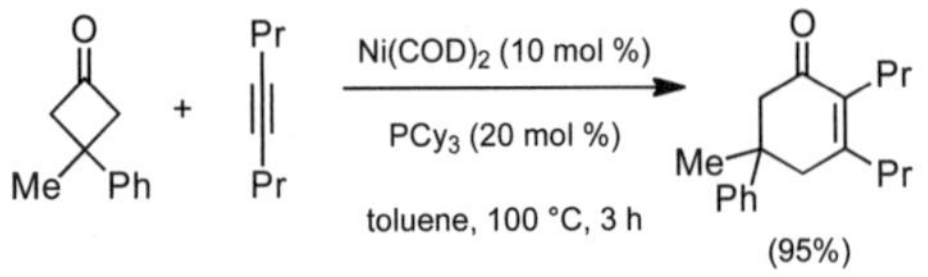

Ni(COD)$_2$ (10 mol %), PCy$_3$ (20 mol %), toluene, 100 °C, 3 h (95%)

To a toluene solution (1.0 mL) of Ni(cod)$_2$ (5.5 mg, 0.02 mmol) and PCy$_3$ (11.2 mg, 0.04 mmol) were added 3-methyl-3-phenylcyclobutanone (32.7 mg, 0.20 mmol) and 4-octyne (33 mg, 0.30 mmol). After being stirred for 3 h at 100 °C, the reaction mixture was concentrated, and the residue was purified by preparative thin layer chromatography of silica gel (hexane:EtOAc = 9:1) to afford the desired product (52.4 mg, 95%).

In another unconventional reaction pathway, formal cycloadditions have been described by a dehydrogenative pathway involving a double C-H activation mechanism. This class of reactions provides an oxidative cycloaddition entry to six-membered rings. In a report by Hiyama, oxidative cycloadditions of simple formamides and alkynes provide unsaturated lactam derivatives by a process involving formation of two carbon–carbon bonds (Scheme 3-24).[97] The proposed mechanism for this unusual process involves consumption of one equivalent of alkyne through hydrogenation as the redox partner to drive the oxidative cycloaddition. The process is initiated by the oxidative addition of Ni(0) to an *N*-formyl functional group activated by AlMe$_3$ coordination, followed by alkyne insertion to the nickel-hydride to generate intermediate **16**. Further C-H activation extrudes the product from alkyne hydrogenation along with five-membered metallacycle **17** derived from coordination of a second equivalent of alkyne. Insertion of the alkyne followed by reductive elimination then provides the oxidative cycloaddition product.

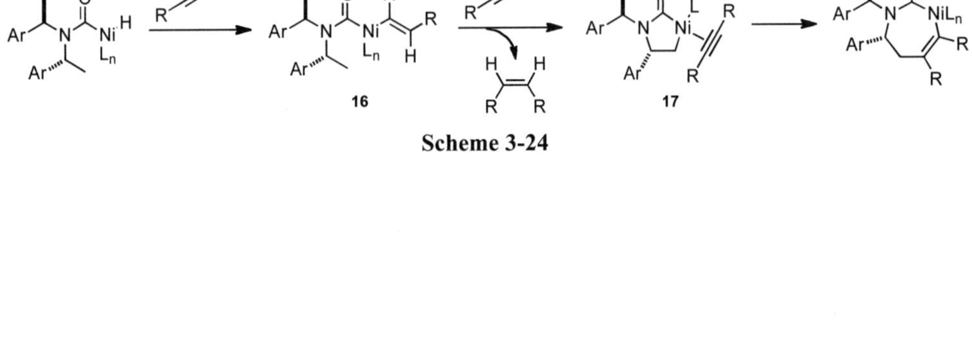

Ni(COD)$_2$ (1 mol %), P(*t*-Bu)$_3$ (4 mol %), AlMe$_3$ (20 mol %), toluene, 80 °C; (1.0 equiv), (2.2 equiv); **16**; **17**

Scheme 3-24

Representative procedure: synthesis of (R)-6-phenyl-1-[(R)-1-phenylethyl]-3,4-dipropyl-5,6-dihydropyridin-2(1H)-one[97]

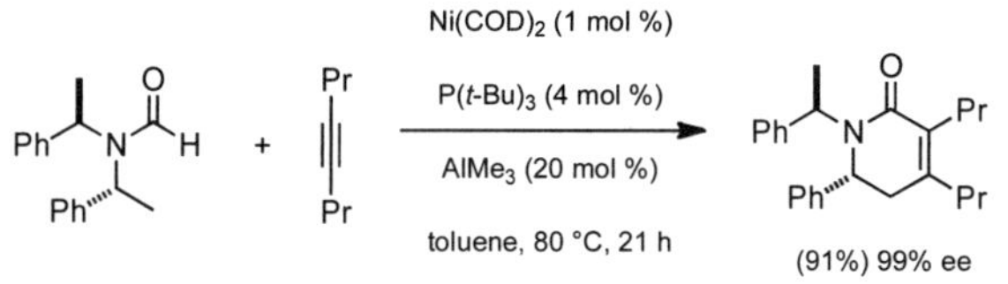

Ni(COD)$_2$ (1 mol %)
P(*t*-Bu)$_3$ (4 mol %)
AlMe$_3$ (20 mol %)
toluene, 80 °C, 21 h
(91%) 99% ee

General procedure: An alkyne (2.2 mmol) was added to a solution of a formamide (1.00 mmol), Ni(cod)$_2$ (2.8–28 mg, 10–100 μmol), P(*t*-Bu)$_3$ (8.1–81 mg, 40–400 μmol), a 1.08-M solution of AlMe$_3$ in hexane (0.19 mL, 0.20 mmol), and dodecane (an internal standard, 57 mg, 0.33 mmol) in toluene prepared in a 15-mL vial under an Ar atmosphere in a dry box. The vial was sealed with a screw cap, taken outside the dry box, and heated at the temperature for the time specified. The resulting mixture was filtered through a silica gel pad, concentrated *in vacuo*, and purified by flash column chromatography on silica gel to give the desired product. Isomers were further separated by preparative recycling silica gel chromatography. Following the general procedure at 80 °C with a 21-h reaction time and 1 mol% Ni(cod)$_2$ and 4 mol% P(*t*-Bu)$_3$, *N,N-bis*-((*R*)-1-phenylethyl)formamide and 4-octyne were used to give the desired product (91%, 99% ee).

A related process from Chatani described a process involving oxidative cycloaddition of aromatic amides and alkynes to produce six-membered lactams (Scheme 3-25).[98] This sequence involves formation of one carbon–carbon and one carbon–nitrogen bond during the cycloaddition, again with alkyne hydrogenation serving as the reduction that accompanies the oxidative cycloaddition. A key requirement of this process is the presence of a pyridyl directing group to drive the complexation of nickel to the amide nitrogen.

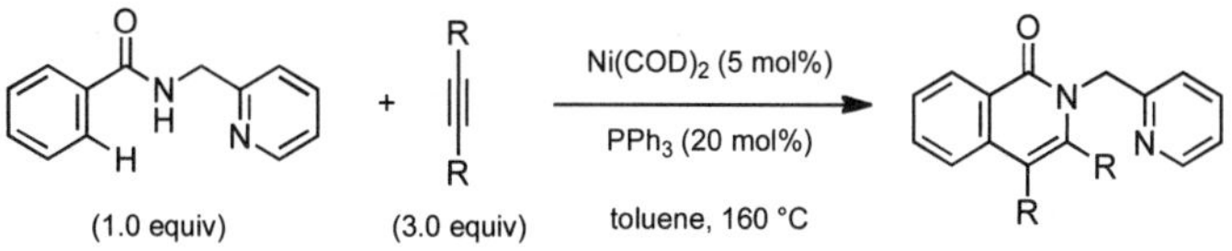

(1.0 equiv) + (3.0 equiv) Ni(COD)$_2$ (5 mol%), PPh$_3$ (20 mol%), toluene, 160 °C

Scheme 3-25

3.3 [2+2+2] Multicomponent Cycloadditions

Review. Chopade, P.R.; Louie, J.; *Adv. Synth. Cat.* **2006,** *348*, 2307 – 2327.

A large number of nickel-catalyzed [2+2+2] cycloadditions have been reported in many different contexts. Exhaustive coverage is well beyond the scope of this chapter, and an excellent review noted above is devoted to the subject. The discussion herein will provide a representative sampling of these processes to demonstrate the types of reactivity and mechanisms that are possible and to highlight the processes that are mechanistically unique or that are of notable synthetic utility.

The trimerization of alkynes to substituted benzenes and the cotrimerization of alkynes with activated alkenes to form cyclohexenes and cyclohexadienes was initially reported in early studies from Reppe.[99] Since that time, many types of [2+2+2] processes have been developed. All carbon frameworks may be prepared by the cyclotrimerization of different alkynes or combinations of alkynes with alkenes, allenes, or ketenes, but in all of these cases, the control of chemoselectivity and regioselectivity remains a substantial challenge. A common strategy to simplify these considerations involves the connection of two or three components *via* tethers, which sometimes allows a single product to be selectively obtained. For example, 1,6- and 1,7-diynes undergo efficient 1:1 cycloaddition with simple alkynes to generate functionalized benzene derivatives (Scheme 3-26).[100, 101] Additionally, a range of heterocycles may be prepared using acetylene by a related procedure, and prochiral triynes may be cyclized in an asymmetric fashion using chiral nickel catalysts.[102]

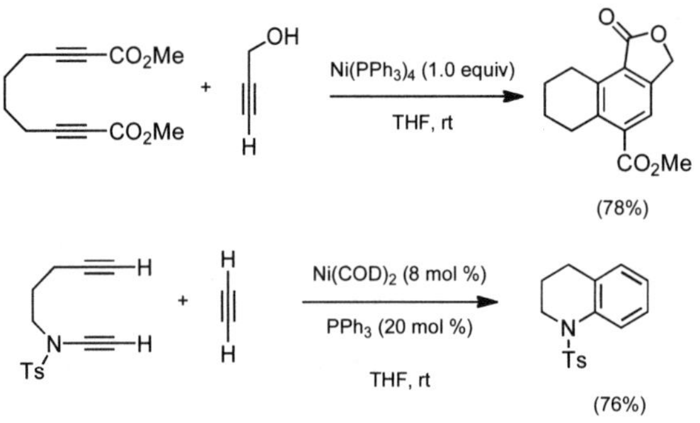

CO_2Me / CO_2Me + OH / H; $Ni(PPh_3)_4$ (1.0 equiv), THF, rt; CO_2Me (78%)

H / N—H, Ts + H / H; $Ni(COD)_2$ (8 mol %), PPh_3 (20 mol %), THF, rt; N, Ts (76%)

Scheme 3-26

Tethered diynes undergo chemoselective 1:1 cycloaddition with simple enones to generate substituted benzenes (after air oxidation of the initially formed cyclohexadiene adduct) or the corresponding cyclohexadiene (if oxidation is blocked by quaternization) (Scheme 3-27).[103] Similarly, tethered alkynyl enones undergo chemoselective cycloadditions with simple enones to generate cyclohexenes.[104] In contrast to other catalysts (*i.e.*, Co and Rh) that display a strong preference for alkynes over alkenes in cyclotrimerizations, the nickel-catalyzed cycloaddition of alkynyl enones illustrated a strong preference for incorporation of two enones with one alkyne. An implication of this finding is the opportunity to generate four contiguous stereocenters during cycloaddition. Furthermore, stereochemistry studies showed that the cycloaddition of two enones with one alkyne proceeds without stereospecificity, as either *E*- or *Z*-enones provide the identical stereochemical outcome, likely because of the involvement of a nickel *O*-enolate intermediate species. Related procedures have been developed for the [2+2+2] cycloaddition of diynes with allenes,[105] of alkenes and alkynes with arynes,[106] and of diynes with ketenes.[107]

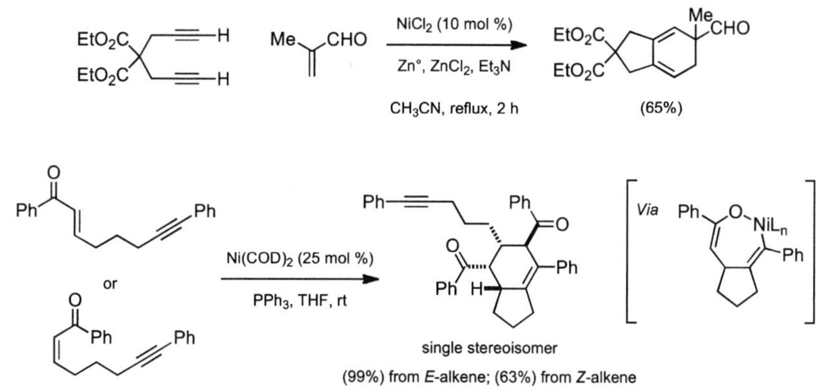

Scheme 3-27

Representative procedure: synthesis of 3aR,4R*,6R*-6-acetyl-4-benzoyl-7-phenyl-1H-indene*[104]

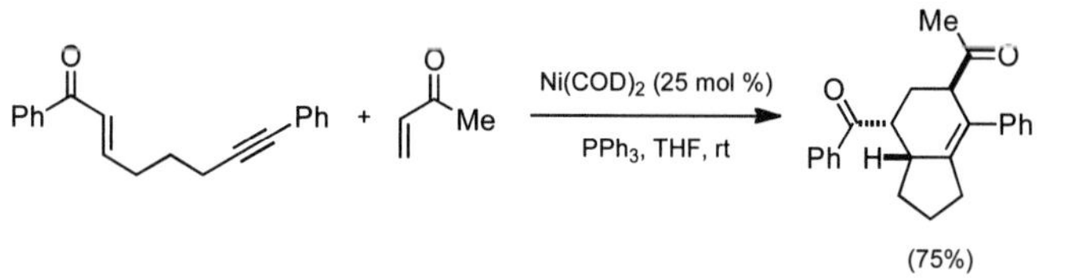

General procedure: A 0.02–0.04-M solution of PPh_3 (0.4–1.0 equiv) in THF was added to $Ni(cod)_2$ (0.20-0.25 equiv) at 0 °C and stirred for 2 min. The nickel solution was transferred by cannula to a 0.4–0.5-M, 0 °C THF solution of the simple enone (5.0 equiv) and the alkynyl enone substrate (1.0 equiv). The reaction was stirred at 0 °C for 5 min and then at 25 °C until the starting material consumption was observed by TLC analysis (generally 1.5–3.0 h). The reaction mixture was subjected to an extractive workup (NH_4Cl/NH_4OH pH = 8 buffer/Et_2O) followed by flash chromatography on SiO_2. Following the general procedure (with stirring at 25 °C for 20 h), (*E*)-1,8-diphenyloct-2-en-7-yn-1-one (50 mg, 0.183 mmol), methyl vinyl ketone (76 mL, 0.915 mmol), $Ni(cod)_2$ (13 mg, 0.046 mmol), and PPh_3 (48 mg, 0.183 mmol) were employed to produce, after flash chromatography (15:1 hexanes:EtOAc), in order of elution, 11.5 mg (23%) of the dimer of the alkynyl enone and 47 mg (75%) of the desired product as a colorless oil.

In addition to the preceding processes that involve the preparation of carbocyclic ring systems, [2+2+2] cycloadditions have also been successfully employed with substrates that enable the incorporation of nitrogen and oxygen into the ring formed by cycloaddition. Early studies from Hoberg illustrated the cycloaddition of two alkynes with an isocyanate,[108] and processes of this type have been extensively developed in intramolecular versions to provide access to an array of substituted pyridone derivatives.[109] More recent studies have illustrated that

N-heterocyclic carbene ligands are generally most effective at promoting cycloadditions of this type (Scheme 3-28). Fully intermolecular versions of a related process are possible by the 2:1 coupling of isocyanates with allenes to produce dihydropyrimidine-2,4-diones.[110]

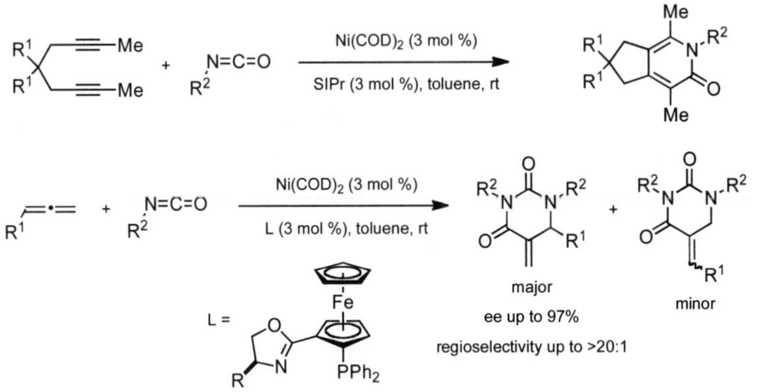

Scheme 3-28

> *Representative procedure: synthesis of dimethyl 1,4-dimethyl-3-oxo-2-phenyl-5,7-dihydro-2H-cyclopenta[c]pyridine-6,6(3H)-dicarboxylate*[109] 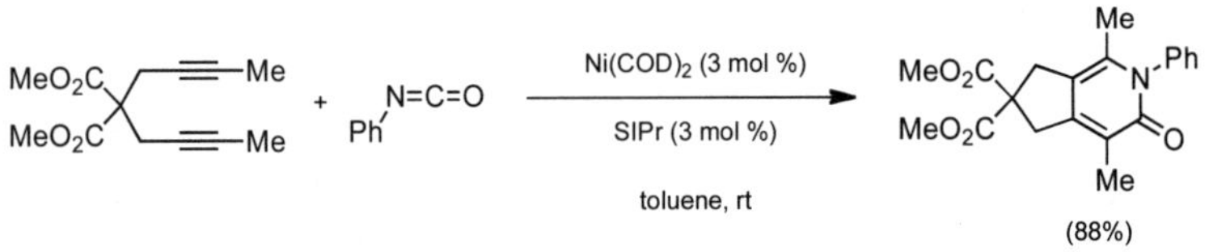
>
> General procedure: A toluene solution of $Ni(cod)_2$ and SIPr was prepared and allowed to equilibrate for at least 6 h. In a glove box, a solution of diyne and isocyanate in toluene was added to an oven-dried vial equipped with a stir bar. To the stirring solution, a solution of $Ni(cod)_2$ and SIPr was added and the reaction was stirred at RT for 30 min (or until complete consumption of starting material was observed as judged by GC). The mixture was then concentrated and purified by silica gel column chromatography. The general procedure was used with dimethyl 2,2-di(but-2-yn-1-yl)malonate (200 mg, 0.85 mmol, 0.1 M), phenyl isocyanate (100 mg, 0.85 mmol, 0.1 M), $Ni(cod)_2$ (7.0 mg, 0.025 mmol, 3 mol%), SIPr (9.8 mg, 0.025 mmol, 3 mol%), and 8.5 mL of toluene. The reaction mixture was purified by column chromatography on silica gel (hexane/ethyl acetate 1:2) to afford the desired product as a white solid (264 mg, 88%).

The related cycloaddition of diynes with carbon dioxide has also been developed as an efficient entry to pyrones (Scheme 3-29). As observed with isocyanate couplings, initial reports described the use of phosphine complexes of Ni(0),[111] but more recent work with *N*-heterocyclic carbene ligands demonstrated considerably milder conditions and broader scope.[112] Additionally, the cycloaddition of diynes with nitriles efficiently proceeds to generate pyridine derivatives under similar conditions.[113, 114]

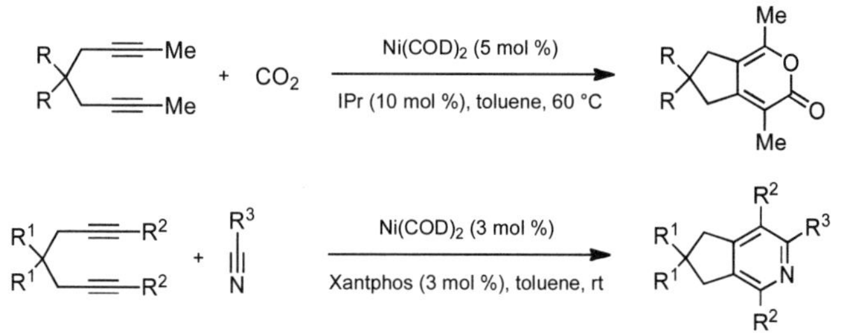

Scheme 3-29

Representative procedure: synthesis of dimethyl 4-methyl-3-phenyl-5H-cyclopenta[c]pyridine-6,6(7H)-dicarboxylate[114]

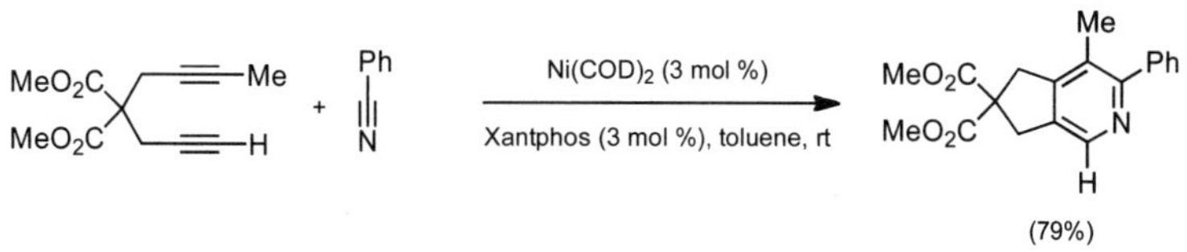

The catalyst was prepared by stirring Ni(cod)$_2$ and xantphos (in a 1:1 molar ratio) in toluene. In nitrogen-filled glove box, a solution of 49.0 mg (0.23 mmol, 0.1 M) of dimethyl 2-(but-2-yn-1-yl)-2-(prop-2-yn-1-yl)malonate in toluene was added dropwise (over a period of 1 h) to a vial containing 36.4 mg (0.35 mmol) of benzonitrile, and 3 mol% of catalyst in toluene. Then, the reaction was stirred at RT for 3 h. The reaction mixture was purified *via* flash column chromatography using first 15% and then 30% ethyl acetate in hexanes to afford the desired product as colorless oil (79%).

In addition to the assembly of various carbocyclic and heterocyclic six-membered ring assembly, nickel-catalyzed [2+2+2] homo-Diels-Alder cycloadditions have been illustrated using norbornadiene derivatives (Scheme 3-30). Electron-deficient dienophiles undergo cycloadditions with norbornadiene using catalysts prepared from Ni(cod)$_2$ and PPh$_3$. Chemical yields and exo-selectivities are both generally high in these processes.

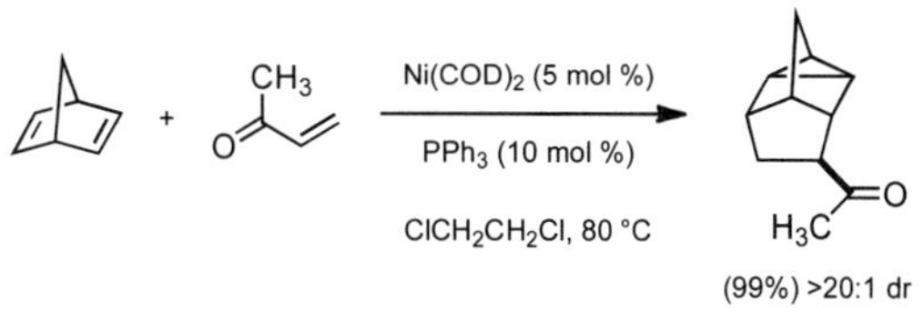

Scheme 3-30

3.4 [3+2] and [2+2+1] Cycloadditions

Several distinct strategies have been devised for the assembly of five-membered rings by nickel-catalyzed cycloaddition. Studies from Noyori[115] and Binger[116] established that Ni(0) complexes effectively catalyze cycloaddition processes of methylene cyclopropanes with acrylates and vinyl ketones to produce five-membered cycloadducts. The process was illustrated not to involve symmetrical trimethylene methane intermediates because methylene cyclopropane **18a** provides product **19a**, and isomer **18b** provides product **19b** (Scheme 3-31). Instead, four-membered nickel metallacycles derived from the oxidative addition into a strained carbon–carbon bond are the likely reactive intermediates in this process.

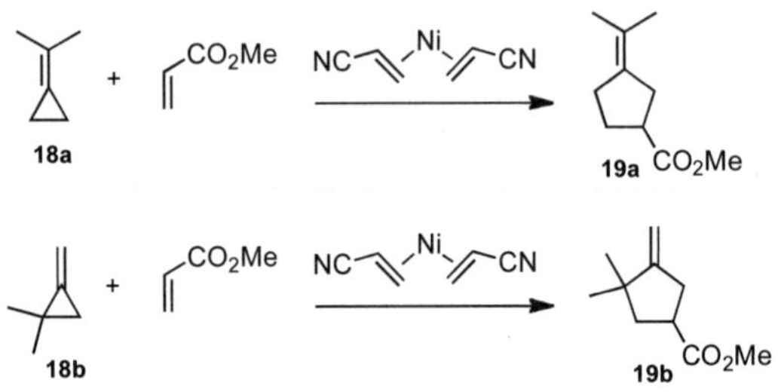

Scheme 3-31

The scope of catalytic ring-opening [3+2] cycloadditions was expanded to include simple cyclopropyl ketones, which combine with vinyl ketones to provide substituted cyclopentane derivatives (Scheme 3-32).[117] Cyclopropyl imines were also found to be effective cycloaddition substrates under similar conditions, as illustrated in the sample procedure below.[118] Oxidative additions of nickel(0) to cyclopropyl ketones were found to produce six-membered metallacyclic nickel *O*-enolates, which were competent species in the [3+2] cycloaddition process.[119, 120]

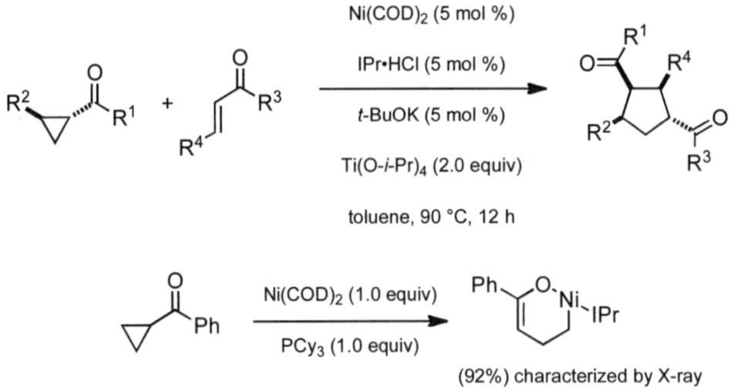

Scheme 3-32

Representative procedure: synthesis of (1S,2R*,3R*)-3-benzoyl-2-phenylcyclopentanecarbaldehyde*[118]

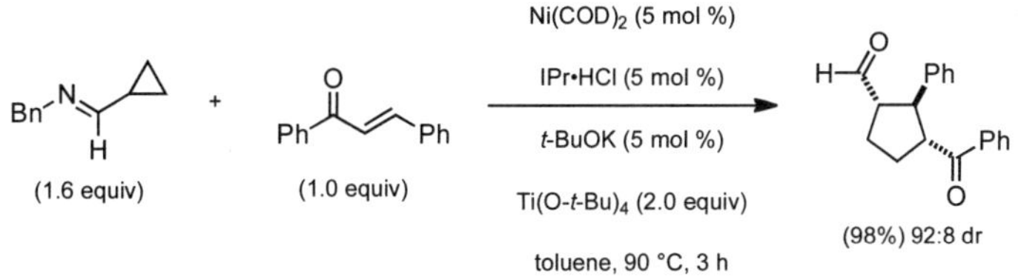

General procedure: Toluene was added to a solid mixture of IPr•HCl, potassium *t*-butoxide, and $Ni(cod)_2$ (to make a 0.1–0.15-M solution) at RT. The resulting solution was stirred for 20–30 min until the color turned to dark red. Then, a 0.5-M toluene solution of the cyclopropyl imine and neat $Ti(O\text{-}t\text{-}Bu)_4$ were added by syringe in sequence. Subsequently, a 0.25-M toluene solution of the enone was prepared. Twenty percent of this solution was added at once, the reaction mixture was heated to 90 °C, and the remaining enone solution was added by syringe pump over 1.5–2 h. The reaction mixture was stirred for 3–6 h. The reaction mixture was allowed to cool to RT, 10% HCl was added, and the mixture was stirred 20 min and was extracted with diethyl ether three times. The combined organic layer was washed with sodium bicarbonate, brine, dried over anhydrous magnesium sulfate, filtered, and concentrated *in vacuo*. The residue was purified *via* column chromatography (SiO_2) to afford the desired product. Following the general procedure, (*E*)-*N*-(cyclopropylmethylene)-1-phenylmethanamine (210 mg, 1.3 mmol), (*E*)-chalcone (166 mg, 0.8 mmol), $Ni(cod)_2$ (22 mg, 0.08 mmol), IPr•HCl (34 mg, 0.08 mmol), *t*-BuOK (9 mg, 0.08 mg), and $Ti(O\text{-}t\text{-}Bu)_4$ (0.62 mL, 1.6 mmol) gave, after column chromatography (10% EtOAc in hexanes), 219 mg of product (98%, dr 92:8) as a white solid.

Whereas the preceding processes involving ring openings of cyclopropanes provide useful [3+2] entries to cyclopentanes, the use of simpler substrates in [3+2] cycloadditions is made possible by the development of reductive cycloaddition pathways. Such reactions were initially developed in the context of stoichiometric processes, where metallacycles were prepared by oxidative cyclizations of enones and alkynes, followed by either protonation or alkylation of the nickel *O*-enolate functionality.[121] Catalytic protocols involving various intramolecular combinations of π-systems include formal [3+2] reductive cycloadditions of bis-enones to form bicyclooctanols (Scheme 3-33),[122] of enones with unsaturated acyl oxazolidinones to form triquinane derivatives,[123] and of enals with alkynes to form bicyclooctenols.[124]

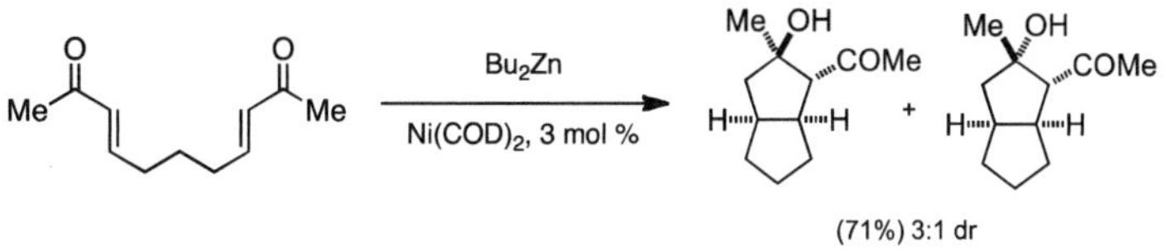

Scheme 3-33

Representative procedure: synthesis of (3aα,5aβ,8aβ)-1,2,3a,5,5a,6,7,8-octahydro-2,2-dimethylcyclopenta[c]pentalene-3,4-dione[123]

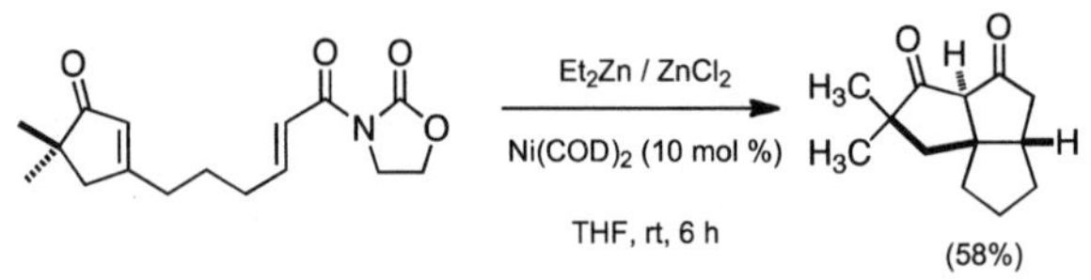

Et$_2$Zn / ZnCl$_2$
Ni(COD)$_2$ (10 mol %)
THF, rt, 6 h
(58%)

Et$_2$Zn (32 μL, 0.309 mmol) was added dropwise to ZnCl$_2$ (14 mg, 0.103 mmol) in 2 mL of THF at 0 °C followed by stirring for 20 min at 0 °C. A 2-mL portion of a THF solution of Ni(cod)$_2$ (3 mg, 0.0103 mmol) was added, and the resultant mixture was immediately transferred by cannula to a solution of the enone substrate (30 mg, 0.103 mmol) in 1 mL of THF at 0 °C and then allowed to warm to 25 °C. After being stirred for 6 h at 25 °C, the reaction mixture was subjected to an extractive workup (NH$_4$Cl/NH$_4$OH pH 8 buffer/Et$_2$O) followed by flash chromatography (5:1 hexanes/EtOAc) on silica gel to produce 12.2 mg (58%) of the desired product.

More recent advances in intermolecular [3+2] reductive cycloadditions have involved combinations of enals or enoates with alkynes (Scheme 3-34).[124–126] The initially developed cycloadditions of enals and alkynes likely proceeds by initial formation of a metallacyclic enolate derivative, followed by enolate protonation and addition of the vinyl nickel unit to the resulting carbonyl to produce the boron alkoxide of the observed cyclopentenol product (Scheme 3-35). The analogous transformation with enoates may also proceed by this mechanism, depicted below by the sequence of initial generation of metallacycle **20**, followed by enolate protonation to form **21** en route to product generation. Alternatively, the collapse of the metallacycle **20** to a ketene intermediate **22** may occur in the enoate variant. The precise pathway followed likely depends on whether protic or aprotic media are used.

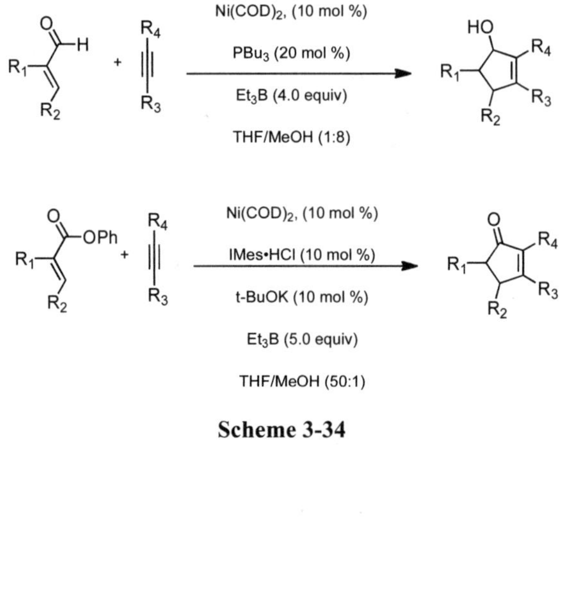

Ni(COD)$_2$, (10 mol %)
PBu$_3$ (20 mol %)
Et$_3$B (4.0 equiv)
THF/MeOH (1:8)

Ni(COD)$_2$, (10 mol %)
IMes•HCl (10 mol %)
t-BuOK (10 mol %)
Et$_3$B (5.0 equiv)
THF/MeOH (50:1)

Scheme 3-34

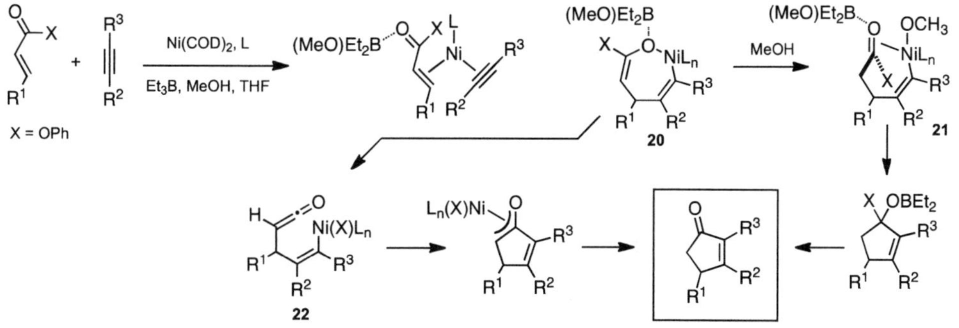

Scheme 3-35

Representative procedure: synthesis of (1R,3aR*,7aS*)-3-methyl-2-phenyl-3a,4,5,6,7,7a-hexahydro-1H-inden-1-ol*[124]

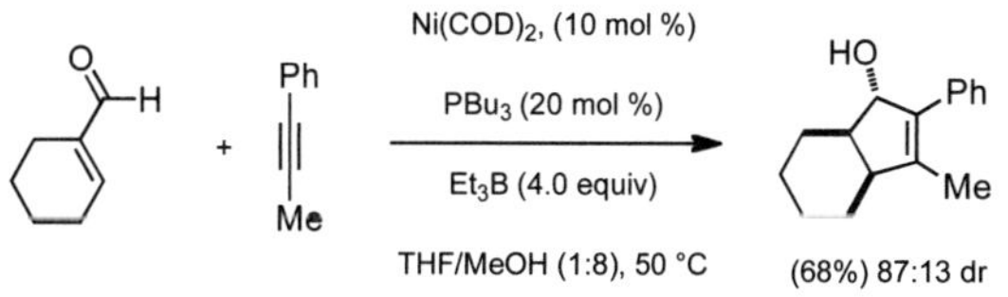

Ni(COD)$_2$, (10 mol %), PBu$_3$ (20 mol %), Et$_3$B (4.0 equiv), THF/MeOH (1:8), 50 °C; (68%) 87:13 dr

General procedure: To a solution of Ni(cod)$_2$ (0.1 equiv) in THF (0.6 mL) was added dropwise tributylphosphine (PBu$_3$) (0.2 equiv) at RT. After stirring for 5–10 min at RT, the reaction mixture became bright yellow. A solution of enal (1.0 equiv) and alkyne (2.0 equiv) at RT in MeOH (4.4 mL) was added, and then Et$_3$B (4.0 equiv) was added. The reaction mixture was stirred at 50 °C until TLC analysis indicated disappearance of the enal. The reaction mixture was concentrated *in vacuo*. The residue was purified *via* flash chromatography (SiO$_2$) to afford the desired product. Following the general procedure, 1-cyclohexene-1-carboxaldehyde (34 μL, 0.30 mmol), 1-phenyl-1-propyne (70 mg, 0.60 mmol), Ni(cod)$_2$ (8 mg, 0.03 mmol), PBu$_3$ (16 μL, 0.06 mmol), and Et$_3$B (174 μL, 1.20 mmol) were stirred for 2 h at 50 °C. Silica gel chromatography using 20% Et$_2$O in hexanes gave the desired product (47 mg, 68%, dr 87:13) as a white solid.

Other classes of nickel-catalyzed [3+2] cycloadditions include the addition of 2-haloacetophenone derivatives to alkynes to produce indenol derivatives (Scheme 3-36).[127] This process likely involves initial oxidative addition to the haloaromatic, followed by alkyne insertion and carbonyl addition. In this case, zinc powder serves as reducing agent to regenerate the active nickel(0) catalyst. A mechanistically intriguing cycloaddition that proceeds without the action of reducing agents is the direct formation of bicyclic products from the addition of α,β,γ,δ-unsaturated ketones with alkynes.[128] This process likely involves initial metallacycle formation followed by unusual rearrangement steps unique to the requisite doubly unsaturated carbonyl component.

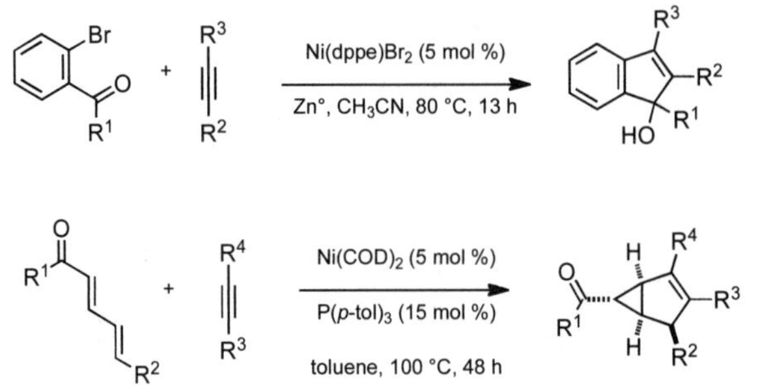

Scheme 3-36

In addition to the above classes of [3+2] cycloadditions, several nickel-catalyzed multicomponent [2+2+1] cycloadditions have been developed as an alternative entry to functionalized five-membered rings. Isocyanides have been used in processes analogous to the Pauson-Khand cycloaddition (Scheme 3-37). Whereas early reports involved stoichiometric processes,[129] catalytic versions were enabled by the use of hindered isocyanides.[130]

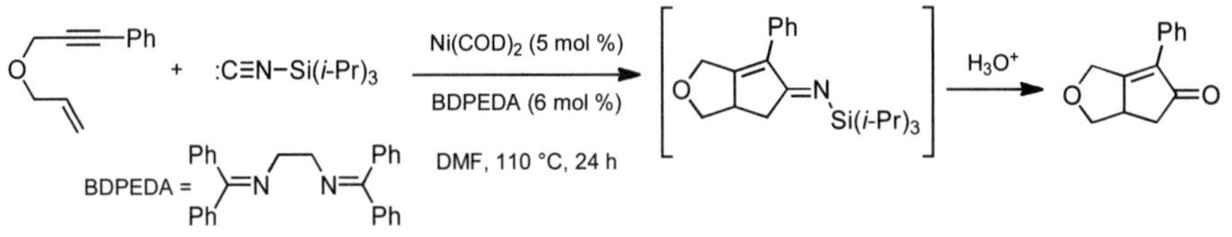

Scheme 3-37

Although nickel-catalyzed carbonylative Pauson-Khand cycloadditions have not been broadly developed, related doubly carbonylative cycloadditions involving allyl halides have been demonstrated as an entry to functionalized cyclopentenones.[131] In recent catalytic versions, iron powder was used as the terminal reductant and dehalogenating agent (Scheme 3-38).[132, 133] [Caution: $Ni(CO)_4$, which could potentially be liberated in this reaction, is a highly toxic gas.]

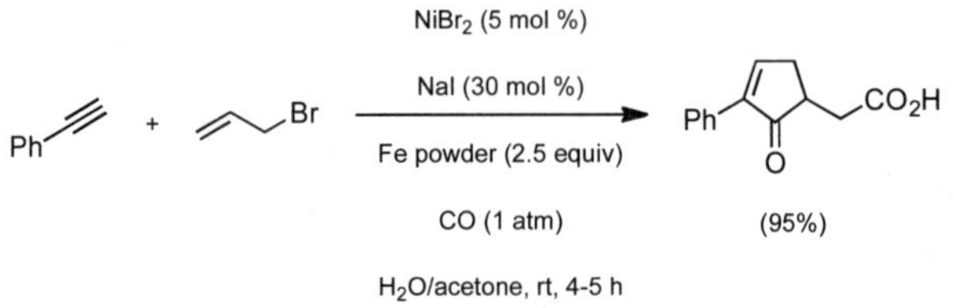

Scheme 3-38

In addition to the syntheses of carbocycles in Scheme 3-38, oxygen-, silicon-, and aluminum-containing heterocycles have also been prepared by [2+2+1] cycloadditions. An interesting tetrahydrofuran synthesis involves stoichiometric metallacycle formation by oxidative cyclization of Ni(0) with two equivalents of norbornadiene, followed by oxygen atom insertion with N_2O (Scheme 3-39).

Tetrahydrofuran formation by reductive elimination is induced by molecular iodine.[134] In a catalytic process, dialkylsilanes undergo three-component cycloadditions with aldehydes and alkynes to provide silacyclic products.[135] The formal transfer of the silylene unit (R_2Si) involves a net two-electron reduction in the process. A related reaction involving the condensation of imines, alkynes, and triorganoaluminum reagents provides an efficient entry to azaaluminacyclopentene products.[136]

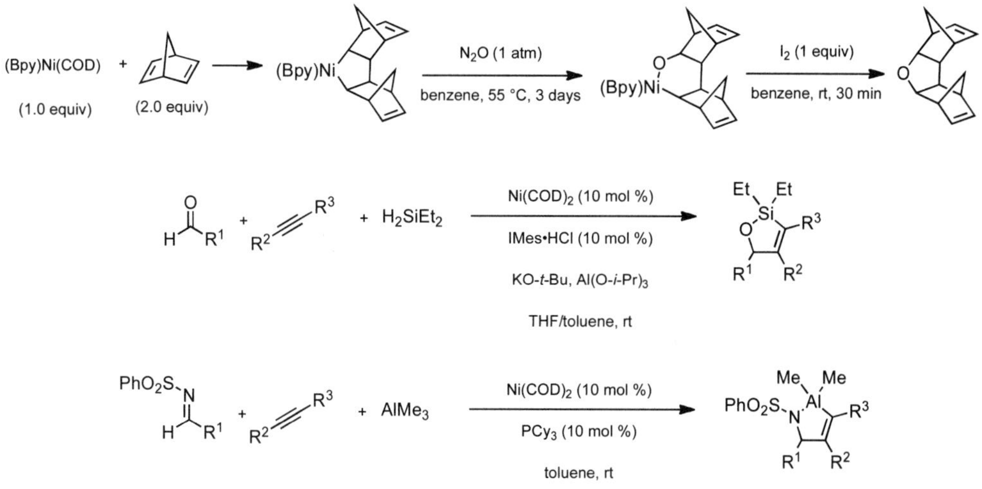

(Bpy)Ni(COD) (1.0 equiv) + (2.0 equiv) → (Bpy)Ni; N_2O (1 atm), benzene, 55 °C, 3 days → (Bpy)Ni–O; I_2 (1 equiv), benzene, rt, 30 min → O

Ni(COD)$_2$ (10 mol %), IMes•HCl (10 mol %), KO-*t*-Bu, Al(O-*i*-Pr)$_3$, THF/toluene, rt; + H_2SiEt_2

Ni(COD)$_2$ (10 mol %), PCy$_3$ (10 mol %), toluene, rt; + $AlMe_3$

Scheme 3-39

In addition to the two- and three-component entries to five-membered rings in Scheme 3-39, the nickel-catalyzed rearrangement of vinyl cyclopropanes to cyclopentenes has also been developed. Whereas various catalysts promote the rearrangement of vinyl cyclopropanes under harsh conditions, nickel-*N*-heterocyclic carbene (NHC) complexes were found to catalyze the rearrangement under mild conditions with relatively broad scope (Scheme 3-40).[137]

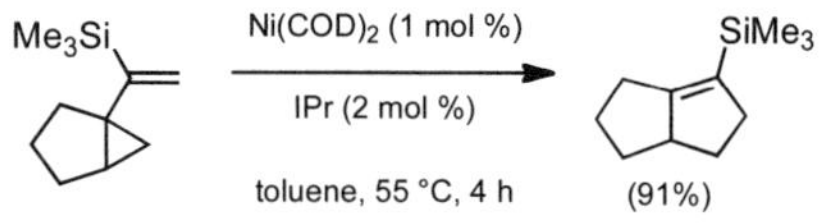

Me_3Si — Ni(COD)$_2$ (1 mol %), IPr (2 mol %), toluene, 55 °C, 4 h → $SiMe_3$ (91%)

Scheme 3-40

3.5 Cycloadditions in Medium-Ring Construction

As illustrated in the early achievements from Reppe[87] on the cyclotetramerization of acetylene and from Wilke on the cyclodimerization and cyclotrimerization of butadienes,[88] nickel-catalyzed cycloadditions possess considerable potential in the construction of medium ring systems. Pioneering studies by Wender have demonstrated many practical implications that have advanced the general utility of medium-ring synthesis by nickel-catalyzed cycloadditions.[138–140] In a series of reports, the efficient assembly of cyclooctadienes by nickel-catalyzed intramolecular [4+4] cycloaddition of bis-dienes was illustrated with a range of accessible skeletal frameworks (Scheme 3-41). Depending on the nature of the tether system employed (type I or II cycloadditions), complex fused ring and bridged bicyclic products may be accessed. The efficient total syntheses of asteriscanolide[141] and salsolene oxide[142] illustrate the utility of nickel-catalyzed [4+4] cycloadditions in complex applications. In the cycloaddition approach to salsolene oxide, starting from a mixture of *E*- and *Z*-isomers of the requisite bis-diene, only the *Z*-isomer shown was reactive in the cycloaddition. However, the unreactive *E*-isomer could be recycled through photoequilibration. A computational study illustrated that diastereoselectivities could best be explained by evaluating the energetics of η^1,η^3 bis-allyl nickel intermediates.[143]

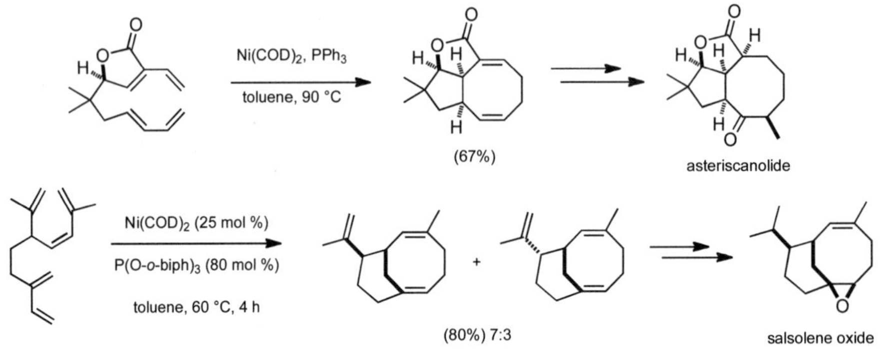

$Ni(COD)_2$, PPh_3
toluene, 90 °C
(67%)
asteriscanolide

$Ni(COD)_2$ (25 mol %)
$P(O\text{-}o\text{-}biph)_3$ (80 mol %)
toluene, 60 °C, 4 h
(80%) 7:3
salsolene oxide

Scheme 3-41

A complementary entry to eight-membered ring products is provided by the [2+2+2+2] cycloaddition of terminal diynes (Scheme 3-42).[144] Treatment of 1,6-diynes with nickel phosphine catalysts can favor the preparation of substituted benzenes by a [2+2+2] process when monodentate phosphines are employed. Alternatively, cyclooctatetraenes are favored by [2+2+2+2] cycloaddition when dimethoxyethane (dme) is employed as the ligand. The dme-based procedure provides a range of functionalized cyclooctatetraenes, which are useful as ligands in a variety of other processes.[145]

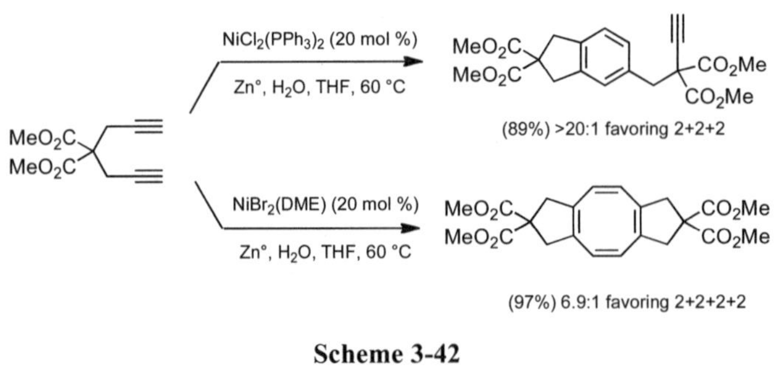

MeO_2C, MeO_2C
$NiCl_2(PPh_3)_2$ (20 mol %)
Zn°, H_2O, THF, 60 °C
MeO_2C, MeO_2C, CO_2Me, CO_2Me
(89%) >20:1 favoring 2+2+2

$NiBr_2(DME)$ (20 mol %)
Zn°, H_2O, THF, 60 °C
MeO_2C, MeO_2C, CO_2Me, CO_2Me
(97%) 6.9:1 favoring 2+2+2+2

Scheme 3-42

Representative procedure: synthesis of (4Z,9Z)-1,3,6,8-tetrahydrocycloocta[1,2-c:5,6-c']difuran[144]

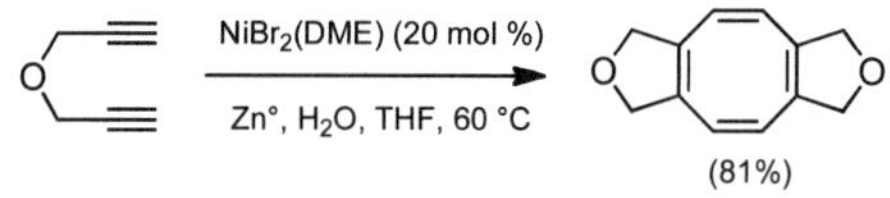

NiBr$_2$(DME) (20 mol %)
Zn°, H$_2$O, THF, 60 °C
(81%)

Under N$_2$, THF (0.75 mL) was added *via* syringe through a septum to an oven-dried test tube containing (dme)NiBr$_2$ (14.2 mg, 0.046 mmol), zinc dust (6.0 mg, 0.092 mmol), and a stir bar, which had been flushed with N$_2$ for 10 min. Water (0.8 μL, 0.046 mmol) was added, and the test tube was placed into a 60 °C oil bath. A solution of 3-(prop-2-yn-1-yloxy)prop-1-yne (48 μL, 0.46 mmol) in THF (0.75 mL) was then added over 3 h *via* a syringe pump. After stirring at this temperature for an additional 1 h, the reaction was allowed to cool to RT and filtered through Celite (Et$_2$O eluent). The solvent was removed under reduced pressure, and the residue was purified *via* column chromatography (20% EtOAc/pentane), yielding the desired product (yellow solid, 35.4 mg, 0.37 mmol, 81%).

The synthesis of seven-membered rings by nickel-catalyzed cycloadditions was made possible by the use of trimethylsilyl (TMS) diazomethane as one of the required reagents. Treatment of TMS diazomethane with 1,3-dienes tethered to isolated alkynes allowed efficient [4+2+1] cycloadditions to prepare cycloheptadiene derivatives (Scheme 3-43).[146, 147] Whereas seven-membered ring products were typically produced without evidence of cyclopropane intermediates, a comparison of stereoisomeric dienes provided direct evidence that divinylcyclopropanes were likely intermediates. In the case of an *E,Z*-diene starting material, $A^{1,3}$-strain introduced in the divinylcyclopropane slowed the resulting [3,3]-sigmatopic rearrangement, allowing the cyclopropane species to be isolated and characterized. Further heating this species in the absence of catalyst allowed its conversion into the expected cycloheptadiene derivative.

+ N$_2$CHSi(CH$_3$)$_3$ — Ni(COD)$_2$ (10 mol %), THF, 60 °C → (49%) >95:5 dr

+ N$_2$CHSi(CH$_3$)$_3$ — Ni(COD)$_2$ (10 mol %), THF, 60 °C → (69%) >95:5 dr — toluene, 105 °C → (75%) >95:5 dr

fast rearrangement

$A^{1,3}$ strain

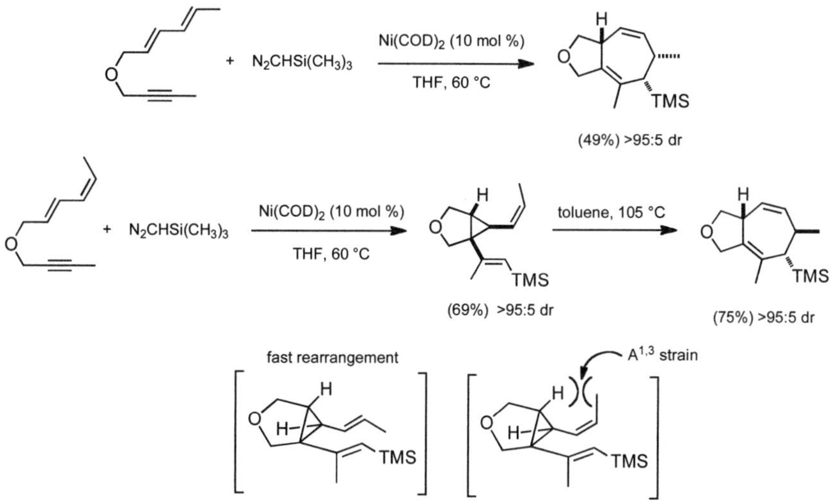

Scheme 3-43

Representative procedure: synthesis of (4,7-dimethyl-1-phenyl-3,5,6,8a-tetrahydro-1H-cyclohepta[c]furan-5-yl)-trimethylsilane[147]

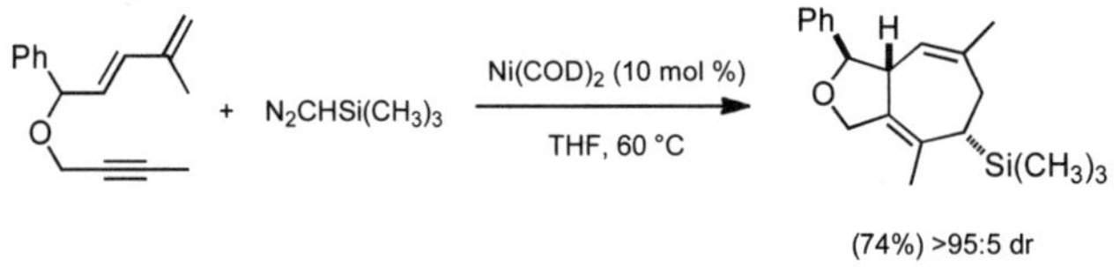

General procedure: To a premixed THF solution of dienyne (1 equiv, 0.1 M) and $TMSCHN_2$ (2 equiv) was added a THF solution of $Ni(cod)_2$ (10 mol%) at 60 °C. The resulting brown solution was stirred at 60 °C for 10–30 min and allowed to cool to RT. The solvent was removed by rotary evaporation, and the residue was absorbed onto silica gel. Flash column chromatography provided the bicyclic products. The general procedure was followed using 140 mg (0.6 mmol) of (*E*)-(1-(but-2-yn-1-yloxy)-4-methylpenta-2,4-dien-1-yl)benzene, 0.6 mL (1.2 mmol, 2.0 M in Et_2O) of $TMSCHN_2$, and 18 mg (0.06 mmol) of $Ni(cod)_2$ to afford 138 mg (74%, >95:5 dr) of the desired product.

In addition to simple [3+2] cycloadditions described previously with methylene cyclopropanes, more complex variations using this same functionality provide efficient access to both seven- and nine-membered rings (Scheme 3-44). The fully intermolecular [3+2+2] cycloaddition of methylene cyclopropanes with two alkynes provides an efficient entry to cycloheptadiene derivatives.[148] By moving to a partially intramolecular system, the efficient assembly of nine-membered ring products was illustrated by the [4+3+2] cycloaddition of methylene cyclopropanes with dienes tethered with an alkyne.[149]

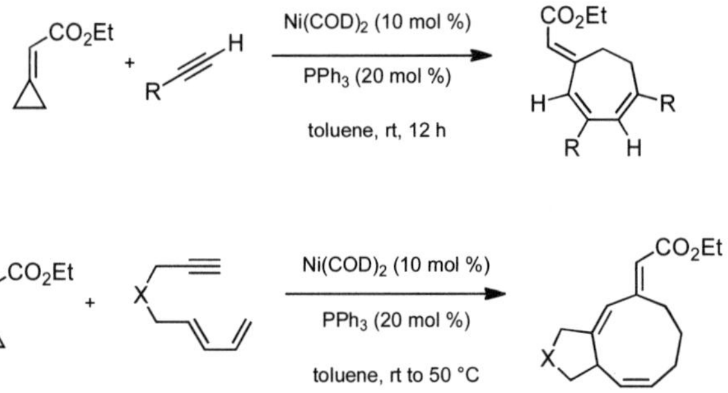

Scheme 3-44

Representative procedure: synthesis of (E)-1-ethoxycarbonylmethylene-3,5-bis(trimethylsilyl)-2,4-cycloheptadiene[148]

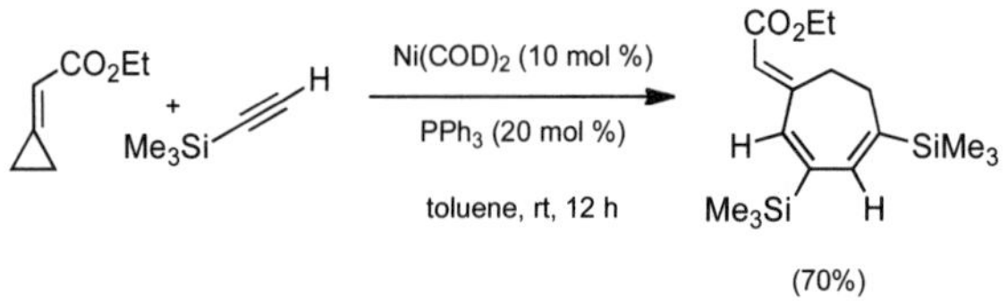

To a dark-red mixture of $Ni(cod)_2$ (27.5 mg, 0.1 mmol) and PPh_3 (52.5 mg 0.2 mmol) in dry toluene (0.5 mL) was added dropwise a solution of ethyl 2-cyclopropylideneacetate (126 mg, 1 mmol) and trimethylsilyl acetylene (5 mmol) in dry toluene (0.5 mL) at RT for 5 h under Ar. The progress of the reaction was monitored by TLC and gas chromatography-mass spectroscopy (GC-MS), and the mixture was stirred until the methylene cyclopropane was consumed (overnight). The mixture was passed through a short alumina column (ether). Evaporation of the solvent gave an oil, which was further purified by a silica gel column chromatography (hexane/EtOAc 40:1) to give the desired product (70%).

Finally, the cycloaddition of 1,6-diynes with cyclobutanones provides a mechanistically intriguing entry to eight-membered ring products (Scheme 3-45).[150] This sequence likely involves metallacycle formation involving addition of the cyclobutanone carbonyl, followed by ring expansion to a nine-membered metallacycle, and then C-C bond-forming reductive elimination.

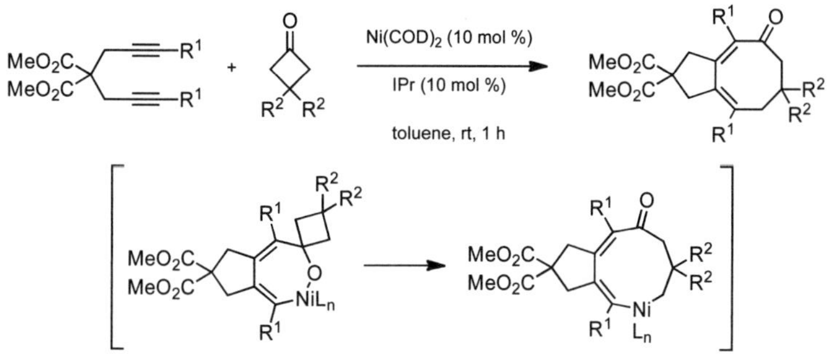

Ni(COD)$_2$ (10 mol %)
IPr (10 mol %)
toluene, rt, 1 h

Scheme 3-45

4. Reductive Coupling Processes

4.1 Early Developments

Reviews. Montgomery, J.; *Acc. Chem. Res.* **2000,** *33*, 467 – 473; Ikeda, S.; *Acc. Chem. Res.* **2000,** *33*, 511 – 519; Montgomery, J.; *Angew. Chem., Int. Ed.* **2004,** *43*, 3890 – 3908; Moslin, R.M.; Miller-Moslin, K.; Jamison, T.F.; *Chem. Commun.* **2007**, 4441 – 4449; Montgomery, J.; Sormunen, G. J. *Top. Curr. Chem.* **2007**, *279*, 1. Malik, H. A.; Baxter, R. D.; Montgomery, J. "Nickel-Catalyzed Reductive Couplings and Cyclizations" in *Catalysis Without Precious Metals*, Bullock, R. M., Ed.; Wiley-VCH: Weinheim, Germany, **2010**, pp. 181 – 212.

An extensive array of nickel-catalyzed reductive coupling processes has been developed, wherein two different π-components are coupled concomitant with a net two-electron reduction of the π-components employed. Union of the two reactive π-components is accompanied by hydrogen atom transfer in processes termed reductive couplings, whereas the analogous processes involving carbon substituent transfer are termed alkylative couplings. A major challenge in the development of heterocouplings of two different π-components is the suppression of competing homocouplings and oligomerizations. However, the highly selective fashion in which nickel catalysts select for the heterocoupling of two different π-components contributes to the considerable synthetic utility of this class of nickel-catalyzed processes. Processes of this type are formally three-component couplings, typically involving a highly polar π-component, a relatively nonpolar π-component, and a reducing agent. Many distinct classes of mechanisms have been proposed, involving fundamentally different mechanistic steps. Although many of the synthetic advances preceded the acquisition of rigorous mechanistic data, recent experimental and computational mechanistic studies have begun to clarify several key mechanistic questions in these reaction classes.

Processes involving the 2:1 coupling of alkynes with silyl hydrides are among the earliest examples of nickel-catalyzed reductive couplings of two π-components. Initial studies from Lappert and Takahashi in 1972 described the propensity of nickel to catalyze the 2:1 addition of 1-pentyne to organosilanes,[151] and the utility of this reactivity was realized by the extensive studies from Tamao and Ito in the mid-to-late 1980s.[152] The utility of nickel-catalyzed reductive couplings involving a polar π-component matched with a nonpolar component was first demonstrated in a series of advances in the mid-1990s, including the reductive coupling of aldehydes and dienes from Mori and Sato, of enones and alkynes from Ikeda and Montgomery, and of aldehydes and alkynes from Montgomery. Following these early advances, extensive developments in many other classes of reductive couplings soon followed, including major programs from Tamaru and Jamison.

4.2 Reductive Couplings of Electron-Deficient Alkenes

The nickel-catalyzed couplings of alkynes with electron-deficient alkenes have been widely explored as alternatives to classic conjugate addition strategies. Couplings of this type can proceed with hydrogen atom transfer (reductive coupling) or carbon substituent transfer (alkylative coupling) using a range of reducing agents or transmetallating agents including organotin, organozinc, organozirconium, or organosilane reagents as well as simple alcohol reductants. Early developments in this area described two possible mechanistic pathways in prototypical enone-based processes, involving reaction initiation by either oxidative addition of nickel(0) to a Lewis acid-activated enone to generate a π-allyl intermediate[153] or by oxidative cyclization of nickel(0) to the enone and alkyne to generate a metallacycle intermediate.[154] Both of these key steps have been demonstrated in stoichiometric studies as demonstrated in Scheme 3-46.

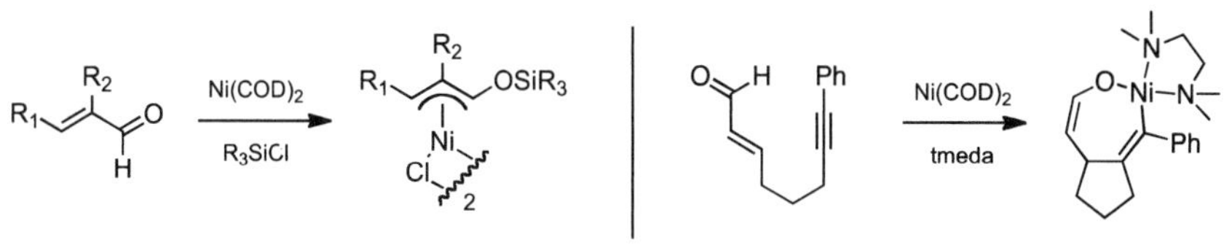

Scheme 3-46

Catalytic couplings of enones and alkynes involving alkyl transfer proceed efficiently with alkynyl tin,[155] organozinc,[122, 156, 157] or alkenylzirconium reagents (Scheme 3-47).[158] Intermolecular and intramolecular versions of the process have been reported.

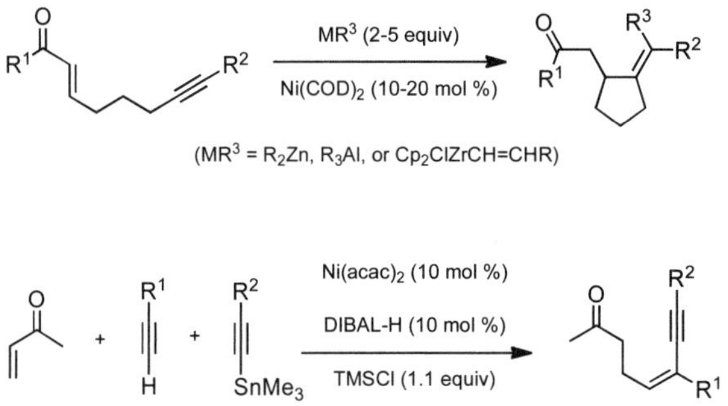

Scheme 3-47

Several total synthesis applications have been developed using alkylative coupling of enoate derivatives with alkynes. The stereoselective preparation of pyrrolidines serves as a key step in the synthesis of members of the kainic acid and domoic acid family of neuroexcitatory amino acids.[159, 160] The syntheses of isodomoic acids G and H make use of the complementary organozinc and organozirconium protocols (Scheme 3-48).[161, 162] By using a simple methyl-substituted alkyne, cyclization using an alkenylzirconium reagent provides stereoselective access to the framework of isodomoic acid G. Alternatively, by first assembling a more complex conjugated enyne precursor, cyclization with dimethylzinc provides stereoselective access to isodomoic acid H. The combination of these methods provides completely selective access to either the *E*- or *Z*-isomer of the challenging tetrasubstituted alkenes characteristic of these natural products.

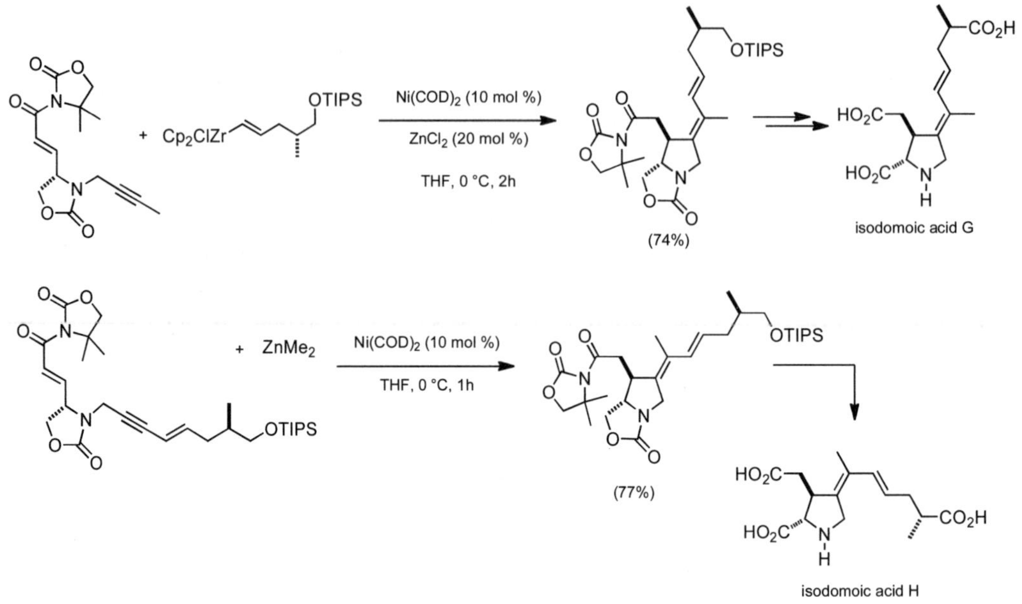

Scheme 3-48

Representative procedure: synthesis of (7S,7aS,E)-7-(2-(4,4-dimethyl-2-oxooxazolidin-3-yl)-2-oxoethyl)-6-((R,E)-6-methyl-7-(triisopropylsilyloxy)hept-3-en-2-ylidene)tetrahydropyrrolo[1,2-c]oxazol-3(1H)-one[162]

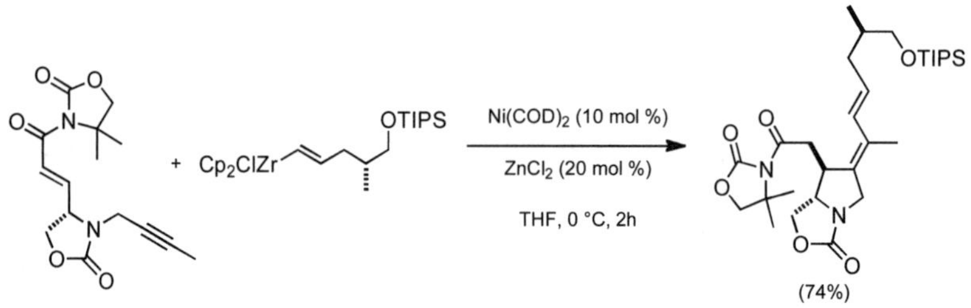

To a suspension of Cp_2ZrHCl (387 mg, 1.5 mmol) in THF (3 mL) was added a 1-mL THF solution of (*R*)-triisopropyl((2-methylpent-4-yn-1-yl)oxy)silane (381 mg, 1.5 mmol) at RT. The resulting suspension was stirred at RT for 30 min, until it was bright yellow and homogeneous, and the solution was then cooled to 0 °C. A 0.5-mL THF solution of $ZnCl_2$ (16 mg, 0.12 mmol) was added slowly to the vinyl zirconium species, and the resulting solution was stirred for 20 min. The mixture was then added quickly to a 0.5-mL THF solution of $Ni(cod)_2$ (16 mg, 0.06 mmol), followed by the addition of the alkynyl oxazolidinone substrate (184 mg, 0.6 mmol) in 1-mL THF. The final solution became dark brown and was stirred at 0 °C for 2 h. Extraction with pH 8 buffered solution and chromatographic purification (hexane:ethyl acetate, 3:2) afforded the desired product (250 mg, 74%) as a pale oil.

Whereas the above methods involve transfer of a carbon substituent from the transmetallating component during cyclization, the analogous reductive couplings involving hydrogen atom transfer have also been reported. While the early demonstrations of enone-alkyne reductive couplings involved organozinc reductants,[157] more efficient protocols involving organoborane[163] and organosilane reductants[164] were subsequently developed (Scheme 3-49).

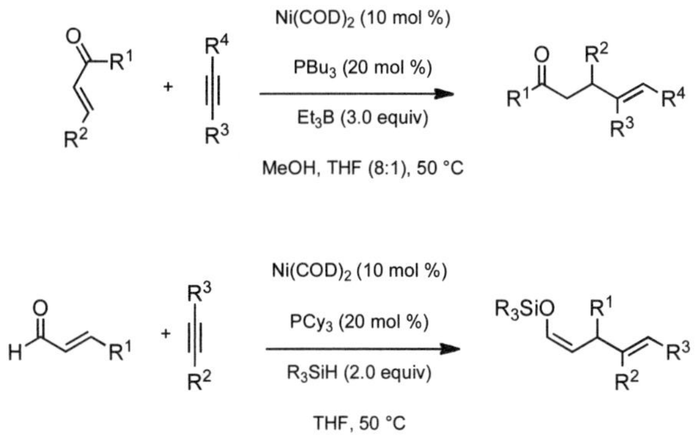

Scheme 3-49

Recent studies have illustrated that both borane and silane reductants may be avoided entirely by the use of methanol as the stoichiometric reductant (Scheme 3-50).[165] Deuterium labeling studies illustrated that the hydroxyl proton and a single methyl-derived proton from methanol are transferred to the observed reductive coupling product. The likely mechanistic pathway for this transformation involves metallacycle formation, followed by nickel enolate protonation by methanol to afford intermediate **23**. β-Hydride elimination of the methoxy ligand extrudes formaldehyde and forms a nickel hydride intermediate that undergoes product-forming reductive elimination.

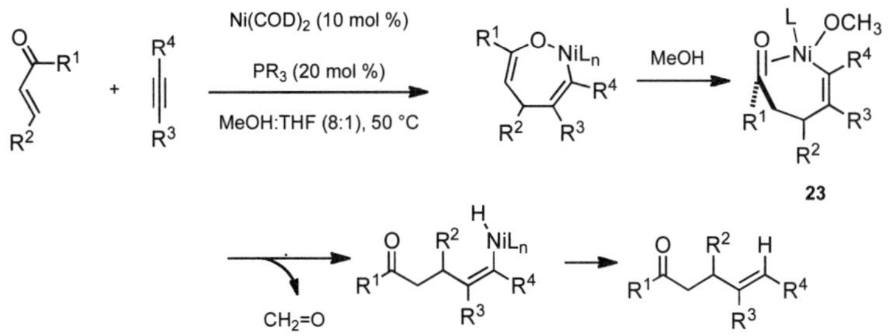

Scheme 3-50

Representative procedure: synthesis of (Z)-5-benzylidene-6-hydroxyheptan-2-one[165]

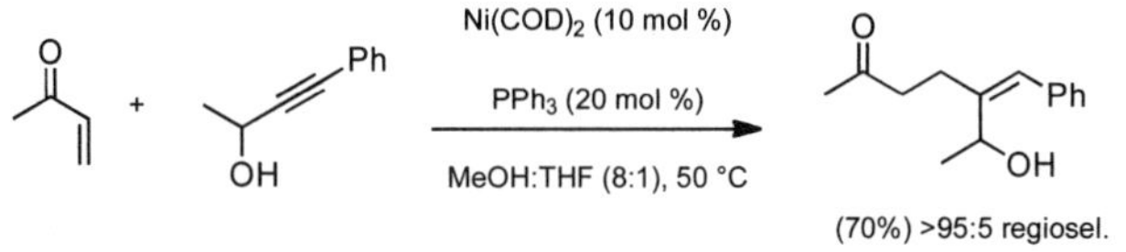

Ni(COD)2 (10 mol %)
PPh3 (20 mol %)
MeOH:THF (8:1), 50 °C
(70%) >95:5 regiosel.

General procedure: THF (0.5 mL) was added to a solid mixture of $Ni(cod)_2$ (0.1 equiv) and ligand (0.2 equiv) at RT. The resulting solution was stirred for 5 min. Then, a solution of enone (1 equiv) and alkyne (1.5 equiv) in methanol (4 mL) was added. With terminal alkynes, a solution of enone (1 equiv) in methanol (2 mL) was added, followed by syringe addition of the alkyne (1.5 equiv) as a solution in methanol (2 mL) at 50 °C. The reaction mixture was stirred at 50 °C until TLC analysis indicated the disappearance of the enone. The reaction was quenched with saturated solution of NH_4Cl, extracted with ethyl acetate, washed with brine, dried over anhydrous $MgSO_4$, and concentrated. The residue was purified *via* flash chromatography (SiO_2) to afford the desired product. Following the general procedure, 3-buten-2-one (21 mg, 0.3 mmol), 4-phenylbut-3-yn-2-ol (66 mg, 0.45 mmol), $Ni(cod)_2$ (8 mg, 0.03 mmol), and PPh_3 (16 mg, 0.06 mmol) were stirred for 3 h at 50 °C. SiO_2 chromatography (25% EtOAc in hexanes) afforded the desired product (46 mg, 70%, >95:5 regioselectivity) as a light yellow oil.

The above classes of alkylative and reductive couplings have been applied in the synthesis of various classes of complex alkaloids, including an alkylative cyclization entry to deformyl isogeissoschizine[166] and a reductive cyclization entry to 11-methoxy mitragynine pseudoindoxyl (Scheme 3-51).[167]

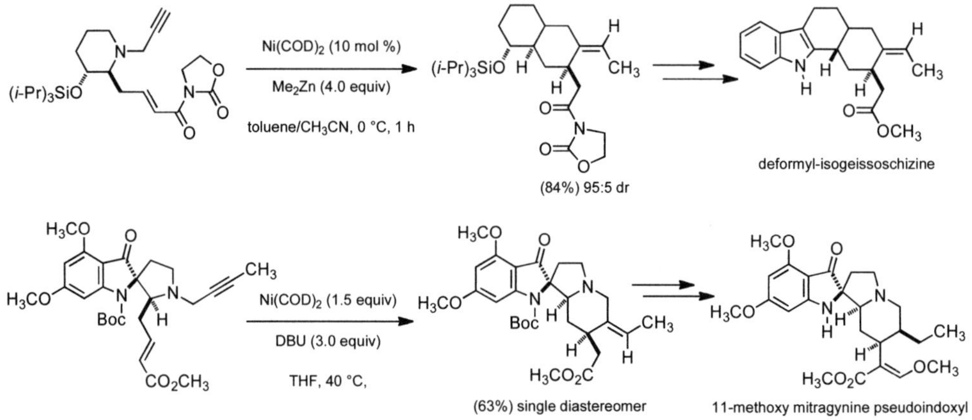

Ni(COD)2 (10 mol %)
Me2Zn (4.0 equiv)
toluene/CH3CN, 0 °C, 1 h
(84%) 95:5 dr
deformyl-isogeissoschizine

Ni(COD)2 (1.5 equiv)
DBU (3.0 equiv)
THF, 40 °C,
(63%) single diastereomer
11-methoxy mitragynine pseudoindoxyl

Scheme 3-51

The reductive coupling of enones and allenes proceeds under conditions related to the more broadly studied alkyne-based processes described previously (Scheme 3-52).[168] By using Et_3SiH as a reductant with the catalyst prepared from $Ni(cod)_2$ and PPh_3, efficient reductive coupling proceeds by addition of the enone β-carbon to the allene central carbon. By using this procedure, couplings of terminal allenes to enones provide efficient access to γ,δ-unsaturated ketone products with a terminal methylene functionality, which is a structural motif not accessible by the corresponding addition reactions of terminal alkynes.

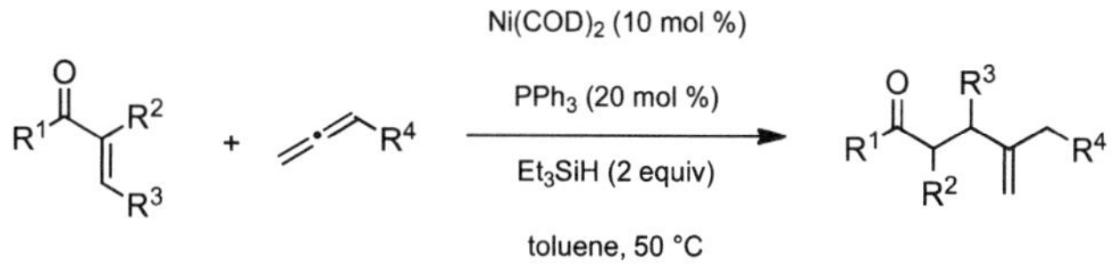

$Ni(COD)_2$ (10 mol %)
PPh_3 (20 mol %)
Et_3SiH (2 equiv)
toluene, 50 °C

Scheme 3-52

4.3 Reductive Couplings of Aldehydes and Dienes

A large number of reductive coupling processes involving aldehydes has been demonstrated. The reductive coupling of aldehydes with 1,3-dienes has been developed primarily with trialkylsilane and trialkylborane reagents. Carefully matching the ligand and reducing agent provides considerable control of regiochemistry and stereochemistry. For example, in intramolecular versions, the use of phosphine ligands with silane reductants provides primarily products from 1,4-addition to the diene (allylation), whereas the same catalyst system with i-$Bu_2Al(acac)$ as reductant provides products from 1,2-addition to the diene (homoallylation) (Scheme 3-53).[169] This study makes the important illustration that using $Ni(cod)_2$ as pre-catalyst can provide very different results than using $Ni(acac)_2$ reduced by i-Bu_2AlH, given the competing reactivity of i-$Bu_2Al(acac)$, which is generated as a by-product in the generation of Ni(0).

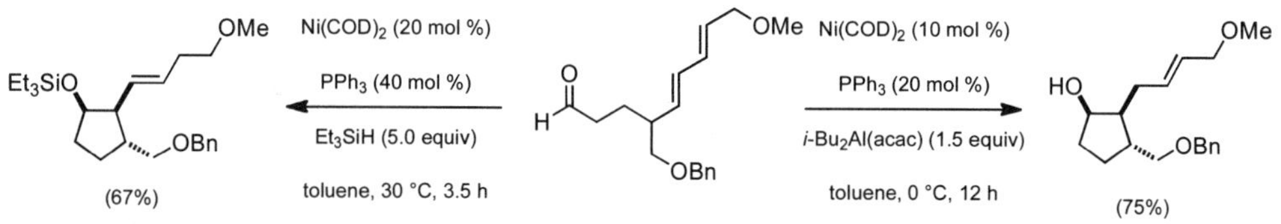

$Ni(COD)_2$ (20 mol %), PPh_3 (40 mol %), Et_3SiH (5.0 equiv), toluene, 30 °C, 3.5 h (67%)
$Ni(COD)_2$ (10 mol %), PPh_3 (20 mol %), i-$Bu_2Al(acac)$ (1.5 equiv), toluene, 0 °C, 12 h (75%)

Scheme 3-53

Intermolecular processes are also subject to control of regiochemistry and stereochemistry. For example, by employing silane reductants, 1,4-addition products (allylation) are obtained with the E-alkene by using PPh_3 as ligand,[170] whereas the 1,4-addition products possessing the Z-alkene are obtained by using N-heterocyclic carbene ligands (Scheme 3-54).[171, 172] Alternatively, the use of triethylborane as the reducing agent with $Ni(acac)_2$ as catalyst provides selective access to 1,2-addition products (homoallylation).[173, 174]

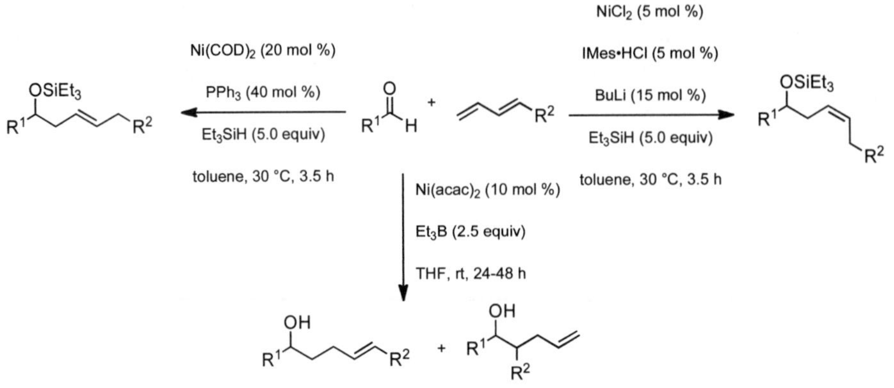

Scheme 3-54

The triethylborane procedure provides exceptional *anti*-diastereoselectivities when 2-substituted dienes are employed and thus provides a highly effective strategy for the homoallylation of carbonyls (Scheme 3-55). The use of diethylzinc as reductant with $Ni(acac)_2$ also provides homoallylation products, and complementary scope with organoboranes and organozincs was illustrated.

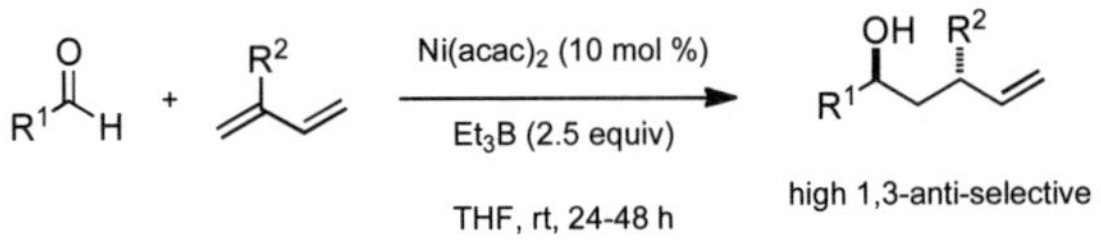

Scheme 3-55

Representative procedure: synthesis of 1,2-anti-4(E)-2-methoxycarbonyl-1-phenyl-4-hexenol[173]

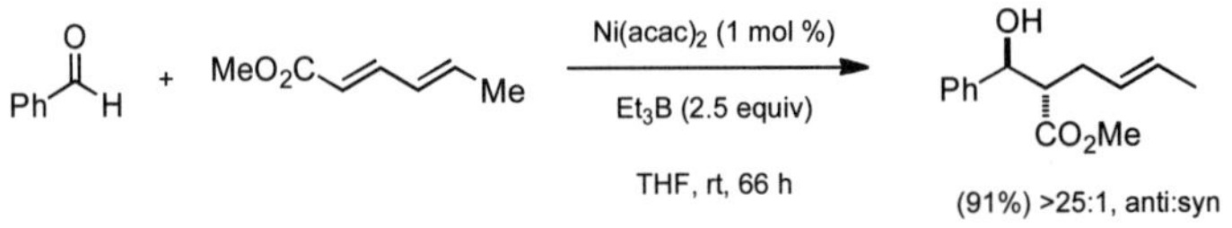

Into a nitrogen-purged flask containing $Ni(acac)_2$ (12.8 mg, 0.05 mmol) are introduced successively freshly dried THF (3 mL), methyl sorbate (2.52 g, 20 mmol), benzaldehyde (530 mg, 5 mmol), and Et_3B (12.0 mL, 1 M in hexane) *via* syringe. The homogeneous mixture is stirred at RT for 66 h until benzaldehyde disappears completely. After dilution with ethyl acetate (50 mL), the mixture is washed successively with 2 M of HCl, sat. $NaHCO_3$, and sat. NaCl, and then dried (anhydrous $MgSO_4$) and concentrated *in vacuo*. The residual oil is subjected to column chromatography over silica gel (hexane/ethyl acetate = 16/1) to give an analytically pure sample of the desired product (1.07 g, 91%).

Several fundamentally different mechanisms have been proposed for aldehyde-diene reductive couplings, including those involving metallacyclic intermediates as well as distinct mechanisms involving reaction initiation by Ni(0) oxidative addition to the reductant. However, a common mechanistic pathway involving metallacyclic intermediates can potentially explain the diverse reactivity patterns observed (Scheme 3-56). Structural evidence for oxametallacycles possessing an η^3-allyl unit has been provided.[175] Related mechanistic questions have been studied in greater detail in the analogous aldehyde-alkyne couplings as described subsequently.

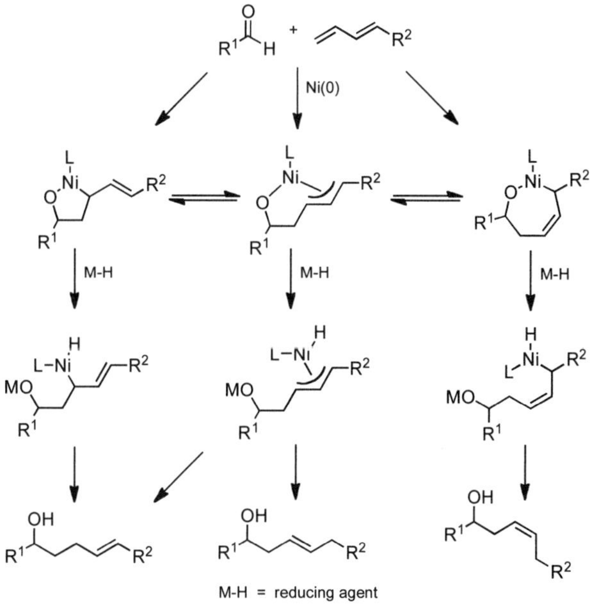

M-H = reducing agent

Scheme 3-56

The use of diboranes provides a highly effective procedure for the synthesis of functionalized allylboranes, which lead to versatile 1,3- or 1,5 diols (Scheme 3-57).[176, 177] Whereas several ligands were effective in producing the 1,5-diol framework, $P(SiMe_3)_3$ displayed unique reactivity in providing access to the synthetically useful 1,3-diol products.

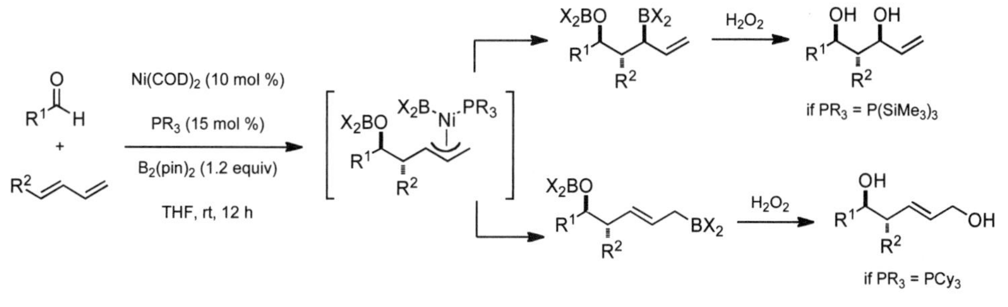

Ni(COD)$_2$ (10 mol %)
PR$_3$ (15 mol %)
B$_2$(pin)$_2$ (1.2 equiv)
THF, rt, 12 h

if PR$_3$ = P(SiMe$_3$)$_3$

if PR$_3$ = PCy$_3$

Scheme 3-57

Representative procedure: synthesis of (1S,2S*,3S*)-2-methyl-1-phenylpent-4-ene-1,3-diol*[177]

Ni(COD)$_2$ (10 mol %)
P(SiMe$_3$)$_3$ (15 mol %)
B$_2$(pin)$_2$ (1.2 equiv)
THF, rt, 12 h
then NaOH, H$_2$O$_2$
(67%)

General procedure: An oven-dried, 20-mL scintillation vial, equipped with a magnetic stir-bar, was charged with Ni(cod)$_2$ (0.10 mmol), P(SiMe$_3$)$_3$ (0.15 mmol), and THF (5 mL, 0.2 M) in a dry box under an Ar atmosphere. After stirring for 5 min, the aldehyde (1.0 mmol), *trans*-1,3-pentadiene (1.1 mmol or 3.0 mmol), and B$_2$(pin)$_2$ (1.2 mmol or 3.0 mmol) were added sequentially. The vial was sealed with a polypropylene cap and removed from the dry box. The reaction mixture was then allowed to stir at RT for 12 h. After this time, the mixture was cooled to 0 °C, and 4 mL of 3 M of NaOH and 3 mL of 30% H$_2$O$_2$ were added dropwise with caution. The mixture was then allowed to stir at RT for 10 h. The resulting solution was cooled to 0 °C and quenched by the addition of 2 mL of saturated aqueous Na$_2$S$_2$O$_3$. The two-phase mixture was extracted with ethyl acetate (3 × 30 mL), and the combined organic layers were dried over anhydrous Na$_2$SO$_4$. The drying agent was removed by filtration, and the solvent was evaporated *in vacuo*. The crude material was purified by silica gel chromatography (hexanes/EtOAc) to afford the title compounds. The reaction was performed according to the general procedure with 27.5 mg (0.10 mmol) of Ni(cod)$_2$, 37.5 mg (0.15 mmol) of P(SiMe$_3$)$_3$, 106 mg (1.0 mmol) of benzaldehyde, 204.4 mg (3.0 mmol) of *trans*-1,3-pentadiene, and 761.8 mg (3.0 mmol) of B$_2$(pin)$_2$ in THF (5.0 mL) for 12 h, followed by oxidation, to afford the desired product as a white solid (128 mg, 67%).

4.4 Reductive Couplings of Aldehydes and Alkynes

The nickel-catalyzed reductive coupling of aldehydes and alkynes has been extensively developed as a stereoselective route to allylic alcohols.[178] Many variants have been developed, and the most widely studied versions include triethylsilane-mediated couplings with either monodentate phosphine or *N*-heterocyclic carbene ligands[179–182] or triethylborane-mediated couplings with monodentate phosphine ligands.[183, 184] Triethylsilane-mediated ynal cyclizations using nickel(0) complexes of PBu$_3$ or PCy$_3$ are highly effective in the formation of five- or six-membered rings with phosphine ligands, and a variety of alkaloid templates has been prepared using this protocol (Scheme 3-58). Complex members of the allopumiliotoxin alkaloid family have been prepared using this procedure.[179, 180] In addition to the more widely developed reductive couplings of aldehydes and alkynes, intermolecular and intramolecular alkylative versions involving three-component addition of aldehydes, alkynes, and organozincs have also been developed.[178, 185]

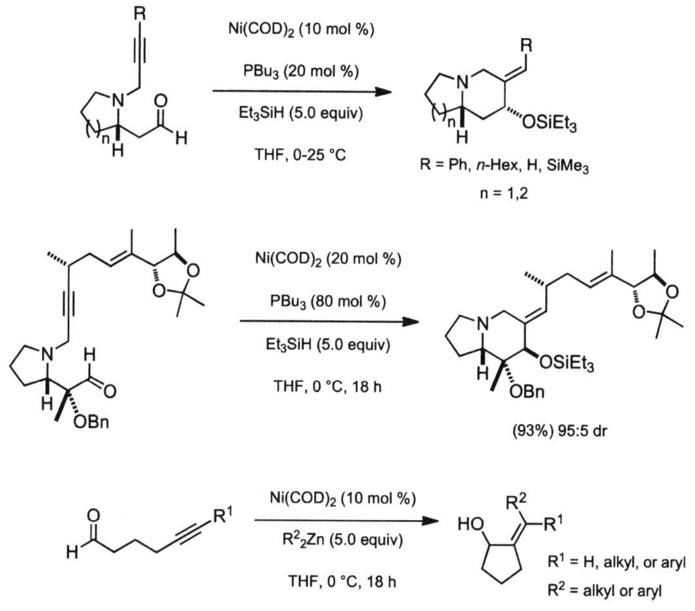

$Ni(COD)_2$ (10 mol %)
PBu_3 (20 mol %)
Et_3SiH (5.0 equiv)
THF, 0-25 °C
R = Ph, *n*-Hex, H, $SiMe_3$
n = 1,2

$Ni(COD)_2$ (20 mol %)
PBu_3 (80 mol %)
Et_3SiH (5.0 equiv)
THF, 0 °C, 18 h
(93%) 95:5 dr

$Ni(COD)_2$ (10 mol %)
$R^2{}_2Zn$ (5.0 equiv)
THF, 0 °C, 18 h
R^1 = H, alkyl, or aryl
R^2 = alkyl or aryl

Scheme 3-58

Representative procedure: synthesis of (2R,9aS*)-3-[(E)-heptylidene]-2-(triethylsilyloxy)octahydroquinolizine*[180]

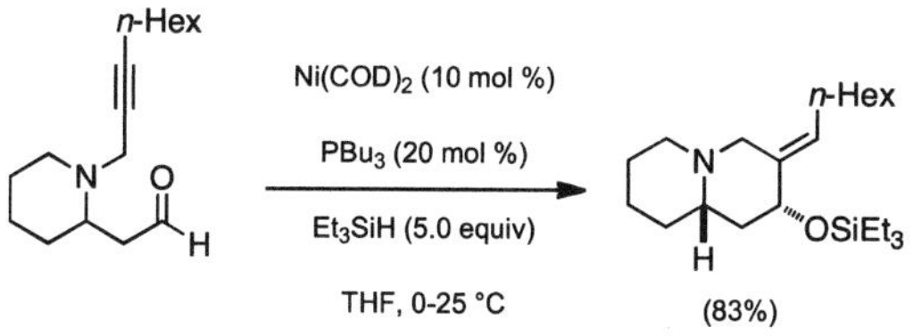

$Ni(COD)_2$ (10 mol %)
PBu_3 (20 mol %)
Et_3SiH (5.0 equiv)
THF, 0-25 °C
(83%)

General procedure: To a solution of $Ni(cod)_2$ (0.1–0.2 equiv) in THF was added dropwise PBu_3 (0.2–0.4 equiv) at RT. After 5 min at RT, the solution was cooled to 0 °C with an ice-bath and Et_3SiH (5 equiv) was added dropwise. Then, a solution of ynal in THF was added dropwise. The reaction mixture was stirred at the indicated temperature until TLC analysis indicated disappearance of the ynal. The reaction mixture was quenched with saturated $NaHCO_3$ at RT and then extracted twice with Et_2O. The combined organic layers were washed with brine, dried over anhydrous Na_2SO_4, filtered, and concentrated, and the residue was purified by column chromatography on silica gel. Following the general procedure at 45 °C, the ynal substrate (175 mg, 0.70 mmol), Et_3SiH (561 μL, 3.51 mmol), $Ni(cod)_2$ (20 mg, 0.073 mmol), and PBu_3 (35 μL, 0.14 mmol) were employed to give (212 mg, 83%) of the desired product as a colorless oil.

The process may be extended to macrocyclizations and intermolecular couplings by using Et_3B as reducing agent with trialkyl phosphine ligands[186, 187] or by using triethylsilane as reducing agent with *N*-heterocyclic carbene ligands (Scheme 3-59).[188] These procedures have been developed in a variety of complex applications, as illustrated by the total synthesis of amphidinolide T1 (Scheme 3-60).[187]

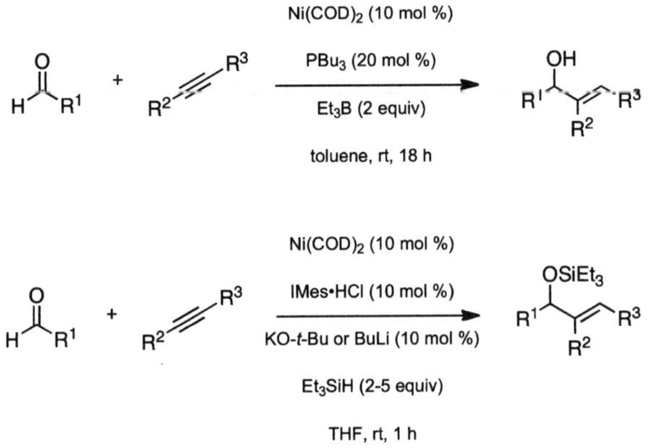

$Ni(COD)_2$ (10 mol %)
PBu_3 (20 mol %)
Et_3B (2 equiv)
toluene, rt, 18 h

$Ni(COD)_2$ (10 mol %)
IMes•HCl (10 mol %)
KO-*t*-Bu or BuLi (10 mol %)
Et_3SiH (2-5 equiv)
THF, rt, 1 h

Scheme 3-59

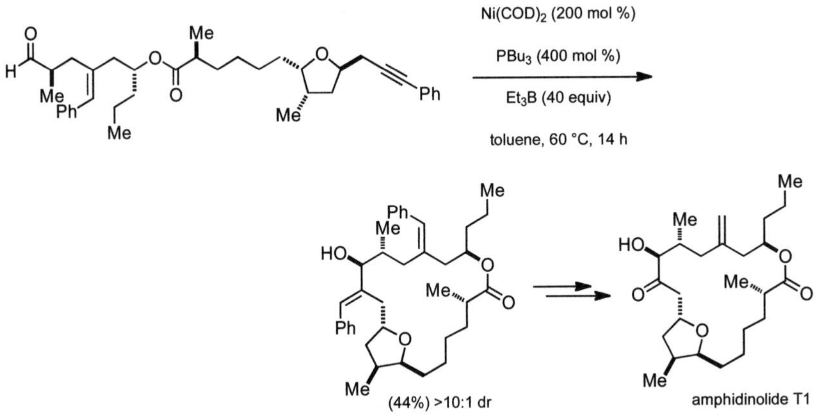

$Ni(COD)_2$ (200 mol %)
PBu_3 (400 mol %)
Et_3B (40 equiv)
toluene, 60 °C, 14 h

(44%) >10:1 dr

amphidinolide T1

Scheme 3-60

Representative procedure: synthesis of (E)-1,3-diphenyl-2-methyl-2-propen-1-ol[183]

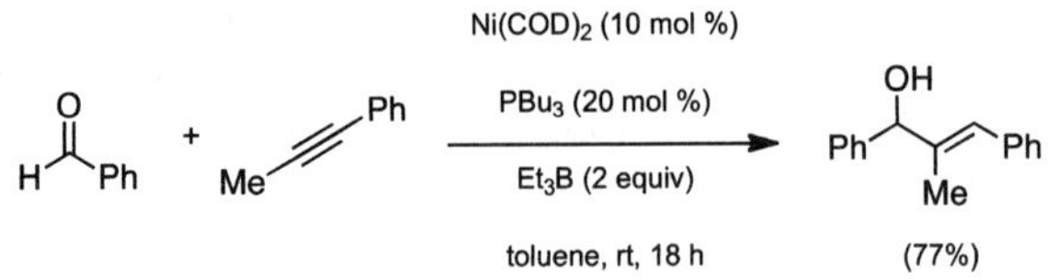

$Ni(COD)_2$ (10 mol %)
PBu_3 (20 mol %)
Et_3B (2 equiv)
toluene, rt, 18 h

(77%)

General procedure: In a glove box, $Ni(cod)_2$ (28 mg, 0.10 mmol) was placed into an oven-dried, one-necked Schlenk flask, which was then sealed with a rubber septum. The flask was removed from the glove box, and toluene (2 mL, degassed with Ar or N_2 by three freeze-pump-thaw cycles) was added *via* syringe. To this yellow solution Bu_3P (0.050 mL, 0.20 mmol) was added *via* syringe, and the resulting solution was stirred 5 min at RT. Et_3B (2.0 mL, 1.0 M in hexanes, 2.0 mmol) was added *via* syringe, and the resulting mixture was stirred 10–15 min. A solution of the alkyne (1.0 mmol) and aldehyde (1.0 mmol) in toluene (2 mL, degassed with Ar or N_2 by three freeze-pump-thaw cycles) was added dropwise *via* syringe over 1 min. The resulting solution turned brown in 5–10 min and was stirred at RT for 18 h unless otherwise indicated. Saturated aqueous NH_4Cl (4 mL) and 1 N HCl (1 mL) were added, and the resulting mixture was diluted with water (5 mL) and EtOAc (5 mL). The aqueous layer was extracted with EtOAc (3 × 10 mL), and the combined organic solutions were washed with saturated aqueous NaCl and then dried over anhydrous Na_2SO_4. After filtration and removal of the solvent *in vacuo*, the allylic alcohol was purified by silica gel chromatography using hexanes:ethyl acetate (15:1 to 5:1) as eluent. The general procedure was followed using 1- phenylpropyne (1.0 mmol, 0.13 mL) and benzaldehyde (1.0 mmol, 0.10 mL) to afford the desired product (173 mg, 77%) as a colorless oil.

Representative procedure: synthesis of (E)-10,10-diisopropyl-2,2,3,3,11-pentamethyl-8-phenyl-4,9-dioxa-3,10-disiladodec-6-ene[189]

H(C=O)Ph + H–C≡C–CH2OSiMe2-t-Bu → [$Ni(COD)_2$ (12 mol %), IMes•HCl (10 mol %), KO-t-Bu (10 mol %), i-Pr_3SiH (2 equiv), THF, rt, 1 h] → Ph–CH(OSi(i-Pr)3)–CH=CH–CH2OSiMe2-t-Bu (88%)

General procedure: To a solid mixture of $Ni(cod)_2$ (12 mol%), IMes·HCl salt (10 mol%), and t-BuOK (10 mol%) was added THF (0.125 M). The resulting solution was stirred for 5 min at RT until the solution turned dark blue in appearance. The alkyne (1.0 equiv), aldehyde (1.0 equiv), and silane (2.0 equiv) were added sequentially, and the reaction mixture was allowed to stir at RT until the starting materials were consumed. The reaction mixture was filtered through silica gel eluting with 50% EtOAc/hexanes. The solvent was removed *in vacuo*, and the crude residue was purified *via* flash chromatography on silica gel to afford the desired product. The general procedure was followed using $Ni(cod)_2$ (20 mg, 0.073 mmol), IMes·HCl salt (20 mg, 0.059 mmol), *t*-BuOK (6.7 mg, 0.059 mmol), triisopropylsilane (0.24 mL, 1.18 mmol), *t*-butyldimethyl(prop-2-yn-1-yloxy)silane (101 mg, 0.59 mmol), and benzaldehyde (63 mg, 0.59 mmol) to afford the desired product (227 mg, 0.52 mmol, 88% yield). With terminal alkynes, syringe drive addition of the alkyne is sometimes beneficial.

The regiocontrol of nickel-catalyzed reductive couplings of aldehydes and alkynes is often dictated by biases of the substrate, particularly the alkyne. However, the procedure involving *N*-heterocyclic carbene ligands with silane reductants is subject to ligand control of regioselectivity (Scheme 3-61).[189] The use of small ligands provides access to formation of the least-hindered carbon–carbon bond, whereas the use of large ligands provides access to formation of the most-hindered carbon–carbon bond. A computational study described a ligand contour analysis to explain the origin of the tunable regiocontrol.[190] The regioselectivity of even relatively complex reductive macrocyclizations may be reversed to provide selective access to 12-membered endocyclizations or 11-membered exocyclizations, depending on the ligand employed (Scheme 3-62).[191]

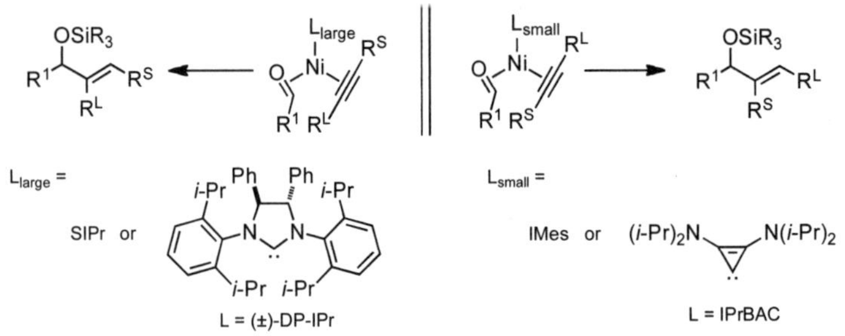

L_{large} = SIPr or L = (±)-DP-IPr

L_{small} = IMes or L = IPrBAC

Scheme 3-61

Ni(COD)$_2$ (30 mol %), (±)-DP-IPr·HCl (30 mol %), KO-*t*-Bu (40 mol %), Et$_3$SiH (2.0 equiv), THF, rt, 15 h: 59 % (single diast)

Ni(COD)$_2$ (30 mol %), IMes·HCl (30 mol %), KO-*t*-Bu (40 mol %), Et$_3$SiH (2.0 equiv), THF, rt, 15 h: 58 % (4:1 dr)

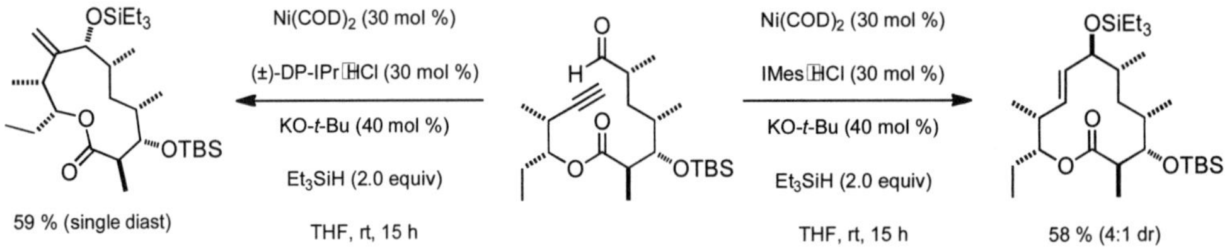

Scheme 3-62

The mechanism of the reductive coupling of aldehydes and alkynes involving either silane or borane reductants appears to involve rate-determining metallacycle formation followed by a fast reductive conversion of the metallacycle into product (Scheme 3-63). A kinetics analysis of a five-membered reductive cyclization involving Et_3SiH as the reducing agent and a catalyst prepared *in situ* from $Ni(cod)_2$ and PBu_3 illustrated a first-order dependence on ynal and catalyst, and a zeroth-order dependence on silane. Additionally, an *in situ* IR study illustrated that the $Ni(cod)_2/PBu_3$-derived catalyst does not undergo oxidative addition to the silane under the reaction conditions, and that silane is consumed only if both the aldehyde and the alkyne are present.[192] In addition to these experimental observations, computational studies of both the borane-mediated and silane-mediated processes found that the processes proceed by rate-limiting metallacycle formation.[190, 193, 194]

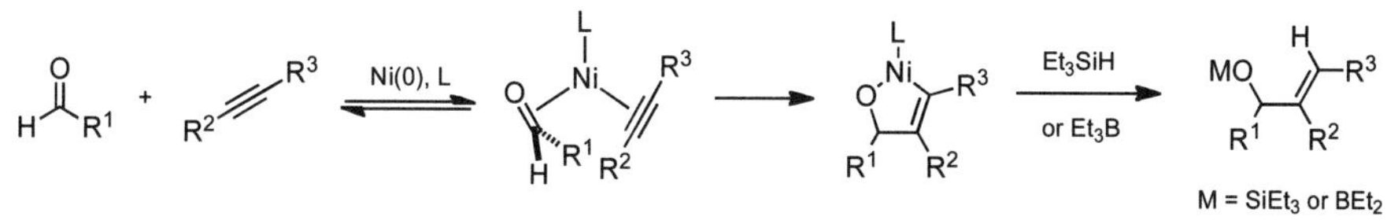

Scheme 3-63

In addition to the more extensively developed aldehyde-alkyne additions noted above, related reductive and alkylative couplings of alkynes with imines,[195–197] of electron-deficient alkenes with imines,[198] and of styrenes[199] or alkynes[200, 201] with carbon dioxide have been developed (Scheme 3-64). The reductive and alkylative couplings of imines were proposed to proceed through metallacycle mechanisms analogous to those depicted above (Scheme 3-63), but the reductive couplings involving carbon dioxide have been proposed to proceed through alternative hydrometallative pathways.

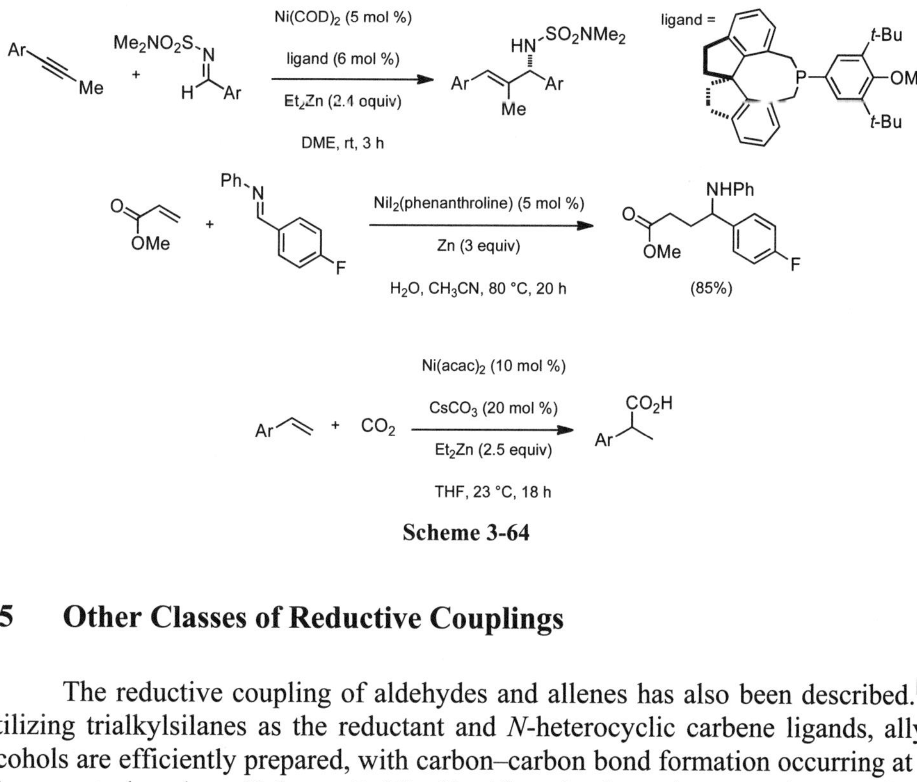

Scheme 3-64

4.5 Other Classes of Reductive Couplings

The reductive coupling of aldehydes and allenes has also been described.[202] Utilizing trialkylsilanes as the reductant and *N*-heterocyclic carbene ligands, allylic alcohols are efficiently prepared, with carbon–carbon bond formation occurring at the allene central carbon (Scheme 3-65). Significantly, by using enantiopure allenes, efficient chirality transfer occurs, and selective access to *Z*-alkenes is provided, which is opposite the stereochemistry provided by the more widely studied alkyne addition processes. The chirality transfer and stereoselectivity likely originates from the conversion of metallacycle **24** into π-allyl complex **25** as depicted subsequently. Labeling experiments with Et_3SiD provided support for this proposal.

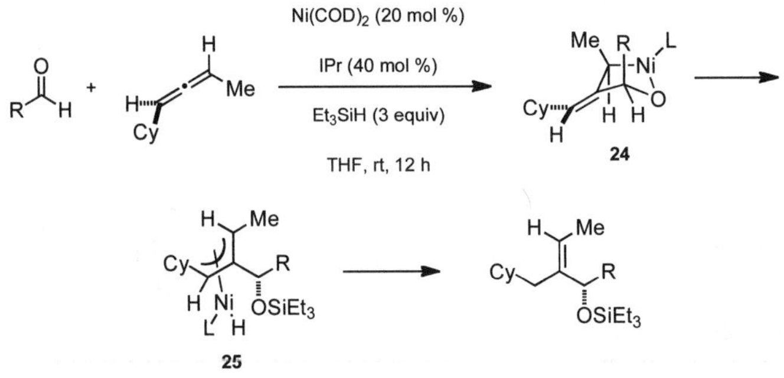

Scheme 3-65

Alkylative versions of nickel-catalyzed cyclizations of aldehydes and allenes employing organozinc reagents have also been developed.[203, 204] This procedure involves coupling of the aldehyde with the internal position of the allene, typically preparing the *cis* ring juncture during coupling (Scheme 3-66). A variety of simple examples has been demonstrated, and a more complex example was used as a key step in the synthesis of testudinariol A, wherein a relatively functionalized cyclopentane structure was stereoselectively prepared by the process.[205]

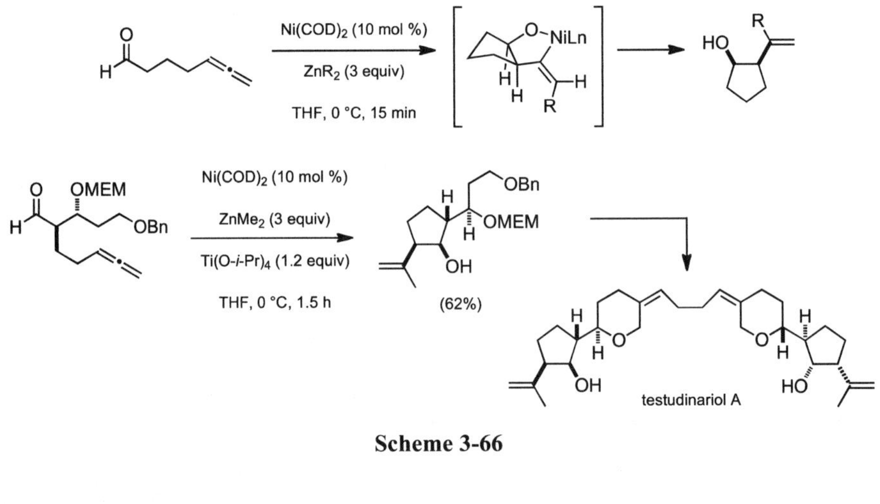

Scheme 3-66

In contrast to the previous methods involving reductive coupling of two π-components, a method involving participation of the sigma bond of an epoxide has been demonstrated. Treatment of a monosubstituted epoxide and alkyne with Et_3B in the presence of a nickel catalyst generated from $Ni(cod)_2$ and PBu_3 affords homoallylic alcohols from reductive coupling (Scheme 3-67).[206] The process proceeds with retention of epoxide stereochemistry, and intramolecular versions are endoselective. These aspects can be explained by initial oxidative addition of Ni(0) to the epoxide, followed by alkyne insertion, and nickel alkoxide reductive cleavage by Et_3B. This method has been applied in several relatively complex fragment couplings and macrocyclizations, as illustrated in the total synthesis of (–)-gloeosporone (Scheme 3-67).[207]

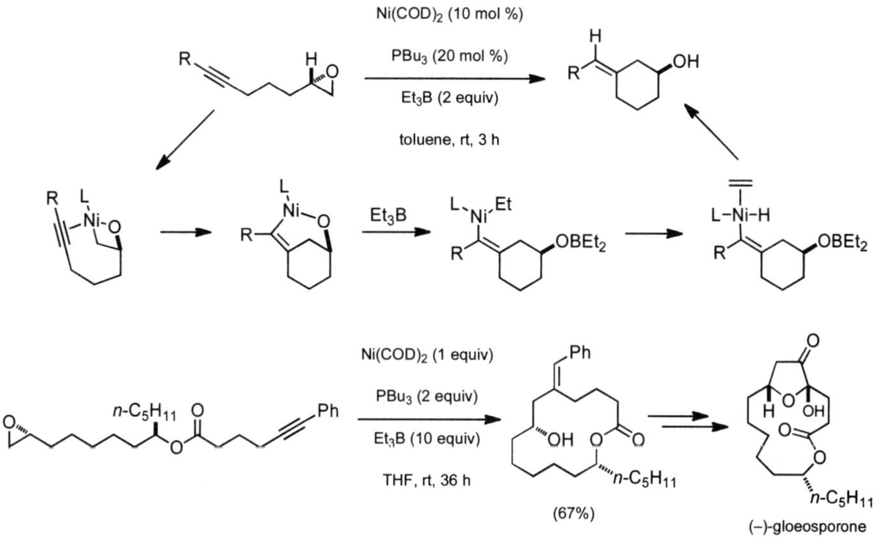

Scheme 3-67

Representative procedure: synthesis of (+) (4E)-4-methyl-5-phenylpent-4-en-2-ol[206]

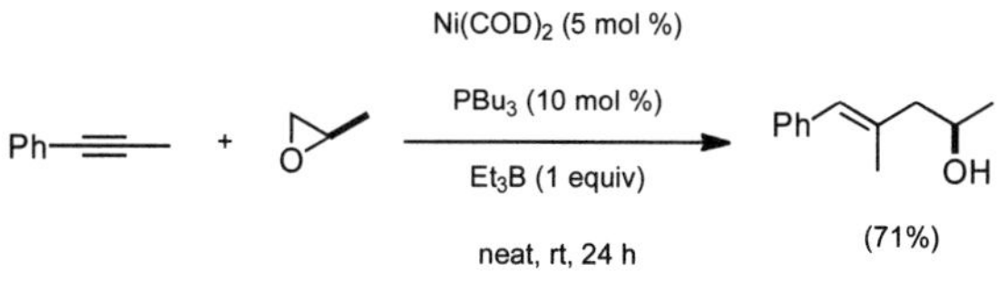

To Ni(cod)$_2$ (137 mg, 0.50 mmol) at RT were added PBu_3 (250 μL, 1.00 mmol), Et_3B (300 μL, 2.07 mmol), (*R*)-propylene oxide (700 μL, 9.99 mmol), 3-phenyl-1-propyne (630 μL, 5.02 mmol), and additional Et_3B (1.20 mL, 8.28 mmol) *via* syringe pump over 4 h. The brownish solution was stirred at RT for 24 h. EtOAc (2.5 mL) was added and the resulting solution was quenched with 2 M of aqueous NaOH (2.5 mL) and 30% aqueous H_2O_2. The resulting biphasic solution was vigorously stirred for 5 min. The layers were separated and the organic phase was washed with saturated aqueous $NaHCO_3$, dried over anhydrous $MgSO_4$, filtered, and concentrated under reduced pressure. The residue was purified by flash chromatography on silica gel (0–20% EtOAc/Hex) to afford the desired product (0.63 g, 71%) as a white solid.

5. Carbonyl Additions and Conjugate Additions

5.1 Nozaki-Hiyama-Kishi Couplings

Reviews. Fürstner, A.; *Chem. Rev.* **1999**, *99*, 991 – 1046.

The nickel-catalyzed, chromium-mediated addition of alkenyl iodides or triflates to aldehydes, known as the Nozaki-Hiyama-Kishi (NHK) coupling, is an exceptionally powerful carbon–carbon bond-forming process (Scheme 3-68). It proceeds with broad scope, extraordinary chemoselectivity, and often excellent stereoselectivity. These features combine to make this process a widely applied transformation in complex synthetic applications. The discovery of this process occurred during the study of processes promoted solely by $CrCl_2$.[208, 209] It was found that the catalytic addition of alkenyl iodides or triflates to aldehydes, promoted by $CrCl_2$, proceeded efficiently in some cases but was highly dependent on the batch of $CrCl_2$ employed. Ultimately, the presence of nickel contaminants in commercial $CrCl_2$ were found to be essential for high-yielding and reproducible additions to occur, and current protocols include the addition of $NiCl_2$ as a beneficial component. It is important to note that related processes with more reactive electrophiles, such as allylic halides, proceed cleanly in the absence of nickel. The addition of nickel salts is often detrimental with highly reactive electrophiles because homocoupling of the electrophile is often promoted by the nickel additive in these cases.

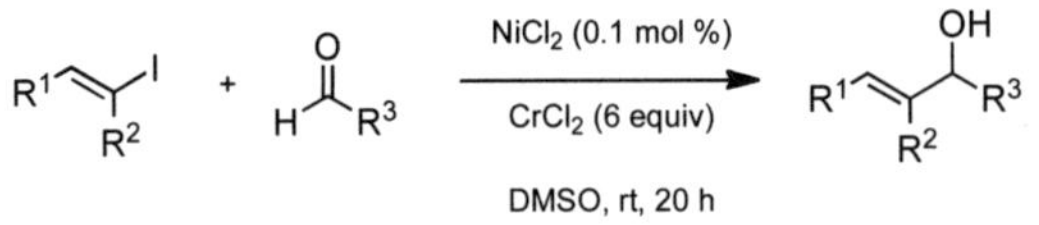

Scheme 3-68

A commonly proposed mechanism of the standard NHK protocol involves reduction of Ni(II) to Ni(0) by $CrCl_2$, then oxidative addition of Ni(0) to the alkenyl iodide to generate an alkenyl Ni(II) species (Scheme 3-69). Transmetallation to $CrCl_3$ (from the initial nickel reduction step) produces a Cr(III) alkenyl species, which undergoes direct addition to the aldehyde to generate product. Alternatively, alkenyl transfer from a vinyl nickel(II) species to $CrCl_2$ may instead occur with release of the Cr(III) alkenyl species and a Ni(I) species.

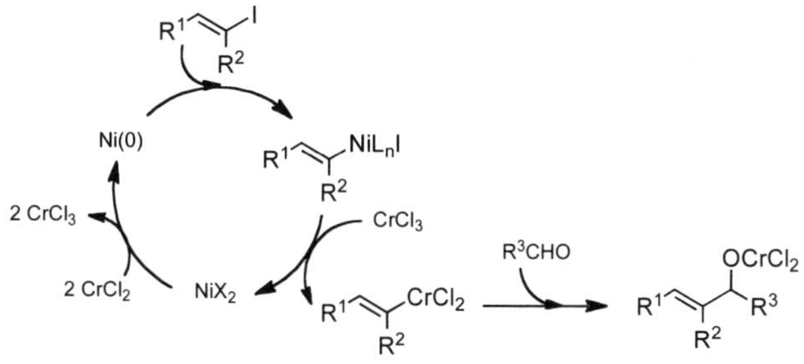

Scheme 3-69

A significant practical advance came in the form of a procedure that is catalytic in both nickel and chromium, using manganese powder as the stoichiometric reductant (Scheme 3-70).[210–212] In this process, manganese powder continually regenerates $CrCl_2$ from $CrCl_3$. Additionally, the stoichiometric use of Me_3SiCl converts the chromium(III) alkoxide product into the corresponding silyl ether while releasing $CrCl_3$ for further reduction to $CrCl_2$ by manganese powder. Cp_2ZrCl_2 may also be used in place of Me_3SiCl and with superior activity to achieve turnover of the chromium additive.[213]

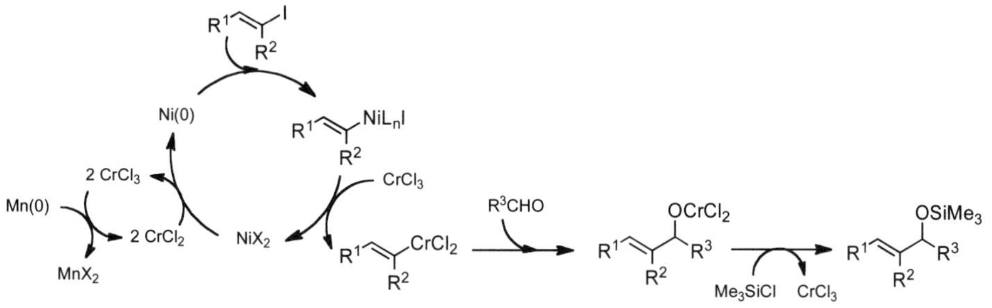

Scheme 3-70

Although the above mechanisms are logical and explain many of the experimental observations, it should be stressed that key questions remain regarding the precise nature of the proposed steps. The involvement of a Ni(I) or Ni(0) species in the oxidative addition to the alkenyl halide and the involvement of a Cr(II) or Cr(III) species in the transmetallation process have not been clarified. The oxidation state of chromium in the step that forms the vinyl chromium species may be different depending on chromium stoichiometry because Cr(III) builds up in the stoichiometric chromium procedure, whereas it is continually depleted by reduction in the procedures that are catalytic in chromium and stoichiometric in manganese. A recent study has articulated many of the remaining mechanistic questions surrounding the precise role of nickel and chromium in the process.[214] For example, evidence was provided that a Ni(I) species may be the active catalytic form of nickel that undergoes addition to the alkenyl iodide. Additionally, the nature of the species generated by interaction of the alkenyl nickel species with $CrCl_3$ is unclear. The formal transmetallation to generate an alkenyl chromium species as depicted previously may occur. Alternatively, a mixed Cr-Ni alkenyl species may instead be involved in the aldehyde addition step. Clarity for these questions awaits further mechanistic investigation.

Representative procedure: synthesis of a palytoxin fragment[208]

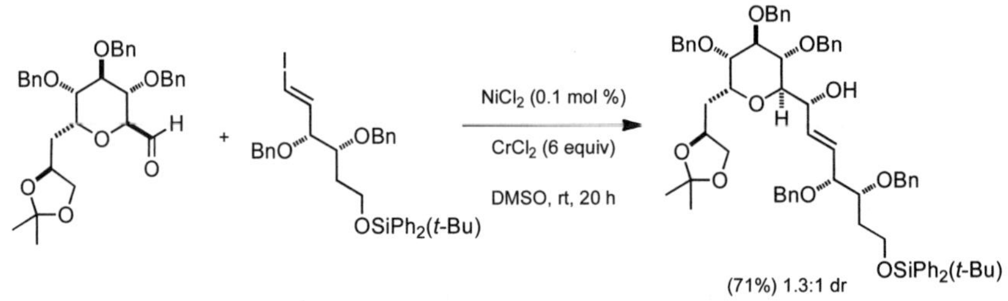

$NiCl_2$ (0.1 mol %)
$CrCl_2$ (6 equiv)
DMSO, rt, 20 h
(71%) 1.3:1 dr

The aldehyde was prepared by oxidation of the corresponding primary alcohol (1.285 g, 2.35 mmol) under Swern conditions. The crude aldehyde was mixed with the iodo olefin (4.766 g, 7.05 mmol) and azeotroped with toluene (two times). The mixture was dissolved in dimethylsulfoxide (DMSO) (60 mL) in a glove box. To this solution was added $CrCl_2$, containing 0.1% $NiCl_2$ (1.75 g, 14.2 mmol) portionwise. The dark-green mixture was stirred in a glove box for 20 h at RT. The reaction mixture was quenched by stirring with saturated NH_4Cl and $CHCl_3$ and then extracted with EtOAc (three times). The combined extracts were washed with brine, dried over anhydrous $MgSO_4$, and evaporated *in vacuo*. The crude products were separated by medium-pressure column chromatography (Merck silica gel 20% EtOAc-hexanes) to give the 16α-allylic alcohol (1.026 g, 39.9% overall yield from the primary alcohol) and the 16β-allylic alcohol (0.802 g, 31.2% overall yield from the primary alcohol).

Representative procedure: synthesis of 2-methylene-1-(4-methoxyphenyl)hexan-1-ol[210]

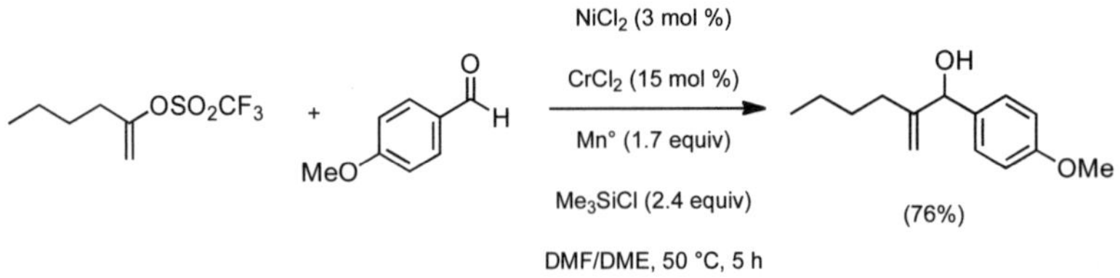

$NiCl_2$ (3 mol %)
$CrCl_2$ (15 mol %)
Mn° (1.7 equiv)
Me_3SiCl (2.4 equiv)
DMF/DME, 50 °C, 5 h
(76%)

A solution of 4-methoxybenzaldehyde (340 mg, 2.5 mmol), 2-trifluoromethylsulfonyloxy-1-hexene (1.06 g, 4.6 mmol) and TMSCl (0.75 mL, 6.0 mmol) in *N*,*N*-dimethylformamide (DMF) (1.5 mL) and DME (5 mL) was dropped into a suspension of Mn powder (230 mg, 4.2 mmol), $CrCl_2$ (46 mg, 0.38 mmol), and $NiCl_2$ (10 mg, 0.07 mmol) in dme (5 mL) at 50 °C. After stirring for 5 h at that temperature, the mixture was quenched with water (15 mL), extracted with ethyl acetate (150 mL in three portions), and the combined organic layers were washed with brine. Aqueous *n*-Bu_4NF (75% w/w) was added, and the solution was stirred at RT until TLC showed complete desilylation of the crude product. Standard work-up followed by flash chromatography with hexane/ethyl acetate (15/1) as eluent afforded the desired product as a colorless syrup (420 mg, 76%).

As noted, many applications of NHK couplings and macrocyclizations have been used in very complex synthetic applications. Macrocyclizations proceed in high yield without the use of high-dilution techniques, as illustrated in an approach to halichondrine (Scheme 3-71).[215] Additionally, highly effective asymmetric additions have been realized by using chromium catalysts derived from chiral sulfonamides (Scheme 3-72).[213, 216] The latter advances display high utility in the catalyst-controlled addition to chiral aldehydes, where the ligand stereochemistry overrides the inherent substrate biases.

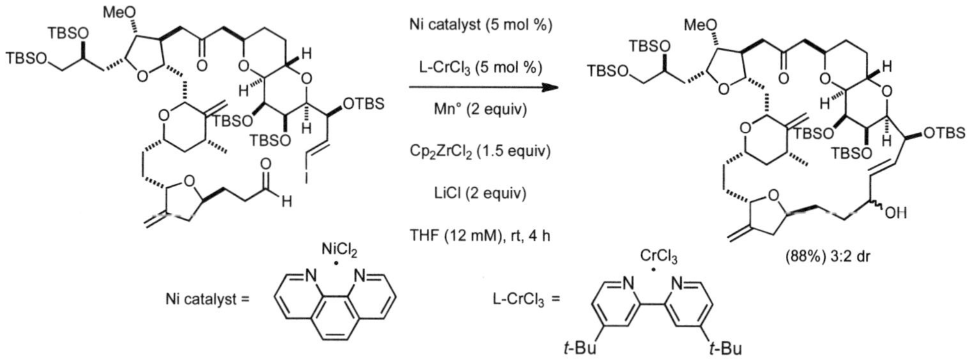

Scheme 3-71

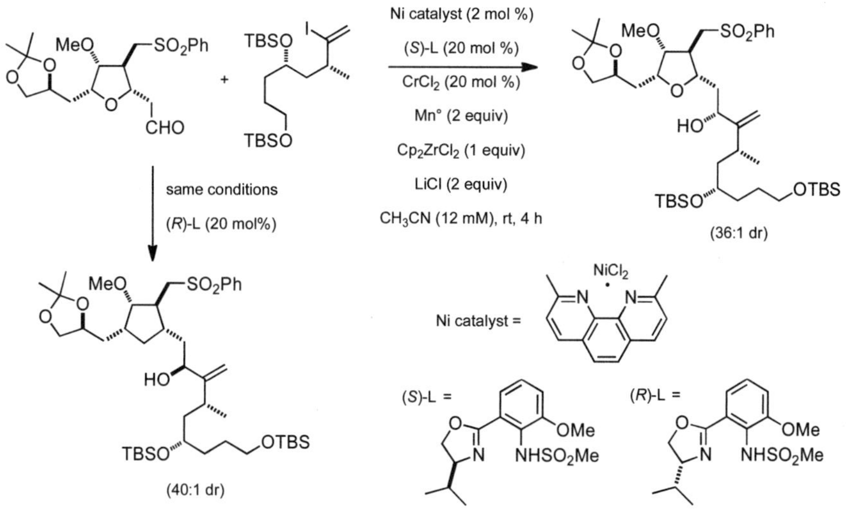

Scheme 3-72

5.2 Other Additions to Aldehydes or Carbon Dioxide

The nickel-catalyzed addition of organozincs to carbon dioxide has been demonstrated for a range of sp^2 and sp^3-hybridized organozinc reagents (Scheme 3-73).[217, 218] These processes complement the reductive additions of alkenes and alkynes to carbon dioxide described in section 4.4. A range of functionalized alkylzinc and arylzinc reagents underwent efficient couplings to CO_2 using $Ni(acac)_2$ or $Ni(cod)_2$-derived catalysts with simple monodentate phosphine ligands.

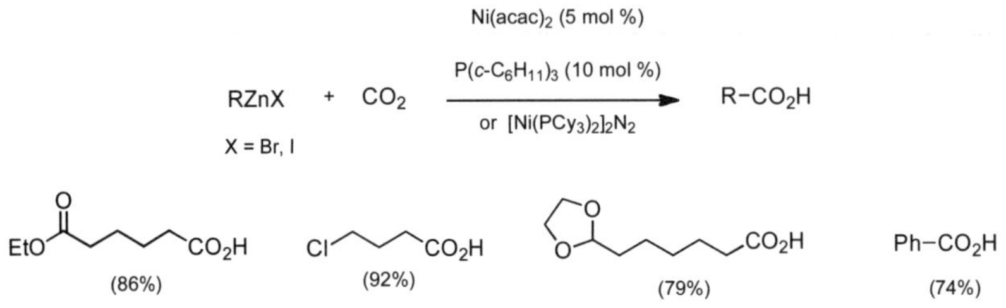

Ni(acac)$_2$ (5 mol %)

RZnX + CO_2 → R–CO_2H [P(c-C_6H_{11})$_3$ (10 mol %) or [Ni(PCy$_3$)$_2$]$_2$N$_2$]

X = Br, I

EtO(C=O)…CO_2H (86%); Cl…CO_2H (92%); (dioxolane)…CO_2H (79%); Ph–CO_2H (74%)

Scheme 3-73

Representative procedure: synthesis of benzyl heptanoate[218]

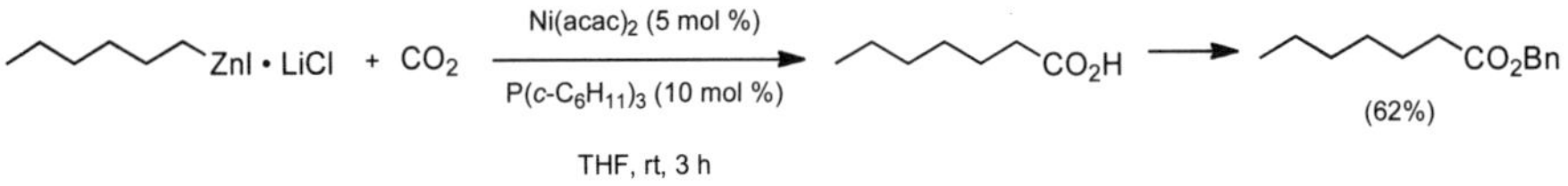

ZnI • LiCl + CO_2 → [Ni(acac)$_2$ (5 mol %), P(c-C_6H_{11})$_3$ (10 mol %), THF, rt, 3 h] → CO_2H → CO_2Bn (62%)

Ni(acac)$_2$ (6.4 mg, 0.025 mmol) was placed in a reaction flask, and the flask was then filled with CO_2 (1 atm, balloon) by using the standard Schlenk technique. THF (5.0 mL) and P(c-C_6H_{11})$_3$ (0.50-M toluene solution, 0.10 mL, 0.050 mmol) were added dropwise, and the solution was stirred for 10 min at RT. Finally, a solution of hexylzinc iodide-lithium chloride complex in THF prepared in advance (0.70 M, 0.71 mL, 0.50 mmol) was added. After being stirred for 3 h at RT, the resulting mixture was poured into 1.0 M of aq. HCl (10 mL). Extraction with ethyl acetate followed by concentration afforded a crude mixture. The crude mixture obtained was dissolved in DMF (3.0 mL), and potassium carbonate (69 mg, 0.50 mmol), sodium iodide (catalytic amount, *ca* 0.05 mmol), and benzyl bromide (0.059 mL, 0.50 mmol) were then added dropwise. After being stirred for 10 h at RT, the mixture was quenched with H_2O (10 mL). Extraction with Et_2O, concentration, and silica gel column purification (hexane:ethyl acetate = 40 : 1) gave the desired product in 62% yield.

In addition to the direct use of alkenyl halides (NHK-type additions) and reactive organometallics (*i.e.*, organozincs) in nickel-catalyzed additions to aldehydes and carbon dioxide, simple alkenes may also be used in related catalytic processes (Scheme 3-74).[219, 220] The catalytic addition of simple alkenes to aldehydes has been developed as an attractive complement to the reductive coupling of aldehydes and alkynes for preparing allylic alcohols.[221–225] In contrast to the various classes of reductive couplings previously described (section 4), the catalytic coupling of aldehydes and alkenes proceeds without the action of a reducing agent, but rather it involves product generation *via* silylation by TMSOTf. The regioselectivity is controlled by the ligand structure, allowing either allylic alcohols or homoallylic alcohols to be obtained selectively in the process.

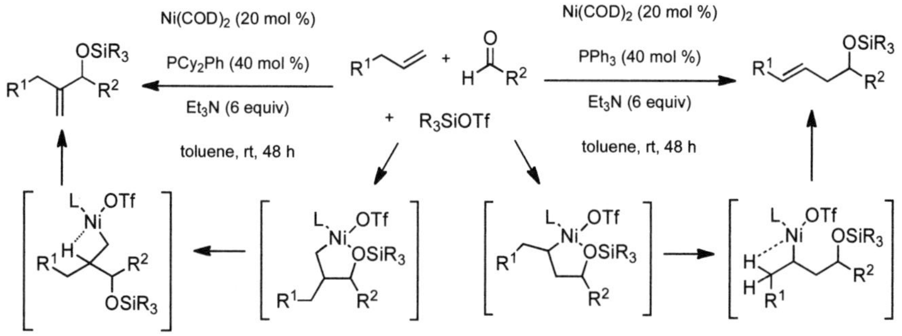

Ni(COD)$_2$ (20 mol %), PCy$_2$Ph (40 mol %), Et$_3$N (6 equiv), toluene, rt, 48 h; Ni(COD)$_2$ (20 mol %), PPh$_3$ (40 mol %), Et$_3$N (6 equiv), toluene, rt, 48 h; + R$_3$SiOTf

Scheme 3-74

The process resembles Prins-type additions and carbonyl ene reactions, but the unique mechanism results in different relative reactivities of substituted alkenes (Scheme 3-75). Simple monosubstituted alkenes are most reactive in the nickel-catalyzed process, whereas more substituted alkenes are more reactive in traditional Lewis acid-catalyzed processes.

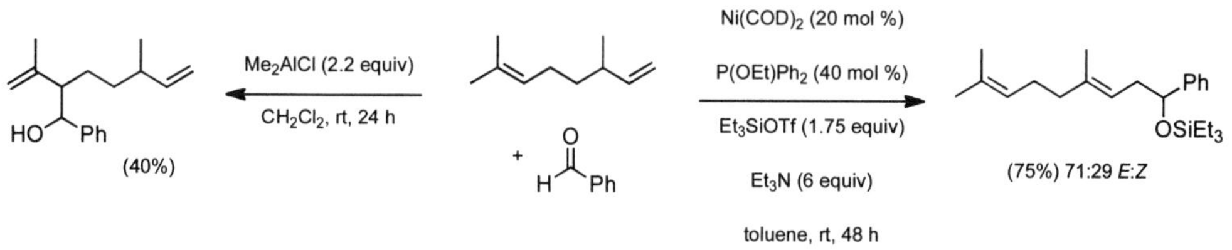

Me$_2$AlCl (2.2 equiv), CH$_2$Cl$_2$, rt, 24 h (40%); Ni(COD)$_2$ (20 mol %), P(OEt)Ph$_2$ (40 mol %), Et$_3$SiOTf (1.75 equiv), Et$_3$N (6 equiv), toluene, rt, 48 h (75%) 71:29 *E*:*Z*

Scheme 3-75

Representative procedure: synthesis of (E)-triethyl((1-phenylnon-3-en-1-yl)oxy)silane or triethyl((2-methylene-1-phenyloctyl)oxy)silane[223]

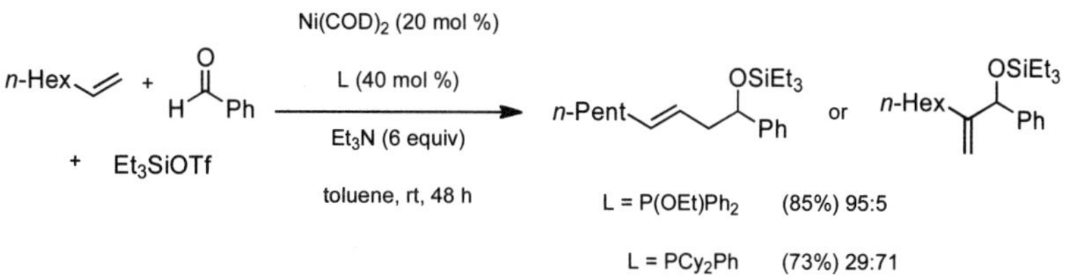

n-Hex alkene + PhCHO + Et$_3$SiOTf; Ni(COD)$_2$ (20 mol %), L (40 mol %), Et$_3$N (6 equiv), toluene, rt, 48 h

L = P(OEt)Ph$_2$ (85%) 95:5

L = PCy$_2$Ph (73%) 29:71

General procedure: A 10-mL test tube and a stir bar were oven dried and brought into a glove box. Ni(cod)$_2$ (27.5 mg, 0.1 mmol, 20 mol%) and ligand (0.2 mmol, 40 mol% as specified) were added to the test tube, the test tube was sealed with a septum, and the sealed tube was brought out of the glove box and connected to an Ar line. The catalyst mixture was dissolved in toluene (2.5 mL) under Ar and stirred 5 min at RT. Alkene (0.5 mL), triethylamine (418 μL, 3 mmol, 600 mol%), and then aldehyde (0.5 mmol, 100 mol%) were added. TESOTf (197 μL, 0.875 mmol, 175 mol%) was added. The mixture was stirred at RT for 48 h. The mixture was filtered through a plug of silica gel. Solvent was removed under reduced pressure, and the crude mixture was diluted in hexane. Purification *via* flash chromatography on silica afforded the coupling product. The reaction of 1-octene and benzaldehyde (51 μL, 0.5 mmol) with $Ni(cod)_2$, $EtOPh_2P$ (43 μL, 0.2 mmol, 40 mol%) TESOTf (197 μL, 0.875 mmol), and triethylamine in toluene following the general procedure afforded a 95:5 ratio of the desired products in 85% yield as a colorless oil. In another experiment, the reaction of 1-octene (1 mL) and benzaldehyde (51 μL, 0.5 mmol) with $Ni(cod)_2$, Cy_2PhP (56 mg, 0.2 mmol, 40 mol%), TESOTf (197 μL, 0.875 mmol), and triethylamine in toluene following the general procedure afforded a 29:71 ratio of the desired products in 73% yield as a colorless oil.

Additionally, a nickel-catalyzed Tishchenko reaction has been developed, where the selective crossed addition of two different aldehydes to generate a single ester product is possible (Scheme 3-76).[226] By employing an aromatic aldehyde with a nonaromatic aldehyde, a metallacycle derived from heterocoupling is formed, followed by its decomposition to the ester product.

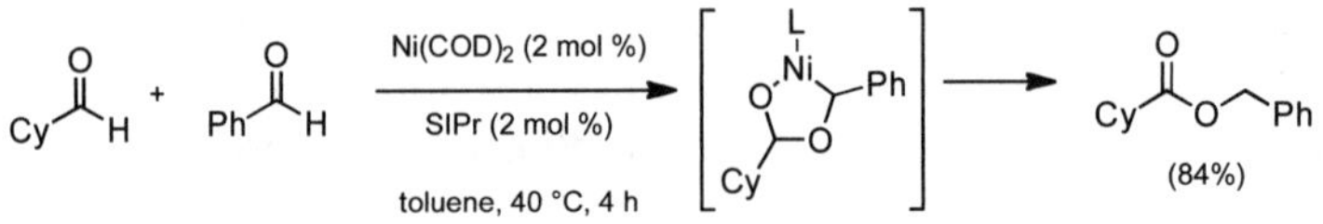

Cy–CHO + Ph–CHO —[$Ni(COD)_2$ (2 mol %), SIPr (2 mol %), toluene, 40 °C, 4 h]→ [L–Ni metallacycle (Ph, Cy)] → Cy–C(O)–O–CH₂Ph (84%)

Scheme 3-76

5.3 Conjugate Addition Processes

Nickel-catalyzed conjugate additions have been developed as versatile tools in synthetic chemistry. The majority of developments have involved the catalyzed addition of organometallic reagents, such as organozirconium, organozinc, organoborane, and organoaluminum reagents to the electrophilic β-carbon of enones and related electron-deficient alkenes. Additionally, reductive methods have been developed for the direct use of organic halides as the reagent that is used to introduce a substituent at an enone β-carbon. These classes of transformations complement the strategies for the reductive coupling of various π-systems to an electron-deficient alkene as described in section 4.2.

The hydrozirconation of alkynes using $Cp_2Zr(H)Cl$ provides alkenyl zirconium species that undergo efficient conjugate additions to enones with catalysis by $Ni(acac)_2$ (Scheme 3-77).[227, 228] Studied to a lesser extent is the $Ni(acac)_2$-catalyzed conjugate addition of alkenylboranes to electron-deficient alkenes.[229] Aluminum acetylides, generated by addition of lithium acetylides to Me_2AlCl, undergo efficient conjugate additions under similar conditions.[230, 231] Arylzinc reagents also undergo efficient nickel-catalyzed conjugate additions, including to challenging hindered enones.[232, 233] Additionally, tandem conjugate addition methods have been developed using arylzinc reagents, where the kinetic enolate generated by conjugate addition undergoes further addition to a tethered enone.[234] Alkenyltin reagents have also been shown to be effective participants in nickel-catalyzed conjugate additions,[235] and the mechanistic findings for this variant are described below. Together, these methods illustrate the general efficiency of conjugate additions of both sp^2- and sp-hybridized carbon groups.

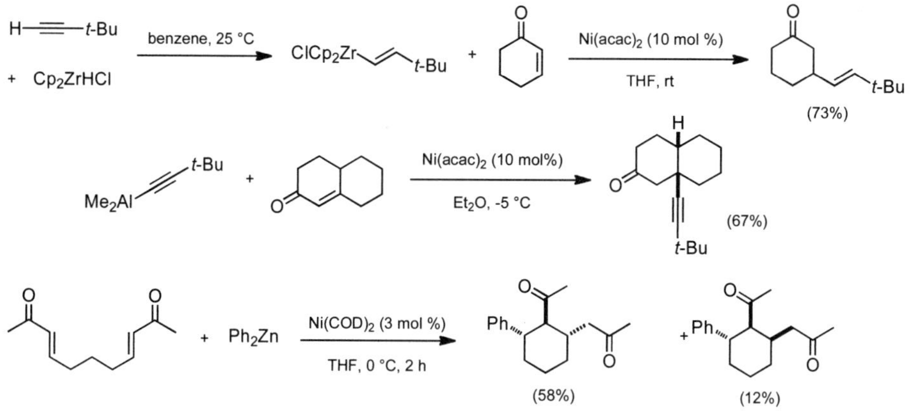

Scheme 3-77

A variety of sp^3-hybridized carbon groups also undergoes efficient nickel-catalyzed conjugate addition. The $Ni(acac)_2$-catalyzed addition of dimethylzinc proceeds very efficiently with challenging, hindered systems (Scheme 3-78).[236] This procedure has been used in contexts where other conjugate additions procedures, including those involving organocuprate formulations, were unsuccessful.[237–139] Additionally, trialkylboranes derived from 9-BBN undergo efficient $Ni(acac)_2$-catalyzed conjugate additions to a variety of enoate derivatives.[240]

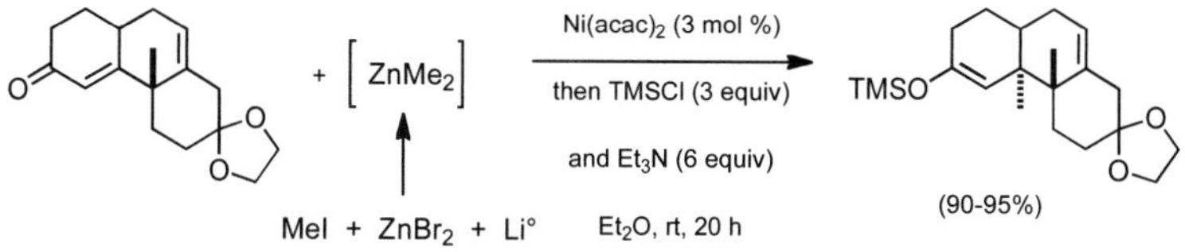

Scheme 3-78

Representative procedure: synthesis of a scopadulcic acid intermediate[238]

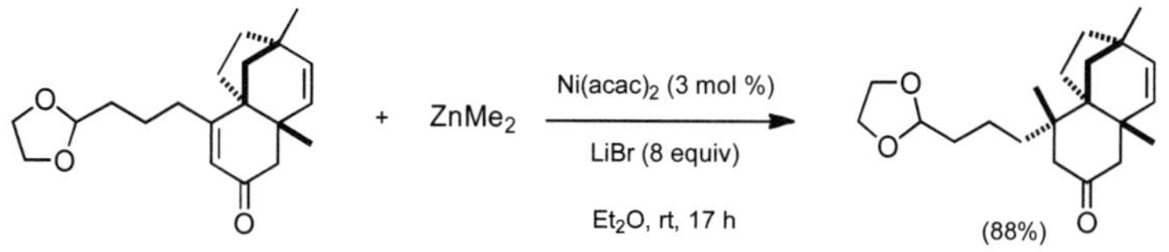

Using a glove box, a flask was charged with dry LiBr (650 mg, 7.5 mmol) and $Ni(acac)_2$ (7 mg, 0.03 mmol) and then flushed with Ar, and dry Et_2O was added. Dimethylzinc (0.26 mL, 3.8 mmol) was added dropwise to this stirred suspension at

0 °C under Ar. The brown heterogeneous mixture was stirred at 0 °C for 10 min, and then a solution of enone (280 mg, 0.93 mmol) and Et_2O (6 mL) was added by cannula. The reaction was allowed to warm to RT and stirred for 17 h. This mixture was cooled to 0 °C and quenched with saturated aqueous NH_4Cl (14 mL). Saturated aqueous ethylenediaminetetraacetic acid (EDTA) (5 mL) was then added, and the organic layer was separated. The aqueous layer was extracted with CH_2Cl_2 (2 × 20 mL), and the combined organic layers were dried (anhydrous Na_2SO_4), filtered, and concentrated. Purification of the residue by flash chromatography (5.7:1 hexanes-EtOAc) afforded 258 mg (88%) of desired product as a colorless oil.

Representative procedure: synthesis of benzyl 14-benzyloxy-3-methyltetradecanoate[240]

Ph⌒O(⌒)$_9$⌒ + H–B(9-BBN) —THF, rt, 15 h→ Ph⌒O(⌒)$_9$⌒B(9-BBN) (6 equiv)

above derivative + crotonate OBn (1 equiv) —$Ni(COD)_2$ (8 mol %), $P(t\text{-}Bu)_3$ (19 mol %), $Cs_2(CO_3)$ (3.0 equiv), toluene, rt, 6.5-10 h→ Ph⌒O(⌒)$_9$⌒CH(CH$_3$)CH$_2$C(O)OBn (96%)

With a glove box filled with Ar, 9-BBN was placed in a reaction flask. A solution of benzyl 11-undecenyl ether (469 mg, 1.8 mmol) in THF (3 mL) was added dropwise. The solution was stirred for 15 h at RT to prepare the alkylborane. Separately, with a glove box filled with Ar, $Ni(cod)_2$ (11 mg, 0.04 mmol) and Cs_2CO_3 (489 mg, 1.5 mmol) were placed in another reaction flask. Toluene (3 mL) and $P(t\text{-}Bu)_3$ (1.0 M, 0.096 mL, 0.096 mmol) were added dropwise. The suspension was stirred for 10 min at 0 °C. Methanol (0.081 mL, 2.0 mmol) was added to the alkylborane, and the solution was then transferred to the mixture of nickel catalyst and Cs_2CO_3 in toluene *via* a syringe under Ar. Finally, a solution of benzyl (*E*)-crotonate (88 mg, 0.5 mmol) in toluene (2 mL) was added. After being stirred for 8.5 h at RT, the resulting mixture was poured into 1.0 M of hydrochloric acid (10 mL). Extraction with hexane/ethyl acetate (20:1) followed by silica gel column purification afforded the desired product (210 mg, 0.48 mmol) in 96% yield.

In addition to the preceding methods that involve catalyzed additions to α,β-unsaturated carbonyls, enantioselective additions to unsaturated acetals have also been developed as a synthetic equivalent to traditional conjugate addition processes. By using chiral bis-phosphine ligands, asymmetic additions of Grignard reagents to unsaturated acetals were realized.[241]

As illustrated by the preceding studies, a wide variety of pre-catalysts, ligands, and transmetallating components has been used in nickel-catalyzed conjugate additions. Based on the range of substrates, catalysts, and reaction conditions, it is likely that different mechanisms are operative between various subclasses of reactions of this type. In early studies from Schwartz, a study of gas evolution during reduction of the Ni(II) pre-catalyst in combination with electrochemical studies suggested that a Ni(I) complex was the active catalyst in alkenylzirconium additions using $Ni(acac)_2$ as a pre-catalyst reduced by diisobutylaluminum hydride (DIBAL).[242, 242] Subsequent studies from Mackenzie described the use of $Ni(cod)_2$ as pre-catayst in the catalytic conjugate addition of alkenylstannanes to enals promoted by trialkylsilyl chlorides (Scheme 3-79).[235] This work involved the synthesis and characterization of Ni(II) π-allyl complexes derived from the oxidative addition of Ni(0) and the demonstration that species of this type are competent catalysts with no accompanying induction period for promoting the alkenyl stannane-enal conjugate additions. Additionally, the Ni(II) π-allyl complexes are directly observable by nuclear magnetic resonance (NMR) as the only nickel-containing species in productive catalytic reactions.

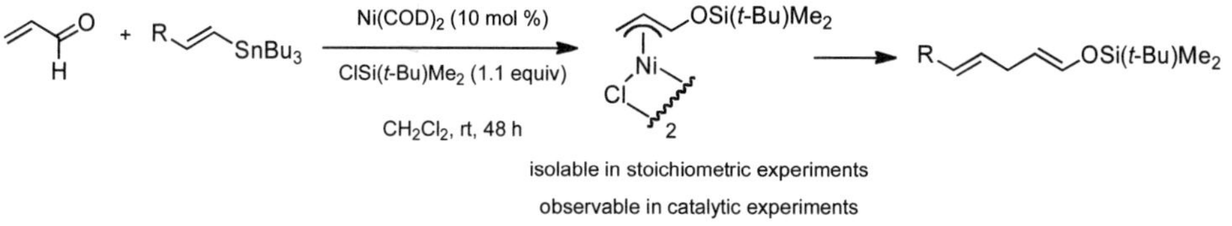

Ni(COD)₂ (10 mol %), ClSi(*t*-Bu)Me₂ (1.1 equiv), CH₂Cl₂, rt, 48 h

isolable in stoichiometric experiments
observable in catalytic experiments

Scheme 3-79

Representative procedure: synthesis of bis(μ-chloro)-[1,2,3-η^3-1-(trimethylsilyloxy)-2-propen-l-yl]nickel(II)[153]

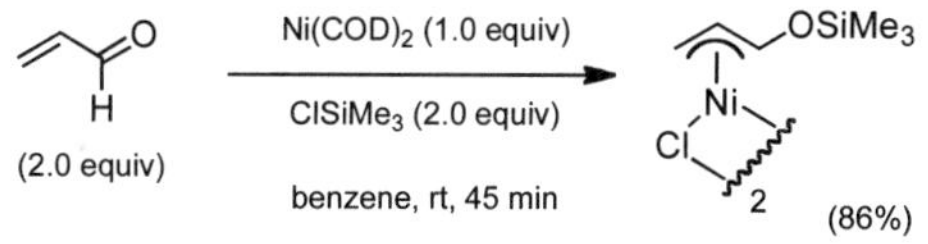

(2.0 equiv) Ni(COD)₂ (1.0 equiv), ClSiMe₃ (2.0 equiv), benzene, rt, 45 min (86%)

In a dry box, a 50-mL Schlenk tube was charged with $Ni(cod)_2$ (500 mg, 1.82 mmol, 1.00 equiv), equipped with a stir bar, and sealed with a septum. A solution of propenal (244 μL, 3.64 mmol, 2.00 equiv) in benzene (8 mL) in a 25-mL Schlenk vessel was treated with $(MeO)(Me_3SiO)C{=}CMe_2$, (185 μL, 0.909 mmol, 0.500 equiv; added as a proton-scavenging reagent), stirred for 5 min, and then transferred by cannula onto the $Ni(cod)_2$. The resultant purple-red slurry was stirred for 10 min and then treated with Me_3SiCl (462 μL, 3.64 mmol, 2.00 equiv) to afford a deep red solution. After 45 min, the volatiles were removed at 0.1 mmHg to obtain an orange-red powder. This was extracted with pentane (20 mL) through a filter paper-tipped cannula. The clear red filtrate was concentrated under vacuum to *ca.* 5 mL and then cooled to –20 °C for 24 h to induce crystallization. The precipitate was isolated by removal of the supernatant through a filter paper-tipped cannula while maintaining the mixture at –20 °C, and washed with pentane (2 × 3 mL) to yield the desired product (0.349 g, 86%) as a burgundy-red crystalline solid.

In addition to the preceding methods that allow conjugate addition of an organometallic reagent to electrophilic alkenes, a variety of methods has also been developed to allow the reductive addition of organohalide reagents to electrophilic alkenes, essentially allowing a polarity reversal strategy for conjugate additions.[244] A key illustration was the development of strategies for the photochemical conversion of enal-derived π-allyl complexes and alkyl halides into conjugate addition products.[153] This stoichiometric process patterns the mechanistic aspects of earlier developments in the photochemical alkylation of simple nickel π-allyl dimers by the involvement of Ni(I)/Ni(III) intermediates in a radical chain process.[245] Catalytic methods for the reductive conjugate addition of alkyl halides have been developed using manganese powder as the terminal reductant,[246] and conjugate additions of aryl halides have been developed in tandem with aldol additions using dimethylzinc as the terminal reductant (Scheme 3-80).[247] These methods hold considerable promise but have not yet been developed with broad scope.

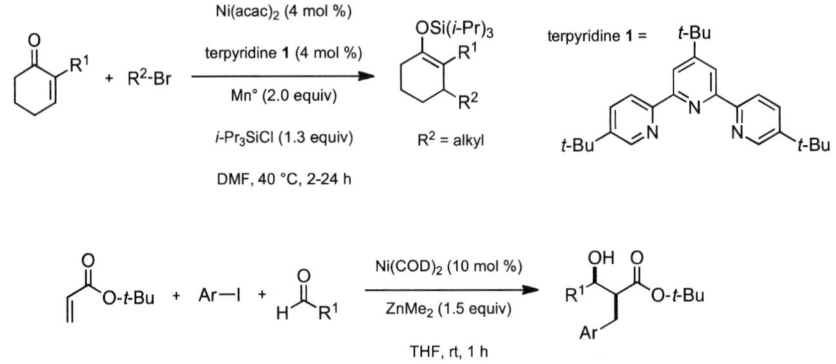

Ni(acac)$_2$ (4 mol %), terpyridine **1** (4 mol %), Mn° (2.0 equiv), *i*-Pr$_3$SiCl (1.3 equiv), DMF, 40 °C, 2-24 h

R^2 = alkyl

terpyridine **1** =

Ni(COD)$_2$ (10 mol %), ZnMe$_2$ (1.5 equiv), THF, rt, 1 h

Scheme 3-80

Representative procedure: synthesis of 2-benzyl-3-hydroxy-3-phenylpropionic acid t-butyl ester[247]

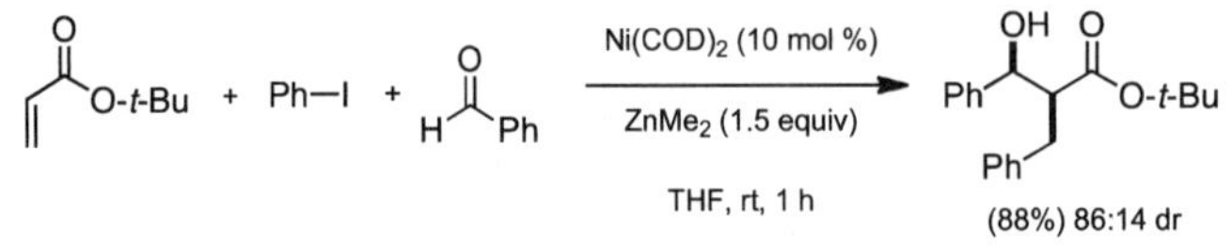

Ni(COD)$_2$ (10 mol %), ZnMe$_2$ (1.5 equiv), THF, rt, 1 h

(88%) 86:14 dr

General procedure: To a solution of Ni(cod)$_2$ (10 mol%) in 8.0 mL of THF at 0 °C was added acrylate (1.0 equiv), aldehyde (2.0 equiv), and aryl halide (2.0 equiv) under Ar atmosphere. To this brown solution, dimethyl zinc (1.5 equiv) was slowly added at 0 °C, the mixture was stirred for 1 h at 0 °C, and the mixture was allowed to warm to RT. After completion of the reaction, as judged by TLC analysis, the mixture was quenched with aqueous NH_4Cl solution at 0 °C, warmed to RT, extracted with ether (3 × 20 mL), washed with water and brine, and concentrated. The crude products were subjected to flash column chromatography on silica gel with hexane:ether (4:1) to give the desired products. Following the general procedure, iodobenzene (224 mL, 2.0 mmol), benzaldehyde (202 mL, 2.0 mmol), *tert*-butyl acrylate (146 mL, 1.0 mmol), dimethyl zinc (750 mL, 2.0-M solution in toluene, 1.5 mmol), and Ni(cod)$_2$ (27 mg, 0.1 mmol) were employed to produce the desired product (276 mg, 88%, dr 86:14) as a colorless thick oil.

In addition to the preceding methods that involve functionalization of an enone or enoate β-carbon with an alkyl or aryl functionality, methods for the nickel-catalyzed reductive aldol functionalization of enoates have also been developed (Scheme 3-81). Using triethylborane as the terminal reductant with $Ni(cod)_2$ as catalyst, reductive aldol reactions proceed to give syn aldol adducts.[248] An unusual role of phenyl iodide as a promoter for the process was found in this study. Intramolecular reductive aldol additions utilizing $Ni(acac)_2$ as catalyst and diethylzinc as the terminal reductant were also described.[249]

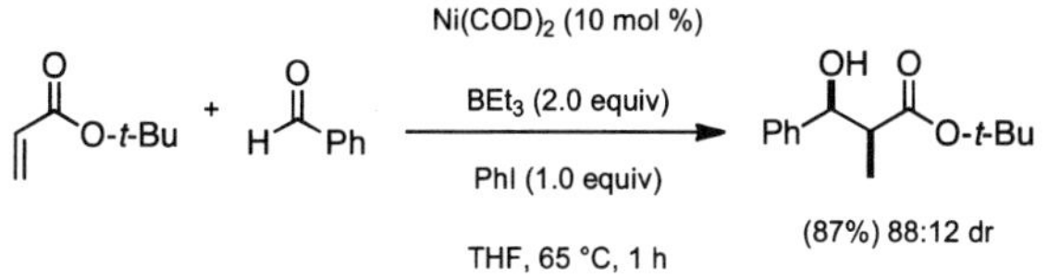

Scheme 3-81

Representative procedure: synthesis of (±)-(3R,4R)-1-benzyl-3-ethyl-4-hydroxy-4-methylpiperidin-2-one[249]

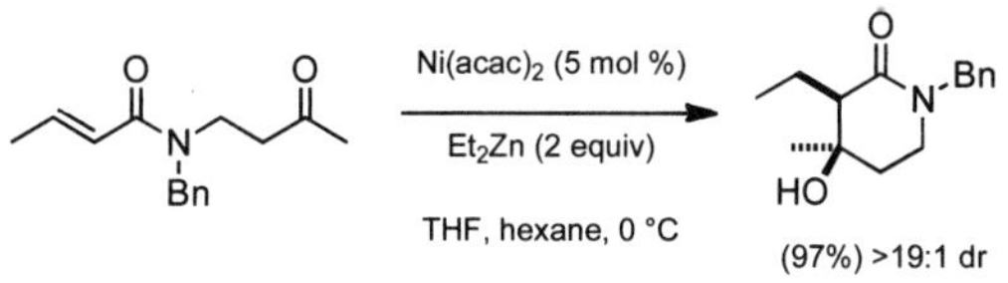

A solution of the (*E*)-*N*-benzyl-*N*-(3-oxobutyl)but-2-enamide (49 mg, 0.20 mmol) and $Ni(acac)_2$ (2.7 mg, 0.01 mmol) in THF (1.5 mL) was stirred at RT for 15 min. The mixture was cooled to 0 °C and Et_2Zn (1.0-M solution in hexane, 0.40 mL, 0.40 mmol) was then added rapidly in one portion. The reaction was stirred at 0 °C for 1 h and then at RT for 14 h. The reaction mixture was filtered through a short plug of SiO_2 using EtOAc as eluent (*ca.* 50 mL), and the filtrate was concentrated *in vacuo*. Purification of the residue by column chromatography (70% EtOAc/petrol) afforded the desired product as a white solid (48 mg, 97%).

6. 1,2- and 1,4-Additions to Alkenes, Alkynes, and Dienes

6.1 Hydrometallation

Many classes of nickel-catalyzed processes involving simple 1,2-additions to alkenes and alkynes or 1,4-additions to 1,3-dienes have been reported. Among these methods, the hydrometallation of simple π-components has been demonstrated with a wide range of main group and transition metal species. Nickel-catalyzed processes involving hydrozincation, hydrosilylation, and hydroalumination have been explored most extensively among the various known nickel-catalyzed hydrometallations.

Hydrosilylations are among the most extensively developed nickel-catalyzed hydrometallations, and examples have been reported with alkenes, alkynes, dienes, aldehydes, and ketones. Early reports described the hydrosilylation of simple alkenes and alkynes using Ni(II) phosphine catalysts (Scheme 3-82).[250–52] With terminal alkenes and alkynes, regioselectivities favoring installation of silicon at the terminal position were observed. More recent work using Ni(0) complexes of NHCs illustrated that the regioselectivity of alkyne hydrosilylation could be tuned through ligand and silane modification (Scheme 3-83).[253]

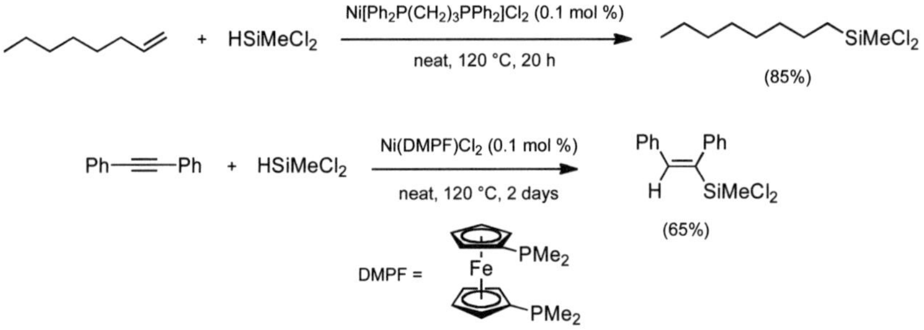

+ $HSiMeCl_2$ — $Ni[Ph_2P(CH_2)_3PPh_2]Cl_2$ (0.1 mol %), neat, 120 °C, 20 h → $SiMeCl_2$ (85%)

Ph—≡—Ph + $HSiMeCl_2$ — $Ni(DMPF)Cl_2$ (0.1 mol %), neat, 120 °C, 2 days → Ph, Ph, H, $SiMeCl_2$ (65%)

DMPF = Fe, PMe_2, PMe_2

Scheme 3-82

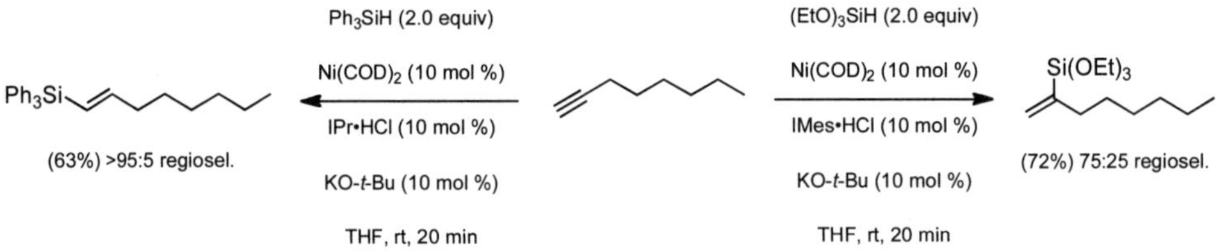

Ph_3SiH (2.0 equiv), $Ni(COD)_2$ (10 mol %), IPr•HCl (10 mol %), KO-*t*-Bu (10 mol %), THF, rt, 20 min → Ph_3Si (63%) >95:5 regiosel.

$(EtO)_3SiH$ (2.0 equiv), $Ni(COD)_2$ (10 mol %), IMes•HCl (10 mol %), KO-*t*-Bu (10 mol %), THF, rt, 20 min → $Si(OEt)_3$ (72%) 75:25 regiosel.

Scheme 3-83

Nickel-catalyzed ketone and aldehyde hydrosilylations have been developed with well-defined Ni(II) hydrides using phosphine anilide ligands (Scheme 3-84).[254] This process is tolerant of a variety of functional groups including aryl halides. In mechanistically distinct processes, nickel(0) complexes of NHCs were shown to catalyze ketone hydrosilylations using carbohydrate-derived silanes in a process that allows reductive glycosylations of ketones.[255] Chemoselectivity of the latter method was optimal in the presence of $Ti(OR)_4$ additives.

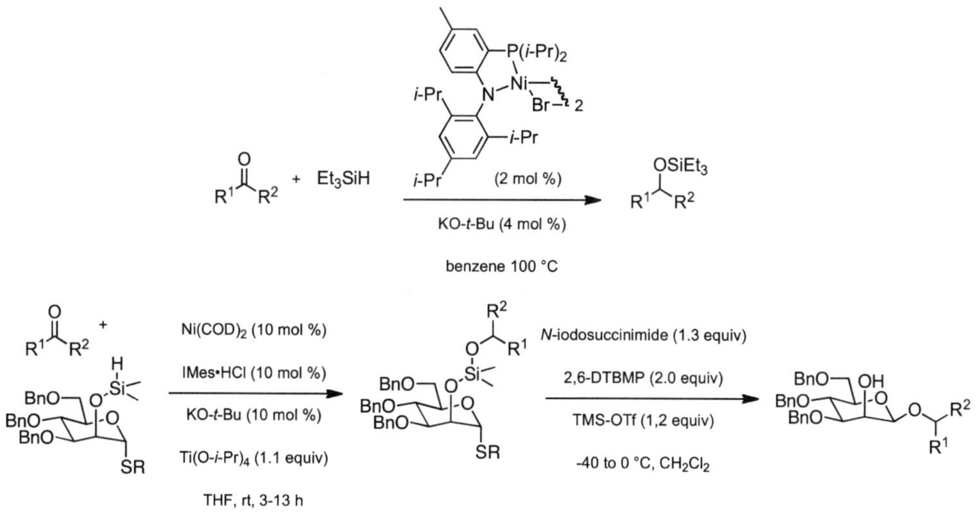

Scheme 3-84

Representative procedure: general conditions for nickel complex-mediated hydrosilylation of carbonyl-containing substrates[254]

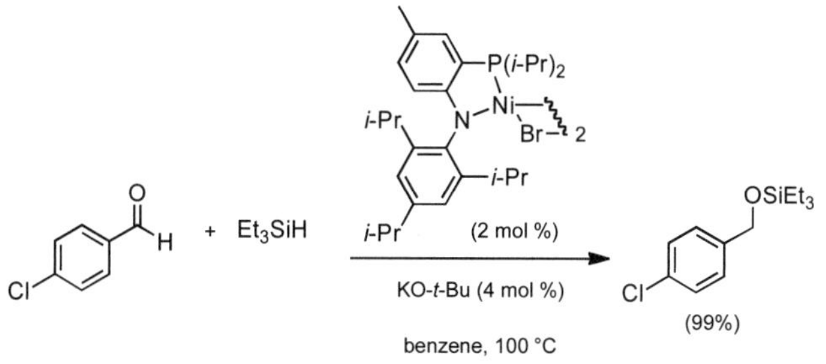

General procedure: In a glove box, a J-Young NMR tube equipped with a Teflon screw cap was charged with 1.0 mL of C_6D_6 solution containing 2 mol% of the nickel catalyst (30.8 mg, 0.027 mmol), 4 mol% of KO-*t*-Bu (6.15 mg, 0.055 mmol), and triethylsilane (262 μL, 1.64 mmol). The J-Young tube was removed from the glove box, and the carbonyl compound (1.37 mmol) was added to the J-Young *via* syringe. The sample was placed into a preheated 100 °C oil bath. The reaction was monitored by ^{1}H NMR spectroscopy. After completion, all volatiles were removed under reduced pressure and the crude product was subjected to flash chromatography on a thin, long silica gel column (Mallinckrodt Baker 60-200 mesh) with pentane as eluent. The oily product was obtained after removal of all pentane *via* a rotary evaporator.

The nickel-catalyzed hydrozincation of alkenes has been developed as a useful entry to functionalized organozincs (Scheme 3-85). Simple terminal olefins, when treated with Et_2Zn and catalytic quantities of $Ni(acac)_2$ and cyclooctadiene, are efficiently converted into the corresponding dialkylzinc reagents from hydrozincation of the alkene.[256, 257] A wide variety of copper-catalyzed alkylations are accessible from the dialkylzincs obtained by this method. The proposed mechanism for the hydrozincation involves generation of a Ni(II) hydride followed by olefin insertion and transmetallation to produce the dialkylzinc product.

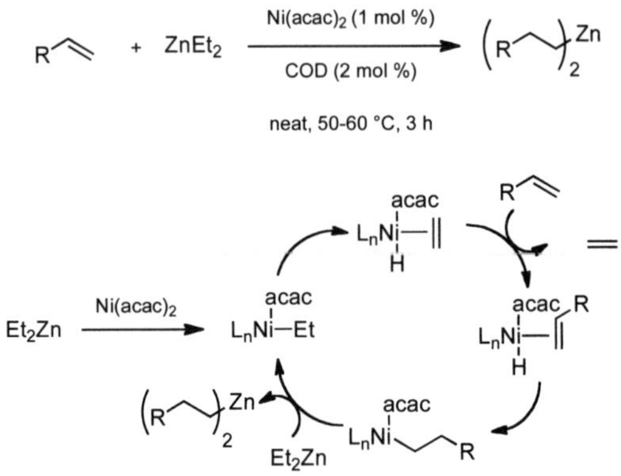

Scheme 3-85

The hydroalumination of oxabicyclic alkenes has been developed as a novel enantioselective entry to cycloalkenols (Scheme 3-86).[258–260] Catalysts derived from $Ni(cod)_2$ and BINAP efficiently catalyze the asymmetric hydroalumination of DIBAL-H to various oxabicyclic alkenes. Initial alkene hydroalumination is followed by ring opening of the intermediate organoaluminum to produce cyclohexenol or cycloheptenol products.

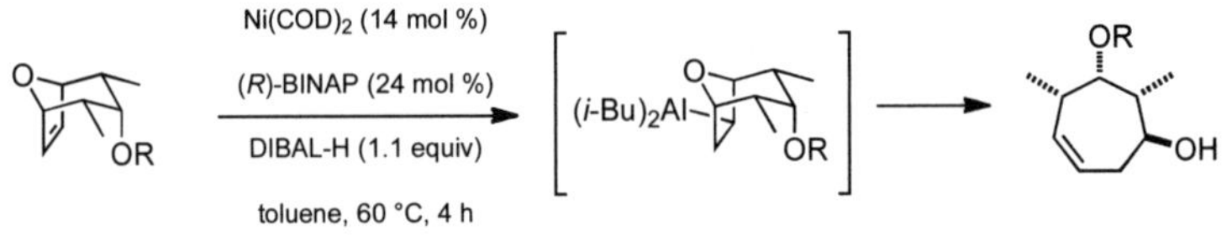

Scheme 3-86

Representative procedure: synthesis of (1S,2R,3S,4S)-3-methoxy-2,4-dimethylcyclohept-5-enol[258]

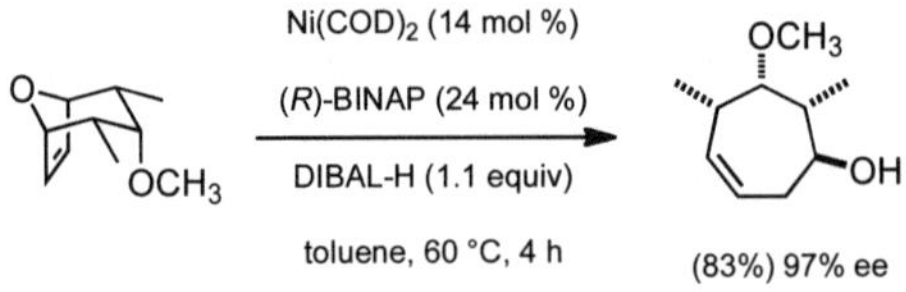

A flame-dried, round-bottom flask was charged with $Ni(cod)_2$ (4.4 mg, 0.016 mmol) in the glove box. $Ni(cod)_2$ was dissolved in freshly distilled toluene (1 mL) and the solution was transferred *via* cannula to a flame-dried round-bottomed flask fitted with a reflux condenser containing *(R)*-BINAP (17.4 mg, 0.028 mmol) under Ar. The resulting yellow solution was allowed to stir for 4 h. The reaction color changed from yellow green to light orange to light red to deep red in the course of about 45 min. After the full 4 h, the reaction color was a dark, deep burgundy. The substrate (25 mg, 0.149 mmol) was dissolved in 1 mL of freshly distilled toluene and transferred to the catalyst solution. There was very little if any color change. The reaction was immediately placed in a preheated 60 °C oil bath followed by addition of DIBAL-H (0.17 mL, 1.0 M in heptane, 0.170 mmol) *via* syringe pump. The addition took about 4 h. The reaction was quenched at RT with Rochelle's salt and allowed to stir open to the air for about 1 h. The organic layer was separated and the aqueous layer was extracted three times with ether. The combined organic layers were washed with brine, dried over anhydrous $MgSO_4$, and concentrated. Bulb-to-bulb distillation yielded 21 mg of the desired product (83%) as a white solid.

6.2 Carbometallation

An interesting series of carbozincation processes has been developed, wherein a halogen-bearing carbon adds to an alkene or alkyne, followed by generation of a carbon–zinc bond at the other terminus of the unsaturation. For example, work from Knochel illustrated that iodoalkenes undergo efficient nickel-catalyzed cyclizations in the presence of diethylzinc to generate a new ring and a functionalized organozinc reagent (Scheme 3-87).[261, 262] Processes of this type likely proceed *via* homolytic cleavage of the C-I bond to initiate cyclization of a carbon-centered radical, followed by capture as the organozinc product. As illustrated, both cis- and trans-configurations of the starting organohalide result in formation of a single cis-fused product, consistent with the configurational instability of the free radical intermediate. The organozinc products are efficiently alkylated by a broad variety of electrophiles.

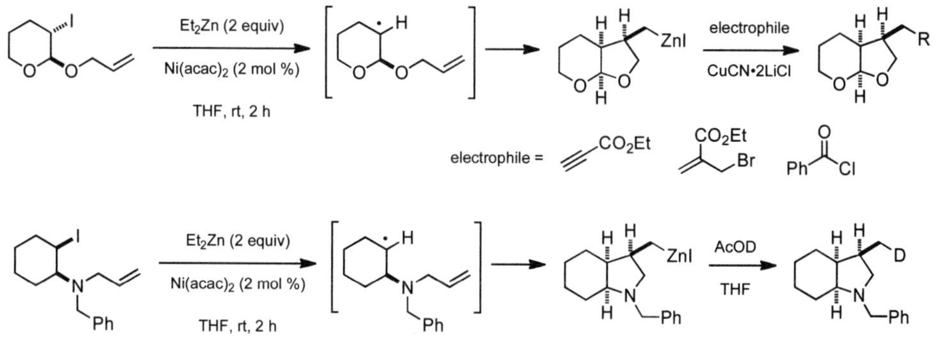

Scheme 3-87

Representative procedure: synthesis of trans-2-butoxy-4-(3-carbethoxy-3-butenyl)tetrahydrofuran[262]

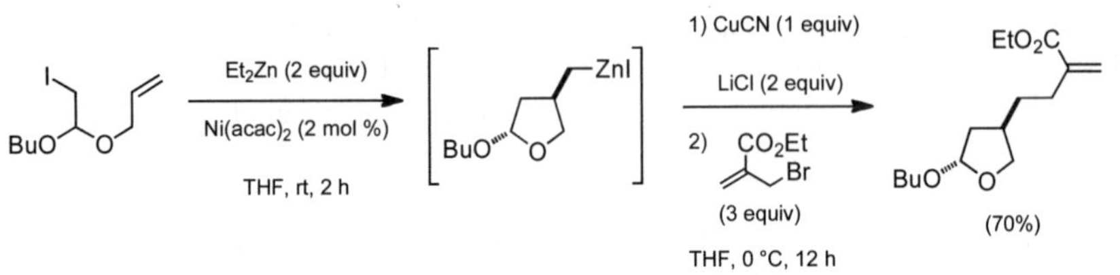

1) CuCN (1 equiv)

Et_2Zn (2 equiv) — $Ni(acac)_2$ (2 mol %) — THF, rt, 2 h — LiCl (2 equiv) — 2) CO_2Et, Br (3 equiv) — THF, 0 °C, 12 h — (70%)

A 50-mL, three-necked flask equipped with an Ar inlet, a magnetic stirring bar, an internal thermometer, and a septum cap was charged with $Ni(acac)_2$ (15 mg, 0.06 mmol, 2 mol%), and a solution of 1-(1-(allyloxy)-2-iodoethoxy)butane (0.85 g, 3.0 mmol, 1 equiv) in THF (5 mL) was added. The resulting green suspension was cooled to −78 °C, and Et_2Zn (0.6 mL, 6.0 mmol, 2 equiv) was added dropwise. The reaction mixture was allowed to warm to 0 °C and stirred at this temperature for 2 h. The excess of Et_2Zn and the solvent were removed *in vacuo* (RT, 2 h). After addition of THF (5 mL) and cooling to −50 °C, CuCN•2LiCl (CuCN: 0.27 g, 3.0 mmol; LiCl: 0.25 g, 6.0 mmol) in THF (5 mL) was added. The reaction mixture was warmed to 0 °C (10 min) and cooled to −78 °C. Ethyl (α-bromomethyl)acrylate (1.78 g, 9.0 mmol, 3 equiv) was added. The reaction mixture was allowed to warm to 0 °C within 12 h and was quenched by addition of saturated aqueous NH_4Cl solution (15 mL). Precipitated copper salts were dissolved by addition of small portions of an aqueous NH_3 solution. The aqueous phase was extracted with ether (3 × 30 mL). The combined organic layer was dried (anhydrous $MgSO_4$) and filtered, and the solvent was evaporated. The residue was purified by flash chromatography (hexanes/ether 6:1), affording the desired product (0.56 g, 2.1 mmol, 70% yield) as a colorless oil.

A related process involves cyclization of iodo alkenes with functionalized organozincs in the presence of a catalyst derived from $Ni(Pyr)_4Cl_2$ and a Pybox ligand.[263] Rather than observing generation of the cyclized functionalized organozinc as shown above (Scheme 3-87), a carbon–carbon bond-forming cross-coupling accompanies the radical cyclization process (Scheme 3-88).

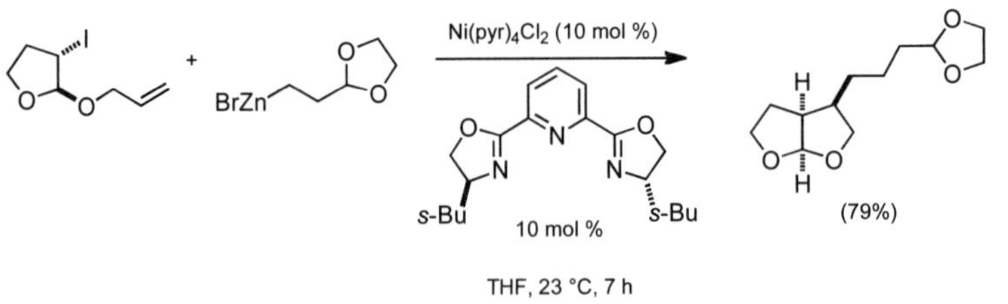

$Ni(pyr)_4Cl_2$ (10 mol %) — s-Bu, s-Bu, 10 mol % — THF, 23 °C, 7 h — (79%)

Scheme 3-88

Cyclizations of iodoalkynes have also been demonstrated to proceed with C-C bond formation (Scheme 3-89). Little mechanistic information is available for this process, but a net syn addition across the alkyne was noted. Additionally, intermolecular carbozincations of alkynes were demonstrated under similar conditions, and one such process formed the basis of an efficient synthesis of tamoxifen.

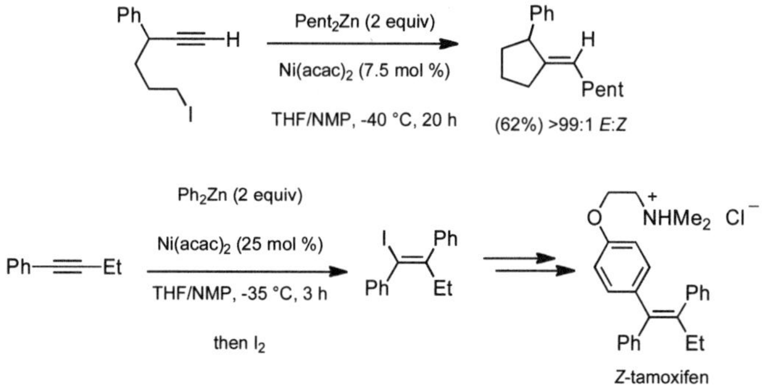

Scheme 3-89

A high-throughput screening effort uncovered a novel hydroarylation process, wherein treatment of aryl bromides and alkynes with a catalyst derived from $Ni(cod)_2$ and PBu_3 in the presence of Et_3SiH generated hydroarylation products with a net trans addition across the alkyne (Scheme 3-90).[264] The origin of the unusual trans addition has not been clarified.

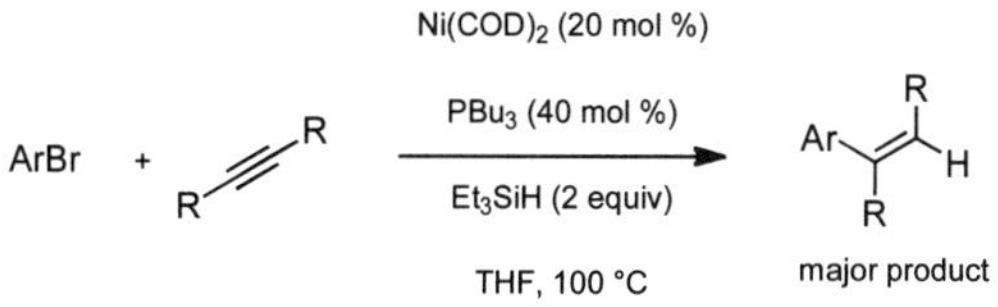

Scheme 3-90

6.3 Hydrovinylation, hydrocyanation, and other hydrocarbations

Review. RajanBabu, T.V.; *Synlett* **2009**, 853 – 885; RajanBabu, T.V.; *Chem. Rev.* **2003,** *103*, 2845 – 2860.

The heterodimerization of alkenes is an exceptionally attractive process given its atom economical nature and utilization of simple feedstocks. Early studies from Wilke demonstrated the nickel-catalyzed homodimerization of simple alkenes[88] and addressed the problems of enantioselective and chemoselective heterocouplings have more recently been addressed, including a major program from Rajanbabu.[265, 266] Ethylene is an especially useful feedstock in the derivatization of vinyl arenes in the hydrovinylation process (Scheme 3-91).[267]

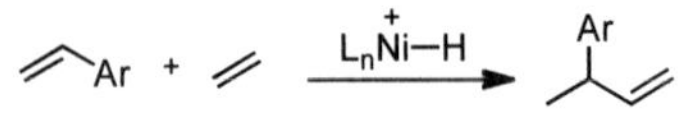

Scheme 3-91

The basic features of the reaction may be summarized as follows (Scheme 3-92). Nickel π-allyl dimers are often used as pre-catalysts. Treatment with monodentate phosphines and Lewis acids likely generates a cationic nickel hydride species, which serves as the active catalyst for the reaction. Regioselective hydrometallation of the styrene generates intermediate **26**, with the regiochemistry being driven by benzylic stabilization. Insertion of ethylene followed by β-hydride elimination produces the hydrovinylation product and regenerates the active cationic nickel hydride species.

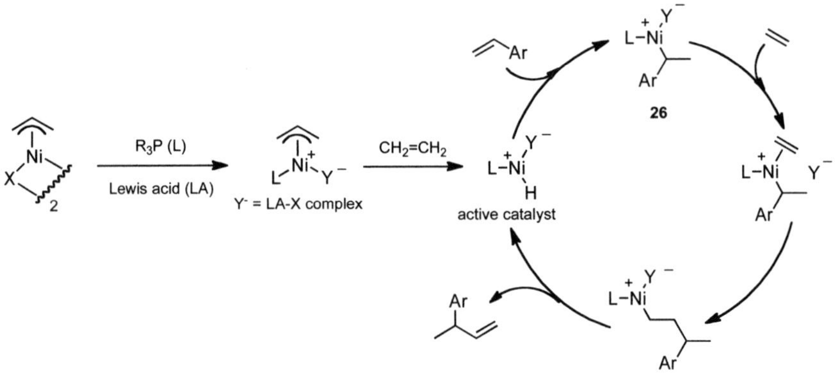

Scheme 3-92

In addition to the extensive studies of hydrovinylations using styrene substrates, 1,3-dienes are also an especially useful substrate class. With phosphoramidite ligands, enantioselective diene hydrovinylation with ethylene provides a practical solution to the difficulties associated with the installation of exocyclic side chain stereochemistry, as the representative procedure below illustrates.[268, 269]

Representative procedure: synthesis of (S)-3-(but-3-en-2-yl)-1H-indene [268]

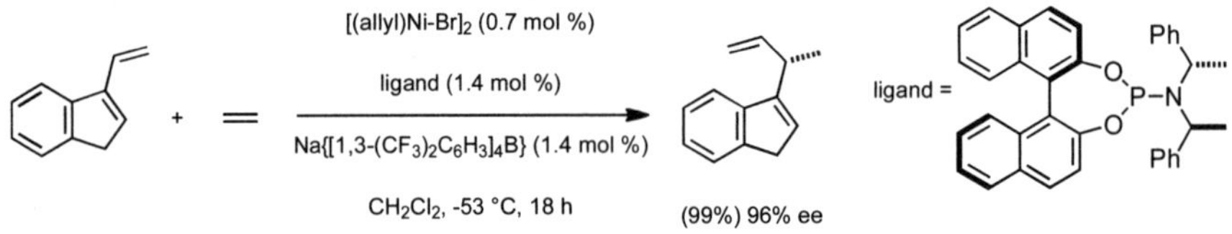

To a solution of $[(allyl)NiBr]_2$ (2.5 mg, 0.007 mmol) in CH_2Cl_2 (0.5 mL) was added a solution of ligand (0.0146 mmol) in CH_2Cl_2 (0.5 mL) at RT in the dry box. The resulting solution was added to a suspension of NaBARF (12.9 mg, 0.0146 mmol)

in CH_2Cl_2 (1 mL) and the mixture was stirred at RT for 1.5 h. The catalyst solution was then filtered through a short pad of Celite into a flame-dried Schlenk flask and 0.5 mL of CH_2Cl_2 was used to rinse the Celite. The flask was taken out of dry box and cooled to –53 °C. After ethylene was introduced to the flask, a solution of 3-vinyl-1*H*-indene (0.25 mmol) in CH_2Cl_2 (1 mL) was added to the catalyst solution. After the mixture was stirred for 18 h, the flask was disconnected from ethylene and 0.5 mL of saturated aqueous ammonium chloride was added to quench the reaction. The flask was allowed to warm to RT, and the mixture was diluted with hexanes. The solution was filtered through a short pad of silica gel using 30 mL of hexanes/ethyl acetate (20/1) to elute the product. The filtrate was collected and concentrated by rotary evaporation to give the desired product (99% conversion, 96% ee).

Catalysts derived from *N*-heterocyclic carbenes have also been demonstrated as highly effective catalysts for hydrovinylations involving heterocoupling of aromatic and aliphatic alkenes (Scheme 3-93).[270] In this case, the catalyst is prepared by coupling an aldehyde and aliphatic alkene with a Ni(0) NHC complex in the presence of a silyl triflate.

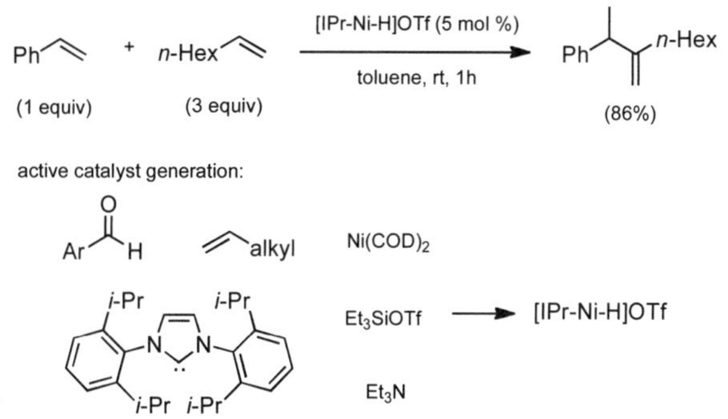

Ph + *n*-Hex (1 equiv) (3 equiv) → [IPr-Ni-H]OTf (5 mol %), toluene, rt, 1h → Ph *n*-Hex (86%)

active catalyst generation:

Ar(C=O)H, alkyl, $Ni(COD)_2$, *i*-Pr *i*-Pr *i*-Pr *i*-Pr (IPr), Et_3SiOTf, Et_3N → [IPr-Ni-H]OTf

Scheme 3-93

Although many notable examples of nickel-catalyzed hydrovinylations have been described, their iterative use in the synthesis of the pseudopterosins provides a clear demonstration of the utility of this strategy (Scheme 3-94).[271] In this impressive illustration, three sequential ligand-controlled hydrovinylation processes enable a highly efficient synthesis of the target molecule.

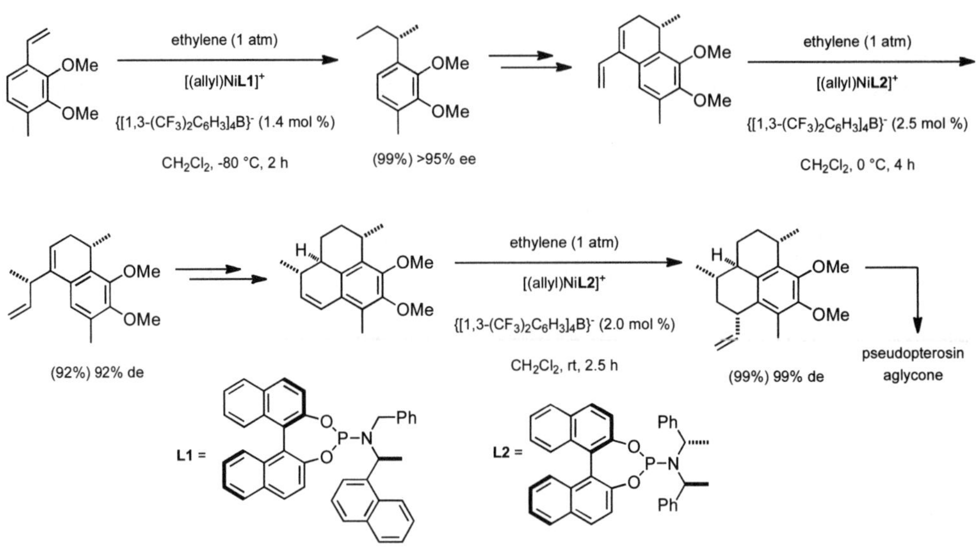

Scheme 3-94

The hydrocyanation of alkenes and dienes has similarly provided an exceptionally useful process for the conversion of simple feedstocks into more complex structures. *[Caution: Hydrogen cyanide is a highly toxic gas.]* The process is best known as a key step in the DuPont adiponitrile process, which involves the dihydrocyanation of 1,3-butadiene (Scheme 3-95). The overall sequence first involves butadiene hydrocyanation to afford a mixture of 3-pentenenitrile and 2-methyl-3-butenenitrile. The unwanted branched isomer 2-methyl-3-butenenitrile is isomerized to 3-pentenenitrile under different conditions, and then 3-pentenenitrile is isomerized to 4-pentenenitrile in a subsequent nickel-catalyzed process in the presence of Lewis acidic additives. Finally, hydrocyanation of the remaining alkene generates the desired product adiponitrile, which serves as a precursor for nylon. A vast number of studies describing the optimization and mechanistic study of this process has appeared, and the interested reader is referred to the many excellent studies describing the details of this process.[272–274]

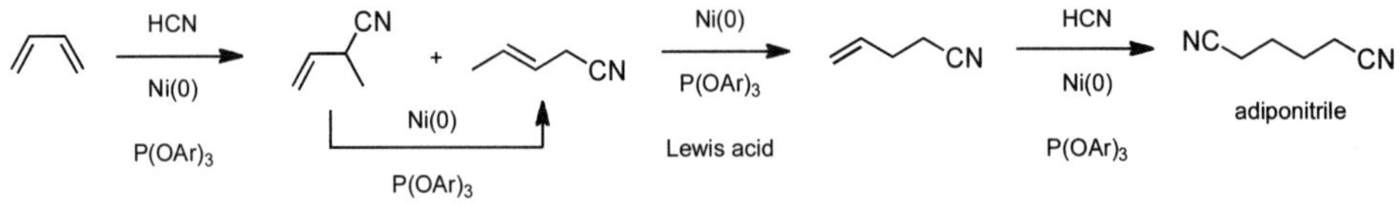

Scheme 3-95

Given the efficiency and utility of the butadiene hydrocyanation process, its potential in more complex settings is underused. Whereas the range of applications has mostly been limited to simple substrates, several important developments in reaction scope and asymmetric induction have appeared. For example, carbohydrate-derived bis-phosphine ligands allowed efficient asymmetric hydrocyanations of styrenes and 1,3-dienes (Scheme 3-96).[275, 276] Similarly, the xantphos ligand class was demonstrated to allow improved conversions in hydrocyanations of simple styrenes, and chiral derivatives of this ligand class allowed asymmetric hydrocyanations of styrenes.[277, 278]

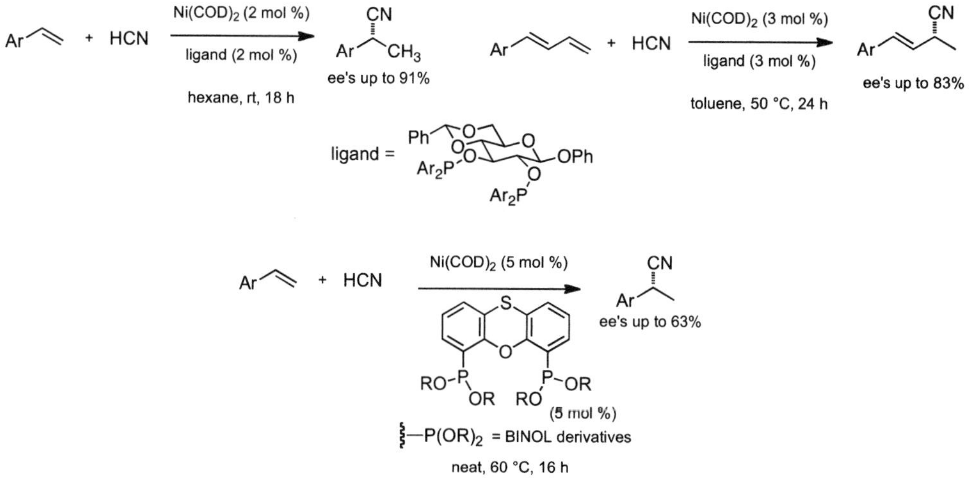

Scheme 3-96

In addition to the best known early developments in hydrovinylations and hydrocyanations, related classes of hydrocarbations (involving formal C-H addition across a unit of unsaturation) have emerged recently with a significantly broadened range of substrate classes. A variety of C-H bonds, including those found in various heterocyclic frameworks, simple formamides, and terminal alkynes, has recently been demonstrated to undergo additions to alkenes and alkynes in work from Nakao and Hiyama (Scheme 3-97). These impressive recent advances greatly expand the scope of the reactivity principles that have long been recognized in hydrocyanation processes and suggest that further advances with other classes of C-H bonds are yet to be discovered.

The classes of C-H bonds within heterocyclic templates that undergo additions to unsaturated functionality include 3-cyanoindoles,[279] indole-3-carboxylates,[280] 2-pyridones,[281] pyridine *N*-oxides,[282] and simple pyridines (Scheme 3-98).[283, 284] Withdrawal of electron density from the heterocycle is often beneficial and may be accomplished by Lewis acidic co-catalysis (as with 2-pyridones and simple pyridines), by substrate oxidation (as with pyridine *N*-oxides), or by installation of electron-withdrawing groups (as with 3-cyanoindoles and indole-3-carboxylates). In all the above cases, it was proposed that direct oxidative addition of Ni(0) to the C-H bond is a key step of the reaction mechanism, as illustrated in a generalized case involving alkyne insertion.

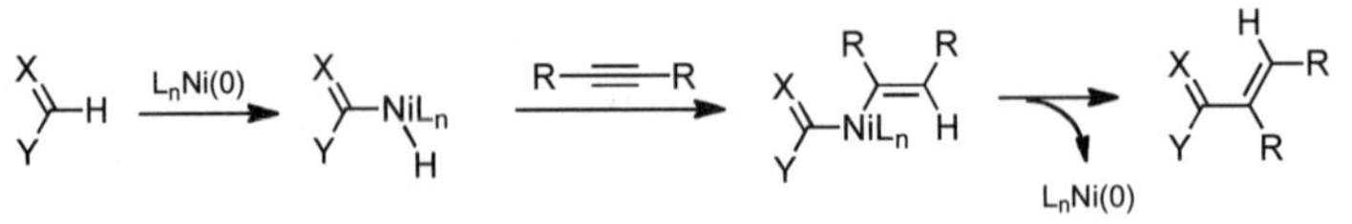
LnNi(0)
R—≡—R
LnNi(0)

Scheme 3-97

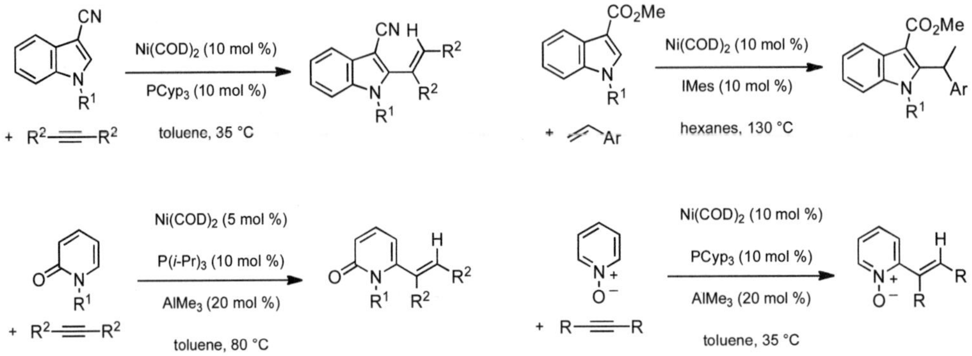
Ni(COD)2 (10 mol %)
PCyp3 (10 mol %)
toluene, 35 °C
Ni(COD)2 (10 mol %)
IMes (10 mol %)
hexanes, 130 °C
Ni(COD)2 (5 mol %)
P(i-Pr)3 (10 mol %)
AlMe3 (20 mol %)
toluene, 80 °C
Ni(COD)2 (10 mol %)
PCyp3 (10 mol %)
AlMe3 (20 mol %)
toluene, 35 °C

Scheme 3-98

Generally, substitution occurs adjacent to a heteroatom, but with simple pyridines, regiochemical reversal was possible by modification of the ligand and Lewis acid structure.[283, 284] Standard phosphine ligands and Lewis acids favor substitution at the pyridine C-2 position in alkenylations with alkyne substrates, whereas bulky NHC ligands paired with bulky Lewis acids favor pyridine C-4 alkylations with alkene substrates on steric grounds (Scheme 3-99). The study with bulky ligands and Lewis acids illustrated that reversible C-H activation occurs at C-2 and C-3, but irreversible reductive elimination is favored at the less hindered C-4 position and governs the regiochemical outcome.

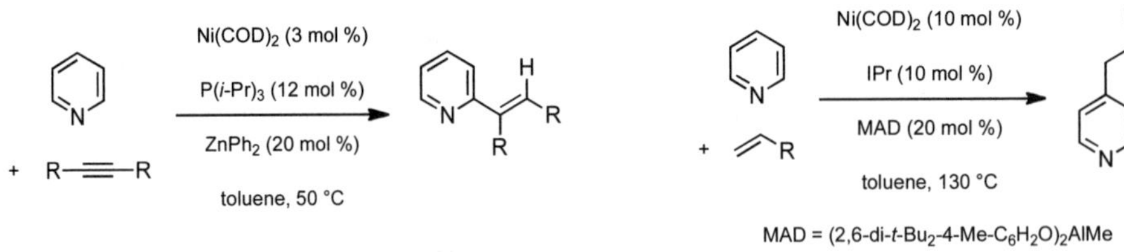
Ni(COD)2 (3 mol %)
P(i-Pr)3 (12 mol %)
ZnPh2 (20 mol %)
toluene, 50 °C
Ni(COD)2 (10 mol %)
IPr (10 mol %)
MAD (20 mol %)
toluene, 130 °C
MAD = (2,6-di-t-Bu2-4-Me-C6H2O)2AlMe

Scheme 3-99

Representative procedure: synthesis of (E)-2-(4-octen-4-yl)quinoline[283]

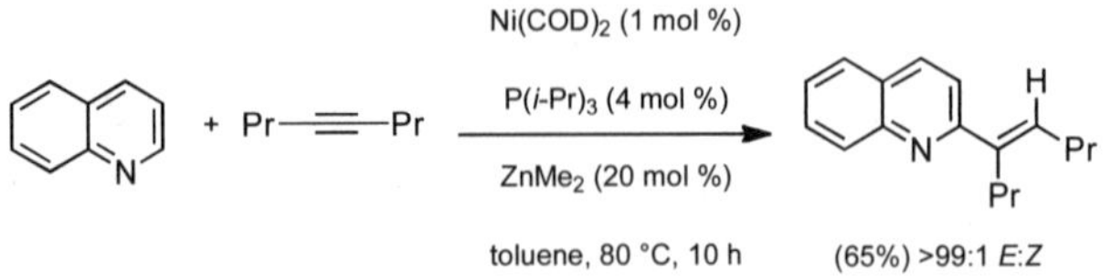
Pr—≡—Pr
Ni(COD)2 (1 mol %)
P(i-Pr)3 (4 mol %)
ZnMe2 (20 mol %)
toluene, 80 °C, 10 h
(65%) >99:1 E:Z

To a solution of quinoline (3.0 mmol), $ZnMe_2$ (60 µmol of a 1.00-M solution in hexane), and 4-octyne (1.0 mmol) in toluene (1.25 mL) was added a solution of $Ni(cod)_2$ (8.2 mg, 30 µmol) and $P(i\text{-}Pr)_3$ (19.2 mg, 0.12 mmol) in toluene (1.25 mL) in a dry box. After addition of undecane (internal standard, 78 mg, 0.50 mmol), the vial was taken outside the dry box and heated 80 °C for 10 h. The resulting mixture was filtered through a silica gel pad, concentrated *in vacuo*, and purified by flush silica gel chromatography to give the desired product in 65% yield (>99:1 *E*:*Z*).

In addition to the preceding advances primarily associated with heterocyclic substrates, similar additions were observed with simple *N*-formyl substrates (Scheme 3-100).[285] Moreover, terminal alkynes were effective substrates in additions to styrenes and 1,3-dienes.[286, 287] Based on this combination of recent developments, the nickel-catalyzed functionalization of a variety of relatively unreactive C-H bonds has now been demonstrated. Given the range of substrates that participates, many other classes of reactions that involve nickel-catalyzed C-H functionalization are to be expected. The special role of the combined effect of nickel and Lewis acid co-catalysis has emerged as an important influence in this growing body of catalytic processes.

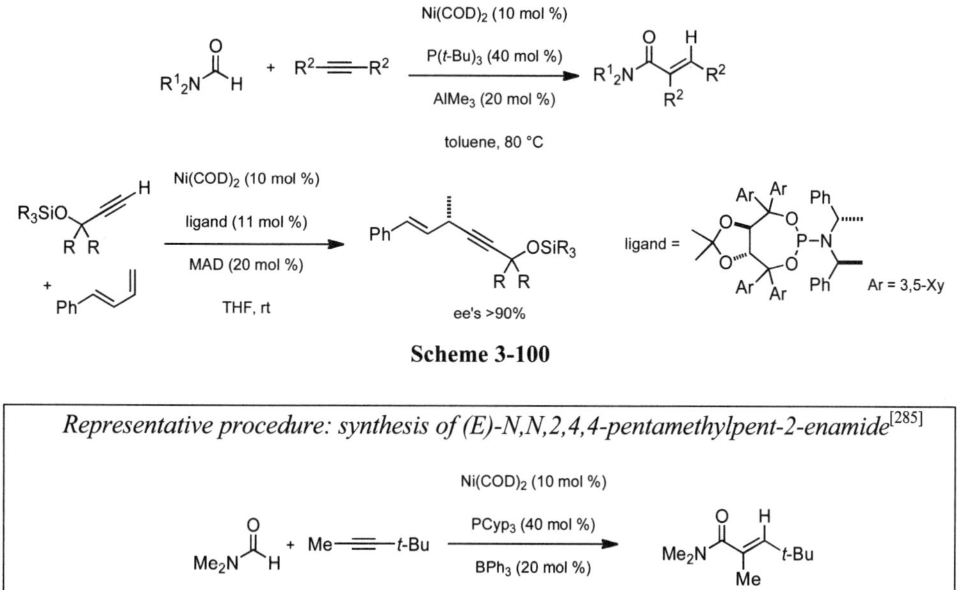

Ni(COD)2 (10 mol %)
P(t-Bu)3 (40 mol %)
AlMe3 (20 mol %)
toluene, 80 °C

Ni(COD)2 (10 mol %)
ligand (11 mol %)
MAD (20 mol %)
THF, rt

ee's >90%

ligand =

Ar = 3,5-Xy

Scheme 3-100

Representative procedure: synthesis of (E)-N,N,2,4,4-pentamethylpent-2-enamide[285]

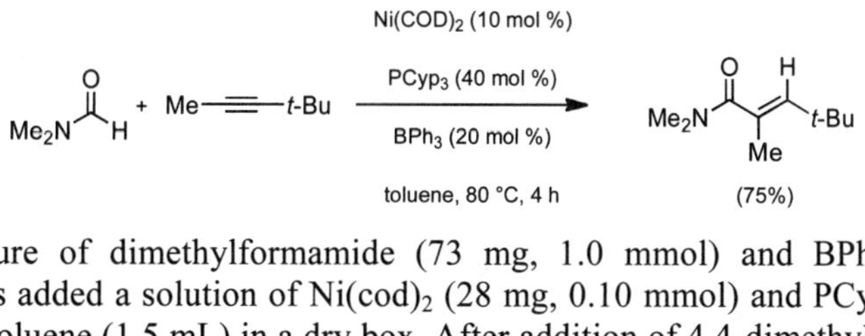

Ni(COD)2 (10 mol %)
PCyp3 (40 mol %)
BPh3 (20 mol %)
toluene, 80 °C, 4 h

(75%)

To a mixture of dimethylformamide (73 mg, 1.0 mmol) and BPh_3 (48 mg, 0.20 mmol) was added a solution of $Ni(cod)_2$ (28 mg, 0.10 mmol) and $PCyp_3$ (95 mg, 0.40 mmol) in toluene (1.5 mL) in a dry box. After addition of 4,4-dimethylpent-2-yne (1.0 mmol) and dodecane (internal standard, 78 mg, 0.50 mmol), the vial was taken outside the dry box and heated at 80 °C for 4 h. The resulting mixture was filtered through a silica gel pad, concentrated *in vacuo*, and purified by flash column chromatography on silica gel to give the desired product (75% yield).

6.4 Carbocyanation

Review: Nakao, Y.; Hiyama, T.; *Pure Appl. Chem.* **2008,** *80*, 1097 – 1107.

As described, recent developments from Nakao and Hiyama have provided impressive achievements in the insertion of unsaturated functionality into various classes of carbon–hydrogen bonds. A major effort from the same investigators has described related processes involving insertions into carbon-cyano bonds. Whereas C-C bonds not embedded within strained rings are typically unreactive toward oxidative addition with low-valent metals, the reactivity of simple nitriles in nickel-catalyzed processes has been attributed to the affinity of cyanide toward nickel, along with the electron-withdrawing ability of the cyano functionality.[288] These aspects have allowed the development of nickel-catalyzed carbocyanation processes involving C-C cleavage of simple nitriles into a general process for the synthesis of functionalized nitriles.

The arylcyanation of alkynes was first reported with nickel catalysts of monodentate phosphines at 100 °C for 24 h,[289] but subsequent studies illustrated a dramatic effect of Lewis acids on the reaction efficiency.[290] Employing BPh_3, $AlMe_3$, or $AlMe_2Cl$ as the Lewis acid, a considerably broader scope was observed in reactions conducted at 50 °C (Scheme 3-101). A variety of aromatic and heteroaromatic nitriles undergoes the alkyne insertion process to generate functionalized cyanoalkenes. Additionally, intramolecular versions involving alkene insertion, including an asymmetric variant employing the TangPHOS bidentate phosphine, allowed the efficient construction of quaternary centers through five-membered ring cyclizations.[291, 292]

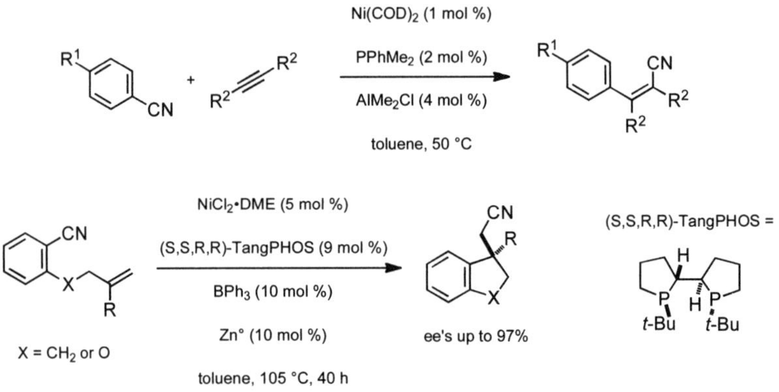

Scheme 3-101

A theoretical investigation of the phenylcyanation of alkynes found that the mechanism involves rate-determining oxidative addition of nickel(0) to the Ph-CN bond, followed by alkyne insertion into the Ph-Ni bond then reductive elimination of the alkenyl-CN bond (Scheme 3-102).[293]

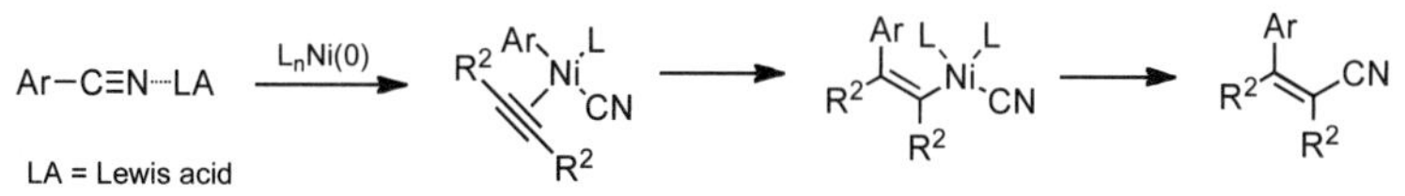

Ar−C≡N···LA → ($L_nNi(0)$) → → →

LA = Lewis acid

Scheme 3-102

Representative procedure: synthesis of (Z)-3-(1-methyl-1H-indol-3-yl)-2-propylhex-2-enenitrile[290]

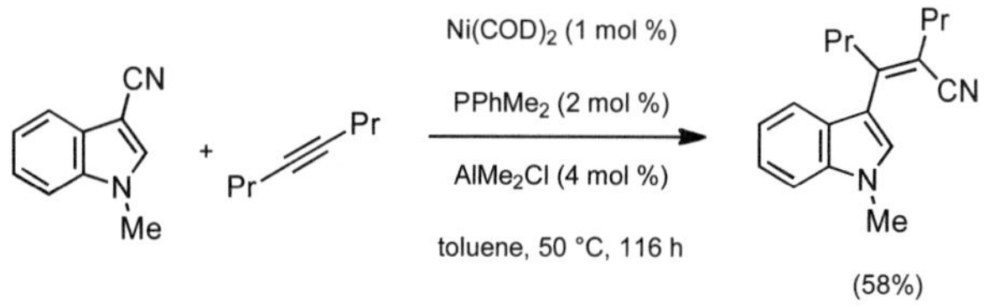

$Ni(COD)_2$ (1 mol %), $PPhMe_2$ (2 mol %), $AlMe_2Cl$ (4 mol %), toluene, 50 °C, 116 h (58%)

1-Methyl-1*H*-indole-3-carbonitrile (1.00 mmol), dimethylaluminum chloride (0.040 mmol), 4-octyne (1.00 mmol), and $C_{12}H_{26}$ (internal standard, 56 mg, 0.33 mmol) were added sequentially to a solution of $Ni(cod)_2$ (2.8 mg, 0.010 mmol) and $PPhMe_2$ (0.020 mmol) in toluene (1.0 mL) in a dry box. The vial was taken outside the dry box and heated at 50 °C for 116 h. The resulting mixture was filtered through a silica gel pad, concentrated *in vacuo*, and purified by flash silica gel column chromatography to give the desired arylcyanation products (58% yield).

Representative procedure: synthesis of (S)-(1-methyl-indan-1-yl)-acetonitrile[291]

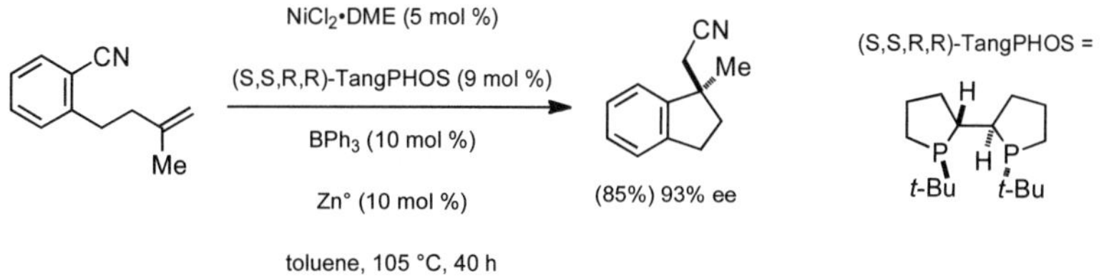

$NiCl_2$•DME (5 mol %), (S,S,R,R)-TangPHOS (9 mol %), BPh_3 (10 mol %), Zn° (10 mol %), toluene, 105 °C, 40 h (85%) 93% ee

(S,S,R,R)-TangPHOS =

General procedure: In a N_2 atmosphere glove box, Zn (3.9 mg, 0.060 mmol, 10 mol%) was weighed into a 1-dram vial. $NiCl_2$·dme (6.6 mg, 0.030 mmol, 5 mol%), (*S,S,R,R*)-TangPHOS (15.5 mg, 0.054 mmol, 9 mol%) and BPh_3 (14.5 mg, 0.060 mmol, 10 mol%) were added. Then, a solution of aryl cyanide substrate (0.600 mmol, 1.0 equiv) and PhMe (600 μL) was added. The vial was capped with a Teflon-lined cap, removed from the glove box, and heated in an aluminum heating block at 105 °C. After cooling to RT, the mixture was diluted with Et_2O (1 mL) and filtered through a plug of silica gel, which was rinsed with Et_2O (10 mL). The filtrate was concentrated *in vacuo* and then purified by silica gel chromatography. The general procedure was followed for 40 h to afford the desired product (87.0 mg, 85%, 93% ee) as a colorless oil.

Other alkyne insertion processes were described employing allyl cyanides as substrates.[294, 295] α-Silyloxyallyl cyanides were especially valuable substrates given the ease of substrate access, providing access to functionalized γ,δ-unsaturated carbonyl products (Scheme 3-103). Additionally, cyanoformates undergo similar alkyne insertions, leading to β-cyanoacrylates and acrylamides.[296] The participation of C(sp)-functionality in a C-CN cleavage process was also illustrated by the catalytic addition of alkynyl cyanides with alkynes to produce cyano-substituted conjugated enynes.[297]

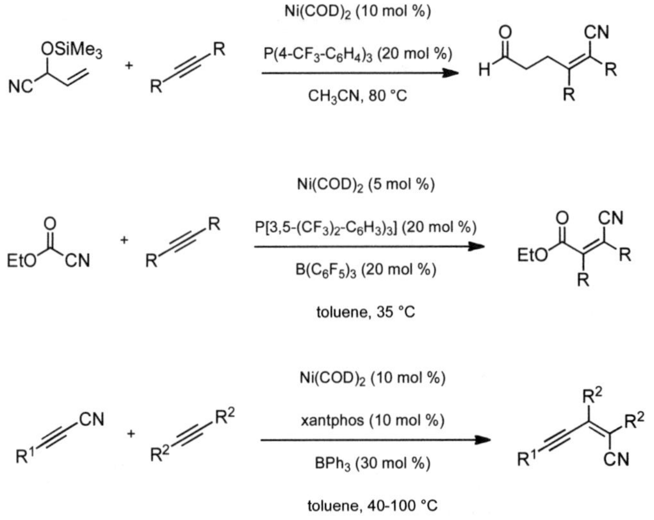

$Ni(COD)_2$ (10 mol %), $P(4\text{-}CF_3\text{-}C_6H_4)_3$ (20 mol %), CH_3CN, 80 °C

$Ni(COD)_2$ (5 mol %), $P[3,5\text{-}(CF_3)_2\text{-}C_6H_3)_3]$ (20 mol %), $B(C_6F_5)_3$ (20 mol %), toluene, 35 °C

$Ni(COD)_2$ (10 mol %), xantphos (10 mol %), BPh_3 (30 mol %), toluene, 40-100 °C

Scheme 3-103

Whereas the above processes illustrate that the functionality directly attached to the C-CN unit can dictate the efficiency of insertion processes, an intriguing use of more remote heteroatom directors, including nitrogen, oxygen, and sulfur groups, facilitates participation of a broader range of alkyl cyanides (Scheme 3-104).[298] Interestingly, insertion processes of 4-aminobutanenitriles proceed cleanly without rearrangement, whereas 5-aminopentanenitriles and 6-aminohexanenitriles undergo alkyne insertions with rearrangement of the carbon framework. This outcome was attributed to the involvement of five-membered azametallacyclic species, which are directly accessed by C-CN cleavage of 4-aminobutanenitriles (*i.e.*, intermediate **27**), or by rearrangement processes involving 5-aminopentanenitriles (*i.e.*, intermediate **28**; Scheme 3-105).

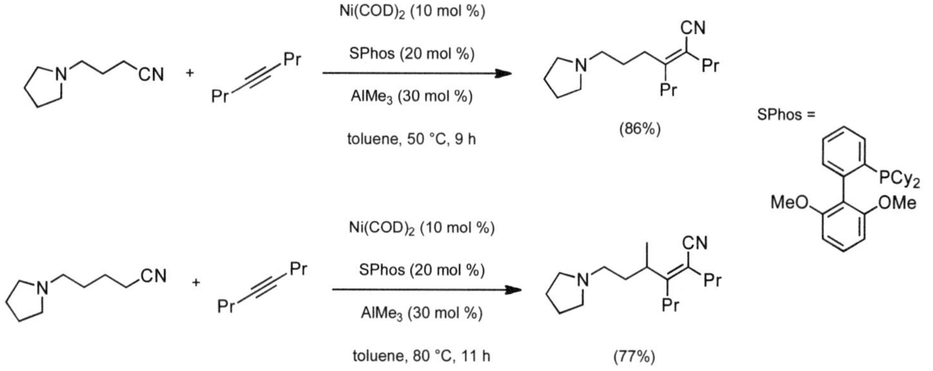

Scheme 3-104

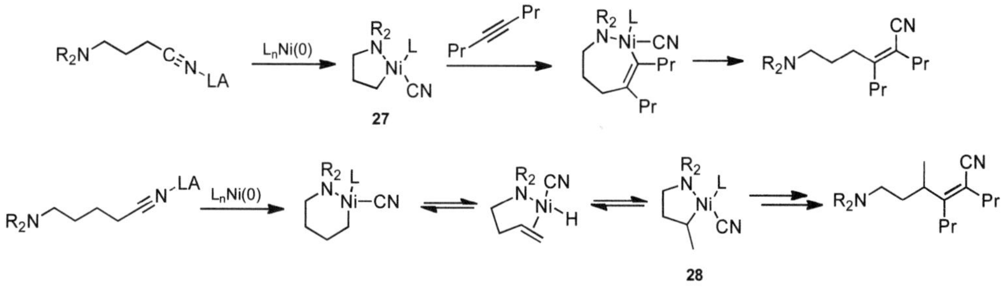

Scheme 3-105

Although the majority of processes of this type involves alkyne insertion into the reactive C-CN species, examples of allene insertions have been demonstrated with cyanoformate and with alkynyl cyanide substrates (Scheme 3-106).[297, 299, 300]

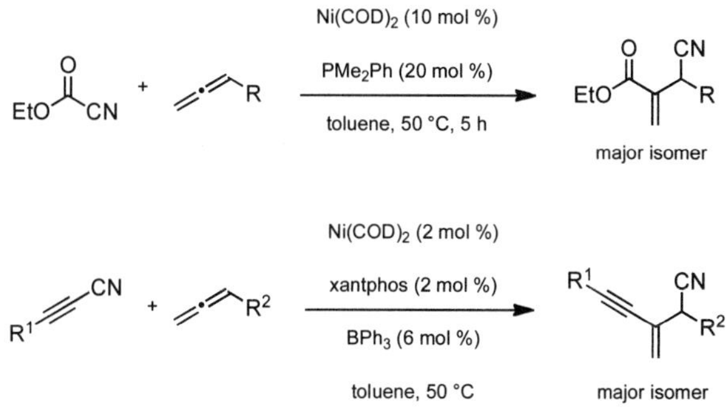

Scheme 3-106

The combination of procedures highlighted above illustrate that the nickel-catalyzed insertion of unsaturated functionality into carbon–cyanide bonds provides a general and versatile entry to more highly functionalized cyano-containing species.

This series of advances provides an important step in demonstrating the utility of C-C cleavage processes that do not require the use of strained ring systems.

6.5 Hydroamination

Review. Hartwig, J.F.; *Pure Appl. Chem.* **2004,** *76*, 507 – 516.

Following early studies of the nickel-catalyzed coupling of dienes with amines to generate mixtures of 1:1 and 2:1 adducts[301] a high-throughput colorimetric assay discovered that the catalyst derived from $Ni(cod)_2$ and dppf efficiently catalyzes the 1:1 hydroamination of 1,3-dienes with primary or secondary amines to produce allylic amine products (Scheme 3-107).[302] A catalytic quantity of trifluoroacetic acid was a key component in the optimized procedure.

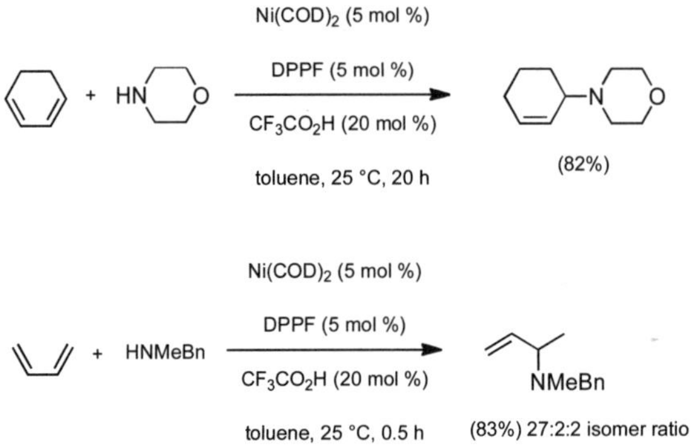

Ni(COD)$_2$ (5 mol %)
DPPF (5 mol %)
CF$_3$CO$_2$H (20 mol %)
toluene, 25 °C, 20 h
(82%)

Ni(COD)$_2$ (5 mol %)
DPPF (5 mol %)
HNMeBn
CF$_3$CO$_2$H (20 mol %)
toluene, 25 °C, 0.5 h
NMeBn
(83%) 27:2:2 isomer ratio

Scheme 3-107

Mechanistic studies suggested that the reaction proceeds by way of trifluoroacetic acid protonation of Ni(0), followed by hydrometallation of the 1,3-diene to generate a π-allyl intermediate (Scheme 3-108). An external attack of the amine to the π-allyl intermediate then generates the product allylic amine. Reaction of the π-allyl intermediate with the amine is thermodynamically favorable only in the presence of additional diene.

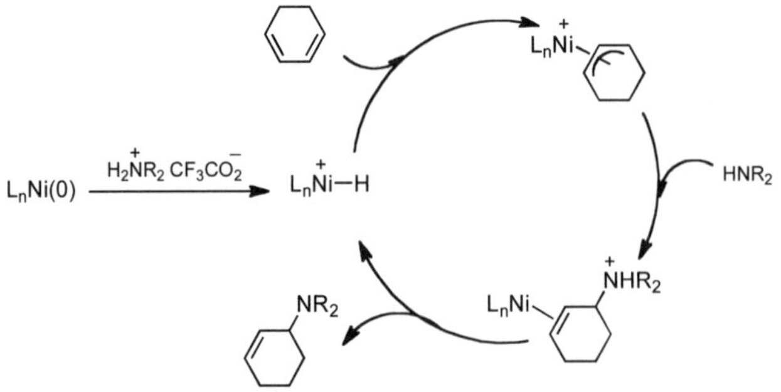

L$_n$Ni(0)
H$_2$N$^+$R$_2$ CF$_3$CO$_2^-$
L$_n$Ni$^+$–H
L$_n$Ni$^+$
HNR$_2$
N$^+$HR$_2$
L$_n$Ni
NR$_2$

Scheme 3-108

Representative procedure: synthesis of (2-cyclohexen-1-yl)diethylamine[302]

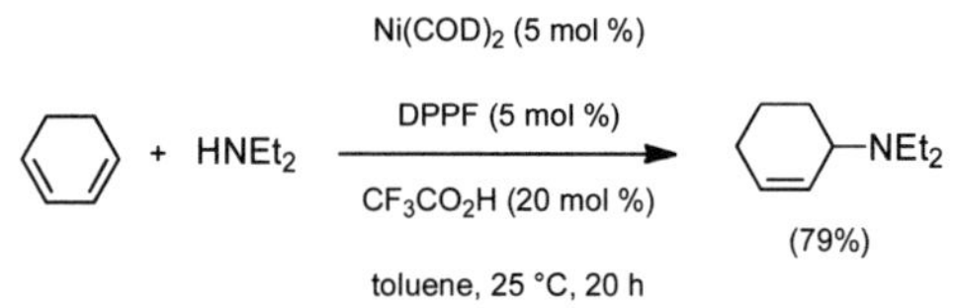

A reaction vial was charged with dppf (55.4 mg, 0.100 mmol, 0.0500 equiv), $Ni(cod)_2$ (27.5 mg, 0.100 mmol, 0.0500 equiv), 1,3-cyclohexadiene (320 mg, 4.00 mmol, 2.00 equiv), diethylamine (146 mg, 2.00 mmol, 1.00 equiv), and 2 mL of toluene under a N_2 atmosphere in a dry box. The reaction vessel was sealed with a PTFE septum and removed from the dry box. TFA (30.8 μL, 0.400 mmol, 0.200 equiv) was added by syringe, and the mixture was stirred at RT for 20 h. The mixture was then filtered through a pad of silica gel and concentrated *in vacuo*, and the product was isolated by Kügelrohr distillation (50 mT, 65 °C) to give 239 mg (79% yield) of the purified product as a colorless oil.

A nickel-catalyzed hydroamination of acrylonitrile using cationic pincer complexes has been described (Scheme 3-109).[303] Mechanistic studies suggested that a simple Lewis acidic role of nickel may be responsible for the catalysis in this group of reactions.

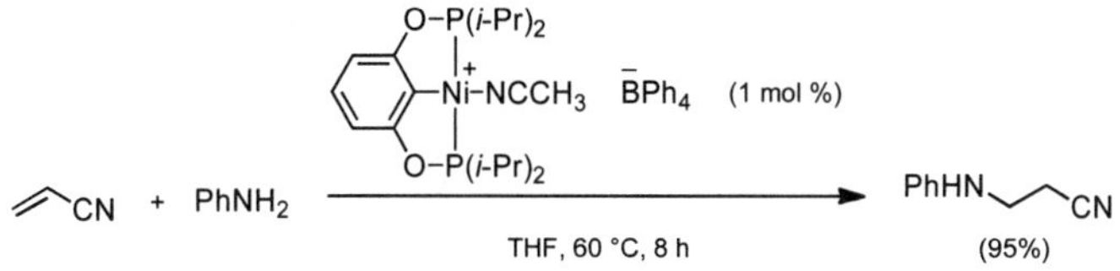

Scheme 3-109

In addition to simple hydroamination processes, nickel-catalyzed alkene diaminations have been developed under oxidative conditions (Scheme 3-110).[304] Treatment of sulfamide substrates with $PhI(OAc)_2$ in the presence of $Ni(acac)_2$ as catalyst directly provides diamination products. This key report suggests that further related developments in oxidative nickel catalysis are likely.

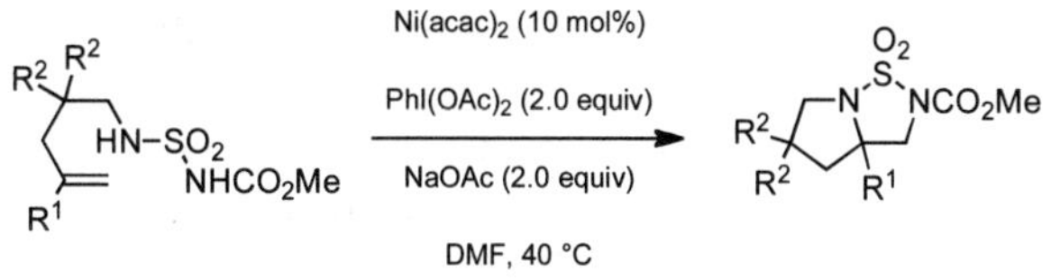

Scheme 3-110

7. Conclusions and Outlook

Although nickel catalysis has played a fundamental role in the advent of organometallics in synthesis, its broader popularity has only recently begun to live up to these early expectations. Previously viewed as a poor sister to palladium in cross-coupling chemistry, nickel exhibits many unique characteristics that are now enabling a broad range of novel and important transformations that in many cases cannot be achieved by any other metal. The increased oxophilic character of nickel compared with palladium ; the easy access to the 0, +1, +2, and +3 oxidation states ; and the efficiency of a versatile range of accessible oxidative addition, reductive elimination, transmetallation, and migratory insertion processes provides some of the basis for the broad and often unique reactivity and capabilities of nickel.

Following the early developments in nickel-catalyzed cycloadditions and cross-coupling processes, continued sophistication has been demonstrated over the years, and recent developments have elevated these classes of transformations into robust and useful catalytic processes. The recent burst of activity in these areas can be attributed to a better appreciation of both ligand effects and mechanistic considerations. Additionally, recent developments in reductive couplings, carbonyl additions, and C-H and C-C functionalization processes have demonstrated new reactivity trends and have suggested many new directions for continued development. Based on the current trajectory and recent fundamental findings, one can expect increased use of the types of reactions described herein as well as the discovery of many types of entirely new processes.

8. Acknowledgments

I would like to thank the many students who contributed over the years to my group's long-running interest in organonickel chemistry. We also thank the National Science Foundation and the National Institutes of Health for funding numerous projects that involve the development of nickel-catalyzed processes. Additionally, I thank my current colleagues at the University of Michigan, my former colleagues at Wayne State University, and my sabbatical colleagues at the Institute of Chemical Research of Catalonia (ICIQ, winter 2011) for many helpful discussions related to the developments from my laboratory described in this chapter.

Montgomery group photo (summer 2012).

9. References

[1] Keim, W.; *Angew. Chem., Int. Ed.* **1990**, *29*, 235 – 244.

[2] Ittel, S.D.; Johnson, L.K.; Brookhart, M.; *Chem. Rev.* **2000**, *100*, 1169 – 1203.

[3] Tamao, K.; Sumitani, K.; Kumada, M.; *J. Am. Chem. Soc.* **1972**, *94*, 4374 – 4376.

[4] Corriu, J.P.; Masse, J.P.; *J. Chem. Soc. Chem. Commun.* **1972**, 144.

[5] Tamao, K.; Kiso, Y.; Sumitani, K.; Kumada, M.; *J. Am. Chem. Soc.* **1972**, *94*, 9268 – 9269.

[6] Hayashi, T.; Tajika, M.; Tamao, K.; Kumada, M.; *J. Am. Chem. Soc.* **1976**, *98*, 3718 – 3719.

[7] Hayashi, T.; Konishi, M.; Fukushima, M.; Mise, T.; Kagotani, M.; Tajika, M.; Kumada, M.; *J. Am. Chem. Soc.* **1982**, *104*, 180 – 186.

[8] Hayashi, T.; Konishi, M.; Kobori, Y.; Kumada, M.; Higuchi, T.; Hirotsu, K.; *J. Am. Chem. Soc.* **1984**, *106*, 158 – 163.

[9] Baba, S.; Negishi, E.; *J. Am. Chem. Soc.* **1976**, *98*, 6729 – 6731.

[10] Negishi, E.; King, A.O.; Okukado, N.; *J. Org. Chem.* **1977**, *42*, 1821 – 1823.

[11] Grushin, V.V.; Alper, H.; *Chem. Rev.* **1994**, *94*, 1047 – 1062.

[12] Kochi, J.K.; *Pure Appl. Chem.* **1980**, *52*, 571 – 605.

[13] Morrell, D.G.; Kochi, J.K.; *J. Am. Chem. Soc.* **1975**, *97*, 7262 – 7270.
[14] Tsou, T.T.; Kochi, J.K.; *J. Am. Chem. Soc.* **1979**, *101*, 6319 – 6332.
[15] Rudolph, A.; Lautens, M.; *Angew. Chem., Int. Ed.* **2009**, *48*, 2656 – 2670.
[16] Frisch, A.C.; Beller, M.; *Angew. Chem., Int. Ed.* **2005**, *44*, 674 – 688.
[17] Lipshutz, B.H.; Mollard, P.; Pfeiffer, S.S.; Chrisman, W.; *J. Am. Chem. Soc.* **2002**, *124*, 14282 – 14283.
[18] Lipshutz, B.H.; Blomgren, P.A.; *J. Am. Chem. Soc.* **1999**, *121*, 5819 – 5820.
[19] Phapale, V.B.; Guisan-Ceinos, M.; Bunuel, E.; Cárdenas, D.J.; *Chem. Eur. J.* **2009**, *15*, 12681 – 12688.
[20] Devasagayaraj, A.; Studemann, T.; Knochel, P.; *Angew. Chem., Int. Ed.* **1995**, *34*, 2723 – 2725.
[21] Jensen, A.E.; Knochel, P.; *J. Org. Chem.* **2002**, *67*, 79 – 85.
[22] Giovannini, R.; Studemann, T.; Dussin, G.; Knochel, P.; *Angew. Chem., Int. Ed.* **1998**, *37*, 2387 – 2390.
[23] Terao, J.; Watanabe, H.; Ikumi, A.; Kuniyasu, H.; Kambe, N.; *J. Am. Chem. Soc.* **2002**, *124*, 4222 – 4223.
[24] Terao, J.; Todo, H.; Watanabe, H.; Ikumi, A.; Kambe, N.; *Angew. Chem., Int. Ed.* **2004**, *43*, 6180 – 6182.
[25] Zhou, J.R.; Fu, G.C.; *J. Am. Chem. Soc.* **2003**, *125*, 14726 – 14727.
[26] Zhou, J.; Fu, G.C.; *J. Am. Chem. Soc.* **2004**, *126*, 1340 – 1341.
[27] Strotman, N.A.; Sommer, S.; Fu, G.C.; *Angew. Chem., Int. Ed.* **2007**, *46*, 3556 – 3558.
[28] Zultanski, S.L.; Fu, G.C.; *J. Am. Chem. Soc.* **2011**, *133*, 15362 – 15364.
[29] Lundin, P.M.; Fu, G.C.; *J. Am. Chem. Soc.* **2010**, *132*, 11027 – 110291.
[30] Lou, S.; Fu, G.C.; *J. Am. Chem. Soc.* **2010**, *132*, 5010 – 5011.
[31] Lundin, P.M.; Esquivias, J.; Fu, G.C.; *Angew. Chem., Int. Ed.* **2009**, *48*, 154 – 156.
[32] Smith, S.W.; Fu, G.C.; *J. Am. Chem. Soc.* **2008**, *130*, 12645 – 12646.
[33] Saito, B.; Fu, G.C.; *J. Am. Chem. Soc.* **2008**, *130*, 6694 – 6695.
[34] Lipshutz, B.H.; Blomgren, P.A.; *J. Am. Chem. Soc.* **1999**, *121*, 5819 – 5820.
[35] Lipshutz, B.H.; Ueda, H.; *Angew. Chem., Int. Ed.* **2000**, *39*, 4492 – 4494.
[36] Lipshutz, B.H.; Tasler, S.; Chrisman, W.; Spliethoff, B.; Tesche, B.; *J. Org. Chem.* **2003**, *68*, 1177 – 1189.
[37] Jana, R.; Pathak, T.P.; Sigman, M.S.; *Chem. Rev.* **2011**, *111*, 1417 – 1492.

[38] Hu, X.L.; *Chem. Sci.* **2011**, *2*, 1867 – 1886.
[39] Phapale, V.B.; Cárdenas, D.J.; *Chem. Soc. Rev.* **2009**, *38*, 1598 – 1607.
[40] Lin, X.F.; Phillips, D.L.; *J. Org. Chem.* **2008**, *73*, 3680 – 3688.
[41] Anderson, T.J.; Jones, G.D.; Vicic, D.A.; *J. Am. Chem. Soc.* **2004**, *126*, 8100 – 8101.
[42] Jones, G.D.; Martin, J.L.; McFarland, C.; Allen, O.R.; Hall, R.E.; Haley, A.D.; Brandon, R.J.; Konovalova, T.; Desrochers, P.J.; Pulay, P.; Vicic, D.A.; *J. Am. Chem. Soc.* **2006**, *128*, 13175 – 13183.
[43] Ren, P.; Vechorkin, O.; von Allmen, K.; Scopelliti, R.; Hu, X.L.; *J. Am. Chem. Soc.* **2011**, *133*, 7084 – 7095.
[44] Zhing, K.N.; Conda-Sheridan, M.; Cooke, S.R.; Louie, J.; *Organometallics* **2011**, *30*, 2546 – 2552.
[45] Lanni, E.L.; McNeil, A.J.; *J. Am. Chem. Soc.* **2009**, *131*, 16573 – 16579.
[46] Lanni, E.L.; McNeil, A.J.; *Macromolecules* **2010**, *43*, 8039 – 8044.
[47] Wenkert, E.; Han, A.L.; Jenny, C.J.; *J. Chem. Soc. Chem. Commun.* **1988**, 975 – 976.
[48] Blakey, S.B.; MacMillan, D.W.C.; *J. Am. Chem. Soc.* **2003**, *125*, 6046 – 6047.
[49] Xie, L.G.; Wang, Z.X.; *Angew. Chem., Int. Ed.* **2011**, *50*, 4901 – 4904.
[50] Rosen, B.M.; Quasdorf, K.W.; Wilson, D.A.; Zhang, N.; Resmerita, A.M.; Garg, N.K.; Percec, V.; *Chem. Rev.* **2011**, *111*, 1346 – 1416.
[51] Wilson, D.A.; Wilson, C.J.; Rosen, B.M.; Percec, V.; *Org. Lett.* **2008**, *10*, 4879 – 4882.
[52] Quasdorf, K.W.; Antoft-Finch, A.; Liu, P.; Silberstein, A.L.; Komaromi, A.; Blackburn, T.; Ramgren, S.D.; Houk, K.N.; Snieckus, V.; Garg, N.K.; *J. Am. Chem. Soc.* **2011**, *133*, 6352 – 6363.
[53] Quasdorf, K.W.; Riener, M.; Petrova, K.V.; Garg, N.K.; *J. Am. Chem. Soc.* **2009**, *131*, 17748 – 17749.
[54] Tobisu, M.; Shimasaki, T.; Chatani, N.; *Angew. Chem., Int. Ed.* **2008**, *47*, 4866 – 4869.
[55] Gooßen, L.J.; Gooßen, K.; Stanciu, C.; *Angew. Chem., Int. Ed.* **2009**, *48*, 3569 – 3571.
[56] Li, B.-J.; Li, Y.-Z.; Lu, X.-Y.; Liu, J.; Guan, B.-T.; Shi, Z.-J.; *Angew. Chem., Int. Ed.* **2008**, *47*, 10124 – 10127.
[57] Li, Z.; Zhang, S.-L.; Fu, Y.; Guo, Q.-X.; Liu, L.; *J. Am. Chem. Soc.* **2009**, *131*, 8815 – 8823.
[58] Alvarez-Bercedo, P.; Martin, R.; *J. Am. Chem. Soc.* **2010**, *132*, 17352 – 17353.
[59] Sergeev, A.G.; Hartwig, J.F.; *Science* **2011**, *332*, 439 – 443.

[60] Hayashi, T.; Konishi, M.; Yokota, K.-I.; Kumada, M.; *J. Organomet. Chem.* **1985**, *285*, 359 – 373.
[61] Bercot, E.A.; Rovis, T.; *J. Am. Chem. Soc.* **2001**, *124*, 174 – 175.
[62] Johnson, J.B.; Bercot, E.A.; Rowley, J.M.; Coates, G.W.; Rovis, T.; *J. Am. Chem. Soc.* **2007**, *129*, 2718 – 2725.
[63] Castano, A.M.; Echavarren, A.M.; *Organometallics* **1994**, *13*, 2262 – 2268.
[64] Wolfe, J.P.; Buchwald, S.L.; *J. Am. Chem. Soc.* **1997**, *119*, 6054 – 6058.
[65] Taylor, B.L.H.; Swift, E.C.; Waetzig, J.D.; Jarvo, E.R.; *J. Am. Chem. Soc.* **2011**, *133*, 389 – 391.
[66] Greene, M.A.; Yonova, I.M.; Williams, F.J.; Jarvo, E.R.; *Org. Lett.* **2012**, *14*, 4293 – 4296.
[67] Nielsen, D.K.; Doyle, A.G.; *Angew. Chem., Int. Ed.* **2011**, *50*, 6056 – 6059.
[68] Shacklady-McAtee, D.M.; Dasgupta, S.; Watson, M.P.; *Org. Lett.* **2011**, *13*, 3490 – 3493.
[69] Graham, T.J.A.; Doyle, A.G.; *Org. Lett.* **2012**, *14*, 1616 – 1619.
[70] Huang, C.-Y.; Doyle, A.G.; *J. Am. Chem. Soc.* **2012**, *134*, 9541 – 9544.
[71] Semmelhack, M.F.; Helquist, P.M.; Jones, L.D.; *J. Am. Chem. Soc.* **1971**, *93*, 5908 – 5910.
[72] Zembayashi, M.; Tamao, K.; Yoshida, J.-I.; Kumada, M.; *Tetrahedron Lett.* **1977**, *18*, 4089 – 4091.
[73] Tsou, T.T.; Kochi, J.K.; *J. Am. Chem. Soc.* **1979**, *101*, 7547 – 7560.
[74] Colon, I.; Kelsey, D.R.; *J. Org. Chem.* **1986**, *51*, 2627 – 2637.
[75] Percec, V.; Bae, J.-Y.; Zhao, M.; Hill, D.H.; *J. Org. Chem.* **1995**, *60*, 176 – 185.
[76] Prinsell, M.R.; Everson, D.A.; Weix, D.J.; *Chem. Commun.* **2010**, *46*, 5743 – 5745.
[77] For a palladium-based approach towards this goal: Krasovskiy, A.; Duplais, C.; Lipshutz, B.H.; *J. Am. Chem. Soc.* **2009**, *131*, 15592 – 15593.
[78] Durandetti, M.; Nédélec, J.-Y.; Périchon, J.; *J. Org. Chem.* **1996**, *61*, 1748 – 1755.
[79] Everson, D.A.; Shrestha, R.; Weix, D.J.; *J. Am. Chem. Soc.* **2010**, *132*, 920 – 921.
[80] Everson, D.A.; Jones, B.A.; Weix, D.J.; *J. Am. Chem. Soc.* **2012**, *134*, 6146 – 6159.
[81] Yu, X.L.; Yang, T.; Wang, S.L.; Xu, H.L.; Gong, H.G.; *Org. Lett.* **2011**, *13*, 2138 – 2141.

[82] Canivet, J.; Yamaguchi, J.; Ban, I.; Itami, K.; *Org. Lett.* **2009**, *11*, 1733 – 1736.
[83] Hachiya, H.; Hirano, K.; Satoh, T.; Miura, M.; *Org. Lett.* **2009**, *11*, 1737 – 1740.
[84] Muto, K.; Yamaguchi, J.; Itami, K.; *J. Am. Chem. Soc.* **2012**, *134*, 169 – 172.
[85] Hachiya, H.; Hirano, K.; Satoh, T.; Miura, M.; *Angew. Chem., Int. Ed.* **2010**, *49*, 2202 – 2205.
[86] Tobisu, M.; Hyodo, I.; Chatani, N.; *J. Am. Chem. Soc.* **2009**, *131*, 12070 – 12071.
[87] Reppe, W.; Schlichting, O.; Klager, K.; Toepel, T.; *Ann. Chem. Justus Liebig* **1948**, *560*, 1 – 92.
[88] Wilke, G.; *Angew. Chem., Int. Ed.* **1988**, *27*, 185 – 206.
[89] Wender, P.A.; Jenkins, T.E.; *J. Am. Chem. Soc.* **1989**, *111*, 6432 – 6434.
[90] Wender, P.A.; Smith, T.E.; *J. Org. Chem.* **1995**, *60*, 2962 – 2963.
[91] Wender, P.A.; Smith, T.E.; *J. Org. Chem.* **1996**, *61*, 824 – 825.
[92] Wender, P.A.; Smith, T.E.; *Tetrahedron* **1998**, *54*, 1255 – 1275.
[93] Koyama, I.; Kurahashi, T.; Matsubara, S.; *J. Am. Chem. Soc.* **2009**, *131*, 1350 – 1351.
[94] Ohashi, M.; Takeda, I.; Ikawa, M.; Ogoshi, S.; *J. Am. Chem. Soc.* **2011**, *133*, 18018 – 18021.
[95] Murakami, M.; Ashida, S.; Matsuda, T.; *J. Am. Chem. Soc.* **2005**, *127*, 6932 – 6933.
[96] Murakami, M.; Ashida, S.; Matsuda, T.; *Tetrahedron* **2006**, *62*, 7540 – 7546.
[97] Nakao, Y.; Morita, E.; Idei, H.; Hiyama, T.; *J. Am. Chem. Soc.* **2011**, *133*, 3264 – 3267.
[98] Shiota, H.; Ano, Y.; Aihara, Y.; Fukumoto, Y.; Chatani, N.; *J. Am. Chem. Soc.* **2011**, *133*, 14952 – 14955.
[99] Reppe, W.; Schweckendiek, W.J.; *Ann. Chem. Justus Liebig* **1948**, *560*, 104 – 116.
[100] Duckworth, D.M.; LeeWong, S.; Slawin, A.M.Z.; Smith, E.H.; Williams, D.J.; *J. Chem. Soc. Perkin Trans. 1* **1996**, 815 – 821.
[101] Bhatarah, P.; Smith, E.H.; *J. Chem. Soc. Perkin Trans. 1* **1992**, 2163 – 2168.
[102] Sato, Y.; Nishimata, T.; Mori, M.; *J. Org. Chem.* **1994**, *59*, 6133 – 6135.
[103] Ikeda, S.; Watanabe, H.; Sato, Y.; *J. Org. Chem.* **1998**, *63*, 7026 – 7029.

[104] Seo, J.; Chui, H.M.P.; Heeg, M.J.; Montgomery, J.; *J. Am. Chem. Soc.* **1999**, *121*, 476 – 477.
[105] Shanmugasundaram, M.; Wu, M.S.; Jeganmohan, M.; Huang, C.W.; Cheng, C.H.; *J. Org. Chem.* **2002**, *67*, 7724 – 7729.
[106] Qiu, Z.Z.; Xie, Z.W.; *Angew. Chem., Int. Ed.* **2009**, *48*, 5729 – 5732.
[107] Kumar, P.; Troast, D.M.; Cella, R.; Louie, J.; *J. Am. Chem. Soc.* **2011**, *133*, 7719 – 7721.
[108] Hoberg, H.; Oster, B.W.; *Synthesis* **1982**, 324 – 325.
[109] Duong, H.A.; Cross, M.J.; Louie, J.; *J. Am. Chem. Soc.* **2004**, *126*, 11438 – 11439.
[110] Miura, T.; Morimoto, M.; Murakami, M.; *J. Am. Chem. Soc.* **2010**, *132*, 15836 – 15838.
[111] Tsuda, T.; Morikawa, S.; Hasegawa, N.; Saegusa, T.; *J. Org. Chem.* **1990**, *55*, 2978 – 2981.
[112] Louie, J.; Gibby, J.E.; Farnworth, M.V.; Tekavec, T.N.; *J. Am. Chem. Soc.* **2002**, *124*, 15188 – 15189.
[113] McCormick, M.M.; Duong, H.A.; Zuo, G.; Louie, J.; *J. Am. Chem. Soc.* **2005**, *127*, 5030 – 5031.
[114] Kumar, P.; Prescher, S.; Louie, J.; *Angew. Chem., Int. Ed.* **2011**, *50*, 10694 – 10698.
[115] Noyori, R.; Odagi, T.; Takaya, H.; *J. Am. Chem. Soc.* **1970**, *92*, 5780 – 5781.
[116] Binger, P.; *Angew. Chem., Int. Ed.* **1972**, *11*, 309 – 310.
[117] Liu, L.; Montgomery, J.; *J. Am. Chem. Soc.* **2006**, *128*, 5348 – 5349.
[118] Liu, L.; Montgomery, J.; *Org. Lett.* **2007**, *9*, 3885 – 3887.
[119] Ogoshi, S.; Nagata, M.; Kurosawa, H.; *J. Am. Chem. Soc.* **2006**, *128*, 5350 – 5351.
[120] Tamaki, T.; Nagata, M.; Ohashi, M.; Ogoshi, S.; *Chem. Eur. J.* **2009**, *15*, 10083 – 10091.
[121] Mahandru, G.M.; Skauge, A.R.L.; Chowdhury, S.K.; Amarasinghe, K.K.D.; Heeg, M.J.; Montgomery, J.; *J. Am. Chem. Soc.* **2003**, *125*, 13481 – 13485.
[122] Montgomery, J.; Oblinger, E.; Savchenko, A.V.; *J. Am. Chem. Soc.* **1997**, *119*, 4911 – 4920.
[123] Seo, J.; Fain, H.; Blanc, J.B.; Montgomery, J.; *J. Org. Chem.* **1999**, *64*, 6060 – 6065.
[124] Herath, A.; Montgomery, J.; *J. Am. Chem. Soc.* **2006**, *128*, 14030 – 14031.
[125] Jenkins, A.D.; Herath, A.; Song, M.; Montgomery, J.; *J. Am. Chem. Soc.* **2011**, *133*, 14460 – 14466.

[126] Ohashi, M.; Taniguchi, T.; Ogoshi, S.; *J. Am. Chem. Soc.* **2011**, *133*, 14900 – 14903.
[127] Rayabarapu, D.K.; Yang, C.H.; Cheng, C.H.; *J. Org. Chem.* **2003**, *68*, 6726 – 6731.
[128] Horie, H.; Kurahashi, T.; Matsubara, S.; *Angew. Chem., Int. Ed.* **2011**, *50*, 8956 – 8959.
[129] Tamao, K.; Kobayashi, K.; Ito, Y.; *J. Am. Chem. Soc.* **1988**, *110*, 1286 – 1288.
[130] Zhang, M.; Buchwald, S.L.; *J. Org. Chem.* **1996**, *61*, 4498 – 4499.
[131] Chiusoli, G.P.; *Acc. Chem. Res.* **1973**, *6*, 422 – 427.
[132] Nadal, M.L.; Bosch, J.; Vila, J.M.; Klein, G.; Ricart, S.; Moretó, J.M.; *J. Am. Chem. Soc.* **2005**, *127*, 10476 – 10477.
[133] del Moral, D.; Ricart, S.; Moretó, J.M.; *Chem. Eur. J.* **2010**, *16*, 9193 – 9202.
[134] Koo, K.; Hillhouse, G.L.; *Organometallics* **1998**, *17*, 2924 – 2925.
[135] Baxter, R.D.; Montgomery, J.; *J. Am. Chem. Soc.* **2008**, *130*, 9662 – 9663.
[136] Ohashi, M.; Kishizaki, O.; Ikeda, H.; Ogoshi, S.; *J. Am. Chem. Soc.* **2009**, *131*, 9160 – 9161.
[137] Zuo, G.; Louie, J.; *Angew. Chem., Int. Ed.* **2004**, *43*, 2277 – 2279.
[138] Wender, P.A.; Ihle, N.C.; *J. Am. Chem. Soc.* **1986**, *108*, 4678 – 4679.
[139] Wender, P.A.; Snapper, M.L.; *Tetrahedron Lett.* **1987**, *28*, 2221 – 2224.
[140] Wender, P.A.; Ihle, N.C.; *Tetrahedron Lett.* **1987**, *28*, 2451 – 2454.
[141] Wender, P.A.; Ihle, N.C.; Correia, C.R.D.; *J. Am. Chem. Soc.* **1988**, *110*, 5904 – 5906.
[142] Wender, P.A.; Croatt, M.P.; Witulski, B.; *Tetrahedron* **2006**, *62*, 7505 – 7511.
[143] Gugelchuk, M.M.; Houk, K.N.; *J. Am. Chem. Soc.* **1994**, *116*, 330 – 339.
[144] Wender, P.A.; Christy, J.P.; *J. Am. Chem. Soc.* **2007**, *129*, 13402 – 13403.
[145] Wender, P.A.; Christy, J.P.; Lesser, A.B.; Gieseler, M.T.; *Angew. Chem., Int. Ed.* **2009**, *48*, 7687 – 7690.
[146] Ni, Y.; Montgomery, J.; *J. Am. Chem. Soc.* **2004**, *126*, 11162 – 11163.
[147] Ni, Y.; Montgomery, J.; *J. Am. Chem. Soc.* **2006**, *128*, 2609 – 2614.
[148] Saito, S.; Masuda, M.; Komagawa, S.; *J. Am. Chem. Soc.* **2004**, *126*, 10540 – 10541.
[149] Saito, S.; Maeda, K.; Yamasaki, R.; Kitamura, T.; Nakagawa, M.; Kato, K.; Azumaya, I.; Masu, H.; *Angew. Chem., Int. Ed.* **2010**, *49*, 1830 – 1833.

[150] Murakami, M.; Ashida, S.; Matsuda, T.; *J. Am. Chem. Soc.* **2006**, *128*, 2166 – 2167.
[151] Lappert, M.F.; Takahashi, S.; *J. Chem. Soc. Chem. Commun.* **1972**, 1272.
[152] Tamao, K.; Kobayashi, K.; Ito, Y.; *J. Am. Chem. Soc.* **1989**, *111*, 6478 – 6480.
[153] Johnson, J.R.; Tully, P.S.; Mackenzie, P.B.; Sabat, M.; *J. Am. Chem. Soc.* **1991**, *113*, 6172 – 6177.
[154] Amarasinghe, K.K.D.; Chowdhury, S.K.; Heeg, M.J.; Montgomery, J.; *Organometallics* **2001**, *20*, 370 – 372.
[155] Ikeda, S.; Sato, Y.; *J. Am. Chem. Soc.* **1994**, *116*, 5975 – 5976.
[156] Ikeda, S.; Yamamoto, H.; Kondo, K.; Sato, Y.; *Organometallics* **1995**, *14*, 5015 – 5016.
[157] Montgomery, J.; Savchenko, A.V.; *J. Am. Chem. Soc.* **1996**, *118*, 2099 – 2100.
[158] Ni, Y.K.; Amarasinghe, K.K.D.; Montgomery, J.; *Org. Lett.* **2002**, *4*, 1743 – 1745.
[159] Chevliakov, M.V.; Montgomery, J.; *J. Am. Chem. Soc.* **1999**, *121*, 11139 – 11143.
[160] ElDouhaibi, A.S.; Kassab, R.M.; Song, M.; Montgomery, J.; *Chem. Eur. J.* **2011**, *17*, 6326 – 6329.
[161] Ni, Y.; Amarasinghe, K.K.D.; Ksebati, B.; Montgomery, J.; *Org. Lett.* **2003**, *5*, 3771 – 3773.
[162] Ni, Y.; Kassab, R.M.; Chevliakov, M.V.; Montgomery, J.; *J. Am. Chem. Soc.* **2009**, *131*, 17714 – 17718.
[163] Herath, A.; Thompson, B.B.; Montgomery, J.; *J. Am. Chem. Soc.* **2007**, *129*, 8712 – 8713.
[164] Herath, A.; Montgomery, J.; *J. Am. Chem. Soc.* **2008**, *130*, 8132 – 8133.
[165] Li, W.; Herath, A.; Montgomery, J.; *J. Am. Chem. Soc.* **2009**, *131*, 17024 – 17029.
[166] Fornicola, R.S.; Subburaj, K.; Montgomery, J.; *Org. Lett.* **2002**, *4*, 615 – 617.
[167] Kim, J.; Schneekloth, J. S.; Sorensen, E. J. *Chem. Sci.* **2012**, *3*, 2849 – 2852.
[168] Li, W.; Chen, N.; Montgomery, J.; *Angew. Chem., Int. Ed.* **2010**, *49*, 8712 – 8716.
[169] Sato, Y.; Takimoto, M.; Mori, M.; *J. Am. Chem. Soc.* **2000**, *122*, 1624 – 1634.

[170] Takimoto, M.; Hiraga, Y.; Sato, Y.; Mori, M.; *Tetrahedron Lett.* **1998**, *39*, 4543 – 4546.
[171] Sato, Y.; Sawaki, R.; Mori, M.; *Organometallics* **2001**, *20*, 5510 – 5512.
[172] Sato, Y.; Hinata, Y.; Seki, R.; Oonishi, Y.; Saito, N.; *Org. Lett.* **2007**, *9*, 5597 – 5599.
[173] Kimura, M.; Ezoe, A.; Shibata, K.; Tamaru, Y.; *J. Am. Chem. Soc.* **1998**, *120*, 4033 – 4034.
[174] Kimura, M.; Fujimatsu, H.; Ezoe, A.; Shibata, K.; Shimizu, M.; Matsumoto, S.; Tamaru, Y.; *Angew. Chem., Int. Ed.* **1999**, *38*, 397 – 400.
[175] Ogoshi, S.; Tonomori, K.-I.; Oka, M.-A.; Kurosawa, H.; *J. Am. Chem. Soc.* **2006**, *128*, 7077 – 7086.
[176] Cho, H.Y.; Morken, J.P.; *J. Am. Chem. Soc.* **2008**, *130*, 16140 – 16141.
[177] Cho, H.Y.; Morken, J.P.; *J. Am. Chem. Soc.* **2010**, *132*, 7576 – 7577.
[178] Oblinger, E.; Montgomery, J.; *J. Am. Chem. Soc.* **1997**, *119*, 9065 – 9066.
[179] Tang, X.Q.; Montgomery, J.; *J. Am. Chem. Soc.* **1999**, *121*, 6098 – 6099.
[180] Tang, X.Q.; Montgomery, J.; *J. Am. Chem. Soc.* **2000**, *122*, 6950 – 6954.
[181] Mahandru, G.M.; Liu, G.; Montgomery, J.; *J. Am. Chem. Soc.* **2004**, *126*, 3698 – 3699.
[182] Chaulagain, M.R.; Sormunen, G.J.; Montgomery, J.; *J. Am. Chem. Soc.* **2007**, *129*, 9568 – 9569.
[183] Huang, W.S.; Chan, J.; Jamison, T.F.; *Org. Lett.* **2000**, *2*, 4221 – 4223.
[184] Miller, K.M.; Huang, W.S.; Jamison, T.F.; *J. Am. Chem. Soc.* **2003**, *125*, 3442 – 3443.
[185] Yang, Y.; Zhu, S.F.; Zhou, C.Y.; Zhou, Q.L.; *J. Am. Chem. Soc.* **2008**, *130*, 14052 – 14053.
[186] Colby, E.A.; O'Brien, K.C.; Jamison, T.F.; *J. Am. Chem. Soc.* **2004**, *126*, 998 – 999.
[187] Colby, E.A.; O'Brien, K.C.; Jamison, T.F.; *J. Am. Chem. Soc.* **2005**, *127*, 4297 – 4307.
[188] Knapp-Reed, B.; Mahandru, G.M.; Montgomery, J.; *J. Am. Chem. Soc.* **2005**, *127*, 13156 – 13157.
[189] Malik, H.A.; Sormunen, G.J.; Montgomery, J.; *J. Am. Chem. Soc.* **2010**, *132*, 6304 – 6305.
[190] Liu, P.; Montgomery, J.; Houk, K.N.; *J. Am. Chem. Soc.* **2011**, *133*, 6956 – 6959.

[191] Shareef, A.R.; Sherman, D.H.; Montgomery, J.; *Chem. Sci.* **2012**, *3*, 892 – 895.
[192] Baxter, R.D.; Montgomery, J.; *J. Am. Chem. Soc.* **2011**, *133*, 5728 – 5731.
[193] McCarren, P.R.; Liu, P.; Cheong, P.H.Y.; Jamison, T.F.; Houk, K.N.; *J. Am. Chem. Soc.* **2009**, *131*, 6654 – 6655.
[194] Liu, P.; McCarren, P.; Cheong, P.H.Y.; Jamison, T.F.; Houk, K.N.; *J. Am. Chem. Soc.* **2010**, *132*, 2050 – 2057.
[195] Patel, S.J.; Jamison, T.F.; *Angew. Chem., Int. Ed.* **2003**, *42*, 1364 – 1367.
[196] Patel, S.J.; Jamison, T.F.; *Angew. Chem., Int. Ed.* **2004**, *43*, 3941 – 3944.
[197] Zhou, C.Y.; Zhu, S.F.; Wang, L.X.; Zhou, Q.L.; *J. Am. Chem. Soc.* **2010**, *132*, 10955 – 10957.
[198] Yeh, C.-H.; Prasad Korivi, R.; Cheng, C.-H.; *Angew. Chem., Int. Ed.* **2008**, *47*, 4892 – 4895.
[199] Williams, C.M.; Johnson, J.B.; Rovis, T.; *J. Am. Chem. Soc.* **2008**, *130*, 14936 – 14937.
[200] Takimoto, M.; Shimizu, K.; Mori, M.; *Org. Lett.* **2001**, *3*, 3345 – 3347.
[201] Li, S.H.; Yuan, W.M.; Ma, S.M.; *Angew. Chem., Int. Ed.* **2011**, *50*, 2578 – 2582.
[202] Ng, S.S.; Jamison, T.F.; *J. Am. Chem. Soc.* **2005**, *127*, 7320 – 7321.
[203] Kang, S.K.; Yoon, S.K.; *Chem. Commun.* **2002**, 2634 – 2635.
[204] Montgomery, J.; Song, M.S.; *Org. Lett.* **2002**, *4*, 4009 – 4011.
[205] Amarasinghe, K.K.D.; Montgomery, J.; *J. Am. Chem. Soc.* **2002**, *124*, 9366 – 9367.
[206] Molinaro, C.; Jamison, T.F.; *J. Am. Chem. Soc.* **2003**, *125*, 8076 – 8077.
[207] Trenkle, J.D.; Jamison, T.F.; *Angew. Chem., Int. Ed.* **2009**, *48*, 5366 – 5368.
[208] Jin, H.; Uenishi, J.; Christ, W.J.; Kishi, Y.; *J. Am. Chem. Soc.* **1986**, *108*, 5644 – 5646.
[209] Takai, K.; Tagashira, M.; Kuroda, T.; Oshima, K.; Utimoto, K.; Nozaki, H.; *J. Am. Chem. Soc.* **1986**, *108*, 6048 – 6050.
[210] Fürstner, A.; Shi, N.; *J. Am. Chem. Soc.* **1996**, *118*, 2533 – 2534.
[211] Fürstner, A.; Shi, N.; *J. Am. Chem. Soc.* **1996**, *118*, 12349 – 12357.
[212] Fürstner, A.; *Chem. Rev.* **1999**, *99*, 991 – 1046.
[213] Guo, H.; Dong, C.-G.; Kim, D.-S.; Urabe, D.; Wang, J.; Kim, J.T.; Liu, X.; Sasaki, T.; Kishi, Y.; *J. Am. Chem. Soc.* **2009**, *131*, 15387 – 15393.

[214] Harnying, W.; Kaiser, A.; Klein, A.; Berkessel, A.; *Chem. Eur. J.* **2011**, *17*, 4765 – 4773.
[215] Namba, K.; Kishi, Y.; *J. Am. Chem. Soc.* **2005**, *127*, 15382 – 15383.
[216] Liu, X.; Henderson, J.A.; Sasaki, T.; Kishi, Y.; *J. Am. Chem. Soc.* **2009**, *131*, 16678 – 16680.
[217] Yeung, C.S.; Dong, V.M.; *J. Am. Chem. Soc.* **2008**, *130*, 7826 – 7827.
[218] Ochiai, H.; Jang, M.; Hirano, K.; Yorimitsu, H.; Oshima, K.; *Org. Lett.* **2008**, *10*, 2681 – 2683.
[219] Ogoshi, S.; Oka, M.-A.; Kurosawa, H.; *J. Am. Chem. Soc.* **2004**, *126*, 11802 – 11803.
[220] Ogoshi, S.; Ueta, M.; Arai, T.; Kurosawa, H.; *J. Am. Chem. Soc.* **2005**, *127*, 12810 – 12811.
[221] Ng, S.S.; Jamison, T.F.; *J. Am. Chem. Soc.* **2005**, *127*, 14194 – 14195.
[222] Ho, C.Y.; Ng, S.S.; Jamison, T.F.; *J. Am. Chem. Soc.* **2006**, *128*, 5362 – 5363.
[223] Ng, S.S.; Ho, C.Y.; Jamison, T.F.; *J. Am. Chem. Soc.* **2006**, *128*, 11513 – 11528.
[224] Ho, C.Y.; Jamison, T.F.; *Angew. Chem., Int. Ed.* **2007**, *46*, 782 – 785.
[225] Ho, C.Y.; Schleicher, K.D.; Chan, C.W.; Jamison, T.F.; *Synlett* **2009**, 2565 – 2582.
[226] Hoshimoto, Y.; Ohashi, M.; Ogoshi, S.; *J. Am. Chem. Soc.* **2011**, *133*, 4668 – 4671.
[227] Loots, M.J.; Schwartz, J.; *J. Am. Chem. Soc.* **1977**, *99*, 8045 – 8046.
[228] Schwartz, J.; Loots, M.J.; Kosugi, H.; *J. Am. Chem. Soc.* **1980**, *102*, 1333 – 1340.
[229] Yanagi, T.; Sasaki, H.; Suzuki, A.; Miyaura, N.; *Synth. Commun.* **1996**, *26*, 2503 – 2509.
[230] Hansen, R.T.; Carr, D.B.; Schwartz, J.; *J. Am. Chem. Soc.* **1978**, *100*, 2242 – 2245.
[231] Schwartz, J.; Carr, D.B.; Hansen, R.T.; Dayrit, F.M.; *J. Org. Chem.* **1980**, *45*, 3053 – 3061.
[232] Greene, A.E.; Lansard, J.P.; Luche, J.L.; Petrier, C.; *J. Org. Chem.* **1984**, *49*, 931 – 932.
[233] Houpis, I.N.; Molina, A.; Dorziotis, I.; Reamer, R.A.; Volante, R.P.; Reider, P.J.; *Tetrahedron Lett.* **1997**, *38*, 7131 – 7134.
[234] Savchenko, A.V.; Montgomery, J.; *J. Org. Chem.* **1996**, *61*, 1562 – 1563.
[235] Grisso, B.A.; Johnson, J.R.; Mackenzie, P.B.; *J. Am. Chem. Soc.* **1992**, *114*, 5160 – 5165.
[236] Petrier, C.; Barbosa, J.C.D.; Dupuy, C.; Luche, J.L.; *J. Org. Chem.* **1985**, *50*, 5761 – 5765.

[237] Kucera, D.J.; Oconnor, S.J.; Overman, L.E.; *J. Org. Chem.* **1993**, *58*, 5304 – 5306.
[238] Fox, M.E.; Li, C.; Marino, J.P.; Overman, L.E.; *J. Am. Chem. Soc.* **1999**, *121*, 5467 – 5480.
[239] Smith, A.B.; Leenay, T.L.; *J. Am. Chem. Soc.* **1989**, *111*, 5761 – 5768.
[240] Hirano, K.; Yorimitsu, H.; Oshima, K.; *Org. Lett.* **2007**, *9*, 1541 – 1544.
[241] Gomez-Bengoa, E.; Heron, N.M.; Didiuk, M.T.; Luchaco, C.A.; Hoveyda, A.H.; *J. Am. Chem. Soc.* **1998**, *120*, 7649 – 7650.
[242] Dayrit, F.M.; Gladkowski, D.E.; Schwartz, J.; *J. Am. Chem. Soc.* **1980**, *102*, 3976 – 3978.
[243] Dayrit, F.M.; Schwartz, J.; *J. Am. Chem. Soc.* **1981**, *103*, 4466 – 4473.
[244] Sustmann, R.; Hopp, P.; Holl, P.; *Tetrahedron Lett.* **1989**, *30*, 689 – 692.
[245] Hegedus, L.S.; Thompson, D.H.P.; *J. Am. Chem. Soc.* **1985**, *107*, 5663 – 5669.
[246] Shrestha, R.; Weix, D.J.; *Org. Lett.* **2011**, *13*, 2766 – 2769.
[247] Subburaj, K.; Montgomery, J.; *J. Am. Chem. Soc.* **2003**, *125*, 11210 – 11211.
[248] Chrovian, C.C.; Montgomery, J.; *Org. Lett.* **2007**, *9*, 537 – 540.
[249] Joensuu, P.M.; Murray, G.J.; Fordyce, E.A.F.; Luebbers, T.; Lam, H.W.; *J. Am. Chem. Soc.* **2008**, *130*, 7328 – 7338.
[250] Kiso, Y.; Kumada, M.; Tamao, K.; Umeno, M.; *J. Organomet. Chem.* **1973**, *50*, 297 – 310.
[251] Kiso, Y.; Kumada, M.; Maeda, K.; Sumitani, K.; Tamao, K.; *J. Organomet. Chem.* **1973**, *50*, 311 – 318.
[252] Lappert, M.F.; Nile, T.A.; Takahashi, S.; *J. Organomet. Chem.* **1974**, *72*, 425 – 439.
[253] Chaulagain, M.R.; Mahandru, G.M.; Montgomery, J.; *Tetrahedron* **2006**, *62*, 7560 – 7566.
[254] Tran, B.L.; Pink, M.; Mindiola, D.J.; *Organometallics* **2009**, *28*, 2234 – 2243.
[255] Buchan, Z.A.; Bader, S.J.; Montgomery, J.; *Angew. Chem., Int. Ed.* **2009**, *48*, 4840 – 4844.
[256] Vettel, S.; Vaupel, A.; Knochel, P.; *Tetrahedron Lett.* **1995**, *36*, 1023 – 1026.
[257] Vettel, S.; Vaupel, A.; Knochel, P.; *J. Org. Chem.* **1996**, *61*, 7473 – 7481.
[258] Lautens, M.; Rovis, T.; *J. Am. Chem. Soc.* **1997**, *119*, 11090 – 11091.
[259] Lautens, M.; Chiu, P.; Ma, S.H.; Rovis, T.; *J. Am. Chem. Soc.* **1995**, *117*, 532 – 533.

[260] Lautens, M.; Ma, S.H.; Chiu, P.; *J. Am. Chem. Soc.* **1997**, *119*, 6478 – 6487.

[261] Vaupel, A.; Knochel, P.; *Tetrahedron Lett.* **1994**, *35*, 8349 – 8352.

[262] Vaupel, A.; Knochel, P.; *J. Org. Chem.* **1996**, *61*, 5743 – 5753.

[263] Phapale, V.B.; Buñuel, E.; García-Iglesias, M.; Cárdenas, D.J.; *Angew. Chem., Int. Ed.* **2007**, *46*, 8790 – 8795.

[264] Robbins, D.W.; Hartwig, J.F.; *Science* **2011**, *333*, 1423 – 1427.

[265] RajanBabu, T.V.; *Synlett* **2009**, 853 – 885.

[266] RajanBabu, T.V.; *Chem. Rev.* **2003**, *103*, 2845 – 2860.

[267] RajanBabu, T.V.; Nomura, N.; Jin, J.; Nandi, M.; Park, H.; Sun, X.F.; *J. Org. Chem.* **2003**, *68*, 8431 – 8446.

[268] Zhang, A.; RajanBabu, T.V.; *J. Am. Chem. Soc.* **2005**, *128*, 54 – 55.

[269] Saha, B.; Smith, C.R.; RajanBabu, T.V.; *J. Am. Chem. Soc.* **2008**, *130*, 9000 – 9005.

[270] Ho, C.-Y.; He, L.; *Angew. Chem., Int. Ed.* **2010**, *49*, 9182 – 9186.

[271] Mans, D.J.; Cox, G.A.; RajanBabu, T.V.; *J. Am. Chem. Soc.* **2011**, *133*, 5776 – 5779.

[272] McKinney, R.J.; Roe, D.C.; *J. Am. Chem. Soc.* **1986**, *108*, 5167 – 5173.

[273] Li, T.; Jones, W.D.; *Organometallics* **2011**, *30*, 547 – 555.

[274] Tauchert, M.E.; Warth, D.C.M.; Braun, S.M.; Gruber, I.; Ziesak, A.; Rominger, F.; Hofmann, P.; *Organometallics* **2011**, *30*, 2790 – 2809.

[275] Casalnuovo, A.L.; Rajanbabu, T.V.; Ayers, T.A.; Warren, T.H.; *J. Am. Chem. Soc.* **1994**, *116*, 9869 – 9882.

[276] Saha, B.; RajanBabu, T.V.; *Org. Lett.* **2006**, *8*, 4657 – 4659.

[277] Kranenburg, M.; Kamer, P.C.J.; Vanleeuwen, P.; Vogt, D.; Keim, W.; *J. Chem. Soc. Chem. Commun.* **1995**, 2177 – 2178.

[278] Goertz, W.; Kamer, P.C.J.; van Leeuwen, P.; Vogt, D.; *Chem. Eur. J.* **2001**, *7*, 1614 – 1618.

[279] Nakao, Y.; Kanyiva, K.S.; Oda, S.; Hiyama, T.; *J. Am. Chem. Soc.* **2006**, *128*, 8146 – 8147.

[280] Nakao, Y.; Kashihara, N.; Kanyiva, K.S.; Hiyama, T.; *Angew. Chem., Int. Ed.* **2010**, *49*, 4451 – 4454.

[281] Nakao, Y.; Idei, H.; Kanyiva, K.S.; Hiyama, T.; *J. Am. Chem. Soc.* **2009**, *131*, 15996 – 15997.

[282] Kanyiva, K.S.; Nakao, Y.; Hiyama, T.; *Angew. Chem., Int. Ed.* **2007**, *46*, 8872 – 8874.

[283] Nakao, Y.; Kanyiva, K.S.; Hiyama, T.; *J. Am. Chem. Soc.* **2008**, *130*, 2448 – 2449.

[284] Nakao, Y.; Yamada, Y.; Kashihara, N.; Hiyama, T.; *J. Am. Chem. Soc.* **2010**, *132*, 13666 – 13668.

[285] Nakao, Y.; Idei, H.; Kanyiva, K.S.; Hiyama, T.; *J. Am. Chem. Soc.* **2009**, *131*, 5070 – 5071.

[286] Shirakura, M.; Suginome, M.; *J. Am. Chem. Soc.* **2008**, *130*, 5410 – 5411.

[287] Shirakura, M.; Suginome, M.; *Angew. Chem., Int. Ed.* **2010**, *49*, 3827 – 3829.

[288] Nakao, Y.; Hiyama, T.; *Pure Appl. Chem.* **2008**, *80*, 1097 – 1107.

[289] Nakao, Y.; Oda, S.; Hiyama, T.; *J. Am. Chem. Soc.* **2004**, *126*, 13904 – 13905.

[290] Nakao, Y.; Yada, A.; Ebata, S.; Hiyama, T.; *J. Am. Chem. Soc.* **2007**, *129*, 2428 – 2429.

[291] Watson, M.P.; Jacobsen, E.N.; *J. Am. Chem. Soc.* **2008**, *130*, 12594 – 12595.

[292] Nakao, Y.; Ebata, S.; Yada, A.; Hiyama, T.; Ikawa, M.; Ogoshi, S.; *J. Am. Chem. Soc.* **2008**, *130*, 12874 – 12875.

[293] Ohnishi, Y.; Nakao, Y.; Sato, H.; Hiyama, T.; Sakaki, S.; *Organometallics* **2009**, *28*, 2583 – 2594.

[294] Nakao, Y.; Yukawa, T.; Hirata, Y.; Oda, S.; Satoh, J.; Hiyama, T.; *J. Am. Chem. Soc.* **2006**, *128*, 7116 – 7117.

[295] Hirata, Y.; Yukawa, T.; Kashihara, N.; Nakao, Y.; Hiyama, T.; *J. Am. Chem. Soc.* **2009**, *131*, 10964 – 10973.

[296] Hirata, Y.; Yada, A.; Morita, E.; Nakao, Y.; Hiyama, T.; Ohashi, M.; Ogoshi, S.; *J. Am. Chem. Soc.* **2010**, *132*, 10070 – 10077.

[297] Nakao, Y.; Hirata, Y.; Tanaka, M.; Hiyama, T.; *Angew. Chem., Int. Ed.* **2008**, *47*, 385 – 387.

[298] Nakao, Y.; Yada, A.; Hiyama, T.; *J. Am. Chem. Soc.* **2010**, *132*, 10024 – 10026.

[299] Nakao, Y.; Hirata, Y.; Hiyama, T.; *J. Am. Chem. Soc.* **2006**, *128*, 7420 – 7421.

[300] Hirata, Y.; Inui, T.; Nakao, Y.; Hiyama, T.; *J. Am. Chem. Soc.* **2009**, *131*, 6624 – 6631.

[301] Kiji, J.; Sasakawa, E.; Yamamoto, K.; Furukawa, J.; *J. Organomet. Chem.* **1974**, *77*, 125 – 130.

[302] Pawlas, J.; Nakao, Y.; Kawatsura, M.; Hartwig, J.F.; *J. Am. Chem. Soc.* **2002**, *124*, 3669 – 3679.

[303] Castonguay, A.; Spasyuk, D.M.; Madern, N.; Beauchamp, A.L.; Zargarian, D.; *Organometallics* **2009**, *28*, 2134 – 2141.

[304] Muñiz, K.; Streuff, J.; Hövelmann, C.H.; Núñez, A.; *Angew. Chem., Int. Ed.* **2007**, *46*, 7125 – 7127.

Chapter Four

Organogold Chemistry

Norbert Krause
Organic Chemistry
Dortmund University of Technology
Dortmund, Germany

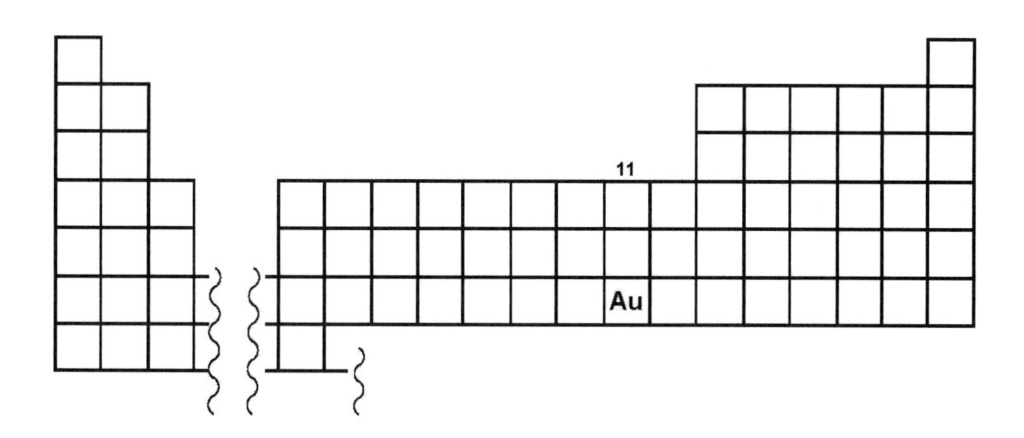

Contents

1. Introduction 431

2. C–H Bond Formation 433

3. C–C Bond Formation 434
- 3.1 Nucleophilic Additions to Alkynes 435
- 3.2 Nucleophilic Additions to Allenes 440
- 3.3 Nucleophilic Additions to Alkenes 444
- 3.4 Cycloisomerizations 449
- 3.5 Cycloadditions 462
- 3.6 Rearrangements 467

4. C–N Bond Formation 469
- 4.1 Nucleophilic Additions to Alkynes 469
- 4.2 Nucleophilic Additions to Allenes 474
- 4.3 Nucleophilic Additions to Alkenes 480

5. C–O Bond Formation 485
- 5.1 Nucleophilic Additions to Alkynes 486
- 5.2 Nucleophilic Additions to Allenes 498
- 5.3 Nucleophilic Additions to Alkenes 514
- 5.4 Rearrangements 516

6. C–S and C–Halogen Bond Formation 522

7. Conclusions and Outlook 525

8. Acknowledgments 526

9. References 527

1. Introduction

Homogeneous gold catalysis began its growth just a decade ago, which is in remarkable contrast to heterogeneous gold catalysis, a field that is much more established and has made crucial contributions to many industrially important processes, such as the hydrochlorination of acetylene and the low-temperature oxidation of CO to CO_2.[1] The long-lasting lack of interest in homogeneous gold catalysts may have persisted for psychological reasons: After all, it was commonly accepted that gold is very expensive, that gold salts are catalytically inactive, and that redox cycles (which are crucial for the catalytic activity of many late transition metals) are impossible. After 10 years of intense research activities resulting in thousands of original publications, 100 or so reviews, and 2 books,[2] it is today safe to say that common knowledge is not always right.

Although it may seem strange to look at such a young field from a historical point of view, it is instructive to identify the milestones that were instrumental in the development of homogeneous gold catalysis. As early as 1986, Ito, Sawamura, and Hayashi reported enantioselective gold-catalyzed aldol-type reactions of aldehydes with isocyanoacetates that afforded substituted, nonracemic oxazolines with high diastereoselectivity and enantioselectivity (Scheme 4-1).[3] These can be opened to the corresponding aminoalcohols that are highly valuable synthetic intermediates. The chiral catalyst was formed *in situ* from cationic gold complex $[Au(c\text{-}HexNC)_2]^+BF_4^-$ and ferrocenyl *bis*-phosphines of type **A**.

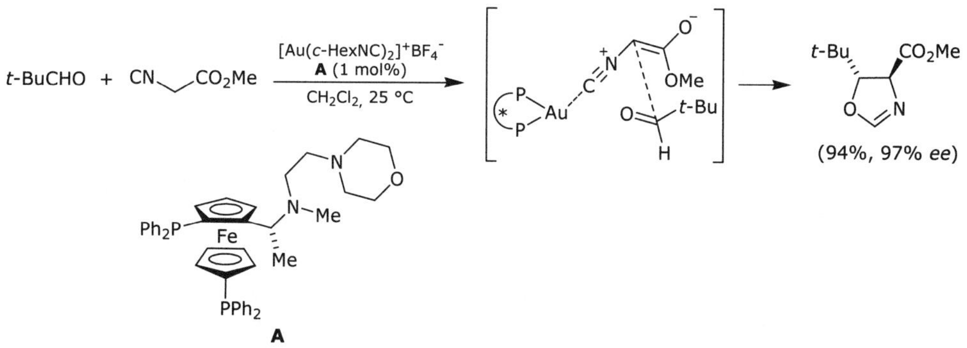

t-BuCHO + CN⌄CO2Me → [Au(c-HexNC)2]+BF4− / A (1 mol%) / CH2Cl2, 25 °C → (94%, 97% ee)

Scheme 4-1

This remarkable transformation was not only the first reaction promoted by a chiral gold catalyst but also the first example of a catalytic enantioselective aldol reaction.[4] It could have been the trigger for the "gold rush" in homogeneous gold catalysis that took place more than a decade later—but it was not. Why? It turned out that the activation mode of the isonitrile by the cationic gold catalyst is unique and cannot be used for other reaction types. Consequently, interest in gold-catalyzed aldol reactions ceased a few years after their discovery.

The actual turning point for homogeneous gold catalysis came in 1998 when a group of industrial chemists at BASF (Ludwigshafen, Germany) led by Teles reported on addition reactions of alcohols to alkynes catalyzed by cationic gold(I) complexes, which were prepared by treatment of Ph_3PAuMe with a Brønsted acid ($MeSO_3H$ or H_2SO_4; Scheme 4-2).[5] This hydroalkoxylation relies on the activation of the triple bond by the gold catalyst that enables nucleophilic attack by the alcohol. Thus, the potential of gold salts to act as soft carbophilic Lewis acids (π-acids) was demonstrated, and it is this particular feature of gold that caught the attention of the scientific community.

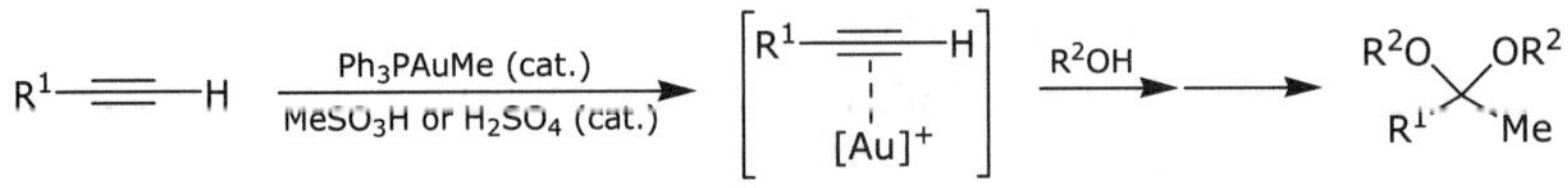

$$R^1{-}{\equiv}{-}H \xrightarrow[MeSO_3H \text{ or } H_2SO_4 \text{ (cat.)}]{Ph_3PAuMe \text{ (cat.)}} \left[R^1{-}{\equiv}{-}H \cdots [Au]^+ \right] \xrightarrow{R^2OH} R^1C(Me)(OR^2)_2$$

Scheme 4-2

Today, we know that the catalytic activity of gold salts is a consequence of relativistic effects that reach a maximum at gold and render not only Au(III) but also Au(I) unique for the selective π-activation of alkynes, allenes, and alkenes in the presence of other hard Lewis-basic sites.[6] Additional modes of reactivity are opened up when enynes are used as substrate; selective activation of the triple bond by the gold catalyst and intramolecular attack of the pendant alkene generate a cyclopropyl gold carbene that can also be viewed as a gold-stabilized cyclopropylmethyl/cyclobutyl/homoallyl cation and that readily participates in cycloisomerizations and cycloadditions. Alternatively, gold carbenes can be formed from propargyl carboxylates by 1,2-carboxylate migration. Many of these transformations can be rendered enantioselective by using chiral phosphines or *N*-heterocyclic carbenes as ligands, even though the two-coordinate linear coordination geometry of gold(I) poses limitations such as the inability to form chelate complexes.

In this chapter, important contributions to the field of homogeneous gold catalysis are summarized for the practitioner. Heterogeneous gold catalysis is covered only briefly in the section on C–H bond formation. The field is far too broad for this chapter to be comprehensive. The reader will notice a preference of the author for cyclic molecules (and, hence, for reaction types that start with "cyclo"), for stereoselective transformations, and for allenes. Moreover, the author prefers individual examples with specific substrates and catalysts over general schemes and long tables containing different substituent patterns. Likewise, representative procedures with specific conditions for a certain substrate (including exact amounts of the reactants and solvents, reaction time, and temperature) are preferred over general procedures. Unfortunately, it is becoming ever more difficult to find this information in the literature, a trend that is alarming. The material is organized according to the bond formed in the gold-catalyzed step (C–H, C–C, C–N, C–O, C–S, C–halogen). More often than not, two or more different bonds are generated in a gold-catalyzed transformation; in those cases, the reaction is classified according to the first (or more important) bond formed. Reference to other transition metal catalysts is made wherever this information is available.[7] It is fully anticipated that the reader will come to the conclusion that GOLD IS THE BEST METAL CATALYST OF THE 21ST CENTURY!

2. C–H Bond Formation

Reviews. Claus, P.; *Appl. Catal. A* **2005**, *291*, 222; Hashmi, A.S.K.; Hutchings, G.J.; *Angew. Chem., Int. Ed.* **2006**, *45*, 7896.

The use of gold as a hydrogenation catalyst has been disfavored by the fact that H_2 does not absorb dissociatively on extended gold surfaces.[8] For this reason, mixed heterogeneous catalysts were used where hydrogen atoms were supplied by a second metal, *e.g.*, Pt or Pd.[9] Alternatively, gold nanoparticles on a suitable solid support can be employed. Even though the activity of these catalysts is low, they exhibit an extraordinarily high selectivity in the partial reduction of buta-1,3-diene.[10] For this purpose, small (<5 nm) gold nanoparticles on Al_2O_3, TiO_2, or SiO_2 surfaces were generated by deposition-precipitation, liquid-phase grafting, or gas-phase grafting. Low temperatures favored the formation of but-1-ene, whereas higher proportions of but-2-enes were obtained above 500 K. Subsequently, it was shown that individual Au^{3+} cations on a ZrO_2 surface exhibit a high reactivity in the partial hydrogenation of buta-1,3-diene.[11]

The selective hydrogenation of α,β-unsaturated aldehydes to allylic alcohols represents another challenge that can be met by heterogeneous gold catalysis. A moderately selective activation of the carbonyl group of acrolein or crotonaldehyde was achieved using gold nanoparticles on a ZnO or ZrO_2 support that was enhanced by doping the catalyst with thiophene.[12] Higher, preparatively useful selectivities were achieved in the gold-catalyzed hydrogenation of structurally more complex enones and enals (*e.g.*, citral) where bulky substituents at the C–C double bond disfavor its adsorption on the catalytic sites (Scheme 4-3).[13] The hydrogenation of aromatic nitro compounds is another interesting application of heterogeneous gold catalysis.[14]

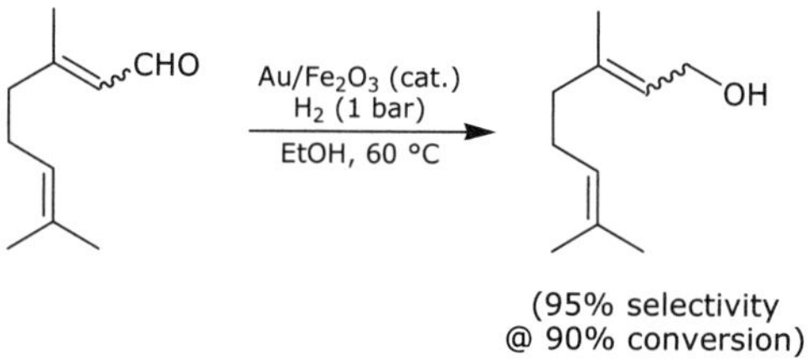

CHO

Au/Fe_2O_3 (cat.)
H_2 (1 bar)
EtOH, 60 °C

OH

(95% selectivity
@ 90% conversion)

Scheme 4-3

The hydrogenation of olefins and imines in the presence of homogeneous gold(I) catalysts was described by Corma and co-workers.[15] By using only 0.1 mol% of AuCl-Me-DUPHOS catalyst **B**, up to 95% *ee* and turnover frequencies (TOFs) between 200 and 4000 h^{-1} were achieved in the hydrogenation of various alkylidenesuccinates (Scheme 4-4). The selectivities are higher than those obtained with the corresponding Pt- and Ir-catalysts. Gold(III)-Schiff base complexes are also active hydrogenation catalysts and seem to operate *via* heterolytic cleavage of H_2.[16] Gold-catalyzed hydroborations[17] and hydrosilylations[18] of various unsaturated substrates have been described as well.

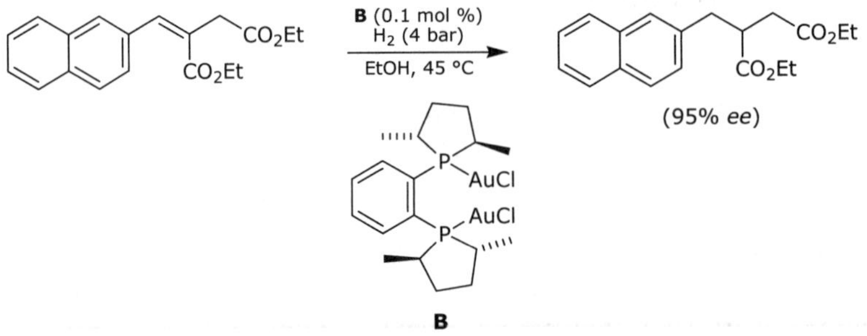

Scheme 4-4

A different approach to gold-catalyzed C–H bond formation involves the protonation of suitable nucleophiles. Cationic gold(I) complexes can generate a so-called Lewis-acid–activated Brønsted acid (LBA) by forming a σ-complex with an alcohol. This strongly increases the acidity of the alcohol and allows the use of chiral LBAs in enantioselective protonation. Toste and co-workers[19] used this strategy for converting various prochiral silyl enol ethers into the corresponding ketones that were obtained with high enantioselectivity in the presence of (*R*)-BINAP$(AuCl)_2$, $AgBF_4$, and ethanol (Scheme 4-5).

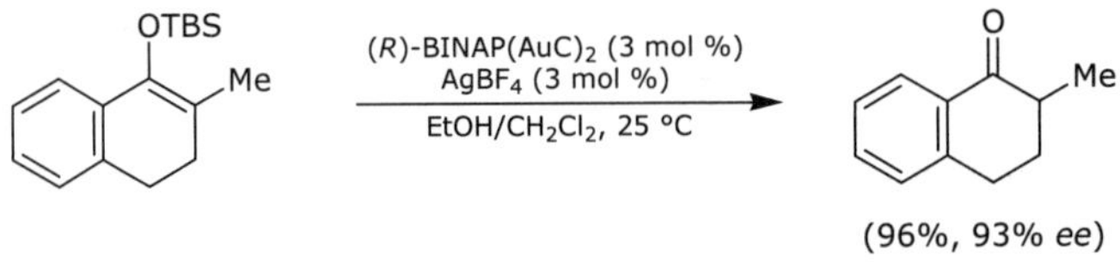

Scheme 4-5

General procedure for the enantioselective protonation of silyl enol ethers[19]

To a test tube equipped with a magnetic stirring bar, a solution of (*R*)-BINAP$(AuCl)_2$ in CH_2Cl_2 (6.0 mM in CH_2Cl_2, 0.50 mL, 3.0 μmol, 0.030 equiv) and a solution of $AgBF_4$ in EtOH (6.0 mM in EtOH, 0.50 mL, 3.0 μmol, 0.030 equiv) are added at room temperature (RT). The reaction mixture is stirred for 30 min at RT. The silyl enol ether (0.10 mmol, 1.0 equiv) is added to the reaction mixture through a septum, and the reaction mixture is monitored by thin layer chromatography (TLC). When the starting silyl enol ether is completely consumed, the reaction is quenched with aqueous sodium bicarbonate ($NaHCO_3$), extracted with Et_2O, dried over anhydrous sodium sulfate (Na_2SO_4), and concentrated. The crude mixture is further purified by preparative TLC using a mixture of EtOAc and hexanes (1:10) as eluent.

3. C–C Bond Formation

Reviews. Hashmi, A.S.K.; *Chem. Rev.* **2007,** *107*, 3180; Fürstner, A.; Davies, P.W.; *Angew. Chem., Int. Ed.* **2007**, *46*, 3410; Li, Z.; Brouwer, C.; He, C.; *Chem. Rev.* **2008,** *108*, 3239; Arcadi, A.; *Chem. Rev.* **2008,** *108*, 3266; Jiménez-Núñez, E.; Echavarren, A.M.; *Chem. Rev.* **2008,** *108*, 3326; Gorin, D.J.; Sherry, B.D.; Toste,

F.D.; *Chem. Rev.* **2008,** *108*, 3351; Shen, H.C.; *Tetrahedron* **2008,** *64*, 3885; Shen, H.C.; *Tetrahedron* **2008,** *64*, 7847; Michelet, V.; Toullec, P.Y.; Genêt, J.-P.; *Angew. Chem., Int. Ed.* **2008,** *47*, 4268; Shapiro, N.D.; Toste, F.D.; *Synlett* **2010**, 675; Hashmi, A.S.K.; *Pure Appl. Chem.* **2010,** *82*, 1517; Aubert, C.; Fensterbank, L.; Garcia, P.; Malacria, M.; Simonneau, A.; *Chem. Rev.* **2011,** *111*, 1954.

3.1 Nucleophilic Additions to Alkynes

Hydroarylation of an alkyne requires activation of the triple bond by a Lewis acid to afford a π-complex, which then reacts with an electron-rich (hetero)arene. Various metal salts, including carbophilic palladium and platinum complexes, have been used for these nucleophilic additions to alkynes, which can also be viewed as Friedel-Crafts reactions.[20] Gold catalysts joined this team only recently. By using $AuCl_3$ together with $AgSbF_6$, Reetz and Sommer[21] achieved intermolecular additions of mesitylene and other benzene derivatives to electron-rich alkynes to give 1,1-disubstituted alkenes as a result of a Markovnikov hydroarylation. The opposite regioselectivity was observed in the hydroarylation of ethyl propiolate and other Michael acceptors with $Ph_3PAuCl/BF_3{\cdot}Et_2O$[21] or $AuCl_3/AgOTf$[22] as catalyst, affording (*Z*)-olefins that in some cases isomerize to the (*E*)-alkene. Spectroscopic and isotope labeling experiments indicate the presence of an arenegold intermediate formed by gold-catalyzed C–H activation,[23] which then undergoes a *cis*-addition to the triple bond.[22] Double hydroarylation products were formed in the gold-catalyzed addition reactions of heteroarenes to alkynes.[24]

Representative procedure for the gold-catalyzed hydroarylation of alkynes. 1-(3-Hydroxy-2,4,6-trimethylphenyl)-1-(4-methoxyphenyl)ethene[21]

$AuCl_3$ (1.5 mol %)
$AgSbF_6$ (3 mol %)
$MeNO_2$, 50 °C, 4 h
OH OMe

2,4,6-Trimethylphenol (13.62 g, 100.0 mmol) was weighed into a 100-mL Schlenk flask with exposure to air. The flask was then evacuated for several minutes and charged with argon (three times). Nitromethane (20 mL) was added. Solutions of 45.5 mg $AuCl_3$ (0.150 mmol), 103 mg $AgSbF_6$ (0.300 mmol) in 8.0 mL of nitromethane and 4-ethynylanisole (10 mmol, concentrated solution in nitromethane) were added rapidly. The reaction mixture was stirred at 50 °C for 4 h. After cooling, the solvent was removed *in vacuo*. Excess 2,4,6-trimethylphenol was removed by sublimation *in vacuo* (ca. 10^{-2} mbar) at 80 °C. The residue was purified by column chromatography (silica gel, hexanes/EtOAc) to afford 1-(3-hydroxy-2,4,6-trimethylphenyl)-1-(4-methoxyphenyl) ethene in 92% yield.

Representative procedure for the gold-catalyzed hydroarylation of alkynes under solvent-free conditions. (Z)-Ethyl 3-(pentamethylphenyl)propenoate[22]

CO_2Et
$AuCl_3$ (2.5 mol %)
AgOTf (7.5 mol %)
25 °C, 1 h
CO_2Et

A mixture of pentamethylbenzene (296.0 mg, 2.0 mmol), ethyl propiolate (196.0 g, 2.0 mmol), $AuCl_3$ (15.2 mg, 0.05 mmol), and AgOTf (38.4 mg, 0.15 mmol) in a test tube was stirred for 60 min until the mixture turned into solid. Then the product was filtered through a short silica gel column with 10/1 hexane/EtOAc as eluent. Removal of solvents yielded white crystals of (*Z*)-ethyl 3-(pentamethylphenyl)propenoate in quantity.

The gold-catalyzed hydroarylation of 1-alkynyl-2-alkenylbenzene derivatives afforded substituted naphthalenes by 6-*endo*-selective cyclization.[25] The same regioselectivity was observed in the cyclization of 2-alkynylbiphenyls to phenanthrenes, which is catalyzed by $PtCl_2$, $AuCl_3$, $GaCl_3$, or $InCl_3$ (Scheme 4-6).[26] In the case of haloalkynes, a 1,2-halogen shift occurs with gold(I) chloride as catalyst; this leads to a gold vinylidene species that cyclizes to the corresponding phenanthrene derivative. This halogen shift was not observed when the cyclization was carried out with (stoichiometric amounts of) $InCl_3$.

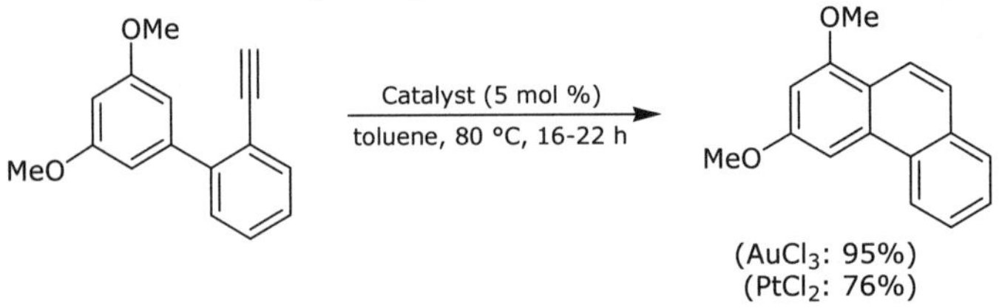

OMe, MeO — Catalyst (5 mol %), toluene, 80 °C, 16-22 h — OMe, MeO

($AuCl_3$: 95%)
($PtCl_2$: 76%)

Scheme 4-6

Representative procedure for the gold-catalyzed cyclization of (2-haloalkynyl)biphenyls. 9-Bromo-1,3-dimethylphenanthrene[26b]

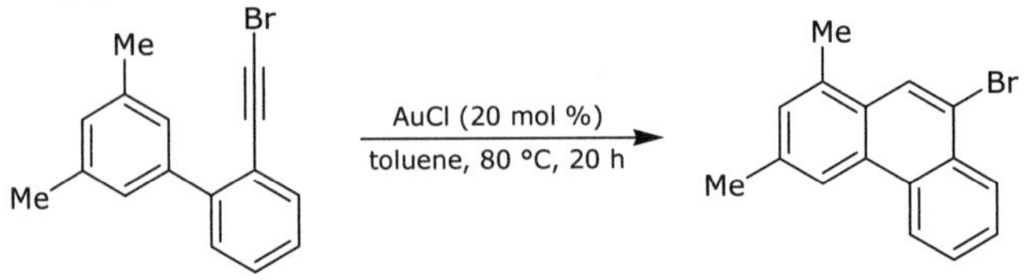

Me, Br, Me — AuCl (20 mol %), toluene, 80 °C, 20 h — Me, Br, Me

A solution of AuCl (4.9 mg, 0.021 mmol) and 2-bromoethynyl-3',5'-dimethylbiphenyl (30 mg, 0.105 mmol) in toluene (1 mL) was stirred at 80 °C for 20 h. The mixture was then adsorbed on silica gel and added on top of a silica gel column. Flash chromatography (hexanes/EtOAc 7:3) afforded 9-bromo-1,3-dimethylphenanthrene as a pale yellow solid (30 mg, quant.).

The intramolecular hydroarylation of aryl propargyl amines or ethers is also catalyzed by $PtCl_2$ or cationic gold(I) complexes.[27] The latter afforded better yields under milder conditions. In most cases, the 6-*endo*-dig cyclization to dihydroquinolines or 2*H*-chromenes prevails over the competing 5-*exo*-dig pathway (Scheme 4-7). Likewise, coumarins were obtained from aryl propiolates,[27b] a reaction that has been used as a fluorescence probe for gold(III) ions.[28] Quinoline and dihydroquinoline derivatives are also accessible from aromatic amines and alkynes by a gold-catalyzed tandem hydroamination/hydroarylation sequence.[29] Aryl homopropargyl amines undergo 6-*endo*-dig-cyclizations in the presence of bulky cationic gold(I) catalysts to afford mixtures of 4-methylenetetrahydroquinolines and 4-methyl-1,2-dihydroquinolines; treatment of the crude product mixtures with *p*-TsOH induced isomerization of the former to the thermodynamically preferred dihydroquinolines.[30] Tricyclic pyrroloquinolines were obtained by heating

N-arylpent-4-ynes or *N*-arylpent-4-ynamides with terminal alkynes in the presence of $AuBr_3$ and $AgSbF_6$.[31] These transformations involve the formation of two C–C bonds and one C–N bond in a one-pot procedure.

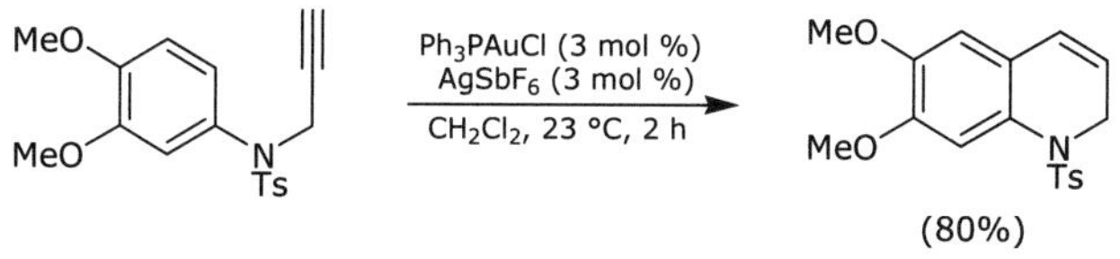

Scheme 4-7

Representative procedure for the gold-catalyzed cyclization of aryl propargyl ethers. 2H-Chromene[27b]

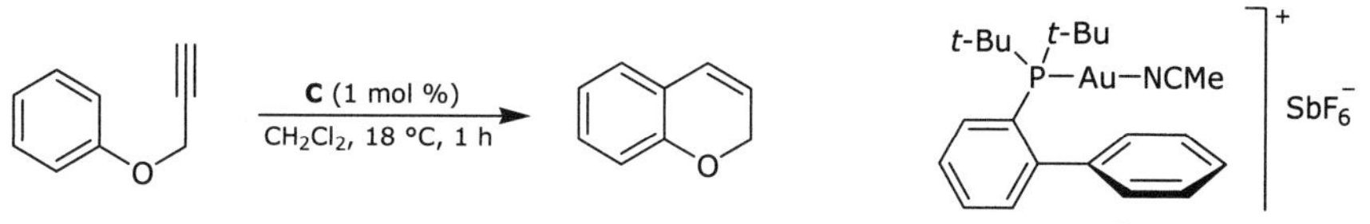

A magnetically stirred solution of phenyl propargyl ether (40 mg, 0.30 mmol) in CH_2Cl_2 (4 mL) maintained at 18 °C was treated with catalyst **C** (2.4 mg, 1 mol%). The resulting solution was stirred at 18 °C for 1 h and then concentrated under reduced pressure, and the ensuing yellow oil was subjected to flash column chromatography (silica gel, hexane) to give, after concentration of the appropriate fractions (R_f 0.3 in 1:49 v/v EtOAc/hexane), 2*H*-chromene (25 mg, 62%) as a clear, colorless oil.

Representative procedure for the gold-catalyzed cyclization of aryl homopropargyl amines. Diethyl 6-bromo-1,4-dimethyl-1H-quinoline-2,2-dicarboxylate[30]

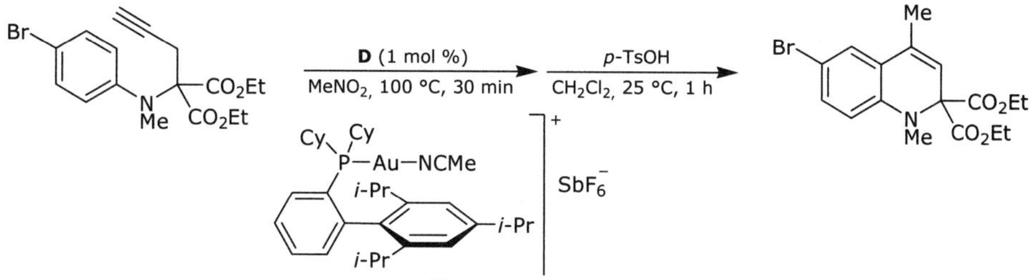

To a solution of 0.5 mmol (190 mg) of diethyl 2-[*N*-(4-bromophenyl)-*N*-methylamino]-2-(prop-2-ynyl)malonate in nitromethane (1 M) was added 5.0 μmol (3.8 mg) of catalyst **D**. The reaction mixture was heated at 100 °C. Upon complete consumption of the substrate (30 min), the mixture was concentrated under reduced pressure. The resulting crude material was dissolved in CH_2Cl_2 (0.1 M), and *p*-TsOH (0.05 equiv) was added to the solution. The reaction mixture was stirred at RT until complete isomerization of the 4-methylenetetrahydroquinoline to the corresponding dihydroquinoline (1 h). A saturated aqueous $NaHCO_3$ solution was then added to the mixture. The organic layer was separated and the aqueous layer was extracted with EtOAc (three times). The combined organic layers were dried over anhydrous $MgSO_4$ and concentrated under reduced pressure to afford 185 mg (97%) of diethyl 6-bromo-1,4-dimethyl-1*H*-quinoline-2,2-dicarboxylate. If necessary, a rapid purification by flash column chromatography (SiO_2)

can be performed to remove impurities. (Note: Most dihydroquinolines were found to be unstable on silica.)

Electron-rich heteroarenes can be used as nucleophiles in gold-catalyzed hydroarylations as well. Thus, *N*-propargylindole-2-carboxamides cyclize to β-carbolines in the presence of gold(III) chloride.[32] This 6-*exo*-dig cyclization with bond formation between C-3 of the indole and the internal carbon atom of the triple bond takes place when the amide nitrogen bears a substituent other than hydrogen. By contrast, unsubstituted amides undergo attack of the carbonyl oxygen at the activated triple bond, which leads to the formation of oxazoles.[32, 33]

Representative procedure for the gold-catalyzed cyclization of N-propargylindole-2-carboxamides. 4,9-Dimethyl-2-tosyl-2H-pyrido[3,4-b]indol-1(9H)-one[32]

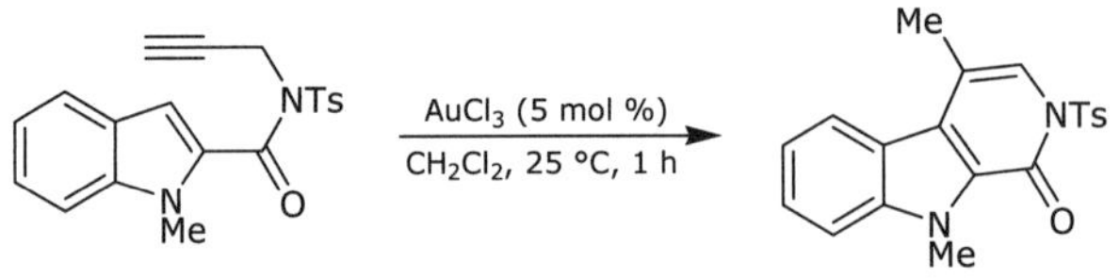

A 0.08-g (0.22 mmol) sample of 1-methyl-*N*-(prop-2-ynyl)-*N*-tosyl-1*H*-indole-2-carboxamide was stirred with $AuCl_3$ (3 mg) in 5 mL of CH_2Cl_2 at RT for 1 h. At the end of this time, the mixture was filtered through a plug of silica gel washed with triethylamine and rinsed with EtOAc, and the filtrate was concentrated under reduced pressure. The crude residue was subjected to flash silica gel chromatography to give 0.07 g (85%) of 4,9-dimethyl-2-tosyl-2*H*-pyrido[3,4-*b*]indol-1(9*H*)-one as an off-white solid.

Whereas other indoles tethered to an alkyne by two or three atoms also undergo six-membered ring annulation in the presence of gold catalysts, rare 7-*exo*- or 8-*endo*-cyclizations can be realized with their counterparts bearing a 4-atom-chain between the reactive subunits.[33] Interestingly, the regioselectivity of these intramolecular hydroarylations can be controlled by the oxidation state of the gold catalyst (Scheme 4-8). Thus, treatment of indoloalkynes with cationic gold(I) catalysts of the type **C** afforded predominantly or exclusively azepino[4,5-*b*]indoles formed by attack of the indole at the internal carbon atom of the activated triple bond. In contrast, 8-*endo*-selective cyclization to indoloazocines prevailed with gold(III) chloride as catalyst. At longer reaction times, allenes or tetracyclic annulated derivatives were also formed as a result of a fragmentation reaction. The method has been applied to the synthesis of the 1*H*-azocino[5,4-*b*]indole skeleton of the lundurine alkaloids.[34]

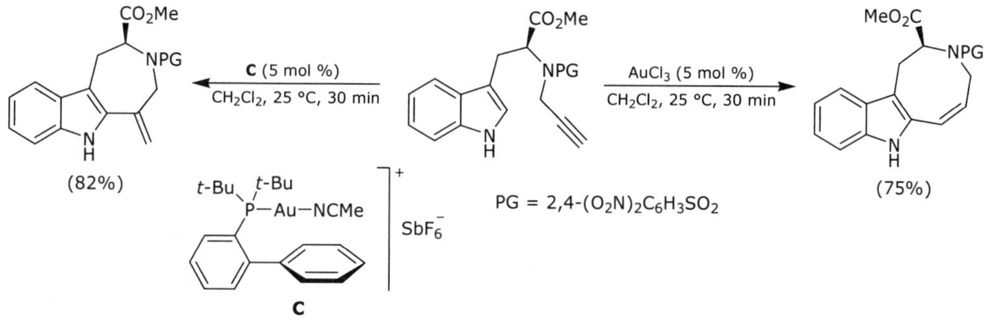

Scheme 4-8

The intermolecular hydroarylation of indoles or pyrroles with alkynes in the presence of cationic gold(I) catalysts leads to 2:1-adducts.[33b] With (*Z*)-2-en-4-yn-1-ols as an unsaturated reaction partner, indoles are converted into dihydrocyclohepta[*b*]indoles in a one-pot reaction sequence that involves a gold-catalyzed intermolecular Friedel-Crafts alkylation followed by an intramolecular hydroarylation (Scheme 4-9).[35]

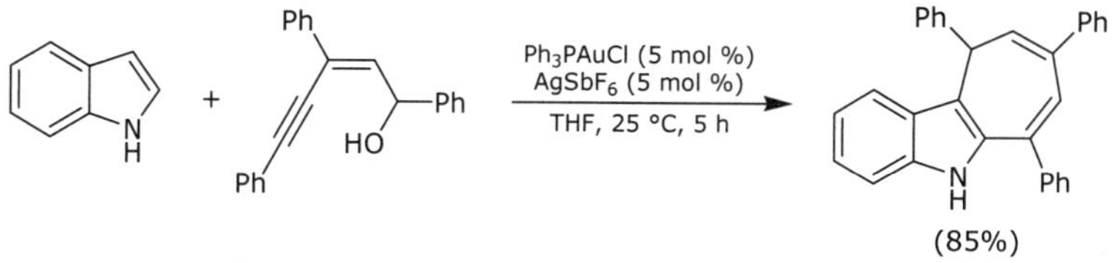

Scheme 4-9

Furans bearing a tethered triple bond in the 2-position afford substituted phenols in the presence of various types of gold catalysts.[36] This transformation is initiated by intramolecular nucleophilic attack of the furan at the activated alkyne, which gives a cyclopropyl gold carbene intermediate. Subsequent cleavage of a C–C and a C–O bond furnish a second gold carbene that cyclizes to an oxepine. This is in equilibrium with the corresponding arene oxide that finally rearranges to the phenol. For furans with a propargyl alcohol side chain, the initial hydroarylation is followed by dehydration, which furnishes benzofurans.[37] These heterocycles are also formed from substrates with two alkynes at opposite sides of the furan ring,[38] whereas furans with two diastereotopic alkynyl groups in the side chain react to diastereomerically pure annulated phenols.[39] Enantiomerically enriched substrates with a center of chirality in the tether undergo phenol formation without racemization.[40] In contrast to the intramolecular phenol synthesis, the intermolecular gold-catalyzed reaction of 2,5-dimethylfuran with phenylacetylene proceeds only sluggishly to afford a mixture of 3,6-dimethyl-2-phenylphenol and 2,5-dimethyl-3-(1-phenylethen-1-yl)furan.[41]

Representative procedure for the gold-catalyzed synthesis of phenols from alkynylfurans. 5-(4-Methoxy-2,6-dimethylphenyl)-2-tosylisoindolin-4-ol[36i]

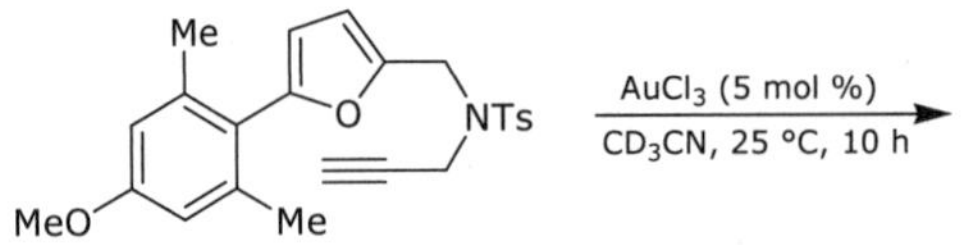

$AuCl_3$ (5 mol %), CD_3CN, 25 °C, 10 h

In a nuclear magnetic resonance (NMR) tube, 58.0 mg (137 μmol) of *N*-((5-(4-methoxy-2,6-dimethylphenyl)furan-2-yl)methyl)-*N*-tosylprop-2-yn-1-amine was dissolved in 600 μL of CD_3CN. Then, 41.5 mg (6.84 μmol) of gold(III) chloride (5 wt% solution in acetonitrile) was added and the reaction was monitored by ^1H NMR. After completion (10 h), the solvent was removed under reduced pressure and the product was purified by column chromatography on silica gel (hexanes/EtOAc, 5:1) to furnish 51.0 mg (88%) of 5-(4-methoxy-2,6-dimethylphenyl)-2-tosylisoindolin-4-ol.

Representative procedure for the gold-catalyzed synthesis of benzofurans from alkynylfurans. 2-Methyl-6-phenylbenzofuran [37]

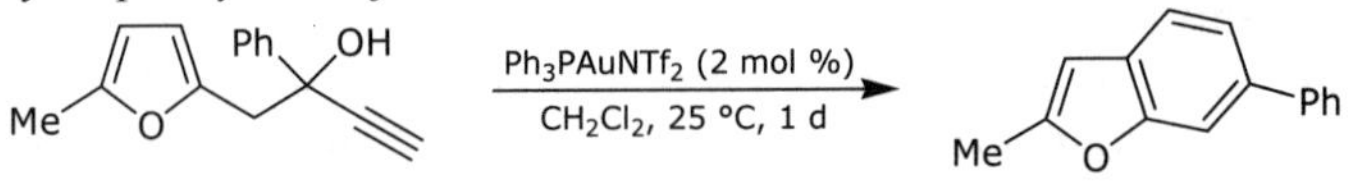

$Ph_3PAuNTf_2$ (2 mol %), CH_2Cl_2, 25 °C, 1 d

To a solution of 1-(5-methylfuran-2-yl)-2-phenylbut-3-yn-2-ol (340 mg, 1.50 mmol) in CH_2Cl_2 (10 mL) was added $Ph_3PAuNTf_2$ (22.2 mg, 30.0 μmol, 2 mol%), and the mixture was stirred for 1 day at RT. Then, the solvent was removed *in vacuo*. Purification of the residue by column chromatography (silica gel, petrol ether/EtOAc, 10:1) afforded 2-methyl-6-phenylbenzofuran (219 mg, 1.05 mmol, 64%) as a colorless oil.

3.2 Nucleophilic Additions to Allenes

The first gold-catalyzed addition reactions of carbon nucleophiles to allenes were only first disclosed in 2006, and the number of examples is still small. Toste and co-workers[42] showed that allenic silyl enol ethers undergo a 5-*endo*-trig cyclization to hexahydroindenone derivatives in the presence of a cationic gold catalyst (Scheme 4-10). In these transformations, water or methanol is used as an external proton source for protodeauration of an intermediate vinylgold species. In an analogous manner, cyclopentenes were obtained in good yields from allenic β-ketoesters.[43] In the presence of a palladium catalyst and an allylic halide, these substrates afford functionalized 2,3-dihydrofurans.

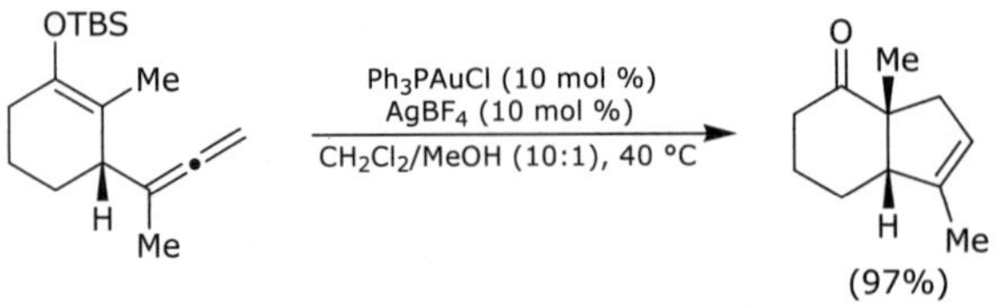

Ph_3PAuCl (10 mol %), $AgBF_4$ (10 mol %), CH_2Cl_2/MeOH (10:1), 40 °C; (97%)

Scheme 4-10

Representative procedure for the gold-catalyzed cyclization of allenic β-ketoesters. 4-Acetyl-4-(ethoxycarbonyl)cyclopentene[43]

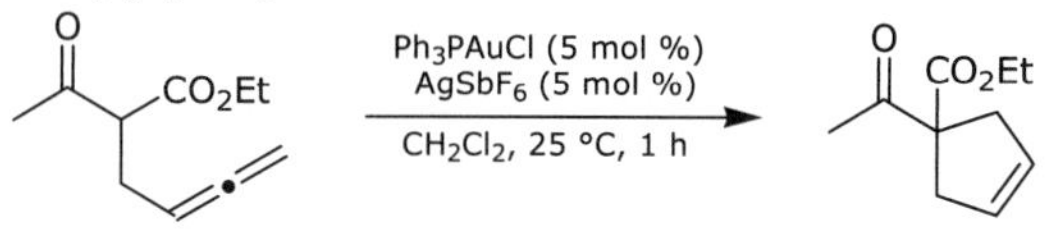

Ph_3PAuCl (5 mol %), $AgSbF_6$ (5 mol %), CH_2Cl_2, 25 °C, 1 h

A mixture of ethyl 2-acetylhexa-4,5-dienoate (47 mg, 0.26 mmol), Ph_3PAuCl (6 mg, 0.0121 mmol), and $AgSbF_6$ (3 mg, 0.0117 mmol) in CH_2Cl_2 (2 mL) was stirred under Ar in a flame-dried Schlenk tube at RT for 1 h. After the reaction was complete (monitored by TLC, petroleum ether/EtOAc 10:1), rotary evaporation and flash chromatography on silica gel (petroleum ether/Et_2O 20:1) afforded 4-acetyl-4-(ethoxycarbonyl)cyclopentene (32 mg, 68%) as a liquid.

The cyclization of malonate-derived allenenes in the presence of (*R*)-xylyl-BINAP$(AuCl)_2$ and silver triflate afforded a mixture of isomeric vinylcyclohexenes (Scheme 4-11).[44] The formation of these products is rationalized by nucleophilic attack of the olefinic double bond at the activated allene, followed by deprotonation

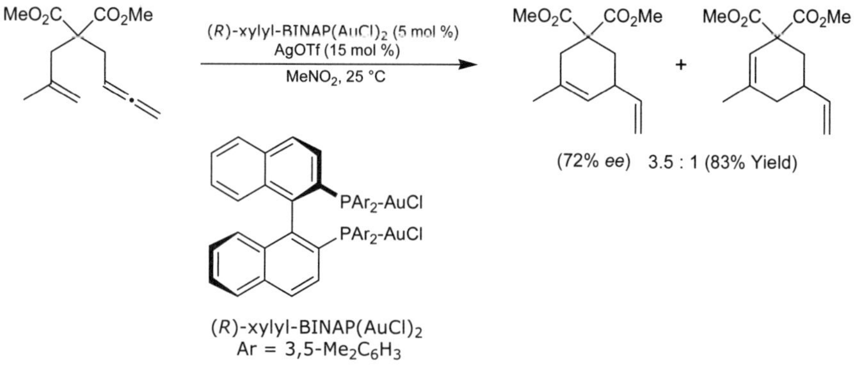

(*R*)-xylyl-BINAP$(AuCl)_2$ (5 mol %), AgOTf (15 mol %), $MeNO_2$, 25 °C

(72% *ee*) 3.5 : 1 (83% Yield)

(*R*)-xylyl-BINAP$(AuCl)_2$
Ar = 3,5-$Me_2C_6H_3$

Scheme 4-11

and proto-deauration of the resulting cyclohexyl cation. Related allenynes afford cyclic ketones by heating with a cationic gold catalyst in wet 1,4-dioxane.[45] This hydrative cyclization probably proceeds *via* 5-*endo*-attack of the triple bond at the activated allene. The resulting vinyl cation is then trapped by water, and subsequent protodeauration closes the catalytic cycle.

Representative procedure for the gold-catalyzed hydrative cyclization of allenynes. (E)-(2-Methyl-2-styrylcyclopentyl)(phenyl)methanone[45]

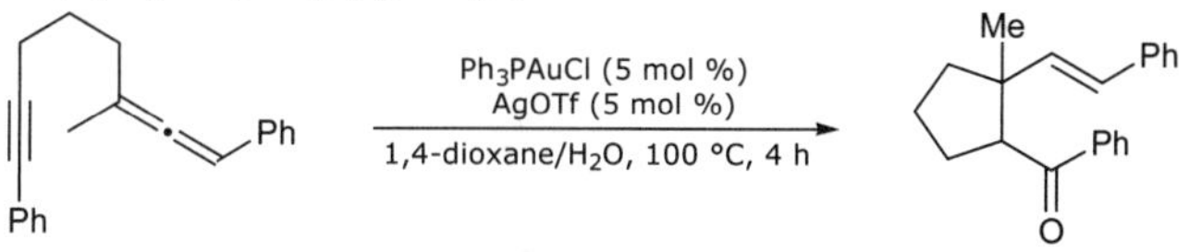

Ph_3PAuCl (5 mol %), AgOTf (5 mol %), 1,4-dioxane/H_2O, 100 °C, 4 h

A solution of $Ph_3PAuOTf$ (5 mol%) was prepared by mixing Ph_3PAuCl (9.1 mg, 0.018 mmol) and AgOTf (4.6 mg, 0.018 mmol) in 1,4-dioxane (1.9 mL). To this solution was added 3-methyl-1,8-diphenylocta-1,2-dien-7-yne (100 mg, 0.37 mmol) at 100 °C, and then the mixture was stirred for 4 h. The resulting solution was filtered through a Celite (Sigma-Aldrich, St. Louis, MO) bed and eluted through a silica gel column

(EtOAc/hexane 1/15) to give (*E*)-(2-methyl-2-styrylcyclopentyl)(phenyl)methanone (91 mg, 0.28 mmol, 85%) as a yellow oil.

Most reports on gold-catalyzed intramolecular C–C bond formation of allenes take advantage of electron-rich aromatics or heteroaromatics as nucleophiles. This hydroarylation has been used in natural product synthesis even before the method was studied extensively. Thus, Nelson and co-workers[46] used a cationic gold catalyst to activate an allene for nucleophilic attack of a tethered pyrrole ring, which delivered a precursor of the alkaloid (–)-rhazinilam with high yield and excellent chirality transfer (Scheme 4-12). It seems reasonable to assume that coordination of the Lewis-acidic gold catalyst to the carbonyl group is key to the high diastereoselectivity. In contrast, palladium or silver catalysts either failed to deliver the desired cyclization product or gave poor stereoselectivities.

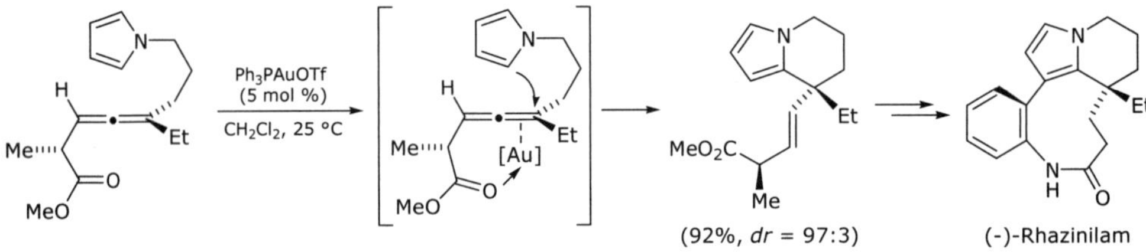

$Ph_3PAuOTf$ (5 mol %), CH_2Cl_2, 25 °C

(92%, *dr* = 97:3)

(-)-Rhazinilam

Scheme 4-12

Representative procedure for the intramolecular gold-catalyzed hydroarylation of allenes. (R,E)-Methyl 4-[(R)-8-ethyl-5,6,7,8-tetrahydroindolizin-8-yl]-2-methylbut-3-enoate[46]

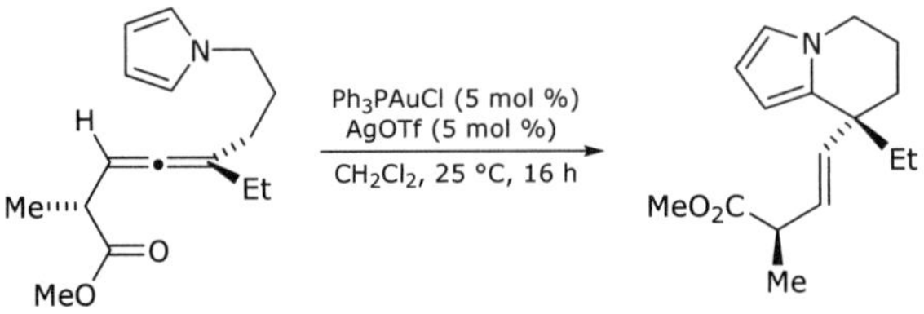

Ph_3PAuCl (5 mol %), AgOTf (5 mol %), CH_2Cl_2, 25 °C, 16 h

To a solution of (2*R*)-methyl 5-ethyl-2-methyl-8-(*1H*-pyrrol-1-yl)octa-3,4-dienoate (1.50 g, 5.75 mmol) in CH_2Cl_2 (480 mL) was added AgOTf (74 mg, 0.29 mmol) and Ph_3PAuCl (142 mg, 0.29 mmol) under N_2. The mixture was stirred at RT for 16 h, quenched with saturated NH_4Cl solution (50 mL), and extracted with EtOAc (3 × 200 mL). The combined organic extracts were washed successively with H_2O and brine, dried (Na_2SO_4), and evaporated. The residue was purified by flash chromatography on silica gel (hexanes/EtOAc, 10:1) to afford (*R,E*)-Methyl 4-[(*R*)-8-ethyl-5,6,7,8-tetrahydroindolizin-8-yl]-2-methylbut-3-enoate as a colorless oil (1.39 g, 92%).

Other electron-rich heteroaromatics like indoles can be used as nucleophiles in gold-catalyzed hydroarylations as well. Widenhoefer et al.[47] reported on the first examples for the formation of tetrahydrocarbazoles and related heterocycles from allenyl indoles. Starting from achiral allenes, the hydroarylation products were obtained with high enantioselectivity by using chiral gold pre-catalyst [Au_2{(*S*)-3,5-*t*-Bu-4-MeO-MeOBIPHEP}Cl_2] together with $AgBF_4$ (Scheme 4-13).[48] Recently, *endo*-selective gold-catalyzed cycloisomerizations of *N*-(2,3-butadienyl)-substituted indole derivatives,[49] as well as intermolecular hydroarylations of allenes with indoles[50] were also described.

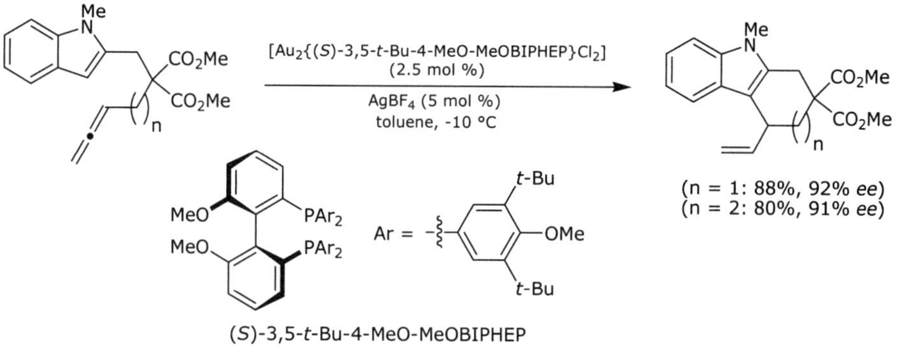

[Au_2\{(*S*)-3,5-*t*-Bu-4-MeO-MeOBIPHEP\}Cl_2]
(2.5 mol %)
$AgBF_4$ (5 mol %)
toluene, -10 °C

(n = 1: 88%, 92% *ee*)
(n = 2: 80%, 91% *ee*)

Ar = 3,5-*t*-Bu-4-OMe-phenyl

(*S*)-3,5-*t*-Bu-4-MeO-MeOBIPHEP

Scheme 4-13

Electron-rich phenyl rings are also suitable nucleophiles for the intramolecular gold-catalyzed hydroarylation of allenes. Thus, dihydroquinoline and chromene derivatives are accessible by treating allenic anilines or allenic arylethers with a cationic gold catalyst (Scheme 4-14).[51] Because of the limited stability of

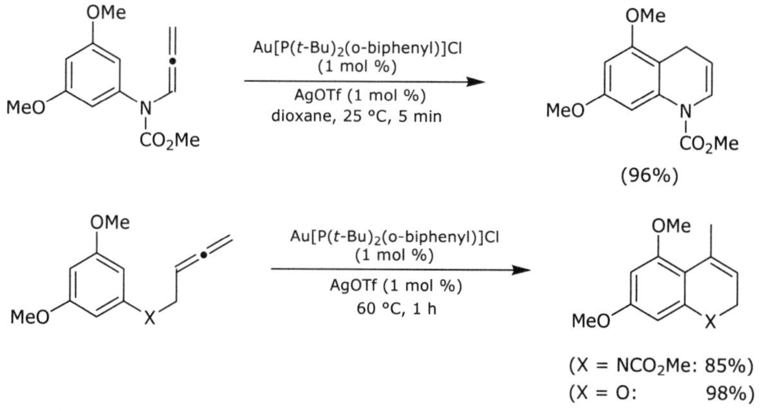

Au[P(*t*-Bu)$_2$(o-biphenyl)]Cl
(1 mol %)
AgOTf (1 mol %)
dioxane, 25 °C, 5 min

(96%)

Au[P(*t*-Bu)$_2$(o-biphenyl)]Cl
(1 mol %)
AgOTf (1 mol %)
60 °C, 1 h

(X = NCO_2Me: 85%)
(X = O: 98%)

Scheme 4-14

dihydroquinolines, they were immediately hydrogenated to the corresponding tetrahydroquinolines. Depending on the structure of the substrate, the C–C bond is formed at the terminal or central allenic carbon atom. Platinum dichloride also induces the cyclization of allenic anilines, albeit at a much slower rate.

This method was applied by Tarselli and Gagné[52] to arylallenes bearing an extended all-carbon linker between the reactive sites. In this case, a 6-*exo*-trig cyclization was induced by a mixture of triphenylphosphite gold(I) chloride (3 mol%) and silver hexafluoroantimonate (5 mol%), giving the corresponding tetralin derivative in high yield. Extensive mechanistic studies have revealed a diaurated species as a catalyst resting state that is activated by the silver salt.[53]

Representative procedure for the gold-catalyzed cyclization of arylallenes. Dimethyl 3,4-dihydro-5,7-dimethoxy-4-vinylnaphthalene-2,2(1H)-dicarboxylate [52a]

MeO_2C, CO_2Me, MeO, OMe; $(PhO)_3PAuCl$ (10 mol %), $AgSbF_6$ (14 mol %), CH_2Cl_2, 25 °C, 6 h

In a 5-mL vial charged with a stirring bar, to $(PhO)_3PAuCl$ (27.2 mg, 0.05 mmol, 0.1 equiv), and $AgSbF_6$ (24.0 mg, 0.07 mmol, 0.14 equiv) was added CH_2Cl_2 (1.0 mL) by syringe, at which point a white-gray suspension formed. After 2 min, dimethyl 2-(3,5-dimethoxybenzyl)-2-(buta-2,3-dienyl)malonate (168 mg, 0.5 mmol, 1.0 equiv) was added by pipette. The suspension turned deep green within 20 min. After 6 h, the reaction mixture was loaded directly onto a silica flash column and purified with 1:7 EtOAc/hexanes. Yield: 85% of dimethyl 3,4-dihydro-5,7-dimethoxy-4-vinylnaphthalene-2,2(1*H*)-dicarboxylate as a clear oil.

Intermolecular hydroarylations of terminal allenes with highly nucleophilic methoxyarenes are catalyzed by cationic phosphite gold(I) complexes.[54] The addition takes place regioselectively at the unsubstituted allene terminus and affords (*E*)-allylated benzene derivatives with moderate-to-high yields.

Representative procedure for the intermolecular gold-catalyzed hydroarylation of allenes. 1,3,5-Trimethoxy-2-(3-methylbut-2-enyl)benzene[54a]

OMe, MeO, OMe, +; $(4\text{-}ClC_6H_4O)_3PAuCl$ (5 mol %), $AgBF_4$ (5 mol %), CH_2Cl_2, 25 °C, 4 h

To a 5-mL vial charged with a stirring bar were added $(4\text{-}ClC_6H_4O)_3PAuCl$ (9.6 mg, 15 μmol) and $AgBF_4$ (3.0 mg, 15 μmol), and CH_2Cl_2 (1.0 mL) by syringe, resulting in a light gray suspension. After stirring for 2 min, 1,3,5-trimethoxybenzene (100 mg, 0.6 mmol) was added, resulting in a color change to light orange. After stirring for additional 2 min, 3-methylbuta-1,2-diene (20.0 mg, 0.3 mmol) was added dropwise by microsyringe. Stirring was continued until gas chromatography (GC)/TLC (product R_f 0.5 in 1:7 EtOAc/hexanes) analysis indicated complete consumption of the allene (4 h). The reaction mixture was concentrated, loaded directly onto a silica flash column, and eluted with 1:10 to 1:8 EtOAc/hexanes to yield 1,3,5-trimethoxy-2-(3-methylbut-2-enyl)benzene (67% yield) as a clear oil.

3.3 Nucleophilic Additions to Alkenes

Several examples of gold-catalyzed addition reactions of carbon nucleophiles to unactivated alkenes have been reported in recent years. Yao and Li obtained adducts with high regioselectivity from β-diketones and various alkenes (styrene derivatives, conjugated dienes, enol ethers) in the presence of cationic gold species prepared *in situ* from $AuCl_3$ and AgOTf (Scheme 4-15).[55] This hydroalkylation is

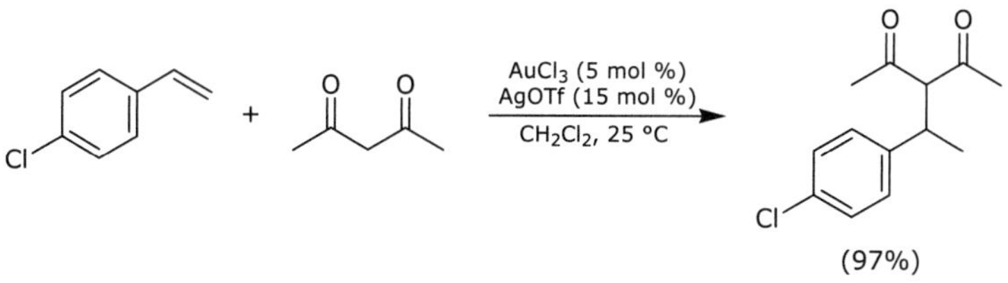

$AuCl_3$ (5 mol %)
AgOTf (15 mol %)
CH_2Cl_2, 25 °C
(97%)

Scheme 4-15

also catalyzed by other transition metals, including palladium and silver, but harsher conditions are often required. For example, with silver triflate as catalyst, the reactants have to be heated to 80–100 °C,[56] whereas the combination $AuCl_3$/AgOTf allows the addition to be carried out at room temperature.

Representative procedure for the gold-catalyzed hydroalkylation of alkenes. 2-Cyclooct-2-enyl-1,3-diphenylpropane-1,3-dione[55b]

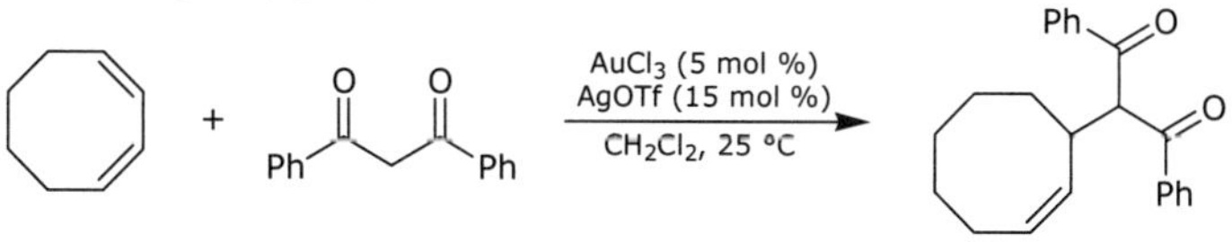

$AuCl_3$ (5 mol %)
AgOTf (15 mol %)
CH_2Cl_2, 25 °C

A solution of $AuCl_3$ (7.6 mg, 0.025 mmol) and AgOTf (19.3 mg, 0.075 mmol) was stirred in CH_2Cl_2 (2 mL) at RT for 2 h. Dibenzoylmethane (112 mg, 0.5 mmol) was then added, which was followed by the addition of cycloocta-1,3-diene (128 µL, 1 mmol) *via* a syringe pump over 5 h, and the resulting solution was stirred overnight. The solvent was removed under reduced pressure, and 2-cyclooct-2-enyl-1,3-diphenylpropane-1,3-dione (65% yield) was isolated *via* column chromatography on silica gel (gradient eluent: hexane: EtOAc = 60:1 to 2:1).

β-Ketoamides[57] and simple ketones[58] can also be used as nucleophiles in the corresponding intramolecular additions, which afforded lactams and cyclic ketones, respectively, with good yields and *trans*-selectivity. Highly active cationic gold(I) catalysts formed *in situ* from Au[P(*t*-Bu)$_2$(*o*-biphenyl)]Cl or IPrAuCl and silver salts were used for these transformations (Scheme 4-16).

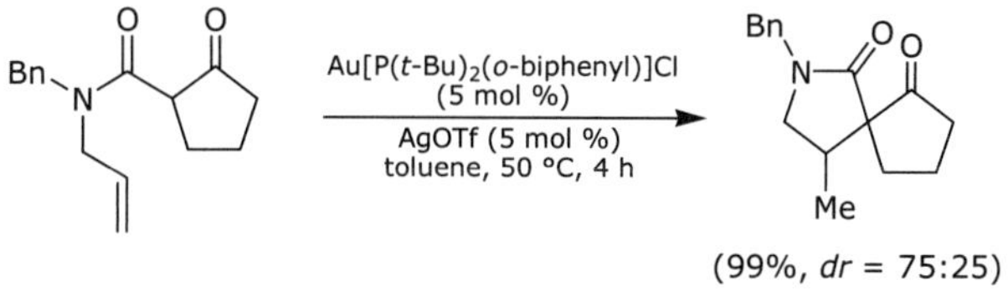

Au[P(*t*-Bu)$_2$(*o*-biphenyl)]Cl (5 mol %)
AgOTf (5 mol %)
toluene, 50 °C, 4 h
(99%, *dr* = 75:25)

Scheme 4-16

Representative procedure for the gold-catalyzed intramolecular hydroalkylation of unsaturated ketones. Dimethyl 3-acetyl-4-methylcyclopentane-1,1-dicarboxylate[58]

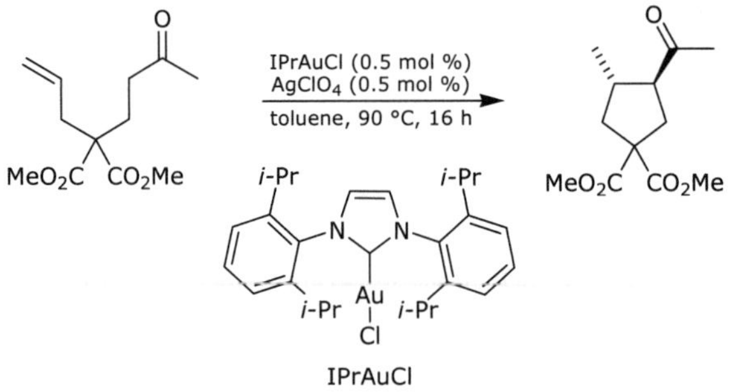

IPrAuCl (0.5 mol %)
$AgClO_4$ (0.5 mol %)
toluene, 90 °C, 16 h

MeO_2C CO_2Me

IPrAuCl

Warning! The perchlorate salt is potentially explosive and should be handled with great caution.

A mixture of IPrAuCl (31.1 mg, 0.05 mmol), $AgClO_4$ (10.4 mg, 0.05 mmol), and dimethyl 2-allyl-2-(3-oxobutyl)malonate (2.45 g, 10 mmol) in toluene (5 mL) was stirred at 90 °C under Ar atmosphere, and the reaction was monitored by TLC. Upon completion (16 h), the solvent was removed under reduced pressure, and then the residue was purified by silica gel column chromatography (eluent: EtOAc/petroleum ether = 1:10) to give dimethyl 3-acetyl-4-methylcyclopentane-1,1-dicarboxylate in 97% yield (2.38 g, 9.7 mmol, *trans*:*cis* = 90:10).

In a similar fashion, indoles can be added to styrenes and other olefins in the presence of Ph_3PAuCl and AgOTf (Scheme 4-17).[59] Heating to 70–85 °C in toluene was required for styrene derivatives and conjugated dienes, whereas hydroarylations of simple olefins were carried out at 130–140 °C under microwave irradiation. Under these conditions, allylbenzenes suffer double-bond migration to the corresponding styrene derivatives that then undergo the gold-catalyzed hydroarylation. Moreover, *E*/*Z*-mixtures of 1-arylbuta-1,3-dienes were isomerized to the *Z*-diene under the influence of the gold catalyst.

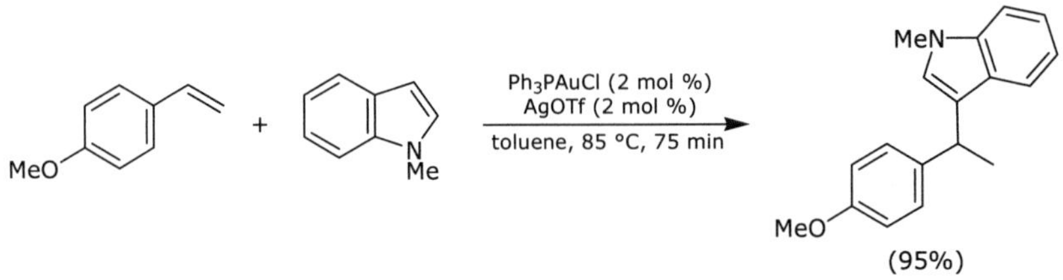

Ph_3PAuCl (2 mol %)
AgOTf (2 mol %)
toluene, 85 °C, 75 min

(95%)

Scheme 4-17

Gold(III) chloride and silver hexafluoroantimonate can also be used as catalyst for addition reactions of electron-rich arenes and heteroarenes to olefins.[60] In a similar fashion, the intramolecular hydroarylation of aryl homoallyl ethers and related substrates takes place upon heating with $AuCl_3$ and AgOTf in dichloroethane to 80 °C and affords dihydrobenzopyrans, tetrahydroquinolines, and tetralins with good yield (Scheme 4-18).[61]

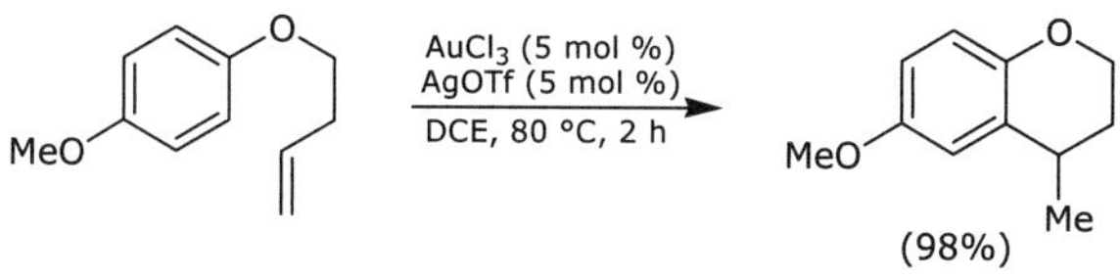

MeO, O; $AuCl_3$ (5 mol %), AgOTf (5 mol %), DCE, 80 °C, 2 h; MeO, O, Me (98%)

Scheme 4-18

The same catalyst was used by Li and co-workers for an efficient annulation of electron-rich phenols or naphthols with cyclic dienes.[62] Tricyclic or tetracyclic benzofuran derivatives were obtained as mixtures of *syn* and *anti* isomers with the former predominating. The reaction probably proceeds by intermolecular C–C bond formation *via* addition of the phenol to the activated diene, followed by protodeauration and intramolecular formation of the C–O bond. In a related transformation, naphthalenes were obtained by treatment of aryl-substituted propargyl esters with cationic gold(I) catalysts.[63] The reaction may proceed *via* consecutive 1,3- and 1,2-rearrangements to afford conjugated dienes that then undergo an intramolecular hydroarylation.

Representative procedure for the gold-catalyzed annulation of phenols with dienes. 1,2,3,4,4a,9b-Hexahydrodibenzofuran[62]

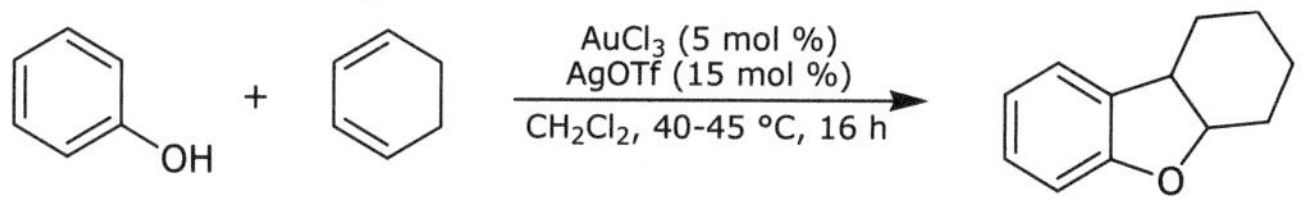

OH + ; $AuCl_3$ (5 mol %), AgOTf (15 mol %), CH_2Cl_2, 40-45 °C, 16 h; O

A solution of $AuCl_3$ (7.6 mg, 0.025 mmol) and AgOTf (19.3 mg, 0.075 mmol) was stirred in dry CH_2Cl_2 (1.5 mL) at RT for 2 h. Phenol (94 mg, 0.5 mmol) was then added, which was followed by the dropwise (ca. 1 drop/s) addition of cyclohexa-1,3-diene (124 μL, 1.0 mmol diluted in 0.5 mL of dry CH_2Cl_2), and the resulting solution was stirred overnight at 40–45 °C. The resulting solution was filtered through a cotton plug, and the solvent was removed under reduced pressure to afford a dark oil. The internal standard (nitromethane) was added, and a crude ^1H NMR was recorded. The product was isolated *via* column chromatography on silica gel (gradient eluent: hexane: CH_2Cl_2 = 40:1 to 2:1) to afford 74% of 1,2,3,4,4a,9b-hexahydrodibenzofuran (*syn:anti* = 75:25) as a colorless oil.

Not only simple, unactivated olefins but also α,β-unsaturated carbonyl compounds undergo gold-catalyzed addition reactions. Electron-rich arenes or heteroarenes (*e.g.*, indoles) are suitable carbon nucleophiles for these Michael additions. For the 1,4-addition of indoles to enones, enals, and related Michael acceptors, simple gold(III) catalysts like $NaAuCl_4 \cdot 2\,H_2O$[64] or $AuCl_3$[24a] can be used (Scheme 4-19). In the case of unsubstituted indoles or 7-azaindoles,[65] C–C bond

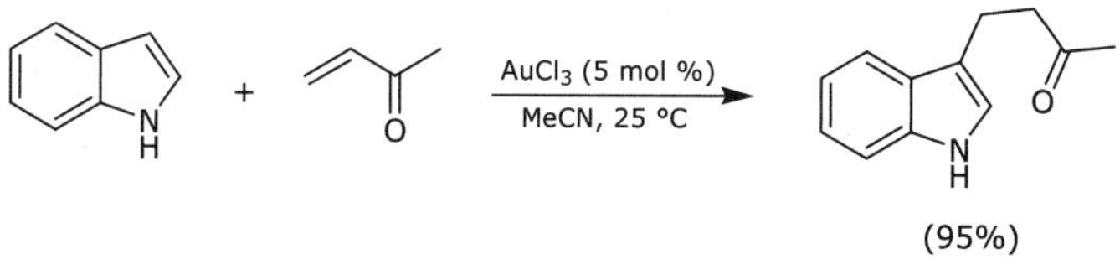

N H + O; $AuCl_3$ (5 mol %), MeCN, 25 °C; O, N H (95%)

Scheme 4-19

formation occurs in the 3-position,whereas 3-methylindole reacts at C-2. Because of the lower reactivity of 7-azaindoles, the reaction mixture has to be heated to 100–140 °C. It is not clear whether C–H activation of the indole by the gold catalyst is involved in these transformations.

General procedure for the gold-catalyzed 1,4-addition of indoles to enones[64]

To a 1:1-mol ratio solution of indole and enone in EtOH is added $NaAuCl_4{\cdot}2\,H_2O$ (5 mol%). The resulting mixture is allowed to react with stirring at RT or at 30 °C while being monitored by TLC or GC-mass spectroscopy (MS). After completion, the solvent is removed by evaporation under reduced pressure. To the residue, acetone (few mL) is added to precipitate the catalyst, which is separated by filtration. The filtrate is concentrated and the crude products are purified by chromatography on silica gel (230–400 mesh) eluting with *n*-hexane/EtOAc mixtures.

General procedure for the gold-catalyzed 1,4-addition of 7-azaindoles to enones[65]

Enone (2.52 mmol) and $NaAuCl_4{\cdot}2\,H_2O$ (0.042 mmol) are added to a screw-top vial (60 × 18 mm) with a solid-top cap charged with a solution of 7-azaindole (0.850 mmol) in absolute ethanol (2 mL). The resulting mixture is then heated while stirring at 100 °C (for more reactive enones) or 140 °C. The reaction is monitored by TLC and GC-MS. After cooling, the mixture is filtered to remove the catalyst and the solvent is concentrated under reduced pressure. The residue is directly purified by flash chromatography (silica gel, *n*-hexane/EtOAc) to give the product.

Michael additions of furans, azulene derivatives, and electron-rich benzenes to enones are also catalyzed by gold(III) salts (Scheme 4-20).[24a, 66] Compared with the corresponding transformations in the presence of Brønsted acids, the gold-catalyzed 1,4-additions are often much faster, which may indicate the formation of arylgold intermediates by auration of the arene.

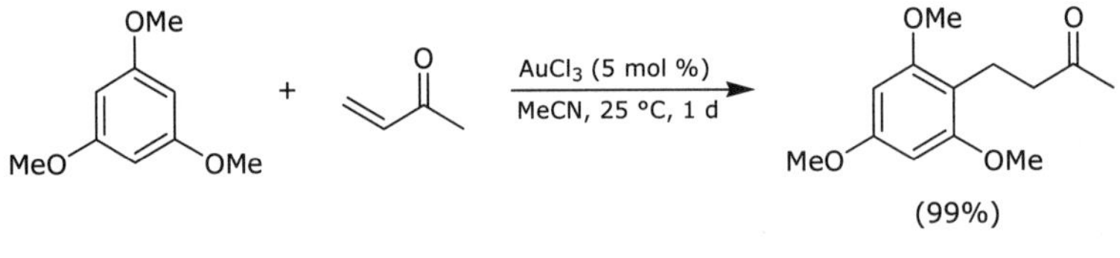

OMe, MeO, OMe + O; $AuCl_3$ (5 mol %), MeCN, 25 °C, 1 d; OMe, O, MeO, OMe (99%)

Scheme 4-20

Representative procedure for the gold-catalyzed 1,4-addition of furans to enones. 3-(5-Methylfuran-2-yl)cyclopentanone[66b]

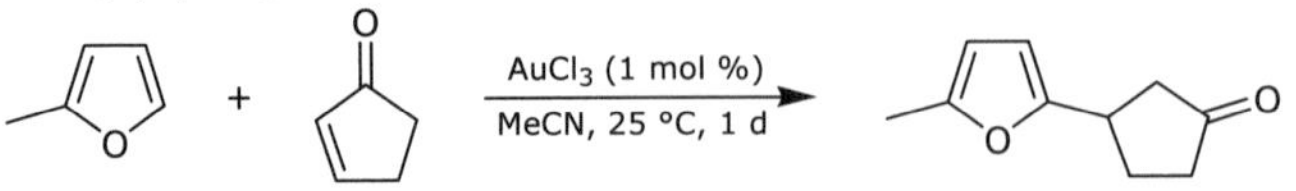

$AuCl_3$ (1 mol %), MeCN, 25 °C, 1 d

To 1.18 g (14.4 mmol) of 2-methylfuran and 1.18 g (14.4 mmol) of cyclopent-2-enone in 10 mL of acetonitrile were added 43.7 mg (144 μmol, 1 mol%) of $AuCl_3$. The reaction was monitored by withdrawing small aliquots, which were analyzed by ^1H NMR. After 1 day, the solvent was removed under vacuum and the crude material was purified by column chromatography on silica gel using hexanes/EtOAc (4:1) to afford 3-(5-methylfuran-2-yl)cyclopentanone; yield: 2.05 g (87%).

Representative procedure for the gold-catalyzed 1,4-addition of azulenes to enones. 1,4-Dimethyl-7-(1-methylethyl)-3-(3-oxobutyl)-azulene[66b]

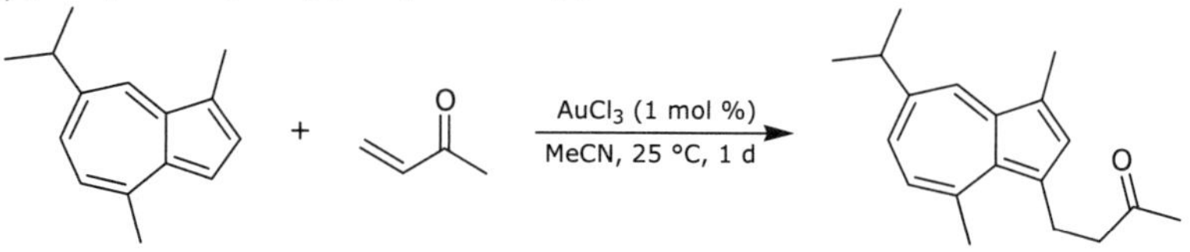

A mixture of 140 mg (2.00 mmol) of methyl vinyl ketone, 950 mg of acetonitrile, and 3.00 mg (10 μmol, 1 mol%) of $AuCl_3$ was slowly added under argon to a vigorously stirred solution of 198 mg (1.00 mmol) of 1,4-dimethyl-7-(1-methylethyl)-azulene in 2 mL of acetonitrile. After 1 day at RT, the reaction mixture was filtered through a small pad of silica (2 g) with an additional 5 mL of acetonitrile as eluent. The solvent was removed under vacuum at 50 °C (from 15 mbar to 0.2 mbar) and the residue purified by flash chromatography on silica gel (petroleum ether/*t*-butyl methyl ether, 2:1; $R_f = 0.50$). Recrystallization from CH_2Cl_2/Et_2O afforded 1,4-dimethyl-7-(1-methylethyl)-3-(3-oxobutyl)-azulene as dark blue crystals; yield: 148 mg (55%), mp 68 °C.

3.4 Cycloisomerizations

The substrates of choice for C–C bond formation by gold-catalyzed cycloisomerization are 1,5-, 1,6-, or 1,7-enynes. Selective activation of the triple bond for an intramolecular attack of the pendant alkene affords cyclopropyl gold carbenes as key intermediates that (in the absence of an external nucleophile) undergo skeletal rearrangement.[67] Thus, conjugated dienes were obtained by treatment of 1,6-enynes with cationic gold catalysts formed *in situ* from Ph_3PAuCl and silver salts (or with stable cationic complexes of the type **C**; Scheme 4-21).[68] Under these conditions, the cycloisomerization occurs even at –60 °C, whereas prolonged heating to 90 °C is required with $FeCl_3$ as catalyst. A 5-*exo*-dig cyclization leads to the gold carbene that

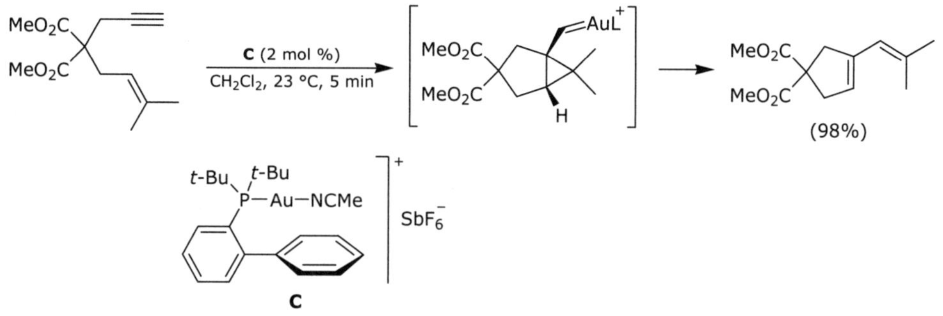

Scheme 4-21

can also be viewed as a gold-stabilized cyclopropylmethyl/cyclobutyl/homoallyl cation. Subsequent cleavage of a cyclopropyl C–C-bond and 1,2-migration of the other leads to the conjugated diene.[68, 69] Interestingly, enynes with strongly electron-donating groups at the alkene afforded dienes with *Z*-configuration at the exocyclic double bond.[70]

General procedure for the gold-catalyzed cycloisomerization of 1,6-enynes to conjugated dienes[68b]

The enyne (0.10–0.50 mmol) in CH_2Cl_2 (1 mL) is added to a mixture of Ph_3PAuCl (2 mol%) and silver(I) salt (2 mol%) (or the corresponding cationic Au(I) complex) in CH_2Cl_2 (2 mL), and the mixture is stirred for the time and at the temperature indicated. The resulting mixture is filtered through SiO_2, and the solvent is evaporated to give the corresponding product.

In contrast to 1,6-enynes with a terminal triple bond, substrates with an internal alkyne suffer cleavage of both multiple bonds (Scheme 4-22).[69, 71] In this case, the ring opening of the cyclopropyl carbene and 1,2-migration of the carbene fragment are followed by a proton shift and removal of the cationic gold catalyst.

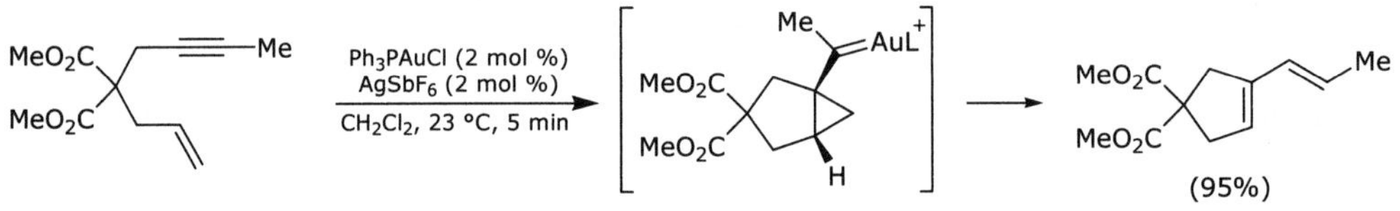

Scheme 4-22

Six-membered dienes with an exocyclic double bond can also be obtained by gold-catalyzed cycloisomerization of 1,6-enynes.[68, 72] This transformation is stereospecific; *i.e.*, substrates with defined double-bond geometries afforded the products as single stereoisomers (Scheme 4-23). Low temperatures and electron-withdrawing

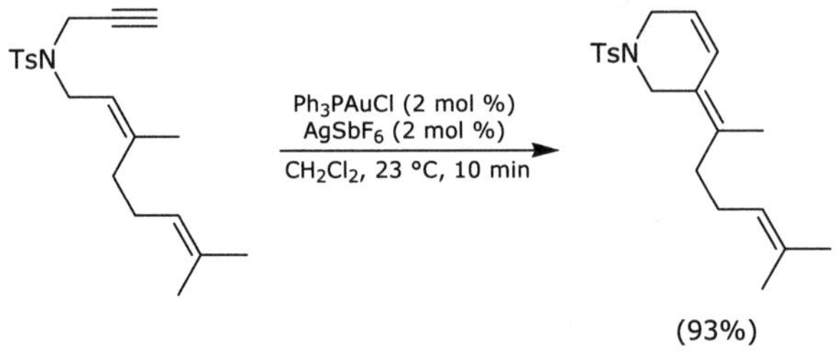

Scheme 4-23

substituents in the tether between double and triple bond favor the formation of six-membered rings.[73] Mechanistically, the cyclization seems to proceed *via* the same cyclopropyl gold carbene intermediates as discussed above; cleavage of the endocyclic cyclopropane bond and loss of the cationic gold catalyst lead to the product.

Another cycloisomerization pathway of 1,6-enynes involves intramolecular cyclopropanation of the alkene by the alkyne (Scheme 4-24).[74] This reaction is favored for enynes bearing a cyclic olefin and probably proceeds by 6-*endo*-dig-cyclization, followed by proton loss and protodeauration of the gold carbene intermediate. In the presence of the cationic gold catalyst $Ph_3PAuSbF_6$, the bicyclic products were obtained with excellent yield within 30 min at room temperature.

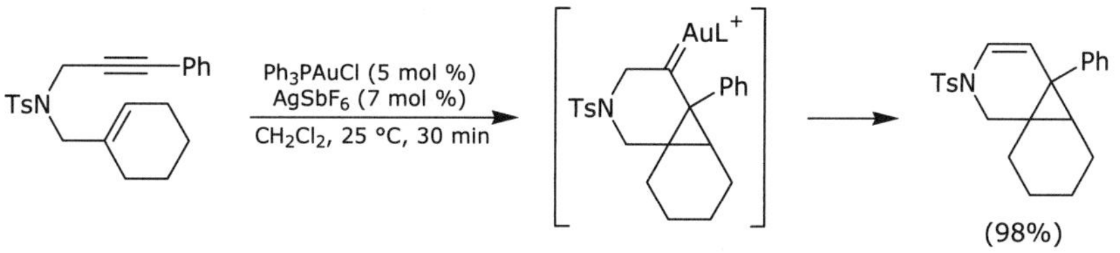

TsN, Ph; Ph_3PAuCl (5 mol %), $AgSbF_6$ (7 mol %), CH_2Cl_2, 25 °C, 30 min; [AuL$^+$, Ph, TsN]; TsN, Ph (98%)

Scheme 4-24

General procedure for the gold-catalyzed cycloisomerization of 1,6-enynes to bicyclo[4.1.0]hept-4-enes[74]

To a flame-dried, 15-mL Schlenk flask capped with a rubber septum is injected 5 mL of CH_2Cl_2 *via* syringe under N_2 flow. Ph_3PAuCl (13 mg, 5 mol%) and $AgSbF_6$ (16 mg, 7 mol%) are added sequentially. The catalyst solution is stirred for 10 min. The enyne (0.52 mmol) is put into the flask under N_2 flow. The reaction is monitored by TLC. After the reactant is consumed, the solvent is removed under reduced pressure. Flash chromatography on silica gel eluting with hexane and EtOAc (v/v, 10:1) gives the product.

The starting materials for the intramolecular cyclopropanation can be obtained from 1,6-diynes by rhodium-catalyzed Diels-Alder reaction. This allows the overall transformation to be performed as a tandem or one-pot process (Scheme 4-25).[74] The relative configuration at the highly substituted cyclopropane ring was determined with X-ray crystallography.

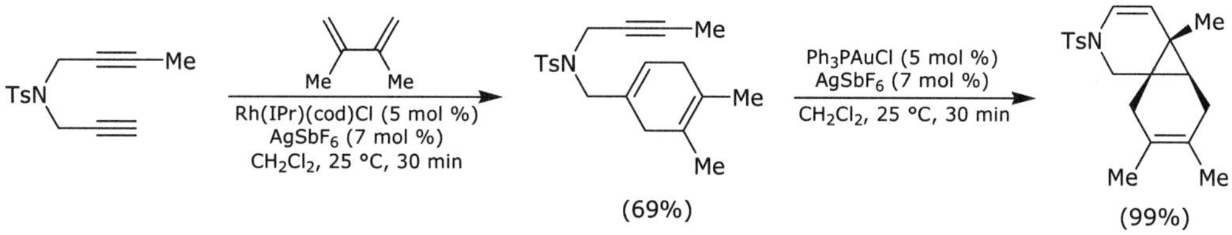

TsN, Me; Me, Me; Rh(IPr)(cod)Cl (5 mol %), $AgSbF_6$ (7 mol %), CH_2Cl_2, 25 °C, 30 min; TsN, Me, Me, Me (69%); Ph_3PAuCl (5 mol %), $AgSbF_6$ (7 mol %), CH_2Cl_2, 25 °C, 30 min; TsN, Me, Me, Me (99%)

Scheme 4-25

The gold-catalyzed intramolecular cyclopropanation can also be performed with allyl propargyl ethers that are prepared by substitution of allyl acetates with propargyl alcohols. These steps can be combined in a tandem synthesis of 3-oxabicyclo[4.1.0]hept-4-ene derivatives using $Ph_3PAuNTf_2$ as catalyst for both steps.[75] The products were obtained as a single diastereomer.

Representative procedure for the gold-catalyzed synthesis of 3-oxabicyclo[4.1.0]hept-4-enes by tandem allylic substitution/cycloisomerization. 2-Phenethyl-6,7-diphenyl-3-oxabicyclo[4.1.0]hept-4-ene[75]

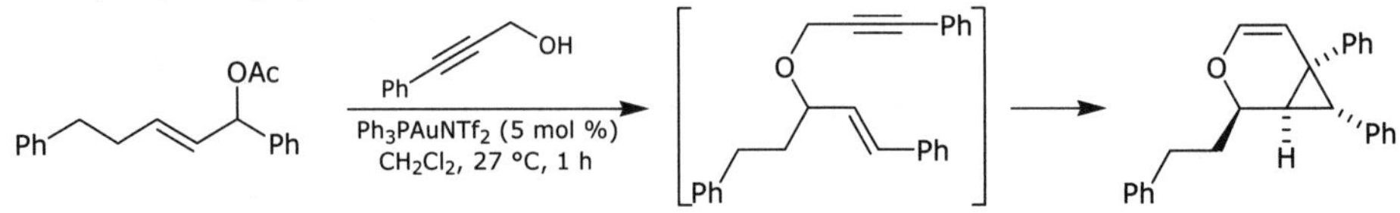

Ph₃PAuNTf₂ (5 mol %)
CH₂Cl₂, 27 °C, 1 h

To a solution of (*E*)-1,5-diphenylpent-2-enyl acetate (28 mg, 0.1 mmol) and 3-phenylprop-2-yn-1-ol (79.2 mg, 0.6 mmol) in CH_2Cl_2 (3 mL) was added 3 mol% of $Ph_3PAuNTf_2$ at 27 °C. Then 30 min later, the second batch of 2 mol% of $Ph_3PAuNTf_2$ was added. The resulting mixture was then stirred for another 30 min. After the complete consumption of starting material (*E*)-1,5-diphenylpent-2-enyl acetate and the enyne intermediate, petrol ether (0.5 mL) was added to quench the reaction. Removal of the solvent followed by column chromatography on silica gel (petrol ether/EtOAc = 100/1) afforded 2-phenethyl-6,7-diphenyl-3-oxabicyclo[4.1.0]hept-4-ene in 86% yield.

An enantioselective version of the gold-catalyzed cyclopropanation was reported by Michelet and co-workers.[76] In the presence of [Au_2{(*R*)-3,5-*t*-Bu-4-MeO-MeOBIPHEP}Cl_2] and AgOTf, 3-oxabicyclo[4.1.0]hept-4-ene derivatives were obtained with excellent enantioselectivities (Scheme 4-26). Both electron-rich and electron-poor aryl groups are tolerated under these conditions. Lower reactivities and stereoselectivities were observed in the corresponding platinum- or rhodium-catalyzed cycloisomerizations.[77]

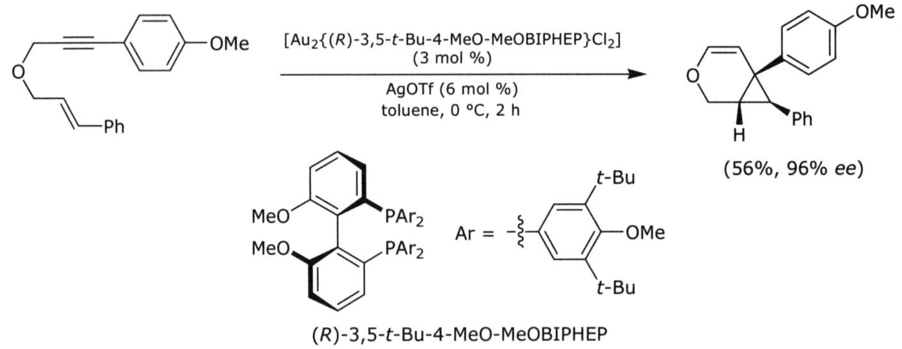

[Au_2{(*R*)-3,5-*t*-Bu-4-MeO-MeOBIPHEP}Cl_2] (3 mol %)
AgOTf (6 mol %)
toluene, 0 °C, 2 h

(56%, 96% *ee*)

(*R*)-3,5-*t*-Bu-4-MeO-MeOBIPHEP

Scheme 4-26

General procedure for the enantioselective gold-catalyzed cycloisomerization of allyl propargyl ethers[76]

A mixture of [Au_2{(*R*)-3,5-*t*-Bu-4-MeO-MeOBIPHEP}Cl_2] (3 mol%) and AgOTf (6 mol%) in distilled toluene (0.5 M) is stirred under an Ar atmosphere at RT for 30 min. The enyne (1 equiv) is then added, and the mixture is stirred until completion. The mixture is then filtered through a short pad of silica to eliminate the catalyst (elution with EtOAc), and the solvents are removed under reduced pressure. The crude product is purified by silica gel flash chromatography (petroleum ether/EtOAc, 98/2 to 80/20 v/v) if necessary.

The gold carbene species formed in the cycloisomerization of 1,6-enynes can be intercepted with olefins to form a cyclopropane in an intra- or intermolecular manner. Thus, treatment of 1,5-dien-10-ynes with gold(I) catalysts afforded tetracyclic products containing two cyclopropyl rings with high yield and excellent stereoselectivity (Scheme 4-27).[78] The cationic gold complex $Ph_3PAuNCMe^+SbF_6^-$ prepared from Ph_3PAuCl by chloride abstraction with $AgSbF_6$ in acetonitrile gave the best results in this transformation and is far superior to silver catalysts.[79] The corresponding cycloisomerization of cyclohexadienyl-substituted alkynes afforded complex polycyclic products with high yields.[80]

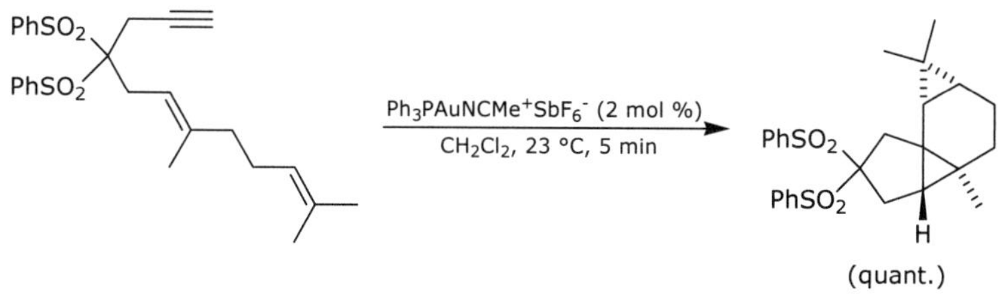

PhSO2, PhSO2; $Ph_3PAuNCMe^+SbF_6^-$ (2 mol %), CH_2Cl_2, 23 °C, 5 min; PhSO2, PhSO2, H (quant.)

Scheme 4-27

Representative procedure for the gold-catalyzed cycloisomerization of cyclohexadienyl-substituted alkynes. 2-Tosyloctahydro-3a,5-methanocyclopropa[4',5']cyclopenta[1',2':1,3]-cyclopropa [1,2-c]pyrrole [80]

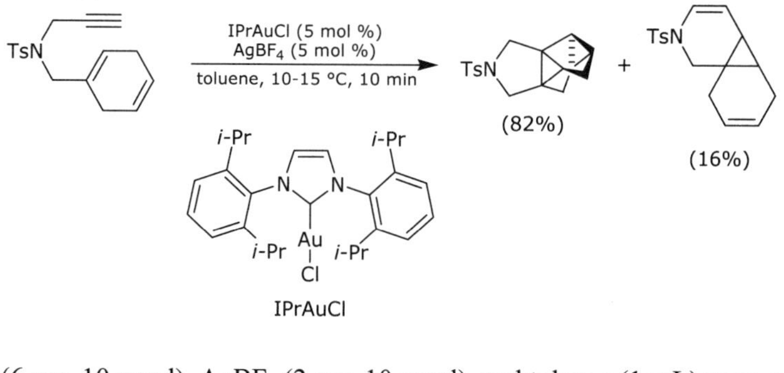

TsN; IPrAuCl (5 mol %), $AgBF_4$ (5 mol %), toluene, 10-15 °C, 10 min; TsN (82%) + TsN (16%)

i-Pr, *i*-Pr, N, N, *i*-Pr, Au, Cl, *i*-Pr

IPrAuCl

IPrAuCl (6 mg, 10 μmol), $AgBF_4$ (2 mg, 10 μmol), and toluene (1 mL) were added to a flame-dried Schlenk tube equipped with a stirring bar. The reaction mixture was stirred at RT for 5 min. Then *N*-(cyclohexa-1,4-dienylmethyl)-4-methyl-*N*-(prop-2-ynyl)benzenesulfonamide (0.5 mmol) and toluene (2 mL) were added to the flask. The resulting mixture was stirred at 10–15 °C until the starting material has completely disappeared (10 min as checked by TLC). The reaction mixture was purified by flash chromatography on a silica gel column eluting with *n*-hexane/EtOAc (10:1) to afford 2-tosyl-1,2,4a,4b,5,8-hexahydrobenzo[1,3]cyclopropa[1,2-*c*]pyridine (24 mg, 16%) as a white solid, and with *n*-hexane/EtOAc (15:2) to afford 2-tosyloctahydro-3a,5-methanocyclopropa-[4',5']cyclopenta[1',2':1,3]cyclopropa[1,2-*c*]pyrrole (123 mg, 0.408 mmol, 82%) as a white solid.

The intermolecular reaction of gold carbene intermediates with alkenes can lead to two different products, depending on the starting material.[81] 1,6-Enynes with a substituent at C-1 afforded cyclopropyl-substituted bicyclo[3.1.0]hexanes *via* the gold carbene formed by 5-*exo*-dig attack at the activated triple bond (Scheme 4-28). If no substituent is present at C-1 of the enyne, then the gold carbene undergoes a rapid skeletal rearrangement, and subsequent trapping by the external olefin provided allylated cyclopropanes (Scheme 4-29).

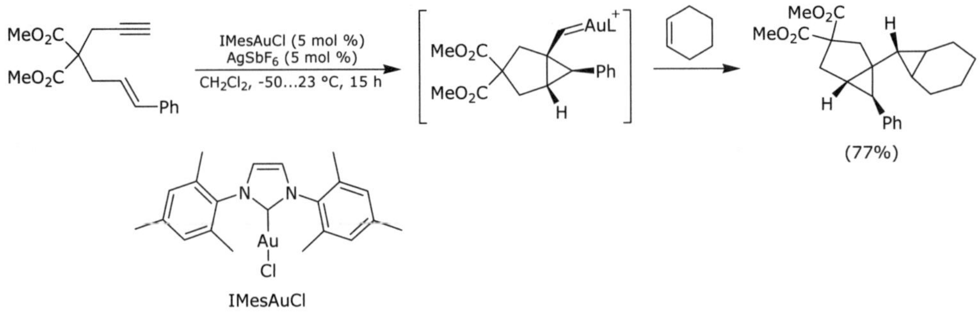

Scheme 4-28

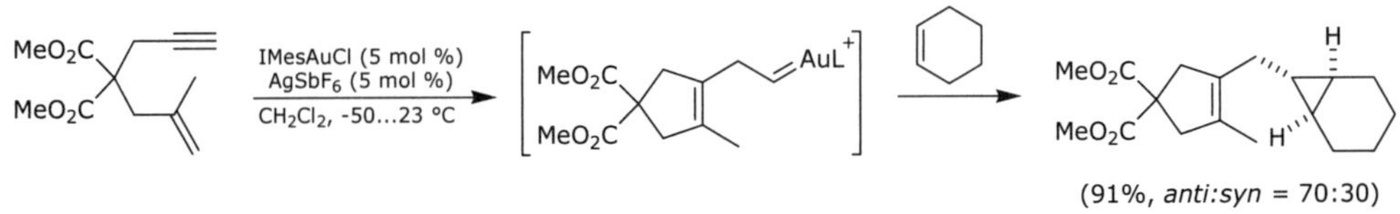

Scheme 4-29

General procedure for the gold-catalyzed intermolecular cyclopropanation of 1,6-enynes with alkenes [81a]

A solution of the enyne (0.15–0.50 mmol) in dry CH_2Cl_2 (1 mL) is added to a mixture of gold(I) complex (5 mol%) and alkene (5 equiv) in dry CH_2Cl_2 (1 mL) at –50 °C (Haake EK 90 immersion cooler). The mixture is then slowly warmed up in the cooling bath by turning the immersion cooler off until the mixture reaches RT (ca. 15 h). The resulting mixture is filtered through Celite and purified by column chromatography (hexanes/EtOAc) to give the corresponding cyclopropane. Solids are additionally purified by trituration in pentane.

In the presence of cationic gold catalysts, 1,6-enynes with a terminal double bond undergo intermolecular addition reactions to carbonyl compounds.[82] Again, a skeletal rearrangement of the gold carbene intermediate is the key to the formation of the tricyclic heterocycles.

Representative procedure for the gold-catalyzed intermolecular addition of 1,6-enynes to carbonyl compounds. Dimethyl 3-phenyltetrahydro-3H-cyclopenta[c]cyclopropa[b]furan-5,5(6H)-dicarboxylate[82b]

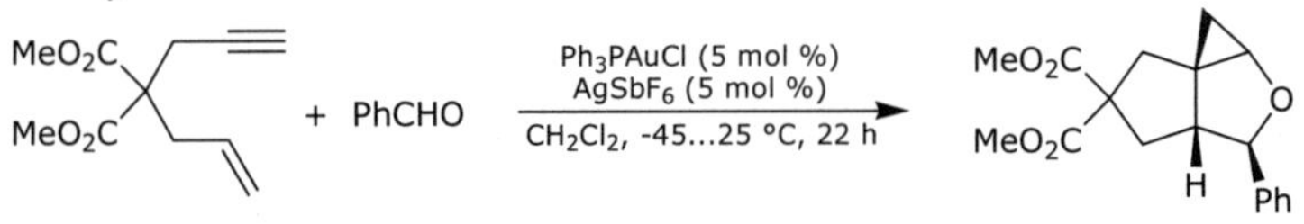

In a flame-dried Schlenk flask under an argon atmosphere, a stirred mixture of benzaldehyde (530 mg, 5.0 mmol), dimethyl 2-allyl-2-(prop-2-ynyl)malonate (210 mg,

1.0 mmol), and $AgSbF_6$ (17 mg, 50 μmol) in anhydrous CH_2Cl_2 (5 mL) was cooled to –45 °C and treated with Ph_3PAuCl (24.7 mg, 50 μmol). The mixture turned orange and was kept at –45 °C for 6 h. The cooling device was switched off, which initiated slow warming to RT. After an overall reaction time of 22 h, a black precipitate had formed and GC-MS indicated complete consumption of the substrate. The mixture was filtered through Celite, which was then washed with CH_2Cl_2. Flash column chromatography on silica (25 g, petroleum ether/EtOAc 10:1) yielded dimethyl 3-phenyltetrahydro-3*H*-cyclopenta[*c*]cyclopropa[*b*]furan-5,5(6*H*)-dicarboxylate (216 mg, 68%) as a colorless oil.

Various other nucleophiles can be employed for the trapping of cyclopropyl gold carbenes formed by cycloisomerization of 1,6-enynes. The intermolecular hydroxycyclization or alkoxycyclization of 1,6-enynes in the presence of water or alcohols is also catalyzed by Pt(II) or Pd(II), but they proceed under milder conditions and more efficiently in the presence of cationic gold(I) catalysts.[68, 83, 84] The catalyst is formed *in situ* either from a phosphinegold(I) chloride and a silver salt, or by reaction of Ph_3PAuMe with a Brønsted acid, *e.g.*, HBF_4 (Scheme 4-30). No cyclization takes place with protic acids in the absence of gold(I).

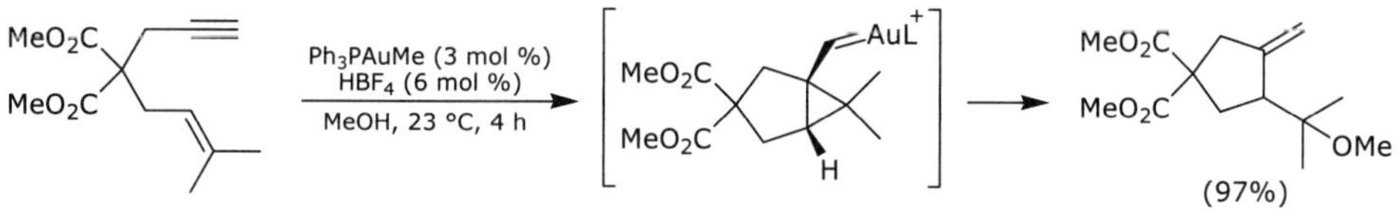

MeO2C MeO2C Ph3PAuMe (3 mol %) HBF4 (6 mol %) MeOH, 23 °C, 4 h MeO2C MeO2C AuL+ H MeO2C MeO2C OMe (97%)

Scheme 4-30

By an analogy with the cyclizations discussed previously, the course of the gold-catalyzed alkoxycyclization depends on the substituent pattern of the starting enyne. Thus, 1,6-enynes with a substituent at C-1 give five-membered ring products (Scheme 4-31).[68] Under these conditions, enol ethers afforded acetals. In contrast to this, enynes with a terminally unsubstituted double bond give six-membered alkoxycyclization products formed by 6-*endo*-dig attack of the alkene at the activated alkyne. Substrates with a defined double-bond geometry react in a stereospecific manner to afford the ethers as a single diastereomer. Various examples for enantioselective alkoxycyclizations of 1,6-enynes using chiral gold catalysts have been reported as well. Unfortunately, in most cases only moderate yields and enantioselectivities were achieved.[85]

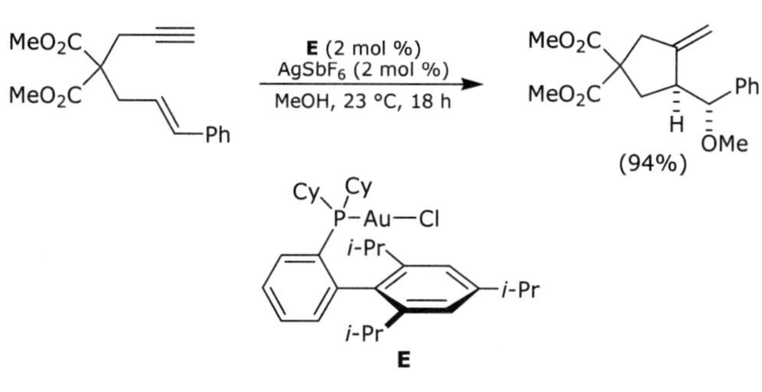

MeO2C MeO2C Ph **E** (2 mol %) AgSbF6 (2 mol %) MeOH, 23 °C, 18 h MeO2C MeO2C H Ph OMe (94%)

Cy Cy P–Au–Cl i-Pr i-Pr i-Pr **E**

Scheme 4-31

Gold-catalyzed intramolecular alkoxycyclizations of suitable 1,6-enynols open up efficient access to cyclic ethers (Scheme 4-32).[68, 86] In the presence of $Ph_3PAuSbF_6$, primary and secondary hydroxy groups participate in the cyclization. Platinum dichloride is much less efficient in this transformation. In a similar manner, lactones are obtained by gold-catalyzed cycloisomerization of 1,6-enynes bearing a pendant carboxylic acid (Scheme 4-33).[86] Both γ- and δ-lactones are accessible by this method in a highly diastereoselective manner. Again, the substitution pattern at the alkene dictates the cyclization mode.

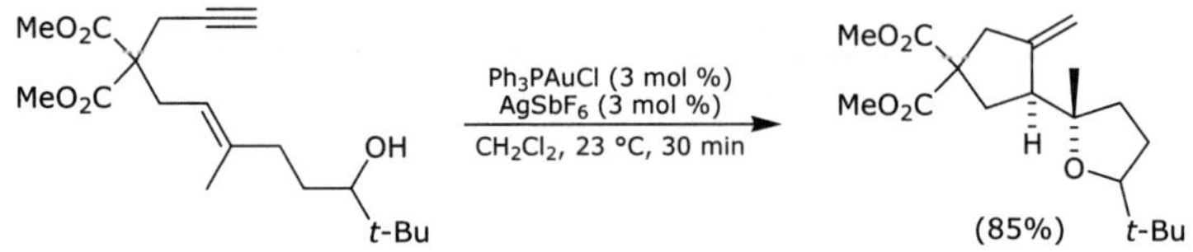

MeO2C MeO2C OH t-Bu Ph3PAuCl (3 mol %) AgSbF6 (3 mol %) CH2Cl2, 23 °C, 30 min MeO2C MeO2C H O (85%) t-Bu

Scheme 4-32

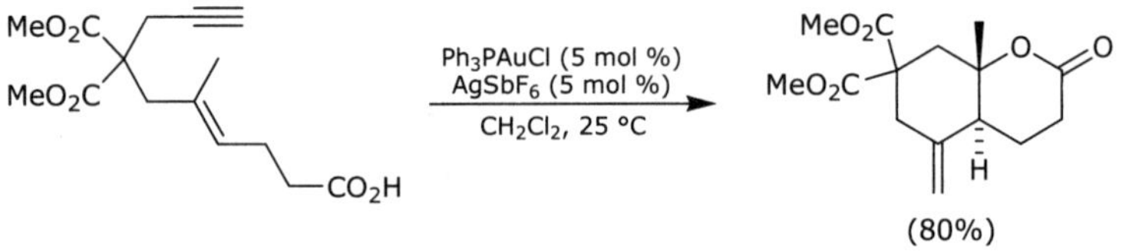

MeO2C MeO2C CO2H Ph3PAuCl (5 mol %) AgSbF6 (5 mol %) CH2Cl2, 25 °C MeO2C MeO2C O O H (80%)

Scheme 4-33

Representative procedure for the gold-catalyzed cycloisomerization of enynoic acids. Dimethyl 3-(tetrahydro-2-methyl-5-oxofuran-2-yl)-4-methylenecyclopentane-1,1-dicarboxylate[86]

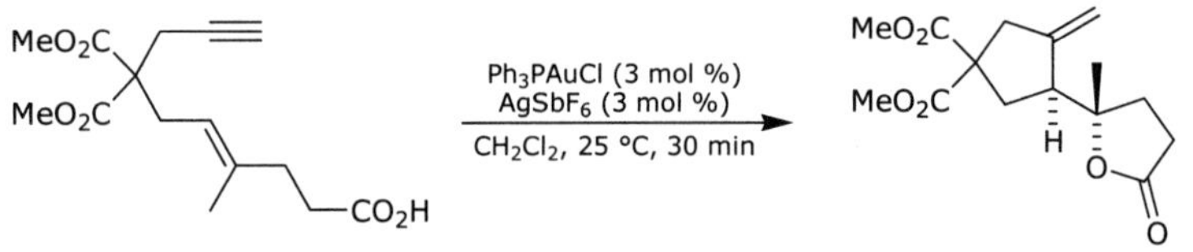

MeO2C MeO2C CO2H Ph3PAuCl (3 mol %) AgSbF6 (3 mol %) CH2Cl2, 25 °C, 30 min MeO2C MeO2C H O O

A solution of Ph_3PAuCl (5 mg, 0.01 mmol) and $AgSbF_6$ (4 mg, 0.01 mmol) in CH_2Cl_2 (7 mL) was added *via* cannula to (*E*)-7,7-di(methoxycarbonyl)-4-methyldec-4-en-9-ynoic acid (112 mg, 0.37 mmol) and the resulting mixture was stirred for 30 min at RT. The reaction mixture was filtered through a plug of SiO_2, the filtrate was evaporated under reduced pressure, and the crude product was purified by flash column chromatography (30% *v/v* EtOAc in hexanes) to afford dimethyl 3-(tetrahydro-2-methyl-5-oxofuran-2-yl)-4-methylenecyclopentane-1,1-dicarboxylate as a colorless oil (92 mg, 82%).

Enantioselective cycloisomerizations of enynoic acids are possible in the presence of BIPHEP-gold catalysts.[87] Under these conditions, 1,6-enynes bearing an intramolecular amide, phenol, or electron-rich arene trap also provide complex polycyclic products with high *ee* (Scheme 4-34).

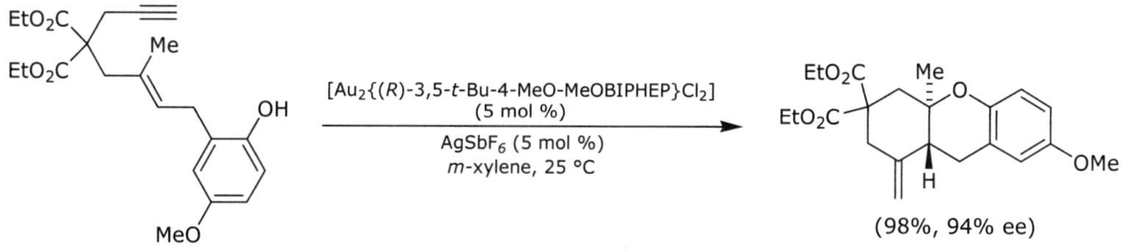

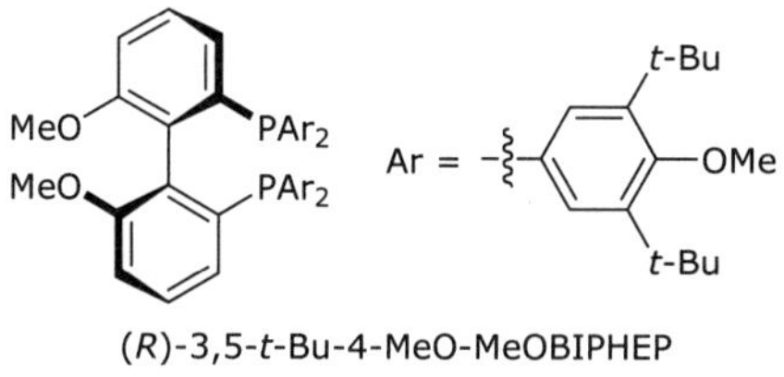

$[Au_2\{(R)\text{-3,5-}t\text{-Bu-4-MeO-MeOBIPHEP}\}Cl_2]$ (5 mol %)
$AgSbF_6$ (5 mol %)
m-xylene, 25 °C

(98%, 94% ee)

Ar =

(*R*)-3,5-*t*-Bu-4-MeO-MeOBIPHEP

Scheme 4-34

Suitable carbon nucleophiles for the intermolecular trapping of cyclopropyl gold carbenes generated by cyclization of 1,6-enynes comprise electron-rich benzene derivatives, indoles, and β-dicarbonyl compounds.[88] Even though the cyclopropane ring can remain intact during the trapping reaction, it is normally opened to give functionalized carbocycles or heterocycles. For the enantioselective version of this transformation, chiral Au(I) or Pt(II) catalysts have been used.[89]

Representative procedure for the gold-catalyzed addition of arenes to 1,6-enynes. 3-((1,2,3,6-Tetrahydro-4-methyl-1-tosylpyridin-3-yl)(phenyl)methyl)-1H-indole[88d]

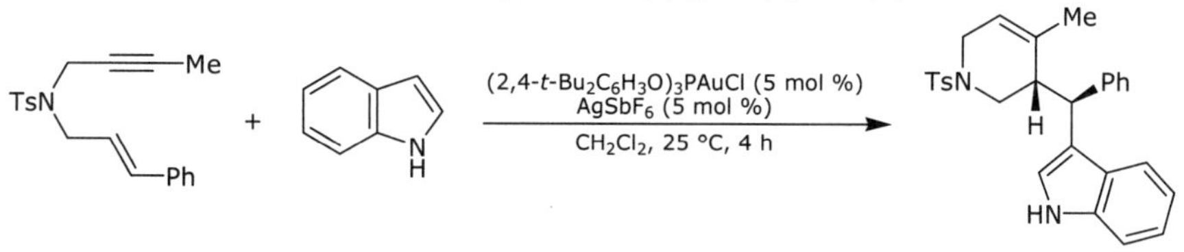

$(2,4\text{-}t\text{-}Bu_2C_6H_3O)_3PAuCl$ (5 mol %)
$AgSbF_6$ (5 mol %)
CH_2Cl_2, 25 °C, 4 h

A solution of *N*-cinnamyl-*N*-tosylbut-2-yn-1-amine (20.0 mg, 0.059 mmol) and indole (7.6 mg, 0.064 mmol) was added to a solution of $(2,4\text{-}t\text{-}Bu_2C_6H_3O)_3PAuCl$ and $AgSbF_6$ (5 mol%) in CH_2Cl_2. The reaction mixture was stirred at RT for 4 h. The reaction mixture was filtered through silica gel with CH_2Cl_2, and the solvents were evaporated under reduced pressure. The residue was purified by chromatography (5:1 hexane/EtOAc) to give 3-((1,2,3,6-tetrahydro-4-methyl-1-tosylpyridin-3-yl)(phenyl)methyl)-1*H*-indole (18.2 mg, 68%) as a light brown solid; mp 84–86 °C.

The cycloisomerization of 1,5-enynes is catalyzed by platinum(II) or gold(I) complexes and affords bicyclo[3.1.0]hexene derivatives (Scheme 4-35).[90] Mechanistically, this transformation is closely related to the corresponding cyclization of 1,6-enynes; thus, activation of the alkyne by the cationic gold catalyst is followed by nucleophilic addition of the pendant alkene to produce a gold carbene intermediate, which affords the product by 1,2-hydrogen shift. Excellent diastereoselectivities were observed in most cases and are explained by half-chair transition states with large groups occupying pseudoequatorial positions. In a similar manner, 1,5-allenynes produce cross-conjugated trienes by nucleophilic addition of an allenic double bond at the activated alkyne, followed by a 1,5-H-shift.[91] A deviating 6-*endo*-cyclization mode leading to aromatic products has been observed for 3-hydroxy-1,5-enynes.[92]

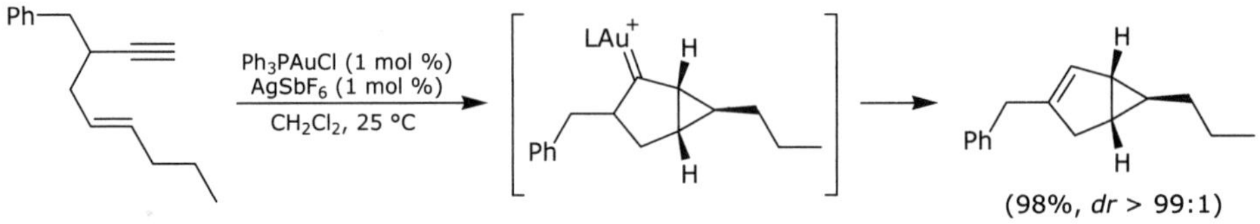

Ph$_3$PAuCl (1 mol %)
AgSbF$_6$ (1 mol %)
CH$_2$Cl$_2$, 25 °C

(98%, *dr* > 99:1)

Scheme 4-35

The 1,5-enynes can be prepared by rhenium-catalyzed coupling of propargyl alcohols with allyl silanes.[93] This step can be conveniently combined with the gold-catalyzed cycloisomerization.

Representative procedure for the one-pot rhenium/gold-catalyzed synthesis of bicyclo[3.1.0]hexane derivatives. 1,3-Diphenylbicyclo[3.1.0]hex-2-ene[90b]

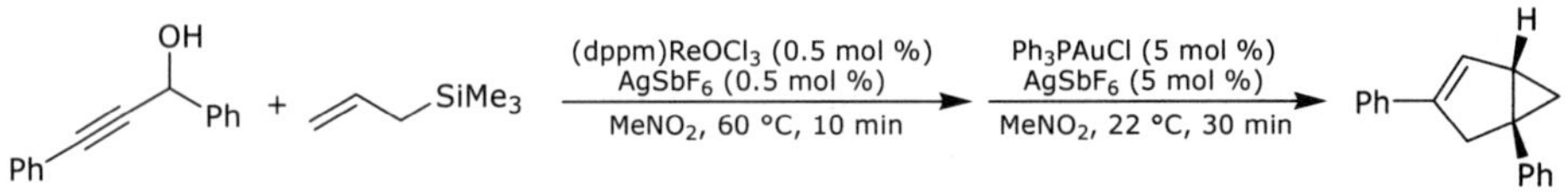

(dppm)ReOCl$_3$ (0.5 mol %)
AgSbF$_6$ (0.5 mol %)
MeNO$_2$, 60 °C, 10 min

Ph$_3$PAuCl (5 mol %)
AgSbF$_6$ (5 mol %)
MeNO$_2$, 22 °C, 30 min

To a stirring solution of (dppm)ReOCl$_3$ (3.4 mg, 5.0 μmol) and AgSbF$_6$ (1.71 mg, 5.0 μmol) in nitromethane (1 mL) in a 1-dram vial was added allyltrimethylsilane (175 μL, 1.10 mmol) and then 1,3-diphenylprop-2-yn-1-ol (208 mg, 1.00 mmol). The vial was submerged in a 60 °C oil bath for 10 min and then allowed to cool to 22 °C. The contents of the vial were transferred *via* pipette—assisted with 0.5 mL of nitromethane—to a 1-dram vial containing Ph$_3$PAuCl (25 mg, 0.050 mmol) and AgSbF$_6$ (17 mg, 0.050 mmol). After stirring for 30 min at 22 °C, the solvent was removed under reduced pressure and silica gel chromatography with hexanes eluent afforded 1,3-diphenylbicyclo[3.1.0]hex-2-ene (209 mg, 89%).

In the absence of a nucleophile, 1,5-enynes with a quarternary carbon atom at the propargylic position undergo a 1,2-alkyl shift. This can be exploited in ring enlargement processes. Thus, enynes with four- or five-membered rings in the tether afforded the corresponding tricyclic products (Scheme 4-36).[90b] In contrast to this, substrates with larger rings react under intramolecular C–H insertion.[94]

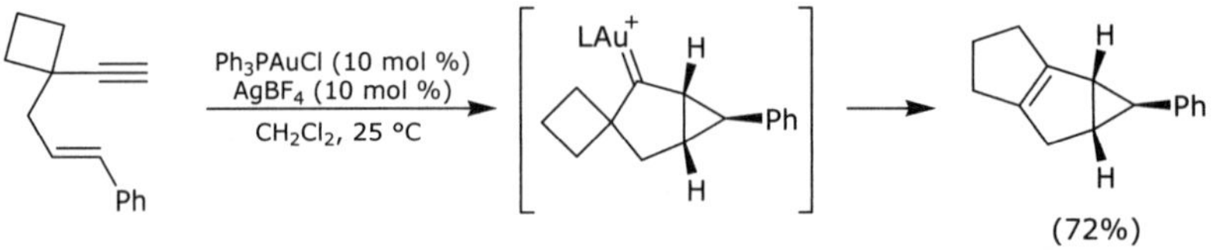

Ph$_3$PAuCl (10 mol %)
AgBF$_4$ (10 mol %)
CH$_2$Cl$_2$, 25 °C

(72%)

Scheme 4-36

In the presence of a nucleophile, the bicyclic gold carbene intermediates can be trapped to afford monocyclic products with opening of the cyclopropyl ring (Scheme 4-37).[90b, 95, 96] Depending on the bond cleaved, functionalized cyclopentenes or cyclohexenes are obtained. Suitable nucleophiles for intermolecular trapping reactions are water, alcohols, and carboxylic acids.

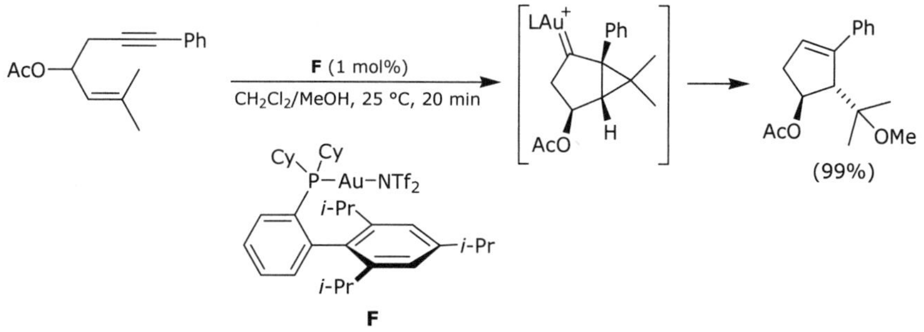

Scheme 4-37

Representative procedure for the gold-catalyzed addition of alcohols to 1,5-enynes. 1-(5-Methoxy-3,3-dimethylcyclohex-1-enyl)benzene[90b]

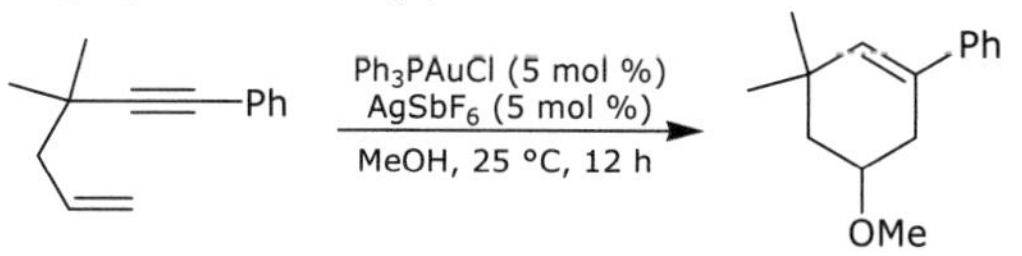

To a filtered solution of Ph_3PAuCl (12 mg, 0.025 mmol) and $AgSbF_6$ (8.6 mg, 0.025 mmol) in methanol (1 mL) was added 1-(3,3-dimethylhex-5-en-1-ynyl)benzene (92 mg, 0.50 mmol). After stirring for 12 h at RT, the methanol was removed under reduced pressure. Flash chromatography eluting with 50:1 hexanes:EtOAc afforded 1-(5-methoxy-3,3-dimethylcyclohex-1-enyl)benzene (92 mg, 85%) as a colorless oil.

The intramolecular trapping of cyclopropyl gold carbene intermediates formed by gold-catalyzed cyclization of 1,5-enynes can be achieved with oxygen or nitrogen nucleophiles.[97] The cycloisomerization of enynols is catalyzed by $Ph_3PAuClO_4$ or $Ph_3PAuNTf_2$ and leads to annulated or spiroethers with high yield. Analogous products were obtained with sulfonamides as internal nucleophile (Scheme 4-38).

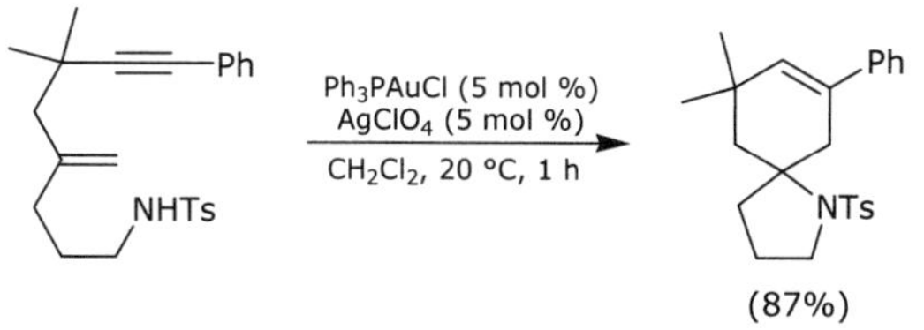

Scheme 4-38

A different cyclization mode was observed for 1,5-enynes bearing a silyloxy group at the triple bond. In the presence of gold(I) chloride, they undergo a double cleavage of both the olefin and the triple bond to afford conjugated or skipped cyclohexadienes (Scheme 4-39).[98] In this transformation, the cyclopropyl gold carbene species formed initially probably undergoes skeletal rearrangement by a series of

1,2-shifts, which finally provide a cyclohexyl gold carbene. The cycloisomerization is also catalyzed by $PtCl_2$, even though this requires heating to 80 °C.

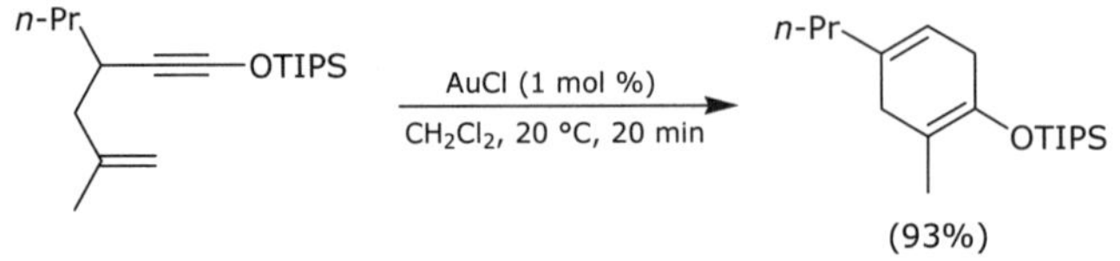

n-Pr —OTIPS AuCl (1 mol %) CH_2Cl_2, 20 °C, 20 min n-Pr OTIPS (93%)

Scheme 4-39

The Conia-ene reaction of β-ketoesters bearing a pendant triple bond involves the cycloisomerization of an enyne formed by enolization. This transformation is efficiently catalyzed by cationic gold(I) complexes and afforded cyclopentane derivatives with excellent yields and moderate to good diastereoselectivities (Scheme 4-40).[99] Acetylenic silyl enol ethers[42, 100] or imines[101] react in an analogous manner. By this method, iodoalkynes were converted into iodocyclopentenes, which are highly useful in natural product synthesis.

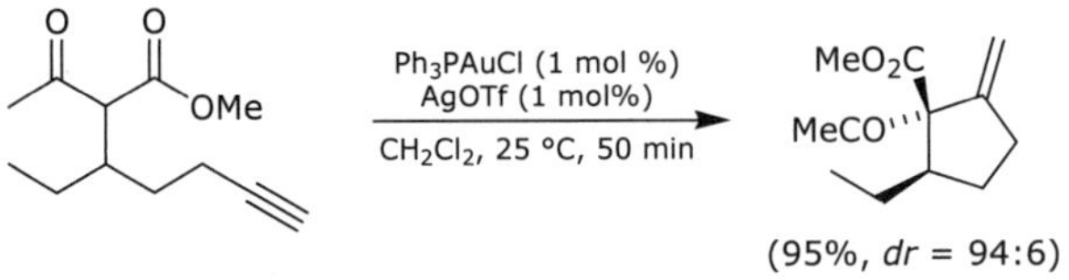

O O OMe Ph_3PAuCl (1 mol %) AgOTf (1 mol%) CH_2Cl_2, 25 °C, 50 min MeO_2C MeCO (95%, *dr* = 94:6)

Scheme 4-40

Representative procedure for the gold-catalyzed Conia-ene reaction of acetylenic silyl enol ethers. (3aR,6R,7aS)-3a-Allyl-5,6,7,7a-tetrahydro-3-iodo-6-methyl-1H-inden-4(3aH)-one[100a]

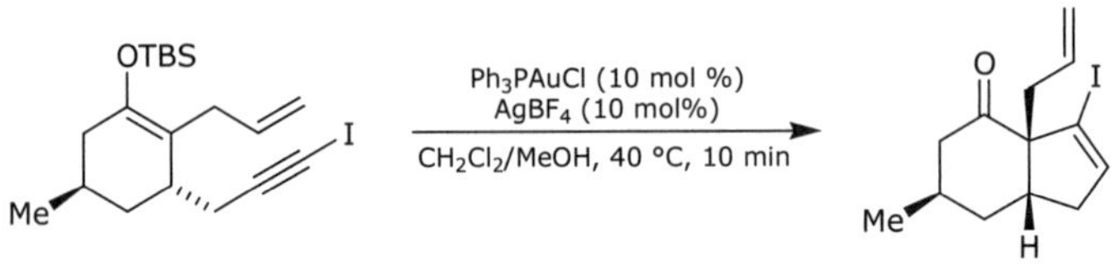

OTBS Me I Ph_3PAuCl (10 mol %) $AgBF_4$ (10 mol%) CH_2Cl_2/MeOH, 40 °C, 10 min O I Me H

To a rapidly stirred solution of ((3*S*,5*R*)-2-allyl-3-(3-iodoprop-2-ynyl)-5-methylcyclohex-1-enyloxy)(*t*-butyl)dimethylsilane (1.57 g, 3.7 mmol) in CH_2Cl_2 (20 mL) was added MeOH (2 mL), followed by Ph_3PAuCl (183 mg, 0.37 mmol) and $AgBF_4$ (72 mg, 0.37 mmol). The reaction mixture was then placed in a preheated 40 °C oil bath and stirred for 10 min; it was then cooled and filtered over Celite to remove gold black. The filtrate was concentrated *in vacuo* and chromatographed directly (18:1 hexanes/ EtOAc) to give (3a*R*,6*R*,7a*S*)-3a-allyl-5,6,7,7a-tetrahydro-3-iodo-6-methyl-*1H*-inden-4(3a*H*)-one (1.02 g, 85%) as a pale yellow liquid.

In the presence of cationic gold(I) catalysts, 3-silyloxy-1,5-enynes undergo a tandem 6-*endo*-dig-cyclization/pinacol rearrangement.[102] Depending on the starting material, monocyclic, spiro, or annulated cyclopentene derivatives were obtained. Isopropanol was used as external proton source for protonation of a σ-vinylgold intermediate and regeneration of the catalyst. In the presence of *N*-iodosuccinimide, the vinylgold species undergoes iododeauration to furnish vinyl iodides.

Representative procedure for the gold-catalyzed tandem cyclization/pinacol rarrangement of 3-silyloxy-1,5-enynes. 3a,4,5,6,7,7a-Hexahydro-1-phenyl-3H-indene-3a-carbaldehyde[102]

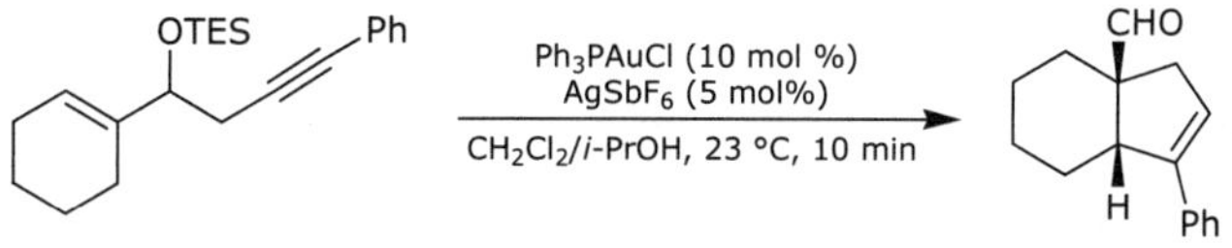

OTES, Ph; Ph_3PAuCl (10 mol %), $AgSbF_6$ (5 mol%), CH_2Cl_2/*i*-PrOH, 23 °C, 10 min; CHO, H, Ph

A solution of Ph_3PAuCl (22.4 mg, 10 mol%) in CH_2Cl_2 (0.3 mL) was added to a solution of $AgSbF_6$ (7.8 mg, 5 mol%) in CH_2Cl_2 (0.3 mL), and the mixture was stirred at RT for 10 min. The resulting suspension was filtered through Celite and concentrated under reduced pressure. To this residue, a solution of (1-cyclohexenyl-4-phenylbut-3-ynyloxy)triethylsilane (156 mg, 0.45 mmol) and *i*-PrOH (0.04 mL, 0.50 mmol) in CH_2Cl_2 (4.5 mL) was added. The pale purple solution was stirred at RT for 10 min. The mixture was concentrated under reduced pressure. Purification of the residue by flash chromatography on silica gel (pentanes/EtOAc 98:2) gave 3a,4,5,6,7,7a-hexahydro-1-phenyl-3*H*-indene-3a-carbaldehyde as a colorless oil (94.7 mg, 0.42 mmol, 93%).

A limited number of examples for the gold-catalyzed cycloisomerization of 1,7-enynes has been disclosed in recent years.[103] In analogy to the corresponding transformation of 1,6-enynes, vinylcyclohexenes are formed by activation of the triple bond, nucleophilic attack of the olefin, and skeletal rearrangement of the cyclopropyl gold carbene thus formed (Scheme 4-41). In the presence of an external nucleophile (H_2O or MeOH), the cyclopropyl carbene undergoes ring opening to afford hydroxy-/alkoxycyclization products. Here, cationic gold(I) catalysts (*e.g.*, **E**) bearing bulky phosphine ligands gave the best results (Scheme 4-42).[103b] A deviating cycloisomerization mode was observed for 1,7-enynes bearing a silyl enol ether. In the presence of a gold(I) complex with an extremely bulky semihollow phosphine ligand, these substrates undergo a 7-*endo*-dig cyclization to afford methylenecycloheptanes.[104] Similar products were also obtained with this catalyst from 1,8-enynes.

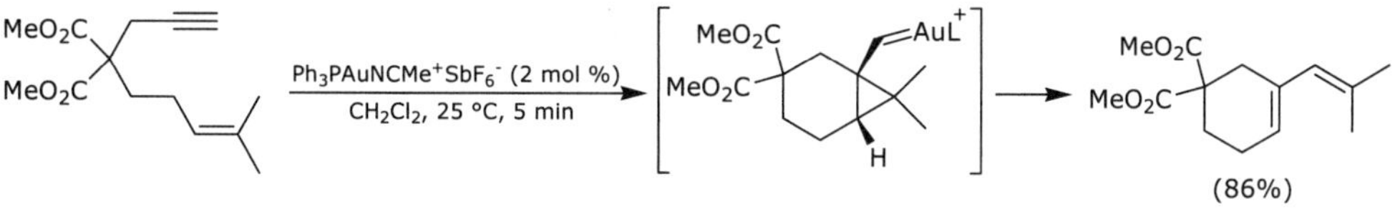

MeO_2C, MeO_2C; $Ph_3PAuNCMe^+SbF_6^-$ (2 mol %), CH_2Cl_2, 25 °C, 5 min; [MeO_2C, MeO_2C, AuL^+, H]; MeO_2C, MeO_2C (86%)

Scheme 4-41

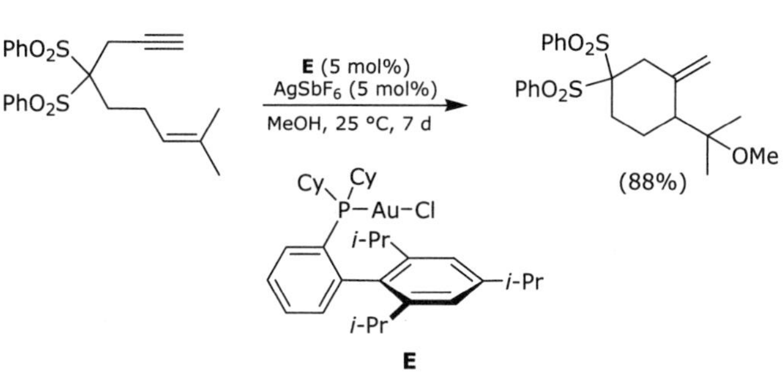

PhO_2S, PhO_2S; **E** (5 mol%), $AgSbF_6$ (5 mol%), MeOH, 25 °C, 7 d; PhO_2S, PhO_2S, OMe (88%)

Cy, Cy, P–Au–Cl, *i*-Pr, *i*-Pr, *i*-Pr

E

Scheme 4-42

3.5 Cycloadditions

Intermolecular [2+2]-cycloadditions of alkenes with alkynes are catalyzed by cationic gold(I) catalysts of type **G** and afford cyclobutenes in high yields and regioselectivities.[105] Because of the steric demand of the catalyst, the alkyne is activated selectively, and competing pathways (including coordination to the alkene) are suppressed. The reaction proceeds well with alkynes bearing either electron-rich or electron-poor substituents. The corresponding *bis*-cyclobutenes were obtained from 3- or 4-diethynylbenzene.

Representative procedure for the gold-catalyzed [2+2]-cycloaddition of alkenes with alkynes. 1,3-Diphenyl-3-methylcyclobutene[105a]

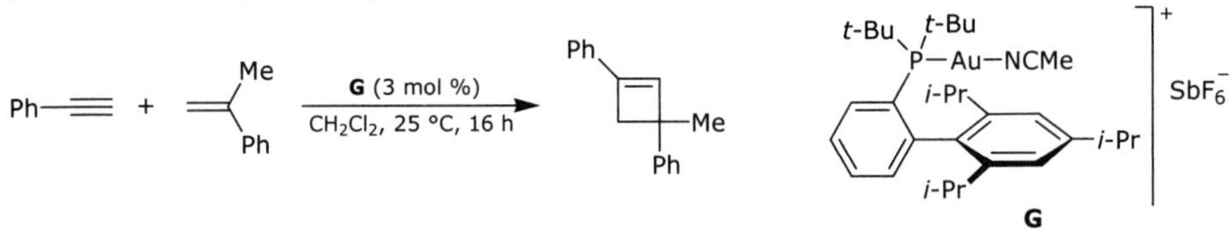

Ph—≡ + Me/Ph-alkene —**G** (3 mol %), CH_2Cl_2, 25 °C, 16 h→ Ph, Me, Ph cyclobutene

t-Bu, t-Bu, P—Au—NCMe, i-Pr, i-Pr, i-Pr, SbF_6^-, **G**

A solution of phenylacetylene (28.8 μL, 0.263 mmol, 1 equiv) and α-methylstyrene (72 μL, 0.526 mmol, 2 equiv) in CH_2Cl_2 (0.53 M) was added to a solution of gold(I) catalyst **G** (7.8 mg, 9 μmol, 3 mol%) in CH_2Cl_2 (0.08 M). The reaction mixture was stirred at RT for 16 h (GC-MS monitoring). Et_3N (0.05 mL) was added and the solvent was evaporated. The residue was purified by preparative TLC (pentane/CH_2Cl_2, 99:1) to give 1,3-diphenyl-3-methyl-cyclobutene as a colorless oil (46.1 mg, 80%).

1,6-Enynes with an amide or ester group in the tether between the multiple bonds undergo (formal) intramolecular [2+2]-cycloadditions in the presence of Ph_3PAuCl and $AgSbF_6$ (Scheme 4-43).[106] The analogous substrates lacking the

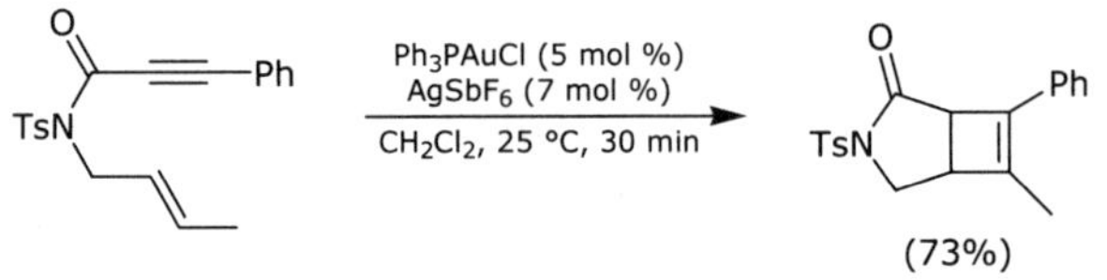

O, TsN, Ph —Ph_3PAuCl (5 mol %), $AgSbF_6$ (7 mol %), CH_2Cl_2, 25 °C, 30 min→ O, TsN, Ph (73%)

Scheme 4-43

carbonyl group afforded conjugated dienes by cycloisomerization (Section 3.4). The reaction probably proceeds by 6-*endo*-dig-cyclization and subsequent skeletal rearrangement of the cyclopropyl gold carbene intermediate. The corresponding gold-catalyzed intramolecular [2+2]-cycloadditions of 1,7- and 1,8-enynes have also been reported.[69, 107]

In the presence of an aryl or alkenyl group at the triple bond, 1,6-enynes undergo (formal) intramolecular [4+2]-cycloadditions catalyzed by various highly reactive cationic gold(I) complexes (*e.g.*, **C**) under mild conditions (Scheme 4-44).[108, 109] These transformations probably occur *via* cyclopropyl gold carbene intermediates formed by 5-*exo*-dig-cyclization, which are transformed into the bicyclic or tricyclic products by Friedel-Crafts-type ring expansion. Various electron-donating or -withdrawing substituents in the substrate are tolerated. Related gold-catalyzed intramolecular cycloadditions of 1,6-diynes[110] and allenynes[111] have been reported as well. Moreover, a gold-catalyzed benzannulation of 2-alkynylbenzaldehydes with alkynes provides substituted naphthalenes *via* a [4+2]/retro-[4+2]-cycloaddition sequence.[112]

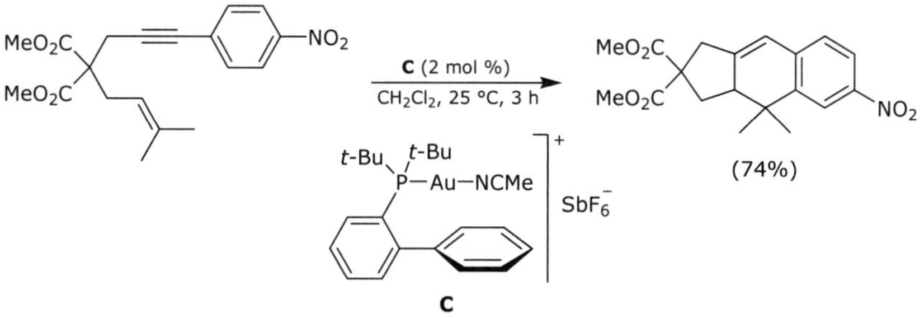

Scheme 4-44

Representative procedure for the gold-catalyzed intramolecular [4+2]-cycloaddition of 1,6-enynes. Dimethyl 3,3a-dihydroacephenanthrylene-4,4(1H,5H,10bH)-dicarboxylate[108c]

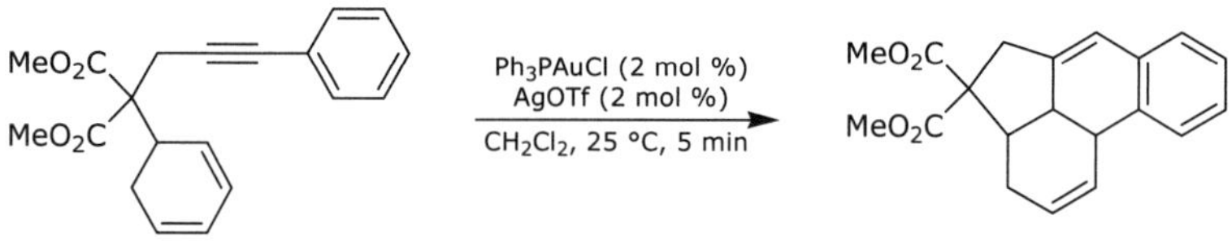

To an oven-dried, 100-mL, round-bottom flask equipped with a stirring bar and capped with a rubber septa was added AgOTf (5.14 mg, 0.02 mmol). The apparatus was evacuated (oil pump) and filled with nitrogen three times. To the reaction mixture were then added *via* syringe dimethyl 2-(cyclohexa-2,4-dienyl)-2-(3-phenylprop-2-ynyl)malonate (0.32 g, 1.0 mmol) and Ph_3PAuCl (0.01 g, 0.02 mmol) in 30 mL of CH_2Cl_2. The resulting mixture was stirred at RT for 5 min. The reaction mixture was filtered through a bed of Celite. The filtrate was concentrated *in vacuo* to give the crude mixture, which was purified by flash column chromatography (silica gel, gradient elution: 10 to 20% EtOAc/hexanes) to give dimethyl 3,3a-dihydroacephenanthrylene-4,4(1*H*,5*H*,10b*H*)-dicarboxylate 0.30 g, 0.93 mmol, 93%) as a yellow solid: mp 66–67 °C.

Representative procedure for the gold-catalyzed intramolecular [3+2]-cycloaddition of 1,6-diynes. 1,3-Dihydro-9-phenylindeno[2,1-c]pyran[110b]

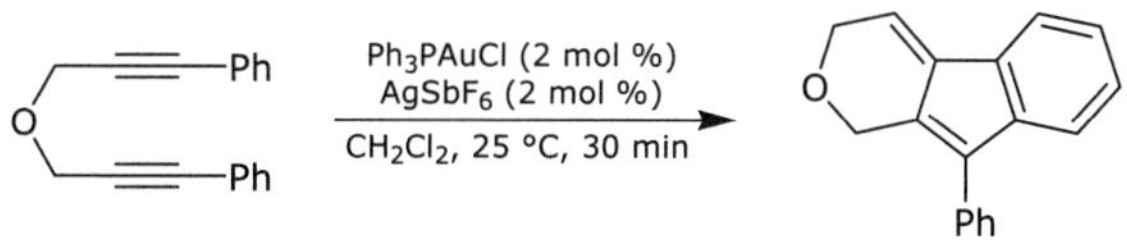

A solution of $Ph_3PAuSbF_6$ (2 mol%) was prepared by mixing Ph_3PAuCl (4.0 mg, 8 μmol) and $AgSbF_6$ (2.1 mg, 8 μmol) in CH_2Cl_2 (1.0 mL). To this solution was added 1-(3-(3-phenylprop-2-ynyloxy)prop-1-ynyl)benzene (100 mg, 0.41 mmol) at 25 °C, and the mixture was stirred for 30 min. The resulting solution was filtered through a Celite bed and eluted through a silica gel column (EtOAc/hexane = 1/15) to give 1,3-dihydro-9-phenylindeno[2,1-*c*]pyran (93.0 mg, 3.78 mmol, 93%) as yellow solid.

Intermolecular[113] or intramolecular [2+2+2]-cycloadditions[114] of 1,6-enynes and carbonyl compounds take place in the presence of cationic gold catalysts. The latter have been used successfully in natural products synthesis.[115] A skeletal rearrangement of a gold carbene intermediate is the key step in these transformations.

Representative procedure for the gold-catalyzed intermolecular [2+2+2]-cycloaddition of 1,6-enynes and aldehydes. Dimethyl 3-(2,4-dimethylphenyl)-1,1-dimethyl-3,5,7,7a-tetrahydrocyclopenta[c]pyran-6,6(1H)-dicarboxylate[113]

MeO_2C
MeO_2C
+
OHC
[IPrAuNCPh][SbF_6] (2 mol %)
CH_2Cl_2, -40 °C, 12 h
MeO_2C
MeO_2C
O
H
i-Pr
i-Pr
N
N
SbF_6^-
Au
i-Pr
i-Pr
NCPh
[IPrAuNCPh][SbF_6]

A solution of dimethyl 2-(3-methylbut-2-enyl)-2-(prop-2-ynyl)malonate (82.6 mg, 0.3467 mmol) and 2,4-dimethylbenzaldehyde (0.098 mL, 0.694 mmol) in CH_2Cl_2 (concentration *ca.* 0.1 M) was cooled to –40 °C and [IPrAuNCPh][SbF_6] (2 mol%) was added after 15 min. The solution was kept at –40 °C for 12 h. A 0.1-M solution of Et_3N in hexane was added, and the solution was filtered through Celite. After evaporation, the crude product was purified by column chromatography (10:1 hexane/EtOAc) to give dimethyl 3-(2,4-dimethylphenyl)-1,1-dimethyl-3,5,7,7a-tetrahydrocyclopenta[*c*]pyran-6,6(1*H*)-dicarboxylate (99.8 mg, 77% yield) as a colorless oil.

Representative procedure for the gold-catalyzed intramolecular [2+2+2]-cycloaddition of ketoenynes. (1S,3aR,4S,5R,7R)-1-(Triethylsilyloxy)-7-isopropyl-1,4-dimethyl-1,2,3,3a,4,5,6,7-octahydro-4,7-epoxyazulen-5-ol[115b]

OH O
OTES
[IPrAuNCPh][SbF_6] (3 mol %)
CH_2Cl_2, MS 4 Å, 23 °C, 5 h
OTES
O
H
OH

[IPrAuNCPh][SbF_6] (11.7 mg, 0.0127 mmol) was added at RT to a solution of (5*R*,10*S*,*E*)-5-hydroxy-10-(triethylsilyloxy)-2,6,10-trimethyldodec-6-en-11-yn-3-one (155 mg, 0.423 mmol) in CH_2Cl_2 (5 mL) containing 4-Å molecular sieves. After stirring the mixture for 5 h, the reaction was stopped by the addition of NEt_3 (0.1 mL). Filtration over SiO_2 and evaporation of the solvent followed by chromatographic purification (3:1 hexane/ Et_2O) of the crude material yielded (1*S*,3a*R*,4*S*,5*R*,7*R*)-1-(triethylsilyloxy)-7-isopropyl-1,4-dimethyl-1,2,3,3a,4,5,6,7-octahydro-4,7-epoxyazulen-5-ol as a colorless oil (90 mg, 58%). The tricycle was obtained as a single diastereoisomer.

Formal intramolecular Diels-Alder reactions catalyzed by gold(I) complexes are also possible with 1,3-dien-8-ynes.[116] Whereas substrates with a terminal triple bond tend to decompose in the presence of a gold catalyst, dienynes with a silyl group at the alkyne afforded the [4+2]-cycloaddition products in high yields in the presence of Ph_3PAuCl and $AgSbF_6$ (Scheme 4-45; the silyl group was lost upon work-up). Copper salts also catalyze this transformation, even though the mechanism is probably different. The method can be applied to dienynes bearing a silyl enol ether moiety as well, furnishing tetrahydroindene derivatives as a single diastereomer.[117]

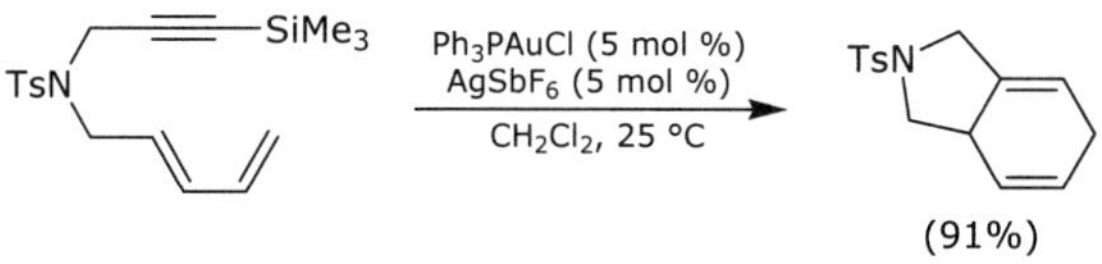

SiMe3 TsN Ph3PAuCl (5 mol %) AgSbF6 (5 mol %) CH2Cl2, 25 °C TsN (91%)

Scheme 4-45

Representative procedure for the gold-catalyzed intramolecular [4+2]-cycloaddition of 1,3-dien-8-ynes. Dimethyl 3,7-diphenyl-5-(triisopropylsiloxy)bicyclo[4.3.0]nona-1,4-dien-8,8-dicarboxylate[117]

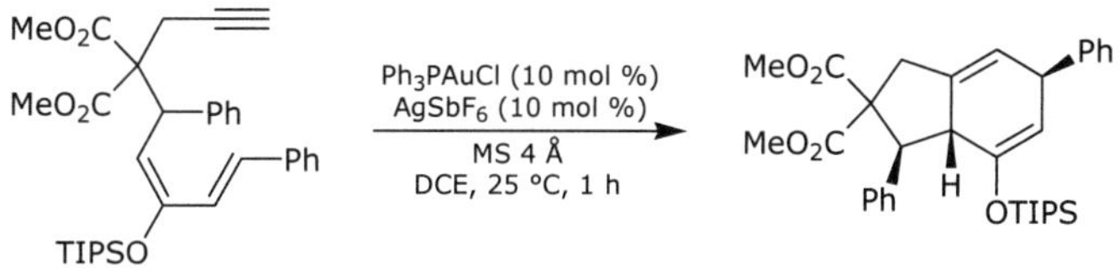

MeO2C MeO2C Ph Ph TIPSO Ph3PAuCl (10 mol %) AgSbF6 (10 mol %) MS 4 Å DCE, 25 °C, 1 h MeO2C MeO2C Ph Ph H OTIPS

Dimethyl 2-((2*E*,4*E*)-3-(triisopropylsiloxy-1,5-diphenylpenta-2,4-dienyl)-2-(prop-2-ynyl)malonate (56 mg, 0.10 mmol) in degassed dichloroethane (1 mL) was added to a mixture of Ph_3PAuCl (4.9 mg, 9.9 μmol, 10 mol%), $AgSbF_6$ (3.4 mg, 9.9 μmol, 10 mol%), and activated MS 4 Å (100 mg), and the resulting mixture was stirred at RT until TLC showed the complete disappearance of the starting material (1 h). The reaction mixture was then filtered through a short pad of silica gel, and the filtrate was concentrated under reduced pressure. The residue was purified by preparative TLC (hexane: EtOAc = 9:1) to give dimethyl 3,7-diphenyl-5-(triisopropylsiloxy)bicycle [4.3.0]nona-1,4-dien-8,8-dicarboxylate (42 mg, 0.075 mmol) in 75% yield as a colorless oil.

High enantioselectivities can be achieved in intramolecular [2+2]-cycloadditions of allenenes using DTBM-SEGPHOS-(5,5′-bis[di(3,5-di-*tert*-butyl-4-methoxyphenyl)phosphino]-4,4′-bi-1,3-benzodioxole)-gold pre-catalyst **H**[118] or TADDOL-(α,α,α′,α′-tetraaryl-1,3-dioxolan-4,5-dimethanol)-derived gold complex **I**[119] in the presence of $AgBF_4$ (Scheme 4-46). By using electron-rich *N*-heterocyclic carbene (NHC)-gold catalysts, an alternative pathway toward [3+2]-cycloaddition products can be favored.[120]

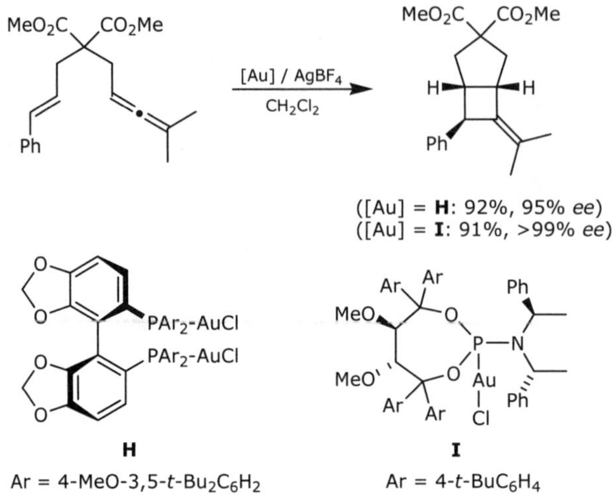

([Au] = **H**: 92%, 95% *ee*)
([Au] = **I**: 91%, >99% *ee*)

H
Ar = 4-MeO-3,5-*t*-$Bu_2C_6H_2$

I
Ar = 4-*t*-BuC_6H_4

Scheme 4-46

Pre-catalyst **I** and related gold phosphoramidite complexes also promote enantioselective intramolecular [4+2]-cycloadditions of allenic dienes.[119a, 121] This transformation was previously studied by Toste *et al.*[122] who obtained the Diels-Alder adduct in the presence of cationic gold species formed from arylphosphite gold chlorides and $AgSbF_6$ (Scheme 4-47). By contrast, [4+3]-cycloaddition products are obtained almost exclusively with gold catalysts bearing electron-rich σ-donor ligands. This is attributed to stabilization of a gold carbenoid intermediate.[120, 122, 123] Intermolecular gold-catalyzed [4+2]-cycloadditions of allenes with conjugated dienes have also been reported recently.[124] Moreover, suitably substituted allenes participate in gold-catalyzed (formal) [3+2]- and [3+3]-cycloadditions.[125]

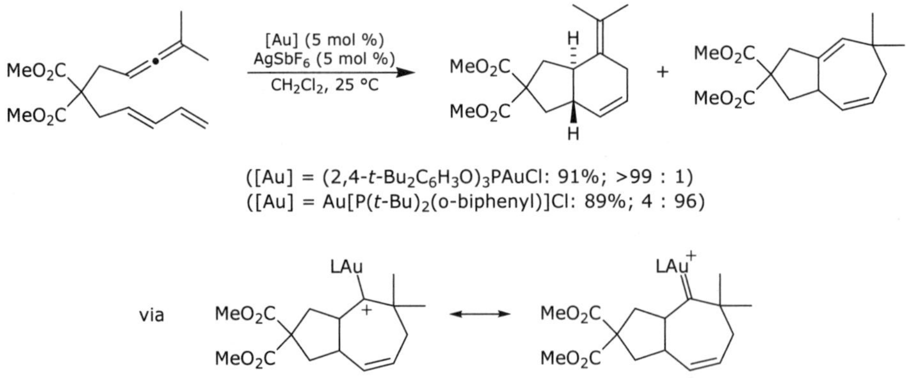

([Au] = (2,4-*t*-$Bu_2C_6H_3O)_3PAuCl$: 91%; >99 : 1)
([Au] = Au[P(*t*-$Bu)_2$(o-biphenyl)]Cl: 89%; 4 : 96)

Scheme 4-47

3.6 Rearrangements

Propargyl vinyl ethers undergo Claisen rearrangements in the presence of gold(I) catalysts to afford β-allenic aldehydes that, because of their limited stability, were immediately reduced to the corresponding alcohols.[126] The gold-oxo complex $[(Ph_3PAu)_3O]BF_4$ gave the best results in this transformation, which tolerates various substituents, including ester groups and silyl ethers. Enantiomerically enriched propargyl vinyl ethers gave the allenes with excellent *center-to-axis chirality transfer* and high diastereoselectivity. The reaction probably proceeds by 6-*endo*-dig addition of the enol ether to the activated triple bond *via* a half-chair transition state and subsequent Grob fragmentation of the cationic vinylgold intermediate. The rearrangement also occurs with propargyl allyl ethers that first suffer double-bond isomerization to propargyl vinyl ethers.[127] In the presence of cationic gold(I)-NHC catalysts, allenyl vinyl ethers undergo a Claisen rearrangement to substituted 1,3-dienes.[128]

Representative procedure for the gold-catalyzed Claisen rearrangement of propargyl vinyl ethers. (R)-3-(2-Phenylvinylidene)heptan-1-ol[126a]

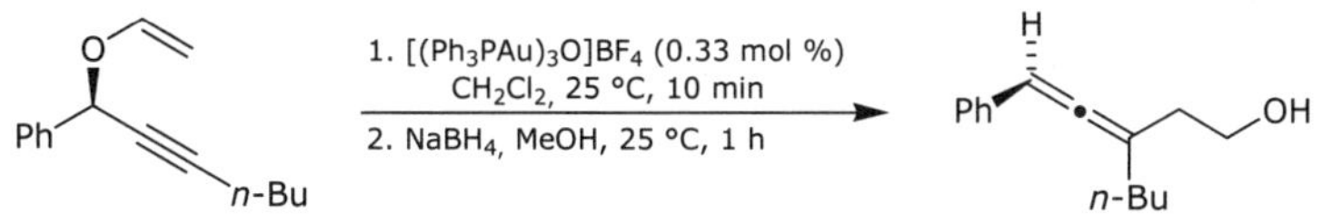

1. $[(Ph_3PAu)_3O]BF_4$ (0.33 mol %), CH_2Cl_2, 25 °C, 10 min
2. $NaBH_4$, MeOH, 25 °C, 1 h

To a solution of 1-((*R*)-1-(vinyloxy)hept-2-ynyl)benzene (0.100 g, 0.467 mmol, 95% *ee*) in CH_2Cl_2 (1.3 mL) was added $[(Ph_3PAu)_3O]BF_4$ (2.3 mg, 1.6 μmol), and the resulting solution was left to stir at RT for 10 min. The reaction mixture was diluted with MeOH (2.6 mL), $NaBH_4$ (17.7 mg, 0.467 mmol) was added, and the solution was left to stir at RT for 1 h. The red solution was concentrated and the crude oil was purified by flash column chromatography (9:1 hexanes: EtOAc eluent) to afford (*R*)-3-(2-phenylvinylidene)heptan-1-ol (92 mg, 91% yield, 90% *ee*, R_f = 0.14 in 9:1 hexanes/EtOAc) as a colorless oil.

In the presence of silver[126a] or gold salts,[129] β-hydroxyallenes undergo a 6-*endo*-trig cyclization to 5,6-dihydro-2*H*-pyrans (see Section 5.2). If the gold-catalyzed Claisen rearrangement of propargyl vinyl ethers was carried out in the presence of an alcohol or water, alkoxy-/hydroxy-substituted dihydropyrans were obtained in high yields (Scheme 4-48).[130] This tandem transformation probably proceeds *via* nucleophilic attack at an intermediate oxocarbenium ion and afforded the heterocycles as a diastereomeric mixture. A related synthesis of dihydrobenzofurans by gold-catalyzed tandem Claisen rearrangement/cyclization of aryl allyl ethers has been described.[131]

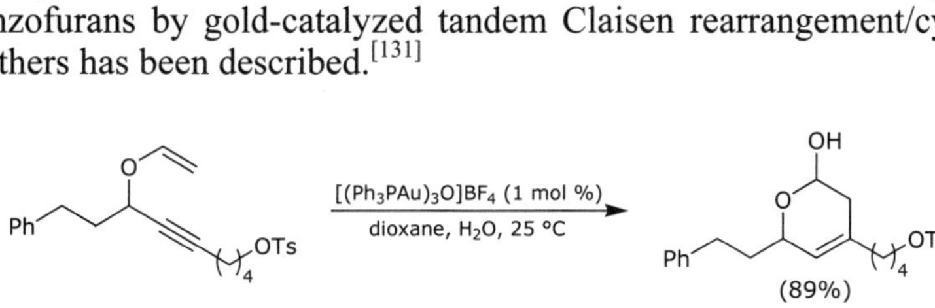

$[(Ph_3PAu)_3O]BF_4$ (1 mol %), dioxane, H_2O, 25 °C

(89%)

Scheme 4-48

General procedure for the gold-catalyzed Claisen rearrangement/cyclization of propargyl vinyl ethers to dihydropyrans[130]

To a solution of propargyl vinyl ether (1 equiv) in 1,4-dioxane (0.30 M) is added water (1 equiv) followed by $[(Ph_3PAu)_3O]BF_4$ (1 mol%). The resulting heterogeneous mixture is left to stir at RT, and dissolution of the catalyst is observed within 10 min. The reaction mixture is monitored until TLC analysis indicates consumption of the starting material. The solution is filtered through a silica gel plug (1:1 hexanes:EtOAc), and the filtrate is concentrated. The diastereoselectivity is ascertained by ^{1}H NMR analysis of the crude reaction mixture, and the resulting residue is purified by flash column chromatography to afford the desired dihydropyran product.

In an analogous manner, propargyl vinyl ethers with a pendant alcohol group as internal nucleophile gave 5,6- or 6,6-spiroketals with excellent diastereoselectivity (Scheme 4-49).[130] This transformation is an example for efficient *center-to-axis-to-center chirality transfer*.

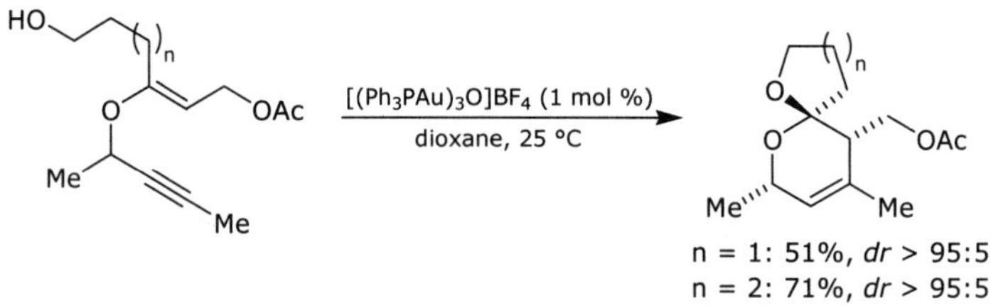

$[(Ph_3PAu)_3O]BF_4$ (1 mol %)
dioxane, 25 °C

n = 1: 51%, *dr* > 95:5
n = 2: 71%, *dr* > 95:5

Scheme 4-49

In the presence of Ph_3PAuCl and $AgBF_4$, the Claisen rearrangement of propargyl vinyl ethers bearing a carbonyl group in the 1-position leads to allenyl β-dicarbonyl compounds that immediately cycloisomerize to the corresponding furan derivatives (Scheme 4-50).[132] Platinum and silver catalysts also promote this tandem transformation, even though their reactivity is much lower.[132, 133] The method can be applied to the synthesis of nitrogen heterocycles as well.[134]

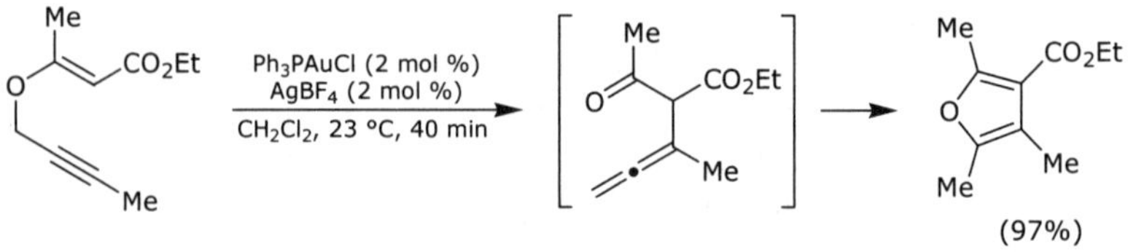

Ph_3PAuCl (2 mol %)
$AgBF_4$ (2 mol %)
CH_2Cl_2, 23 °C, 40 min

(97%)

Scheme 4-50

The related Gosteli-Claisen rearrangement of 2-alkoxycarbonyl-substituted propargyl vinyl ethers can also be catalyzed by cationic gold(I) complexes and provides highly substituted furan derivatives.[135]

Representative procedure for the gold-catalyzed Gosteli-Claisen rearrangement/ cycloisomerization of alkoxycarbonyl-substituted propargyl vinyl ethers. Isopropyl 4,5-dimethyl-3-(4-nitrophenyl)furan-2-carboxylate[135]

CO_2i-Pr
O
Me
NO_2
Ph_3PAuCl (5 mol %)
$AgSbF_6$ (5 mol %)
CH_2Cl_2, 25 °C, 20 min
NO_2
i-PrO_2C
O
Me
Me

To a solution of $AgSbF_6$ (6 mg, 0.016 mmol, 0.05 equiv) in CH_2Cl_2 (1 mL) was added Ph_3PAuCl (8 mg, 0.016 mmol, 0.05 equiv) in the dark. The white suspension was stirred in the dark at RT for 30 min. In the meantime, the solution changed color to light purple. Subsequently, (*Z*)-isopropyl 2-(but-2-ynyloxy)-3-(4-nitrophenyl)acrylate (0.103 g, 0.33 mmol, 1 equiv) was added with CH_2Cl_2 (0.5 mL). The color of the solution changed immediately to dark brown, and heat generation could be observed. The reaction mixture was stirred at RT for 20 min in the dark. The reaction was stopped by the addition of silica gel. The solvent was removed under reduced pressure. Purification by flash chromatography (cyclohexane/EtOAc 100/1) afforded isopropyl 4,5-dimethyl-3-(4-nitrophenyl)furan-2-carboxylate (0.098 g, 0.32 mmol, 96%) as a white solid.

4. C–N Bond Formation

Reviews. Widenhoefer, R.A.; Han, X.; *Eur. J. Org. Chem.* **2006**, 4555; Hashmi, A.S.K.; *Chem. Rev.* **2007,** *107*, 3180; Li, Z.; Brouwer, C.; He, C.; *Chem. Rev.* **2008,** *108*, 3239; Arcadi, A.; *Chem. Rev.* **2008,** *108*, 3266; Gorin, D.J.; Sherry, B.D.; Toste, F.D.; *Chem. Rev.* **2008,** *108*, 3351; Patil, N.T.; Yamamoto, Y.; *Chem. Rev.* **2008,** *108*, 3395; Shen, H.C.; *Tetrahedron* **2008,** *64*, 3885; Corma, A.; Leyva-Pérez, A.; Sabater, M.J.; *Chem. Rev.* **2011,** *111*, 1657; Krause, N.; Winter, C.; *Chem. Rev.* **2011,** *111*, 1994.

4.1 Nucleophilic Additions to Alkynes

Amines and nitrogen heterocycles are highly useful intermediates and target molecules in academia and industry. Among numerous synthetic approaches, the intermolecular or intramolecular hydroamination of alkynes is particularly powerful.[136] Because of their high carbophilicity, late transition metals are the catalysts of choice for the activation of an alkyne in the presence of an amine, even though competing coordination to the nitrogen can lead to considerable deactivation of the catalyst. Nevertheless, gold catalysts have been used for the hydroamination of alkynes as early as 1987 when Utimoto *et al.* described the synthesis of 2,3,4,5-tetrahydropyridines from 5-alkynylamines in the presence of sodium tetrachloroaurate (Scheme 4-51).[137, 138] High dilution conditions were necessary to

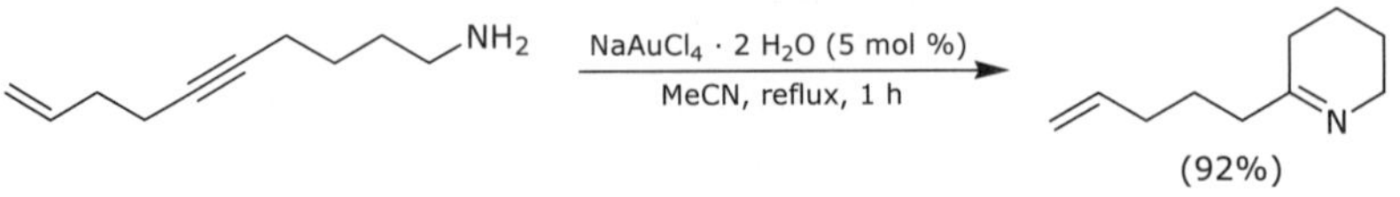

NaAuCl$_4$ · 2 H$_2$O (5 mol %)
MeCN, reflux, 1 h

(92%)

Scheme 4-51

prevent catalyst decomposition. The transformation proceeds by 6-*exo*-selective cyclization and subsequent double-bond migration. The presence of an additional olefin in the substrate does not influence product formation. Likewise, 3- or 4-alkynylamines afforded 3,4-dihydro-2*H*-pyrroles by 5-*endo*- or 5-*exo*-cyclization, whereas indoles were obtained by heating 2-alkynylanilines in the presence of NaAuCl$_4$·2 H$_2$O or PdCl$_2$ (Scheme 4-52).[139] With ethanol or ethanol/water mixtures as the solvent, these cyclization can be performed at room temperature.[140] In the presence of NaAuCl$_4$·2 H$_2$O and *N*-bromosuccinimide (NBS), Br$_2$, or I$_2$, the corresponding 3-bromo- or 3-iodoindoles were obtained in a one-pot hydroamination/ halogenation,[140] whereas with an external alkyne, a double hydroamination to afford

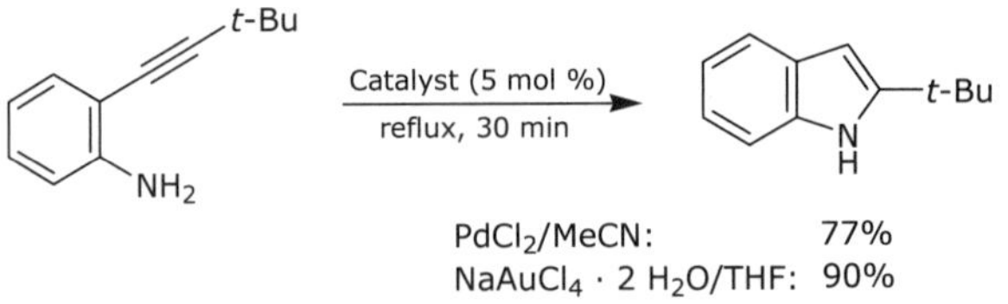

Catalyst (5 mol %)
reflux, 30 min

PdCl$_2$/MeCN: 77%
NaAuCl$_4$ · 2 H$_2$O/THF: 90%

Scheme 4-52

N-vinylindoles took place.[141] Moreover, the gold-catalyzed cyclization of 2-alkynylanilines can be combined with a gold-catalyzed Michael addition to enones,[24a, 66] which afforded 3-alkylated indoles with good yield.[142] Gold-catalyzed hydroamination/hydroarylation cascades of polyenyne-substituted anilines leading to condensed aromatics have been described recently.[143]

General procedure for the gold-catalyzed intramolecular hydroamination of alkynylamines[137b]

The 5-alkynylamine (1 mmol) or 3-/4-alkynylamine (0.66 mmol) is dissolved in MeCN or tetrahydrofuran (THF) (40 mL), and NaAuCl$_4$·2 H$_2$O (20 mg, 5 mol%) is added to the solution. Soon after the addition of the aurate, the color of the solution turns to dark red. The color disappears after several minutes, and the mixture is refluxed for 1–2 h. After evaporation of the solvent *in vacuo*, the residue is diluted with Et$_2$O (30 mL) and washed with a 1:1 mixture of aq. ammonia and brine (30 mL). The etheral layer is dried (anhydrous Na$_2$SO$_4$) and concentrated under reduced pressure, and the crude product is purified by distillation.

Representative procedure for the gold-catalyzed cyclization of 2-alkynylanilines. 5,7-Dichloro-1H-indole[140]

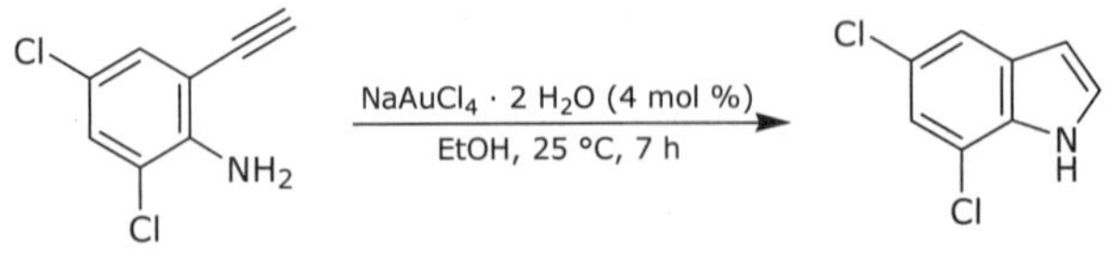

NaAuCl$_4$ · 2 H$_2$O (4 mol %)
EtOH, 25 °C, 7 h

To a solution of 2,4-dichloro-6-ethynylaniline (0.140 g, 0.75 mmol) in EtOH (6 mL) was added $NaAuCl_4 \cdot 2\ H_2O$ (0.012 g, 0.030 mmol). The mixture was stirred at RT for 7 h under N_2 and then purified by chromatography on silica gel (hexane–EtOAc, 98:2) to give 0.110 g (79%) of 5,7-dichloro-1*H*-indole.

Representative procedure for the one-pot cyclization/halogenation of 2-alkynylanilines. 2-Butyl-3-iodo-1H-indole[140]

1. $NaAuCl_4 \cdot 2\ H_2O$ (4 mol %) EtOH, 25 °C, 3.5 h
2. KOH, I_2, EtOH, 25 °C, 2 h

To a solution of 2-hex-1-ynylaniline (0.070 g, 0.40 mmol) in EtOH (1.5 mL) was added $NaAuCl_4 \cdot 2\ H_2O$ (0.006 g, 0.015 mmol). The mixture was stirred under N_2 at RT for 3.5 h, and then KOH (0.057 g, 1.02 mmol) and a solution of I_2 (0.104 g, 0.41mmol) in EtOH (1 mL) were added. The mixture was stirred for a further 2 h and then extracted with 5% sodium thiosulfate (100 mL) and EtOAc (2 × 50 mL). The organic phase was dried (anhydrous Na_2SO_4) and evaporated under reduced pressure. Column chromatographic purification on silica gel (hexane–EtOAc, 95:5) afforded 0.098 g (81%) of 2-butyl-3-iodo-1*H*-indole.

Pyrroles can be synthesized by gold-catalyzed intramolecular hydroamination of various acetylenic substrates; for example, by treatment of acetylenic ketones with primary amines in the presence of gold(I) or gold(III) catalysts.[144] Under these conditions, the initially formed enamine undergoes a gold-catalyzed 5-*exo*-selective hydroamination, which affords aromatic heterocycles after double-bond isomerization. Various other gold-catalyzed syntheses of nitrogen heterocycles involve nucleophilic additions to carbonyl groups or imines.[145]

Representative procedure for the gold-catalyzed synthesis of pyrroles. 1-(1-Benzyl-2,5-dimethyl-1H-pyrrol-3-yl)ethanone[144a]

$BnNH_2$, $NaAuCl_4 \cdot 2\ H_2O$ (5 mol %), EtOH, 40 °C, 2 h

To a solution of 3-(prop-2-ynyl)pentane-2,4-dione (70 mg, 0.51 mmol) and benzylamine (81 mg, 0.76 mmol) in ethanol (3 mL) $NaAuCl_4 \cdot 2\ H_2O$ (10 mg, 0025 mmol) was added at RT under a nitrogen atmosphere. The reaction mixture was heated at 40 °C and stirred for 2 h. The solvent was removed under reduced pressure, and the residue was purified by flash chromatography on silica gel (hexane/EtOAc, 95/5) to give 1-(1-benzyl-2,5-dimethyl-1*H*-pyrrol-3-yl)ethanone; yield: 120 mg (quantitative).

In a related study, 1-en-4-yn-3-ols were treated with tosylamide in the presence of tetrachloroauric acid to yield 1-amino-1-en-3-ynes, which cyclize under the reaction conditions to the corresponding pyrroles (Scheme 4-53).[146] For this tandem amination/intramolecular hydroamination, a rather high catalyst loading of

20 mol% was required. A stepwise process was possible by using 10 mol% $HAuCl_4$. Similar gold-catalyzed cyclizations leading to six-membered nitrogen heterocycles have also been reported.[147]

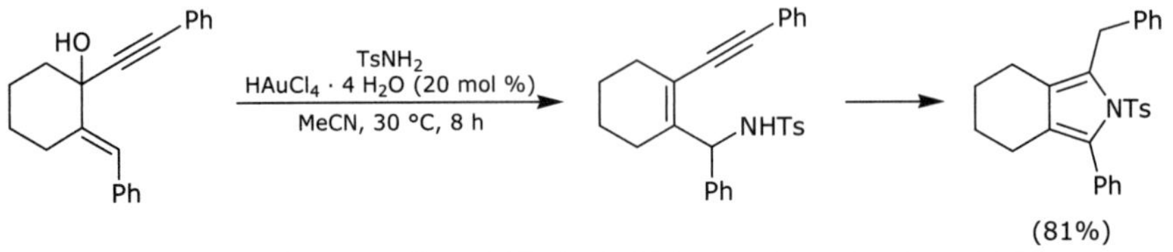

Scheme 4-53

Monosubstituted pyrroles were formed in excellent yields by dehydrative cyclization of acetylenic amino alcohols in the presence of Au[P(*t*-Bu)$_2$(*o*-biphenyl)]Cl and AgOTf (Scheme 4-54).[148] The transformation proceeds not only with tosylamides, but also with basic amino groups, even though longer reaction times were required. Homopropargyl azides are also transformed to substituted pyrroles in the presence of cationic gold(I) catalysts.[149]

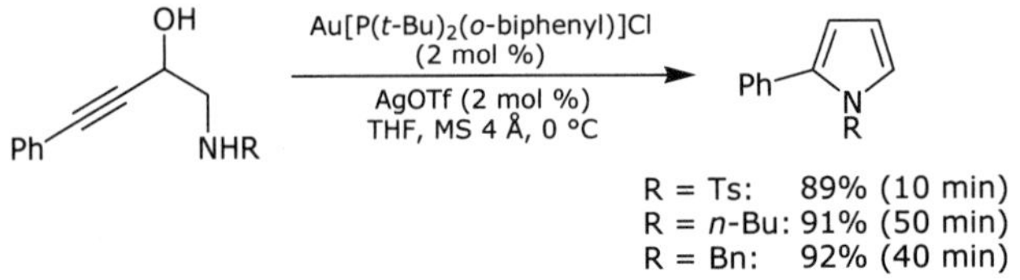

Scheme 4-54

By analogy to propargyl oxiranes (Section 5.1), suitably substituted propargyl aziridines undergo a cycloisomerization to the corresponding pyrroles in the presence of gold(I) catalysts (Scheme 4-55).[150] An interesting counterion effect was observed in these transformations. Use of $Ph_3PAuOTs$ afforded 2,5-disubstituted pyrroles originating from sequential intramolecular attack of the aziridine nitrogen at the activated alkyne, deprotonation, and protodeauration. By contrast, the isomeric 2,4-disubstituted pyrroles were obtained in the presence of $Ph_3PAuOTf$. In the latter case, deprotonation of the intermediate σ-vinylgold species seems to be disfavored because of the absence of a sufficiently basic anion, so that a 1,2-aryl shift takes place instead.

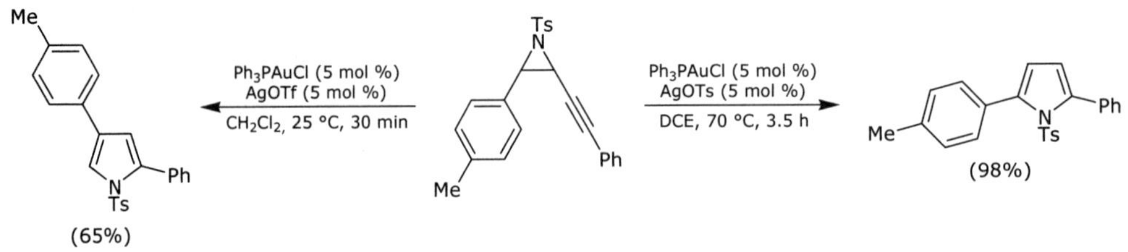

Scheme 4-55

Acyloxylated alkynylaziridines react with an external nucleophile (methanol or ethanol) in the presence of $Ph_3PAuSbF_6$ to afford pyrroles by tandem rearrangement/nucleophilic substitution (Scheme 4-56).[151]

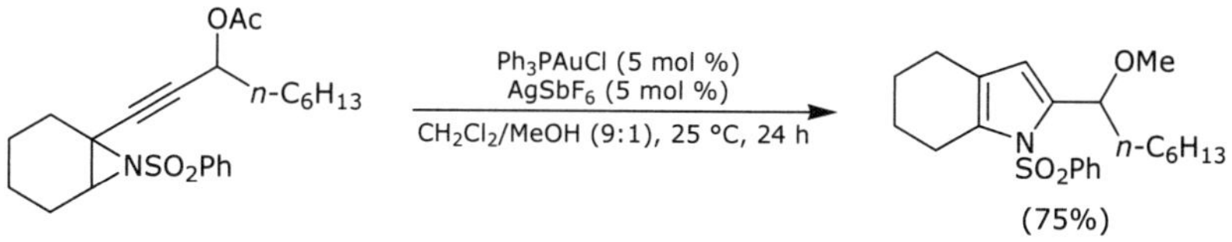

OAc; n-C$_6$H$_{13}$; NSO$_2$Ph → Ph$_3$PAuCl (5 mol %), AgSbF$_6$ (5 mol %), CH$_2$Cl$_2$/MeOH (9:1), 25 °C, 24 h → OMe; n-C$_6$H$_{13}$; N SO$_2$Ph (75%)

Scheme 4-56

Substituted indolizines and related heterocycles are accessible from propargylpyridines in the presence of $AuBr_3$ or AuCl (Scheme 4-57).[152] This remarkable transformation is initiated by a 1,2-migration of a silyl group to afford a gold vinylidene intermediate, which then suffers an intramolecular attack of the nitrogen atom. The resulting zwitterion delivers the aromatic heterocycle by a series of 1,2-hydride shifts. Stannyl or germyl groups, as well as protons, can also be used for the initial migration.

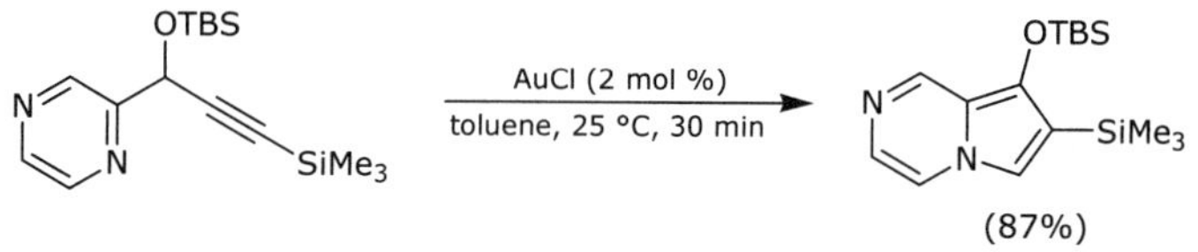

OTBS; N; N; SiMe$_3$ → AuCl (2 mol %), toluene, 25 °C, 30 min → OTBS; N; N; SiMe$_3$ (87%)

Scheme 4-57

The intermolecular hydroamination of terminal or internal alkynes with anilines to form imines takes place at elevated temperatures in the presence of a cationic gold(I) catalyst formed *in situ* from Ph_3PAuMe and $H_3PW_{12}O_{40}$.[153] Other acidic promoters gave inferior results. Electron-withdrawing substituents at the amine, as well as electron-donating groups at the alkyne led to increased rates, allowing turnover numbers close to 9000. The authors assume coordination of both the alkyne and the amine at the gold catalyst, followed by an innersphere C–N bond formation and proton transfer. Alternatively, the intermolecular hydroamination of terminal alkynes with aromatic amines can be catalyzed by $AuCl_3$ or $AuCl_3$/AgOTf,[154] as well as by various other cationic gold(I) complexes.[155] Chiral secondary amines are accessible with high *ee* by combining the gold-catalyzed hydroamination of alkynes with an enantioselective transfer hydrogenation.[156]

Representative procedure for the intermolecular gold-catalyzed hydroamination of alkynes. (E)-4-Bromo-N-(1-phenylethylidene)aniline[153]

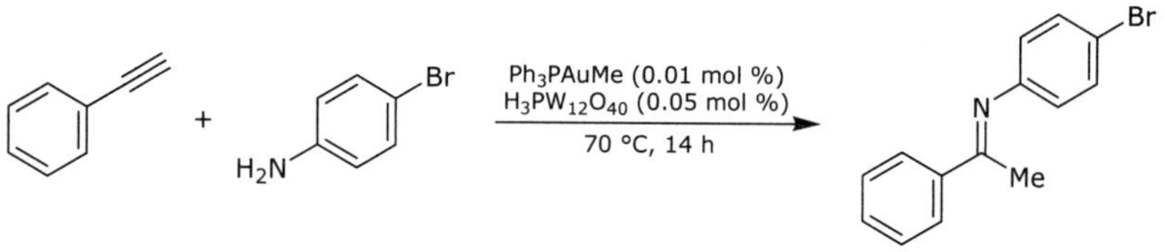

+ H$_2$N; Br → Ph$_3$PAuMe (0.01 mol %), H$_3$PW$_{12}$O$_{40}$ (0.05 mol %), 70 °C, 14 h → Br; N; Me

Under a nitrogen atmosphere, a mixture of phenylacetylene (2.17 g, 21 mmol), 4-bromoaniline (3.61 g, 21 mmol), Ph_3PAuMe (1.0 mg, 0.0021 mmol), and $H_3PW_{12}O_{40}$

(30 mg, 0.01 mmol) was stirred at 70 °C for 14 h. After the mixture was cooled, hexane (25 mL) was added to the reaction mixture to precipitate out a white powder. The powder was recrystallized from hexane-CH_2Cl_2 to give (*E*)-4-bromo-*N*-(1-phenylethylidene)aniline (4.9 g, 18 mmol, 85% isolated yield).

4.2 Nucleophilic Additions to Allenes

The first gold-catalyzed intramolecular hydroamination of allenes was reported by Morita and Krause in 2004.[157] With $AuCl_3$ various α-aminoallenes were converted into the corresponding 3-pyrrolines with high levels of chirality transfer (Scheme 4-58). Isomers observed spectroscopically for Ac- or Boc-protected pyrrolines are rotamers with regard to the amide bond. Whereas short reaction times (30 min) were observed for protected aminoallenes, 5 days at room temperature were required for full conversion of the corresponding unprotected aminoallene. The diminished reactivity is probably caused by the deactivation of the gold catalyst by the Lewis-basic amino group. With gold(I) chloride instead of $AuCl_3$, the reaction time was decreased to several hours at room temperature.[158] Compared with the corresponding cyclizations in the presence of $AgNO_3$ or $Ru_3(CO)_{12}$,[159] gold catalysis is advantageous in terms of lower catalyst loadings and milder reaction conditions. The method was applied to the synthesis of bicyclic β-lactams[160] and tricyclic pyrroloisoindolones.[161] Analogously, β-aminoallenes undergo a slow gold-catalyzed 6-*endo*-cycloisomerizion to the corresponding tetrahydropyridines.[129]

R2 R1 R3 H NHPG [Au] R2 R1 N R3 PG

PG = Ms, Ts, Ac, Boc, H

Scheme 4-58

Representative procedure for the gold-catalyzed cycloisomerization of protected α-aminoallenes. 2-Benzyloxymethyl-5-isopropyl-3-methyl-1-(toluene-4-sulfonyl)-2,5-dihydro-1H-pyrrole[157]

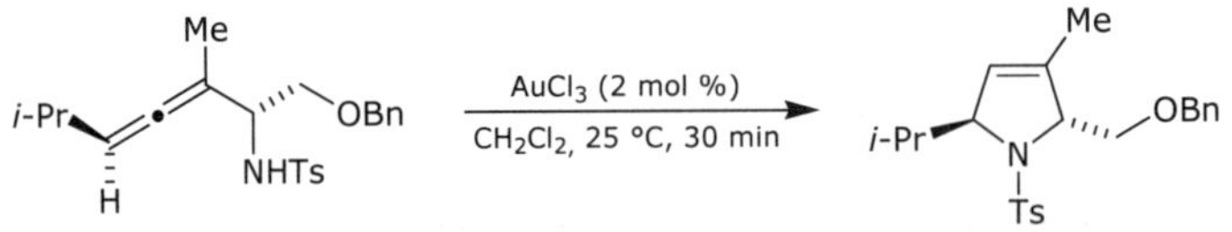

Me *i*-Pr OBn H NHTs $AuCl_3$ (2 mol %) CH_2Cl_2, 25 °C, 30 min Me *i*-Pr OBn N Ts

A 0.165-M solution of $AuCl_3$ in CH_3CN (18 μL, 2 mol%) was added to a solution of *N*-(1-benzyloxymethyl-2,5-dimethylhexa-2,3-dienyl)-4-methylbenzenesulfonamide (58 mg, 0.15 mmol) in dry CH_2Cl_2 (4 mL) at RT under argon. The reaction was monitored by TLC; after 30 min, the solvent was removed *in vacuo* and the crude product was submitted to SiO_2 column chromatography (CH_2Cl_2/MeOH, 10/1) to give 2-benzyloxymethyl-5-isopropyl-3-methyl-1-(toluene-4-sulfonyl)-2,5-dihydro-1*H*-pyrrole (54 mg, 93%, *dr* = 95:5 according to GC and NMR analysis) as a colorless oil.

Representative procedure for the gold-catalyzed cycloisomerization of unprotected α-aminoallenes. 2-(t-Butyldimethylsilyloxymethyl)-3-methyl-5-phenyl-2,5-dihydro-1H-pyrrole[157]

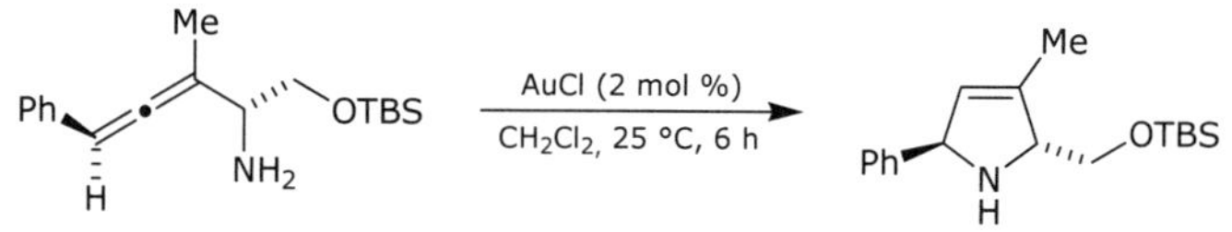

AuCl (2 mol %)
CH_2Cl_2, 25 °C, 6 h

AuCl (0.5 mg, 2 mol%) was added to a solution of 1-(*t*-butyldimethylsilyloxymethyl)-2-methyl-4-phenylbuta-2,3-dienylamine (29 mg, 0.10 mmol) in dry CH_2Cl_2 (3 mL) at RT under argon. After complete conversion (6 h), the solvent was removed *in vacuo* and the crude product was submitted to SiO_2 column chromatography (CH_2Cl_2/MeOH, 10/1) to give 2-(*t*-butyldimethylsilyloxymethyl)-3-methyl-5-phenyl-2,5-dihydro-1*H*-pyrrole (20 mg, 69%) as a colorless oil.

Application of the method to allenic hydroxylamine derivatives proved to be particularly useful because three different chiral heterocycles can be obtained with high regioselectivity and stereoselectivity, depending on the starting material, the gold catalyst, and the protecting group at nitrogen (Scheme 4-59).[162] In all cases, the nitrogen atom acts as the nucleophile and attacks the allene in a 5- or 6-*endo* cyclization. Thus, *N*-hydroxy-α-aminoallenes afforded *N*-hydroxypyrrolines with complete *axis-to-center chirality transfer* in the presence of gold(I) chloride, whereas treatment of unprotected allenic hydroxylamine ethers (with exchanged positions of the heteroatoms) with cationic gold(I) complexes yielded 4,5-dihydroisoxazoles. By contrast, *N*-Boc-protected hydroxylamine ethers underwent a selective 6-*endo*-cyclization to 3,6-dihydro-1,2-oxazines catalyzed by AuCl.

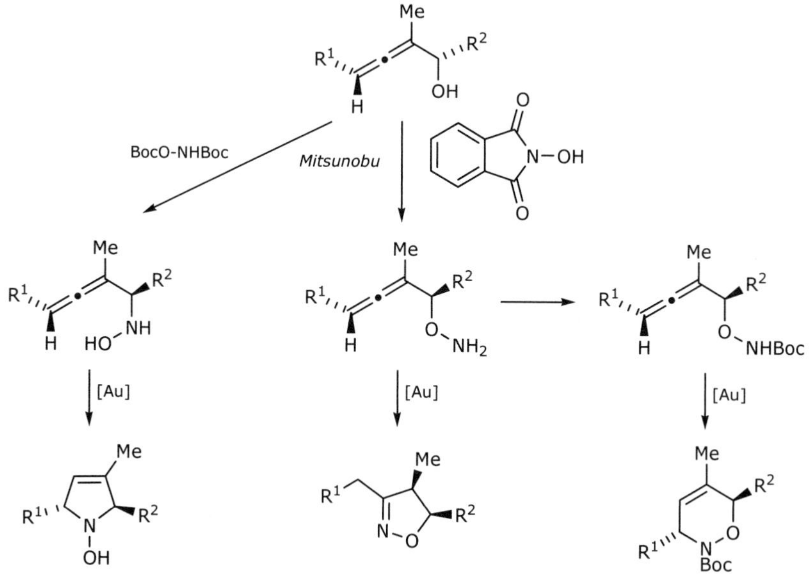

BocO-NHBoc
Mitsunobu
[Au]
[Au]
[Au]

Scheme 4-59

Representative procedure for the gold-catalyzed cycloisomerization of N-hydroxy-α-aminoallenes. N-Hydroxy-2-((benzyloxy)methyl)-2,5-dihydro-5-isopropyl-3-methyl-1H-pyrrole[162]

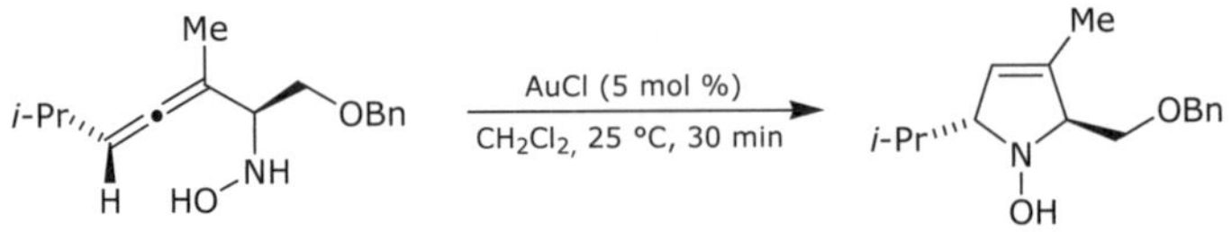

AuCl (5 mol %)
CH_2Cl_2, 25 °C, 30 min

In an oven-dried Schlenk tube, 1-(benzyloxy)-*N*-hydroxy-3,6-dimethylhepta-3,4-dien-2-amine (60.0 mg, 230 μmol) was dissolved in dry CH_2Cl_2 (4 mL) and treated with AuCl (2.7 mg, 11.5 μmol). After complete conversion (30 min, monitored by TLC), the solvent was removed under reduced pressure and the crude product was purified by flash column chromatography (SiO_2, cyclohexane/EtOAc/Et_3N = 91:6:3), affording 56.5 mg (94%) of *N*-hydroxy-2-((benzyloxy)methyl)-2,5-dihydro-5-isopropyl-3-methyl-1*H*-pyrrole as a yellow oil.

Allenic hydrazones undergo a gold(I)-catalyzed cycloisomerization to substituted pyrroles, which are obtained with excellent yield (Scheme 4-60).[163] This transformation requires heating of the allene with $Ph_3PAuNTf$-toluene complex in dichloromethane to 100 °C for 20 min. A 1,2-alkyl or aryl shift is key to the formation of the aromatic heterocycle.

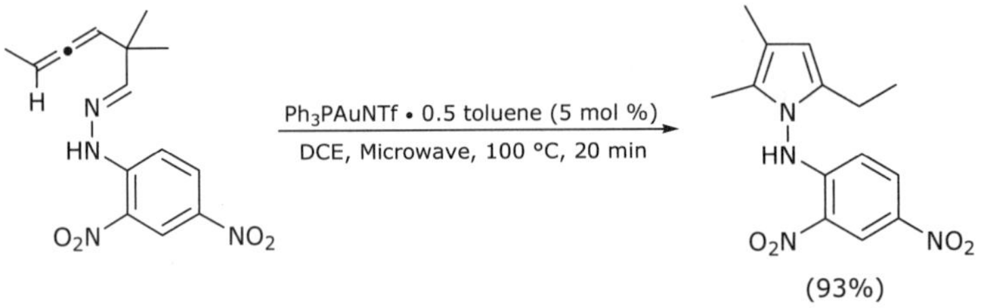

$Ph_3PAuNTf$ • 0.5 toluene (5 mol %)
DCE, Microwave, 100 °C, 20 min

(93%)

Scheme 4-60

In contrast to α- or β-aminoallenes, substrates with an amino group in the γ- or δ-position undergo *exo*-selective hydroaminations in the presence of gold catalysts. Yamamoto *et al.*[164] obtained 2-vinylpyrrolidines or 2-vinylpiperidines by treatment of the corresponding aminoallene with gold(I) chloride (Scheme 4-61). The low catalyst loading of 1 mol% in the case of the γ-aminoallene (n = 1) is revealing a high reactivity of these substrates towards the gold catalyst, whereas the formation of the six-membered ring (n = 2) required 5 mol% and longer reaction times (24 h instead of 3 h). In the case of the enantiomerically enriched γ-aminoallene, efficient *axis-to-center*

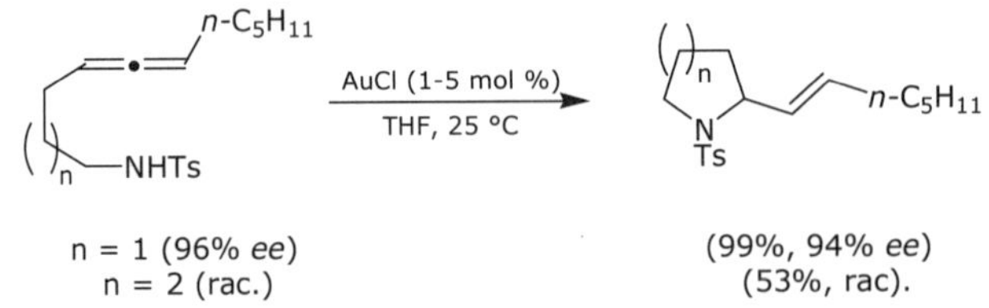

AuCl (1-5 mol %)
THF, 25 °C

n = 1 (96% *ee*) — (99%, 94% *ee*)
n = 2 (rac.) — (53%, rac).

Scheme 4-61

chirality transfer was observed. Analogous results were obtained by Widenhoefer *et al.*[47a] using a phosphine-stabilized gold complex and silver triflate in dioxane at room temperature. The method was applied in natural product synthesis[165] and also employed for the cycloisomerization of allenic ureas[166] or allenamides[167] to the corresponding imidazolidinones.

Representative procedure for the gold-catalyzed cycloisomerization of γ-aminoallenes. Benzyl 4,4-diphenyl-2-vinylpyrrolidine-1-carboxylate[47a]

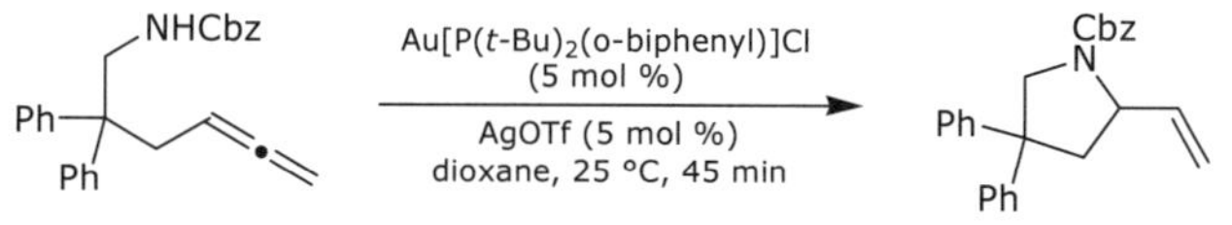

Au[P(*t*-Bu)$_2$(o-biphenyl)]Cl (5 mol %), AgOTf (5 mol %), dioxane, 25 °C, 45 min

A suspension of benzyl 2,2-diphenylhexa-4,5-dienylcarbamate (96 mg, 0.25 mmol), Au[P(*t*-Bu)$_2$(*o*-biphenyl)]Cl (6.9 mg, 13 μmol), and AgOTf (3.3 mg, 13 μmol) in dioxane (2 mL) was stirred at 25 °C for 45 min. The crude reaction mixture was chromatographed (SiO_2; hexanes-EtOAc, 4:1) to give benzyl 4,4-diphenyl-2-vinylpyrrolidine-1-carboxylate (91 mg, 95%) as a viscous, colorless oil.

The Widenhoefer group extended the method to the synthesis of enantiomerically enriched 2-vinylpyrrolidines from monosubstituted or trisubstituted γ-aminoallenes, as well as allenic ureas, in the presence of chiral gold catalysts derived from [Au$_2${(*S*)-3,5-*t*-Bu-4-MeO-MeOBIPHEP}Cl$_2$] and a silver salt (Scheme 4-62).[168]

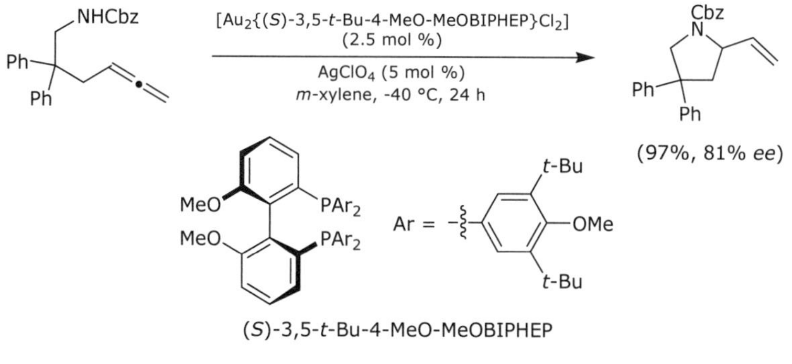

[Au$_2${(*S*)-3,5-*t*-Bu-4-MeO-MeOBIPHEP}Cl$_2$] (2.5 mol %), $AgClO_4$ (5 mol %), *m*-xylene, -40 °C, 24 h

(97%, 81% *ee*)

(*S*)-3,5-*t*-Bu-4-MeO-MeOBIPHEP

Scheme 4-62

Higher levels of chiral induction were achieved with (*R*)-xylyl-BINAP(Au-*p*-nitrobenzoate)$_2$ or (*R*)-ClMeOBIPHEP(Au-*p*-nitrobenzoate)$_2$ as catalyst (Scheme 4-63).[169] These allow the smooth formation of chiral pyrrolidines or piperidines with up to 99% *ee* and high chemical yield from the corresponding trisubstituted tosyl-protected γ- or δ-aminoallene. Gold catalysts with a chiral counterion can also be employed for the highly enantioselective intramolecular *exo*-hydroamination of aminoallenes.[170]

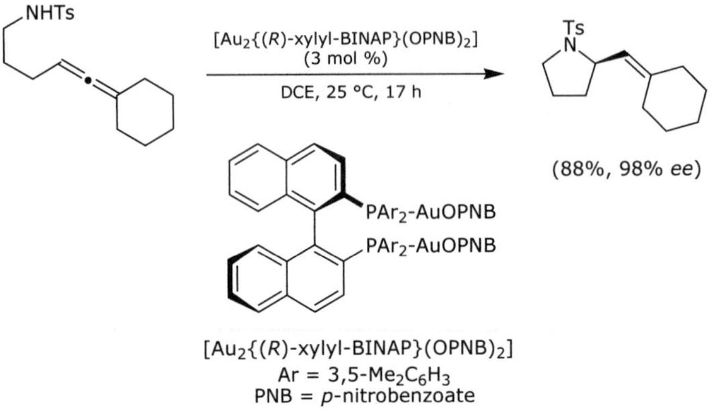

NHTs

$[Au_2\{(R)\text{-xylyl-BINAP}\}(OPNB)_2]$ (3 mol %)

DCE, 25 °C, 17 h

Ts N

(88%, 98% *ee*)

PAr_2-AuOPNB

PAr_2-AuOPNB

$[Au_2\{(R)\text{-xylyl-BINAP}\}(OPNB)_2]$
Ar = 3,5-$Me_2C_6H_3$
PNB = *p*-nitrobenzoate

Scheme 4-63

Recently, the scope of gold-catalyzed intramolecular *exo*-selective hydroaminations was expanded to allenic hydrazines and hydroxylamines.[171] The former substrates afforded pyrazolidines in the presence of DTBM-SEGPHOS-Au complex **J**, whereas $[Au_2\{(R)\text{-xylyl-BINAP}\}(OPNB)_2]$ gave the best results in the cyclization of hydroxylamine derivatives to isoxazolidines. Excellent chemical yields and enantioselectivities were obtained in most cases, and the method was also applied to the synthesis of chiral tetrahydrooxazines.

Representative procedure for the gold-catalyzed cycloisomerization of allenic hydrazines. t-Butyl 2-(mesitylsulfonyl)-3-(2-methylprop-1-enyl)pyrazolidine-1-carboxylate[171]

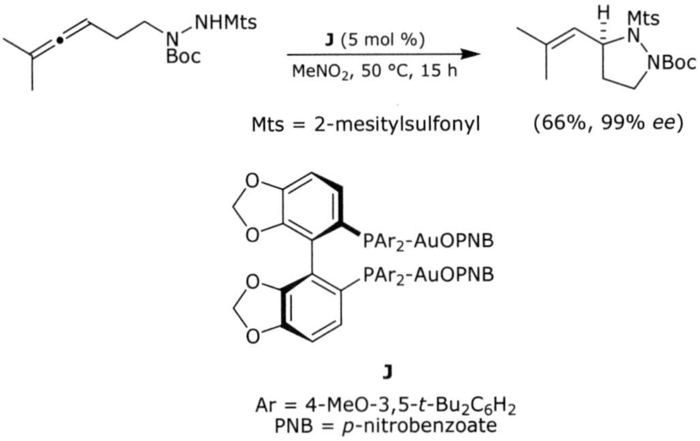

NHMts N Boc

J (5 mol %)

$MeNO_2$, 50 °C, 15 h

H Mts N NBoc

Mts = 2-mesitylsulfonyl

(66%, 99% *ee*)

PAr_2-AuOPNB

PAr_2-AuOPNB

J

Ar = 4-MeO-3,5-*t*-$Bu_2C_6H_2$
PNB = *p*-nitrobenzoate

To a solution of *t*-butyl 2-(mesitylsulfonyl)-1-(5-methylhexa-3,4-dienyl)hydrazine-carboxylate (15 mg, 34 μmol) in $MeNO_2$ (0.3 mL) was added catalyst **J** (3.2 mg, 1.7 μmol). The resulting homogeneous mixture was protected from ambient light and stirred at 50 °C for 15 h. The crude mixture was purified by flash column chromatography (0–6% EtOAc/hexanes) to afford *t*-butyl 2-(mesitylsulfonyl)-3-(2-methylprop-1-enyl)pyrazolidine-1-carboxylate as a clear oil (10 mg, 66% yield, 98% *ee*).

Representative procedure for the gold-catalyzed cycloisomerization of allenic hydroxylamines. t-Butyl 3-(cyclohexylidenemethyl)isoxazolidine-2-carboxylate[171]

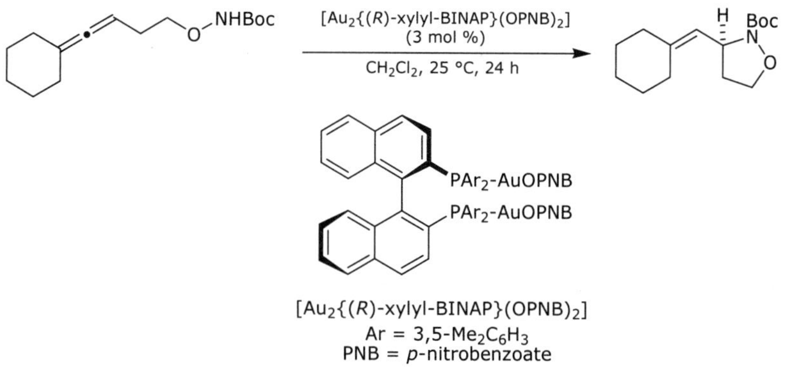

NHBoc — $[Au_2\{(R)\text{-xylyl-BINAP}\}(OPNB)_2]$ (3 mol %), CH_2Cl_2, 25 °C, 24 h → H Boc N O

PAr₂-AuOPNB
PAr₂-AuOPNB

$[Au_2\{(R)\text{-xylyl-BINAP}\}(OPNB)_2]$
Ar = 3,5-$Me_2C_6H_3$
PNB = *p*-nitrobenzoate

To a solution of *t*-butyl 4-cyclohexylidenebut-3-enyloxycarbamate (53.4 mg, 0.20 mmol) in CH_2Cl_2 (0.6 mL) was added $[Au_2\{(R)\text{-xylyl-BINAP}\}(OPNB)_2]$ (8.8 mg, 6 µmol). The resulting homogeneous mixture was protected from ambient light and stirred at 25 °C for 24 h. The crude oil was purified by column chromatography (1:24 EtOAc:hexanes) to yield *t*-butyl 3-(cyclohexylidenemethyl)isoxazolidine-2-carboxylate as a white solid (49.7 mg, 93% yield, 99% *ee*).

In contrast to intramolecular nucleophilic additions, the corresponding gold-catalyzed intermolecular hydroaminations of allenes have been described much less frequently. The first report by Nishina and Yamamoto focused on the reaction of various mono- or 1,3-disubstituted allenes with aniline in the presence of 10 mol% of $AuBr_3$ (Scheme 4-64).[172] The addition took place regioselectively at the less hindered allene terminus and afforded secondary amines with good chemical yield. By using a cationic gold catalyst, morpholine can also be added, affording allylamines in moderate-to-good yields upon heating of the reaction mixture to 80 °C for 9–24 h.[172b, 173] High levels of chirality transfer were observed in the intermolecular hydroamination of chiral allenes. Similar results were obtained by Bertrand *et al.*[174] using gold catalysts bearing bulky cyclic alkylaminocarbene (CAAC) ligands; these even allowed the use of ammonia and hydrazine for the hydroamination. In contrast, the opposite regioselectivity was observed in the addition of benzyl carbamate to 1,1-disubstituted allenes in the presence of a cationic gold-NHC catalyst.[175]

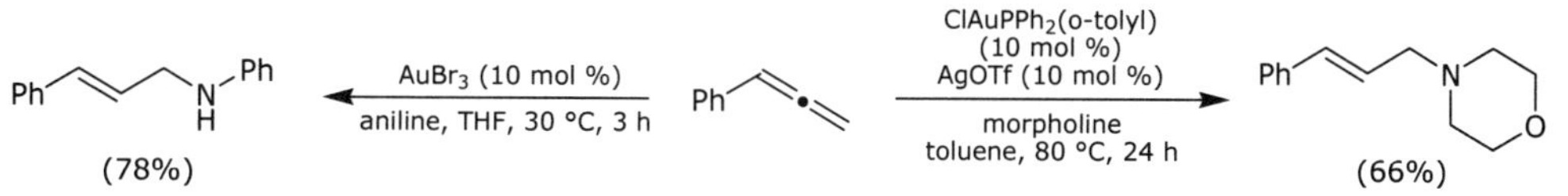

Ph N(H)Ph (78%) ← $AuBr_3$ (10 mol %), aniline, THF, 30 °C, 3 h — Ph (allene) — $ClAuPPh_2$(o-tolyl) (10 mol %), AgOTf (10 mol %), morpholine, toluene, 80 °C, 24 h → Ph (morpholinyl) (66%)

Scheme 4-64

Representative procedure for the gold-catalyzed intermolecular addition of amines to allenes. (E)-(3-(4-Methoxyphenyl)allyl)morpholine[172b]

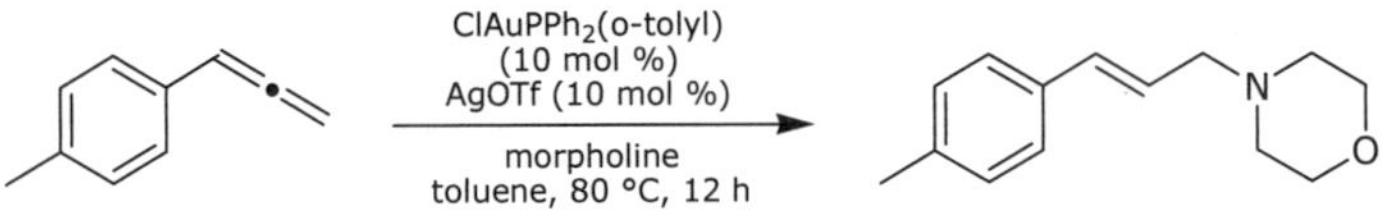

ClAuPPh₂(o-tolyl) (10 mol %), AgOTf (10 mol %); morpholine, toluene, 80 °C, 12 h

To a suspension of $ClAuPPh_2$(*o*-tolyl) (25.4 mg, 0.05 mmol) and AgOTf (12.8 mg, 0.05 mmol) in toluene (0.5 mL) was added morpholine (43.7 mg, 0.502 mmol). To the reaction mixture was added *p*-tolylallene (79.3 mg, 0.6 mmol), and the resulting mixture was stirred at 80 °C under an Ar atmosphere. The reaction mixture was colorless and heterogeneous at the beginning, but it turned yellow to brown as the reaction progressed. After the reaction was completed (12 h), the reaction mixture was filtered through a short basic silica gel pad with EtOAc as eluent, and the resulting filtered solution was concentrated *in vacuo*. The product was purified quickly by short column chromatography (basic silica gel, hexane/EtOAc = 100:1 to 10:1) to give (*E*)-(3-(4-methoxyphenyl)allyl)morpholine in 83% yield (90.8 mg).

4.3 Nucleophilic Additions to Alkenes

The gold-catalyzed hydroamination of alkenes is an important route to acyclic and cyclic amines. He and co-workers reported the first examples for this transformation, using $Ph_3PAuOTf$ at elevated temperatures (Scheme 4-65).[176] *p*-Toluenesulfonamide gave the best results in the intermolecular hydroamination of various acyclic and cyclic olefins, which occurred with Markovnikov regioselectivity. Likewise, tosylated γ-aminoalkenes were converted into the corresponding pyrrolidines under these conditions. 2-Nitrobenzenesulfonamides can also be used as nucleophiles, and microwave heating leads to drastically reduced reaction times.[177] Addition reactions of aniline to styrene derivatives are possible upon heating to 150 °C in the presence of $AuCl_3$ and $CuCl_2$.[178]

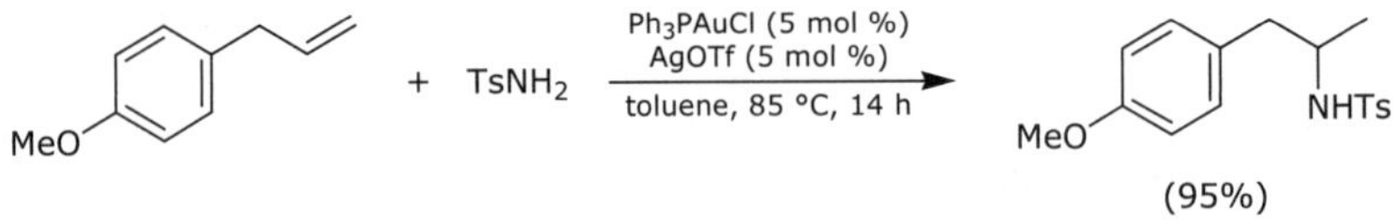

MeO (allylbenzene) + $TsNH_2$ → Ph₃PAuCl (5 mol %), AgOTf (5 mol %); toluene, 85 °C, 14 h → MeO ... NHTs (95%)

Scheme 4-65

The use of carbamates as the nucleophile allowed extension of the method to the gold-catalyzed intermolecular hydroamination of conjugated dienes at room temperature.[179] The reaction took place regioselectively as 1,2-addition at the less substituted double bond.

Representative procedure for the gold-catalyzed hydroamination of conjugated dienes. Benzyl 3-methylpent-3-en-2-ylcarbamate[179]

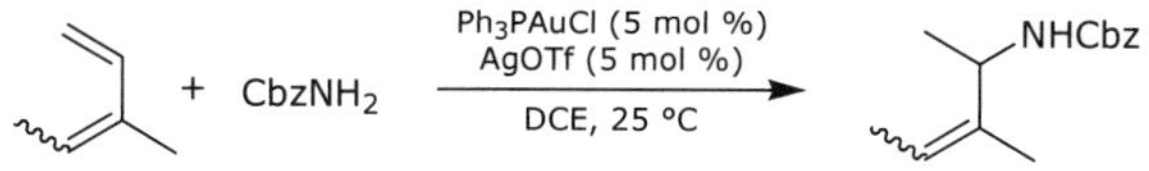

Ph_3PAuCl (5 mol %), AgOTf (5 mol %), DCE, 25 °C; + $CbzNH_2$ → NHCbz

Ph_3PAuCl (25 mg, 0.050 mmol) and AgOTf (13 mg, 0.050 mmol) were allowed to react in a minivial loaded with benzyl carbamate (0.15 g, 1.0 mmol) in anhydrous 1,2-dichloroethane (2.0 mL) for 5 min. 3-Methylpenta-1,3-diene (0.14 mL, 1.2 mmol) was then added, and the reaction mixture was stirred overnight under nitrogen at RT in the dark. The solvent was evaporated *in vacuo*, and the residue was purified by flash column chromatography on silica gel (hexanes/EtOAc 1:0→9:1). Benzyl 3-methylpent-3-en-2-ylcarbamate was isolated in 86% yield (0.20 g, 0.86 mmol) after removal of solvent.

Various transition metals catalyze the aza-Michael addition of carbamates to enones.[180] Among the most efficient are gold(I) and gold(III) salts (Scheme 4-66). The catalyst of the stoichiometry $AuCl_3 \cdot 2\,H_2O$ used by the authors is probably tetrachloroauric acid hydrate ($HAuCl_4 \cdot x\,H_2O$). In the corresponding reaction of

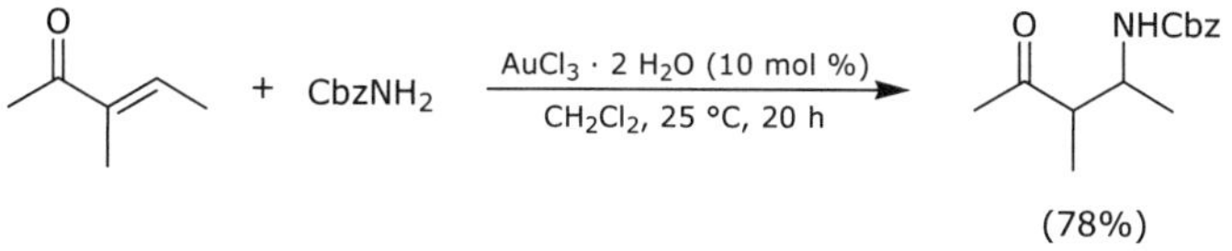

+ $CbzNH_2$; $AuCl_3 \cdot 2\,H_2O$ (10 mol %), CH_2Cl_2, 25 °C, 20 h; O NHCbz (78%)

Scheme 4-66

enones with allylamides in the presence of Au[P(*t*-Bu)$_2$(*o*-biphenyl)]Cl and $AgClO_4$, the aza-1,4-addition was followed by an intramolecular hydroalkylation which afforded various pyrrolidines with moderate to good yield and diastereoselectivity (Scheme 4-67).[181] Control experiments revealed that no conversion took place

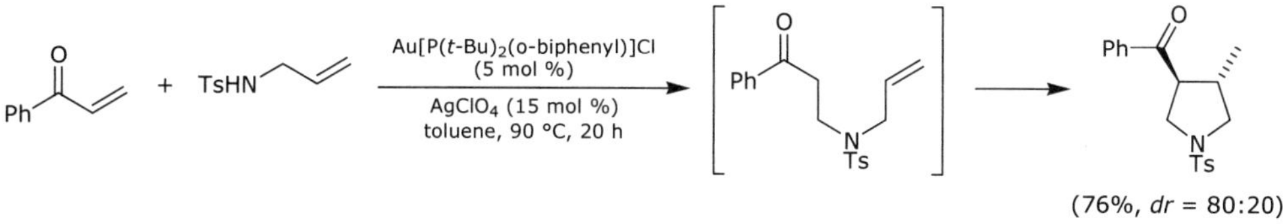

Ph, O + TsHN; Au[P(*t*-Bu)$_2$(o-biphenyl)]Cl (5 mol %), $AgClO_4$ (15 mol %), toluene, 90 °C, 20 h; [Ph, O, N, Ts] → Ph, O, N, Ts (76%, *dr* = 80:20)

Scheme 4-67

with Au[P(*t*-Bu)$_2$(*o*-biphenyl)]ClO_4, whereas with silver perchlorate alone only the Michael adduct was formed. Thus, the silver salt plays a dual role in this transformation, *i.e.*, as catalyst for the intermolecular aza-Michael addition, and as activator of the gold pre-catalyst.

An intramolecular gold-catalyzed hydroamination took place upon heating of cyclic dienes bearing a sulfonamide side chain with Ph_3PAuCl and AgOTf.[182] In this case, 1,4-addition of the amide leads to the formation of hexahydroindoles and related heterocycles. The presence of the bulky phenyl groups in the tether is crucial for regioselectivity. An enantioselective version of this reaction has been reported recently.[183]

Representative procedure for the gold-catalyzed intramolecular hydroamination of cyclic dienes. Cis-3,3-diphenyl-1-(phenylsulfonyl)-1,3,3a,4,5,7a-hexahydro-1H-indole[182]

NHSO$_2$Ph, Ph, Ph → Ph$_3$PAuCl (5 mol %), AgOTf (5 mol %), toluene, 80 °C, 18 h → H, SO$_2$Ph, N, H, Ph, Ph

To an oven-dried, 50-mL, round-bottom flask equipped with a stirring bar and capped with a rubber septa were added 2-(cyclohexa-2,4-dienyl)-2,2-diphenyl-*N*-(phenylsulfonyl)ethanamine (415.5 mg, 1.0 mmol) and AgOTf (12.5 mg, 0.05 mmol). The apparatus was evacuated (oil pump) and filled with nitrogen three times. To the reaction mixture was then added *via* syringe Ph$_3$PAuCl (24.7 mg, 0.05 mmol) in 20 mL of toluene. The resulting mixture was stirred at 80 °C until all starting material was consumed (typically 18 h). The reaction mixture was filtered through a bed of Celite. The filtrate was concentrated *in vacuo* to give the crude mixture, which was purified by flash column chromatography (silica gel, gradient elution: 10–30% EtOAc/hexanes) to give *cis*-3,3-diphenyl-1-(phenylsulfonyl)-1,3,3a,4,5,7a-hexahydro-1*H*-indole (370 mg, 0.89 mmol, 89%) as a yellow solid.

The *exo*-selective gold-catalyzed cyclization of alkenyl carbamates, amides, or ammonium salts affords the corresponding pyrrolidine or piperidine derivatives.[184] Here, cationic gold(I) catalysts bearing sterically hindered, electron-rich phosphine ligands such as P(*t*-Bu)$_2$(*o*-biphenyl) were much more reactive than Ph$_3$PAuOTf.

Representative procedure for the gold-catalyzed intramolecular hydroamination of alkenyl carbamates. Benzyl 2-methyl-4,4-diphenylpyrrolidine-1-carboxylate[184a]

NHCbz, Ph, Ph → Au[P(*t*-Bu)$_2$(o-biphenyl)]Cl (5 mol %), AgOTf (5 mol %), dioxane, 60 °C, 18 h → Cbz, N, Me, Ph, Ph

A mixture of benzyl 2,2-diphenylpent-4-enylcarbamate (0.17 g, 0.45 mmol), Au[P(*t*-Bu)$_2$(*o*-biphenyl)]Cl (12 mg, 0.022 mmol), and AgOTf (6 mg, 0.02 mmol) in dioxane (0.45 mL) was degassed by means of one freeze-pump-thaw cycle, pressurized with nitrogen, and stirred at 60 °C for 18 h. The crude reaction mixture was purified by chromatography (hexanes/EtOAc 20:1) to give benzyl 2-methyl-4,4-diphenylpyrrolidine-1-carboxylate (160 mg, 97%) as a viscous, colorless oil.

Highly reactive cationic Au(I)-NHC complexes allow the efficient cyclization of alkenyl ureas to pyrrolidines or imidazolidinones at room temperature.[185]

Representative procedure for the gold-catalyzed intramolecular hydroamination of alkenyl ureas. Phenyl 2-methyl-N,4,4-triphenylpyrrolidine-1-carboxlamide[185a]

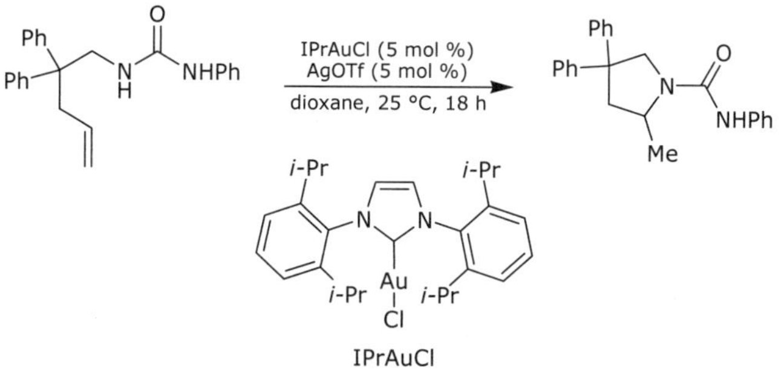

IPrAuCl (5 mol %)
AgOTf (5 mol %)
dioxane, 25 °C, 18 h

IPrAuCl

Dioxane (0.50 mL) was added to a mixture of 1-phenyl-3-(2,2-diphenylpent-4-enyl)urea (90 mg, 0.25 mmol), IPrAuCl (7.5 mg, 0.012 mmol), and AgOTf (3.4 mg, 0.013 mmol), and the resulting suspension was stirred at RT for 18 h. The resulting mixture was concentrated and chromatographed (Et_2O/CH_2Cl_2 = 1:30) to give phenyl 2-methyl-*N*,4,4-triphenylpyrrolidine-1-carboxamide (86 mg, 96%) as white microcrystals.

The corresponding intermolecular addition of cyclic ureas to terminal alkenes is catalyzed by Au[P(*t*-Bu)$_2$(*o*-biphenyl)]Cl and AgOTf, and it provided alkylated ureas in high yield at 100 °C.[186] The reaction even works with ethylene and other gaseous olefins. In the presence of chiral gold(I) pre-catalyst [Au$_2${(*S*)-3,5-*t*-Bu-4-MeO-MeOBIPHEP}Cl$_2$] and silver triflate, the hydroamination products were obtained with 71-76% *ee* (Scheme 4-68). Treatment of cyclic ureas with allylic alcohols in the presence of cationic gold(I) catalysts afforded allylated ureas with

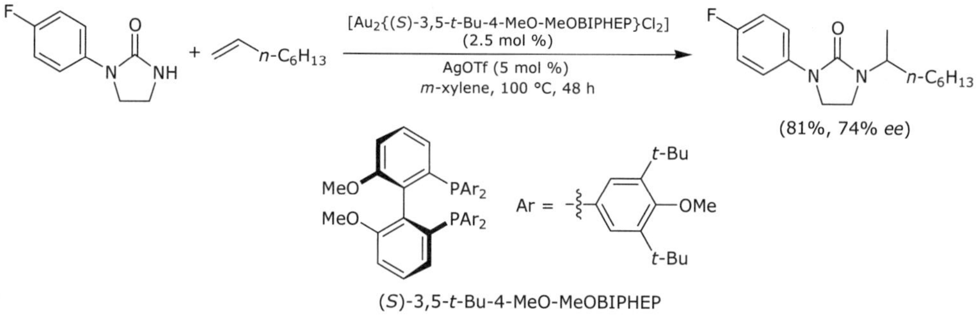

[Au$_2${(*S*)-3,5-*t*-Bu-4-MeO-MeOBIPHEP}Cl$_2$] (2.5 mol %)
AgOTf (5 mol %)
m-xylene, 100 °C, 48 h

(81%, 74% *ee*)

(*S*)-3,5-*t*-Bu-4-MeO-MeOBIPHEP

Scheme 4-68

high γ-regioselectivity and *syn*-stereoselectivity.[187] For the intramolecular version of this transformation, Fmoc-protected amino allylic alcohols worked best and gave the corresponding vinyl-substituted pyrrolidines or piperidines with high yield and enantioselectivity in the presence of [Au$_2${(*S*)-3,5-*t*-Bu-4-MeO-MeOBIPHEP}Cl$_2$] and $AgClO_4$ (Scheme 4-69).[188]

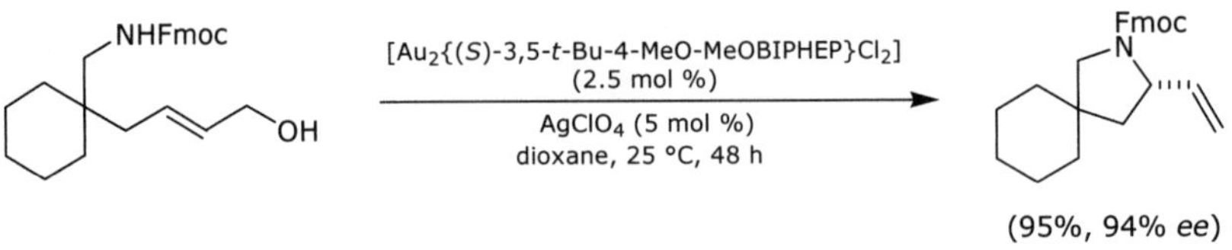

NHFmoc
OH
[Au_2{(*S*)-3,5-*t*-Bu-4-MeO-MeOBIPHEP}Cl_2]
(2.5 mol %)
$AgClO_4$ (5 mol %)
dioxane, 25 °C, 48 h
Fmoc
N
(95%, 94% *ee*)

Scheme 4-69

Olefins bearing an amino group in the γ- or δ-position undergo a gold-catalyzed intramolecular oxidative carboamination to afford pyrrolidines or piperidines in the presence of arylboronic acids and Selectfluor (Air Products and Chemicals, Detroit, MI) (Scheme 4-70).[189] The reaction may proceed *via* an arylgold(III) species that activates the olefin for an *anti*-selective attack of the nitrogen nucleophile. Reductive elimination affords the functionalized heterocycle and gold(I) which is reoxidized to gold(III) with Selectfluor. An alternative mechanistic model invokes binuclear Au(II)–Au(II) complexes as key intermediates for this transformation.[190]

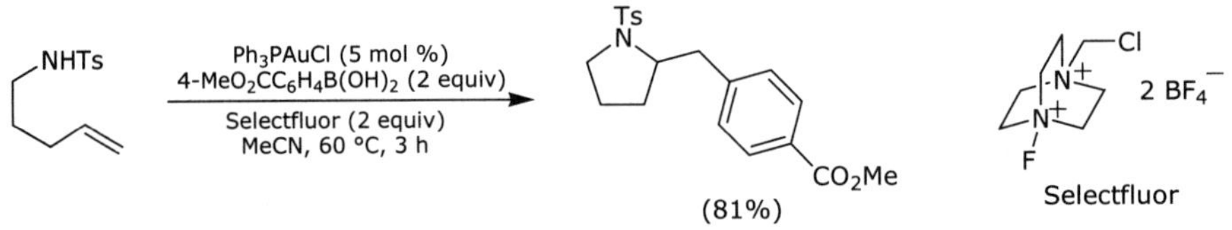

NHTs
Ph_3PAuCl (5 mol %)
4-$MeO_2CC_6H_4B(OH)_2$ (2 equiv)
Selectfluor (2 equiv)
MeCN, 60 °C, 3 h
Ts
N
CO_2Me
(81%)
Cl
N+
N+
F
2 BF_4^-
Selectfluor

Scheme 4-70

In the presence of water, the reaction of tosylated γ-aminoalkenes with $Ph_3PAuSbF_6$ and Selectfluor afforded mixtures of 3-hydroxypiperidines and 2-hydroxymethylpyrrolidines.[191] In this oxidative difunctionalization of the olefin, the 6-*endo*- cyclization products were formed predominantly (Scheme 4-71). Likewise,

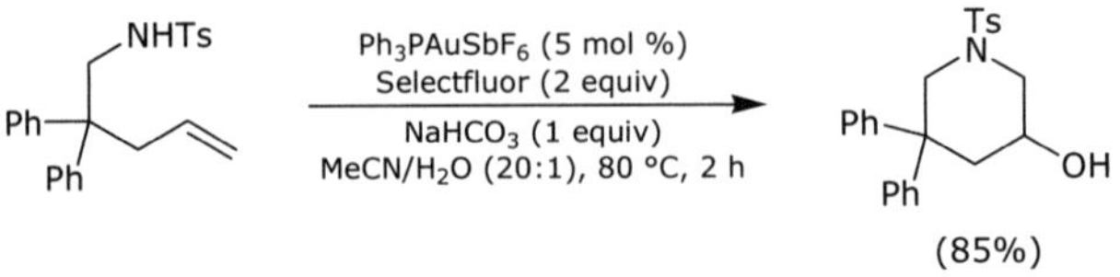

NHTs
Ph
Ph
$Ph_3PAuSbF_6$ (5 mol %)
Selectfluor (2 equiv)
$NaHCO_3$ (1 equiv)
MeCN/H_2O (20:1), 80 °C, 2 h
Ts
N
Ph
Ph
OH
(85%)

Scheme 4-71

3-methoxy-piperidines were obtained in the presence of methanol, whereas the corresponding acetates were formed with $PhI(OAc)_2$. If the amount of water was reduced to two equivalents, the nitrile used as solvent acts as nucleophile; subsequent hydrolysis afforded aminoamidation products (Scheme 4-72).

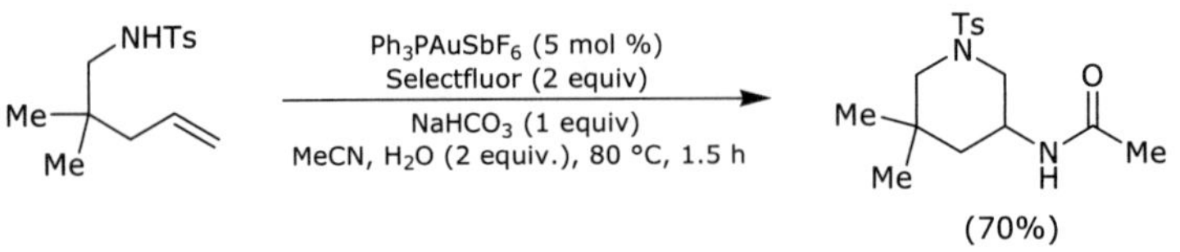

NHTs
Me
Me
$Ph_3PAuSbF_6$ (5 mol %)
Selectfluor (2 equiv)
$NaHCO_3$ (1 equiv)
MeCN, H_2O (2 equiv.), 80 °C, 1.5 h
Ts
N
Me
Me
O
N
H
Me
(70%)

Scheme 4-72

General procedure for the gold-catalyzed aminoamidation of alkenes[191]

A mixture of the γ-aminoalkene (1 equiv), Selectfluor (2 equiv), and $NaHCO_3$ (1 equiv) is dissolved in acetonitrile or propionitrile or butyronitrile (0.02 M) and H_2O (2 equiv), followed by addition of $Ph_3PAuSbF_6$ (0.05 equiv). The mixture is stirred for 1.5 h at 80 °C, and the reaction is monitored by TLC. Upon consumption of the starting material, the mixture is diluted with CH_2Cl_2 (5 mL) and filtered over Celite (washing with CH_2Cl_2 (2.5 mL) and EtOAc (2.5 mL)). The solvent is evaporated under reduced pressure and the residue is purified through silica gel flash column chromatography (hexane/EtOAc = 1/1 to 0/100) to yield the desired aminoamides.

In a related study, alkenyl ureas were found to undergo a gold-catalyzed diamination of the double bond in the presence of $PhI(OAc)_2$ as oxidant (Scheme 4-73).[192]

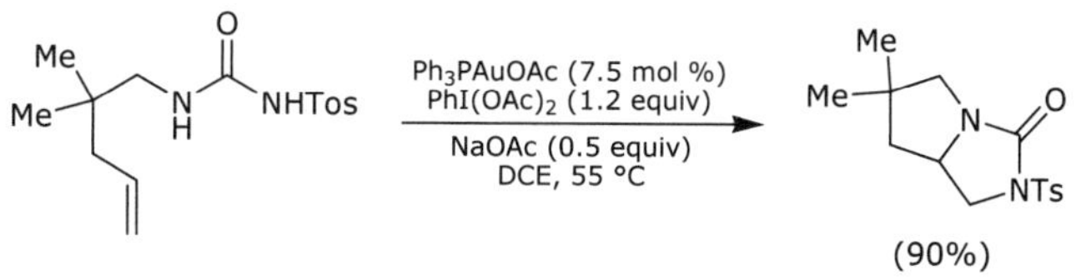

Me Me N H O NHTos

$Ph_3PAuOAc$ (7.5 mol %)
$PhI(OAc)_2$ (1.2 equiv)
NaOAc (0.5 equiv)
DCE, 55 °C

Me Me N O NTs

(90%)

Scheme 4-73

General procedure for the gold-catalyzed diamination of alkenes[192]

The urea starting material (0.2 mmol, 1.0 equiv), iodosobenzene diacetate (0.24 mmol, 1.2 equiv) and NaOAc (0.08 mmol, 0.4 equiv) are weighed into an oven-dried Pyrex tube equipped with a magnetic stirring bar and set under N_2 atmosphere. Preformed $Ph_3PAuOAc$ (0.015 mmol, 7.5 mol%) or a combination of Ph_3PAuCl (0.015 mmol) and AgOAc (0.015 mmol) are added in one portion, and the combined solids are dissolved in dichloroethane (2 mL). The resulting mixture is stirred for 12 h at 50 °C oil bath temperature. The reaction is quenched by addition of sat. aq. $Na_2S_2O_3$ solution, and a small additional amount of H_2O is added. The organic phase is separated, and the aqueous phase is extracted with EtOAc several times. The combined organic layers are dried over anhydrous $MgSO_4$, and the solvent is removed under reduced pressure. All products are purified by flash chromatography.

5. C–O Bond Formation

Reviews. Hashmi, A.S.K.; *Chem. Rev.* **2007,** *107*, 3180; Li, Z.; Brouwer, C.; He, C.; *Chem. Rev.* **2008,** *108*, 3239; Arcadi, A.; *Chem. Rev.* **2008,** *108*, 3266; Gorin, D.J.; Sherry, B.D.; Toste, F.D.; *Chem. Rev.* **2008,** *108*, 3351; Patil, N.T.; Yamamoto, Y.; *Chem. Rev.* **2008,** *108*, 3395; Shen, H.C.; *Tetrahedron* **2008,** *64*, 3885; Muzart, J.; *Tetrahedron* **2008,** *64*, 5815; Corma, A.; Leyva-Pérez, A.; Sabater, M.J.; *Chem. Rev.* **2011,** *111*, 1657; Krause, N.; Winter, C.; *Chem. Rev.* **2011**, *111*, 1994; Alcaide, B.; Almendros, P.; Alonso, J. M.; *Org. Biomol. Chem.* **2011**, 9, 4405.

5.1 Nucleophilic Additions to Alkynes

The addition of water or alcohols to alkynes is a benchmark reaction for testing the efficiency of various types of gold catalysts. Whereas classic reagents for this transformation (sulfuric acid and mercury salts) have been known for 150 years,[193] the application of gold catalysis commenced much more recently. After some early examples by Utimoto and co-workers,[194] it was the Teles group at BASF who, in their landmark contribution, demonstrated the extraordinary activity of cationic gold complexes in the hydroalkoxylation of alkynes (Scheme 4-74).[5] These catalysts were prepared by treatment of Ph_3PAuMe with a Brønsted acid ($MeSO_3H$ or H_2SO_4) and achieved total turnover numbers of up to 10^5 with turnover frequencies of up to 5400 h^{-1}. They are neither air nor water sensitive and allow the reactions to be performed without solvent. Both terminal and internal alkynes afforded acetals with high regioselectivity, the latter as a result of nucleophilic attack at the less sterically hindered position of the triple bond. Occasionally, the corresponding enol ether was formed as a side or main product. Primary and secondary propargyl alcohols also react smoothly to form cyclic *bis*-ketals. With diols and terminal alkynes, cyclic acetals were formed more efficiently in the presence of Ph_3PAuCl and $AgBF_4$ than with Brønsted acids.[195]

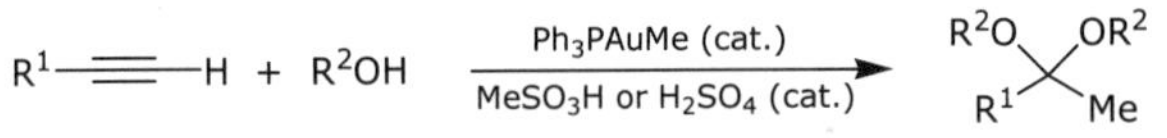

R^1—≡—H + R^2OH → (Ph_3PAuMe (cat.); $MeSO_3H$ or H_2SO_4 (cat.)) → R^2O, OR^2, R^1, Me

Scheme 4-74

The nature of the ligand in the cationic gold catalyst has a strong influence on the activity, in the following order: L = $Ph_3As < Et_3P < Ph_3P < (4\text{-}FC_6H_4)_3P < (MeO)_3P < (PhO)_3P$. Hard anions afforded particularly active catalysts. The reaction is probably initiated by activation of the alkyne by the cationic gold catalyst, followed by attack of the alcohol at the gold and intramolecular transfer to the coordinated alkyne. This leads to a σ-vinylgold intermediate, which adds a second molecule of the alcohol and finally delivers the product by protodeauration.

Representative procedure for the gold-catalyzed hydroalkoxylation of alkynes. Trans-2,5-dimethyl-2,5-dimethoxy-1,4-dioxane[5]

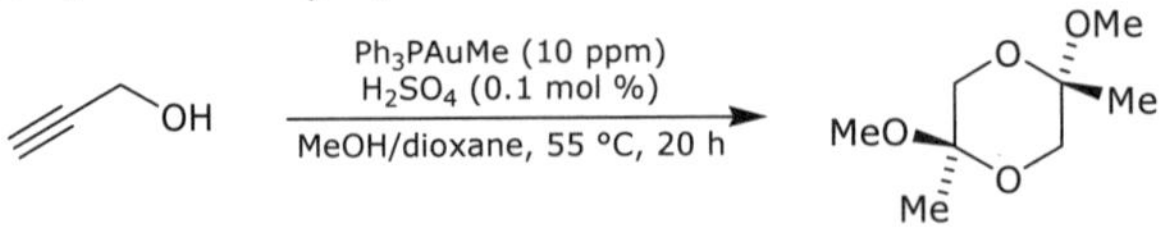

OH → (Ph_3PAuMe (10 ppm); H_2SO_4 (0.1 mol %); MeOH/dioxane, 55 °C, 20 h) → MeO, Me, O, OMe, Me, O

Propargyl alcohol (15.1 mol, 848 g, freshly distilled), methanol (59.6 mol, 1908 g), and concentrated sulfuric acid (15 mmol, 1.47 g) were mixed and heated to 55 °C. A solution of Ph_3PAuMe (148 μmol, 70.4 mg) in 125 mL of dioxane was then added within 10 h. The mixture was stirred for an additional 10 h, and then most of the remaining methanol was removed by distillation at reduced pressure. The residual solution was neutralized with 30% sodium methanolate in methanol and cooled in an ice bath. Precipitated product was collected by filtration and dried. A second batch of product can be obtained by further concentration of the mother liquor. Total yield of isolated product: 1238 g (93%).

The intramolecular hydroalkoxylation of suitable acetylenic alcohols in the presence of AuCl and K_2CO_3 afforded 2-alkylidenetetrahydrofurans or -pyrans by *exo*-selective cyclization (Scheme 4-75).[196, 197] Exclusive *Z*-stereoselectivity was observed with substrates bearing an internal triple bond. This was explained by an *anti*-attack of the hydroxy group at the activated triple bond and subsequent stereoselective protodeauration. Interestingly, even rather sensitive bromoalkynes participate in the cycloisomerization with high efficiency. The analogous reaction of 2-(prop-2-yn-1-ylamino)phenols furnished substituted benzoxazines,[198] whereas propargyl esters with a pendant hydroxy group underwent cycloetherification with concomitant 1,3-acetoxy transfer.[199] Likewise, alkylidene-substituted γ- or δ-lactones were obtained from alkynoic acids,[200] and Boc-protected propargyl alcohols or amines were cyclized to 4-alkylidene-1,3-dioxolan-2-ones or the corresponding oxazolidinones in the presence of $Ph_3PAuNTf_2$.[201] 2,3-Dihydroisoxazoles were obtained in a one-pot procedure from propargyl alcohols and protected hydroxylamines by iron-catalyzed formation of propargyl hydroxylamines and subsequent cyclization in the presence of sodium tetrachloroaurate.[202]

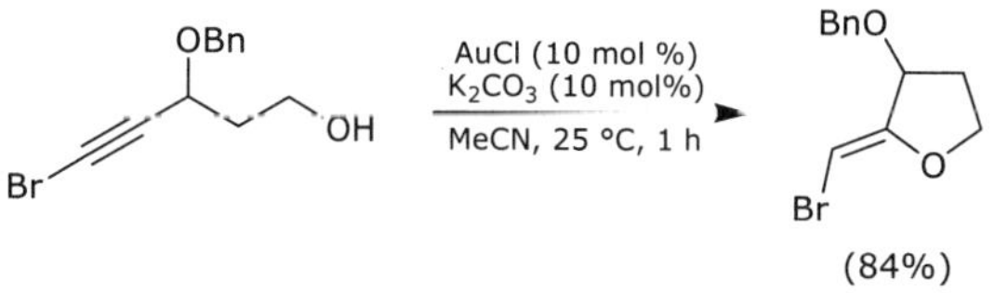

OBn
Br
OH
AuCl (10 mol %)
K_2CO_3 (10 mol%)
MeCN, 25 °C, 1 h
BnO
O
Br
(84%)

Scheme 4-75

The corresponding reaction of but-3-yn-1-ols or pent-4-yn-1-ols with primary or secondary alcohols in the presence of catalytic amounts of Ph_3PAuBF_4 and *p*-TsOH afforded tetrahydrofuranyl ethers (Scheme 4-76).[203] This tandem 5-*endo*-cycloisomerization/hydroalkoxylation proceeds *via* 2,3-dihydrofurans,[204] which then undergo an intermolecular Brønsted acid-catalyzed addition of the external alcohol. The transformation is not restricted to internal alkynols but can be applied to terminal acetylenes as well. Application of the method to the synthesis of bicyclic heterocycles with a β-lactam structure was reported recently.[205] Under the same conditions, epoxyalkynes undergo a sequence of epoxide opening, 6-*exo*-cycloisomerization, and nucleophilic addition to afford tetrahydropyranyl ethers.[206] In a closely related transformation, cyclic acetals were obtained from alk-2-ynoates bearing a hydroxy group in 6- or 7-position by treatment with $AuCl_3$ and MeOH.[207]

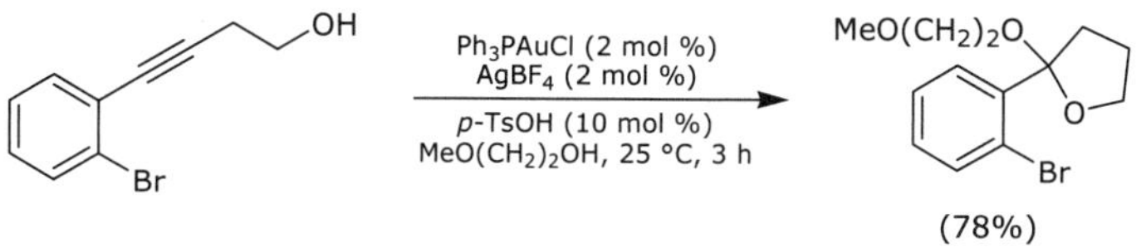

OH
Br
Ph_3PAuCl (2 mol %)
$AgBF_4$ (2 mol %)
p-TsOH (10 mol %)
$MeO(CH_2)_2OH$, 25 °C, 3 h
$MeO(CH_2)_2O$
O
Br
(78%)

Scheme 4-76

General procedure for the gold-catalyzed tandem cycloisomerization/hydroalkoxylation of homopropargyl alcohols[203]

To a solution of the homopropargyl alcohol (0.5 mmol) in 4 mL of the alcohol under argon is added 2 mol% each of Ph_3PAuCl and $AgBF_4$, as well as 10 mol% of *p*-TsOH. The mixture is stirred at RT, and the reaction is monitored by TLC. After completion, 3 mL of sat. $NaHCO_3$ solution is added, the organic layer is separated, and the aqueous layer is washed with Et_2O (2 × 5 mL). The combined organic layers are dried over anhydrous $MgSO_4$, and the solvent is removed under vacuum. The crude product is purified by flash column chromatography on alumina with cyclohexane/EtOAc (20:1).

Spiroketals are accessible from suitable linear alkynediols[208] or epoxy-alkynols[209] in the presence of platinum or gold catalysts and a Brønsted acid. The twofold intramolecular hydroalkoxylation was applied to the synthesis of the spiroketal core of rubromycins,[210] as well as to the marine polyether okadaic acid.[211] A closely related variation of this transformation was the key step in a synthesis of the A–D domain of the marine toxin azaspiracid.[212] In an analogous fashion, diyndiols were converted into *bis*-spiroketals bearing different combinations of five- and six-membered rings.[213] Linear acetylenic triols cyclize to monounsaturated five- or six-membered ring spiroketals in the presence of cationic gold(I) catalysts and molecular sieves.[214] This dehydrative cyclization, which has also found application in Forsyth's okadaic acid synthesis,[211] occurs already at 0 °C and does not take place with AgOTf or TfOH as catalyst.

Representative procedure for the gold-catalyzed dehydrative cyclization of acetylenic triols. 3,3-Diphenyl-1,6-dioxaspiro[4.5]dec-9-ene[214]

HO, Ph, Ph, OH, OH → Au[P(*t*-Bu)$_2$(o-biphenyl)]Cl (2 mol %), AgOTf (2 mol %), THF, MS 4 Å, 0 °C, 35 min → Ph, Ph, O, O

Dry THF (0.5 mL) was added to an aluminum-foil-covered test tube containing Au[P(*t*-Bu)$_2$(*o*-biphenyl)]Cl (2.2 mg, 0.004 mmol), AgOTf (1.0 mg, 0.004 mmol), and activated MS 4 Å (17 mg). After stirring for 10 min, the mixture was cooled to 0 °C and a solution of 7,7-diphenyloct-4-yne-1,3,8-triol (62.0 mg, 0.2 mmol) in dry THF (1.5 mL) was added. After TLC analysis showed the reaction to be complete (35 min), the mixture was filtered through a short plug of silica with Et_2O (40 mL). The solution of crude product was concentrated and then purified by flash chromatography (5% Et_2O/pentanes) to give 3,3-diphenyl-1,6-dioxaspiro[4.5]dec-9-ene as a colorless oil (57.5 mg, 99%).

A twofold intramolecular *exo*-selective C–O bond formation takes place in the gold-catalyzed cycloisomerization of branched acetylenic diols to strained bicyclic ketals (Scheme 4-77).[215] Both gold(I) and gold(III) chloride can be employed with similar efficiency. Even though methanol was used as solvent, no intermolecular hydroalkoxylation was observed. In a similar fashion, treatment of 2-alkynylpentan-1,5-diols with AuCl induces a rapid cyclization to dioxabicyclo[4.2.1]nonanes, which then undergo slow ring opening to tetrahydropyrans.[216] The method was used in the synthesis of a structural mimic of platensimycin.[217] If one of the two pendant hydroxy groups is replaced by an alkene, then the gold-catalyzed cyclization is followed by a Prins reaction to generate 9-oxabicyclo[3.3.1]nonane derivatives.[218]

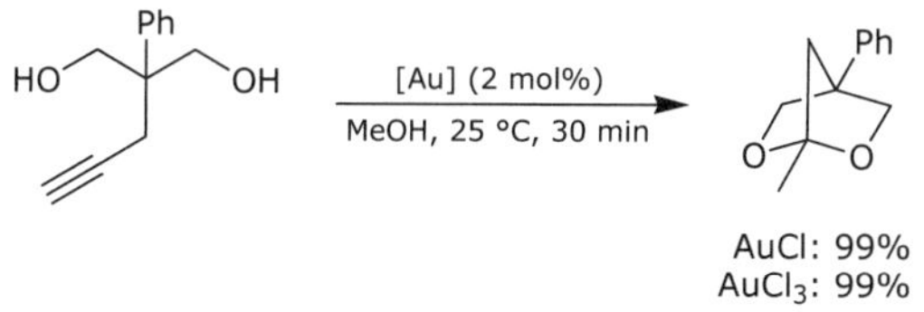

[Au] (2 mol%)
MeOH, 25 °C, 30 min

AuCl: 99%
$AuCl_3$: 99%

Scheme 4-77

Competition between *endo*- and *exo*-selective cyclization modes was studied by Czekelius and co-workers using 1,4-diynes with a pendant hydroxy group (Scheme 4-78).[219] Whereas diynes bearing an alkoxy group in the tether undergo a 7-*endo*-cycloisomerization to dioxepine derivatives in the presence of cationic gold(I) catalysts, the corresponding substrates with an all-carbon chain between the diyne and the hydroxy group afforded enol ethers by *exo*-selective hydroalkoxylation. A mechanistic rationale for the regioselectivity difference invokes a stereoelectronic interaction between the oxygen atom in the tether and the gold catalyst.

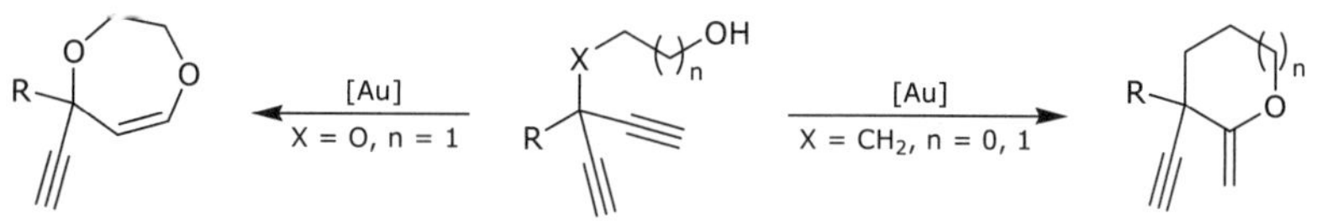

[Au]
X = O, n = 1

[Au]
X = CH_2, n = 0, 1

Scheme 4-78

Representative procedure for the gold-catalyzed cycloisomerization of 1,4-diynes. 3-Ethynyl-2-methylene-3-phenyltetrahydrofuran[219b]

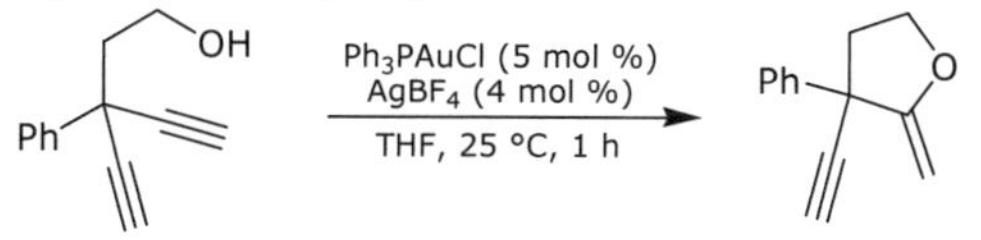

Ph_3PAuCl (5 mol %)
$AgBF_4$ (4 mol %)
THF, 25 °C, 1 h

In a glove box were added Ph_3PAuCl (93.0 mg, 0.187 mmol, 5.0 mol%) and $AgBF_4$ (29.0 mg, 0.149 mmol, 4.0 mol%). This mixture was dissolved in dry THF (0.05 M) and stirred for 10 min. 3-Ethynyl-3-phenylpent-4-yn-1-ol (690 mg, 3.74 mmol, 1.0 equiv) in dry THF (0.05 M) was added to this solution at RT. The mixture was stirred at RT for 1 h. After complete conversion as indicated by TLC analysis, the mixture was directly purified by column chromatography (hexanes/EtOAc 50:1) to afford 3-ethynyl-2-methylene-3-phenyltetrahydrofuran (686 mg, 3.72 mmol, 99%) as colorless oil.

In the presence of gold(III) chloride, conjugated enynes bearing a tethered hydroxy group at the triple bond cyclize to conjugated dienes that undergo a Diels-Alder reaction with a suitable dienophile (Scheme 4-79).[220] Depending on the length of the tether, this tandem cycloisomerization/cycloaddition leads to the formation of annulated or spiro bicyclic heterocycles. Thus, hex-5-en-3-yn-1-ols react in a 5-*endo*-cycloisomerization to 2-alkenyl-substituted 2,3-dihydrofurans, which afford hexahydrobenzofurans in the subsequent cycloaddition. In contrast to this, enynes with a two or three carbon linker between the triple bond and the hydroxy group undergo an *exo*-selective cyclization furnishing spiro bicyclic Diels-Alder adducts.

Whereas the former transformation occurs at room temperature, heating to reflux in dichloroethane is required for the latter.

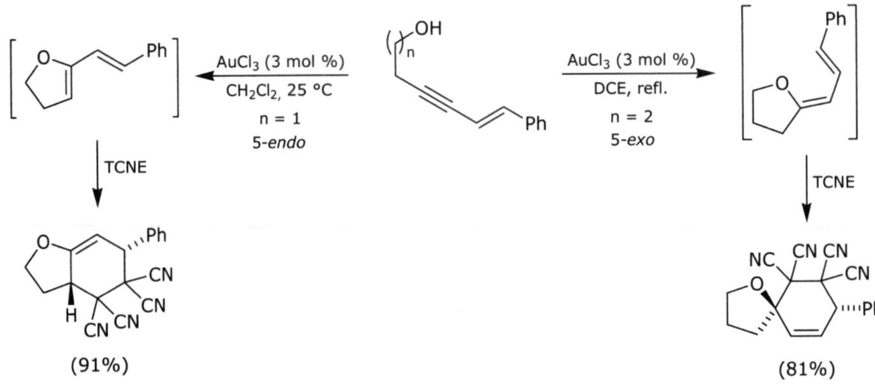

Scheme 4-79

(*Z*)-2-En-4-yn-1-ols bearing a primary or secondary hydroxy group undergo cycloisomerization to substituted furans (Scheme 4-80).[66a, 221] In this transformation, gold was observed to be much more efficient than Ru, Pd, or Ag catalysts. The isomeric 2-methylenebut-3-yn-1-ols also cyclize to substituted furans in the presence

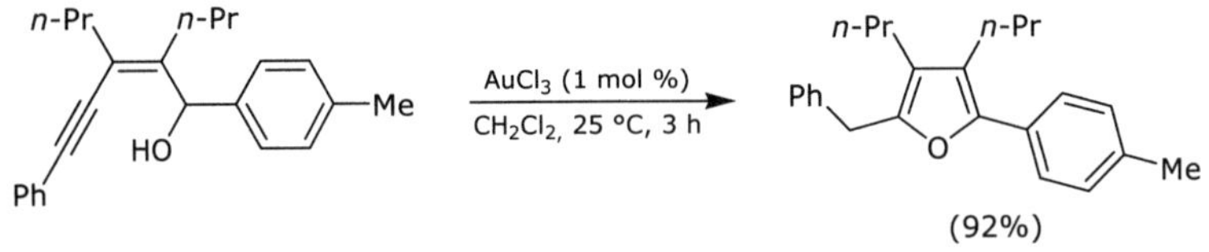

Scheme 4-80

of cationic gold(I)-NHC catalysts.[222] Likewise, benzofurans were formed by treatment of 2-alkynylphenols with gold(I) or gold(III) catalysts.[38, 203] With $AuCl_3$ or $Ph_3PAuOTf$ as catalyst, (*Z*)-2-en-4-yn-1-ols with a tertiary alcohol group cyclize to 2-alkylidene-2,5-dihydrofurans. In all cases, the reaction proceeds by 5-*exo*-dig attack of the hydroxy group at the activated alkyne and subsequent protodeauration. Aliphatic substrates with a primary or secondary alcohol then undergo aromatization to the furan derivative by proton transfer.

Various other acetylenic substrates can also be used for the synthesis of furans by intramolecular gold-catalyzed C–O bond formation. Thus, acetylenic diols gave a variety of monosubstituted, disubstituted, or trisubstituted furans by dehydrative cyclization in the presence of 2 mol% Au[P(*t*-Bu)$_2$(*o*-biphenyl)]Cl/AgOTf or AuCl at 0 °C (Scheme 4-81).[148] In the latter case, the catalyst loading can be decreased to 0.05 mol% by scavenging water with activated molecular sieves and conducting the reaction in refluxing THF.

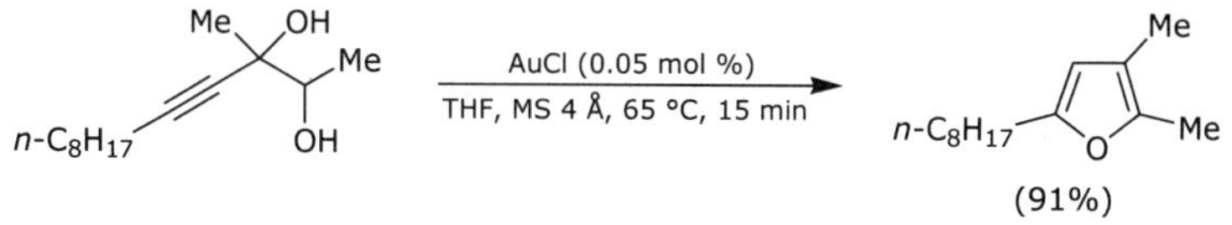

$n\text{-}C_8H_{17}$ / Me / OH / Me / OH — AuCl (0.05 mol %), THF, MS 4 Å, 65 °C, 15 min → Me / $n\text{-}C_8H_{17}$ / O / Me (91%)

Scheme 4-81

Treatment of suitable propargyl oxiranes with $AuCl_3$ or $Ph_3PAuOTf$ also afforded furans in moderate to high yields (Scheme 4-82).[223] Free hydroxy groups or aryl bromides were tolerated under the reaction conditions. The transformation is accelerated in the presence of an alcohol (*e.g.*, MeOH), which opens the epoxide with

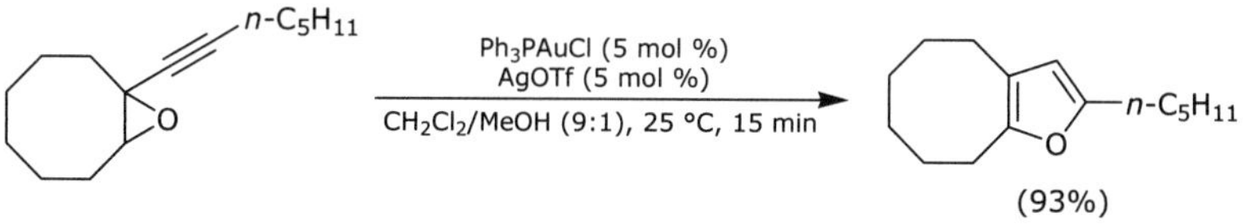

$n\text{-}C_5H_{11}$ / O — Ph_3PAuCl (5 mol %), AgOTf (5 mol %), CH_2Cl_2/MeOH (9:1), 25 °C, 15 min → O / $n\text{-}C_5H_{11}$ (93%)

Scheme 4-82

formation of regioisomeric propargyl/homopropargyl alcohols; these then undergo cyclization and elimination of the alcohol. Silver, platinum, or mercury catalysts also promote this transformation, albeit with lower reactivity.[223b, 224]

Representative procedure for the gold-catalyzed cycloisomerization of propargyl oxiranes to furans. 4-(4-Methylfuran-2-yl)butan-1-ol[223a]

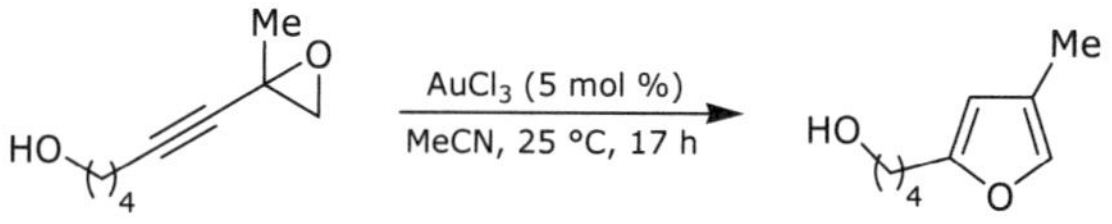

HO / 4 / Me / O — $AuCl_3$ (5 mol %), MeCN, 25 °C, 17 h → HO / 4 / Me / O

A solution of $AuCl_3$ in MeCN (255 mg; 5% w/w, 12.8 mg, 42.0 μmol $AuCl_3$) was added to a solution of 6-(2-methyloxiran-2-yl)hex-5-yn-1-ol (130 mg, 843 μmol) in MeCN (1 mL) at 25 °C. After 17 h (the reaction was monitored by TLC), the solvent was evaporated under reduced pressure and residue was purified by column chromatography (silica gel, 10% EtOAC-hexane). Pure 4-(4-methylfuran-2-yl)butan-1-ol was obtained in as a colorless viscous liquid; yield: 109 mg (84%).

Acetylenic oxiranes bearing an additional acetoxy group cyclize to functionalized furans in the presence of a gold catalyst and a nucleophile. Thus, 2,5-disubstituted furans were obtained by treating 1-oxiranylpropargyl acetates with an alcohol or an electron-rich arene and a gold(III) catalyst.[225] This transformation is probably initiated by nucleophilic attack of the epoxide oxygen atom at the activated triple bond, followed by opening of the resulting oxonium ion with the nucleophile, and acetate elimination. The regioisomeric 3-oxiranylpropargyl acetates afforded the same products in the presence of $Ph_3PAuSbF_6$ and an alcohol or thiol.[151]

Representative procedure for the gold-catalyzed cyclization of oxiranylpropargyl acetates to furans. 4,5,6,7-Tetrahydro-2-(1-methoxyheptyl)benzofuran[151]

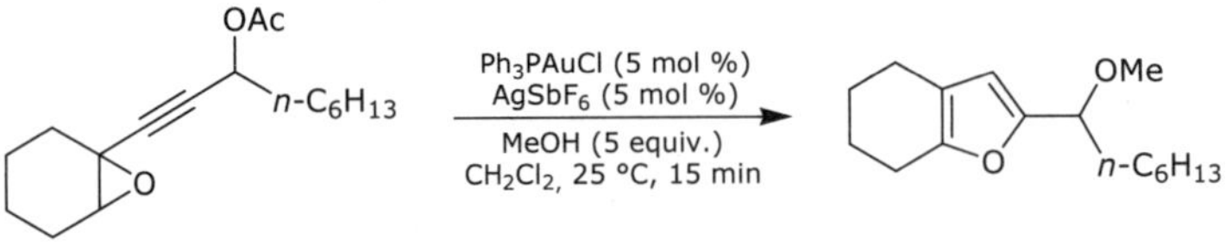

OAc; n-C_6H_{13}; O — Ph_3PAuCl (5 mol %), $AgSbF_6$ (5 mol %) / MeOH (5 equiv.), CH_2Cl_2, 25 °C, 15 min → OMe; O; n-C_6H_{13}

1-(7-Oxabicyclo[4.1.0]heptan-1-yl)non-1-yn-3-yl acetate (56 mg, 0.2 mmol) in 1 mL of CH_2Cl_2 was added to a stirred solution of premixed Ph_3PAuCl (0.01 mmol) and $AgSbF_6$ (0.01 mmol) in 1 mL of CH_2Cl_2 containing 5 equiv of MeOH at RT. The reaction was monitored by TLC until completion (15 min). The reaction mixture was filtered through a pad of silica gel with CH_2Cl_2. Solvents were removed *in vacuo*, and the crude residue was purified by flash chromatography (pentane/Et_2O). Yield: 95% (48 mg, 0.191 mmol) of 4,5,6,7-tetrahydro-2-(1-methoxyheptyl)benzofuran as pale yellow oil.

Acetylenic ketones are viable substrates as well for gold-catalyzed cyclizations to furans. Whereas alk-3-yn-1-ones readily cyclize to substituted furans in the presence of gold(I) or gold(III) chloride[66a, 226] (possibly *via* isomerization to α-allenyl ketones; see Section 5.2), the corresponding transformation of alk-4-yn-1-ones takes place in the presence of the cationic gold catalyst generated *in situ* from Ph_3PAuCl and AgOTf in toluene (Scheme 4-83).[227] The cyclization is accelerated by

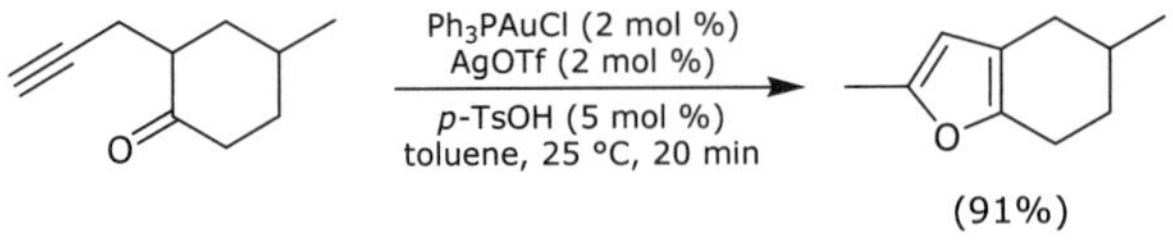

Ph_3PAuCl (2 mol %), AgOTf (2 mol %) / p-TsOH (5 mol %), toluene, 25 °C, 20 min → (91%)

Scheme 4-83

p-toluenesulfonic acid and may proceed by nucleophilic attack of the enol oxygen atom at the activated triple bond. In a similar fashion, *N*-propargylcarboxamides can be cyclized to oxazoles.[32, 33, 228] In contrast to this, alkylidene-substituted tetrahydrofuranyl ethers were formed when the cyclization of alk-4-yn-1-ones was carried out in alcoholic solvents. Alk-4-yn-1-ones bearing two substituents at C-3 cannot afford furans but rather undergo a 6-*endo*-dig cyclization to 4*H*-pyrans instead of the usual 5-*exo*-dig cycloisomerization.

In a closely related reaction, cross-conjugated enynones are converted into furans in the presence of a nucleophile and gold(III) chloride.[229] Alcohols, β-dicarbonyl compounds, and electron-rich aromatics are suitable nucleophiles for this transformation. A variety of other transition metals also catalyzes the reaction but at a much slower rate. Even though the order of events could not be established, it seems reasonable to assume that the reaction proceeds *via* a (gold-catalyzed?) Michael addition of the external nucleophile to the enone and subsequent nucleophilic attack of the carbonyl oxygen atom at the activated triple bond. This leads to a σ-vinylgold species that can be trapped with various electrophiles (proton sources, I_2, NIS, or PhSeCl). The transformation can also be performed with Bu_4NAuCl_4 in the ionic liquid [BMIM][BF_4], a catalyst that is air-stable and recyclable.[230]

Representative procedure for the gold-catalyzed synthesis of furans from cross-conjugated enynones. 4,5,6,7-Tetrahydro-4-methoxy-2-(4-methoxyphenyl)benzofuran[229b]

$AuCl_3$ (1 mol %)
MeOH (1.5 equiv)
CH_2Cl_2/MeCN, 25 °C, 1 h

A solution of $AuCl_3$ (30.3 mg) in MeCN (970 mg) was prepared. To 2-(4-methoxyphenylethynyl)cyclohex-2-enone (45.2 mg, 0.2 mmol) and MeOH (9.6 mg, 1.5 equiv) in CH_2Cl_2 (1.0 mL) was added a portion of the above $AuCl_3$ solution (20 mg, 1 mol%). The mixture was stirred at RT for 1 h. The solvent was removed under reduced pressure and the residue was chromatographed (silica gel, 7:1 hexane/EtOAc) to afford 45.5 mg (88%) of 4,5,6,7-tetrahydro-4-methoxy-2-(4-methoxyphenyl)benzofuran as a colorless oil.

A variation of this method uses alkynylcyclopropylketones that undergo ring expansion and furan formation in the presence of a nucleophile and $Ph_3PAuOTf$.[231] The substrates can be readily prepared from the corresponding enynones by cyclopropanation. With AgOTf as catalyst, comparable yields were achieved at higher catalyst loading (5 mol%) and longer reaction time (1 h at room temperature).

Representative procedure for the gold-catalyzed ring expansion/cyclization of alkynylcyclopropylketones. 5,6,7,8-Tetrahydro-5-methoxy-2-phenyl-4H-cyclohepta[b]furan[231]

$Ph_3PAuOTf$ (1 mol %)
MeOH (2 equiv)
CH_2Cl_2, 25 °C, 15 min

A solution of 1-phenylethynylbicyclo[4.1.0]heptan-2-one (105 mg, 0.5 mmol) and MeOH (32 mg, 1.0 mmol) in dry CH_2Cl_2 (2 mL) was added to a solution of $Ph_3PAuOTf$ (generated by mixing equal equivalents of Ph_3PAuCl and AgOTf, with AgCl filtered off) in CH_2Cl_2 (6.25 mM, 0.8 mL, 5 μmol). The mixture was stirred for 15 min at RT. After evaporation of the solvent, the residue was purified by column chromatography on silica gel (cyclohexane/ EtOAc = 10:1) to afford 110 mg of 5,6,7,8-tetrahydro-5-methoxy-2-phenyl-4*H*-cyclohepta[*b*]furan (91%) as a light yellow oil.

In the presence of the highly electrophilic gold(I) complex $(4\text{-}CF_3C_6H_4)_3PAuCl$ and AgOTf, 4-hydroxyalky-2-yn-1-ones afforded substituted 3(2*H*)-furanones with excellent yield.[232] The cyclization may proceed by intramolecular Michael addition of the hydroxy group at the activated ynone to generate an unsaturated oxirane, which then undergoes ring-opening by nucleophilic attack of the pendant carbonyl oxygen atom. An alternative pathway *via* hydration of the alkyne seems unlikely because the reaction is inhibited in the presence of water.

Representative procedure for the gold-catalyzed cyclization of 4-hydroxyalky-2-yn-1-ones to 3(2H)-furanones. 2,2-Dimethyl-5-(2-phenylethyl)-3(2H)-furanone[232a]

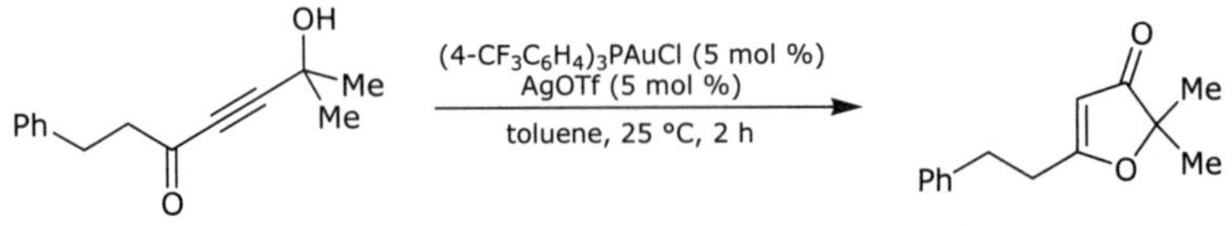

$(4\text{-}CF_3C_6H_4)_3PAuCl$ (5 mol %), AgOTf (5 mol %), toluene, 25 °C, 2 h

To a solution of 6-hydroxy-6-methyl-1-phenylhept-4-yn-3-one (111 mg, 0.51 mmol) in toluene (2.6 mL, 0.20 M) were added $(4\text{-}CF_3C_6H_4)_3PAuCl$ (17.9 mg, 0.026 mmol) and AgOTf (6.6 mg, 0.026 mmol) in this order at RT. The reaction mixture was stirred at RT for 2.0 h and then quenched with saturated aqueous NH_4Cl. The organic materials were extracted with EtOAc, and the combined organic extracts were washed with brine, dried over anhydrous Na_2SO_4, and evaporated *in vacuo*. The residue was purified by column chromatography (silica gel, hexanes/EtOAc = 4:1) to give 2,2-dimethyl-5-(2-phenylethyl)-3(2*H*)-furanone (101 mg, 91%) as a pale yellow solid.

The corresponding saturated heterocycles are accessible from homopropargylic alcohols in the presence of $Ph_3PAuNTf_2$, MsOH, and a pyridine *N*-oxide as stoichiometric oxidant (Scheme 4-84).[233] It is assumed that an α-oxogold carbene is formed under the reaction conditions, which provides the product by intramolecular O–H insertion. Various reactive functionalities in the substrate are tolerated, including halides, azide, and acid-sensitive protecting groups. The analogous reaction of propargyl alcohols afforded oxetan-3-ones.[234] Enantiomerically enriched propargyl alcohols gave the product without any loss of

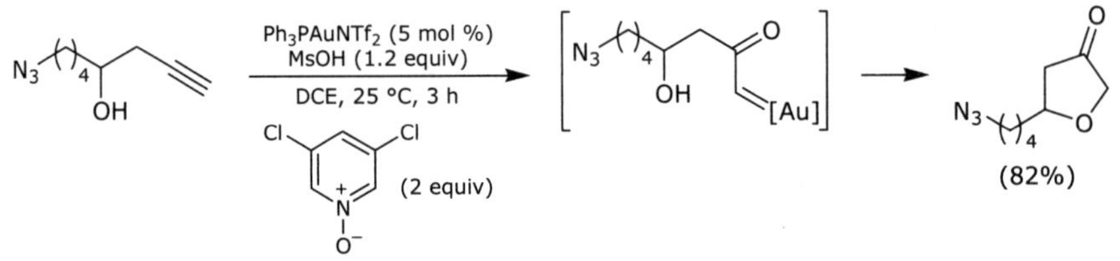

$Ph_3PAuNTf_2$ (5 mol %), MsOH (1.2 equiv), DCE, 25 °C, 3 h, 3,5-dichloropyridine N-oxide (2 equiv); [Au]; (82%)

Scheme 4-84

stereochemical information. The reaction also works well for the parent compound, which is difficult to prepare by other methods. Coupled together with a subsequent Wittig or Strecker reaction, highly useful oxetane derivatives are accessible in an efficient and straightforward manner.

Representative procedure for the gold-catalyzed oxidative cyclization of propargyl alcohols to oxetan-3-ones. Ethyl 2-(oxetan-3-ylidene)acetate[234]

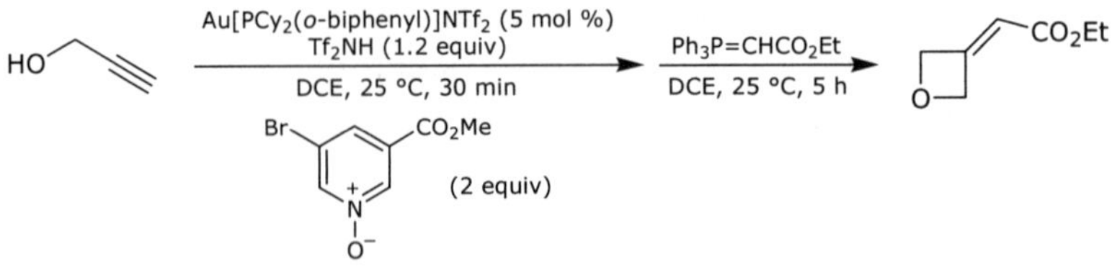

$Au[PCy_2(o\text{-}biphenyl)]NTf_2$ (5 mol %), Tf_2NH (1.2 equiv), DCE, 25 °C, 30 min, 3-bromo-5-(methoxycarbonyl)pyridine N-oxide (2 equiv); $Ph_3P{=}CHCO_2Et$, DCE, 25 °C, 5 h; CO_2Et

3-Methoxycarbonyl-5-bromopyridine *N*-oxide (1.0 mmol), Tf_2NH (3.0 mL, 0.20 M in 1,2-dichloroethane [DCE]), and Au[PCy_2(*o*-biphenyl)]NTf_2 (20.7 mg, 0.025 mmol) were added sequentially to a solution of propargyl alcohol (28.0 mg, 0.50 mmol) in DCE (7.0 mL) at RT. The reaction mixture was stirred at RT for 0.5 h; after which, it was treated with saturated aqueous $NaHCO_3$ (5 mL). The resulting solution was then extracted with CH_2Cl_2 (1 × 10 mL), and the combined organic layers were dried over anhydrous $MgSO_4$. The mixture was filtrated and the mixture was used directly for the next step.

To the above solution was added carboethoxymethylene triphenylphosphorane (192 mg, 0.55 mmol, 1.5 equiv) at RT. After stirring for 5 h, the mixture was concentrated *in vacuo* and the residue was purified by silica gel flash chromatography (eluent: hexanes/EtOAc) to afford ethyl 2-(oxetan-3-ylidene)acetate with 62% overall yield.

The gold-catalyzed tandem nucleophilic addition/cyclization of 2-alkynylbenzaldehydes affords 1*H*-isochromene derivatives. The reaction is promoted by palladium,[235] copper,[235b, 236] silver,[237] and gold catalysts[237a, 238] In most cases, the 6-*endo*-dig cyclization is preferred over the competing 5-*exo*-dig cyclization mode.[239] With terminal alkynes as the nucleophiles, the reaction works best with Me_3PAuCl as catalyst in the presence of water and Hünig's base (*i*-Pr_2NEt). Under these conditions, a gold acetylide is probably formed that may generate a chelate complex with the 2-alkynylbenzaldehyde. Acetylide addition to the aldehyde and subsequent attack of the oxygen atom at the activated triple bond provides a σ-vinylgold intermediate that undergoes protodeauration.

Representative procedure for the gold-catalyzed nucleophilic addition/cyclization of 2-alkynylbenzaldehydes. 3-Phenyl-1-(phenylethynyl)-1H-isochromene[238a]

CHO; Ph; + ≡—Ph; Me_3PAuCl (5 mol %); *i*-Pr_2NEt (20 mol %); toluene/H_2O, 70 °C, 1 d; Ph; O; Ph

Me_3PAuCl (1.9 mg, 6.25 μmol, 5 mol%) was mixed with 0.5 mL distilled water, 0.5 mL of toluene, 0.028 mL of phenylacetylene (0.25 mmol, 2 equiv), and 5 μL of *i*-Pr_2NEt
(0.025 mmol, 20 mol%). The mixture was stirred at RT until all solids disappeared. 2-(Phenylethynyl)benzaldehyde (25.8 mg, 0.125 mmol) was added, and the mixture was stirred at RT for 10 min and then at 70 °C for 1 day. The reaction mixture was extracted with Et_2O (3 × 5 mL). The extracts were dried with anhydrous Na_2SO_4 and concentrated under reduced pressure. The residue was purified by flash column (30–40:1 hexanes/EtOAc). 3-Phenyl-1-(phenylethynyl)-1*H*-isochromene was obtained in 81% yield (31.1 mg).

An enantioselective version of this reaction has been described recently by Handa and Slaughter using gold(I) catalysts of the type **K** bearing chiral acyclic diaminocarbene ligands (ADCs; Scheme 4-85).[240] Compared with NHCs, their N–C–N angles are wider so that chiral substituents are closer to the metal center. $LiNTf_2$

was identified as a necessary additive for the reaction with a 0.9:1 ratio to **K**; higher ratios resulted in racemization of the 1*H*-isochromenes.

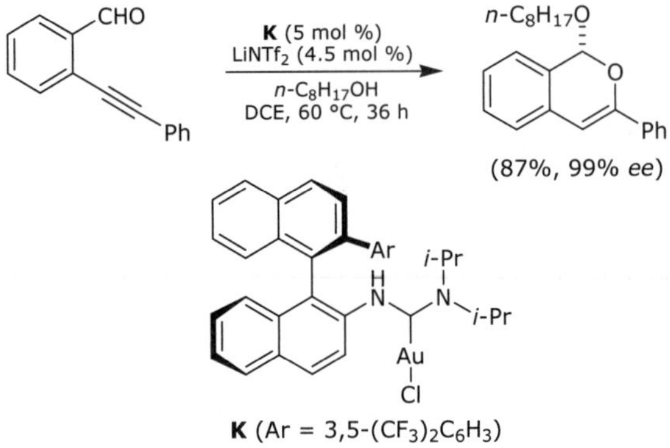

Scheme 4-85

General procedure for the enantioselective gold-catalyzed tandem acetalization/cyclization of 2-alkynylbenzaldehydes[240]

Under dry and oxygen-free conditions, enantiomerically pure chiral gold catalyst of the type **K** (5 mol%) and $LiNTf_2$ (4.5 mol%) are dissolved in 1.0 mL of dry DCE in a 4-mL screw-cap reaction vial fitted with a polytetrafluoroethylene (PTFE)/silicone septum. The vial is placed in an aluminum heating block at 45 °C, and the solution is stirred for 15 min. The mixture is then allowed to cool to RT or heated to 60 °C if a preliminary trial indicates a slow reaction at RT. A solution of the 2-alkynylbenzaldehyde (0.5 mmol) and alcohol (1.0 equiv; 0.55 equiv in the case of MeOH) in 1.0 mL of dry DCE is then added *via* syringe. After 8 h (6 h in the case of MeOH), another 1.0 equiv of alcohol (0.55 equiv for MeOH) is added. Reaction progress is monitored by TLC using Et_2O:hexanes as the eluent. Products are purified by flash chromatography and then analyzed by polarimetry, chiral high-performance liquid chromatography (HPLC), and NMR within 1–2 h.

In a related study, Uemura and co-workers obtained planar-chiral isochromene chromium complexes by gold-catalyzed enantioselective cyclization of prochiral (1,3-dihydroxymethyl-2-alkynyl)benzene chromium complexes.[241] The highest selectivities in this 6-*endo*-dig-hydroalkoxylation were observed in the presence of a chiral cationic gold catalyst prepared *in situ* from (*R*)-xylyl-BINAP$(AuCl)_2$ and $AgSbF_6$. With (*R*)-SEGPHOS$(AuCl)_2$ and $AgBF_4$, the enantiomeric product was obtained, albeit with low enantioselectivity.

Representative procedure for the enantioselective gold-catalyzed cyclization of (1,3-dihydroxymethyl-2-alkynyl)benzene chromium complexes. (R_p)-Tricarbonyl-(3-methyl-5-hydroxymethyl-1H-isochromene)chromium[241]

OH Me OH $Cr(CO)_3$

(R)-xylyl-BINAP$(AuCl)_2$ (10 mol %)
$AgSbF_6$ (20 mol %)
CH_2Cl_2, 25 °C, 5 min

OH Me O $Cr(CO)_3$

PAr_2-AuCl
PAr_2-AuCl

(R)-xylyl-BINAP$(AuCl)_2$
Ar = 3,5-$Me_2C_6H_3$

A mixture of (R)-xylyl-BINAP$(AuCl)_2$ (5.8 mg, 4.8 μmol) and $AgSbF_6$ (3.3 mg, 9.6 μmol) in CH_2Cl_2 (1 mL) was stirred at RT for 10 min under nitrogen. AgCl was precipitated during that time. A solution of [1,3-dihydroxymethyl-2-(propy-1-nyl)benzene]tricarbonylchromium (15.0 mg, 0.048 mmol) in CH_2Cl_2 (2.0 mL) was added to the reaction mixture by syringe at RT. After 5 min of stirring, Et_2O (5 mL) was added, and the reaction mixture was filtered through a short layer of silica gel. Concentration of the organic layer under reduced pressure and purification by silica gel column chromatography (5% Et_2O in hexane) gave 12.8 mg of (R_p)-tricarbonyl-(3-methyl-5-hydroxymethyl-1*H*-isochromene)chromium (85% yield, 99% *ee*) as yellow crystals.

Like the hydroalkoxylation of alkynes, the hydration is also catalyzed by Ph_3PAuMe/H_2SO_4,[242] as well as, by various gold(III) catalysts.[243] However, the cationic gold complexes $R_3PAuNTf_2$[244] and $IPrAuSbF_6$[245] are more efficient catalysts for this transformation. In the latter case, turnover numbers up to 84,000 and turnover frequencies of 4700 h^{-1} were achieved upon heating terminal or internal alkynes to 120 °C in a dioxane-water mixture, so that catalyst loadings as low as 10 ppm are possible under optimized conditions. The key to this tremendous efficiency is the pronounced stability of gold-NHC catalysts that allows for their prolonged use at elevated temperatures. The analogous hydration of nitriles catalyzed by $IPrAuNTf_2$ generates amides,[246] whereas homopropargyl ethers containing a pendant oxygen or nitrogen nucleophile give saturated heterocyclic ketones formed by a sequence of alkyne hydration, retro-Michael addition, and intramolecular conjugate addition.[247] These transformations proceed with good to excellent diastereoselectivity and have been applied to the synthesis of the natural product andrachcinidine.

Representative procedure for the gold-catalyzed hydration of alkynes. Octan-4-one[245a]

IPrAuCl (100 ppm)
$AgSbF_6$ (100 ppm)
dioxane/H_2O, 120 °C, 18 h

i-Pr i-Pr N N i-Pr Au i-Pr Cl
IPrAuCl

In a sealed 50-mL reaction vial equipped with a magnetic stirring bar, IPrAuCl (1.0 mg, 2 μmol, 100 ppm) was added to 1,4-dioxane (13 mL). $AgSbF_6$ (0.7 mg, 2 μmol, 100 ppm) was added and the solution was stirred for 1 min. Oct-4-yne (20 mmol, 2.20 g, 1 equiv) was added, followed by the addition of distilled H_2O (6.6 mL). The reaction mixture was then heated for 18 h at 120 °C. Volatile components were then removed under reduced pressure and the residue dissolved in CH_2Cl_2. The solution was filtered over a plug of silica (~1 cm) and the solvent was removed under reduced pressure, affording 2.15 g (16.8 mmol, 84%) of octan-4-one.

Representative procedure for the gold-catalyzed transformation of homopropargyl ethers to heterocyclic ketones. 1-((2R,6S)-6-((R)-2-Hydroxypentyl)-1-(2-nitrophenylsulfonyl)-piperidin-2-yl)propan-2-one[247b]

OMe OH NsHN
Ph_3PAuCl (5 mol %)
$AgSbF_6$ (10 mol %)
toluene/H_2O, 40 °C, 24 h
O OH N Ns

To a solution of *N*-((4*R*,6*S*)-4-hydroxy-10-methoxytridec-12-yn-6-yl)-2-nitrobenzene-sulfonmide (70.0 mg, 0.164 mmol) in water-saturated toluene (6.6 mL, 0.025 M) were added Ph_3PAuCl (4.1 mg, 0.008 mmol) and $AgSbF_6$ (5.6 mg, 0.016 mmol). The reaction mixture was stirred at 40 °C for 24 h. The reaction solution was directly purified by chromatography on silica gel (10:1, CH_2Cl_2/EtOAc) to afford 1-((2*R*,6*S*)-6-((*R*)-2-hydroxypentyl)-1-(2-nitrophenylsulfonyl)-piperidin-2-yl)propan-2-one (60.5 mg, 0.147 mmol, 89%) as a colorless oil.

5.2 Nucleophilic Additions to Allenes

The cycloisomerization of α-allenyl ketones to the corresponding substituted furans[66a] was the first example of a gold-catalyzed addition of an oxygen nucleophile to an allene (Scheme 4-86). Traditionally, silver[248] or palladium[249] catalysts were employed for cyclizations of this type; advantages of gold catalysis include shorter reaction times, milder conditions, and/or lower catalyst loadings. Variable amounts of

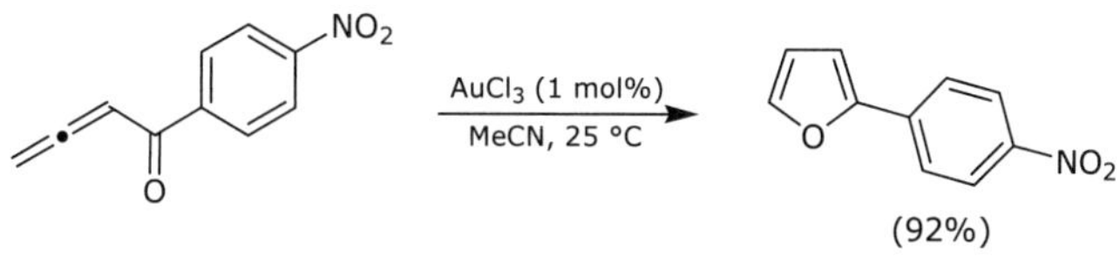

Scheme 4-86

a side product formed by Michael addition of the furan to unconsumed allenic ketone were obtained with gold(III) chloride as catalyst. This side reaction can be prevented by using a gold(III)-porphyrin complex.[250] Under these conditions, no dimer could be detected, but the presence of trifluoroacetic acid and elevated temperatures are essential, which can lead to problems with acid-labile substrates. [Au(TPP)]Cl is highly reactive (the catalyst loading can be decreased to 0.1 mol%, corresponding to turnover numbers of up to 850) and can be recovered and reused in up to nine consecutive runs with no appreciable loss of reactivity or decrease of yield. Mechanistically, the cycloisomerization proceeds by coordination of the active gold species to the "distal" double bond of the allene. This process induces a nucleophilic attack of the oxygen atom to furnish a cationic intermediate, which after deprotonation and protodeauration (which is facilitated in the presence of an external proton donor like CF_3CO_2H) leads to the furan and releases the gold catalyst into the catalytic cycle.

Representative procedure for the gold-catalyzed cycloisomerization of allenic ketones. 2-Phenylfuran[250]

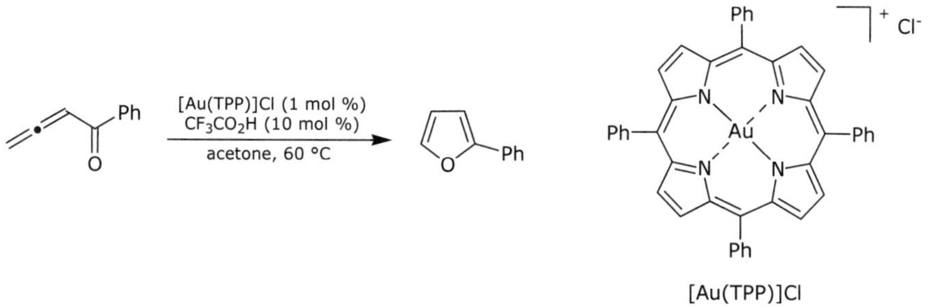

To an acetone solution (30 mL) containing 1-phenylbuta-2,3-dien-1-one (15 mmol) was added [Au(TPP)]Cl (15 μmol) and CF_3CO_2H (1.5 mmol), and the resultant mixture was stirred at 60 °C for 2 h. About 27 mL of solvent was removed under reduced pressure, and *n*-hexane was added to the mixture to precipitate out the gold(III)-porphyrin catalyst, which was filtered and washed with *n*-hexane (3 × 40 mL). Excess solvent was removed under reduced pressure, and the product yield was determined by ^1H NMR analysis (80–85%). After each reaction, the catalyst recovered by filtration was dried under reduced pressure. The cycloisomerization of 1-phenylbuta-2,3-dien-1-one was repeated nine more times using the recovered catalyst under the same reaction conditions.

An extension of the method to haloallenones afforded halogenated furans. Gevorgyan and co-workers[251] demonstrated that the structure of the product is highly dependent on the gold catalyst (Scheme 4-87). When a carbophilic Au(I) species (*e.g.*, Et_3PAuCl) was used, the cycloisomerization led to the expected 5-halofuran, whereas the more oxophilic $AuCl_3$ preferentially generated the 4-halofuran. The latter finding can be explained by formation of a bromoirenium ion arising from coordination of the oxophilic gold(III) catalyst to the carbonyl oxygen atom.

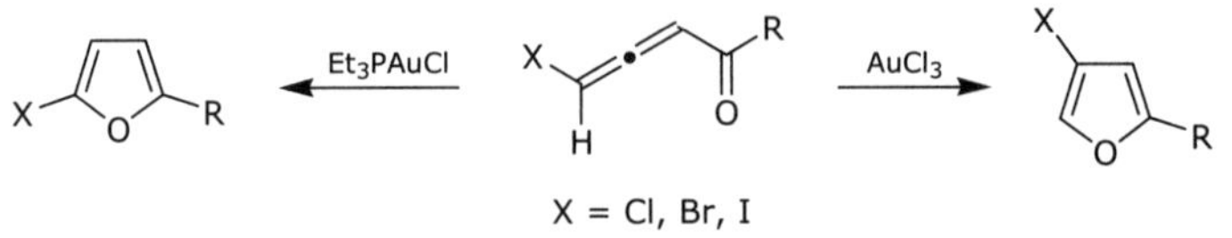

X = Cl, Br, I

Scheme 4-87

Representative procedure for the gold-catalyzed cycloisomerization of haloallenones. 3-bromo-2-butyl-4,5-diphenylfuran[251]

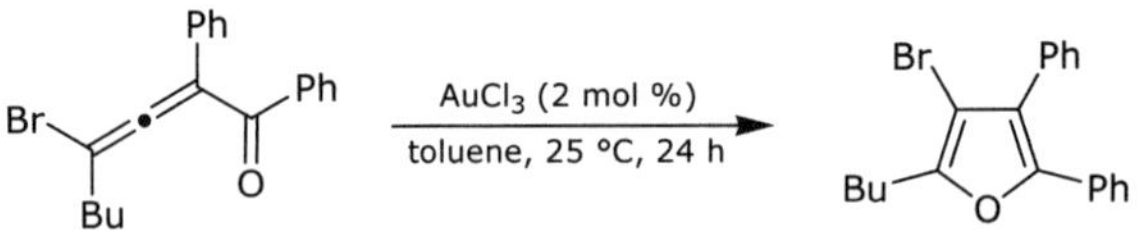

In a glove box under nitrogen atmosphere, to a 3.0-mL Wheaton microreactor equipped with a spin vane and screw cap with a PTFE faced silicone septum was added 6.2 mg (0.02 mmol, 2 mol%) of $AuCl_3$. The microreactor was removed from the glove box, 1 mL of anhydrous toluene and 360 mg (1.02 mmol) of 4-bromo-1,2-diphenylocta-2,3-dien-1-one were sequentially added, and the mixture was stirred for 24 h. After the reaction was judged complete by TLC, it was quenched by filtration through a pad of alumina with CH_2Cl_2, the filtrate was concentrated, and the residue was purified over 30 mL of silica gel using hexanes as eluent to afford 270 mg (0.75 mmol, 75%) of 3-bromo-2-butyl-4,5-diphenylfuran as a colorless oil that solidifies on storage.

In an analogous manner, silylfurans, thiofurans, or selenofurans can be obtained from the corresponding allenes by a 1,2-Si, 1,2-S, or 1,2-Se shift (Scheme 4-88).[251b, 252] In the latter two cases, the allene is formed *in situ* from the corresponding thio-/selenoalkynone by 1,2-migration of the chalcogenide.

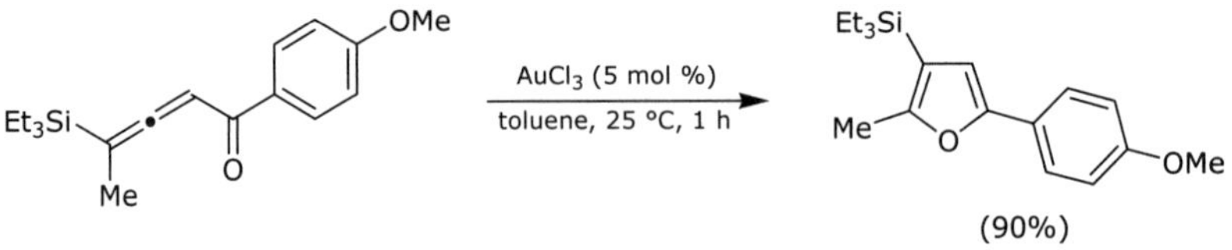

Scheme 4-88

The driving force for the cycloisomerization of allenones is the formation of the aromatic heterocycle. This is even possible with allenones bearing two carbon substituents in the γ-position if at least one of the substituents has a pronounced tendency to migrate.[251b, 253] For example, heating of aromatic allenones with the cationic gold catalyst $Ph_3PAuOTf$ afforded furans *via* cationic intermediates, *i.e.*, by ring closure and subsequent [1,5]-phenyl shift (Scheme 4-89). Even though gold(I) seems to be the most active catalyst, this transformation can also be performed in the presence of various other Lewis acids (*e.g.*, $Sn(OTf)_2$, $In(OTf)_3$, AgOTf, $Cu(OTf)_2$, and Me_3SiOTf) as well as Brønsted acids (TfOH).

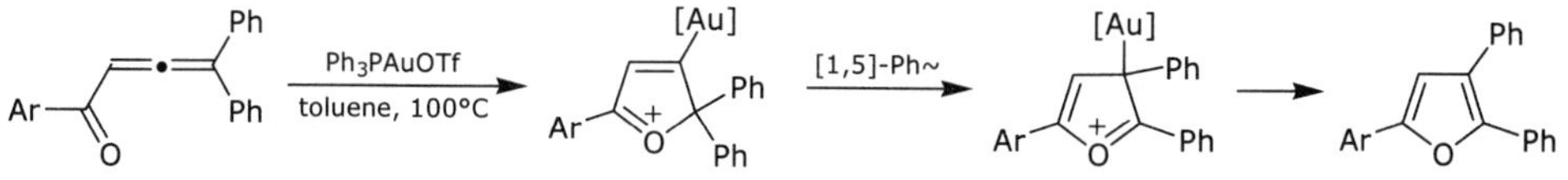

Scheme 4-89

Whereas the cycloisomerization of allenic ketones affords achiral products, replacing the keto with a hydroxy group leads to the formation of chiral heterocycles. In 2001, the synthesis of chiral 2,5-dihydrofurans by treatment of α-hydroxyallenes with catalytic amounts of $AuCl_3$ in unpolar solvents was reported (Scheme 4-90).[254] Many functionalities (*e.g.*, carbonyl groups, additional free alcohols, and acid-sensitive

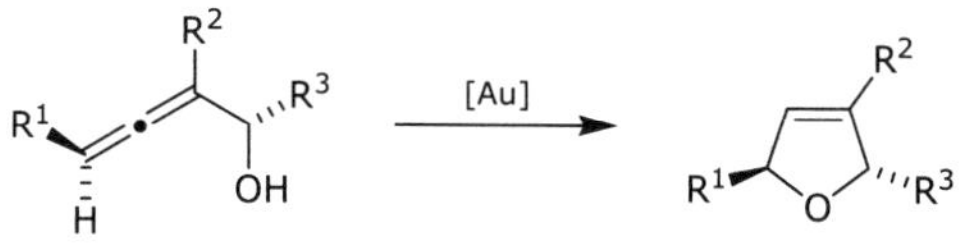

Scheme 4-90

protecting groups) are tolerated under these conditions.[254, 255] Besides gold(III) salts, various neutral or cationic gold(I) catalysts (*e.g.*, AuCl and Ph_3PAuBF_4) can also be used, and the catalyst loading can be as low as 0.05 mol%, so that this transformation is among the most efficient in homogeneous gold catalysis. In the case of chiral allenes bearing alkyl substituents, the stereochemical information of the chirality axis is completely transferred to the new stereogenic center.[256] Mechanistically, the gold-catalyzed cycloisomerization of α-allenols proceeds *via* activation of the allenic double bond distal to the hydroxy group, nucleophilic attack of the oxygen atom at the activated double bond, and protodeauration of the σ-gold species thus formed. The cyclization is accelerated in the presence of external proton donors (water and methanol); this suggests that the protodeauration is the rate-limiting step. Various applications in target-oriented synthesis underline the utility of the method.[257]

Representative procedure for the gold-catalyzed cycloisomerization of α-hydroxyallenes. (2S,5S)-2-(5-(2-(Benzyloxy)ethyl)-2,5-dihydro-5-methylfuran-2-yl)propan-2-ol [257b, 257c]

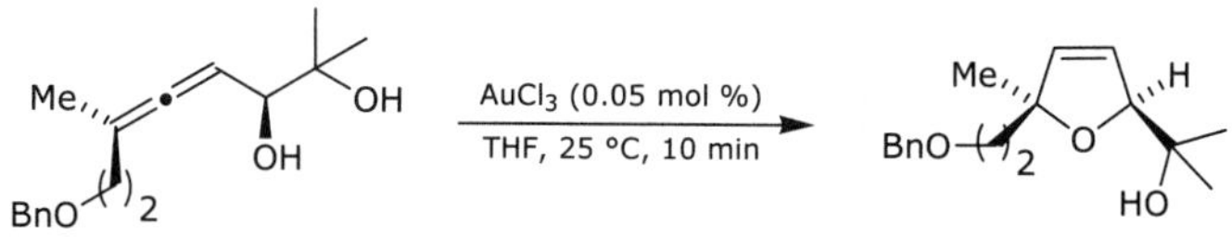

To a solution of (3*S*,5*S*)-8-(benzyloxy)-2,6-dimethylocta-4,5-diene-2,3-diol (1.40 g, 5.1 mmol) in dry THF (51 mL) was added at RT $AuCl_3$ (15 μL, 2.54 μmol, 0.1664 M in MeCN). After 10 min stirring at RT, the solvent was removed *in vacuo* and the crude product was purified by column chromatography on silica gel (cyclohexane/EtOAc, 7:3) to afford (2*S*,5*S*)-2-(5-(2-(benzyloxy)ethyl)-2,5-dihydro-5-methylfuran-2-yl)propan-2-ol (1.36 g, 97%, *dr* = 98:2, *ee* > 98% by GC and NMR analysis) as a colorless oil.

In contrast to alkyl-substituted allenes, substrates with phenyl or electron-rich aromatic substituents undergo epimerization when treated with gold catalysts. Whereas $AuCl_3$ in CH_2Cl_2 epimerizes both the allene and the dihydrofuran, only the allene but not the dihydrofuran is epimerized in the presence of gold(I) chloride in dichloromethane.[258] The epimerization probably proceeds *via* zwitterionic intermediates comprising a benzyl cation substructure and an anionic aurate moiety. The stereochemical integrity of substrate and product can be preserved by decreasing the Lewis acidity of the gold catalyst.[258] This is achieved by using σ-donor ligands to gold, *e.g.*, 2,2'-bipyridine, and/or weakly coordinating solvents like THF. Another possibility is to conduct the cycloisomerization at low temperature (–30 °C) instead of at room temperature, taking advantage of the fact that the epimerization is slower than cyclization.

Representative procedure for the gold-catalyzed cycloisomerization of aryl-substituted α-hydroxyallenes. Cis-2-(t-butyldimethylsilyloxy)-3-methyl-5-phenyl-2,5-dihydrofuran[258]

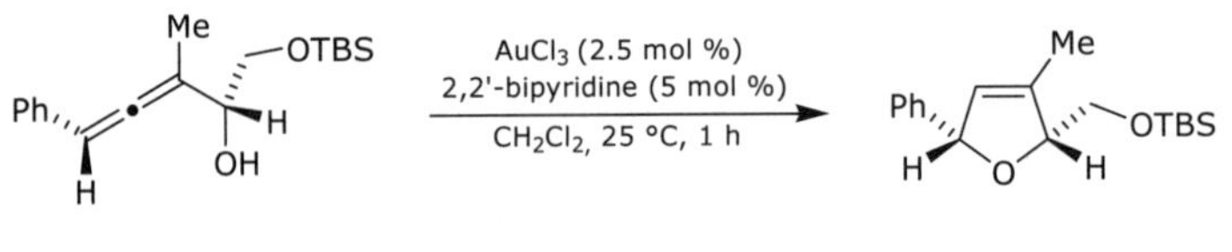

$AuCl_3$ (2.5 mol %)
2,2'-bipyridine (5 mol %)
CH_2Cl_2, 25 °C, 1 h

To a solution of 1-(*t*-butyldimethylsilyloxy)-3-methyl-5-phenylpenta-3,4-dien-2-ol (80 mg, 0.26 mmol, *dr* > 97:3) and 2,2'-bipyridine (2.0 mg, 12.8 μmol) in 3 mL of dry CH_2Cl_2 was added $AuCl_3$ (2.0 mg, 6.6 μmol). After stirring for 1 h at RT, the solvent was evaporated, and the residue was purified by column chromatography (SiO_2, EtOAc/cyclohexane, 1:10) furnishing 56 mg (70%) of *cis*-2-(*t*-butyldimethylsilyloxy)-3-methyl-5-phenyl-2,5-dihydrofuran as a slightly yellow oil (*dr* > 97:3 according to NMR analysis).

The gold-catalyzed cycloisomerization of α,β-dihydroxyallenes is not only stereoselective but also highly chemoselective because no product resulting from nucleophilic attack of the β-hydroxy group was observed. Nevertheless, β-hydroxyallenes undergo a 6-*endo*-trig cyclization to the corresponding 5,6-dihydro-2*H*-pyrans in the presence of a cationic gold catalyst formed *in situ* from Ph_3PAuCl and $AgBF_4$ (Scheme 4-91).[129, 259] These cyclizations usually occur with complete *axis-to-center chirality transfer*; for phenyl-substituted β-hydroxyallenes, the stereoselectivity can be improved by adding pyridine to the reaction mixture.[258]

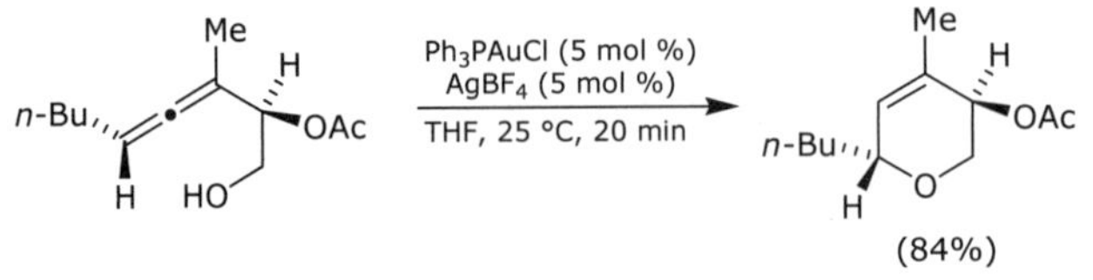

Ph_3PAuCl (5 mol %)
$AgBF_4$ (5 mol %)
THF, 25 °C, 20 min

(84%)

Scheme 4-91

Representative procedure for the gold-catalyzed cycloisomerization of β-hydroxyallenes. (2R)-2-Methyl-2-[(2R)-4-methyl-6-(2-methylpropenyl)-5,6-dihydro-2H-pyran-2-ylmethyl]-1,4-dioxaspiro[4.5]decane[259]

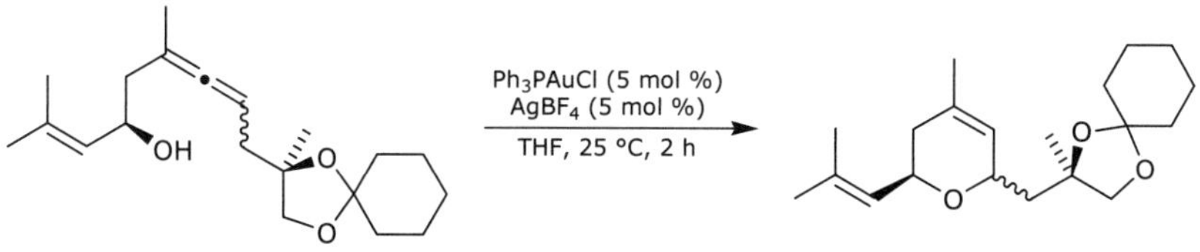

Ph$_3$PAuCl (5 mol %)
AgBF$_4$ (5 mol %)
THF, 25 °C, 2 h

To a solution of (4*R*)-2,6-dimethyl-9-[(2*R*)-2-methyl-1,4-dioxaspiro[4.5]dec-2-yl]-nona-2,6,7-trien-4-ol (52 mg, 0.162 mmol, *dr* = 60:40) in dry THF (2.5 mL) was added Ph$_3$PAuCl (4.0 mg, 8 μmol) and AgBF$_4$ (1.6 mg, 8 μmol) at RT under argon. After stirring for 2 h, the reaction mixture was filtered through Celite and the filtrate was concentrated under vacuum. The residue was purified by column chromatography using cyclohexane/EtOAc (50:1) to give (2*R*)-2-methyl-2-[(2*R*,6*R*)-4-methyl-6-(2-methylpropenyl)-5,6-dihydro-2*H*-pyran-2ylmethyl]-1,4-dioxaspiro[4.5]decane (26 mg, 0.081 mmol, 50%) and (2*R*)-2-methyl-2-[(2*R*,6*S*)-4-methyl-6-(2-methylpropenyl)-5,6-dihydro 2*H* pyran-2-yl-methyl]-1,4-dioxaspiro[4.5]decane (18 mg, 0.057 mmol, 35%) as colorless oils.

The cycloisomerization of β-hydroxyallenes is often very slow, resulting in reaction times of several days. However, addition of *N*-iodosuccinimide (NIS) to the reaction mixture induces a tremendous acceleration, leading to the formation of the corresponding iodinated dihydropyran within 1 min at room temperature (Scheme 4-92).[260] This effect is probably caused by a very rapid iododeauration of a σ-gold intermediate by NIS, which is activated by the gold catalyst.[261]

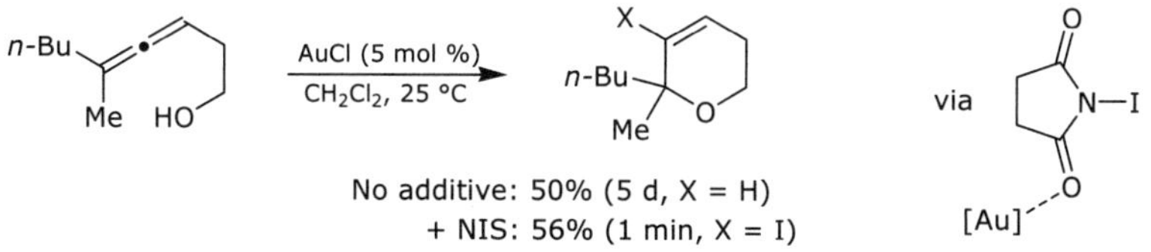

n-Bu Me HO
AuCl (5 mol %)
CH$_2$Cl$_2$, 25 °C
X *n*-Bu Me O
via N—I O O [Au]
No additive: 50% (5 d, X = H)
+ NIS: 56% (1 min, X = I)

Scheme 4-92

Substrates bearing both an allenol and a homopropargyl alcohol moiety can be activated in a chemoselective manner with coinage metal catalysts. Whereas AuCl$_3$ coordinates preferably to the allene and gives the corresponding dihydrofuran as major product, reaction with silver triflate yields the furan by activation of the triple bond (Scheme 4-93).[262] This is a rare example for the formation of different products using gold and silver catalysts.[263]

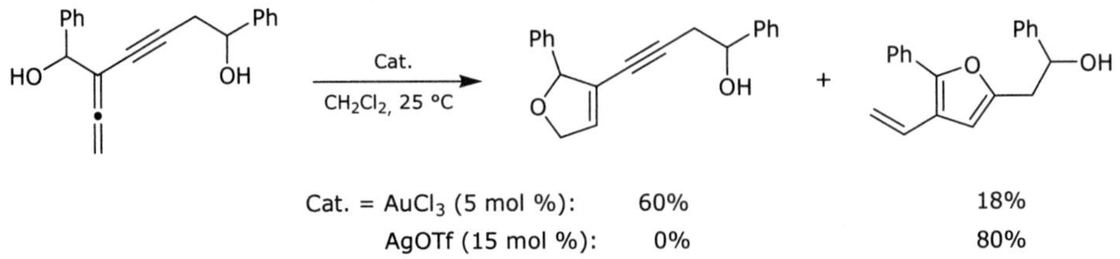

Cat. = $AuCl_3$ (5 mol %): 60% 18%
AgOTf (15 mol %): 0% 80%

Scheme 4-93

A serious drawback of homogeneous gold catalysis is the (seemingly unavoidable) reduction of the catalyst to metallic gold after the reaction or upon workup, which renders recycling of the catalyst impossible. This problem can be overcome by using ionic liquids as reaction medium, for example $AuBr_3$ in the imidazolium-derived medium [BMIM][PF_6] (Scheme 4-94).[264] This catalyst system is not only stable to water and air but also can be recycled easily by extraction of the product. Over five runs, only 0.03% of the original catalyst loading is lost during extraction of the product. This almost negligible leaching makes the method attractive for the synthesis of pharmacologically active target molecules and indicates that the solution of $AuBr_3$ in [BMIM][PF_6] is potentially recyclable several thousand times.

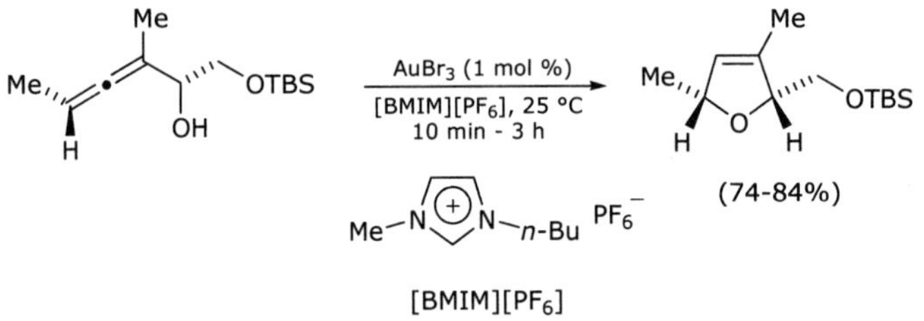

$AuBr_3$ (1 mol %)
[BMIM][PF_6], 25 °C
10 min - 3 h
(74-84%)
[BMIM][PF_6]

Scheme 4-94

Recently, the first example of gold catalysis in micellar systems was reported, using the vitamin-E-*derived amphiphiles polyoxyethanyl α-tocopheryl sebacate (PTS) or D-α-tocopherol-polyethylenglycol-750-succinate monomethylether (TPGS-750-M).[265] With gold(III) bromide, these form air-stable aqueous gold catalyst solutions that show excellent reactivity and recyclability, thereby allowing smooth and efficient cycloisomerization of various α-functionalized allenes (Scheme 4-95). The addition of NaCl to the reaction mixture affords larger micelles and therefore induces faster reactions, allowing catalyst loadings as low as 1 mol%.[265, 266] Recycling of the catalyst solution is possible by extraction of the product from the reaction flask with hexane, with a loss of only 0.29% of the original catalyst loading over four runs.

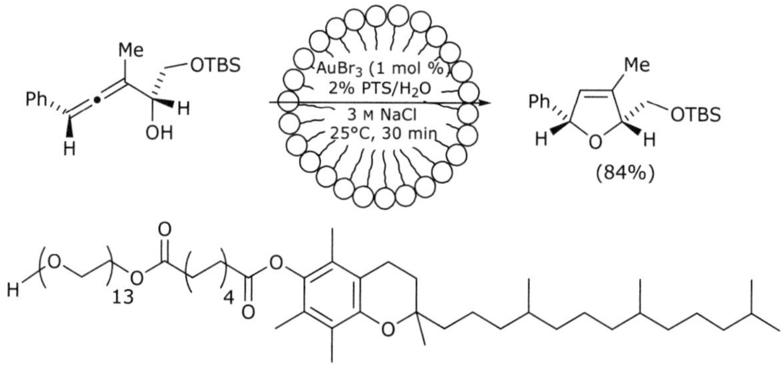

Scheme 4-95

Representative procedure for the gold-catalyzed cycloisomerization of α-hydroxyallenes in micelles. ((Cis-2,5-dihydro-5-isopropyl-3-methylfuran-2-yl)methoxy)t-butyldimethylsilane[265]

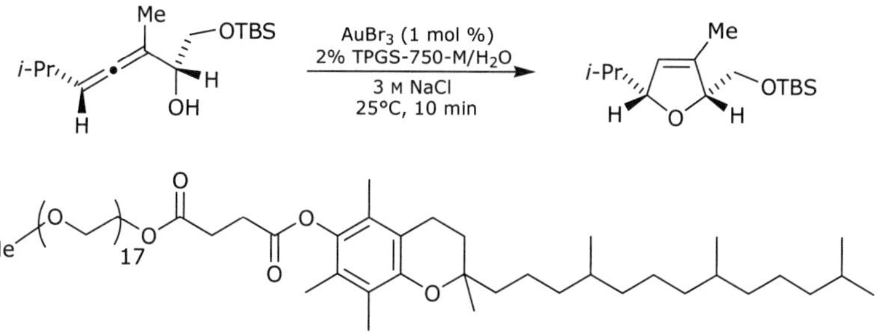

In a vial, 1-(*t*-butyldimethylsilyloxy)-3,6-dimethylhepta-3,4-dien-2-ol (50.0 mg, 185 μmol) was dissolved in a 2% aqueous TPGS-750-M solution (1.5 mL) containing NaCl (263 mg, 3 M) and treated with $AuBr_3$ (0.8 mg, 1.85 μmol) under air. After complete conversion (10 min, monitored by TLC), the reaction mixture was extracted with *n*-hexane. The solvent was removed under reduced pressure and the crude product was purified by flash column chromatography (SiO_2, cyclohexane/EtOAc, 10:1) to afford ((*cis*-2,5-dihydro-5-isopropyl-3-methylfuran-2-yl)methoxy)*t*-butyldimethylsilane (46.2 mg, 92%) as a yellow oil.

The cycloisomerization of α- or β-hydroxyallenes can also be carried out in water, for example with tetrachloroauric acid as catalyst.[267] This system was used for the first example of a tandem lipase/gold-catalyzed transformation. The one-pot kinetic resolution/cycloisomerization of racemic allenic acetates with *Burkholderia cepacia* lipase (PS Amano SD; Amano Enzyme USA Co., Ltd., Elgin, IL) and $HAuCl_4$ afforded 2,5-dihydrofurans as well as unreacted starting material with 28–50% isolated yield and 86–98% *ee* (Scheme 4-96).[268] The mutual tolerance of the Lewis-acidic gold catalyst with the Lewis-basic lipase is maintained as long as low amounts of the former are used.

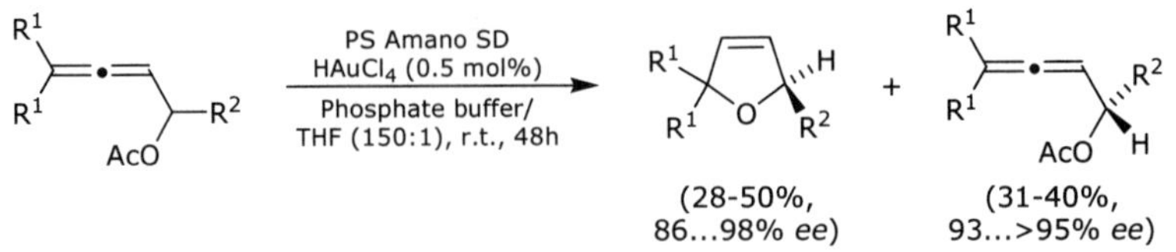

Scheme 4-96

Besides *endo*-cyclizations, functionalized allenes can also undergo gold-catalyzed *exo*-selective attack, in particular if the distance between the nucleophilic group and the allene moiety is large. Alcaide *et al.*[269] examined gold-catalyzed transformations of β-lactams containing an α,γ-dihydroxyallene structure and observed the formation of different cyclization products (Scheme 4-97). Whereas treatment of the tert-butyldimethylsilyl (TBS)-protected substrate with $AuCl_3$ afforded the tetrahydrofuran derivative resulting from 5-*exo*-attack of the γ–hydroxy group, the corresponding methoxymethyl-(MOM)-protected starting material underwent deprotection and 5-*endo*-trig cyclization to the 2,5-dihydrofuran under these conditions. Various other transition metal catalysts (*e.g.*, $AgNO_3$, $[PtCl_2(CH_2{=}CH_2)]_2$, $FeCl_3$) can also be used for cyclizations of this type.

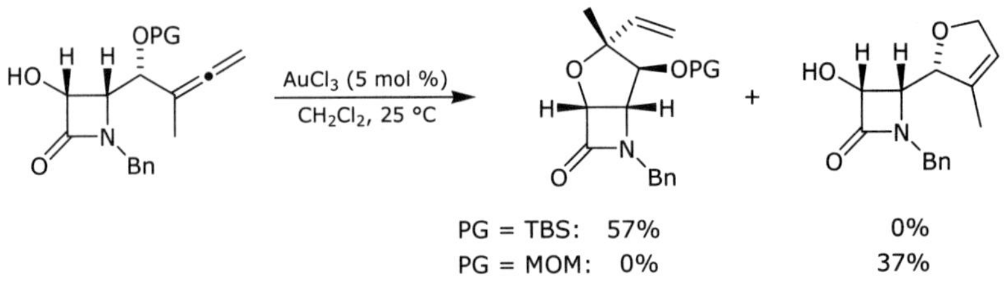

Scheme 4-97

Moving the MOM group to the γ-position leads to the formation of bicyclic tetrahydrooxepines by gold-catalyzed deprotection and (rather unusual) 7-*endo*-trig cycloisomerization (Scheme 4-98).

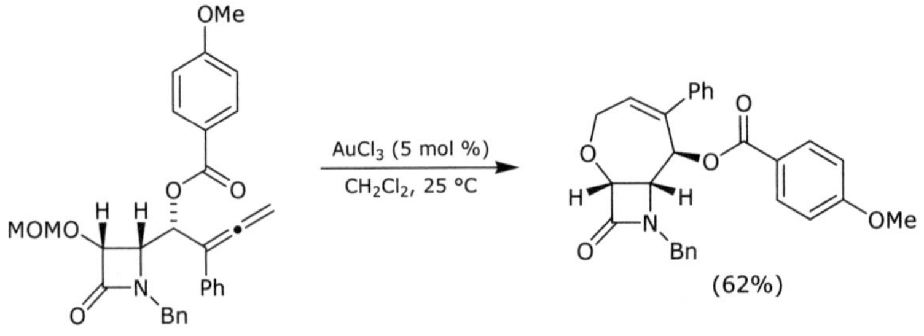

Scheme 4-98

General procedure for the gold-catalyzed synthesis of fused tetrahydrooxepines by 7-endo-trig cycloisomerization[269c]

$AuCl_3$ (0.025 mmol) is added under Ar to a stirred solution of the corresponding methoxymethyl-substituted allene (0.5 mmol) in CH_2Cl_2 (0.5 mL). The resulting mixture is stirred at RT until the starting material disappears (TLC control). The reaction is then quenched with brine (0.5 mL), the mixture is extracted with EtOAc (3 × 3 mL), and the combined extracts are washed twice with brine. The organic layer is dried (anhydrous $MgSO_4$) and concentrated under reduced pressure. Chromatography of the residue with elution with EtOAc/hexanes mixtures gives analytically pure tetrahydrooxepines.

Intriguing variable cyclization modes were also observed in the reaction of allenic epoxyalcohols with cationic gold catalysts (Scheme 4-99).[270] Whereas substrates bearing a *bis*-(phenylsulfonyl)methyl or sulfonamide group in the tether afforded mainly or exclusively the 7-*exo*-trig cyclization product, a very unusual 9-*endo*-trig cycloisomerization was observed with the corresponding malonate derivative.

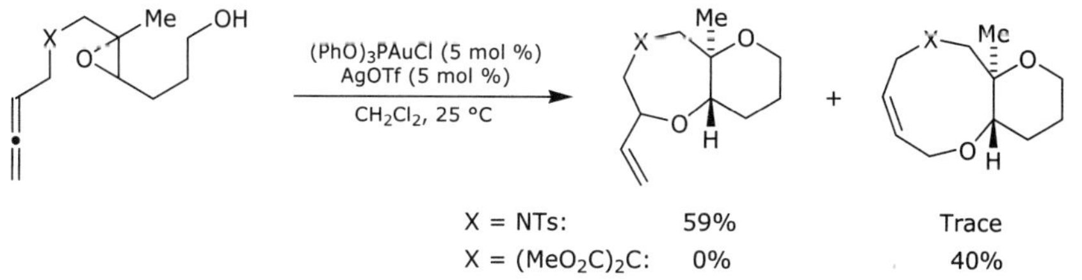

$(PhO)_3PAuCl$ (5 mol %), AgOTf (5 mol %), CH_2Cl_2, 25 °C

X = NTs:	59%	Trace
X = $(MeO_2C)_2C$:	0%	40%

Scheme 4-99

Application of the method to the gold-catalyzed cascade cyclization of an allenic diepoxyalcohol proceeded with formation of three C–O bonds in one pot to furnish a tricyclic polyether (Scheme 4-100).

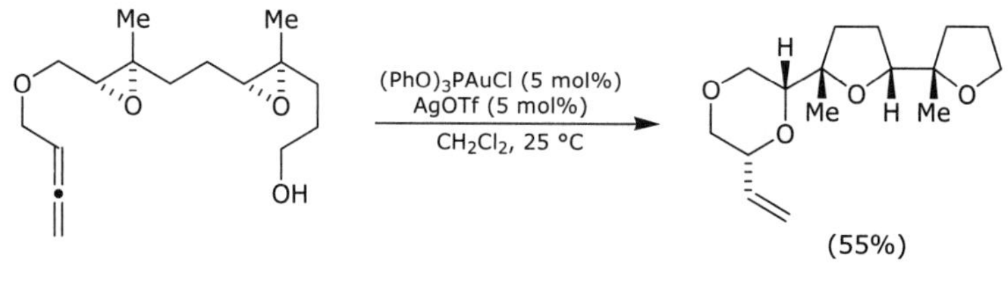

$(PhO)_3PAuCl$ (5 mol%), AgOTf (5 mol%), CH_2Cl_2, 25 °C

(55%)

Scheme 4-100

Additional examples for *exo*-selective cyclizations of allenols with an extended tether between the hydroxy group and the allene have been reported using cationic gold(I) catalysts. Thus, treatment of γ-hydroxyallenes with Au[P(*t*-Bu)$_2$(*o*-biphenyl)]Cl and AgOTs afforded vinyltetrahydrofurans by 5-*exo*-trig cycloisomerization.[47a] The corresponding 6-*exo*-dig cyclization leading to dihydropyrans did not take place under these conditions but rather in the presence of a platinum catalyst. Application to axially chiral allenols provided the corresponding tetrahydrofurans with high levels of chirality transfer.

Representative procedure for the gold-catalyzed cycloisomerization of γ-hydroxyallenes. 4,4-Diphenyl-2-vinyltetrahydrofuran[47a]

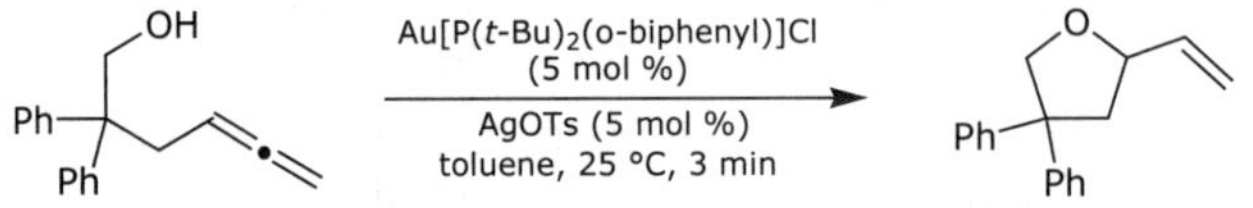

Au[P(*t*-Bu)$_2$(o-biphenyl)]Cl (5 mol %), AgOTs (5 mol %), toluene, 25 °C, 3 min

A mixture of Au[P(*t*-Bu)$_2$(*o*-biphenyl)]Cl (3.3 mg, 6.25 μmol) and AgOTs (1.7 mg, 6.25 μmol) in toluene (0.4 mL) was stirred at RT for 10 min and then treated with a solution of 2,2-diphenylhexa-4,5-dien-1-ol (31.3 mg, 0.125 mmol) in toluene (0.6 mL). The resulting suspension was stirred at RT for 3 min and then chromatographed (SiO_2; hexanes/EtOAc, 50:1 to 20:1) to give 4,4-diphenyl-2-vinyltetrahydrofuran (28.4 mg, 91%) as a colorless oil.

Extension of the method to the synthesis of enantiomerically enriched tetrahydrofurans or tetrahydropyrans was possible by using chiral phosphine ligands on the gold catalyst.[271] In the presence of chiral gold pre-catalyst [Au$_2${(*S*)-3,5-*t*-Bu-4-MeO-MeOBIPHEP}Cl$_2$] and AgOTs, prochiral allenes afforded the desired heterocycles with high enantioselectivity (Scheme 4-101).

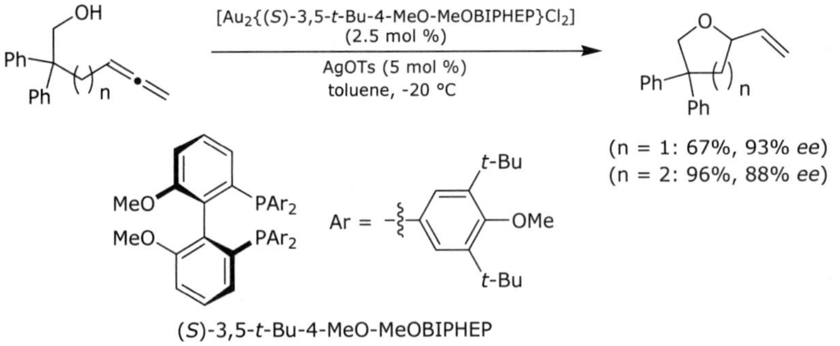

[Au$_2${(*S*)-3,5-*t*-Bu-4-MeO-MeOBIPHEP}Cl$_2$] (2.5 mol %), AgOTs (5 mol %), toluene, -20 °C

(n = 1: 67%, 93% *ee*)
(n = 2: 96%, 88% *ee*)

(*S*)-3,5-*t*-Bu-4-MeO-MeOBIPHEP

Scheme 4-101

Representative procedure for the enantioselective gold-catalyzed cycloisomerization of γ-hydroxyallenes. 4,4-Diphenyl-2-vinyltetrahydrofuran[271]

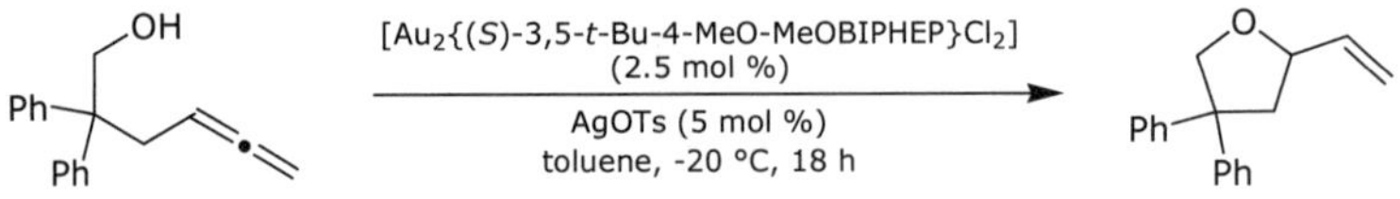

[Au$_2${(*S*)-3,5-*t*-Bu-4-MeO-MeOBIPHEP}Cl$_2$] (2.5 mol %), AgOTs (5 mol %), toluene, -20 °C, 18 h

A mixture of [Au$_2${(*S*)-3,5-*t*-Bu-4-MeO-MeOBIPHEP}Cl$_2$] (5.1 mg, 3.1 μmol) and AgOTs (1.7 mg, 6.3 μmol) in toluene (0.4 mL) was stirred at RT for 10 min. The mixture was then cooled to –20 °C and treated with a solution of 2,2-diphenylhexa-4,5-dien-1-ol (31.3 mg, 0.125 mmol) in toluene (0.6 mL). The resulting suspension was stirred at –20 °C for 18 h. Column chromatography of the reaction mixture (hexanes/EtOAc, 50:1 to 20:1) gave 4,4-diphenyl-2-vinyltetrahydrofuran (20.9 mg, 67%, 93% *ee*) as a colorless oil.

With chiral racemic allenes, the cyclization products were formed as a mixture of *E*/*Z*-isomers with high enantioselectivities in most cases (Scheme 4-102).[272] Hydrogenation of the mixture afforded 2-heptyl-4,4-diphenyltetrahydrofuran with 90% *ee*, which indicates that the stereogenic centers of both cyclization products have the same absolute configuration. The corresponding cyclization of the enantiomerically enriched (*R*)-allene gave diastereo- and enantiomerically pure 2-((*Z*)-hept-1-enyl)-tetrahydro-4,4-diphenylfuran. Thus, the absolute configuration of the catalyst determines the configuration of the chirality center of the product, whereas both the substrate and the catalyst determine the configuration of the double bond.

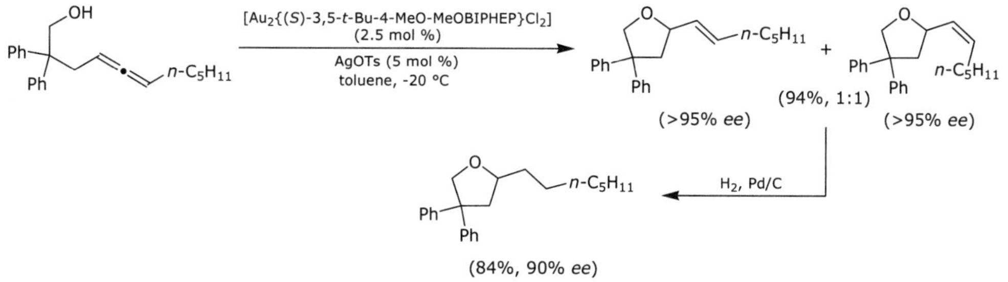

Scheme 4-102

An alternative catalytic system for *exo*-selective cycloisomerizations of hydroxy-allenes was reported by Toste *et al.*[170a] and takes advantage of a chiral counterion, which was introduced into the catalyst by a silver salt (Scheme 4-103). Thus, treatment of γ- or δ-allenols with catalytic amounts the of achiral gold pre-catalyst dppm$(AuCl)_2$ and the chiral silver salt **L** afforded tetrahydrofurans or tetrahydropyrans in high yield and excellent enantioselectivity in most cases. It is also possible to use a chiral gold catalyst together with the chiral silver salt. Application of this method to allenic carboxylic acids furnished chiral γ-lactones with up to 82% *ee*.[170]

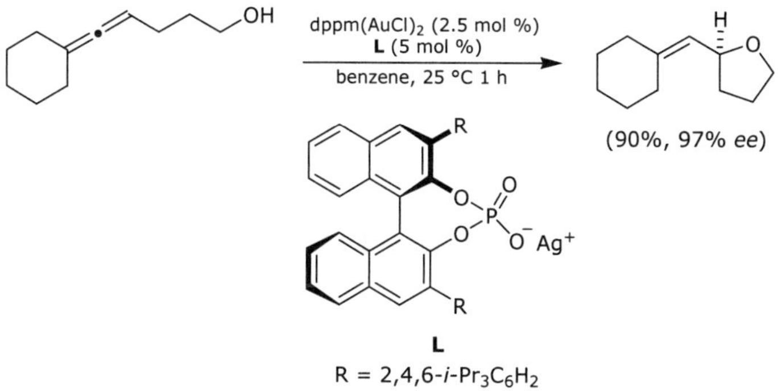

Scheme 4-103

Instead of allenic carboxylic acids, the corresponding esters can also be used as substrates for gold-catalyzed cyclization reactions. Thus, heating *t*-butyl allenoates with gold(III) chloride in dichloromethane afforded butenolides in high yield (Scheme 4-104).[273] In a similar way, Bäckvall *et al.*[274] obtained δ-lactones by 6-*endo*-trig-cyclization of allenic malonate esters with $AuCl_3$ and $AgSbF_6$ in acetic acid.

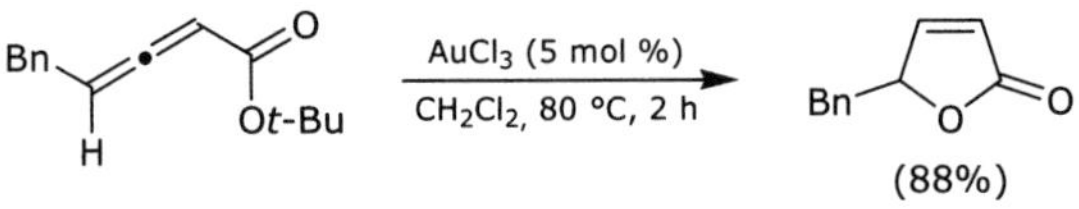

Scheme 4-104

Representative procedure for the gold-catalyzed cyclization of allenic malonate esters. Methyl 3-(cyclohex-2-enyl)-3,6-dihydro-6,6-dimethyl-2-oxo-2H-pyran-3-carboxylate[274]

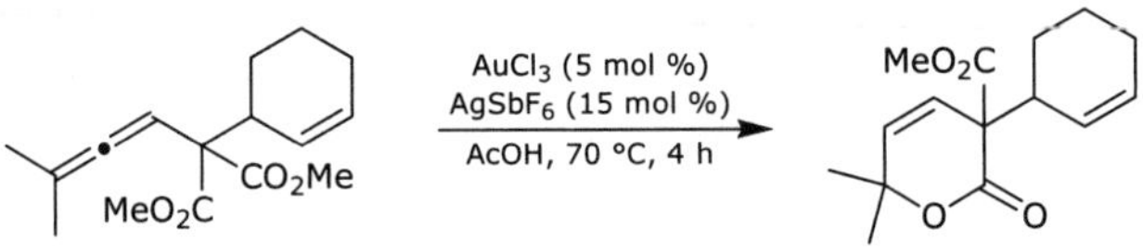

To a stirred solution of dimethyl 2-(cyclohex-2-enyl)-2-(3-methylbuta-1,2-dienyl)malonate (100 mg, 0.36 mmol) in 2 mL of acetic acid at 70 °C, $AuCl_3$ (5.5 mg, 0.018 mmol) and $AgSbF_6$ (16.4 mg, 0.054 mmol) were added portionwise. The mixture was then stirred for 4 h until total disappearance of the starting material (TLC analysis). The reaction mixture was filtered through a small amount of silica gel and the silica gel was washed with Et_2O (5 mL). The combined organic phases were concentrated under vacuum to give the crude product. Purification was carried out by flash chromatography (*n*-pentane-Et_2O, 4:1) on silica gel to yield 75 mg (79%) of pure methyl 3-(cyclohex-2-enyl)-3,6-dihydro-6,6-dimethyl-2-oxo-2*H*-pyran-3-carboxylate.

The gold-catalyzed cyclization of allenic esters proceeds *via* σ-vinylgold intermediates that can be isolated when stoichiometric amounts of gold (*e.g.*, $Ph_3PAuOTf$) are used (Scheme 4-105).[275] They were characterized by NMR spectroscopy and X-ray crystallography. In the presence of triflic acid or iodine, vinylgold compounds undergo protodeauration or iododeauration to the corresponding butenolides.

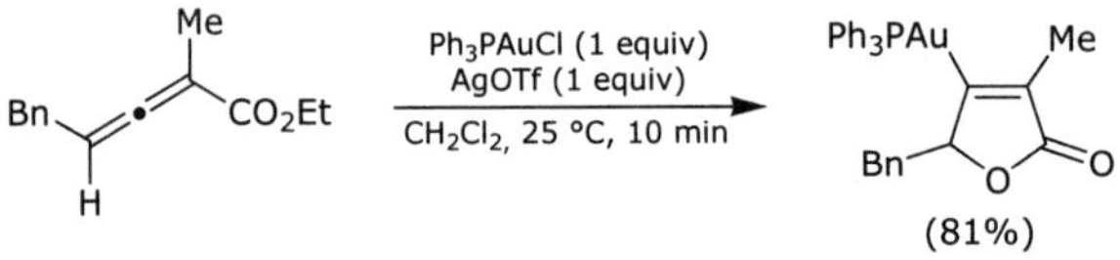

Scheme 4-105

Representative procedure for the iododeauration of vinylgold compounds. 5-Hexyl-4-iodo-3-methylfuran-2(5H)-one[275]

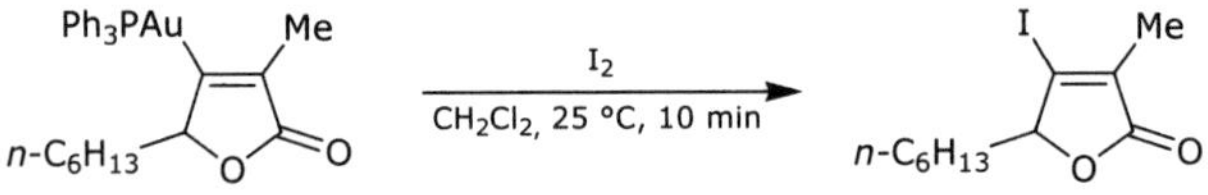

To a solution of 5-hexyl-4-(triphenylphosphinegold)-3-methylfuran-2(5*H*)-one (64 mg, 0.10 mmol) in CH_2Cl_2 (1.0 mL) was added I_2 (25 mg, 0.10 mmol). The mixture was stirred for 10 min at RT; afterward, the solvent was removed under reduced pressure and the residue was subjected to a flash column chromatography (eluent: EtOAc/petroleum ether = 1:10) to give 5-hexyl-4-iodo-3-methylfuran-2(5*H*)-one (22 mg, 72%) as a white solid.

Vinylgold compounds readily participate in palladium-catalyzed cross-couplings, *e.g.*, with aryl halides (Scheme 4-106).[276]

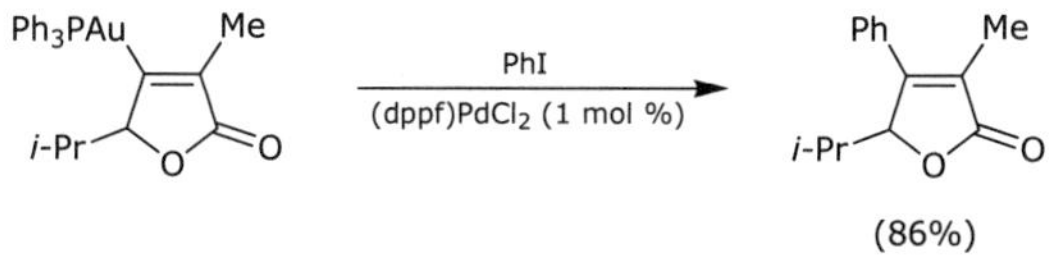

Ph3PAu Me i-Pr O O → PhI (dppf)PdCl2 (1 mol %) → Ph Me i-Pr O O (86%)

Scheme 4-106

Whereas these transformations require stoichiometric gold compounds, catalytic amounts of both gold and palladium are sufficient for the cycloisomerization of allyl allenoates to allyl-substituted butenolides.[277] Blum and co-workers reported this tandem C–O/C–C bond formation, which is initiated by activation of the distal allenic double bond with $Ph_3PAuOTf$ (Scheme 4-107). This induces cyclization to an allyl oxonium intermediate, which undergoes deallylation in the presence of Pd_2dba_3. Nucleophilic attack of the resulting σ-vinylgold intermediate at the π-allylpalladium species and reductive elimination furnish the allylated butenolide and regenerate both catalysts.

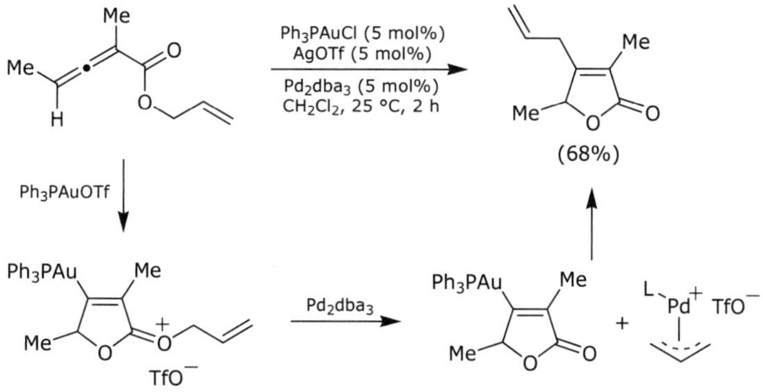

Ph3PAuCl (5 mol%) AgOTf (5 mol%) Pd2dba3 (5 mol%) CH2Cl2, 25 °C, 2 h (68%) Ph3PAuOTf Pd2dba3 TfO⁻ L Pd+ TfO⁻

Scheme 4-107

Representative procedure for the gold- and palladium-catalyzed cycloisomerization of allyl allenoates. 4-Allyl-5-(4-bromophenyl)-3-methylfuran-2(5H)-one[277]

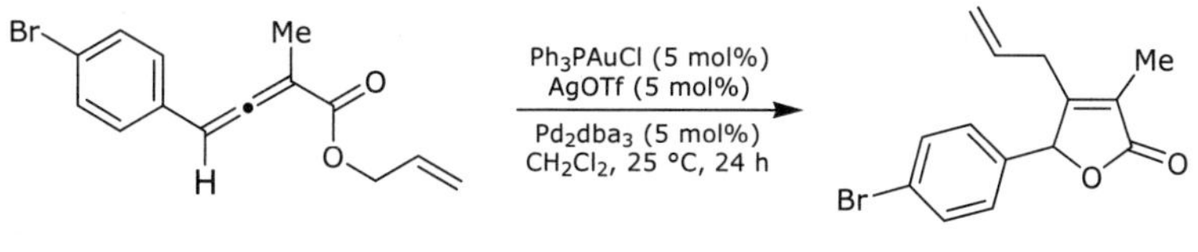

Ph3PAuCl (5 mol%) AgOTf (5 mol%) Pd2dba3 (5 mol%) CH2Cl2, 25 °C, 24 h

In a glove box, Ph_3PAuCl (1.2 mg, 2.5 μmol), AgOTf (0.6 mg, 2.5 μmol), $Pd_2(dba)_3$ (2.3 mg, 2.5 μmol), and allyl 4-(4-bromophenyl)-2-methylbuta-2,3-dienoate (14.6 mg, 0.05 mmol) were weighed in separate dram vials. Dry CH_2Cl_2 (0.5 mL) was added *via* syringe to the dram vial containing Ph_3PAuCl, and this solution was then added to the dram vial containing AgOTf. The Ph_3PAuCl/AgOTf solution was added to the dram vial containing the substrate. To the dram vial containing $Pd_2(dba)_3$ was added the solution containing substrate, Ph_3PAuCl, and AgOTf in CH_2Cl_2. The vial was capped and allowed to stir for 24 h. Purification by flash chromatography (hexanes/EtOAc, 4:1) gave 4-allyl-5-(4-bromophenyl)-3-methylfuran-2(5*H*)-one (10.0 mg, 68% yield) as a clear colorless oil.

Normally, homogeneous gold catalysis does not involve a change of oxidation state by oxidative addition/reductive elimination. In the presence of suitable oxidizing agents such as Selectfluor, however, oxidative C–C coupling of σ-gold intermediates involving a gold(I)/gold(III) redox cycle becomes possible. Gouverneur and co-workers[278] have developed this powerful synthetic method for the cycloisomerization of aryl-substituted allenoates to indenofuranones and further applied it to the formation of the alkynylated lactones from allenic esters and terminal alkynes (Scheme 4-108).

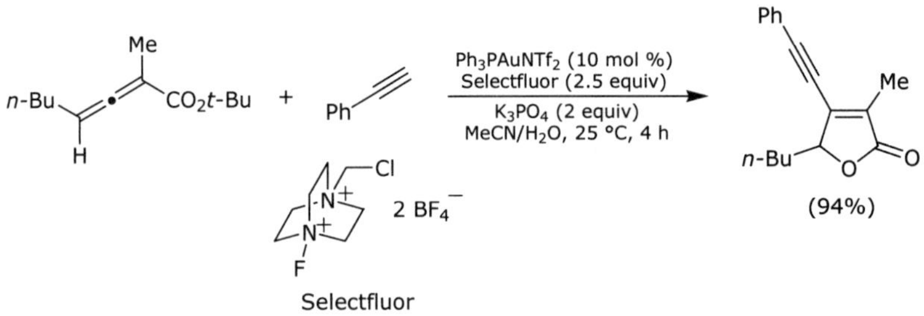

n-Bu, Me, H, CO_2t-Bu + Ph; $Ph_3PAuNTf_2$ (10 mol %), Selectfluor (2.5 equiv), K_3PO_4 (2 equiv), $MeCN/H_2O$, 25 °C, 4 h; Ph, Me, n-Bu, O, O (94%); Cl, N+, N+, F, 2 BF_4^-, Selectfluor

Scheme 4-108

Representative procedure for the gold-catalyzed cascade cyclization/oxidative alkylation of allenoates. 5-Butyl-4-[(4-fluorophenyl)ethynyl]-3-methylfuran-2(5H)-one[278b]

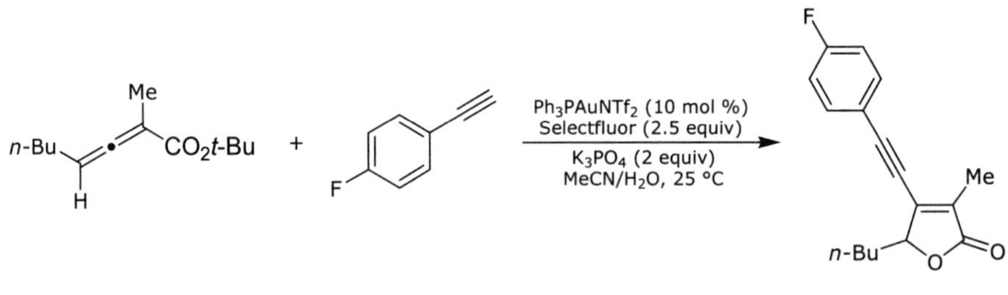

n-Bu, Me, H, CO_2t-Bu + F; $Ph_3PAuNTf_2$ (10 mol %), Selectfluor (2.5 equiv), K_3PO_4 (2 equiv), $MeCN/H_2O$, 25 °C; F, Me, n-Bu, O, O

To a stirring solution of *t*-butyl 2-methylocta-2,3-dienoate (100 mg, 0.48 mmol, 1 equiv) in anhydrous acetonitrile (5 mL) and water (86 μL, 4.8 mmol, 10 equiv) was added $Ph_3PAuNTf_2$ (36 mg, 0.050 mmol, 10 mol%), Selectfluor (430 mg, 1.2 mmol, 2.5 equiv), potassium phosphate (tribasic) (200 mg, 0.96 mmol, 2 equiv), and 1-ethynyl-4-fluorobenzene (78 μL, 0.53 mmol, 1.5 equiv). The reaction mixture was stirred at RT until TLC showed complete consumption of the allenoate (4 to 48 h). The reaction mixture was diluted with CH_2Cl_2 (20 mL), filtered through Celite, dried over anhydrous Na_2SO_4, and filtered again; then, the solvent was removed *in vacuo*. The resulting crude mixture was purified by flash column chromatography on silica gel (10:1, petrol ether 40–60/Et_2O) to afford 5-butyl-4-[(4-fluorophenyl)ethynyl]-3-methylfuran-2(5*H*)-one as a yellow oil (102 mg, 78%).

Gold-catalyzed intermolecular hydroalkoxylations of allenes have received much less attention than their intramolecular counterpart. Nishina and Yamamoto[172b, 279] performed the addition of primary or secondary alcohols using 5 mol% each of Ph_3PAuCl and AgOTf without any solvent and obtained the allylic ethers resulting from attack of the alcohol at the less substituted allenic terminus with moderate to high yield. The best results were obtained for aryl-substituted allenes. Unfortunately, enantiomerically enriched allenes provided only racemic addition products under these conditions. In the presence of a gold(I) catalyst and *N*-iodosuccinimide, 2-iodoallylic ethers were obtained in a regioselective and stereoselective manner.[280]

Representative procedure for the gold-catalyzed intermolecular hydroalkoxylation of allenes. 1-((E)-3-Isopropoxyprop-1-enyl)-4-methylbenzene[172b, 279a]

Ph_3PAuCl (5 mol %)
AgOTf (5 mol %)
i-PrOH, 30 °C, 2 h

To a suspension of Ph_3PAuCl (12.4 mg, 0.025 mmol) and AgOTf (6.4 mg, 0.025 mmol) in *i*-PrOH (57 µL, 0.75 mmol; neat conditions) was added *p*-tolylallene (64.9 mg, 0.50 mmol) and the mixture was stirred at 30 °C under an Ar atmosphere. After the reaction was complete (2 h), the reaction mixture was filtered through a short silica gel pad with Et_2O as eluent. The product was purified by column chromatography (silica gel and pentane) to give 1-((*E*)-3-isopropoxyprop-1-enyl)-4-methylbenzene in 98% yield (93.4 mg).

The opposite regioselectivity, *i.e.*, addition at the more substituted end of the allene, can also be obtained.[281]

Representative procedure for the gold-catalyzed intermolecular hydroalkoxylation of allenes. (4-(2-Methylbut-3-en-2-yloxy)butyl)benzene[281c]

\+ Ph⁄⁄OH
IPrAuCl (10 mol %)
AgOTf (10 mol %)
DMF, 0 °C, 20 h
O Ph

i-Pr N N *i*-Pr
i-Pr Au *i*-Pr
Cl
IPrAuCl

To a stirred solution of 3-methylbuta-1,2-diene (19.1 mg, 0.28 mmol) and 4-phenyl-1-butanol (0.43 mL, 2.8 mmol) in anhydrous DMF (0.28 mL) at 0 °C was added IPrAuCl (17.4 mg, 0.028 mmol) and AgOTf (7.2 mg, 0.028 mmol). This was left to stir under N_2 atmosphere at 0 °C for 20 h. The reaction mixture was then flushed through two plugs of silica (Et_2O), washed with water and brine, and then dried over anhydrous $MgSO_4$. The product, (4-(2-methylbut-3-en-2-yloxy)butyl)benzene, was obtained by flash column chromatography (eluent: 98:2 *n*-pentane:Et_2O) as a clear colorless liquid (49.9 mg, 0.23 mmol, 82%).

Zhang and Widenhoefer[282] also employed gold(I)-NHC complex IPrAuCl together with silver triflate for the regioselective and stereoselective addition of various alcohols to substituted allenes (Scheme 4-109).[283] The efficiency of chirality transfer was remarkably influenced by the concentration of the alcohol: A 0.44-M solution of benzylic alcohol led to an enantiomeric excess of 64% after 30 min, whereas the use of a 1.76 M solution delivered the allyl ether with 79% *ee* after 20 min. Treatment of the allene with the catalyst in the absence of an alcohol led to complete racemization within 30 min. This is probably caused by formation of a π-allylgold intermediate.[282, 284]

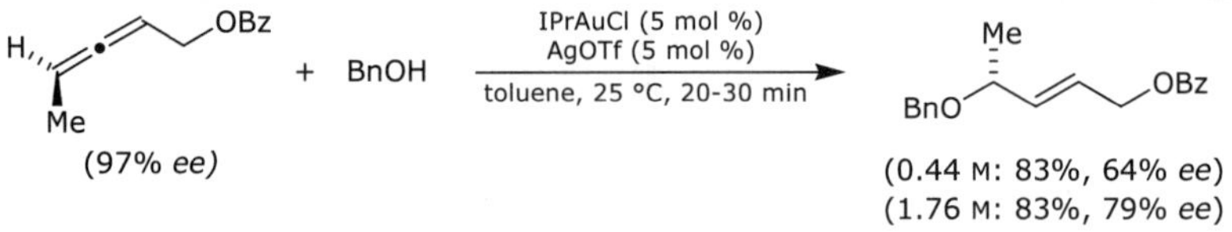

Scheme 4-109

5.3 Nucleophilic Additions to Alkenes

Because of the low reactivity toward gold catalysts, the number of examples for gold-catalyzed nucleophilic additions of oxygen nucleophiles to alkenes is small. In 2005, Yang and He reported that heating of olefins with phenols or carboxylic acids in the presence of $Ph_3PAuOTf$ generated ethers or esters with moderate to good yield (Scheme 4-110).[285] Nucleophilic attack at terminal alkenes occurred with Markovnikov regioselectivity to afford secondary ethers or esters. Occasionally, side products resulting from migration of the double bond were observed.

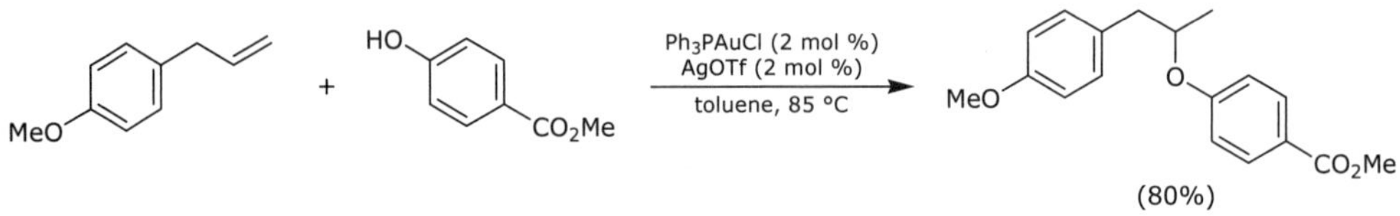

Scheme 4-110

The corresponding addition reaction of aliphatic alcohols to alkenes takes place after strong heating (without solvent) in the presence of $AuCl_3$ and $CuCl_2$ (Scheme 4-111).[178, 286] Markovnikov regioselectivity was observed as well in these transformations. Primary and secondary alcohols can be used; the desired ethers were accompanied by variable amounts of side products resulting from chlorination, hydration, or oxidation of the alkene. The copper(II) chloride stabilizes the gold catalyst, probably by reoxidizing it to the catalytically active Au(III) species.[287]

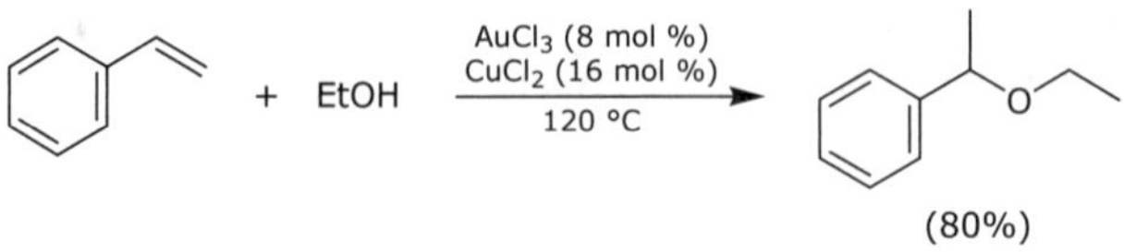

Scheme 4-111

Interestingly, aliphatic alcohols bearing a coordinating substituent (*e.g.*, halogens and alkoxy groups) in the β-position are more reactive in gold-catalyzed addition reactions to olefins than are their simpler counterparts.[288] Here, a cationic gold(I) catalyst prepared *in situ* from $(C_6F_5)_3PAuCl$ and AgOTf in the presence of additional $(C_6F_5)_3P$ gave the best results (Scheme 4-112). Again, side products formed by double-bond isomerization and subsequent hydroalkoxylation were also isolated in some cases.

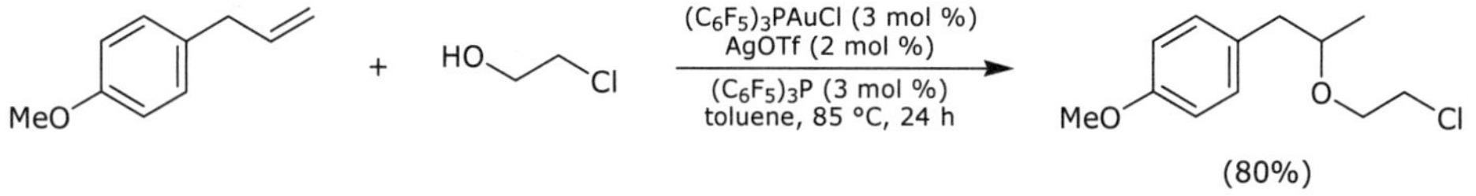

MeO + HO Cl; $(C_6F_5)_3PAuCl$ (3 mol %), AgOTf (2 mol %), $(C_6F_5)_3P$ (3 mol %), toluene, 85 °C, 24 h; MeO O Cl (80%)

Scheme 4-112

Similar to the corresponding aminoalkenes, olefins bearing an OH group in the γ- or δ-position undergo a gold-catalyzed intramolecular oxidative carboalkoxylation to afford tetrahydrofurans or tetrahydropyrans in the presence of arylboronic acids and Selectfluor (Scheme 4-113).[189]

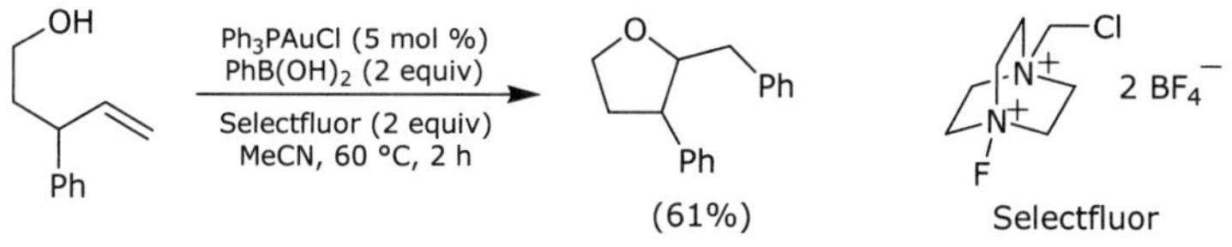

OH Ph; Ph_3PAuCl (5 mol %), $PhB(OH)_2$ (2 equiv), Selectfluor (2 equiv), MeCN, 60 °C, 2 h; O Ph Ph (61%); Cl N+ N+ F 2 BF_4^- Selectfluor

Scheme 4-113

A related gold-catalyzed cyclization of allyl acetates bearing a malonate as internal nucleophile leads to γ-vinyl butyrolactones (Scheme 4-114).[289] $AuBr_3$ and *in situ* formed $Ph_3PAuSbF_6$ are the best catalysts for this transformation, which occurred with good yield and excellent diastereoselectivity in most cases. Intermolecular C–O bond

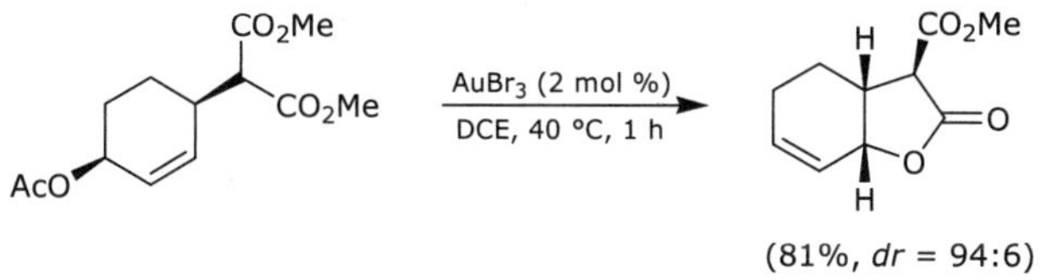

CO_2Me CO_2Me AcO; $AuBr_3$ (2 mol %), DCE, 40 °C, 1 h; H CO_2Me O O H (81%, *dr* = 94:6)

Scheme 4-114

formation is also possible under these conditions, albeit with lower efficiency. Enantiomerically pure allyl acetates afforded nearly racemic products; this can be explained by formation of an allyl cation that is then attacked by the ester.

Even allylic alcohols participate in gold-catalyzed cyclizations of this type. Aponick and co-workers obtained substituted tetrahydrofurans and tetrapyrans by dehydrative cyclization of allylic diols in the presence of Ph_3PAuCl and AgOTf (Scheme 4-115).[290] High *cis*-diastereoselectivities were observed when the reaction was carried out at low temperature (−10…−78 °C). Bandini and co-workers[291] applied the method to the synthesis of *cis*-disubstituted morpholines in the presence of dppf$(AuCl)_2$ and AgOTf. Moreover, enantioselectivities of up to 95% *ee* were achieved using (*R*)-DTBM-SEGPHOS as chiral ligand.

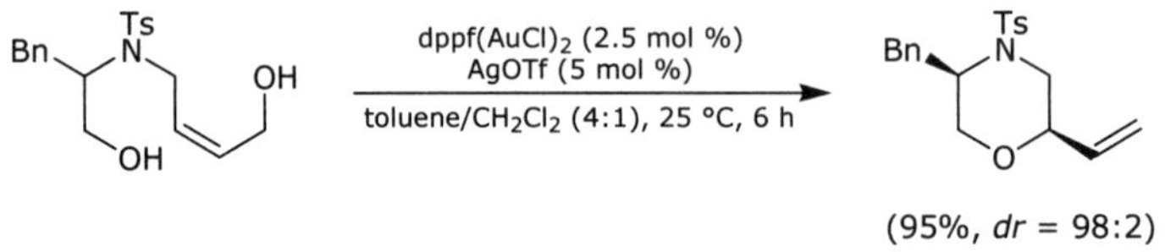

dppf$(AuCl)_2$ (2.5 mol %)
AgOTf (5 mol %)
toluene/CH_2Cl_2 (4:1), 25 °C, 6 h

(95%, *dr* = 98:2)

Scheme 4-115

General procedure for the synthesis of morpholines by gold-catalyzed cyclization of diols[291]

In a 10-mL, one-necked, round-bottom flask, 0.8 mL of reagent grade toluene is added, followed by AuCl·SMe_2 (2.2 mg, 7.5 µmol, 0.05 equiv) and 1,1'-bis(diphenylphosphino)-ferrocene dppf (2 mg, 3.75 µmol, 0.025 equiv). The mixture is stirred at RT for 15 min. Then, the flask is covered with aluminum foil and 2 mg of AgOTf (7.5 µmol, 0.05 equiv) is added. The suspension is stirred for another 30 min, and a solution of the aminodiol (0.15 mmol, 1 equiv) in CH_2Cl_2 (200 µL) is added. The mixture is stirred in the dark at the same temperature; then after completion as indicated by TLC, a small aliquot is withdrawn for GC-MS and ^{1}H-NMR analysis. The reaction mixture is directly poured onto a plug of silica gel for flash chromatography purification.

5.4 Rearrangements

The Meyer-Schuster rearrangement of propargyl alcohols or carboxylates to α,β-unsaturated carbonyl compounds is catalyzed by various gold(I) and gold(III) complexes. Initially observed as an undesired side reaction in gold(III)-catalyzed nucleophilic substitution,[292] the method has evolved as a powerful access to enones, enals, and enoates.[293] Among different catalysts, cationic gold(I)-NHC complexes are the most efficient. The presence of water in the reaction mixture turned out to be crucial and may be responsible for formation of a catalytically active hydroxo-gold species. For the Meyer-Schuster rearrangement of propargyl alcohols, the combination of molybdenum and cationic gold catalysts gave the best results.[284] In the presence of gold(I) or platinum(II) catalysts, suitable acetylenic diols undergo a tandem Meyer-Schuster rearrangement/oxa-Michael addition to afford tetrahydrofurans or tetrapyrans.[295] Performing the gold-catalyzed Meyer-Schuster rearrangement in the presence of *N*-bromosuccinimide or *N*-iodosuccinimide led to the formation of α-bromoenones or α-iodoenones.[296]

Representative procedure for the gold-catalyzed Meyer-Schuster-rearrangement of propargyl acetates. (E)-1-Phenyloct-2-en-4-one[293e]

ItBuAuCl (2 mol %)
$AgSbF_6$ (2 mol %)
THF/H_2O

It-BuAuCl

Conventional heating: $AgSbF_6$ (6.8 mg, 0.02 mmol, 2 mol%) was added to a THF solution (5 mL) of I*t*-BuAuCl (8.2 mg, 0.02 mmol, 2 mol%) in a screw-cap vial. The solution instantly became cloudy and distilled water (1 mL) was added. The reaction mixture was stirred for 1 min before a THF solution (5 mL) of 1-phenyloct-3-yn-2-yl acetate (244 mg, 1 mmol, 1 equiv) was added. The vial was then placed in an oil bath at 60 °C, and the reaction mixture was stirred for 8 h and then allowed to cool to RT. The resulting mixture was dissolved in pentane, filtered through Celite, and evaporated. The crude product was purified by flash chromatography on silica gel (pentane/Et_2O, 98:2); yield 176 mg (87%) of (*E*)-1-phenyloct-2-en-4-one.

Microwave heating: $AgSbF_6$ (6.8 mg, 0.02 mmol, 2 mol%) was added to a THF solution (5 mL) of I*t*BuAuCl (8.2 mg, 0.02 mmol, 2 mol%) in a microwave-designed vial. The solution instantly became cloudy and distilled water (1 mL) was added. The reaction mixture was stirred for 1 min before a THF solution (5 mL) of 1-phenyloct-3-yn-2-yl acetate (244 mg, 1 mmol, 1 equiv) was added. The vial was then placed in a microwave reactor and heated at 80 °C for 12 min. The resulting mixture was dissolved in pentane, filtered through Celite, and evaporated. The crude product was purified by flash chromatography on silica gel (pentane/Et_2O, 98:2); yield 188 mg (93%) of (*E*)-1-phenyloct-2-en-4-one.

Representative procedure for the molybdenum/gold-catalyzed Meyer-Schuster-rearrangement of propargyl alcohols. 5-Phenylpent-1-en-3-one[294]

$MoO_2(acac)_2$ (1 mol %)
Ph_3PAuCl (1 mol %)
AgOTf (1 mol %)
toluene, 25 °C, 30 min

$MoO_2(acac)_2$ (2.1 mg, 6.3 μmol), Ph_3PAuCl (3.1 mg, 6.3 μmol), and AgOTf (1.6 mg, 6.3 μmol) were added in this order to a solution of 5-phenylpent-2-yn-1-ol (0.10 g, 0.63 mmol) in toluene (1.6 mL) at RT. The reaction mixture was stirred for 30 min and then quenched with saturated aqueous NH_4Cl. The organic materials were extracted with Et_2O, and the combined organic layer was washed with brine, dried over anhydrous $MgSO_4$, and evaporated *in vacuo*. The residue was purified by column chromatography (silica gel, hexanes/Et_2O 10:1) to give 5-phenylpent-1-en-3-one (93 mg, 93%) as a colorless oil.

In a related transformation, propargyl pivalates rearrange to (1*Z*,3*E*)-pivaloyloxy-1,3-dienes after treatment with the cationic gold(I)-NHC catalyst $IPrAuNTf_2$ (Scheme 4-116).[297] Here, a 1,2-acyloxy-migration to the activated triple bond is followed by a 1,2-C–H-insertion of the intermediate gold carbene. The competing enone formation is minimized under anhydrous conditions. The same

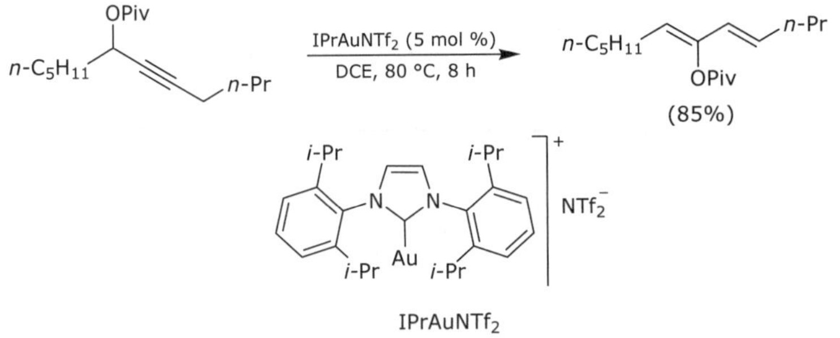

OPiv; n-C5H11; n-Pr → IPrAuNTf2 (5 mol %), DCE, 80 °C, 8 h → n-C5H11; n-Pr; OPiv (85%)

i-Pr; i-Pr; N; N; Au; i-Pr; i-Pr; NTf2−; +

IPrAuNTf2

Scheme 4-116

products were obtained from trimethylsilylmethyl-substituted propargyl esters by 1,3-acyloxy transfer and subsequent desilylation.[298] The dienes formed by these methods are excellent substrates for Diels-Alder reactions; both steps can be performed sequentially without isolation of the diene. A gold-catalyzed rearrangement of β-alkynylpropiolactones to substituted α-pyrones[299] may proceed *via* a similar mechanism.

Representative procedure for the gold-catalyzed rearrangement/Diels-Alder reaction of propargyl pivatales. 4-Formyl-4,6-dimethyl-3-propylcyclohex-1-enyl pivalate[297]

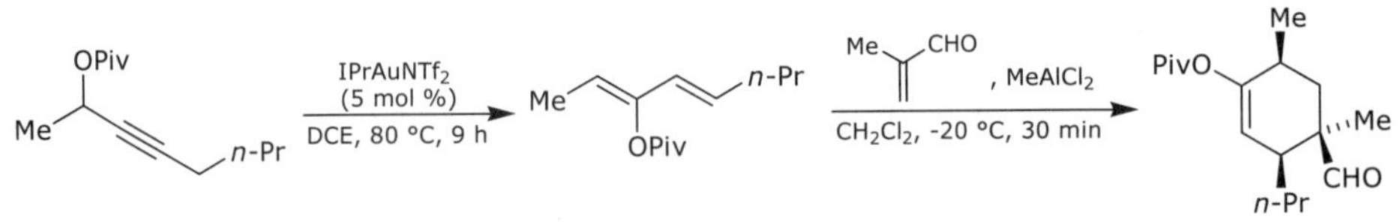

OPiv; Me; n-Pr → IPrAuNTf2 (5 mol %), DCE, 80 °C, 9 h → Me; n-Pr; OPiv → Me, CHO, MeAlCl2; CH2Cl2, -20 °C, 30 min → PivO; Me; Me; CHO; n-Pr

To a solution of oct-3-yn-2-yl pivalate (0.249 g, 1.19 mmol) in anhydrous 1,2-dichloroethane (2.4 mL) in a flame-dried, 2-dram vial was added $IPrAuNTf_2$ (51 mg, 0.06 mmol). The vial was sealed tightly with Teflon tape inside and outside the cap and heated up to 80 °C for 9 h. The reaction mixture was transferred to a 50-mL, round-bottom flask and CH_2Cl_2 (10 mL) was added. The resulting mixture was cooled to –20 °C and treated with methacrolein (167 mg, 2.38 mmol) and $MeAlCl_2$ (2.4 mL, 1 M solution in hexanes) for 30 min. Then, the reaction was quenched with Et_3N (1 mL) and diluted with saturated aqueous NH_4Cl solution (20 mL). Extraction was carried out with Et_2O (3 × 20 mL), and the combined organic layers were dried (anhydrous $MgSO_4$), filtered, and concentrated. The resulting residue was purified *via* silica gel flash column chromatography (eluent: EtOAc/hexanes = 1:8) to yield an inseparable mixture of 4-formyl-4,6-dimethyl-3-propylcyclohex-1-enyl pivalate and the regioisomer 5-formyl-5,6-dimethyl-3-propylcyclohex-1-enyl pivalate (ratio: 88:12) in a combined 82% yield.

In the presence of platinum or gold catalysts, 1,5- or 1,6-enynes bearing an acyloxy group in the propargylic position undergo the Ohloff-Rautenstrauch rearrangement to bicyclic products.[90a, 300] The transformation probably proceeds by 1,2-migration of the acetoxy group to the activated alkyne, affording a vinylgold carbene intermediate that then undergoes an intramolecular cyclopropanation.[301] Alternatively, the alkene can act as the primary nucleophile to give a cyclopropyl gold carbene intermediate, which is then attacked by the adjacent acetoxy group. Whereas cationic gold(I) catalysts induce a rapid rearrangement at room temperature, prolonged heating is required in the presence of $PtCl_2$. The enol ester moiety of the rearrangement products can be easily cleaved to provide bicyclic ketones. The method has found application in the synthesis of various terpenoids.[302]

Representative procedure for the gold-catalyzed Ohloff-Rautenstrauch rearrangement of 4-acetoxy-1-en-5-ynes. 1-Phenylbicyclo[3.1.0]hexan-2-one[90a]

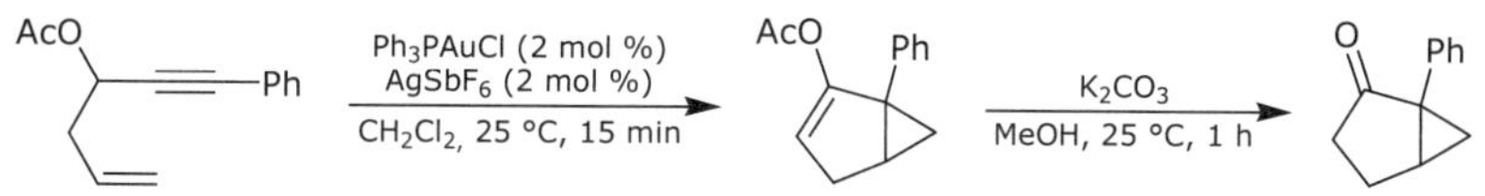

AcO, Ph; Ph_3PAuCl (2 mol %), $AgSbF_6$ (2 mol %), CH_2Cl_2, 25 °C, 15 min → AcO, Ph; K_2CO_3, MeOH, 25 °C, 1 h → O, Ph

A solution of 1-phenylhex-5-en-1-yn-3-yl acetate (0.200 g, 0.933 mmol) in CH_2Cl_2 (2.4 mL) was added to a suspension of Ph_3PAuCl (9.25 mg, 18.7 μmol) and $AgSbF_6$ (6.43 mg, 18.7 μmol) in CH_2Cl_2 (21 mL). After stirring at RT for 15 min, the solvent was evaporated and the crude product was dissolved in methanol (8 mL). K_2CO_3 (12.9 mg) was added and the suspension was stirred for 1 h before the reaction was quenched with water and the aqueous phase was extracted with *t*-butyl methyl ether. The combined organic layers were dried (anhydrous $MgSO_4$), filtered, and evaporated, and the residue was purified by flash chromatography (hexane/EtOAc, 4:1 + 1% triethylamine, v/v) to give 1-phenylbicyclo[3.1.0]hexan-2-one as a yellow liquid (0.119 g, 74%).

The gold-catalyzed Ohloff-Rautenstrauch rearrangement of 1-en-4-yn-3-yl pivalates furnished cyclopent-2-enones with high yield (Scheme 4-117).[303] The method can be applied to the synthesis of annulated enones bearing sulfonamide or ether linkages. Enantiomerically enriched enyne pivalates react with excellent chirality transfer. The transformation was rationalized in terms of an intramolecular 1,2-rearrange-ment of the ester group to the activated alkyne, followed by a Nazarov cyclization of the cationic pentadienylgold intermediate. Here, the chiral information of the starting material was transferred to the helically chiral pentadienyl cation and then further to the product.[304]

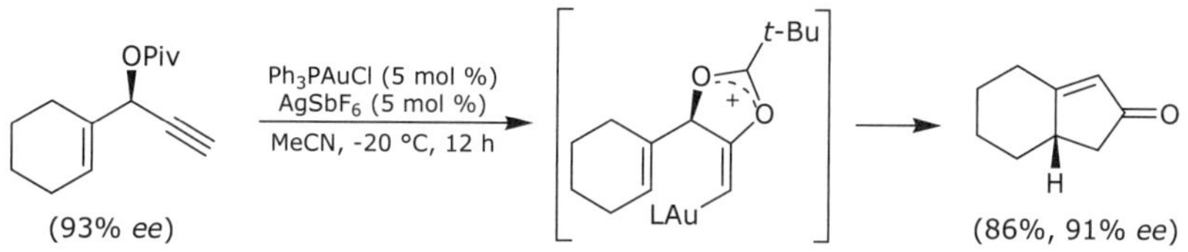

OPiv (93% *ee*); Ph_3PAuCl (5 mol %), $AgSbF_6$ (5 mol %), MeCN, -20 °C, 12 h → [*t*-Bu, O, +, O, LAu] → O, H (86%, 91% *ee*)

Scheme 4-117

In a closely related study, 4-en-2-yn-1-yl acetates also afforded cyclopent-2-enones after treatment with Ph_3PAuCl and $AgSbF_6$.[305] In this case, a 1,3-acetoxy migration leads to a pentadienyl cation that undergoes the Nazarov cyclization. A subsequent 1,2-hydride shift is the rate-determining step of the transformation; this is accelerated in the presence of water, so that the reaction should be conducted in wet dichloromethane.[306] The cyclopentadienyl acetate formed is hydrolyzed to the corresponding cyclopentenone under the reaction conditions in most cases; in the presence of aryl substituents, however, the hydrolysis is achieved by addition of Tf_2NH. In the presence of cationic gold(I) catalysts, 5-en-2-yn-1-yl acetates also undergo a 1,3-acetoxy shift, and the acetoxyallenes[307] formed are converted into bicyclo[3.1.0]hexenyl acetates which can be hydrolyzed to cyclohex-2-enones.[308]

Representative procedure for the gold-catalyzed cyclization of 5-acetoxy-1-en-3-ynes to cyclopent-2-enones. 3-(4-(Trifluoromethyl)phenyl)-5-methylcyclopent-2-enone[305a]

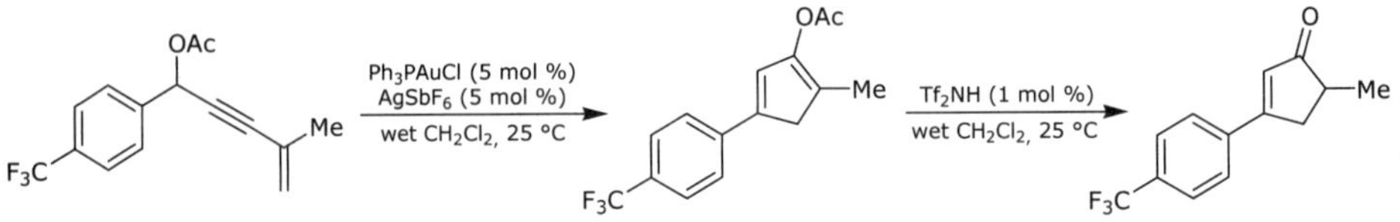

To a solution of 1-(4-(trifluoromethyl)phenyl)-4-methylpent-4-en-2-ynyl acetate (57 mg, 0.2 mmol) in wet CH_2Cl_2 (20.0 mL; generated by shaking distilled CH_2Cl_2 with deionized water in a separatory funnel) was added $Ph_3PAuCl/AgSbF_6$ (pre-generated as 0.01-M solution in CH_2Cl_2 by mixing Ph_3PAuCl with $AgSbF_6$, 1.0 mL) at RT and the reaction mixture was stirred overnight. To the reaction mixture was added 0.20 mL of a solution of Tf_2NH (pre-generated as 0.01-M solution in CH_2Cl_2) and the mixture was stirred for another 4.0 h. The resulting solution was concentrated and the residue was purified by silica gel flash column chromatography (hexanes/EtOAc = 10/1) to give 3-(4-(trifluoromethyl)phenyl)-5-methylcyclopent-2-enone (46 mg, 96% yield).

Simple aromatic propargyl acetates can be efficiently rearranged to substituted indenes in the presence of cationic gold(I)-NHC catalysts.[309] Also in this transformation, acetoxyallenes formed by 1,3-acetoxy transfer are probable intermediates that then undergo an intramolecular electrophilic aromatic substitution. Indeed, exposure of independently prepared acetoxyallenes to the gold catalyst also afforded the corresponding indenes, albeit with slightly longer reaction times. This might be a consequence of a higher affinity of the catalyst toward alkynes.

Representative procedure for the gold-catalyzed rearrangement of aromatic propargyl acetates to indenes. 5-Butyl-5H-indeno[5,6-d][1,3]dioxol-5-yl acetate [309]

IPrAuCl (2 mol %), $AgBF_4$ (2 mol %), CH_2Cl_2, 25 °C, 5 min

IPrAuCl

To a solution of IPrAuCl (12.4 mg, 0.02 mmol) in anhydrous CH_2Cl_2 (35 mL) in a round-bottom flask equipped with a septum, $AgBF_4$ (2 mg, 0.02 mmol) was added in the absence of light. The solution instantly became cloudy. A 5-mL solution of 1-(benzo[*d*][1,3]dioxol-6-yl)hept-2-ynyl acetate (274 mg, 1 mmol) in anhydrous CH_2Cl_2 was then injected through the septum. When TLC analysis showed total consumption of the starting material (5 min), the solvent was removed. The resulting mixture was dissolved in pentane, filtered through Celite· and evaporated. The crude oil was purified by flash chromatography on silica gel (pentane/Et_2O, 95/5) to afford 247 mg (90%) of 5-butyl-5*H*-indeno[5,6-*d*][1,3]dioxol-5-yl acetate.

The isomerization of simple allyl acetates proceeds by 1,3-migration *via* a six-membered cyclic acetoxonium intermediate and is most efficiently catalyzed by cationic gold(I)-NHC complexes.[310] The rearrangement is carried out either by refluxing in dichloroethane or with microwave heating. By contrast, the allylic rearrangement of Baylis-Hillman acetates already occurs at room temperature in the presence of AuCl and AgOTf.[311]

Representative procedure for the gold-catalyzed rearrangement of allyl acetates. 4-Nitrocinnamyl acetate[310a]

IPrAuCl (3 mol %), $AgBF_4$ (2 mol %), DCE

Conventional heating: To a DCE solution (15 mL) of IPrAuCl (18 mg, 0.03 mmol, 3 mol%) in an oven-dried, round-bottom flask equipped with a condenser, $AgBF_4$ (3.9 mg, 0.02 mmol, 2 mol%) was added. The solution instantly became cloudy, and the reaction mixture was stirred for 1 min before a DCE solution (5 mL) of 1-(4-nitrophenyl)allyl acetate (221 mg, 1 mmol, 1 equiv) was added. The flask was then placed in an oil bath at 85 °C, and the reaction mixture was stirred for 12 h and allowed to cool to RT. The resulting mixture was dissolved in pentane, filtered through Celite, and evaporated. The crude product was purified by flash chromatography on silica gel (gradient hexane/Et_2O, from 90/10 to 80/20); yield: 212 mg (96%, *E*:*Z* = 92:8) of 4-nitrocinnamyl acetate.

Microwave heating: To a DCE solution (15 mL) of IPrAuCl (18 mg, 0.03 mmol, 3 mol%) in a microwave-designed vial, $AgBF_4$ (3.9 mg, 0.02 mmol, 2 mol%) was added.

The solution instantly became cloudy, and the reaction mixture was stirred for 1 min before a DCE solution (5 mL) of 1-(4-nitrophenyl)allyl acetate (221 mg, 1 mmol, 1 equiv) was added. The vial was then placed in a microwave reactor and heated at 80 °C for 12 min. The resulting mixture was dissolved in pentane, filtered through Celite, and evaporated. The crude product was purified by flash chromatography on silica gel (gradient hexane/Et_2O, from 90/10 to 80/20); yield: 217 mg (98%, *E*:*Z* = 95:5) of 4-nitrocinnamyl acetate.

6. C–S and C–Halogen Bond Formation

The gold-catalyzed formation of a carbon-sulfur bond is not an easy task, given the high strength of Au–S bonds that renders sulfides potent poisons for gold catalysts. Nevertheless, the first example for a gold-catalyzed C–S bond formation was reported in 2006 when Morita and Krause[312] converted various α-thioallenes into the corresponding 2,5-dihydrothiophenes (Scheme 4-118). In this transformation,

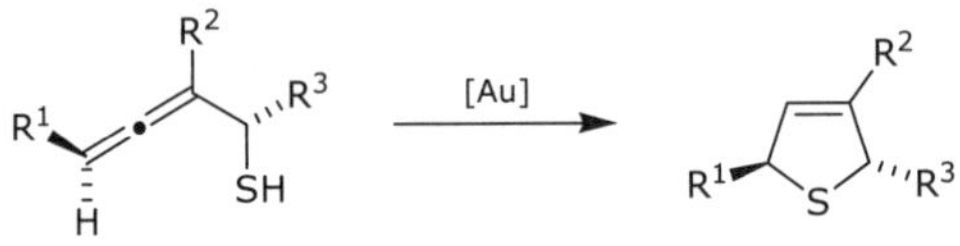

Scheme 4-118

gold(I) catalysts showed a higher reactivity and gave better yields than $AuCl_3$. In the latter case, the disulfide formed by oxidative coupling of the thioallene was isolated as side product, indicating that gold(III) is reduced to gold(I) under the reaction conditions. Independent of the catalyst used, complete *axis-to-center chirality transfer* was observed in the cycloisomerization of α-thioallenes.[313]

Representative procedure for the gold-catalyzed cycloisomerization of α-thioallenes. Trans-2,5-dihydro-5-isopropyl-2-(methoxymethyl)-3-methylthiophene[312]

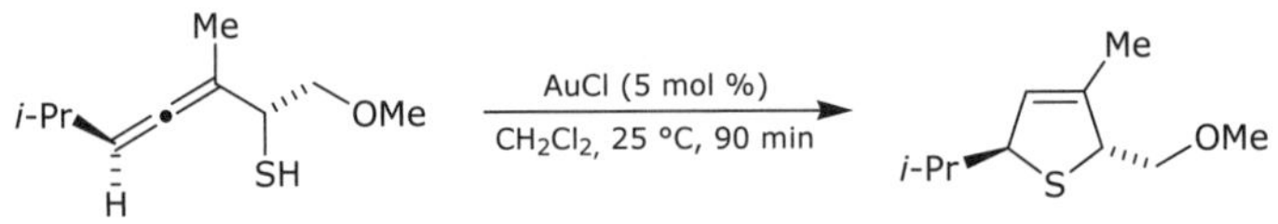

AuCl (5 mol %), CH_2Cl_2, 25 °C, 90 min

AuCl (3 mg, 5 mol%) was added to a stirred solution of 1-methoxy-3,6-dimethylhepta-3,4-diene-2-thiol (50 mg, 0.27 mmol) in CH_2Cl_2 (5 mL) at RT under argon, and the mixture was stirred for 90 min at RT (the reaction was monitored by TLC). The reaction mixture was concentrated under reduced pressure. The residue was purified by column chromatography on silica gel with cyclohexane/EtOAc (30:1) as eluent to afford *trans*-2,5-dihydro-5-isopropyl-2-(methoxymethyl)-3-methylthiophene (44 mg, 88%) as a yellow oil.

A single case of thiophene synthesis by gold-catalyzed dehydrative cyclization of a thiolated propargyl alcohol has been reported.[148a] Compared with the corresponding transformation of propargyl diols or aminoalcohols (Sections 4.1 and

5.1), higher reaction temperature and catalyst loading had to be used to reach an excellent yield of 90%.

Representative procedure for the gold-catalyzed dehydrative cyclization of thiolated propargyl alcohols. 2-Phenylthiophene[148a]

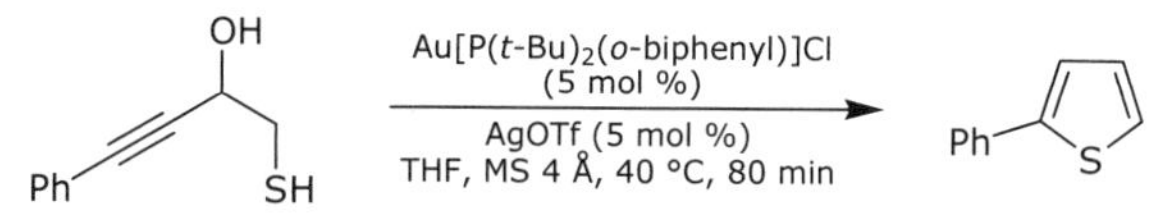

Au[P(*t*-Bu)$_2$(*o*-biphenyl)]Cl (5 mol %), AgOTf (5 mol %), THF, MS 4 Å, 40 °C, 80 min

Dry THF (1.0 mL) was added to an aluminum-foil–covered test tube containing Au[P(*t*-Bu)$_2$(*o*-biphenyl)]Cl (10.6 mg, 0.02 mmol), AgOTf (5.1 mg, 0.02 mmol), and activated MS 4 Å (25 mg), and the mixture was heated to 40 °C. A solution of 1-mercapto-4-phenylbut-3-yn-2-ol (71.2 mg, 0.4 mmol) in dry THF (3.0 mL) was added. After TLC analysis showed the reaction to be complete (80 min), the mixture was cooled to RT and filtered through a short plug of silica with Et_2O (40 mL). The solution of crude product was concentrated and then purified by flash chromatography to give 2-phenylthiophene (57.9 mg, 90%).

An intramolecular carbothiolation of alkynes takes place after treatment of suitable (2-alkynylphenyl) sulfides with gold(I) chloride. For example, reaction of substrates bearing an alkoxyalkyl group at sulfur afforded 2,3-disubstituted benzothiophenes, which are formed by attack of the sulfide at the activated the triple bond and subsequent nucleophilic 1,3-migration of the alkoxyalkyl fragment (Scheme 4-119).[314] Analogous products were obtained with allyl, benzyl, and silyl[315] migrating groups. Substrates bearing a chiral 1-arylethyl group at the sulfide undergo the intramolecular carbothiolation with up to 91% chirality transfer and retention of the configuration, suggesting that the 1,3-migration occurs *via* a contact ion pair.[316] Intermolecular addition reactions of dithiols to phenylacetylene in the presence of Ph_3PAuCl and $AgBF_4$ provided the corresponding thioacetals with high yield and regioselectivity.[195]

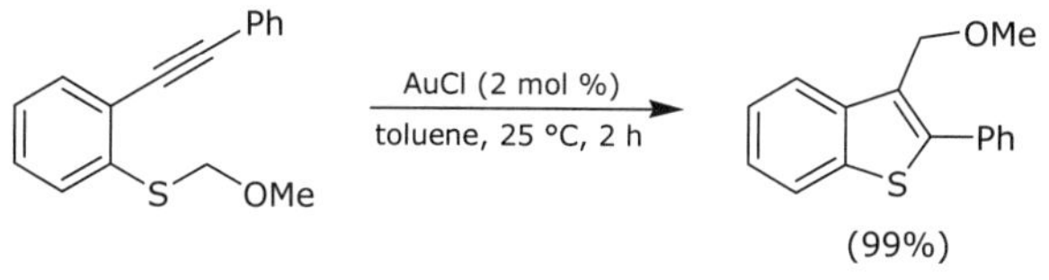

AuCl (2 mol %), toluene, 25 °C, 2 h (99%)

Scheme 4-119

In the presence of gold(I) or gold(III) chloride, aromatic propargyl sulfides undergo a 1,2-sulfide shift to the activated triple bond, and the gold carbene species thus generated reacts with the adjacent aryl ring to afford substituted indenes.[317] Application of the method to dithioacetals gave rise to the formation of dithiofluorene derivatives (Scheme 4-120). The transformation tolerates the presence of halogen or oxygen atoms in the starting material. In a related transformation, dithio-substituted cycloocta-1,2-dienes were obtained by treatment of propargyl 1,3-dithianes with $Ph_3PAuCl/AgSbF_6$.[318]

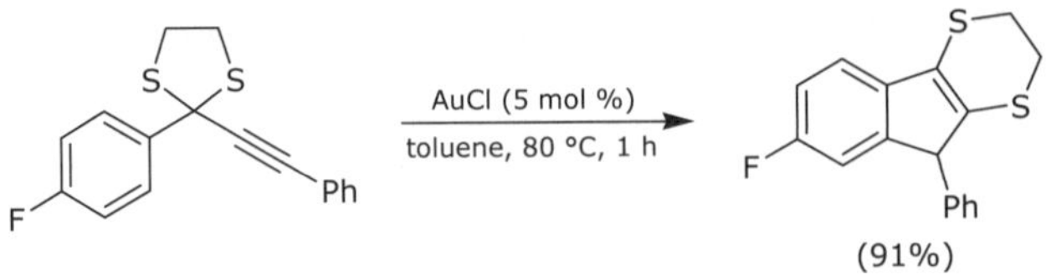

Scheme 4-120

The gold(III)-catalyzed direct nucleophilic substitution of allyl or propargyl alcohols is possible with many nucleophiles, including allyl silanes, electron-rich aromatics, β-dicarbonyl compounds, alcohols, and sulfonamides.[292, 319] With thiols, the corresponding propargyl thioethers were obtained with moderate yield (Scheme 4-121).

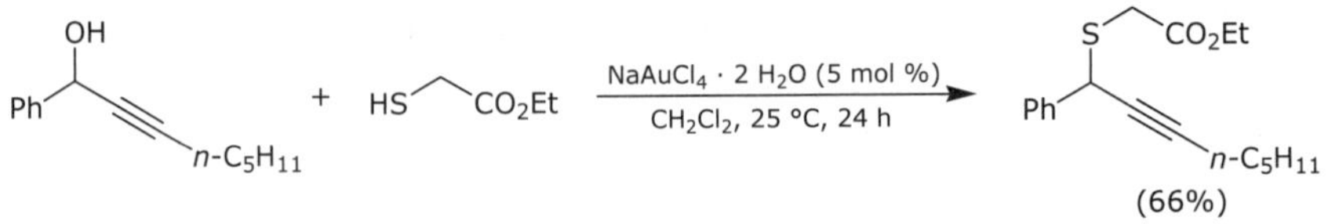

Scheme 4-121

Conjugated dienes and activated olefins like indene undergo intermolecular gold-catalyzed additions of sulfur nucleophiles.[320] Aliphatic and aromatic thiols, as well as thioacids, afforded the hydrothiolation products with good yield in the presence of Ph_3PAuBF_4 (Scheme 4-122).

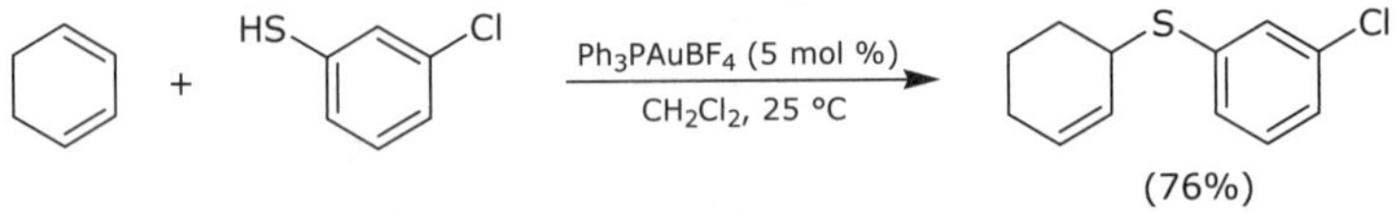

Scheme 4-122

Representative procedure for the gold-catalyzed addition of thiols to conjugated dienes. (3-Methylpent-3-en-2-yl)(phenyl)sulfane[320]

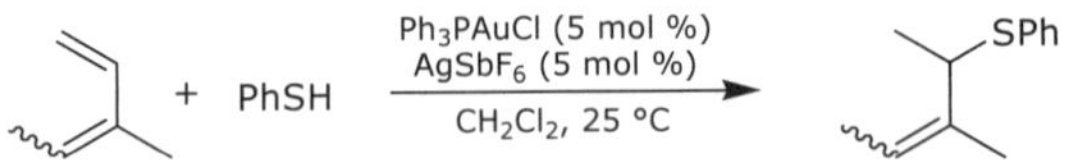

An oven-dried amber vial was charged with Ph_3PAuCl (25 mg, 0.050 mmol) and $AgBF_4$ (9.7 mg, 0.050 mmol) under an atmosphere of N_2. The mixture was stirred in anhydrous CH_2Cl_2 (1.0 mL) for several minutes. To this was added benzenethiol (0.10 mL, 1.0 mmol), then 3-methylpenta-1,3-diene (*cis/trans* mixture, 0.14 mL, 1.2 mmol), and the mixture was stirred overnight. The mix was then filtered through a Celite 545 pad and the solvent removed *via* rotavap. The crude material was then purified by column chromatography on silica gel (hexane), and the solvent was removed to yield (3-methylpent-3-en-2-yl)(phenyl)sulfane as an off-white oil (0.17 g, 90%).

The halogenation of aromatic compounds with *N*-bromosuccinimide can be realized after activation of the electrophile with a Lewis or Brønsted acid. Among many different Lewis acids examined for this purpose, gold(III) chloride shows an extraordinarily high activity.[261] Various activated and slightly deactivated aromatics can be brominated under these conditions, and the catalyst loading can be as low as 0.01 mol%. A kinetic isotope effect observed in the bromination of benzene-d_6 suggests that the transformation proceeds *via* dual activation of both NBS and the aromatic substrate. The $AuCl_3$-catalyzed halogenation can be extended to NCS and NIS, respectively; because of lower reactivity of NCS, the chlorination is limited to activated substrates. The aryl halides thus obtained can be used for subsequent Pd-catalyzed coupling reactions in a one-pot procedure.

Representative procedure for the gold-catalyzed bromination of aromatics with NBS. Methyl 5-bromo-2-methoxybenzoate[261]

MeO, CO_2Me (benzene) → NBS (1 equiv), $AuCl_3$ (0.1 mol %), DCE, 80 °C, 23 h → MeO, CO_2Me, Br (benzene)

NBS (3 mmol, 540 mg) and $AuCl_3$ (3 μmol, 0.1 mol%, 1 mg) were weighed into a 25-mL, round-bottom flask. DCE (2 mL) and methyl 2-methoxybenzoate (3 mmol, 498 mg) were then added to the flask in succession. The resulting reaction mixture was stirred for about 23 h at 80 °C (monitored by GC-MS). The solution was then concentrated under reduced pressure and the residue was purified by flash column chromatography to give (703 mg, 96%).

7. Conclusions and Outlook

The investigations summarized in this chapter demonstrate that homogeneous gold catalysis has reached a remarkable state of maturity in just one decade. In particular, it appears that every conceivable combination of acetylenic substrate and nucleophilic reagent has been activated with gold catalysts for intermolecular or intramolecular (cyclo)addition, cycloisomerization, or rearrangement. So where do we go from here? In the author's personal view, progress will be made in the following areas:

- Improvement of the activity of homogeneous gold catalysts that will make it possible to activate alkenes and other less reactive substrates at ambient temperatures
- Development of homogeneous gold catalysts with improved stability and recyclability that will render gold catalysis more sustainable and allow its use in industrial applications
- Combinations of gold with other transition metals, organocatalysts, or biocatalysts in tandem transformations
- Development of new methods for realizing gold(I)–gold(III) redox cycles that will make it possible to combine the traditional π-carbophilicity of gold catalysts with C–C and C–heteroatom cross-couplings

- Design of new chiral ligand scaffolds for enantioselective transformations of gold(I) and gold(III) catalysts

The reader is invited to amend this list with his or her favorite subjects. Even though nobody can be sure which advances will be made sooner and which later, it seems certain that the future of homogeneous gold catalysis will be bright and full of surprises!

8. Acknowledgments

The author wishes to express his sincere appreciation to all the graduate students and postdocs who have contributed to the research program on stereoselective gold catalysis with allenes. Their names are listed in the reference section. Funding by the European Community *via* COST and Marie-Curie initiatives was particularly helpful.

9. References

[1] (a) Hashmi, A.S.K.; Hutchings, G.J.; *Angew. Chem., Int. Ed.* **2006**, *45*, 7896; (b) Hutchings, G.J.; *Chem. Commun.* **2008**, 1148.

[2] (a) Toste, F.D.; Michelet, V., in *Gold Catalysis: An Homogeneous Approach*, Imperial College Press, London, **2012**; (b) Hashmi, A.S.K.; Toste, F.D., in *Modern Gold Catalyzed Synthesis*, Wiley-VCH, Weinheim, **2012**.

[3] (a) Ito, Y.; Sawamura, M.; Hayashi, T.; *J. Am. Chem. Soc.* **1986**, *108*, 6405; (b) Ito, Y.; Sawamura, M.; Hayashi, T.; *Tetrahedron Lett.* **1987**, *28*, 6215; (c) Hayashi, T.; Ito, Y.; Sawamura, M.; *Tetrahedron* **1992**, *48*, 1999.

[4] (a) Hayashi, T.; *Pure Appl. Chem.* **1988**, *60*, 7; (b) Sawamura, M.; Ito, Y.; *Chem. Rev.* **1992**, *92*, 857; (c) Gulevich, A.V.; Zhdanko, A.G.; Orru, R.V.A.; Nenajdenko, V.G.; *Chem. Rev.* **2010**, *110*, 5235.

[5] Teles, J.H.; Brode, S.; Chabanas, M.; *Angew. Chem., Int. Ed.* **1998**, *37*, 1415.

[6] (a) Pyykkö, P.; *Angew. Chem., Int. Ed.* **2004**, *43*, 4412; (b) Gorin, D.J.; Toste, F.D.; *Nature* **2007**, *446*, 395.

[7] Leyva-Pérez, A.; Corma, A.; *Angew Chem., Int. Ed.* **2012**, *51*, 614.

[8] Hammer, B.; Nørskov, J. K.; *Nature* **1995**, *376*, 238.

[9] Wood, B.J.; Wise, H.; *J. Catal.* **1966**, *5*, 135.

[10] Okumura, M.; Akita, T.; Haruta, M.; *Catal. Today* **2002**, *74*, 265.

[11] Zhang, X.; Shi, H.; Xu, B.Q.; *Angew. Chem., Int. Ed.* **2005**, *44*, 7132.

[12] (a) Bailie, J.E.; Hutchings, G.J.; *Chem. Commun.* **1999**, 2151; (b) Bailie, J.E.; Abdullah, H.A.; Anderson, J.A.; Rochester, C.H.; Richardson, N.V.; Hodge, N.; Zhang, J.G.; Burrows, A.; Kiely, C.; Hutchings, G.; *Phys. Chem. Chem. Phys.* **2001**, *3*, 4113; (c) Mohr, C.; Hofmeister, H.; Lucas, M.; Claus, P.; *Chem. Eng. Tech.* **2000**, *23*, 324; (d) Mohr, C.; Hofmeister, H.; Radnik, J.; Claus, P.; *J. Am. Chem. Soc.* **2003**, *125*, 1905.

[13] (a) Milone, C.; Tropeano, M.L.; Gulino, G.; Neri, G.; Ingoglia, R.; Galvagno, S.; *Chem. Commun.* **2002**, 868; (b) Milone, C.; Ingoglia, R.; Tropeano, M.L.; Neri, G.; Galvagno, S.; *Chem. Commun.* **2003**, 868; (c) Milone, C.; Ingoglia, R.; Pistone, A.; Neri, G.; Frusteri, F.; Galvagno, S.; *J. Catal.* **2004**, *222*, 348.

[14] (a) Corma, A.; Serna, P.; *Science* **2006**, *313*, 332; (b) Corma, A.; Concepcion, P.; Serna, P.; *Angew. Chem., Int. Ed.* **2007**, *46*, 7266.

[15] González-Arellano, C.; Corma, A.; Iglesias, M.; Sánchez. F.; *Chem. Commun.* **2005**, 3451.

[16] Comas-Vives, A.; González-Arellano, C.; Corma, A.; Iglesias, M.; Sánchez, F.; Ujaque, G.; *J. Am. Chem. Soc.* **2006**, *128*, 4756.

[17] (a) Baker, R.T.; Calabrese, J.C.; Westcott, S.A.; *J. Organomet. Chem.* **1995**, *498*, 109; (b) Leyva, A.; Zhang, X.; Corma, A.; *Chem. Commun.* **2009**, 4947.

[18] (a) Ito, H.; Yajima, T.; Tateiwa, J.; Hosomi, A.; *Chem. Commun.* **2000**, 981; (b) Corma, A.; González-Arellano, C.; Iglesias, M.; Sánchez, F.; *Angew. Chem., Int. Ed.* **2007**, *46*, 7820.

[19] Cheon, C.H.; Kanno, O.; Toste, F.D.; *J. Am. Chem. Soc.* **2011**, *133*, 13248.

[20] (a) Nevado, C.; Echavarren, A.M.; *Synthesis* **2005**, 167; (b) De Mendoza, P.; Echavarren, A.M.; *Pure Appl. Chem.* **2010**, *82*, 801; (c) Kitamura, T.; *Eur. J. Org. Chem.* **2009**, 1111.

[21] Reetz, M.T.; Sommer, K.; *Eur. J. Org. Chem.* **2003**, 3485.

[22] Shi, Z.; He, C.; *J. Org. Chem.* **2004**, *69*, 3669.

[23] (a) Skouta, R.; Li, C.-J.; *Tetrahedron* **2008**, *64*, 4917; (b) De Haro, T.; Nevado, C.; *Synthesis* **2011**, 2530.

[24] (a) Li, Z.; Shi, Z.; He, C.L.; *J. Organomet. Chem.* **2005**, *690*, 5049; (b) Hashmi, A.S.K.; Blanco, M.C.; *Eur. J. Org. Chem.* **2006**, 4340.

[25] (a) Dankwardt, J.W.; *Tetrahedron Lett.* **2001**, *42*, 5809; (b) Shibata, T.; Ueno, Y.; Kanda, K.; *Synlett* **2006**, 411; (c) Oh, C.H.; Kim, A.; Park, W.; Park, D.I.; Kim, N.; *Synlett* **2006**, 2781.

[26] (a) Fürstner, A.; Mamane, V.; *J. Org. Chem.* **2002**, *67*, 6264; (b) Mamane, V.; Hannen, P.; Fürstner, A.; *Chem. Eur. J.* **2004**, *10*, 4556.

[27] (a) Nevado, C.; Echavarren, A.M.; *Chem. Eur. J.* **2005**, *11*, 3155; (b) Menon, R.S.; Findlay, A.D.; Bissember, A.C.; Banwell, M.G.; *J. Org. Chem.* **2009**, *74*, 8901.

[28] (a) Do, J.H.; Kim, H.N.; Yoon, J.; Kim, J.S.; Kim, H.-J.; *Org. Lett.* **2010**, *12*, 932; (b) Yang, Y.-K.; Lee, S.; Tae, J.; *Org. Lett.* **2009**, *11*, 5610.

[29] (a) Liu, X.-Y.; Ding, P.; Huang, J.-S.; Che, C.-M.; *Org. Lett.* **2007**, *9*, 2645; (b) Zeng, X.; Frey, G.D.; Kinjo, R.; Donnadieu, B.; Bertrand, G.; *J. Am. Chem. Soc.* **2009**, *131*, 8690.

[30] Gronnier, C.; Odabachian, Y.; Gagosz, F.; *Chem. Commun.* **2011**, 218.

[31] (a) Liu, X.-Y.; Che, C.-M.; *Angew. Chem., Int. Ed.* **2008**, *47*, 3805; (b) Zhou, Y.; Feng, E.; Liu, G.; Ye, D.; Li, J.; Jiang, H.; Liu, H.; *J. Org. Chem.* **2009**, *74*, 7344.

[32] England, D.B.; Padwa, A.; *Org. Lett.* **2008**, *10*, 3631.

[33] (a) Ferrer, C.; Echavarren, A.M. *Angew. Chem., Int. Ed.* **2006**, *45*, 1105; (b) Ferrer, C.; Amijs, C.H.M.; Echavarren, A.M.; *Chem. Eur. J.* **2007**, *13*, 1358.

[34] Ferrer, C.; Escribano-Cuesta, A.; Echavarren, A.M.; *Tetrahdron* **2009**, *65*, 9015.

[35] Lu, Y.; Du, X.; Jia, X.; Liu, Y.; *Adv. Synth. Catal.* **2009**, *351*, 1517.

[36] (a) Hashmi, A.S.K.; Frost, T.M.; Bats, J.W.; *J. Am. Chem. Soc.* **2000**, *122*, 11553; (b) Hashmi, A.S.K.; Frost, T.M.; Bats, J.W.; *Catal. Today* **2002**, *72*, 19; (c) Hashmi, A.S.K.; Ding, L.; Bats, J.W.; Fischer, P.; Frey, W.; *Chem. Eur. J.* **2003**, *9*, 4339; (d) Hashmi, A. S.K.; Weyrauch, J.P.; Rudolph, M.; Kurpejović, E.; *Angew. Chem., Int. Ed.* **2004**, *43*, 6545; (e) Hashmi, A.S.K.; Rudolph, M.; Weyrauch, J.P.; Wölfle, M.; Frey, W.; Bats, J.W.; *Angew. Chem., Int. Ed.* **2005**, *44*, 2798; (f) Carrettin, S.; Blanco, M.C.; Corma, A.; Hashmi, A.S.K.; *Adv. Synth. Catal.* **2006**, *348*, 1283; (g) Hashmi, A.S.K.; Haufe, P.; Schmid, C.; Rivas Nass, A.; Frey, W.; *Chem. Eur. J.* **2006**, *12*, 5376; (h) Hashmi, A.S.K.; Weyrauch, J.P.; Kurpejović, E.; Frost, T.M.; Miehlich, B.; Frey, W.; Bats, J.W.; *Chem. Eur. J.* **2006**, *12*, 5806; (i) Hashmi, A.S.K.; Salathé, R.; Frey, W.; *Chem. Eur. J.* **2006**, *12*, 6991; (j) Hashmi, A.S.K.; Rudolph, M.; Bats, J.W.; Frey, W.; Rominger, F.; Oeser, T.; *Chem. Eur. J.* **2008**, *14*, 6672.

[37] Hashmi, A.S.K.; Wölfle, M.; *Tetrahedron* **2009**, *65*, 9021.

[38] (a) Hashmi, A.S.K.; Frost, T.M.; Bats, J.W.; *Org. Lett.* **2001**, *3*, 3769; (b) Hashmi, A.S.K.; Enns, E.; Frost, T.M.; Schäfer, S.; Frey, W.; Rominger, F.; *Synthesis* **2008**, 2707.

[39] Hashmi, A.S.K.; Hamzić, M.; Rudolph, M.; Ackermann, M.; Rominger, F.; *Adv. Synth. Catal.* **2009**, *351*, 2469.

[40] (a) Hashmi, A.S.K.; Ata, F.; Kurpejovic, E.; Huck, J.; Rudolph, M.; *Top. Catal.* **2007**, *44*, 245; (b) Hashmi, A.S.K.; Schäfer, S.; Bats, J.W.; Frey, W.; Rominger, F.; *Eur. J. Org. Chem.* **2008**, 4891; (c) Hashmi, A.S.K.; Ata, F.; Haufe, P.; Rominger, F.; *Tetrahedron* **2009**, *65*, 1919.

[41] Hashmi, A.S.K.; Blanco, M.C.; Kurpejović, E.; Frey, W.; Bats, J.W.; *Adv. Synth. Catal.* **2006**, *348*, 709.

[42] Staben, S.T.; Kennedy-Smith, J.J.; Huang, D.; Corkey, B.K.; LaLonde, R.L.; Toste, F.D.; *Angew. Chem., Int. Ed.* **2006**, *45*, 5991.

[43] Jiang, X.; Ma, X.; Zheng, Z.; Ma, S.; *Chem. Eur. J.* **2008**, *14*, 8572.

[44] Tarselli, M.A.; Chianese, A.R.; Lee, S.J.; Gagné, M.R.; *Angew. Chem., Int. Ed.* **2007**, *46*, 6670.

[45] Yang, C.-Y.; Lin, G.-Y.; Liao, H.-Y.; Datta, S.; Liu, R.-S.; *J. Org. Chem.* **2008**, *73*, 4907.

[46] Liu, Z.; Wasmuth, A.S.; Nelson, S.G.; *J. Am. Chem. Soc.* **2006**, *128*, 10352.

[47] (a) Zhang, Z.; Liu, C.; Kinder, R.E.; Han, X.; Quian, H.; Widenhoefer, R.A.; *J. Am. Chem. Soc.* **2006**, *128*, 9066; (b) Kong, W.; Fu, C.; Ma, S.; *Chem. Eur. J.* **2011**, *17*, 13134.

[48] Liu, C.; Widenhoefer, R.A.; *Org. Lett.* **2007**, *9*, 1935.

[49] Barluenga, J.; Piedrafita, M.; Ballesteros, A.; Suárez-Sobrino, A.L.; González, J.M.; *Chem. Eur. J.* **2010**, *16*, 11827.

[50] (a) Toups, K.L.; Liu, G.T.; Widenhoefer, R.A.; *J. Organomet. Chem.* **2009**, *694*, 571; (b) Wang, M.-Z.; Zhou, C.-Y.; Guo, Z.; Wong, E.L.-M.; Wong, M.-K.; Che, C.-M.; *Chem. Asian J.* **2011**, *6*, 812.

[51] Watanabe, T.; Oishi, S.; Fujii, N.; Ohno, H.; *Org. Lett.* **2007**, *9*, 4821.

[52] (a) Tarselli, M.A.; Gagné, M.R.; *J. Org. Chem.* **2008**, *73*, 2439; (b) Park, C.; Lee, P.H.; *Org. Lett.* **2008**, *10*, 3359.

[53] (a) Weber, D.; Tarselli, M.A.; Gagné, M.R.; *Angew. Chem., Int. Ed.* **2009**, *48*, 5733; (b) Weber, D.; Gagné, M.R.; *Org. Lett.* **2009**, *11*, 4962; (c) Seidel, G.; Lehmann, C.W.; Fürstner, A.; *Angew. Chem., Int. Ed.* **2010**, *49*, 8466.

[54] (a) Tarselli, M.A.; Liu, A.; Gagné, M.R.; *Tetrahedron* **2009**, *65*, 1785; (b) Kimber, M.C.; *Org. Lett.* **2010**, *12*, 1128.

[55] (a) Yao, X.; Li, C.-J.; *J. Am. Chem. Soc.* **2004**, *126*, 6884; (b) Nguyen, R.-V.; Yao, X.-Q.; Bohle, D.S.; Li, C.-J.; *Org. Lett.* **2005**, *7*, 673.

[56] Yao, X.; Li, C.-J.; *J. Org. Chem.* **2005**, *70*, 5752.

[57] Zhou, C.-Y.; Che, C.-M.; *J. Am. Chem. Soc.* **2007**, *129*, 5828.

[58] Xiao, Y.-P.; Liu, X.-Y.; Che, C.-M.; *Angew. Chem., Int. Ed.* **2011**, *50*, 4937.

[59] (a) Rozenman, M.M.; Kanan, M.W.; Liu, D.R.; *J. Am. Chem. Soc.* **2007**, *129*, 14933; (b) Wang, M.-Z.; Wong, M.-K.; Che, C.-M.; *Chem. Eur. J.* **2008**, *14*, 8353.

[60] Xiao, P-Y.; Liu, X-Y.; Che, C-M.; *J. Organomet. Chem.* **2009**, *694*, 494.

[61] Jean, M.; Van de Weghe, P.; *Tetrahedron Lett.* **2011**, *52*, 3509.

[62] Nguyen, R.-V.; Yao, X.; Li, C.-J.; *Org. Lett.* **2006**, *8*, 2397.

[63] (a) Dudnik, A.S.; Schwier, T.; Gevorgyan, V.; *Org. Lett.* **2008**, *10*, 1465; (b) Dudnik, A.S.; Schwier, T.; Gevorgyan, V.; *Tetrahedron* **2009**, *65*, 1859; (c) Kong, W.; Fu, C.; Ma, S.; *Eur. J. Org. Chem.* **2010**, *2010*, 6545.

[64] Arcadi, A.; Bianchi, G.; Chiarini, M.; D'Anniballe, G.; Marinelli, F.; *Synlett* **2004**, 944.

[65] Alfonsi, M.; Arcardi, A.; Bianchi, G.; Marinelli, F.; Nardini, A.; *Eur. J. Org. Chem.* **2006**, 2393.

[66] (a) Hashmi, A.S.K.; Schwarz, L.; Choi, J.-H.; Frost, T.M.; *Angew. Chem., Int. Ed.* **2000**, *39*, 2285; (b) Dyker, G.; Muth, E.; Hashmi, A.S.K.; Ding, L.; *Adv. Synth. Catal.* **2003**, *345*, 1247; (c) Aguilar, D.; Contel, M.; Navarro, R.; Soler, T.; Urriolabeitia, E.P.; *J. Organomet. Chem.* **2009**, *694*, 486.

[67] (a) Zhang, L.; Sun, J.; Kozmin, S.A.; *Adv. Synth. Catal.* **2006**, *348*, 2271; (b) Soriano, E.; Marco-Contelles, J.; *Acc. Chem. Res.* **2009**, *42*, 1026.

[68] (a) Nieto-Oberhuber, C.; Muñoz, M.P.; Buñuel, E.; Nevado, C.; Cárdenas, D.J.; Echavarren, A.M.; *Angew. Chem., Int. Ed.* **2004**, *43*, 2402; (b) Nieto-Oberhuber, C.; Muñoz, M.P.; López, S.; Jiménez-Núñez, E.; Nevado, C.; Herrero-Gómez, E.; Raducan, M.; Echavarren, A.M.; *Chem. Eur. J.* **2006**, *12*, 1677.

[69] Nieto-Oberhuber, C.; López, S.; Muñoz, M.P.; Cárdenas, D.J.; Buñuel, E.; Nevado, C.; Echavarren, A.M.; *Angew. Chem., Int. Ed.* **2005**, *44*, 6146.
[70] Jiménez-Núñez, E.; Claverie, C.K.; Bour, C.; Cárdenas, D.J.; Echavarren, A.M.; *Angew. Chem., Int. Ed.* **2008**, *47*, 7892.
[71] Nieto-Oberhuber, C.; López, S.; Jiménez-Núñez, E.; Echavarren, A.M.; *Chem. Eur. J.* **2006**, *11*, 5916.
[72] Lee, J.C.H.; Hall, D.G.; *Tetrahedron Lett.* **2011**, *52*, 321.
[73] Cabello, N.; Jiménez-Núñez, E.; Buñuel, E.; Cárdenas, D.J.; Echavarren, A.M.; *Eur. J. Org. Chem.* **2007**, 4217.
[74] Lee, S.I.; Kim, S.M.; Choi, M.R.; Kim, S.Y.; Chung, Y.K.; Han, W.-S.; Kang, S.O.; *J. Org. Chem.* **2006**, *71*, 9366.
[75] Chen, Z.; Zhang, Y.-X.; Wang, Y.-H.; Zhu, L.-L.; Liu, H.; Li, X.-X.; Guo, L.; *Org. Lett.* **2010**, *12*, 3468.
[76] Chao, C.-M.; Beltrami, D.; Toullec, P.Y.; Michelet, V.; *Chem. Commun.* **2009**, 6988.
[77] (a) Brissy, D.; Skander, M.; Retailleau, P.; Marinetti, A.; *Organometallics* **2007**, *26*, 5782; (b) Nishimura, T.; Kawamoto, T.; Nagaosa, M.; Kumamoto, H.; Hayashi, T.; *Angew. Chem., Int. Ed.* **2010**, *49*, 1638.
[78] Nieto-Oberhuber, C.; López, S.; Muñoz, M.P.; Jiménez-Núñez, E.; Buñuel, E.; Cárdenas D.J.; Echavarren, A.M.; *Chem. Eur. J.* **2006**, *12*, 1694.
[79] Porcel, S.; Echavarren, A.M.; *Angew. Chem., Int. Ed.* **2007**, *46*, 2672.
[80] Kim, S.M.; Park, J.H.; Choi, S.Y.; Chung, Y.K.; *Angew. Chem., Int. Ed.* **2007**, *46*, 6172.
[81] (a) López, S.; Herrero-Gómez, E.; Pérez-Galán, P.; Nieto-Oberhuber, C.; Echavarren, A.M. *Angew. Chem., Int. Ed.* **2006**, *45*, 6029; (b) Pérez-Galán, P.; Herrero-Gómez, E.; Hog, D.T.; Martin, N.J.A.; Maseras, F.; Echavarren, A.M.; *Chem. Sci.* **2011**, *2*, 141.
[82] (a) Schelwies, M.; Dempwolff, A.L.; Rominger, F.; Helmchen, G.; *Angew. Chem., Int. Ed.* **2007**, *46*, 5598; (b) Schelwies, M.; Moser, R.; Dempwolff, A.L.; Rominger, F.; Helmchen, G.; *Chem. Eur. J.* **2009**, *15*, 10888.
[83] Genin, E.; Leseurre, L.; Toullec, P.Y.; Genêt, J.-P.; Michelet, V.; *Synlett* **2007**, 1780.
[84] Gold-catalyzed aminocyclizations of 1,6-enynes have been reported as well: Leseurre, L.; Toullec, P.Y.; Genêt, J.-P.; Michelet, V.; *Org. Lett.* **2007**, *9*, 4049.
[85] (a) Muñoz, M.P.; Adrio, J.; Carretero, J.C.; Echavarren, A.M.; *Organometallics* **2005**, *24*, 1293; (b) Chao, C.-M.; Genin, E.; Toullec, P.Y.; Genêt, J.-P.; Michelet, V.; *J. Organomet. Chem.* **2009**, *694*, 538; (c) Matsumoto, Y.; Selim, K.B.; Nakanishi, H.; Yamada, K.; Yamamoto, Y.; Tomioka, K.; *Tetrahedron Lett.* **2010**, *51*, 404.
[86] Fürstner, A.; Morency, L.; *Angew. Chem., Int. Ed.* **2008**, *47*, 5030.
[87] Sethofer, S.G.; Mayer, T.; Toste, F.D.; *J. Am. Chem. Soc.* **2010**, *132*, 8276.
[88] (a) Toullec, P.Y.; Genin, E.; Leseurre, L.; Genêt J.-P.; Michelet, V.; *Angew. Chem., Int. Ed.* **2006**, *45*, 7427; (b) Leseurre, L.; Chao, C.-M.; Seki, T.; Genin, E.; Toullec, P.Y.; Genêt J.-P.; Michelet, V.; *Tetrahedron* **2009**, *65*, 1911; (c) Amijs, C.H.M.; Ferrer, C.; Echavarren, A.M.; *Chem. Commun.* **2007**, 698; (d) Amijs, C.H.M.; López-Carrillo, V.; Raducan, M.; Pérez-Galán, P.; Ferrer, C.; Echavarren, A.M.; *J. Org. Chem.* **2008**, *73*, 7721; (e) Seo, H.; Roberts, B.P.; Abboud, K.A.; Merz, K.M.; Hong, S.; *Org. Lett.* **2010**, *12*, 4860.
[89] (a) Toullec, P.Y.; Chao, C.-M.; Chen, Q.; Gladiali, S.; Genêt, J.-P.; Michelet, V.; *Adv. Synth. Catal.* **2008**, *350*, 2401; (b) Chao, C.-M.; Vitale, M.R.; Toullec, P.Y.; Genêt, J.-P.; Michelet, V.; *Chem.Eur. J.* **2009**, *15*, 1319.
[90] (a) Mamane, V.; Gress, T.; Krause, H.; Fürstner, A.; *J. Am. Chem. Soc.* **2004**, *126*, 8654; (b) Luzung, M.R.; Markham, J.P.; Toste, F.D.; *J. Am. Chem. Soc.* **2004**, *126*,

10858; (c) Gagosz, F.; *Org. Lett.* **2005**, 7, 4129; (d) Ma, S.; Yu, S.; Gu, Z.; *Angew. Chem., Int. Ed.* **2006**, *45*, 200.

[91] Cheong, P.H.-Y.; Morganelli, P.; Luzung, M.; Houk, K.N.; Toste, F.D.; *J. Am. Chem. Soc.* **2008**, *130*, 4517.

[92] (a) Grisé, C.M.; Barriault, L.; *Org. Lett.* **2006**, *8*, 5905; (b) Grisé, C.M.; Rodrigue, E.M.; Barriault, L.; *Tetrahedron* **2008**, *64*, 797; (c) Bhunia, S.; Sohel, S.M.A.; Yang, C.-C.; Lush, S.-F.; Shen, F.-M.; Liu, R.-S.; *J. Organomet. Chem.* **2009**, *694*, 566.

[93] Luzung, M.R.; Toste, F.D.; *J. Am. Chem. Soc.* **2003**, *125*, 15760.

[94] Horino, Y.; Yamamoto, T.; Ueda, K.; Kuroda, S.; Toste, F.D.; *J. Am. Chem. Soc.* **2009**, *131*, 2809.

[95] Buzas, A.K.; Istrate, F.M.; Gagosz, F.; *Angew. Chem., Int. Ed.* **2007**, *46*, 1141.

[96] (a) Park, S.; Lee, D.; *J. Am. Chem. Soc.* **2006**, *128*, 10664; (b) Horino, Y.; Luzung, M.R.; Toste, F.D.; *J. Am. Chem. Soc.* **2006**, *128*, 11364.

[97] (a) Zhang, L.; Kozmin, S.A.; *J. Am. Chem. Soc.* **2005**, *127*, 6962; (b) Toullec, P.Y.; Blarre, T.; Michelet, V.; *Org. Lett.* **2009**, *11*, 2888.

[98] (a) Zhang, L.; Kozmin, S.A.; *J. Am. Chem. Soc.* **2004**, *126*, 11806; (b) Sun, J.; Conley, M.P.; Zhang, L.; Kozmin, S.A.; *J. Am. Chem. Soc.* **2006**, *128*, 9705; (c) Li, C.; Zeng, Y.; Zhang, H.; Feng, J.; Zhang, Y.; Wang, J.; *Angew. Chem., Int. Ed.* **2010**, *49*, 6413.

[99] (a) Kennedy-Smith, J.J.; Staben S.T.; Toste, F.D.; *J. Am. Chem. Soc.* **2004**, *126*, 4526; (b) Staben, S.T.; Kennedy-Smith, J.J.; Toste, F.D.; *Angew. Chem., Int. Ed.* **2004**, *43*, 5350; (c) Pan, J.-H.; Yang, M.; Gao, Q.; Zhu, N.-Y.; Yang, D.; *Synthesis* **2007**, 2539.

[100] (a) Linghu, X.; Kennedy-Smith, J.J.; Toste, F.D.; *Angew. Chem., Int. Ed.* **2007**, *46*, 7671; (b) Lee, K.; Lee, P.H. *Adv. Synth. Catal.* **2007**, *349*, 2092; (c) Barabé, F.; Bétournay, G.; Bellavance, G.; Barriault, L.; *Org. Lett.* **2009**, *11*, 4236.

[101] Binder, J.T.; Crone, B.; Haug, T.T.; Menz, H.; Kirsch, S.F.; *Org. Lett.* **2008**, *10*, 1025.

[102] (a) Kirsch, S.F.; Binder, J.T.; Crone, B.; Duschek, A.; Haug, T.T.; Liébert, C.; Menz, H.; *Angew. Chem., Int. Ed.* **2007**, *46*, 2310; (b) Baskar, B.; Bae, H.J.; An, S.E.; Cheong, J.Y.; Rhee, Y.H.; Duschek, A.; Kirsch, S.F.; *Org. Lett.* **2008**, *10*, 2605; (c) Menz, H.; Binder, J.T.; Crone, B.; Duschek, A.; Haug, T.T.; Kirsch, S.F.; Klahn, P.; Liébert, C.; *Tetrahedron* **2009**, *65*, 1880.

[103] (a) Ochida, A.; Ito, H.; Sawamura, M.; *J. Am. Chem. Soc.* **2006**, *128*, 16486; (b) Cabello, N.; Rodríguez, C.; Echavarren, A.M.; *Synlett* **2007**, 1753.

[104] Ito, H.; Ohmiya, H.; Sawamura, M.; *Org. Lett.* **2010**, *12*, 4380.

[105] (a) López-Carrillo, V.; Echavarren, A.M.; *J. Am. Chem. Soc.* **2010**, *132*, 9292; (b) Li, X.-X.; Zhu, L.-L.; Zhou, W.; Chen, Z.; *Org. Lett.* **2012**, *14*, 436.

[106] (a) Lee, Y.T.; Kang, Y.K; Chung, Y.K.; *J. Org. Chem.* **2009**, *74*, 7922; (b) Oh, C.H.; Kim, A.; *Synlett* **2008**, 777.

[107] Odabachian, Y.; Gagosz, F.; *Adv. Synth. Catal.* **2009**, *351*, 379.

[108] (a) Nieto-Oberhuber, C.; López S.; Echavarren, A.M.; *J. Am. Chem. Soc.* **2005**, *127*, 6178; (b) Nieto-Oberhuber, C.; Pérez-Galán, P.; Herrero-Gómez, E.; Lauterbach, T.; Rodríguez, C.; López, S.; Bour, C.; Rosellón, A.; Cárdenas, D.J.; Echavarren, A.M.; *J. Am. Chem. Soc.* **2008**, *130*, 269; (c) Pérez-Galán, P.; Martin, N.J.A.; Campaña, A.G.; Cárdenas, D.J.; Echavarren, A.M. *Chem. Asian J.* **2011**, *6*, 482; (d) Yeh, M.-C. P.; Tsao, W.-C.; Lee, B.-J.; Lin, T.-L.; *Organometallics* **2008**, *27*, 5326.

[109] Kozak, J.A.; Patrick, B.O.; Dake, G.R.; *J. Org. Chem.* **2010**, *75*, 8585.

[110] (a) Shibata, T.; Fujiwara, R.; Takano, D.; *Synlett* **2005**, 2062; Erratum: *Synlett* **2007**, 2766; (b) Lian, J.-J.; Chen, P.-C.; Lin, Y.-P.; Ting, H.-C.; Liu, R.-S.; *J. Am. Chem. Soc.* **2006**, *128*, 11372.
[111] (a) Lemière, G.; Gandon, V.; Agenet, N.; Goddard, J.-P.; de Kozak, A.; Aubert, C.; Fensterbank, L.; Malacria, M.; *Angew. Chem., Int. Ed.* **2006**, *45*, 7596; (b) Lin, G.-Y.; Yang, C.-Y.; Liu, R.-S.; *J. Org. Chem.* **2007**, *72*, 6753.
[112] Asao, N.; *Synlett* **2006**, 1645.
[113] Escribano-Cuesta, A.; López-Carrillo, V.; Janssen, D.; Echavarren, A.M.; *Chem. Eur. J.* **2009**, *15*, 5646.
[114] Jiménez-Núñez, E.; Claverie, C.K.; Nieto-Oberhuber, C.; Echavarren, A.M.; *Angew. Chem., Int. Ed.* **2006**, *45*, 5452.
[115] (a) Jiménez-Núñez, E.; Molawi, K.; Echavarren, A.M.; *Chem. Commun.* **2009**, 7327; (b) Molawi, K.; Delpont, N.; Echavarren, A.M.; *Angew. Chem., Int. Ed.* **2010**, *49*, 3517; (c) Zhou, Q.; Chen, X.; Ma, D.; *Angew. Chem., Int. Ed.* **2010**, *49*, 3513.
[116] Fürstner, A.; Stimson, C.C.; *Angew. Chem., Int. Ed.* **2007**, *46*, 8845.
[117] Kusama, H.; Karibe, Y.; Onizawa, Y.; Iwasawa, N.; *Angew. Chem., Int. Ed.* **2010**, *49*, 4269.
[118] Luzung, M.R.; Mauleón, P.; Toste, F.D.; *J. Am. Chem. Soc.* **2007**, *129*, 12402.
[119] (a) Teller, H.; Flügge, S.; Goddard, R.; Fürstner, A.; *Angew. Chem., Int. Ed.* **2010**, *49*, 1949; (b) Gonzàlez, A.Z.; Benitez, D.; Tkatchouk, E.; Goddard, W.A.; Toste, F.D.; *J. Am. Chem. Soc.* **2011**, *133*, 5500.
[120] Alcarazo, M.; Stork, T.; Anoop, A.; Thiel, W.; Fürstner, A.; *Angew. Chem., Int. Ed.* **2010**, *49*, 2542.
[121] (a) Alonso, I.; Trillo, B.; López, F.; Montserrat, S.; Ujaque, G.; Castedo, L.; Lledós, A.; Mascareñas, J.L.; *J. Am. Chem. Soc.* **2009**, *131*, 13020; (b) González, A.Z.; Toste, F.D.; *Org. Lett.* **2010**, *12*, 200.
[122] Mauleón, P.; Zeldin, R.M.; González, A.Z.; Toste, F.D.; *J. Am. Chem. Soc.* **2009**, *131*, 6348.
[123] (a) Benitez, D.; Tkatchouk, E.; González, A.Z.; Goddard, W.A., III; Toste, F.D.; *Org. Lett.* **2009**, *11*, 4798; (b) Trillo, B.; López, F.; Montserrat, S.; Ujaque, G.; Castedo, L.; Lledós, A.; Mascareñas, J.L.; *Chem. Eur. J.* **2009**, *15*, 3336; (c) Gung, B.W.; Craft, D.T.; Bailey, L.N.; Kirschbaum, K.; *Chem. Eur. J.* **2010**, *16*, 639; (d) Alonso, I.; Faustino, H.; López, F.; Mascareñas, J.L.; *Angew. Chem., Int. Ed.* **2011**, *50*, 11496.
[124] Wang, G.; Zou, Y.; Li, Z.; Wang, Q.; Goeke, A.; *Adv. Synth. Catal.* **2011**, *353*, 550.
[125] (a) Huang, X.; Zhang, L.; *J. Am. Chem. Soc.* **2007**, *129*, 6398; (b) Huang, X.; Zhang, L.; *Org. Lett.* **2007**, *9*, 4627; (c) Chaudhuri, R.; Liao, H.-Y.; Liu, R.-S.; *Chem. Eur. J.* **2009**, *15*, 8895; (d) Teng, T.-M.; Lin, M.-S.; Vasu, D.; Bhunia, S.; Liu, T.-A.; Liu, R.-S.; *Chem. Eur. J.* **2010**, *16*, 4744.
[126] (a) Sherry B.D.; Toste, F.D.; *J. Am. Chem. Soc.* **2004**, *126*, 15978; (b) Mauleón, P.; Krinsky, J.L.; Toste, F.D.; *J. Am. Chem. Soc.* **2009**, *131*, 4513; (c) Wang, D.; Gautam, L.N.S.; Bollinger, C.; Harris, A.; Li, M.; Shi, X.; *Org. Lett.* **2011**, *13*, 2618.
[127] Ferrer, C.; Raducan, M.; Nevado, C.; Claverie, C.K.; Echavarren, A.M.; *Tetrahedron* **2007**, *63*, 6306.
[128] Krafft, M.E.; Hallal, K.M.; Vidhani, D.V.; Cran, J.W.; *Org. Biomol. Chem.* **2011**, *9*, 7535.
[129] Gockel, B.; Krause, N.; *Org. Lett.* **2006**, *8*, 4485.
[130] Sherry, B.D.; Maus, L.; Laforteza, B.N.; Toste, F.D.; *J. Am. Chem. Soc.* **2006**, *128*, 8132.
[131] Reich, N.W.; Yang, C-G.; Shi, Z.; He, C.; *Synlett* **2006**, 1278.

[132] Suhre, M.H.; Reif, M.; Kirsch, S.F.; *Org. Lett.* **2005**, *7*, 3925.
[133] Cao, H.; Jiang, H.; Mai, R.; Zhu, S.; Qi, C.; *Adv. Synth. Catal.* **2010**, *352*, 143.
[134] (a) Binder, J.T.; Kirsch, S.F.; *Org. Lett.* **2006**, 8, 2151; (b) Saito, A.; Konishi, T.; Hanzawa, Y.; *Org. Lett.* **2010**, *12*, 372; (c) Wei, H.; Wang, Y.; Yue, B.; Xua, P.-F.; *Adv. Synth. Catal.* **2010**, *352*, 2450.
[135] Gille, A.; Rehbein, J.; Hiersemann, M.; *Org. Lett.* **2011**, *13*, 2122.
[136] (a) Müller, T.E.; Hultzsch, K.C.; Yus, M.; Foubelo, F.; Tada, M.; *Chem. Rev.* **2008**, *108*, 3795; (b) Chemler, S.R.; *Org. Biomol. Chem.* **2009**, *7*, 3009.
[137] (a) Fukuda, Y.; Utimoto, K.; Nozaki, H.; *Heterocycles* **1987**, *25*, 297; (b) Fukuda, Y.; Utimoto, K.; *Synthesis* **1991**, 975.
[138] (a) Müller, T.E.; *Tetrahedron Lett.* **1998**, *39*, 5961; (b) Müller, T.E.; Grosche, M.; Herdtweck, E.; Pleier, A.-K.; Walter, E.; Yan, Y.-K.; *Organometallics* **2000**, *19*, 170; (c) Kadzimirsz, D.; Hildebrandt, D.; Merz, K.; Dyker, G.; *Chem. Commun.* **2006**, 661; (d) Yeom, H.-S.; Lee, E.-S.; Shin, S.; *Synlett* **2007**, 2292; (e) Enomoto, T.; Girard, A.-L.; Yasui, Y.; Takemoto, Y.; *J. Org. Chem.* **2009**, *74*, 9158; (f) Han, J.; Xu, B.; Hammond, G.B.; *Org. Lett.* **2011**, *13*, 3450.
[139] Iritani, K.; Matsubara, S.; Utimoto, K.; *Tetrahedron Lett.* **1988**, *29*, 1799.
[140] Arcadi, A.; Bianchi, G.; Marinelli, F.; *Synthesis* **2004**, 610.
[141] Zhang, Y.; Donahue, J.P.; Li, C.-J.; *Org. Lett.* **2007**, *9*, 627.
[142] (a) Alfonsi, M.; Arcadi, A.; Aschi, M.; Bianchi, G.; Marinelli, F.; *J. Org. Chem.* **2005**, *70*, 2265; (b) Ambrogio, I.; Arcadi, A.; Cacchi, S.; Fabrizi, G.; Marinelli, F.; *Synlett* **2007**, 1775.
[143] Hirano, K.; Inaba, Y.; Takasu, K.; Oishi, S.; Takemoto, Y.; Fujii, N.; Ohno, H.; *J. Org. Chem.* **2011**, *76*, 9068.
[144] (a) Arcadi, A.; Di Giuseppe, S.; Marinelli, F.; Rossi, E.; *Adv. Synth. Catal.* **2001**, *343*, 443; (b) Arcadi, A.; Di Giuseppe, S.; Marinelli, F.; Rossi, E.; *Tetrahedron: Asymmetry* **2001**, *12*, 2715; (c) Harrison, T.J.; Kozak, J.A.; Corbella-Pane, M.; Dake, G.R.; *J. Org. Chem.* **2006**, *71*, 4525.
[145] (a) Arcadi, A.; Bianchi, G.; Di Giuseppe, S.; Marinelli, F.; *Green Chem.* **2003**, *5*, 64; (b) Arcadi, A.; Chiarini, M.; Di Giuseppe, S.; Marinelli, F.; *Synlett* **2003**, 203; (c) Abbiati, G.; Arcadi, A.; Bianchi, G.; Di Giuseppe, S.; Marinelli, F.; Rossi, E.; *J. Org. Chem.* **2003**, *68*, 6959; (d) Kusama, H.; Miyashita, Y.; Takaya, J.; Iwasawa, N.; *Org. Lett.* **2006**, *8*, 289; (e) Yuon, S.W.; *J. Org. Chem.* **2006**, *71*, 2521; (f) Qian, J.; Liu, Y.; Zhu, J.; Jiang, B.; Xu, Z.; *Org. Lett.* **2011**, *13*, 4220.
[146] Shu, X.-Z.; Liu, X.-Y.; Xiao, H.-Q.; Ji, K.-G.; Guo, L.-N.; Liang, Y.-M.; *Adv. Synth. Catal.* **2008**, *350*, 243.
[147] (a) Lok, R.; Leone, R.E.; Williams, A.J.; *J. Org. Chem.* **1996**, *61*, 3289; (b) Obika, S.; Kono, H.; Yasui, Y.; Yanada, R.; Takemoto, Y.; *J. Org. Chem.* **2007**, *72*, 4462.
[148] (a) Aponick, A.; Li, C.-Y.; Malinge, J.; Marques, E.F.; *Org. Lett.* **2009**, *11*, 4624; (b) Egi, M.; Azechi, K.; Akai, S.; *Org. Lett.* **2009**, *11*, 5002; (c) Egi, M.; Azechi, K.; Akai, S.; *Adv. Synth. Catal.* **2011**, *353*, 287; (d) Surmont, R.; Verniest, G.; De Kimpe, N.; *Org. Lett.* **2009**, *11*, 2920.
[149] Gorin, D.J.; Davis, N.R.; Toste, F.D.; *J. Am. Chem. Soc.* **2005**, *127*, 11260.
[150] (a) Davies, P.W.; Martin, N.; *Org. Lett.* **2009**, *11*, 2293; (b) Davies, P.W.; Martin, N.; *J. Organomet. Chem.* **2011**, *696*, 159; (c) Du, X.; Xie, X.; Liu, Y.; *J. Org. Chem.* **2010**, *75*, 510.
[151] Blanc, A.; Alix, A.; Weibel, J.-M.; Pale, P.; *Eur. J. Org. Chem.* **2010**, 1644.
[152] Seregin, I.V.; Gevorgyan, V.; *J. Am. Chem. Soc.* **2006**, *128*, 12050.
[153] Mizushima, E.; Hayashi, T.; Tanaka, M.; *Org. Lett.* **2003**, *5*, 3349.
[154] Luo, Y.; Li, Z.; Li, C.-J.; *Org. Lett.* **2005**, *7*, 2675.

[155] (a) Leyva, A.; Corma, A.; *Adv. Synth. Catal.* **2009**, *351*, 2876; (b) Kramer, S.; Dooleweerdt, K.; Lindhardt, A.T.; Rottländer, M.; Skrydstrup, T.; *Org. Lett.* **2009**, *11*, 4208; (c) Duan, H.; Sengupta, S.; Petersen, J.L.; Akhmedov, N.G.; Shi, X.; *J. Am. Chem. Soc.* **2009**, *131*, 12100.
[156] (a) Liu, X.-Y.; Che, C.-M.; *Org. Lett.* **2009**, *11*, 4204; (b) Han, Z.-Y.; Xiao, H.; Chen, X.-H.; Gong, L.-Z.; *J. Am. Chem. Soc.* **2009**, *131*, 9182.
[157] (a) Morita, N.; Krause, N.; *Org. Lett.* **2004**, *6*, 4121; (b) Morita, N.; Krause, N.; *Eur. J. Org. Chem.* **2006**, 4634.
[158] Zhu, R.-X.; Zhang, D.-J.; Guo, J.-X.; Mu, J.-L.; Duan, C.-G.; Liu, C.-B.; *J. Phys. Chem. A* **2010**, *114*, 4689.
[159] Dieter, R.K.; Chen, N.; Yu, H.; Nice, L.E.; Gore, V.K.; *J. Org. Chem.* **2005**, *70*, 2109.
[160] (a) Lee, P.H.; Kim, H.; Lee, K.; Kim, M.; Noh, K.; Kim, H.; Seomoon, D.; *Angew. Chem., Int. Ed.* **2005**, *44*, 1840; (b) Breman, A.C.; Dijkink, J.; Van Maarseveen, J.H.; Kinderman, S.S.; Hiemstra, H.; *J. Org. Chem.* **2009**, *74*, 6327.
[161] Kaden, S.; Reissig, H.-U.; Brüdgam, I.; Hartl, H.; *Synthesis* **2006**, 1351.
[162] Winter, C.; Krause, N.; *Angew. Chem., Int. Ed.* **2009**, *48*, 6339.
[163] Benedetti, E.; Lemière, G.; Chapellet, L.-L.; Penoni, A.; Palmisano, G.; Malacria, M.; Goddard, J.-P.; Fensterbank, L.; *Org. Lett.* **2010**, *12*, 4396.
[164] Patil, N.T.; Lutete, L.M.; Nishina, N.; Yamamoto, Y.; *Tetrahedron Lett.* **2006**, *47*, 4749.
[165] (a) Bates, R.W.; Dewey, M.R.; *Org. Lett.* **2009**, *11*, 3706; (b) Bates, R.W.; Shuyi Ng, P.; *Tetrahedron Lett.* **2011**, *52*, 2969.
[166] Li, H.; Widenhoefer, R.A.; *Org. Lett.* **2009**, *11*, 2671.
[167] Manzo, A.M.; Perboni, A.D.; Broggini, G.; Rigamonti, M.; *Tetrahedron Lett.* **2009**, *50*, 4696.
[168] (a) Zhang, Z.; Bender, C.F.; Widenhoefer, R.A.; *Org. Lett.* **2007**, *9*, 2887; (b) Zhang, Z.; Bender, C.F.; Widenhoefer, R.A. *J. Am. Chem. Soc.* **2007**, *129*, 14148; (c) Li, H.; Du Lee, S.; Widenhoefer, R.A.; *J. Organomet. Chem.* **2011**, *696*, 316; (c) Aikawa, K.; Kojima, M.; Mikami, K.; *Angew. Chem., Int. Ed.* **2009**, *48*, 6073.
[169] LaLonde, R.L.; Sherry, B.D.; Kang, E.J.; Toste, F.D.; *J. Am. Chem. Soc.* **2007**, *129*, 2452.
[170] (a) Hamilton, G.L.; Kang, E.J.; Mba, M.; Toste, F.D.; *Science* **2007**, *317*, 496; (b) Aikawa, K.; Kojima, M.; Mikami, K.; *Adv. Synth. Catal.* **2010**, *352*, 3131.
[171] LaLonde, R.L.; Wang, Z.J.; Mba, M.; Lackner, A.D.; Toste, F.D.; *Angew. Chem., Int. Ed.* **2010**, *49*, 598.
[172] (a) Nishina, N.; Yamamoto, Y.; *Angew. Chem., Int. Ed.* **2006**, *45*, 3314; (b) Nishina, N.; Yamamoto, Y.; *Tetrahedron* **2009**, *65*, 1799; (c) Duncan, A.N.; Widenhoefer, R.A.; *Synlett* **2010**, 419; (d) Hill, A.W.; Elsegood, M.R.J.; Kimber, M.C.; *J. Org. Chem.* **2010**, *75*, 5406.
[173] (a) Nishina, N.; Yamamoto, Y.; *Synlett* **2007**, 1767.
[174] (a) Lavallo, V.; Frey, G.D.; Donnadieu, B.; Soleilhavoup, M.; Bertrand, G.; *Angew. Chem., Int. Ed.* **2008**, *47*, 5224; (b) Zeng, X.; Soleilhavoup, M.; Bertrand, G.; *Org. Lett.* **2009**, *11*, 3166; (c) Kinjo, R.; Donnadieu, B.; Bertrand, G.; *Angew. Chem., Int. Ed.* **2011**, *50*, 5560.
[175] Kinder, R.E.; Zhang, Z.; Widenhoefer, R.A.; *Org. Lett.* **2008**, *10*, 3157.
[176] Zhang, J.; Yang, C.-G.; He, C.; *J. Am. Chem. Soc.* **2006**, *128*, 1798.
[177] (a) Liu, X.-Y.; Li, C.-H.; Che, C.-M. *Org. Lett.* **2006**, *8*, 2707; (b) Giner, X.; Najera, C. *Org. Lett.* **2008**, *10*, 2919.
[178] Zhang, X.; Corma, A.; *Dalton Trans.* **2008**, 397.

[179] Brouwer, C.; He, C.; *Angew. Chem., Int. Ed.* **2006**, *45*, 1744.
[180] Kobayashi, S.; Kakumoto, K.; Sugiura, M.; *Org. Lett.* **2002**, *4*, 1319.
[181] Xiao, Y.-P.; Liu, X.-Y.; Che, C.-M.; *Beilstein J. Org. Chem.* **2011**, *7*, 1100.
[182] Yeh, M.-C. P.; Pai, H.-F.; Lin, Z.-J.; Lee, B.-R.; *Tetrahedron* **2009**, *65*, 4789.
[183] Kanno, O.; Kuriyama, W.; Wang, Z.J.; Toste, F.D.; *Angew. Chem., Int. Ed.* **2011**, *50*, 9919.
[184] (a) Han, X.; Widenhoefer, R.A.; *Angew. Chem., Int. Ed.* **2006**, *45*, 1747; (b) Bender, C.F.; Widenhoefer, R.A.; *Chem. Commun.* **2006**, 4143; (c) Bender, C.F.; Widenhoefer, R.A.; *Chem. Commun.* **2008**, 2741.
[185] (a) Bender, C.F.; Widenhoefer, R.A.; *Org. Lett.* **2006**, *8*, 5303; (b) Li, H.; Song, F.; Widenhoefer, R.A.; *Adv. Synth. Catal.* **2011**, *353*, 955.
[186] Zhang, Z.; Lee, S.D.; Widenhoefer, R.A.; *J. Am. Chem. Soc.* **2009**, *131*, 5372.
[187] Mukherjee, P.; Widenhoefer, R.A.; *Org. Lett.* **2010**, *12*, 1184.
[188] Mukherjee, P.; Widenhoefer, R.A.; *Angew. Chem., Int. Ed.* **2012**, *51*, 1405.
[189] Zhang, G.; Cui, L.; Wang, Y.; Zhang, L.; *J. Am. Chem. Soc.* **2010**, *132*, 1474.
[190] Tkatchouk, E.; Mankad, N.P.; Benitez, D.; Goddard, W.A.; Toste, F.D.; *J. Am. Chem. Soc.* **2011**, *133*, 14293.
[191] De Haro, T.; Nevado, C.; *Angew. Chem., Int. Ed.* **2011**, *50*, 906.
[192] Iglesias, A.; Muñiz, K.; *Chem. Eur. J.* **2009**, *15*, 10563.
[193] Hintermann, L.; Labonne, A.; *Synthesis* **2007**, 1121.
[194] (a) Imi, K.; Imai, K.; Utimoto, K.; *Tetrahedron Lett.* **1987**, *28*, 3127; (b) Fukuda, Y.; Utimoto, K.; *Bull. Chem. Soc. Jpn.* **1991**, *64*, 2013; (c) Fukuda, Y.; Utimoto, K.; *J. Org. Chem.* **1991**, *56*, 3729.
[195] Santos, L.L.; Ruiz, V.R.; Sabater, M.J.; Corma, A.; *Tetrahedron* **2008**, *64*, 7902.
[196] (a) Harkat, H.; Weibel, J.-M.; Pale, P.; *Tetrahedron Lett.* **2007**, *48*, 1439; (b) Harkat, H.; Blanc, A.; Weibel, J.-M.; Pale, P.; *J. Org. Chem.* **2008**, *73*, 1620.
[197] Trost, B.M.; Dong, G.; *J. Am. Chem. Soc.* **2010**, *132*, 16403.
[198] Manzo, A.M.; Perboni, A.; Broggini, G.; Rigamonti, M.; *Synthesis* **2011**, 127.
[199] De Brabander, J.K.; Liu, B.; Qian, M.; *Org. Lett.* **2008**, *10*, 2533.
[200] (a) Genin, E.; Toullec, P.-Y.; Antoniotti, S.; Brancour, C.; Genêt, J.-P.; Michelet, V.; *J. Am. Chem. Soc.* **2006**, *128*, 3112; (b) Toullec, P.Y.; Genin, E.; Antoniotti, S.; Genêt, J.-P.; Michelet, V.; *Synlett* **2008**, 707; (c) Neatu, F.; Parvulescu, V.I.; Michelet, V.; Genêt, J.-P.; Goguet, A.; Hardacre, C.; *New. J. Chem.* **2009**, *33*, 102; (d) Harkat, H.; Weibel, J.-M.; Pale, P.; *Tetrahedron Lett.* **2006**, *47*, 6273; (e) Harkat, H.; Dembele, A.Y.; Weibel, J.-M.; Blanc, A.; Pale, P.; *Tetrahedron* **2009**, *65*, 1871; (f) Marchal, E.; Uriac, P.; Legouin, B.; Toupet, T.; Van de Weghe, P.; *Tetrahedron* **2007**, *63*, 9979.
[201] (a) Buzas, A.; Gagosz, F.; *Org. Lett.* **2006**, *8*, 515; (b) Buzas, A.; Gagosz, F.; *Synlett* **2006**, 2727; (c) Buzas, A.K.; Istrate, F.; Gagosz, F.; *Tetrahedron* **2009**, *65*, 1889; (d) Robles-Machin, R.; Adrio, J.; Carretero, J.C.; *J. Org. Chem.* **2006**, *71*, 5023; (e) Lee, E.-S.; Yeom, H.-S.; Hwang, J.-H.; Shin, S.; *Eur. J. Org. Chem.* **2007**, 3503; (f) Kang, J.-E.; Shin, S.; *Synlett* **2006**, 717; (g) Kang, J.-E.; Kim, H.-B.; Lee, J.-W.; Shin, S.; *Org. Lett.* **2006**, *8*, 3537; (h) Ritter, S.; Horino, Y.; Lex, J.; Schmalz, H.-G.; *Synlett* **2006**, 3309; (i) Hashmi, A.S.K.; Rudolph, M.; Schymura, S.; Visus, J.; Frey, W.; *Eur. J. Org. Chem.* **2006**, 4905; (j) Gouault, N.; Le Roch, M.; Cornée, C.; David, M.; Uriac, P.; *J. Org. Chem.* **2009**, *74*, 5614; (k) Ueda, M.; Sato, A.; Ikeda, Y.; Miyoshi, T.; Naito, T.; Miyata, O.; *Org. Lett.* **2010**, *12*, 2594.
[202] (a) Debleds, O.; Dal Zotto, C.; Vrancken, E.; Campagne, J.-M.; Retailleau, P.; *Adv. Synth. Catal.* **2009**, *351*, 1991; (b) Praveen, C.; Kalyanasundaram, A.; Perumal, P.T. *Synlett* **2010**, 777.

[203] Belting, V.; Krause, N.; *Org. Lett.* **2006**, *8*, 4489.

[204] (a) Jury, J.C.; Swamy, N.K.; Yazici, A.; Willis, A.C.; Pyne, S.G.; *J. Org. Chem.* **2009**, *74*, 5523; (b) Laroche, C.; Kerwin, M.; *J. Org. Chem.* **2009**, *74*, 9229.

[205] Alcaide, B.; Almendros, P.; Martínez del Campo, T.; Carrascosa, R.; *Eur. J. Org. Chem.* **2010**, 4912.

[206] Dai, L.-Z.; Qi, M.-J.; Shi, Y.-L.; Liu, X.-G.; Shi, M.; *Org. Lett.* **2007**, *9*, 3191.

[207] Diéguez-Vázquez, A.; Tzschucke, C.C.; Crecente-Campo, J.; McGrath, S.; Ley, S.V.; *Eur. J. Org. Chem.* **2009**, 1698.

[208] Liu, B.; De Brabander, J.K.; *Org. Lett.* **2006**, *8*, 4907.

[209] Dai, L.-Z.; Shi, M.; *Chem.Eur. J.* **2008**, *14*, 7011.

[210] Zhang, Y.; Xue, J.; Xin, Z.; Xie, Z.; Li, Y.; *Synlett* **2008**, 940.

[211] Fang, C.; Pang, Y.; Forsyth, C.J.; *Org. Lett.* **2010**, *12*, 4528.

[212] Li, Y.; Zhou, F.; Forsyth, C.J.; *Angew. Chem., Int. Ed.* **2007**, *46*, 279.

[213] Volchkov, I.; Sharma, K.; Cho, E.J.; Lee, D.; *Chem. Asian J.* **2011**, *6*, 1961.

[214] Aponick, A.; Li, C.L.; Palmes, J.A.; *Org. Lett.* **2009**, *11*, 121.

[215] Antoniotti, S.; Genin, E.; Michelet, V.; Genêt, J.-P.; *J. Am. Chem. Soc.* **2005**, *127*, 9976.

[216] Liu, L.-P.; Hammond, G.B.; *Org. Lett.* **2009**, *11*, 5090.

[217] Yeung, Y.-Y.; Corey, E.J.; *Org. Lett.* **2008**, *10*, 3877.

[218] (a) Barluenga, J.; Diéguez, A.; Fernández, A.; Rodríguez, F.; Fañanás, F.J.; *Angew. Chem., Int. Ed.* **2006**, *45*, 2091; (b) Barluenga, J.; Fernández, A.; Satrustegui, A.; Diéguez, A.; Rodríguez, F.; Fañanás, F.J.; *Chem. Eur. J.* **2008**, *14*, 4153; (c) Barluenga, J.; Fernández, A.; Diéguez, A.; Rodríguez, F.; Fañanás, F.J.; *Chem. Eur. J.* **2009**, *15*, 11660.

[219] (a) Wilckens, K.; Uhlemann, M.; Czekelius, C.; *Chem. Eur. J.* **2009**, *15*, 13323; (b) Rüttinger, R.; Leutzow, J.; Wilsdorf, M.; Wilckens, K.; Czekelius, C.; *Org. Lett.* **2011**, *13*, 224.

[220] Barluenga, J.; Calleja, J.; Mendoza, A.; Rodríguez, F.; Fañanás, F.J.; *Chem. Eur. J.* **2010**, *16*, 7110.

[221] (a) Liu, Y.; Song, F.; Song, Z.; Liu, M.; Yan, B.; *Org. Lett.* **2005**, *7*, 5409; (b) Du, X.; Song, F.; Lu, Y.; Chen, H.; Liu, Y.; *Tetrahedron* **2009**, *65*, 1839; (c) Zhang, X.; Lu, Z.; Fu, C.; Ma, S.; *J. Org. Chem.* **2010**, *75*, 2589.

[222] (a) Praveen, C.; Kiruthiga, P.; Perumal, P.T.; *Synlett* **2009**, 1990; (b) Kim, S.; Kang, D.; Shin, S.; Lee, P.H.; *Tetrahedron Lett.* **2010**, *51*, 1899; (c) Hashmi, A.S.K.; Häffner, T.; Rudolph, M.; Rominger, F.; *Eur. J. Org. Chem.* **2011**, 667.

[223] (a) Hashmi, A.S.K.; Sinha, P.; *Adv. Synth. Catal.* **2004**, *346*, 432; (b) Blanc, A.; Tenbrink, K.; Weibel, J.-M.; Pale, P.; *J. Org. Chem.* **2009**, *74*, 5342.

[224] (a) Yoshida, M.; Al-Amin, M.; Matsuda, K.; Shishido, K.; *Tetrahedron Lett.* **2008**, *49*, 5021; (b) Yoshida, M.; Al-Amin, M.; Shishido, K.; *Synthesis* **2009**, 2454; (c) Blanc, A.; Tenbrink, K.; Weibel, J.-M.; Pale, P.; *J. Org. Chem.* **2009**, *74*, 4360.

[225] (a) Shu, X.-Z.; Liu, X.-Y.; Xiao, H.-Q.; Ji, K.-G.; Guo, L.-N.; Qi, C.-Z.; Liang, Y.-M.; *Adv. Synth. Catal.* **2007**, *349*, 2493; (b) Ji, K.-G.; Shu, X.-Z.; Chen, J.; Zhao, S.-C.; Zheng, Z.-J.; Liu, X.-Y.; Liang, Y.-M.; *Org. Biomol. Chem.* **2009**, *7*, 2501.

[226] (a) Li, Y.; Wheeler, K.A.; Dembinski, R.; *Adv. Synth. Catal.* **2010**, *352*, 2761; (b) Li, Y.; Wheeler, K.A.; Dembinski, R.; *Eur. J. Org. Chem.* **2011**, *2011*, 2767.

[227] Belting, V.; Krause, N.; *Org. Biomol. Chem.* **2009**, *7*, 1221.

[228] (a) Hashmi, A.S.K.; Weyrauch, J.P.; Frey, W.; Bats, J.W.; *Org. Lett.* **2004**, *6*, 4391; (b) Weyrauch, J.P.; Hashmi, A.S.K.; Schuster, A.; Hengst, T.; Schetter, S.; Littmann, A.; Rudolph, M.; Hamzic, M.; Visus, J.; Rominger, F.; Frey, W.; Bats, J.W.; *Chem. Eur. J.* **2010**, *16*, 956.

[229] (a) Yao, T.; Zhang, X.; Larock, R.C.; *J. Am. Chem. Soc.* **2004**, *126*, 11164; (b) Yao, T.; Zhang, X.; Larock, R.C.; *J. Org. Chem.* **2005**, *70*, 7679.

[230] Liu, X.; Pan, Z.; Shu, X.; Duan, X.; Liang, Y.; *Synlett* **2006**, 1962.

[231] Zhang, J.; Schmalz, H.-G.; *Angew. Chem., Int. Ed.* **2006**, *45*, 6704.

[232] (a) Egi, M.; Azechi, K.; Saneto, M.; Shimizu, K.; Akai, S.; *J. Org. Chem.* **2010**, *75*, 2123; (b) Liu, Y.; Liu, M.; Guo, S.; Tu, H.; Zhou, Y.; Gao, H.; *Org. Lett.* **2006**, *8*, 3445; (c) Kirsch, S.F.; Binder, J.T.; Liebert, C.; Menz, H.; *Angew. Chem., Int. Ed.* **2006**, *45*, 5878; (d) Binder, J.T.; Crone, B.; Kirsch, S.F.; Liebert, C.; Menz, H.; *Eur. J. Org. Chem.* **2007**, 1636.

[233] Ye, L.; Cui, L.; Zhang, G.; Zhang, L.; *J. Am. Chem. Soc.* **2010**, *132*, 3258.

[234] Ye, L.; He, W.; Zhang, L.; *J. Am. Chem. Soc.* **2010**, *132*, 8550.

[235] (a) Asao, N.; Nogami, T.; Takahashi, K.; Yamamoto, Y.; *J. Am. Chem. Soc.* **2002**, *124*, 764; (b) Asao, N.; Chan, C.S.; Takahashi, K.; Yamamoto, Y.; *Tetrahedron* **2005**, *61*, 11322.

[236] Patil, N.T.; Yamamoto, Y.; *J. Org. Chem.* **2004**, *69*, 5139.

[237] (a) Godet, T.; Vaxelaire, C.; Michel, C.; Milet, A.; Belmont, P. *Chem. Eur. J.* **2007**, *13*, 5632; (b) Yu, X.; Ding, Q.; Wang, W.; Wu, J.; *Tetrahedron Lett.* **2008**, *49*, 4390.

[238] (a) Yao, X.; Li, C.-J.; *Org. Lett.* **2006**, *8*, 1953; (b) Dell'Acqua, M.; Facoetti, D.; Abbiatti, G.; Rossi, E.; *Synthesis* **2010**, 2367; (c) Zhou, L.; Liu, Y.; Zhang, Y.; Wang, J.; *Beilstein J. Org. Chem.* **2011**, *7*, 631; (d) Asao, N.; Aikawa, H.; Tago, S.; Umetsu, K. *Org. Lett.* **2007**, *9*, 4299; (e) Aikawa, H.; Tago, S.; Umetsu, K.; Haginiwa, N.; Asao, N.; *Tetrahedron* **2009**, *65*, 1774.

[239] Liu, L.-P.; Hammond, G.B.; *Org. Lett.* **2010**, *12*, 4640.

[240] Handa, S.; Slaughter, L.M.; *Angew. Chem., Int. Ed.* **2012**, *51*, 2912.

[241] Murai, M.; Uenishi, J.; Uemura, M.; *Org. Lett.* **2010**, *12*, 4788.

[242] (a) Mizushima, E.; Sato, K.; Hayashi, T.; Tanaka, M.; *Angew. Chem., Int. Ed.* **2002**, *41*, 4563; (b) Mizushima, E.; Cui, D.-M.; Nath, D.C.D.; Hayashi, T.; Tanaka, M.; Danheiser, R.L.; Lam, T.Y. *Org. Synth.* **2006**, *83*, 55; (c) Roembke, P.; Schmidbaur, H.; Cronje, S.; Raubenheimer, H.; *J. Mol. Catal. A* **2004**, *212*, 35.

[243] (a) Casado, R.; Contel, M.; Laguna, M.; Romero, P.; Sanz, S.; *J. Am. Chem. Soc.* **2003**, *125*, 11925; (b) Wang, W.; Jasinski, J.; Hammond, G.B.; Xu, B.; *Angew. Chem., Int. Ed.* **2010**, *49*, 7247.

[244] Leyva, A.; Corma, A.; *J. Org. Chem.* **2009**, *74*, 2067.

[245] (a) Marion, N.; Ramon, R.S.; Nolan, S.P.; *J. Am. Chem. Soc.* **2009**, *131*, 448; (b) Frémont, P.; Singh, R.; Stevens, E.D.; Petersen, J.L.; Nolan, S.P.; *Organometallics* **2007**, *26*, 1376.

[246] Ramon, R.S.; Marion, N.; Nolan, S.P.; *Chem. Eur. J.* **2009**, *15*, 8695.

[247] (a) Jung, H.H.; Floreancig, P.E.; *Org. Lett.* **2006**, *8*, 1949; (b) Jung, H.H.; Floreancig, P.E.; *J. Org. Chem.* **2007**, *72*, 7359.

[248] (a) Marshall, J.A.; Robinson, E.D.; *J. Org. Chem.* **1990**, *55*, 3450; (b) Marshall, J.A.; Bartley, G.S.; *J. Org. Chem.* **1994**, *59*, 7169; (c) Marshall, J.A.; Sehon, C.A.; *J. Org. Chem.* **1995**, *60*, 5966.

[249] (a) Hashmi, A.S.K. *Angew. Chem., Int. Ed.* **1995**, *34*, 1581; (b) Hashmi, A.S.K.; Ruppert, T.L.; Knöfel, T.; Bats, J.W.; *J. Org. Chem.* **1997**, *62*, 7295.

[250] Zhou, C.-Y.; Chan, P.W.H.; Che, C.-M.; *Org. Lett.* **2006**, *8*, 325.

[251] (a) Sromek, A.W.; Rubina, M.; Gevorgyan, V.; *J. Am. Chem. Soc.* **2005**, *127*, 10500; (b) Dudnik, A.S.; Sromek, A.W.; Rubina, M.; Kim, J.T.; Kel'in, A.V.; Gevorgyan, V.; *J. Am. Chem. Soc.* **2008**, *130*, 1440; (c) Xia, Y.; Dudnik, A.S.; Gevorgyan, V.; Li, Y.; *J. Am. Chem. Soc.* **2008**, *130*, 6940.

[252] Dudnik, A.S.; Xia, Y.; Li, Y.; Gevorgyan, V.; *J. Am. Chem. Soc.* **2010**, *132*, 7645.

[253] Dudnik, A.S.; Gevorgyan, V.; *Angew. Chem., Int. Ed.* **2007**, *46*, 5195.

[254] (a) Hoffmann-Röder, A.; Krause, N.; *Org. Lett.* **2001**, *3*, 2537; (b) Krause, N.; Hoffmann-Röder, A.; Canisius, J.; *Synthesis* **2002**, 1759; (c) Park, J.; Kim, S.H.; Lee, P.H. *Org. Lett.* **2008**, *10*, 5067; (d) Eom, D.; Kang, D.; Lee, P.H.; *J. Org. Chem.* **2010**, *75*, 7447.

[255] (a) Buzas, A.; Istrate, F.; Gagosz, F.; *Org. Lett.* **2006**, *8*, 1957; (b) Hyland, C.J.T.; Hegedus, L.S.; *J. Org. Chem.* **2006**, *71*, 8658; (c) Deutsch, C.; Hoffmann-Röder, A.; Domke, A.; Krause, N.; *Synlett* **2007**, 737; (d) Deutsch, C.; Lipshutz, B.H.; Krause, N.; *Angew. Chem., Int. Ed.* **2007**, *46*, 1650; (e) Brasholz, M.; Reissig, H.-U.; *Synlett* **2007**, 1294; (f) Dugovič, B.; Reissig, H.-U.; *Synlett* **2008**, 769; (g) Brasholz, M.; Reissig, H.-U.; Zimmer, R.; *Acc. Chem. Res.* **2009**, *42*, 45; (h) Lechel, T.; Reissig, H.-U.; *Pure Appl. Chem.* **2010**, *82*, 1835; (i) Brasholz, M.; Dugovič, B.; Reissig, H.-U.; *Synthesis* **2010**, 3855; (j) Alcaide, B.; Almendros, P.; Martínez del Campo, T.; Redondo, M.C.; Fernández, I.; *Chem. Eur. J.* **2011**, *17*, 15005.

[256] (a) Bongers, N.; Krause, N.; *Angew. Chem., Int. Ed.* **2008**, *47*, 2178; (b) Gandon, V.; Lemière, G.; Hours, A.; Fensterbank, L.; Malacria, M.; *Angew. Chem., Int. Ed.* **2008**, *47*, 7534.

[257] (a) Erdsack, J.; Krause, N.; *Synthesis* **2007**, 3741; (b) Volz, F.; Krause, N.; *Org. Biomol. Chem.* **2007**, *5*, 1519; (c) Volz, F.; Wadman, S.H.; Hoffmann-Röder, A.; Krause, N.; *Tetrahedron* **2009**, *65*, 1902; (d) Gao, Z.; Li, Y.; Cooksey, J.P.; Snaddon, T.N.; Schunk, S.; Viseux, E.M.E.; McAteer, S.M.; Kocienski, P.J.; *Angew. Chem., Int. Ed.* **2009**, *48*, 5022.

[258] Deutsch, C.; Gockel, B.; Hoffmann-Röder, A.; Krause, N.; *Synlett* **2007**, 1790.

[259] Sawama, Y.; Sawama, Y.; Krause, N.; *Org. Biomol. Chem.* **2008**, *6*, 3573.

[260] (a) Gockel, B.; Krause, N.; *Eur. J. Org. Chem.* **2010**, 311; (b) Poonoth, M.; Krause, N.; *Adv. Synth. Catal.* **2009**, *351*, 117.

[261] Mo, F.; Yan, J.M.; Qiu, D.; Li, F.; Zhang, Y.; Wang, J.; *Angew. Chem., Int. Ed.* **2010**, *49*, 2028.

[262] Kim, S.; Lee, P. H. *Adv. Synth. Catal.* **2008**, *350*, 547.

[263] Zriba, R.; Gandon, V.; Aubert, C.; Fensterbank, L.; Malacria, M.; *Chem. Eur. J.* **2008**, *14*, 1482.

[264] Aksin, Ö.; Krause, N.; *Adv. Synth. Catal.* **2008**, *350*, 1106.

[265] Minkler, S.R.K.; Lipshutz, B.H.; Krause, N.; *Angew. Chem., Int. Ed.* **2011**, *50*, 7820.

[266] (a) Lipshutz, B.H.; Ghorai, S.; *Aldrichim. Acta* **2008**, *41*, 59; (b) Lipshutz, B.H.; Ghorai, S.; Abela, A.R.; Moser, R.; Nishikata, T.; Duplais, C.; Krasovskiy, A.; Gaston, R.D.; Gadwood, R.C.; *J. Org. Chem.* **2011**, *76*, 4379; (c) Lipshutz, B.H.; Ghorai, S.; Leong, W.W.Y.; Taft, B.R.; Krogstad, D.V.; *J. Org. Chem.* **2011**, *76*, 5061.

[267] Winter, C.; Krause, N.; *Green Chem.* **2009**, *11*, 1309.

[268] Asikainen, M.; Krause, N.; *Adv. Synth. Catal.* **2009**, *351*, 2305.

[269] (a) Alcaide, B.; Almendros, P.; Martínez del Campo, T.; *Angew. Chem., Int. Ed.* **2007**, *46*, 6684; (b) Alcaide, B.; Almendros, P.; Martínez del Campo, T.; *Chem. Eur. J.* **2008**, *14*, 7756; (c) Alcaide, B.; Almendros, P.; Martínez del Campo, T.; Soriano, E.; Marco-Contelles, J.L.; *Chem. Eur. J.* **2009**, *15*, 1901; (d) Alcaide, B.; Almendros, P.; Martínez del Campo, T.; Soriano, E.; Marco-Contelles, J.L.; *Chem. Eur. J.* **2009**, *15*, 1909 (e) Alcaide, B.; Almendros, P.; Martínez del Campo, T.; Soriano, E.; Marco-Contelles, J.L.; *Chem. Eur. J.* **2009**, *15*, 9127; (f) Alcaide, B.; Almendros, P.; Carrascosa, R.; Martínez del Campo, T.; *Chem. Eur. J.* **2010**, *16*, 13243.

[270] Tarselli, M.A.; Zuccarello, J.L.; Lee, S.J.; Gagné, M.R.; *Org. Lett.* **2009**, *11*, 3490.

[271] Zhang, Z.; Widenhoefer, R.A.; *Angew. Chem., Int. Ed.* **2007**, *46*, 283.

[272] (a) Widenhoefer, R.A.; *Chem. Eur. J.* **2008**, *14*, 5382; (b) Sengupta, S.; Shi, X. *ChemCatChem* **2010**, *2*, 609. (c) Pradal, A.; Toullec, P.Y.; Michelet, V.; *Synthesis* **2011**, 1501.

[273] (a) Kang, J.-E.; Lee, E.-S.; Park, S.-I.; Shin, S.; *Tetrahedron Lett.* **2005**, *46*, 7431; (b) Liu, H.; Leow, D.; Huang, K.-W.; Tan, C.-H.; *J. Am. Chem. Soc.* **2009**, *131*, 7212.

[274] Piera, J.; Krumlinde, P.; Strübing, D.; Bäckvall, J.-E.; *Org. Lett.* **2007**, *9*, 2235.

[275] (a) Liu, L.-P.; Xu, B.; Mashuta, M.S.; Hammond, G.B.; *J. Am. Chem. Soc.* **2008**, *130*, 17642; (b) Liu, L.-P.; Hammond, G.B.; *Chem. Asian J.* **2009**, *4*, 1230.

[276] (a) Hashmi, A.S.K.; Lothschütz, C.; Döpp, R.; Rudolph, M.; Ramamurthi, T.D.; Rominger, F.; *Angew. Chem., Int. Ed.* **2009**, *48*, 824; (b) Hashmi, A.S.K.; Döpp, R.; Lothschütz, C.; Rudolph, M.; Riedel, D.; Rominger, F.; *Adv. Synth. Catal.* **2010**, *352*, 1307.

[277] Shi, Y.; Roth, K.E.; Ramgren, S.D.; Blum, S.A.; *J. Am. Chem. Soc.* **2009**, *131*, 18022.

[278] (a) Hopkinson, M.N.; Tessier, A.; Salisbury, A.; Giuffredi, G.T.; Combettes, L.E.; Gee, A.D.; Gouverneur, V.; *Chem. Eur. J.* **2010**, *16*, 4739; (b) Hopkinson, M.N.; Ross, J.E.; Giuffredi, G.T.; Gee, A.D.; Gouverneur, V.; *Org. Lett.* **2010**, *12*, 4904; (c) Hopkinson, M.N.; Gee, A.D.; Gouverneur, V.; *Chem. Eur. J.* **2011**, *17*, 8248.

[279] (a) Nishina, N.; Yamamoto, Y.; *Tetrahedron Lett.* **2008**, *49*, 4908; (b) Cui, D.-M.; Yu, K.-R.; Zhang, C.; *Synlett* **2009**, 1103.

[280] Heuer-Jungemann, A.; McLaren, R.G.; Hadfield, M.S.; Lee, A.-L.; *Tetrahedron* **2011**, *67*, 1609.

[281] (a) Horino, Y.; Takata, Y.; Hashimoto, K.; Kuroda, S.; Kimura, M.; Tamaru, Y.; *Org. Biomol. Chem.* **2008**, *6*, 4105; (b) Cui, D.-M.; Zheng, Z.-L.; Zhang, C.; *J. Org. Chem.* **2009**, *74*, 1426; (c) Hadfield, M.S.; Lee, A.-L.; *Org. Lett.* **2010**, *12*, 484.

[282] Zhang, Z.; Widenhoefer, R.A.; *Org. Lett.* **2008**, *10*, 2079.

[283] Zhang, Z.; Lee, S.D.; Fisher, A.S.; Widenhoefer, R.A.; *Tetrahedron* **2009**, *65*, 1794.

[284] Paton, R.S.; Maseras, F.; *Org. Lett.* **2009**, *11*, 2237.

[285] Yang, C.-G.; He, C.; *J. Am. Chem. Soc.* **2005**, *127*, 6966.

[286] Zhang, X.; Corma, A.; *Chem. Commun.* **2007**, 3080.

[287] Graf, T.A.; Anderson, T.K.; Bowden, N.B.; *Adv. Synth. Catal.* **2011**, *353*, 1033.

[288] Hirai, T.; Hamasaki, A.; Nakamura, A.; Tokunaga, M.; *Org. Lett.* **2009**, *11*, 5510.

[289] Wang, Y.-H.; Zhu, L.-L.; Zhang, Y.-X.; Chen, Z.; *Chem. Commun.* **2010**, 577.

[290] (a) Aponick, A.; Li, C.-Y.; Biannic, B.; *Org. Lett.* **2008**, *10*, 669; (b) Aponick, A.; Biannic, B.; *Synthesis* **2008**, 3356.

[291] Bandini, M.; Monari, M.; Romaniello, A.; Tragni, M.; *Chem. Eur. J.* **2010**, *16*, 14272.

[292] Georgy, M.; Boucard, V.; Campagne, J.-M.; *J. Am. Chem. Soc.* **2005**, *127*, 14180.

[293] (a) Engel, D.A.; Dudley, G.B.; *Org. Lett.* **2006**, *8*, 4027; (b) Lopez, S.S.; Engel, D.A.; Dudley, G.B.; *Synlett* **2007**, 949; (c) Yu, M.; Li, G.; Wang, S.; Zhang, L.; *Adv. Synth. Catal.* **2007**, *349*, 871; (d) Lee, S.I.; Baek, J.Y.; Sim, S.H.; Chung, Y.K. *Synthesis* **2007**, 2107; (e) Marion, N.; Carlqvist, P.; Gealageas, R.; De Frémont, P.; Maseras, F.; Nolan, S.P.; *Chem. Eur. J.* **2007**, *13*, 6437; (f) Ramón, R.S.; Marion, N.; Nolan, S.P. *Tetrahedron* **2009**, *65*, 1767; (g) Wang, D.; Zhang, Y.; Harris, A.; Gautam, L.N.S.; Chen, Y.; Shi, X.; *Adv. Synth. Catal.* **2011**, *353*, 2584.

[294] Egi, M.; Yamaguchi, Y.; Fujiwara, N.; Akai, S.; *Org. Lett.* **2008**, *10*, 1867.

[295] (a) Schwehm, C.; Wohland, M.; Maier, M.E.; *Synlett* **2010**, 1789; (b) Wohland, M.; Maier, M.E.; *Synlett* **2011**, 1523.

[296] (a) Yu, M.; Zhang, G.; Zhang, L.; *Org. Lett.* **2007**, *9*, 2147; (b) Yu, M.; Zhang, G.; Zhang, L.; *Tetrahedron* **2009**, *65*, 1846; (c) Ye, L.; Zhang, L.; *Org. Lett.* **2009**, *11*, 3646; (d) Wang, D.; Ye, X.; Shi, X; *Org. Lett.* **2010**, *12*, 2088.

[297] (a) Li, G.; Zhang, G.; Zhang, L.; *J. Am. Chem. Soc.* **2008**, *130*, 3740; (b) Buzas, A.; Istrate, F.; Gagosz, F.; *Org. Lett.* **2007**, *9*, 985; (c) Huang, X.; De Haro, T.; Nevado, C.; *Chem. Eur. J.* **2009**, *15*, 5904.

[298] (a) Wang, S.; Zhang, L.; *Org. Lett.* **2006**, *8*, 4585; (b) Wang, S.; Zhang, G.; Zhang, L.; *Synlett* **2010**, 692.

[299] Dombray, T.; Blanc, A.; Weibel, J.-M.; Pale, P.; *Org. Lett.* **2010**, *12*, 5362.

[300] Harrak, Y.; Blaszykowski, C.; Bernard, M.; Cariou, K.; Mainetti, E.; Mouriès, V.; Dhimane, A.-L.; Fensterbank, L.; Malacria, M.; *J. Am. Chem. Soc.* **2004**, *126*, 8656.

[301] Johansson, M.J.; Gorin, D.J.; Staben, S.T.; Toste, F.D.; *J. Am. Chem. Soc.* **2005**, *127*, 18002.

[302] (a) Fürstner, A.; Hannen, P.; *Chem. Commun.* **2004**, 2546; (b) Fürstner, A.; Hannen, P.; *Chem. Eur. J.* **2006**, *12*, 3006; (c) Fehr, C.; Galindo, J.; *Angew. Chem., Int. Ed.* **2006**, *45*, 2901; (d) Fehr, C.; Winter, B.; Magpantay, I.; *Chem. Eur. J.* **2009**, *15*, 9773.

[303] Shi, X.; Gorin, D.J.; Toste, F.D.; *J. Am. Chem. Soc.* **2005**, *127*, 5802.

[304] Nieto Faza, O.; Silva López, C.; Álvarez; R.; De Lera, A.R.; *J. Am. Chem. Soc.* **2006**, *128*, 2434.

[305] (a) Zhang L.; Wang, S.; *J. Am. Chem. Soc.* **2006**, *128*, 1442; (b) Shi, F.-Q.; Li, X.; Xia, Y.; Zhang, L.; Yu, Z.-X.; *J. Am. Chem. Soc.* **2007**, *129*, 15503; (c) Kato, K.; Kobayashi, T.; Fujinami, T.; Motodate, S.; Kusakabe, T.; Mochida, T.; Akita, H.; *Synlett* **2008**, 1081; (d) Karmakar, S.; Kim, A.; Oh, C.H.; *Synthesis* **2009**, 194.

[306] Lee, J.H.; Toste, F.D.; *Angew. Chem., Int. Ed.* **2007**, *46*, 912.

[307] Nun, P.; Gaillard, S.; Slawin, A.M.Z.; Nolan, S.P.; *Chem. Commun* **2010**, *46*, 9113.

[308] Buzas, A.; Gagosz, F.; *J. Am. Chem. Soc.* **2006**, *128*, 12614.

[309] Marion, N.; Diez-Gonzalez, S.; de Fremont, P.; Noble, A.R.; Nolan, S.P.; *Angew. Chem., Int. Ed.* **2006**, *45*, 3647.

[310] (a) Marion, N.; Gealageas, R.; Nolan, S.P.; *Org. Lett.* **2007**, *9*, 2653; (b) Gourlaouen, C.; Marion, N.; Nolan, S.P.; Maseras, F.; *Org. Lett.* **2009**, *11*, 81.

[311] Liu, Y.; Mao, D.; Qian, J.; Lou, S.; Xu, Z.; Zhang, Y.; *Synthesis* **2009**, 1170.

[312] Morita, N.; Krause, N.; *Angew. Chem., Int. Ed.* **2006**, *45*, 1897.

[313] Ando, K.; *J. Org. Chem.* **2010**, *75*, 8516.

[314] Nakamura, I.; Sato, T.; Yamamoto, Y.; *Angew. Chem., Int. Ed.* **2006**, *45*, 4473.

[315] Nakamura, I.; Sato, T.; Terada, M.; Yamamoto, Y.; *Org. Lett.* **2007**, *9*, 4081.

[316] Nakamura, I.; Sato, T.; Terada, M.; Yamamoto, Y.; *Org. Lett.* **2008**, *10*, 2649.

[317] Peng, L.; Zhang, X.; Zhang, S.; Wang, J.; *J. Org. Chem.* **2007**, *72*, 1192.

[318] Zhao, X.; Zhong, Z.; Peng, L.; Zhang, W.; Wang, J.; *Chem. Commun.* **2009**, 2535.

[319] (a) Georgy, M.; Boucard, V.; Debleds, O.; Dal Zotto, C.; Campagne, J.-M.; *Tetrahedron* **2009**, *65*, 1758; (b) Guo, S.; Song, F.; Liu, Y.; *Synlett* **2007**, 964; (c) Rao, W.; Chan, P.W.H.; *Org. Biomol. Chem.* **2008**, *6*, 2426; (d) Kothandaraman, P.; Rao, W.; Zhang, X.; Chan, P.W.H.; *Tetrahedron* **2009**, *65*, 1833.

[320] Brouwer, C.; Rahaman, R.; He, C.; *Synlett* **2007**, 1785.

INDEX

bis(acetonitrile)(1,5-cyclooctadiene)rhodium tetrafluoroborate $[Rh(CN)_2(COD)]BF_4$, preparation 142
activation of isonitrile 431
acetoxonium intermediate 521
acetoxyallenes 520
1,3-acetoxy transfer 487, 520
(*S*)-(-)-acromelobic acid 178
acyclic diaminocarbene ligands (ADC's) 495
1,2-acyloxy-migration 518
additives, to generate cationic Rh complexes
silver hexafluoroantimonate ($AgSbF_6$) 139
silver tetrafluoroborate ($AgBF_4$) 139
silver triflate (AgOTf) 139
aldehydes, via hydroformylation 157
Alder-ene reaction, intramolecular, Rh-catalyzed 261, 262, 263
with amide linker 270
aldol cycloreduction, Rh-catalyzed 230
alkoxycyclization 457, 464
alkylative coupling 368
alkylidenecyclopropene 117
4-alkylidene-1,3-dioxolan-2-ones 487
2-alkylidene-2,5-dihydrofurans 490
2-alkylidenetetrahydrofurans/pyrans 487
alkyl halides, in couplings 345
alkynyl cyanides, reactions with alkynes 410
alkynylfurans 440
1,3-alkynyl rearrangement 215
allenamides 477
allenenes 441, 465
allenic hydrazines 478

allenic hydrazones 476
allenic hydroxylamine derivatives 475
allenic β-ketoesters 440
allenic silyl enol ethers 440
allenic ureas 477
allenyl indoles 442
allenynes 441, 463
allopumiliotoxins 376
allylamines 479
allylboranes, synthesis of 375
allyl cyanides, reactions with alkynes 410
π-allylgold intermediate 514
allylic alcohol, silyl ethers
  in couplings 339
  formation of 389
allylic boronates 54, 55
allylic phosphates 89, 90, 91, 93
allylic substitution/cycloisomerization 452
π-allylnickel complexes 393
allyl oxonium intermediate 511
π-allylpalladium species 511
alternative sources of CO, in Rh-catalyzed reactions 235
amidation/hydrolysis 25
aminations, of aryl chlorides 340
amination/intramolecular hydroamination 471
α-amino acid derivatives 176
aminoalcohols 431, 522
α-aminoboronates 57
2-aminopyridine ligand 39
ampenidone 35
amphidinolide T1 378
andrachcinidine 497
angeloylgomisin R 60
anilines 23, 32
annulation 438, 447
anti-selective aldol 110
aqueous ammonia 23, 24
N-arylation 20, 29, 34
arylation, *meta*-selective 114
arylative cyclization, Rh-catalyzed 212
arylboronates 90
arylboronic acids 484, 515
aryl halide, homocoupling, 343
ascorbate 5
aspidosperma alkaloids 201
asteriscanolide 364
asymmetric couplings, benzylic ethers 341
asymmetric hydroboration, Rh-catalyzed
  styrenes 217, 218, 219
asymmetric hydroformylation 152
asymmetric hydrogenation, 173
  2-acetamidoacrylic acids 176
  β-(acylamino)acrylates 178, 179
  mixtures of *E*- and *Z*- isomers 180
  β,β-disubstituted α-dehydroamino acids 181
  enamides 182
  itaconic acid & derivatives 184
  enol esters 186
  unsaturated acids, esters 187
  ketones 189, 191
  N-acylhydrazones 191
asymmetric hydrosilylations 96, 98, 107
  dialkyl ketones 108
Au-catalyzed
  aldol reactions 431
  aminoamidation 485
  annulation of phenols 447
  benzannulation 463
  bromination 525

C–H activation 435, 448
Claisen rearrangement 467, 468, 469
cyclization of arylallenes 444
cyclopropanation 452, 454
diamination 485
dehydrative cyclization 472, 488, 490, 515, 522, 523
hydration of alkynes 498
hydroalkylation of alkenes 445
hydroalkylation of unsaturated ketones 446
hydroarylation 435, 436, 438, 439, 442, 443, 444, 446,447
hydroborations 433
hydrosilylations 433
hydroalkoxylation of alkynes 486, 497
intermolecular Friedel-Crafts alkylation 439
intramolecular alkoxycyclizations 456
intramolecular [2+2+2]-cycloaddition of ketoenynes 464
intramolecular [3+2]-cycloaddition of 1,6-diynes 463
intramolecular [4+2]-cycloaddition of 1,3-dien-8-ynes 465
intramolecular hydroamination 469, 470, 471, 474, 482, 483
intramolecular oxidative carboalkoxylation 515
intramolecular oxidative carboamination 484
nucleophilic substitution 516, 524
oxidative cyclization 494
ring expansion/cyclization of alkynylcyclopropylketones 493
tandem acetalization/cyclization of 2-alkynylbenzaldehydes 496
tandem Claisen rearrangement/cyclization 467
tandem C–O/C–C bond formation 511
tandem cycloisomerization/ cycloaddition 489
tandem cycloisomerization/ hydroalkoxylation of homopropargyl alcohols 488
tandem hydroamination/ hydroarylation 436
tandem lipase/gold-catalyzed transformation 505
tandem Meyer-Schuster rearrangement/oxa-Michael addition 516
tandem nucleophilic addition/cyclization 495

arene oxide 439
(aza)Henry reaction 67
7-azaindoles 447
azametallacycle 342
aza-Michael addition 481
azepino[4,5-*b*]indoles 438
aziridination, intramolecular 199
aziridines 166
in cross-coupling 342
1*H*-azocino[5,4-*b*]indole 438

Baylis-Hillman acetates 521
base metal 3
benzenes, substituted, from vinylallenes 279

benzenes, substituted, from
vinylallenes (*continued*)
formation 354
benzylic ethers, in couplings with
Grignards 341
benzpyrazole 29
benzylamines, nonracemic 219
benzofurans 439, 490
benzoxazines 487
benzoxazoles, in cross-couplings, 347
benzothiophenes 523
biaryls, nonracemic, from
cyclotrimerization, Rh-catalyzed
248
bicyclooctanols 359
binaphthyl couplings 61
BINAP 61, 104, 108
binuclear Au(II)–Au(II) complexes
484
BIPHEP ligand 52, 61, 96, 97, 101,
107, 108, 109
bortezomib (Velcade) 57
bouvardin 39
B_2Pin_2 49, 50, 53
brevetoxin A 94
bromoalkynes 487
bromoirenium ion 500
Brønsted acid 432, 448, 455, 486,
487, 488, 500, 524
N-t-butanesulfinylimine derivatives 57
t-butanol 9, 100
butenolides 87, 509-511

calcination 34
calphostin A 60
carbolines 438
carbazole 26
syn-carbocupration 70
carbocyanation, Ni-catalyzed 408
of alkynes 408
carbometallation
organozinc formation, Ni-
catalyzed 399
carbon dioxide
in cycloadditions 356
reactions with zinc reagents 388
reductive couplings 381
β-carbon elimination 351
carbothiolation of alkynes 523
carbozincation 116
cascade cyclization 507, 512
cesium carbonate 41
cesium phenoxide 44
C-H activation 9, 352, 406
asymmetric, intermolecular 196
borylation, aromatics & benzylic
positions 204
by metal carbenoid & nitrenoid
195
into *N*-Boc-pyrrolidine 198
secondary amines, carbonylation
204
C-H functionalization 347
C–H insertion 458, 518
chiral counterion 477, 509
chirality transfer 442, 47, 468, 474,
475, 477, 479, 502, 507, 514, 519,
522, 523
chlorocarbonyl-*bis*(triphenylphosphine)rhodium(I),
$RhCl(CO)(PPh_3)$ 140
di-μ-chloro-*bis*(η4-1,5-
cyclooctadiene)-dirhodium(I),
$[Rh(C_8H_{12})Cl]_2$, preparation 141
2*H*-chromenes 436
Chxn-Py-Al 39
cinnamyl phosphates 92
Claisen rearrangements 467, 468
clickphine 7

1,4-conjugate addition, boronic acids 208
NMR study 208
1-pot hydroboration/addition 209, 210
regioselectivity, in 4-oxobutenamides 215
to enoates 211
to unsaturated nitro compounds 211, 212
then aldol reaction 220, 222
with trifluoroborates 210
coenzyme Q_{10} 327
computational studies, mechanism of cross-coupling 338, 380
phenylcyanation of alkynes 408
Conia-ene reaction 460
conjugate additions, Ni-catalyzed, with zinc reagents 390-392
contiguous asymmetric centers 109
copper(I) acetylides 5
copper-in-charcoal (Cu/C) 14, 43
copper powder 22
copper sulfate 8
Corriu 322
coumarins 436
crispine A 170
cross-conjugated trienes,
from Rh-catalyzed tandem Alder-ene/Diels-Alder 271
in tandem Diels-Alder reactions 271, 273
CuH/C 106
(*R,R,R,R*)-CuPhEt 110
3-cyanoindoles, C-H activation of 405
cyclic alkylaminocarbene (CAAC) ligands 479
cyclic *bis*-ketals 486
cyclizations, *exo, endo*, macro 378, 380
intramolecular, aldehyde-allene 382
of iodoalkenes 399
of iodoalkynes 400
cycloadditions & cyclizations, Rh-catalyzed 230
enals + alkynes 359
mechanism, general 230
Ni-catalyzed 348, 363
of unsaturated ketones + alkynes 361
reductive 231, 285
with 1,6-enynes 286, 287
with silanes, and mechanism 284, 286
with heteroatoms 351
[2+2]-cycloadditions 462, 465
[2+2+1] cycloaddition 230, 357
asymmetric 234
[2+2+2] cyclotrimerization 248, 251
intermolecular, chemo-, regioselective 248
diyne-alkyne cyclotromerizations 251, 22
of 1,4-dienynes 255, 256, 257
of a 1,6-enyne, asymmetric 255
of a 1,6,11-triyne 253
formation of pyridones 258, 259
multicomponent 353
[2+2+2+1] cycloadditions 282
intramolecular, and mechanism 283, 284, 285
[2+2+2+2] of terminal 1,6-diynes 364

cycloadditions & cyclizations, Rh-catalyzed (*continued*)
[3+2] cycloaddition, cyclopropenones + alkynes 247
multicomponent 357, 358
[3+2+2] cycloadditions 366
[3+3]-cycloadditions 466
[4+1] cycloaddition, vinylallenes 245, 246
[4+2] cycloadditions 273, 274, 349, 350, 462, 463, 465, 466
of dienynes 274, 275, 276, 277
of trienes, intramolecular 276, 277
of allene-diene, tethered 278, 279
[4+3]-cycloaddition 466
[4+4] cycloadditions 364
[5+2] cycloadditions 280
asymmetric versions 282
cyclobutanones, cyclization 351
cyclobutenes 462
cycloetherification 487
cycloheptadiene 365
cyclohydrocarbonylation (CHC) 168-171
cycloisomerizations 449
alkynols & alkynyl anilines 263, 264
alkynylphenols 263
allenic acetates 505
allenic hydrazines 478
α-allenols 501
allenyne 268
allyl allenoates 511
allyl propargyl ethers 451
γ-aminoallenes 474
branched acetylenic diols 488
N-(2,3-butadienyl)-substituted indole derivatives 442
cyclohexadienyl-substituted alkynes 453
α,β-dihydroxyallenes 502
1,4-diynes 489
1,5-enynes 457
1,6-enynes 449, 457, 461
1,7-enynes 461
enynes, and mechanism 262, 264
enynoic acids 456
enynols 459
haloallenones 499
α- or β-hydroxyallenes 505
γ-hydroxyallenes 507, 508
N-hydroxy-α-aminoallenes 475
propargyl aziridines 472
propargyl oxiranes 491
protected α-aminoallenes 474
α-thioallenes 522
unprotected α-aminoallenes 475
bis(1,5-cyclooctadiene)rhodium tetrafluoroborate $[Rh(COD)_2]BF_4$, preparation 142
bis(1,5-cyclooctadiene)rhodium trifluoromethanesulfonate 189
cyclooctatetraenes 364
cyclopentadienones, Rh-catalyzed, 247
cyclopeptide alkaloid 36
cyclopropanation 164
asymmetric 165, 166
intramolecular 168, 450, 451
cyclopropenylcarbinols 117
cyclopropyl gold carbene 432, 439, 449, 450, 455, 457, 459, 461, 462, 519
cyclopropylzinc 117

(-)- deoxyprosophyline 168
diaminations, of alkenes, Ni-catalyzed 413

diarylamines 32
dichlorotetracarbonyldirhodium(I), $Rh_2(CO)_4Cl_2$], preparation 141
Diels-Alder reaction 451, 465, 489, 518
dienes, hydroformylation
conjugated 147
non-conjugated 148
dienynes 465
diethoxymethylsilane (DEMS) 96, 102, 108
N,N-diethylsalicylamide 20
DIFLUORPHOS ligand 81, 82
dihydrobenzopyrans 446
dihydrocyclohepta[*b*]indoles 439
2,3-dihydrofurans 440, 487, 489
2,5-dihydrofurans 490, 501, 505
2,3-dihydroisoxazoles 487
4,5-dihydroisoxazoles 475
3,6-dihydro-1,2-oxazines 475
dihydropyrans 351
5,6-dihydro-2*H*-pyrans 467, 502
dihydropyranones 71
dihydropyrimidine-2,4-diones 356
dihydroquinolines 436, 443
2,5-dihydrothiophenes 522
trans-N,N'-dimethyl-1,2-cyclohexanediamine 35
N,N-dimethylethylenediamine (DMEDA) 25
diorganozinc 31
dioxepines 489
diynes, tethered 354
1,4- 489
1,6- 450, 462
with cyclobutanones 367
domino reactions, Rh-catalyzed 298
allylic substitution, then Pauson-Khand 299
allylic substitution, then carbocyclization 300
Pd- then Rh-catalyzed 299
1,2-*bis*(diphenylphosphino)ethane (1,5-cyclooctadiene)rhodium tetrafluoroborate [$Rh(dppe)(COD)]BF_4$, preparation 142
1,3-dipolar cycloadditions 202
dirhodium carboxamidate, C-H activation 196
dirhodium carboxylate, C-H activation 196
dirhodium phosphate, C-H activation 196
dithioacetals 523
dithiofluorene derivatives 523
L-DOPA 177
(S)-duloxetine 191
duocarmycins 30
dynamic kinetic asymmetric transformatiuon (DYKAT) 89
dynamic kinetic resolution 325, 331, 334

electron transfer, from Ni(0) 326, 343
enals, 1,4-additions to 395
enol ethers 434, 440, 444, 455, 460, 498
enantioselective protonation, silyl enol ethers 434
enantioselective 1,2-reduction 102
5-*endo-dig* cyclization 36
enoates 98
enolate trapping 109
1,5-enynes 449, 457, 458, 459, 519
1,6-enynes 449, 450, 451, 453-457, 461-464, 519

1,7-enynes 449, 461, 462
1,8-enynes 461, 462
1,6-enynols 456
epimerization, allenes 502
epoxides, in cross-couplings 342
Fesulphos ligand 67
Fleming's silane ($PhMe_2SiH$) 96
fluorapatite (CuFAP) 33
fluorescence probe, gold(III) ions 436
(*S*)-fluoxetine 191
formylvinylsilanes 163
Friedel-Crafts-type ring expansion 462
furanones 83, 172
    3(2*H*)- 493
furans 439, 448, 490- 493, 498, 499, 500
    halogenated 499

(–)-gloeosporone 383
D-glucosamine 24
gold acetylide 495
Goldberg reaction 34
gold carbene 432, 439, 449, 450, 453, 454, 455, 457, 458, 459, 460, 461, 462, 464, 518, 523
gold carbenoid intermediate 466
gold(I)/gold(III) redox cycle 512
gold nanoparticles 433
gold(I)-NHC catalysts 467, 476, 490, 514, 516, 518, 520, 521
gold(III)-porphyrin complex 499
gold(III)-Schiff base complexes 433
gold-stabilized cyclopropylmethyl/cyclobutyl/homoallyl cation 449
gold vinylidene intermediate 473
gomisins O and E 60
Gosteli-Claisen rearrangement 468
granules 34
Grob fragmentation 467

halichondrine 387
haloalkynes 436
haloallenones 499
halocrotonates 88
harmicine 170
heteroaromatics 33, 91
heteroatom directors, in alkyne insertions 410, 411
hetero-allylic asymmetric alkylations 87
heterocouplings 323
N-heterocyclic carbene 6, 335, 356, 363, 373, 376, 378, 380, 381, 403
hetero-Diels-Alder 71
heterogeneous catalysis 4, 38, 333
    gold catalysis 431-433
hexahydrobenzofurans 489
hexahydroindenone derivatives 440
hexahydroindoles 481
higher order cyanocuprate 58
homoallylic silyl ethers, formation of 389
Huisgen cycloaddition 5, 11
Hünig's base 495
hydrative cyclization 441
hydration 493, 497
    of alkyne 493, 498
    of nitriles 497
β-hydride elimination 323, 327, 328, 371, 402
hydroacylation 191
    of alkynes/alkenes with aldehydes 193
    of alkynals 194
hydroalkoxylation 432, 486, 487, 488, 489, 496, 497, 513, 515
hydroalkylation 444, 481

hydroamination 412, 469, 473, 479
of acrylonitrile 413
of alkenes 480
of alkenyl ureas 483
of alkynes 469, 473
of allenes 474
of aminoallenes 477
of conjugated dienes 481
of 1,3-dienes 412
mechanistic studies 412
hydroamination/halogenation 470
hydroamination/hydroarylation 470
hydroarylation, 435, 436, 438, 439, 442, 443, 444, 446, 447, 470
of alkynes, *trans* 401, 435
hydroboration 54
hydrocyanation
of alkenes 404
hydroformylation 142
of unsaturated alcohols 150
of alkenyl esters 150
'oxo' process 143
regioselective, alkenes 148
then Wittig olefination 156, 157
then Mannich reaction 160
hydrogenation,
of β-(acylamino)acrylates 178
itaconic acid 184
alkylidenesuccinates 433
aromatic nitro compounds 433
olefins and imines 433
hydrogenation catalyst, gold 433
hydrogenolysis, of ethers 338
hydrometallation 395
hydrosilylations, Ni-catalyzed
alkenes 396
alkynes 396
ketones 396
hydrovinylation 401, 402
heterocouplings 403
hydroxo-gold species 516
α-hydroxyallene 111
hydroxylamines 478, 487
N-hydroxypyrrolines 475
8-hydroxyquinoline 39, 41
hydrozincation, of alkenes, Ni-catalyzed 396, 397
hyellazole 251

imidazole(s) 16, 26, 29, 33
imidazolidinones 477, 482
imidazolium salt 110
immobilized copper 43
indenes 520, 521, 523
indenofuranones 512
indenols 361
indoles 26, 28, 438, 439, 442, 443, 446, 447, 448, 457, 470
indole-3-carboxylates, C-H activation of 405
indolizines 473
indoloalkynes 438
indoloazocines 438
Inductively Coupled Plasma-Atomic Emission Spectroscopy (ICP-AES) 15
interiotherin A 60
intramolecular alkylative coupling 376
intramolecular aminations 19
intramolecular hydroarylation 436, 438, 439, 446, 447
intramolecular silylformylation, alkynes 162
iodoalkynes 460
iododeauration 460, 503, 510
ionic liquid 492, 504
iron/copper catalysts 28
isochromene chromium complexes 496

1*H*-isochromene derivatives 495
isocyanides, in cycloadditions 362
isodomoic acids G and H 369
isogeissoschizine, deformyl 372

kinetic isotope effect 525
kinetic resolution,
    of alkynal hydroacylation 194
    then cycloisomerization 505
Kumada 322, 325
(±)-kumausyne 94

lactams 445, 474, 506,
lactams, unsaturated 352, 353
lactones 456, 487, 509, 512
(+)-lasubine II 260, 261
Lewis acid-activated Brønsted acid
    (LBA) 434
Lewis acidic co-catalysis 405
    in carbocyanation 408
ligand abbreviations 324
ligands
    BenzP* 153, 154
    BICP 182, 264
    (*R,R,R,R*)-BICPO 264
    BISBI 145
    BINOL-based phosphonite 153,
        154
    BINOL-based phosphoramidite
        205
    BINAP 176, 193, 208, 209, 210,
        223, 268
    H_8-BINAP 249
    BINAPHOS 153, 155
    BINAPINE 153, 154
    BINAPO 176
    BIPHEP 176
    BIPHEPHOS 145, 148, 149, 159,
        169, 170, 171
    Cl,MeO-BIPHEP 223
    Chiraphite 153, 154
    3,5-diMeC_6H_4-BINAP 234, 235
    DIPHOS 274
    DPAMP 176, 177
    DMTPPN 151
    DPPB 151
    DTBM-SEGPHOS 259
    DuanPhos 178
    EBTHI 234
    esp 200
    ESPHOS 153, 154
    Et-DuPhos 176, 184, 191
    Et-FerroTANE 184
    Josiphos 49, 52, 74, 77-79, 86,
        88, 96, 100, 101-105, 107,
        176, 210
    Kelliphite 153, 154
    MANDYPHOS 108
    (S,S)-MCCPM 190
    Me-BPE 181
    Me-DuPhos 52, 178, 181, 188,
        193, 264
    Me-DUPHOS(O) 68, 104, 105
    MeO-BIPHEP 210
    Me-PennPhos 186
    MonoPhos 176, 177
    NAPHOS 145
        1-naphthoic acid 39
    NHC 92, 93, 95
    NOBIN-related 153, 154
    PHANEPHOS 223
    (*S,S*)-Ph-bod* 216
    Ph-BPE 182, 184
    phosphazene P_4-*t*-Bu ligand 39
    phosphoramidite ligand 54, 69,
        76, 83, 86, 89
    PINAP 65
    PyBOX ligand 66
    (S)-Quinap 65, 218, 219
    Quinazolinap 50

quinolone-*N*-oxide ligand 24
QuinoxP* 49, 55, 153, 154
reversed JOSIPHOS 88
SEGPHOS 96, 104, 109
DTBM- 68, 97, 99, 100, 101, 104, 106
SIPHOS 179, 180, 182, 234
TADDOL 75, 86
Tangphos 153, 154, 178, 179, 182, 184, 186
Taniaphos 49, 86-88, 108
TBPM 21
o-TDPP 151
TPPTS 143
Tol-BINAP 52, 77, 78, 79, 80, 228
XANTPHOS 145
ligands based on cod and nbd 213, 276
ligands, directing effects 149
ligandless 37
linear selective hydroformylation of alkenes 145
lundurine alkaloids 438

macrocyclizations 378, 380, 383, 387
manganese, in MnF_2/CuI 29
Markovnikov hydroarylation 435
Markovnikov regioselectivity 480, 514
mechanism, of Rh-catalyzed hydroformylation 143, 144
Meldrum's acid 83
4-methyl-1,2-dihydroquinolines 436
2-methylTHF 75
metallacycle 322, 348, 349, 351, 352, 358-362, 367, 368, 371, 380, 381, 390
methylene cyclopropanes, in cycloadditions 358
4-methylenetetrahydroquinolines 436
Meyer-Schuster-rearrangement 517
micellar systems 504
Michael additions 447, 448, 470, 492, 493, 499
Michael adduct 481
microwave 7, 14, 44, 115
C-H activation 207
morpholines 516
multi-component coupling 10

nanoparticles 13, 16, 34, 45
nanowires 34
naphthalenes 436, 447, 463
naphthols 447
Nazarov cyclization 519, 520
Negishi 325
Negishi couplings 330, 342
neocuproine 39
Ni(I) – Ni(III) intermediates 394
nickel *O*-enolates 354, 359
nicotinamides 30
Nozaki-Hiyama-Kishi couplings 384
macrocyclizations 387
mechanism 384, 385
nucleophilic additions to alkenes 444, 480, 514
nucleophilic additions to alkynes 435, 469, 486
nucleophilic additions to allenes 440, 474, 498
Ohloff-Rautenstrauch rearrangement 519
oligomerization 322, 323, 349, 368
okadaic acid 488
on water 9
oxametallacycle 342
oxazoles, 438, 492
in cross-couplings 347
oxazolidinones 487

oxazolines 431
oxepine 439
oxetan-3-ones 494
oxidative addition 326
  allylic alcohol derivatives 339
  anhydrides 339
  phenol derivatives 336, 339
oxidative C–C coupling 512
oxidative cyclization 329
oxidative difunctionalization 484
oxidative sulfamidation, aldehydes 200
oxidizing agent 58
α-oxogold carbene 494
oxycupration 114

palladium-catalyzed cross couplings 511
papulacadin D 251
partial reduction of buta-1,3-diene 433
Pauson-Khand reactions, of 232, 233, 234, 362
  allene-dienes 243, 244
  allenynes 240, 241, 242
  bis-allenes 244
  diene-alkene 239
  dienynes 238
  diynes 236
Pd-Cu bimetallic 11
perovskite 45, 46
phenanthrenes 436
1,10-phenanthroline 18, 35, 41, 44
  4,7-dimethoxy 35
phenols 439, 440, 447, 514
phenylsilane ($PhSiH_3$) 96, 104
phosphoramidite complexes 466
phthalimides, synthesis 207, 208
(+)-pilocarpine 266
pinacol rearrangement 460
pinacolborane (pinBH) 49, 52, 108
(+)-pinnatoxin A 94
piperidines 477, 483, 484
piperazine, ligand 23
platensimycin 488
polymerization 335
polymethylhydrosiloxane (PMHS) 95-97, 100, 103, 107, 109
polyoxyethanyl α-tocopheryl sebacate (PTS) 504
Prins reaction 489
propargyl aziridines 472
propargylic amines 65, 66
propargylic carbonates 112
L-proline 18, 26, 28
(S)-propanolol 190
propargylic epoxide 112
(+)-prosopinine 168, 169
protease inhibitor 57
protodeauration 440, 441, 447, 450, 472, 486, 487, 490, 495, 499, 501, 510
protodesilylation 84
pseudopterosins 403
(+/-)-ptilocaulin 275
Pybox ligand 400
4*H*-pyrans 492
pyranones 83
pyrazole 26, 28
pyridine *N*-oxides, C-H activation of 405
pyridones,
  C-H activation of 405
  via Rh-catalyzed cycloaddition 258, 259
2-pyridylsulfones 79, 80
pyrimidinediol ligand 27
α-pyrones 518
pyrroles 26, 439, 471-473, 476
pyrrolidines 477, 480-484

pyrrolinones 171
3-pyrrolines 474
pyrroloisoindolones 474
pyrroloquinolines 437

quaternary centers, 55, 91
 via cyclobutanones 228, 229
 via enolboration/
  hydroformylation/aldol 156
quinolines 436

racemization 439, 496, 514
rearrangements, 447, 449, 454, 459,
 460, 461, 462, 464, 467, 516
 Meyer-Schuster rearrangement
  516
 Ni-catalyzed, vinylcyclopropanes
  365
 rearrangement/nucleophilic
  substitution 473
GPIIb/IIIa receptor antagonist 30
reductants, in Ni-catalyzed couplings
 368
 i-Bu_2Al(acac) 373
 trialkylsilanes 381
reductive aldol, Ni-catalyzed 395
reductive coupling, Rh-catalyzed 220
 enoate/aldehyde 221, 222
 intramolecular 222
 of dienes with aldehydes, and
  mechanism 225
 of 1,6-diynes, 1,6-enynes 223
reductive elimination 27, 58, 326,
 328, 334, 336
reductive hydrogenation, Rh-catalyzed
 223
reductive Mannich reaction 108
regiocontrol, in Ni-catalyzed
 couplings 379
relativistic effects 432
retro-Michael addition 497
(–)-rhazinilam 442
ring enlargement 458, 493
rhodium carbenoid,
 formation of oxazolidinone 199
 insertion at benzylic position 198
 rhodium catalyst, water soluble,
  $HRh(CO)(TPPTS)_3$ [TPPTS =
  $P(C_6H_4SO_3Na\text{-}m])_3$, 143
rhodium(III) chloride 139
rhodium(III) chloride trihydrate
 ($RhCl_3 \cdot 3H_2O$) 139
rubromycins 488

Salox 39
salsolene oxide 364
Schlenk equilibrium 86
Schweisinger's base 47
Selectfluor® 484, 485, 512, 515
[3,3]-sigmatopic rearrangement 365
silylcarbocyclization,
 cascade, endiyne, and mechanism
  291, 292
 mechanism 288
  of allenynes 287, 288, 289
  of 1,6-diynes, carbonylative
   290
  of triynes 293
silyl enol ethers 434, 440, 460
silylformylation, olefins and alkynes
 160
silyl hydrides 368
sodium *t*-butoxide (NaO-*t*-Bu) 51
solid phase click chemistry 12
spiroethers 459
spiroketals, 5,6- or 6,6- 468, 488
stereoconvergent, to allylic boronates
 56
stereodivergent, to allylic boronates
 55

Stryker's reagent 94, 99
styrene derivatives 444, 446, 480
styrenyl aziridines, in Negishi couplings 342
sulfonates, aryl, in couplings 345
sulfonylimines 66
sulfonyloxime 31
Suzuki couplings 330, 331, 334
syn-elimination 117

tamoxifen 401
tandem 1,4-addition-aldol reactions 294
tandem hydroformylation reactions 155
  acetalization 159
  cyclization 155, 156, 158
  Mannich reaction 160
tandem reductive aldols, Rh-catalyzed 294-297
  mechanism 297, 298
tandem silylformylation reactions
  olefination 162
TangPHOS ligand, asymmetric intramolecular cyclization 408
testudinariol A 382
tetraalkylphosphonium salts 21
2,2,6,6-tetramethylheptane-3,5-dione ligand 35, 41, 46
tetrahydrocarbazoles 442
tetrahydrofurans 362, 506, 508, 509, 515, 516
  yl ethers 487, 492
tetrahydroindene 465
tetrahydrooxepines 506
  bicyclic
tetrahydropyrans 488, 508, 509, 515
  yl ethers 487
tetrahydropyridines 469, 474
tetrahydroquinolines 443, 446
tetralins 443, 446
tetramethyldisiloxane (TMDS) 96
thioacetals 524
thiazoles, in cross-couplings 347
2-thienylaldimine derivatives 67
thioesters 52, 79
N-thioimides 48
thiophene, synthesis 522
2-thiophenecarboxylate (TC) 72, 85, 86
three-component couplings 368
TIPS-acetylene 11
Tishchenko reaction, Ni-catalyzed 390
TMS-Cl 80
TMS diazomethane, 7-membered ring formation 365
α-tocopherol-polyethylenglycol-750-succinate monomethylether (TPGS-750-M) 504
transmetallation 329, 334, 336
Transmission Electron Microscopy (TEM) 13, 45
transfer hydrogenation 473
1,2,3-triazoles 5, 7, 9
tri-μ-carbonyl-nonacarbonyltetrarhodium [$Rh_4(CO)_{12}$], preparation 139
trifluoromethanesulfonylazide 7
trienes, cross-conjugated 268
trimethylsilylazide 10

ultrasonication 14
Ullmann couplings 16, 29, 34, 39, 61
umpolung, to aryl amines 31
α,β,-unsaturated ketones, 1,4-additions with zinc reagents 391
α,β,γ,δ-unsaturated ketones, cyclizations 361
unsaturated lactams 352

γ,δ-unsaturated ketones, from allenes
and enones 373
unsaturated sulfones 104
ureas 477, 482, 483, 485

vancomycin 39
verbenachalcone 39
vineomycinone B 251
vinylalanes 92
α-vinylboronates 54
vinylcyclohexenes 441, 461
vinylgold species/intermediate 440, 460, 467, 472, 486, 492, 495, 510, 511, 519
N-vinylindoles 470
vinyl iodides 36
2-vinylpiperidines 476
2-vinylpyrrolidines 476, 477
vinyltrifluoroborate salts 37
vitamin D analogues 350

WALPHOS 108
Wilkinson's catalyst $RhCl(PPh_3)_3$,
preparation 140
isomerization of propargylic alcohols 226

Xantphos ligand 112

(+)-yatakemycin 30
yohimban alkaloid 350

zaragozic acids 202
zinc, in Ni-catalyzed reactions 343, 361
zwitterionic intermediates 502